SPECIFIC HEAT

Metallic Elements and Alloys

THERMOPHYSICAL PROPERTIES OF MATTER
The TPRC Data Series

A Comprehensive Compilation of Data by the
Thermophysical Properties Research Center (TPRC), Purdue University

Y. S. Touloukian, Series Editor
C. Y. Ho, Series Technical Editor

Volume 1. Thermal Conductivity–Metallic Elements and Alloys
Volume 2. Thermal Conductivity–Nonmetallic Solids
Volume 3. Thermal Conductivity–Nonmetallic Liquids and Gases
Volume 4. Specific Heat–Metallic Elements and Alloys
Volume 5. Specific Heat–Nonmetallic Solids
Volume 6. Specific Heat–Nonmetallic Liquids and Gases
Volume 7. Thermal Radiative Properties–Metallic Elements and Alloys
Volume 8. Thermal Radiative Properties–Nonmetallic Solids
Volume 9. Thermal Radiative Properties–Coatings
Volume 10. Thermal Diffusivity
Volume 11. Viscosity
Volume 12. Thermal Expansion–Metallic Elements and Alloys
Volume 13. Thermal Expansion–Nonmetallic Solids

New data on thermophysical properties are being constantly accumulated at TPRC. Contact TPRC and use its interim updating services for the most current information.

THERMOPHYSICAL PROPERTIES OF MATTER
VOLUME 4

SPECIFIC HEAT

Metallic Elements and Alloys

Y. S. Touloukian
Director
Thermophysical Properties Research Center
and
Distinguished Atkins Professor of Engineering
School of Mechanical Engineering
Purdue University
and
Visiting Professor of Mechanical Engineering
Auburn University

E. H. Buyco
Assistant Professor of Engineering
Purdue University
Calumet Campus
Formerly
Assistant Senior Researcher
Thermophysical Properties Research Center
Purdue University

IFI/PLENUM • NEW YORK-WASHINGTON • 1970

Library of Congress Catalog Card Number 73-129616

SBN (13-Volume Set) 306-67020-8

SBN (Volume 4) 306-67024-0

IFI/Plenum Data Corporation is a subsidiary of
Plenum Publishing Corporation
227 West 17th Street, New York, N.Y. 10011

Distributed in Europe by Heyden & Son, Ltd.
Spectrum House, Alderton Crescent
London N.W. 4, England

Printed in the United States of America

"In this work, when it shall be found that much is omitted, let it not be forgotten that much likewise is performed..."

SAMUEL JOHNSON, A.M.

From last paragraph of Preface to his two-volume *Dictionary of the English Language*, Vol. I, page 5, 1755, London, Printed by Strahan.

Foreword

In 1957, the Thermophysical Properties Research Center (TPRC) of Purdue University, under the leadership of its founder, Professor Y. S. Touloukian, began to develop a coordinated experimental, theoretical, and literature review program covering a set of properties of great importance to science and technology. Over the years, this program has grown steadily, producing bibliographies, data compilations and recommendations, experimental measurements, and other output. The series of volumes for which these remarks constitute a foreword is one of these many important products. These volumes are a monumental accomplishment in themselves, requiring for their production the combined knowledge and skills of dozens of dedicated specialists. The Thermophysical Properties Research Center deserves the gratitude of every scientist and engineer who uses these compiled data.

The individual nontechnical citizen of the United States has a stake in this work also, for much of the science and technology that contributes to his well-being relies on the use of these data. Indeed, recognition of this importance is indicated by a mere reading of the list of the financial sponsors of the Thermophysical Properties Research Center; leaders of the technical industry of the United States and agencies of the Federal Government are well represented.

Experimental measurements made in a laboratory have many potential applications. They might be used, for example, to check a theory, or to help design a chemical manufacturing plant, or to compute the characteristics of a heat exchanger in a nuclear power plant. The progress of science and technology demands that results be published in the open literature so that others may use them. Fortunately for progress, the useful data in any single field are not scattered throughout the tens of thousands of technical journals published throughout the world. In most fields, fifty percent of the useful work appears in no more than thirty or forty journals. However, in the case of TPRC, its field is so broad that about 100 journals are required to yield fifty percent. But that other fifty percent! It is scattered through more than 3500 journals and other documents, often items not readily identifiable or obtainable. Nearly 50,000 references are now in the files.

Thus, the man who wants to use existing data, rather than make new measurements himself, faces a long and costly task if he wants to assure himself that he has found all the relevant results. More often than not, a search for data stops after one or two results are found—or after the searcher decides he has spent enough time looking. Now with the appearance of these volumes, the scientist or engineer who needs these kinds of data can consider himself very fortunate. He has a single source to turn to; thousands of hours of search time will be saved, innumerable repetitions of measurements will be avoided, and several billions of dollars of investment in research work will have been preserved.

However, the task is not ended with the generation of these volumes. A critical evaluation of much of the data is still needed. Why are discrepant results obtained by different experimentalists? What undetected sources of systematic error may affect some or even all measurements? What value can be derived as a "recommended" figure from the various conflicting values that may be reported? These questions are difficult to answer, requiring the most sophisticated judgment of a specialist in the field. While a number of the volumes in this Series do contain critically evaluated and recommended data, these are still in the minority. The data are now being more intensively evaluated by the staff of TPRC as an integral part of the effort of the National Standard Reference Data System (NSRDS). The task of the National Standard Reference Data System is to organize and operate a comprehensive program to prepare compilations of critically evaluated data on the properties of substances. The NSRDS is administered by the National Bureau of Standards under a directive from the Federal Council for Science and Technology, augmented by special legislation of the Congress of the United States. TPRC is one of the national resources participating in the National Standard Reference Data System in a united effort to satisfy the needs of the technical community for readily accessible, critically evaluated data.

As a representative of the NBS Office of Standard Reference Data, I want to congratulate Professor Touloukian and his colleagues on the accomplishments represented by this Series of reference data books. Scientists and engineers the world over are indebted to them. The task ahead is still an awesome one and I urge the nation's private industries and all concerned Federal agencies to participate in fulfilling this national need of assuring the availability of standard numerical reference data for science and technology.

EDWARD L. BRADY
Associate Director for Information Programs
National Bureau of Standards

Preface

Thermophysical Properties of Matter, the TPRC Data Series, is the culmination of twelve years of pioneering effort in the generation of tables of numerical data for science and technology. It constitutes the restructuring, accompanied by extensive revision and expansion of coverage, of the original *TPRC Data Book*, first released in 1960 in loose-leaf format, 11″ × 17″ in size, and issued in June and December annually in the form of supplements. The original loose-leaf *Data Book* was organized in three volumes: (1) metallic elements and alloys, (2) nonmetallic elements, compounds, and mixtures which are solid at N.T.P., and (3) nonmetallic elements, compounds, and mixtures which are liquid or gaseous at N.T.P. Within each volume, each property constituted a chapter.

Because of the vast proportions the *Data Book* began to assume over the years of its growth and the greatly increased effort necessary in its maintenance by the user, it was decided in 1967 to change from the loose-leaf format to a conventional publication. Thus, the December 1966 supplement of the original *Data Book* was the last supplement disseminated by TPRC.

While the manifold physical, logistic, and economic advantages of the bound volume over the loose-leaf oversize format are obvious and welcome to all who have used the unwieldy original volumes, the assumption that this work will no longer be kept on a current basis because of its bound format would not be correct. Fully recognizing the need of many important research and development programs which require the latest available information, TPRC has instituted a *Data Update Plan* enabling the subscriber to inquire, by telephone if necessary, for specific information and receive, in many instances, same-day response on any new data processed or revision of published data since the latest edition. In this context, the TPRC Data Series departs drastically from the conventional handbook and giant multivolume classical works, which are no longer adequate media for the dissemination of numerical data of science and technology without a continuing activity on contemporary coverage. The loose-leaf arrangements of many works fully recognize this fact and attempt to develop a combination of bound volumes and loose-leaf supplement arrangements as the work becomes increasingly large. TPRC's *Data Update Plan* is indeed unique in this sense since it maintains the contents of the TPRC Data Series current and live on a day-to-day basis between editions. In this spirit, I strongly urge all purchasers of these volumes to complete in detail and return the *Volume Registration Certificate* which accompanies each volume in order to assure themselves of the continuous receipt of annual listing of corrigenda during the life of the edition.

The TPRC Data Series consists initially of 13 independent volumes. The initial ten volumes will be published in 1970, and the remaining three by 1972. It is also contemplated that subsequent to the first edition, each volume will be revised, updated, and reissued in a new edition approximately every fifth year. The organization of the TPRC Data Series makes each volume a self-contained entity available individually without the need to purchase the entire Series.

The coverage of the specific thermophysical properties represented by this Series constitutes the most comprehensive and authoritative collection of numerical data of its kind for science and technology.

Whenever possible, a uniform format has been used in all volumes, except when variations in presentation were necessitated by the nature of the property or the physical state concerned. In spite of the wealth of data reported in these volumes, it should be recognized that all volumes are not of the same degree of completeness. However, as additional data are processed at TPRC on a continuing basis, subsequent editions will become increasingly more complete and up to date. Each volume in the Series basically comprises three sections, consisting of a text, the body of numerical data with source references, and a material index.

The aim of the textual material is to provide a complementary or supporting role to the body of numerical data rather than to present a treatise on the subject of the property. The user will find a basic theoretical treatment, a comprehensive presentation of selected works which constitute reviews, or compendia of empirical relations useful in estimation of the property when there exists a paucity of data or when data are completely lacking. Established major experimental techniques are also briefly reviewed.

The body of data is the core of each volume and is presented in both graphical and tabular format for convenience of the user. Every single point of numerical data is fully referenced as to its original source and no secondary sources of information are used in data extraction. In general, it has not been possible to critically scrutinize all the original data presented in these volumes, except to eliminate perpetuation of gross errors. However, in a significant number of cases, such as for the properties of liquids and gases and the thermal conductivity of all the elements, the task of full evaluation, synthesis, and correlation has been completed. It is hoped that in subsequent editions of this continuing work, not only new information will be reported but the critical evaluation will be extended to increasingly broader classes of materials and properties.

The third and final major section of each volume is the material index. This is the key to the volume, enabling the user to exercise full freedom of access to its contents by any choice of substance name or detailed alloy and mixture composition, trade name, synonym, etc. Of particular interest here is the fact that in the case of those properties which are reported in separate companion volumes, the material index in each of the volumes also reports the contents of the other companion volumes.* The sets of companion volumes are as follows:

Thermal conductivity:	Volumes 1, 2, 3
Specific heat:	Volumes 4, 5, 6
Radiative properties:	Volumes 7, 8, 9
Thermal expansion:	Volumes 12, 13

The ultimate aims and functions of TPRC's Data Tables Division are to extract, evaluate, reconcile, correlate, and synthesize all available data for the thermophysical properties of materials with the result of obtaining internally consistent sets of property values, termed the "recommended reference values." In such work, gaps in the data often occur, for ranges of temperature, composition, etc. Whenever feasible, various techniques are used to fill in such missing information, ranging from empirical procedures to detailed theoretical calculations. Such studies are resulting in valuable new estimation methods being developed which have made it possible to estimate values for substances and/or physical conditions presently unmeasured or not amenable to laboratory investigation. Depending on the available information for a particular property and substance, the end product may vary from simple tabulations of isolated values to detailed tabulations with generating equations, plots showing the concordance of the different values, and, in some cases, over a range of parameters presently unexplored in the laboratory.

*For the first edition of the Series, this arrangement was not feasible for Volume 7 due to the sequence and the schedule of its publication. This situation will be resolved in subsequent editions.

The TPRC Data Series constitutes a permanent and valuable contribution to science and technology. These constantly growing volumes are invaluable sources of data to engineers and scientists, sources in which a wealth of information heretofore unknown or not readily available has been made accessible. We look forward to continued improvement of both format and contents so that TPRC may serve the scientific and technological community with ever-increasing excellence in the years to come. In this connection, the staff of TPRC is most anxious to receive comments, suggestions, and criticisms from all users of these volumes. An increasing number of colleagues are making available at the earliest possible moment reprints of their papers and reports as well as pertinent information on the more obscure publications. I wish to renew my earnest request that this procedure become a universal practice since it will prove to be most helpful in making TPRC's continuing effort more complete and up to date.

It is indeed a pleasure to acknowledge with gratitude the multisource financial assistance received from over fifty of TPRC's sponsors which has made the continued generation of these tables possible. In particular, I wish to single out the sustained major support being received from the Air Force Materials Laboratory–Air Force Systems Command, the Office of Standard Reference Data–National Bureau of Standards, and the Office of Advanced Research and Technology–National Aeronautics and Space Administration. TPRC is indeed proud to have been designated as a National Information Analysis Center for the Department of Defense as well as a component of the National

Standard Reference Data System under the cognizance of the National Bureau of Standards.

While the preparation and continued maintenance of this work is the responsibility of TPRC's Data Tables Division, it would not have been possible without the direct input of TPRC's Scientific Documentation Division and, to a lesser degree, the Theoretical and Experimental Research Divisions. The authors of the various volumes are the senior staff members in responsible charge of the work. It should be clearly understood, however, that many have contributed over the years and their contributions are specifically acknowledged in each volume. I wish to take this opportunity to personally thank those members of the staff, research assistants, graduate research assistants, and supporting graphics and technical typing personnel without whose diligent and painstaking efforts this work could not have materialized.

Y. S. Touloukian

Director
Thermophysical Properties Research Center
Distinguished Atkins Professor of Engineering

Purdue University
Lafayette, Indiana
July 1969

Introduction to Volume 4

This volume of *Thermophysical Properties of Matter*, the TPRC Data Series, was initiated in recent years and follows the general format of the Center's work on thermal conductivity.

The volume comprises three major sections: the front text material together with its bibliography, the main body of numerical data and its references, and the material index.

The text material is intended to assume a role complementary to the main body of numerical data, the presentation of which is the primary purpose of this volume. It is felt that a concise discussion of the theoretical nature of the property under consideration together with a review of predictive procedures and recognized experimental techniques will be appropriate in a major reference work of this kind. The extensive reference citations given in the text should lead the interested reader to a highly comprehensive literature for a detailed study. It is hoped, however, that enough detail is presented for this volume to be self-contained for the practical user.

The main body of the volume consists of the presentation of numerical data compiled over the years in a most comprehensive and meticulous manner. The scope of coverage includes the metallic elements and most metallic alloys of engineering importance. The extraction of all data directly from their original sources ensures freedom from errors of transcription. Furthermore, some gross errors appearing in the original source documents have been corrected. The organization and presentation of the data together with other pertinent information in the use of the tables and figures are discussed in detail in the text of the section entitled *Numerical Data.*

It is regrettable that the authors have not yet had the time to review and evaluate critically the extensive data compiled in this volume. However, it is hoped that the user will be able to exercise proper selectivity and discretion among conflicting sets of data based on the extensive information reported for each set in the accompanying specification tables.

As stated earlier, all data have been obtained from their original sources and each data set is so referenced. TPRC has in its files all documents cited in this volume. Those that cannot be readily obtained elsewhere are available from TPRC in microfiche form.

The material index at the end of the volume covers the contents of all three companion volumes (Volumes 4, 5, and 6) on specific heat. It is hoped that the user will find these comprehensive indices helpful.

This volume has grown out of activities made possible principally through the support of the Air Force Materials Laboratory–Air Force Systems Command, under the monitorship of Mr. John H. Charlesworth. In the preparation of Volume 4 we have drawn most heavily upon the scientific literature and hence we feel a debt of gratitude to the authors of the referenced articles.

While this volume is primarily intended as a reference work for the designer, researcher, experimentalist, and theoretician, the teacher at the graduate level may also use it as a teaching tool to point out to his students the topography of the state of knowledge on the specific heat of metals. We believe there is also much food for reflection by the specialist and the academician concerning the meaning of "original" investigation and its "information content."

The authors are keenly aware of the possibility of many weaknesses in a work of this scope. We hope that we will not be judged too harshly and that we will receive the benefit of suggestions regarding references omitted, additional material groups needing more detailed treatment, improvements in presentation, and, most important, any inadvertent errors. If the *Volume Registration Certificate* accompanying this volume is returned, the reader will assure himself of receiving annually a list of corrigenda as possible errors come to our attention.

Lafayette, Indiana
July 1969

Y. S. Touloukian
E. H. Buyco

Contents

Material Index

GROUPING OF MATERIALS AND LIST OF FIGURES AND TABLES

1. ELEMENTS

1. ELEMENTS (continued)

2. NONFERROUS BINARY ALLOYS

2. NONFERROUS BINARY ALLOYS (continued)

2. NONFERROUS BINARY ALLOYS (continued)

3. NONFERROUS MULTIPLE ALLOYS

3. NONFERROUS MULTIPLE ALLOYS (continued)

4. FERROUS ALLOYS

A. CARBON STEELS GROUP I

B. CARBON STEELS GROUP II

C. ALLOY STEELS GROUP I

D. ALLOY STEELS GROUP II

4. FERROUS ALLOYS (continued)

D. ALLOY STEELS GROUP II (continued)

Theory, Estimation, and Measurement

Notation

A Grüneisen constant; Cross-sectional area
a Lattice constant: Empirical constant
b Empirical constant
c, C Heat capacity of mass m, specific heat per unit mass
C_a, C_f Constant which depends on particular type of lattice and on crystal structure, respectively
C_e Electronic specific heat
C_p, C_v Specific heat at constant pressure and constant volume, respectively
d Density
e Base of natural logarithm, 2.71828
E Total energy of an oscillator, particle, or system; Internal energy; Voltage
H Enthalpy
$(\Delta H)_f$ Heat of fusion
h Planck constant, 6.6262×10^{-27} erg sec
I Electrical current
J, J' Quantum mechanical exchange constants
K Calibration factor in ice drop calorimeter
k Boltzmann constant, 1.3806×10^{-16} erg K^{-1}
L Linear dimension
m Mass of a particle, system, or specimen
m_e Mass of an electron
n Integer, 0, 1, 2, 3, . . .
N_A Avogadro's number, 6.0222×10^{23} g-mol^{-1}
N_e Number of electrons per gram atom
p Momentum of a particle; Pressure of a gas
q Direction coordinate from equilibrium position
Q Amount of heat absorbed or removed from the system
R Gas constant, 8.3143 J K^{-1} g-mol^{-1}
s Spin vector
T Temperature, K
t Time
V Volume
v Specific volume
W Work done on or by the system
x, x_m $h\nu/kT$ and $h\nu_D/kT$, respectively, as used in equation (17)
X_i Atomic mole or mass fraction of ith component in an alloy or mixture
α, α_f Coefficient of thermal linear expansion, and a constant which depends on crystal structure, respectively
β Coefficient of isobaric volumetric expansion; Constant in Debye cube law
γ Constant in the electronic specific heat relation (26)
θ_D, θ_E Characteristic Debye temperature and Einstein temperature, $h\nu_D/k$ and $h\nu/k$, respectively
ν Frequency of oscillation of a particle
ν_D Debye frequency
ω Natural angular frequency
ρ Electrical resistivity
ρ_e Number of free electrons per unit volume
ϵ Energy of an oscillation
π Mathematical constant, 3.14159 . . .
κ_T Isothermal compressibility, as used in equation (36)

Notation

Theory of Specific Heat of Solids

1. INTRODUCTION

Rapid advances in the frontiers of science and technology have brought about a general realization of the fact that the present limitations in many technical developments are a direct result of inadequate knowledge of the thermophysical properties of materials. In the high-temperature range ($T > 1000$ K), interest in the determination of specific heats of materials has been hastened because of the requirements in space programs as well as industrial applications. The need for data at high temperatures has advanced our knowledge in many areas of solid state studies such as lattice vibrations, energy levels in magnetic solids, electronic distributions, and many other atomic and molecular phenomena.

The measurement of specific heat at cryogenic temperatures ($C_p \cong C_v$ for $T \leq 4$ K) provides us with a direct means to test theoretical models of a system. For instance, precise specific heat measurements were needed to test the validity of Debye's and Einstein's theory for specific heat of solids at low temperatures. Finally, knowledge of accurate specific heat data at low temperature is very useful in studies of cryogenic techniques.

2. DEFINITIONS

When a quantity of heat Q is added to a system so that there is a change in temperature, $T_2 - T_1$, then the mean heat capacity of the mass m of the substance is defined by

$$\bar{c} = \frac{Q}{T_2 - T_1} \tag{1}$$

The limiting value of the above ratio as the temperature changes by dT is defined as the true heat capacity, i.e.,

$$c = \frac{đQ^*}{dT} \tag{2}$$

*$đQ$ is used instead of dQ to indicate that it is not an exact differential.

In order to obtain a quantity that is independent of the mass, m, of a substance, equation (2) is divided by m; i.e.,

$$C = \frac{c}{m} = \frac{đQ}{m\,dT} \tag{3}$$

The quantity q represents the amount of heat per unit mass, so that equation (3) may also be written as

$$C = \frac{đq}{dT} \tag{4}$$

Raising the temperature of a unit mass of a substance by an amount dT, however, does not define the process in a thermodynamic sense; for instance, it will take a different amount of heat $đq$ if the process is at constant pressure than when the process is at constant volume. As a matter of fact there are an infinite number of different processes for a system at temperature T to change to a temperature $T + dT$. It is clear, therefore, that an infinite number of specific heats could also be defined for a substance. The two processes that are most commonly used in thermodynamics are those at constant volume and constant pressure. For these two processes equation (4) may be written

$$C_p = \left(\frac{đq}{dT}\right)_p \tag{5}$$

and

$$C_v = \left(\frac{đq}{dT}\right)_v \tag{6}$$

Experimentally, the values of the specific heat measured are either at constant pressure, C_p, or at constant volume, C_v. The units most commonly used for specific heat are cal g^{-1} K^{-1}, Btu lb^{-1} F^{-1}, joules kg^{-1} K^{-1}. The units for molar or atomic specific heat are cal g-mol^{-1} K^{-1}, Btu lb-mol^{-1} F^{-1}, joules kg-mol^{-1} K^{-1}, cal g-atom^{-1} K^{-1}, joules kg-atom^{-1} K^{-1}, etc.

3. DULONG AND PETIT'S LAW

In 1819 Dulong and Petit [9] published the results of their measurements on the specific heat at constant pressure of thirteen solid elements at room temperature. From these measurements, they observed that the product of the specific heat at constant pressure and the atomic weight was approximately a constant, about 6 cal g-atom^{-1} K^{-1}. Subsequent researches, extending from 1840 to 1862, revealed the general applicability of the Dulong and Petit's law to several metallic elements, when the specific heat at constant pressure was determined at temperatures sufficiently below their melting point but not far below room temperature. During the same period an important extension of Dulong and Petit's law was applied to chemical compounds, i.e., the molar specific heat of a compound is equal to the sum of the atomic specific heats of its constituent elements. This law which is generally referred to as the Kopp–Neumann law [32] has also been applied to predict the atomic specific heat of alloys. For alloys, the atomic specific heat is equal to the sum of the product of the atomic specific heat of each constituent element and its atomic fraction. If an alloy consists of elements 1, 2, 3, . . . , n, with atomic fraction $X_1, X_2, X_3, \ldots, X_n$ and atomic specific heat $C_{p1}, C_{p2}, C_{p3}, \ldots, C_{pn}$, then the atomic specific heat of the alloy is

$$C_p = \sum_{i=1}^{n} X_i C_{p_i} \tag{7}$$

Equation (7) should be applied with caution for alloys especially near magnetic and phase transitions. Bottema and Jaeger [5] have applied the Kopp–Neumann law to the alloy Ag_3Au and they found that the experimental data on the specific heat at constant pressure of this alloy agree closely with the calculated values between 0 C to 400 C. Between 400 C and 800 C, the values obtained from the Kopp–Neumann law were 0.5 percent to 1.8 percent higher than the experimental results. Buyco [46] calculated the specific heat of the alloys of aluminum, beryllium, nickel, and iron between 300 K to 1000 K and found the calculated values agree with the experimental data to within 5 percent.

The theoretical justification of the law of Dulong and Petit was demonstrated by Boltzmann in 1871. The results obtained previously by Dulong and Petit also follow from Boltzmann's equipartition of energy theorem. Complete and detailed derivation of this theorem is discussed elsewhere [15, 20, 21, 33, and 43].

The following is a brief exposition. The energy of a linear harmonic oscillator consists of kinetic and potential energies, i.e.,

$$E = \frac{p^2}{2m} + \frac{m\omega^2 q^2}{2} \tag{8}$$

where p is the momentum, m is the mass, ω is the natural angular frequency, q is the distance from equilibrium position, and E is the total energy of an oscillator. From the theorem of equipartition of energy [15, 20, 21, 31], each degree of freedom contributes $(kT/2)$ to the energy of a particle in equilibrium. A three-dimensional oscillator which has six degrees of freedom will therefore have an internal energy of 3 kT at thermal equilibrium. A gram-atom of an element has N_A atoms; hence, the internal energy is $3N_AkT$. The specific heat at constant volume is obtained by differentiating the internal energy with respect to temperature at constant volume, i.e.,

$$\left(\frac{\partial E}{\partial T}\right)_v = C_v = 3N_A k \tag{9}$$

where N_A is the Avogadro constant and k is the Boltzmann constant. The product of Avogadro constant and Boltzmann constant is equal to the gas constant R. Therefore:

$$C_v = 3R \cong 5.96 \text{ cal mol}^{-1}\text{ K}^{-1}$$

Hence, the Dulong and Petit value of about 6 cal mol^{-1} deg^{-1} for the specific heat of metallic solids can be accounted for on the basis of classical statistical mechanics. However, the observation of Dulong and Petit was short lived. In 1875 Weber [48] showed that the atomic specific heat of silicon, boron, and carbon are considerably lower than the values predicted by Dulong and Petit. For example, the atomic specific heat of crystalline silicon, boron, and diamond were found to be 4.8, 2.7, and 1.8 cal mol^{-1} deg^{-1}, respectively, at room temperature. Subsequent specific heat measurements at low temperatures ($T < 300$ K) revealed that the specific heat of solids increased rapidly with temperature and almost leveled off about their Debye temperature. Classical theory does not explain this behavior for solids. It should also be noted that classical theory encounters the same difficulty in the behavior of molar specific heats.

4. EINSTEIN'S SPECIFIC HEAT THEORY

Einstein [10] proposed a simple model to account

for the decrease in the specific heat at low temperatures below the value $3R$ per mole which was obtained at elevated temperatures. His oversimplified physical model considers the thermal properties of the vibrations of a lattice of N_A atoms as a set of $3N_A$ independent harmonic oscillators in one dimension, each with the same frequency, ν. He then quantized the energy of the oscillators in accordance with the results obtained by Planck. According to Planck, a harmonic oscillator does not have a continuous energy spectrum but can accept energy values equal to an integer times $h\nu$, where ν is the frequency of oscillations and h is the Planck constant. Hence the possible energy levels of an oscillator may be given by

$$\epsilon = nh\nu \qquad n = 0, 1, 2, 3, \ldots$$

The average energy of an oscillator at temperature T, according to the well known Planck formula [7, 20, 21, 32], is

$$\bar{\epsilon} = \frac{h\nu}{\exp(h\nu/kT) - 1} \tag{10}$$

In Einstein's model the vibrational energy of a solid element containing N_A atoms is $3N_A$ times the average energy of an oscillator, i.e.,

$$\bar{E} = 3N_A \frac{h\nu}{\exp(h\nu/kT) - 1} \tag{11}$$

The results obtained from quantum mechanics however showed that the average energy of an oscillator [7, 15] should be written as

$$\bar{\epsilon} = \frac{h\nu}{2} + \frac{h\nu}{\exp(h\nu/kT) - 1} \tag{12}$$

instead of as in equation (10).

The result obtained for the specific heat by differentiating equation (10) is the same as that obtained from equation (12). In any case the specific heat for one atom of an element is

$$\left(\frac{\partial E}{\partial T}\right)_v = C_v = \frac{3N_A k (h\nu/kT)^2 \exp(h\nu/kT)}{[\exp(h\nu/kT) - 1]^2} \tag{13}$$

For convenience, the characteristic Einstein temperature defined by $\theta_E = h\nu/k$ may be introduced in equation (13) to obtain

$$C_v = \frac{3R(\theta_E/T)^2 \exp(\theta_E/T)}{[\exp(\theta_E/T) - 1]^2} \tag{14}$$

In the high-temperature range with $T \gg \theta_E$ [15, 20, 21, 32], equation (14) upon expansion in power series becomes

$$C_v \cong 3R\left[1 - \frac{1}{12}\left(\frac{\theta_E}{T}\right)^2\right] \tag{15}$$

When the value of $[(\theta_E/T)^2/12]$ is such that it is very much smaller than 1, then Einstein's theory yields the classical Dulong and Petit value of 6 cal mol^{-1} deg^{-1}.

In the low-temperature region $T \ll \theta_E$, equation (14) may be written approximately as

$$C_v \cong 3R\left(\frac{\theta_E}{T}\right)^2 \exp(-\theta_E/T) \tag{16}$$

According to equation (16), the low-temperature specific heat of solids should approach zero exponentially. Experimental evidence indicates that C_v approaches zero more slowly than this. The reason for the discrepancy between Einstein's theoretical prediction and the experimental results may be explained on the basis of the assumption made in the theory that each atom in a solid vibrates independently of the others but with precisely the same frequency. However, in spite of the weakness in Einstein's theory, his pioneering work opened the way for the application of quantum theory to the specific heat of solids.

5. DEBYE'S SPECIFIC HEAT THEORY

From the point of view of the wave whose wavelength is large compared with the interatomic distances, a crystal may appear like a continuum. The fundamental assumption of Debye [6] is that the continuum model may be employed for all possible vibrational modes of the crystal. Debye has given a limit to the total number of vibrational modes equal to $3N_A$, where N_A is the number of atoms in a gram atom of an element. In this case, the frequency spectrum which corresponds to an ideal continuum is cut off in order to comply with a total of $3N_A$ modes. This procedure should provide a maximum frequency ν_D (Debye frequency) which is common to both the longitudinal and transverse modes. By associating with each vibrational mode a harmonic oscillator of the same frequency, Debye obtained the following expression [7, 15, 20, 21, 32] for the vibrational energy:

$$\bar{E} = 9N_A h\nu_D \left(\frac{kT}{h\nu_D}\right)^4 \int_0^{x_m} \frac{x^3\,dx}{e^x - 1} \tag{17}$$

where

$$x = h\nu/kT \qquad x_m = h\nu_D/kT$$

Clearly, when $T \gg \theta_D$, x_m is small compared with unity for the whole integration range. In this case $e^x - 1 \cong x$ so that equation (17) could easily be integrated to obtain the expression

$$\bar{E} \cong 3N_A kT \tag{18}$$

Then

$$\left(\frac{\partial \bar{E}}{\partial T}\right)_v = C_v = 3N_A k = 3R \cong 6\,\text{cal mol}^{-1}\,\text{deg}^{-1}$$

a result agreeing with classical theory.

At very low temperatures, $T \ll \theta_D$, the upper limit of integration in equation (17) may be replaced by infinity since $h\nu/kT \to \infty$ as $T \to 0$. It is now possible to integrate equation (17) as follows [51]

$$\int_0^\infty \frac{x^3\,dx}{e^x - 1} = 6\sum_1^\infty \frac{1}{n^4} = \frac{\pi^4}{15} \tag{19}$$

Hence

$$\bar{E} = \frac{3}{5}\pi^4 N_A kT\left(\frac{T}{\theta_D}\right)^3 \tag{20}$$

and

$$C_v = \left(\frac{\partial \bar{E}}{\partial T}\right)_v = \frac{12}{5}\pi^4 N_A k\left(\frac{T}{\theta_D}\right)^3 \tag{21}$$

or

$$C_v = \frac{12}{5}\pi^4 R\left(\frac{T}{\theta_D}\right)^3 \tag{22}$$

For one atom or one mole of a substance, $R = 1.987$ cal mol^{-1} deg^{-1} so that equation (22) may be written as

$$C_v = 464.5\left(\frac{T}{\theta_D}\right)^3 \text{cal mol}^{-1}\,\text{deg}^{-1} \qquad T < \left(\frac{\theta_D}{50}\right) \tag{23}$$

Debye's theory predicts a cube law dependence of the specific heat of the elements for temperatures $T < (\theta_D/10)$. The range of validity of this law [15] has now been restricted to $T < (\theta_D/50)$ as a result of more recent theoretical work on specific heat studies. The predictions of Debye's theory agree quite well with experimental values of the specific heat of solids and is a definite improvement over Einstein's work.

Due to improved calorimetric measurements at low temperatures ($T < 5$ K), in recent years accurate specific heat values revealed that Debye's equation for C_v does not fit the experimental results precisely. Furthermore, it was observed that θ_D, which according to Debye's theory is a constant, did in fact vary with temperature. The deficiency of the Debye theory may be explained on the basis of the approximation made in treating solids as a continuous elastic media and neglecting the discreteness of the atoms.

Further improvements on Debye's theory was developed by Born and Karman [4]. They calculated the frequency spectrum by considering the lattice modes of vibration for a particular crystal structure under investigation. The method is involved so that one is referred to the original work [4] for detailed discussion.

6. ELECTRONIC SPECIFIC HEAT

In 1900, Drude [8] suggested a model for a free-electron theory of metals. He assumed that metals contain free electrons in thermal equilibrium with the atoms of the solid. He further assumed that the potential energy of the free electrons is equal to the product of the number of electrons per unit volume and the average energy of an electron. The essential feature in the problem is the determination of the number of electrons with energy between E and $E + dE$. Classical theory using Maxwell–Boltzmann statistics [2, 8, 15, 20, 21, 32, 43], would give an expression for the electronic specific heat as

$$C_e = \tfrac{3}{2}N_e k \tag{24}$$

Using Fermi–Dirac statistics [7, 15, 19, 20, 21, 31, 32], the following expression for the electronic specific heat may be obtained at low temperatures:

$$C_e = \pi^2 R(2m_e k/h^2)\left(\frac{\pi}{3\rho_e}\right)^{2/3} T \tag{25}$$

or simply

$$C_e = \gamma T \tag{26}$$

where ρ_e is the number of free electrons per unit volume, γ is the proportionality constant, T is the absolute temperature, N_e is the number of electrons per gram atom, m_e is the mass of an electron, k is the Boltzmann constant, h is the Planck constant, R is the gas constant, and C_e is the electronic specific heat.

The specific heat of metals below the Debye temperature and "very much" below the Fermi temperature [15, 19, 20, 21, 32] may be expressed as the sum of the electronic specific heat and the lattice specific heat, i.e.,

$$C_v = \gamma T + \beta T^3 \tag{27}$$

Indeed, this relationship has been verified by accurate low temperature specific heat measurements. At sufficiently low temperature ($T < 1$ K) the electronic specific heat is dominant, while at high temperatures the lattice contribution is predominant.

7. MAGNETIC SPECIFIC HEAT

There are two types of materials that exhibit a magnetic contribution to the total specific heat: namely, the ferromagnetic and the ferrimagnetic materials.

A ferromagnet is a material [7, 15, 20, 21, 32] that contains a spontaneous magnetic moment. This means that this material possesses a magnetic moment even in the absence of an external magnetic field. This type of material exhibits a magnetic ordering with parallel alignment of adjacent spins. A ferromagnetic material has a Curie temperature, T_c, which is defined as the temperature above which magnetization disappears, and the material becomes paramagnetic. The Curie temperature separates the ordered ferromagnetic phase from the disordered paramagnetic phase.

An antiferromagnet is a material [7, 15, 20, 21, 32], that has spins which are ordered in an antiparallel arrangement. There is no net magnetic moment at temperatures below the Néel temperature. Hysteresis is usually observed and a sharp maximum in the susceptibility curve is exhibited. Above the Néel temperature, the spins are said to be free, and the material becomes paramagnetic. In some ways ferrimagnetic materials are similar to the ferromagnetic materials except that in the former the adjacent spins are unequal and antiparallel. The Néel temperature may be defined for ferrimagnetic material as the temperature separating the ordered ferrimagnetic phase from the disordered paramagnetic phase.

For ferri- and ferromagnets, the internal energy [7, 15, 20, 21, 32], is given by the expression

$$\bar{E} = 4\pi V(2\alpha_f Jsa^2)\left(\frac{kT}{2\alpha_f Jsa^2}\right)^{5/2} \int_0^x \frac{x^4\,dx}{e^{x^2} - 1} \tag{28}$$

At low temperatures the upper limit for x may be taken equal to infinity and hence the integral may be easily determined. Differentiating equation (28) gives the magnetic specific heat [15]

$$C_M = \frac{d\bar{E}}{dT} = C_f N_A k \left(\frac{kT}{2Js}\right)^{3/2} \tag{29}$$

where α_f and C_f are constants which depend upon crystal structure, a is the lattice constant, J is the quantum mechanical exchange constant, k is the Boltzmann constant, N_A is the Avogadro number, s is the magnitude of the spin vector, and V is the volume of the material.

Equation (29) shows that at low temperatures the ferromagnetic contribution to the specific heat is proportional to the three-halves power of the absolute temperature. For metals which are ferromagnetic [15], the total specific heat is equal to the sum of the electronic, lattice, and magnetic terms, i.e.,

$$C_v = \gamma T + \beta T^3 + \delta T^{3/2} \tag{30}$$

For ferrimagnets, which are electrical insulators, [15], the electronic term is negligible compared with the other terms, so that the total specific heat may be given by the expression

$$C_v = \beta T^3 + \delta T^{3/2} \tag{31}$$

Both sides of equation (31) may be divided by $T^{3/2}$ to give

$$C_v / T^{3/2} = \beta T^{3/2} + \delta \tag{32}$$

A plot of $C_v/T^{3/2}$ versus $T^{3/2}$ should give a straight line with slope β and intercept δ.

For the case of antiferromagnetic materials [15], the expressions for the mean internal energy is

$$\bar{E} = 4\pi V(2\alpha_a J' sa^2)\left(\frac{kT}{2\alpha_a J' sa^2}\right)^4 \int_0^x \frac{x^3\,dx}{e^x - 1} \tag{33}$$

The upper limit for integration may be taken as equal to infinity at low temperatures so that differentiation of equation (33) gives the magnetic specific heat [15, 28]

$$C_M = C_a N_A k \left(\frac{kT}{2J's}\right)^3 \tag{34}$$

where C_a is a constant which depends upon the type of lattice and J' is the magnitude of the exchange constant.

The striking difference between the contributions to the specific heat exhibited by ferromagnets and ferrimagnets is the $T^{3/2}$ dependence in the former and T^3 dependence in the latter. Hence for antiferromagnetic materials, the temperature dependence is of the same form as the Debye's T^3 formula. The separation of the spin wave contribution from the lattice specific heat in antiferromagnetic materials is indeed very difficult.

8. LOW-TEMPERATURE SPECIFIC HEAT

The specific heat of solids is ordinarily measured at constant pressure. The specific heat at constant volume is that which is obtained if the interatomic distance is kept constant as the temperature changes. The specific heat at constant volume, C_v, may be assumed to be approximately equal to the specific heat at constant pressure, C_p, at cryogenic temperatures. At high temperatures, $C_p > C_v$. This difference is obtained from the classical thermodynamic relations

$$C_p - C_v = -T\left(\frac{\partial V}{\partial T}\right)_p^2 \Big/ \left(\frac{\partial V}{\partial p}\right)_T \quad (35)$$

From the definition of the isothermal compressibility

$$\kappa_T = -\left(\frac{\partial V}{\partial p}\right)_T \Big/ V \quad (36)$$

and the isobaric coefficient of volumetric expansion

$$\beta = \left(\frac{\partial V}{\partial T}\right)_p \Big/ V \quad (37)$$

Using equations (36) and (37), equation (35) may be written as

$$C_p - C_v = \frac{TV\beta^2}{\kappa_T} \quad (38)$$

By rearranging equation (38), this may also be written as

$$C_p - C_v = \left(\frac{V\beta^2}{\kappa_T C_p^2}\right) C_p^2 T = AC_p^2 T \quad (39)$$

where

$$A = \frac{V\beta^2}{\kappa_T C_p^2}$$

The parameter A is called the Grüneisen constant, which is actually only approximately constant [15] over a wide range of temperature. If A is calculated at any one temperature from values of V, β, and κ_T, it may be used [15, 20, 21, 32] to calculate $C_p - C_v$ over a wide range of temperature without introducting a serious error.

For isotropic substances, the isothermal coefficient of volumetric expansion may be written in terms of the coefficient of linear expansion

$$\beta = \left(\frac{\partial V}{\partial T}\right)_p \Big/ V = 3\left[\left(\frac{\partial L}{\partial T}\right)_p \Big/ L\right] = 3\alpha \quad (40)$$

Hence, from equation (38)

$$C_p - C_v = \frac{9\alpha^2}{\kappa_T} TV = \left(\frac{9V\alpha^2}{\kappa_T C_p^2}\right) C_p^2 T \quad (41)$$

where

$$A = \frac{9V\alpha^2}{\kappa_T C_p^2}$$

In the absence of contributions from magnetic and nuclear specific heat, the expression for C_v for most metals has been shown [15, 20, 21, 32] to be

$$C_v = \gamma T + \beta T^3 \quad (27)$$

where γT is the electronic contribution and βT^3 is the lattice contribution. For nonmetals, the electronic contribution may be very small compared with the lattice term so that

$$C_v = \beta T^3 \quad (42)$$

When the nuclear quadrupole moment interacts with the electronic field gradient of the lattice and the electron, then the total specific heat of the substance is given as

$$C_v = \gamma T + \beta T^3 + \alpha T^{-2} \quad (43)$$

where αT^{-2} is the nuclear contribution to the total specific heat.

9. NORMAL AND SUPERCONDUCTING MATERIALS

At a certain critical temperature (superconducting temperature), several materials exhibit superconducting behavior [15, 20, 21, 32]. Below this temperature, the specific heat of a superconducting material is found to depart significantly from the values obtained for a normally behaving material. It is also found that if an external magnetic field of sufficient strength is applied while the specific heat of the material is being measured, the values obtained correspond to what the normal values would be. Hence, the specific heat values obtained experimentally in the presence of sufficient external magnetic field below the superconducting critical temperature are referred to as the normal specific heat (C_N) while the values obtained in the absence of a magnetic field are referred to as superconducting specific heat (C_S). For example, the critical superconducting temperatures of aluminum and niobium are approximately 1.196 K and 9.22 K, respectively.

Other Major Sources of Data

There exists in the literature a number of reference sources which, while less extensive in scope, may nevertheless prove valuable to the reader. While it is not the intent here to cite every available review, it is felt that the following works, listed in chronological order, are of particular significance. One should note that most of the citations do not present critical evaluation of the data they report.

Furukawa, Saba, and Reilly [12] report on the critical analysis of the thermodynamic properties of copper, silver, and gold between 0 and 300 K. A tabulation is given for the values of specific heat C_p, enthalpy $H - H_0^0$, entropy S^0, Gibbs energy $G - H_0^0$, enthalpy function $(H - H_0^0)/T$ and Gibbs energy function $(G - H_0^0)/T$. The report also contains a comparison of the values of the electronic coefficient of the specific heat and the 0 K limiting Debye characteristic temperature with their selected values. An appraisal of low-temperature calorimetry is also given.

Touloukian [44] edited a handbook entitled *Thermophysical Properties of High Temperature Solid Materials* consisting of nine books totaling more than 8500 pages. The properties covered in the handbook are density, melting point, heat of fusion, heat of vaporization, heat of sublimation, electrical resistivity, specific heat at constant pressure, thermal conductivity, thermal diffusivity, thermal linear expansion, thermal radiative properties (absorptance, emittance, reflectance, and transmittance), and vapor pressure. Generally, only materials with melting points above 800 K are included, except for materials within the categories of polymers, plastics, and composites.

Touloukian, Gerritsen, and Moore [45], *Thermophysical Properties Research Literature Retrieval Guide*, consisting of a set of three books, contains references for 33,700 research documents on thermophysical properties of matter. The properties covered are thermal conductivity, specific heat at constant pressure, viscosity, thermal radiative properties (emissivity, absorptivity, reflectivity, transmissivity), optical constants (total and spectral), diffusion coefficient, thermal diffusivity, and Prandtl number. This publication supersedes the earlier works of this series (Volume I, 1960 and Volume II, 1963), and constitutes an enlarged and consolidated definitive work reporting the total literature through June 1964.

Schick [29] edited a comprehensive work entitled *Thermodynamics of Certain Refractory Compounds*. Volume 2 of this work includes thermodynamic properties of borides, carbides, nitrides, and oxides of 31 elements in the temperature range from 0 to 6000 K. Over 160 thermodynamic tables, together with comprehensive discussions, are presented.

Moeller et al.'s [24] compilation on *Thermophysical Properties of Thermal Insulating Materials* should prove useful in cryogenic and high temperature applications. The properties included in this compilation are thermal conductivity, thermal linear expansion, specific heat, total normal emittance, thermal diffusivity, compressive strength, density, melting point, and modulus of elasticity. Various experimental methods for determining thermal properties are described and their accuracies are indicated.

Wood and Deem [52] report on the compilation of specific heat, thermal linear expansion, and thermal conductivity data for materials of possible structural usefulness above 1500 K. Data are presented graphically with notations as to measurement methods and test conditions.

Hultgren, Orr, Anderson, and Kelley [16] published their book on the *Selected Values of Thermodynamic Properties of Metals and Alloys* in 1963. This book presents in tabular form heat capacity, enthalpy, entropy, free energy function, and vapor pressure. In some cases the heat of fusion, melting point, and other transition temperatures are also given. For the binary alloys, phase diagrams are included.

Eldridge and Deem [11] issued a report under the auspices of the Data and Publication Panel of

ASTM–ASME joint committee on effects of temperature on the properties of metals. The metals covered are Al, Co, Fe, Mg, Mo, Ni, and their alloys. The properties included are thermal conductivity, thermal linear expansion, specific heat, electrical resistivity, density, emissivity, diffusivity, and magnetic permeability. Emphasis is given to data over a range from cryogenic (2 K) to elevated temperatures (2800 K).

Johnson [17] edited a compendium of the properties of materials at low temperatures. The first phase of the compendium covers properties of ten fluids (Part I), properties of solids (Part II), and an extensive bibliography of references (Part III). The properties covered are density, expansivity, thermal conductivity, specific heat, enthalpy, heats of transition, phase equilibria, dielectric constants, adsorption, surface tension, and viscosity for solid, liquid, and gas phases of He, H_2, Ne, N_2, O_2, air, CO, F_2, A, and NH_3. Data sheets, primarily in graphic form, are presented for "best values" of data collected. The sources of the materials used, other references, and tables of selected values with appropriate comments are furnished with each data sheet.

Kelley's [18] bulletin contains the then-available high-temperature specific heat data for the elements and inorganic compounds. The thermodynamic properties are listed in tables and algebraic expressions for their representations are also given.

Stull and Sinke [40] published their well-known reference work on the *Thermodynamic Properties of the Elements* in 1956. This book reports specific heat as well as thermodynamic property values for the elements in their condensed and gaseous state. A search of the literature was made by the authors through 1955. Whenever experimental data were not available, reasonable estimates were made in order to fill the gaps in information. A tabulation of thermodynamic values from 298.15 K to 3000 K is given for the elements.

Methods for the Measurement of the Specific Heat of Solids

1. INTRODUCTION

There are few methods for the practical and precise determination of the specific heat of solids. Although many variants and minor modifications or improvements are reported in the various references cited in this section, the most important ones are described in detail in reference [54]. References [55] to [61] also constitute major works on calorimetry including various specialized applications.

The primary methods for the measurement of the specific heat of solids which are commonly used are the method of mixtures or drop method, adiabatic method, comparative method, pulse-heating method, and modifications of these. A number of specific calorimetric techniques are briefly described in this section.

The method of mixtures [14, 37, 50] is widely employed for measuring specific heats of solids above room temperature. This method frequently gives accurate results in a temperature range where no phase transition exists. The usual method consists of dropping the substance under investigation from a furnace temperature into a calorimeter (at room or ice temperature) and the quantity that is obtained directly is the change in enthalpy. Heat capacities are obtained from these values by differentiation, i.e., $C_p = (\partial H/\partial T)_p$. This method is inherently not suitable for use with substances which undergo phase transitions over the temperature range of interest or whose specific heat is highly temperature sensitive.

Various methods of obtaining directly the true specific heat based on the Nernst calorimeter [38, 42, 47, 49] have been used successfully in obtaining precise data in the temperature range below room temperature. Attempts to use this method at moderately high temperatures have not produced accurate results because of heat exchange with the surroundings. This method involves the measurement of energy required to raise the temperature of the substance over small temperature intervals from a fraction of a degree to a few degrees.

2. NERNST-TYPE ADIABATIC VACUUM CALORIMETER

A typical adiabatic vacuum calorimeter consists of a block over which an insulated coil of platinum wire is wound. The block may be either a solid sample under investigation or a container for the solid sample. The block is suspended by leads in a vacuum-tight container. The container is cooled in a dewar containing liquid air, hydrogen, or helium, depending on the temperature range involved. At the start of the operation, the vacuum-tight container is filled with helium gas at very low pressure while the block is cooled to the bath temperature by heat transfer through the helium gas. After the block has been cooled, the gas is removed by pumping and a known amount of heat is applied to the platinum coil by means of electric current for a given time interval. The temperature rise of the block is measured by means of a suitable resistance thermometer. The specific heat is then determined from the measured heat input and temperature change of the sample. Improved versions of the Nernst-type adiabatic calorimeter are described by Taylor and Smith [42], Wallace *et al.* [47], and Westrum [49].

The calorimeter assembly which is discussed by Wallace *et al.* [47] consisted of the sample container, the thermal shields, the outer jacket with associated radiation shields, and the vacuum system. Figure 1 presents a schematic diagram of the calorimeter.

3. MODIFIED ADIABATIC CALORIMETER

A modification of the direct method has been applied successfully by Schmidt and Leidenfrost [30] to obtain the specific heats of powders and granular materials from 273 K to 773 K. The determination of specific heats was carried out for Mond Nickel (99.85% Ni) with an accuracy of 0.6 percent.

The theory of the method as employed for a continuously heated adiabatic calorimeter for measuring powders and granular materials is discussed in detail in reference [30].

Consider a calorimeter and sample system with negligible heat loss to the surroundings, then the heat input may be expressed as

$$\frac{dQ}{dt} = mC_p\frac{dT}{dt} + W_c\frac{dT}{dt} \tag{44}$$

where dQ/dt is the heat input per unit time, T is the temperature, t is the time, m is the mass of the specimen, W_c is the thermal constant of calorimeter body and heater element, energy per degree, and C_p is the specific heat of specimen. From equation (44),

$$C_p = \frac{1}{m}\left[\frac{dQ/dt}{dT/dt} - W_c\right] \tag{45}$$

It is desirable to achieve as small a temperature variation as possible if the specific heat is assumed

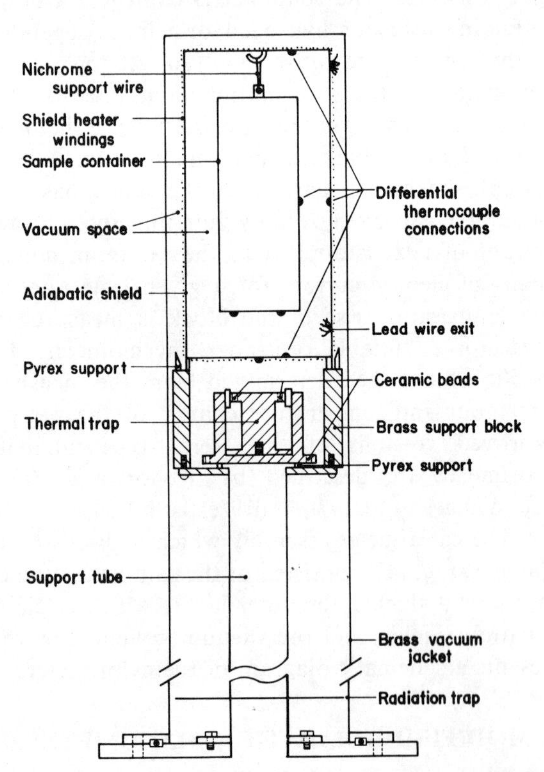

Fig. 1. Schematic diagram for adiabatic specific heat calorimeter [47].

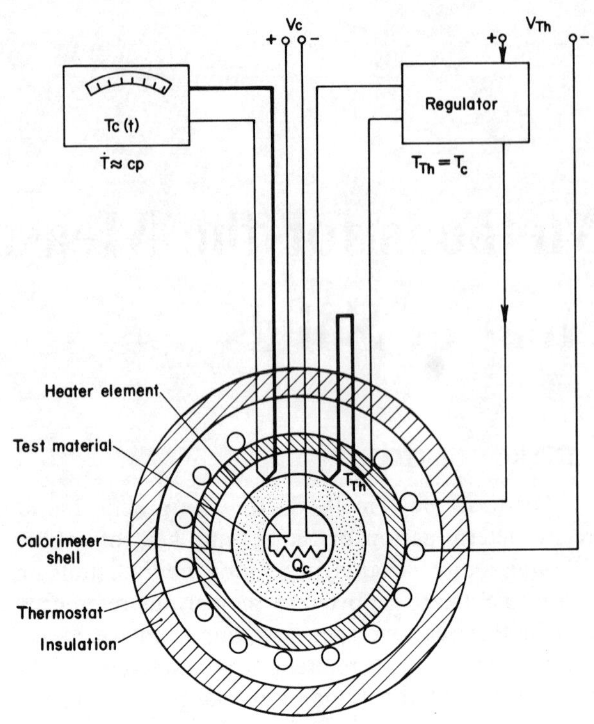

Fig. 2. Schematic diagram for spherical adiabatic calorimeter [30].

constant during each measurement interval. On the other hand, this temperature variation must be large enough to lend itself to precision measurement. The heating must be such that steady-state condition is reached within a reasonable length of time. Schmidt and Leidenfrost [30] have shown that for powders or granular materials of low thermal diffusivity, the following assumptions can be satisfied well enough to yield accurate measurements:

1. The temperature field is dependent only on time and the radial coordinate.
2. The sample is uniformly homogeneous, and its properties are constant over small temperature differences.
3. The sum of the heat capacities of the calorimeter body and its inside heater is small compared with the heat capacity of the sample mass.

The experimental arrangement of the apparatus is shown in schematic form in Fig. 2.

4. DROP ICE CALORIMETER

In this method [13] the heat given off by the sample is used to melt a portion of the ice in an equilibrium ice-water bath and the resulting change

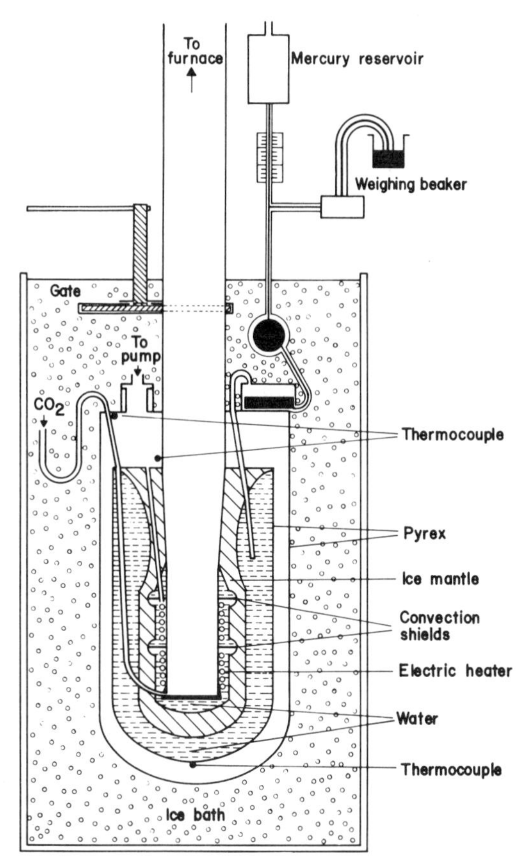

Fig. 3. Schematic diagram for drop ice calorimeter [13].

in volume of the bath is measured by the change in height of a mercury column. The calibration factor for a particular calorimeter (ratio of heat input to mass of mercury displaced by melted ice) is determined from the following expression:

$$K = \Delta H_f/(v_i - v_w)d_m \tag{46}$$

where K is the calibration factor, ΔH_f is the heat of fusion of ice, v_i is the specific volume of ice, v_w is the specific volume of water, and d_m is the density of mercury.

The calibration factor K relates the enthalpy change of the specimen to the height of the mercury column. Values of $(H_T - H_{273.15})$ are then determined for various initial specimen temperatures. These data are either represented graphically or by a suitable empirical relation. The specific heat curve is either derived from the graphically smoothed enthalpy data or from the equation

$$C_p = \frac{d}{dT}(H_T - H_{273.15})_p$$

A schematic drawing of the ice calorimeter is shown in Fig. 3. A central well is provided to receive the specimen whose enthalpy is to be determined. An electric heater, sheathed in a metal tube, is soldered on the outside of the well in order to introduce known amounts of heat for calibration purposes. The lower portion of the well is surrounded by two coaxial glass vessels which provide an insulating space between the inner ice-water system and the surrounding ice bath. Any volume change resulting from the melting of ice in the inner vessel displaces an equivalent volume of mercury and is collected in a beaker and weighed to account for the change in mercury in the calorimeter. A special gate prevents heat transfer from above to the calorimeter along the central well.

5. DROP ISOTHERMAL WATER CALORIMETER

In the drop water calorimeter a sample is heated in the furnace and dropped into the calorimeter

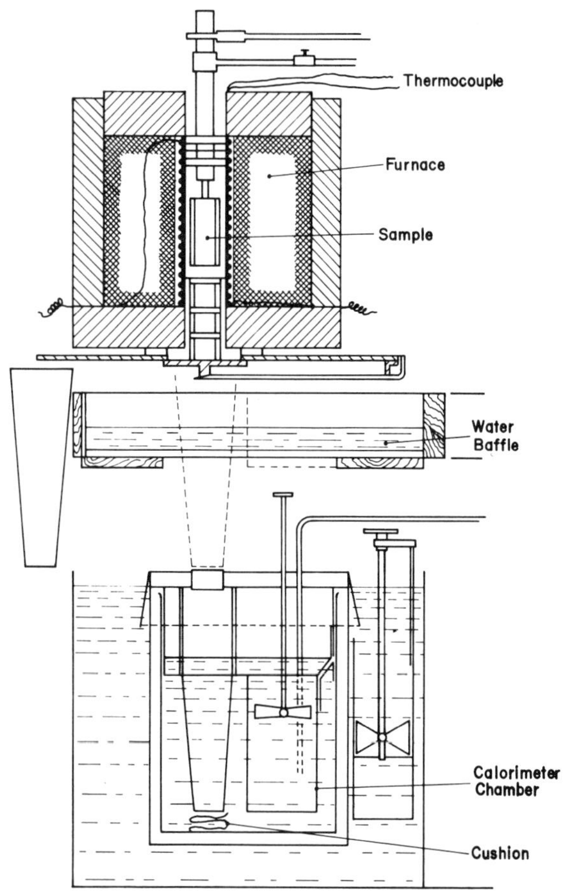

Fig. 4. Schematic diagram for drop isothermal water calorimeter [50].

proper, which consists of a water bath with free air space above. The water in the bath is stirred to assure uniform temperature. The calorimeter is enclosed by an isothermal jacket and the top is covered with copper plates which have a constant temperature because of their high thermal conductivity. The rise in the temperature of the calorimeter is measured with great accuracy by using a Beckmann thermometer or a sensitive thermopile. The enthalpy change of the specimen is determined from the known heat capacity of the calorimeter and its temperature rise. The enthalpy change may be referred to either 273.15 K or 298.15 K. In either case the specific heat is obtained from the smoothed enthalpy data by either graphical or analytical differentiation, i.e.,

$$C_p = \frac{d(H_T - H_{298.15})_p}{dT}$$

A schematic drawing [50] is shown in Fig. 4 to illustrate the details of the apparatus.

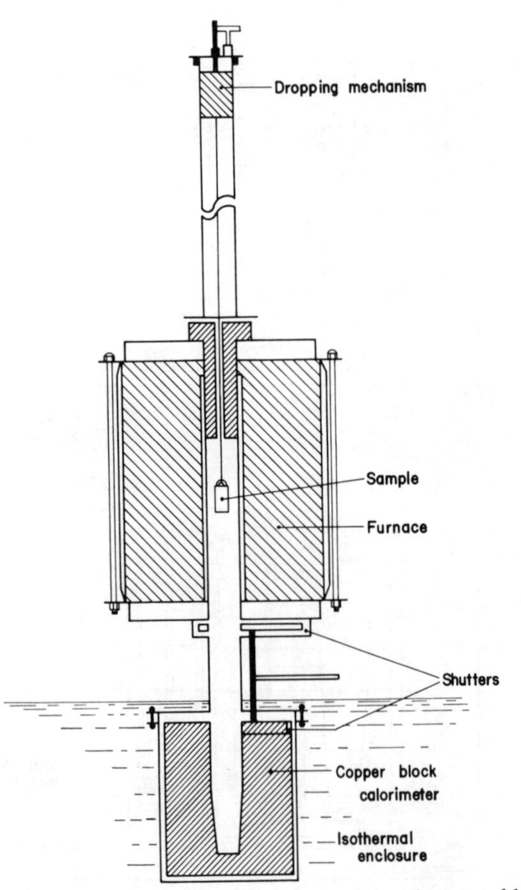

Fig. 5. Schematic diagram for drop isothermal copper block calorimeter [37].

6. DROP COPPER BLOCK CALORIMETER

This drop calorimeter employs a copper block which is submerged in an isothermal oil bath. The temperature of the calorimeter is measured using a special bridge network of copper and manganin resistances. The heat released from the sample is distributed to the copper block because of its high thermal conductivity. Generally it takes some time to achieve uniform heat distribution. The change in enthalpy of the specimen is measured in terms of the amount of heat absorbed by the copper block in changing from its initial temperature to its final temperature. This value is then corrected to 298.15 K so that the tabulated enthalpy values of the specimen are referred to 298.15 K, that is, $H_T - H_{298.15}$. The specific heat as a function of temperature may then be derived from the smoothed enthalpy data obtained either graphically or from the equation

$$C_p = \frac{d}{dT}(H_T - H_{298.15})_p$$

A schematic diagram according to Southard [37] is shown in Fig. 5.

7. PULSE-HEATING METHOD

The pulse-heating method of measuring specific heat is very attractive, particularly for materials that are electrical conductors. This method was first discussed by Avramescu [1] and later modified by other investigators [2, 25, 39, 41]. The method involves the rapid heating of small samples in vacuum. Voltage probes are attached across the central portion of the sample wire which is then mounted in a high-vacuum system. The sample is connected to an electrical circuit consisting of a large storage battery, a variable resistor, a fixed resistor, and a high-current relay controlled by a timing circuit which determines the duration of the pulse. A schematic diagram of a typical circuit [41] for the measurement of specific heat is shown in Fig. 6. The current flowing through the specimen and the voltage drop across the central portion are measured simultaneously as a function of time. The specific resistance at each time interval is calculated from the relationship $\rho = AE/LI$, where A is the cross-sectional area of sample, E is the voltage, I is the current, and L is the distance between voltage probes. This specific electrical resistance is then plotted as a function of time. The specific heat at any temperature T is given by the equation

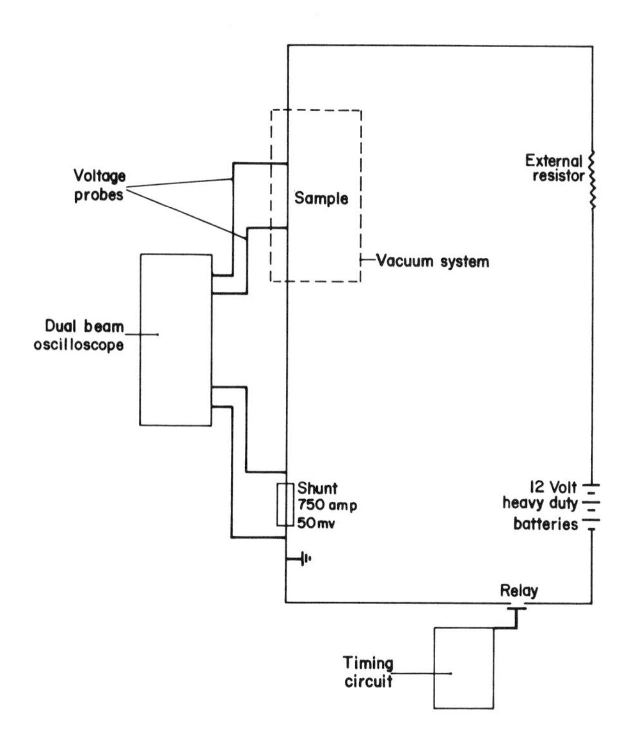

Fig. 6. Schematic diagram of circuit for specific heat measurement using pulse-heating method [41].

$$C_p = \frac{EI(d\rho/dT)}{Jm(d\rho/dt)} \tag{47}$$

where C_p is the specific heat, cal g^{-1} K^{-1}, J is the conversion factor, 4.184 joules cal^{-1}, m is the mass of sample between voltage probes, grams, $d\rho/dT$ is the temperature coefficient of the resistance at temperature T, $d\rho/dt$ is the time rate of change of resistivity at temperature T, and ρ is the electrical resistivity of sample.

8. COMPARATIVE METHOD

The method consists of placing a specimen with its temperature-monitoring thermocouple in a refractory container of low thermal conductivity and in turn placing this in a furnace whose temperature is maintained constant above or below the specimen temperature. The container is calibrated by determining its heating rate when empty and then with a reference sample of known specific heat. Separate electrical heating circuits are usually provided for the specimen and the shield so that their temperature will rise equally and simultaneously in order to reduce heat losses. The specific heat C_{p2} of the unknown specimen is calculated from the following relation:

$$\frac{C_{p2}W_2}{C_{p1}W_1} = \frac{\Delta t_2/\Delta T_2 - \Delta t_r/\Delta T_r}{\Delta t_1/\Delta T_1 - \Delta t_r/\Delta T_r} \tag{48}$$

where $(\Delta t/\Delta T)$ is the slope of a time–temperature curve, and the subscripts r, 1, and 2 represent the empty container, the container with specimen 1, and the container with specimen 2, respectively. The papers by Boggs and Wiebelt [3] and Smith [34] give excellent accounts in the use of this method.

Irreproducible heating or cooling conditions and differences in thermal conductivity between the unknown and reference specimen usually account for the inaccuracies encountered in this method.

References to Text

1. Avramescu, A., "Temperature Variation of the True Specific Heat of Conductivity Copper and Conductivity Aluminum up to the Melting Point," *Z. Tech. Physik* **20**, 213–17, 1939.
2. Baxter, H., "Determination of Specific Heat of Metals," *Nature* **153**, 316, 1944.
3. Boggs, J. H. and Wiebelt, J. A., "An Investigation of a Particular Comparative Method of Specific Heat Determination in the Temperature Range of 1500 F to 2600 F," USAEC TID–5734, 1–91, 1960.
4. Born, M. and Karman, T., "Vibrations in Space Lattices," *Physik Z.* **13**, 297–309, 1912.
5. Bottema, J. A. and Jaeger, F. M., "The Law of Additive Atomic Heats in Intermetallic Compounds," *Proc. Acad. Sci. Amsterdam* **35**, 928–31, 1932.
6. Debye, P., "The Theory of Specific Heat," *Ann. Physik* **39** (4), 789–839, 1912.
7. Dekker, A. J., *Solid State Physics*, Prentice-Hall, Inc., 1–525, 1961.
8. Drude, P., "The Electronic Theory of Metal," *Ann. Physik* **1**, 566–613, 1900.
9. Dulong, P. L. and Petit, A. T., *Ann. Chim.* **10**, 395–413, 1819.
10. Einstein, A., "The Planck's Theory of Radiation and the Theory of Specific Heat," *Ann. Physik* **22** (4), 180–90, 1907.
11. Eldridge, E. A. and Deem, H. W., "Report on Physical Properties of Metals and Alloys from Cryogenic to Elevated Temperatures," ASTM–STP–296, 1–206, 1961.
12. Furukawa, G. T., Saba, W. G., and Reilly, M. L., "Critical Analysis of the Heat-Capacity Data of the Literature and Evaluation of Thermodynamic Properties of Copper, Silver, and Gold from 0 to 300 K," NSRDS–NBS 18, 1–49, 1968.
13. Ginnings, D. C. and Corruccini, R. J., "An Improved Ice Calorimeter—The Determination of its Calibration Factor and the Density of Ice at 0 C," *J. Res. Natl. Bur. Std.* **38**, 583–91, 1947.
14. Ginnings, D. C. and Furukawa, G. T., "Heat Capacity Standards for the Range 14 to 1200 K," *J. Am. Chem. Soc.* **75**, 522–7, 1953.
15. Gopal, E. S. R., *Specific Heats at Low Temperatures*, Plenum Press, 1–111, 1966.
16. Hultgren, R., Orr, R. L., Anderson, P. D., and Kelley, K. K., *Selected Values of Thermodynamic Properties of Metals and Alloys*, John Wiley and Sons, Inc., 1–963, 1963.
17. Johnson, V. J. (Editor), "A Compendium of the Properties of Materials at Low Temperature," WADD–TR–60–56, Pt. 2, 1–333, 1960. [AD 249 786]
18. Kelley, K. K., "Data on Theoretical Metallurgy. XIII. High-Temperature Heat Capacity and Entropy Data for the Elements and Inorganic Compounds," U.S. Bur. Mines Bull. 584, 1–232, 1960.
19. King, A. L., *Thermophysics*, W. H. Freeman and Company, 1–369, 1962.
20. Kittel, C., *Introduction to Solid State Physics*, John Wiley and Sons, Inc., 122–5, 1963.
21. Kittel, C., *Elementary Solid State Physics*, John Wiley and Sons, Inc., 49–52, 1962.
22. Lehman, G. W., "Thermal Properties of Refractory Materials," WADD–TR–60–581, 1–19, 1960. [AD 247 411], [PB 160 804]
23. Levinson, L. S., "High Temperature Drop Calorimeter," *Rev. Sci. Instr.* **36** (6), 639–42, 1962.
24. Moeller, C. E., Loser, J. B., Thompson, M. B., Snyder, W. E., and Hopkins, V., "Thermophysical Properties of Thermal Insulating Materials," ASD–TDR–64–5, 1–362, 1964. [AD 601 535], [N64 22689]
25. Nathan, A. M., "A Dynamic Method for Measuring the Specific Heat of Metals," *J. Appl. Phys.* **22**, 234–5, 1951.
26. Parker, W. J., Jenkins, R. J., Butler, C. P., and Abbott, G. L., "Flash Method of Determining Thermal Diffusivity, Heat Capacity and Thermal Conductivity," *J. Appl. Phys.* **32**, 1679–84, 1961.
27. Reif, F., *Fundamentals of Statistical and Thermal Physics*, McGraw-Hill Book Co., Inc., 1–651, 1965.
28. Sachs, M., *Solid State Theory*, McGraw-Hill Book Co., Inc., 143–68, 1963.
29. Schick, H. L. (Editor), *Thermodynamics of Certain Refractory Compounds, Vol.* 2, Academic Press, 1–775, 1966.
30. Schmidt, E. O. and Leidenfrost, W., "Adiabatic Calorimeter for Measurements of Specific Heats of Powder and Granular Materials at 0 C to 500 C," *ASME 2nd Symp. Thermophysical Properties*, Princeton, N.J., 178–84, 1962.
31. Sears, F. W., *An Introduction to Thermodynamics, The Kinetic Theory of Gases, and Statistical Mechanics*, Addison-Wesley Publishing Co., Inc., 1–373, 1964.
32. Seitz, F., *The Modern Theory of Solids*, McGraw-Hill Book Co., Inc., 38–9, 1940.
33. Slater, J. C., *Introduction to Chemical Physics*, McGraw-Hill Book Co., Inc., 1939.
34. Smith, C. S., "A Simple Method for Thermal Analysis Permitting Quantitative Measurements of Specific and Latent Heats," *Trans. AIME* **137**, 1936, 1940.
35. Smith, D. F., Kaylor, C. E., Walden, G. E., Taylor, A. R., and Gayle, J. B., "Construction, Calibration and Operation of Ice Calorimeter," U.S. Bur. Mines Rept. Invest. 5832, 1–20, 1961.

36. Sommerfeld, A., "The Electronic Theory of Metals," *Naturwiss* **15**, 825–32, 1927.
37. Southard, J. C., "A Modified Calorimeter for High Temperatures. The Heat Content of Silica, Wollastonite, and Thorium Dioxide above 25°," *J. Am. Chem. Soc.* **63**, 3142–6, 1941.
38. Sterrett, K. F., Blackburn, D. A., Bestul, A. B., Chang, S. S., and Horman, J., "An Adiabatic Calorimeter for the Range 10 K to 360 K," *J. Res. Natl. Bur. Std.* **69C**, 19–26, 1965.
39. Strittmater, R. C., Pearson, G. J., and Danielson, G. C., "Measurements of Specific Heats by a Pulse Method," *Proc. Iowa Acad. Sci.* **64**, 466–70, 1957.
40. Stull, D. R. and Sinke, G. C., *Thermodynamic Properties of the Elements*, Am. Chem. Soc., 1–234, 1956.
41. Taylor, R. E. and Finch, R. A., "The Specific Heats and Resistivities of Molybdenum, Tantalum, and Rhenium," *J. Less-Common Metals* **6**, 283–94, 1964.
42. Taylor, A. R. and Smith, D. F., "Construction, Calibration, and Operation of a Low-Temperature Adiabatic Calorimeter," U.S. Bur. Mines Rept. Invest. 5974, 1–17, 1962.
43. Tolman, R. C., *Principles of Statistical Mechanics*, Oxford Univ. Press, London, 1938.
44. Touloukian, Y. S. (Editor), *Thermophysical Properties of High Temperature Solid Materials*, MacMillan Co., Vols. 1, 2, 3, 4, 5, and 6, 1–8500, 1967.
45. Touloukian, Y. S., Gerritsen, J. K., and Moore, N. Y., *Thermophysical Properties Research Literature Retrieval Guide*, Plenum Press, 2nd Ed., Books 1, 2, and 3, 1967.
46. Touloukian, Y. S. (Editor), "Recommended Values of the Thermophysical Properties of Eight Alloys, Major Constituents and Their Oxides," TPRC Rept. 16, 323–46, 1966.
47. Wallace, W. E., Craig, R. S., and Johnston, W. V., "An Adiabatic Calorimeter for the Range 15 C to 290 C," U.S. At. Energy Comm., NYD–6328, 1–16, 1966.
48. Weber, H. F., "The Specific Heat of Elements Carbon, Boron, and Silicon," *Phil. Mag.* **49**, 161–301, 1875.
49. Westrum, E. F., Jr., "Cryogenic Calorimetric Contributions to Chemical Thermodynamics," *J. Chem. Educ.* **39** (9), 443–54, 1962.
50. White, W. P., "Specific Heat Determination at Higher Temperatures," *Am. J. Sci.* **47** (4), 1–59, 1919.
51. Whittaker, E. T. and Watson, G. N., *Modern Analysis*, Cambridge Univ. Press, 4th edition, 1938.
52. Wood, W. D. and Deem, H. W., "Thermal Properties of High-Temperature Materials," RSIC–202, 1–399, 1964. [AD 455 069]
53. Zemansky, M. W., *Heat and Thermodynamics*, McGraw-Hill Book Co., Inc., 1–484, 1957.
54. McCullough, J. P. and Scott, D. W. (Editors), *Experimental Thermodynamics, Volume I, Calorimetry of Non-Reacting Systems*, Plenum Press (New York)/Butterworths (London), 1968.
55. White, W. P., *The Modern Calorimeter*, Chemical Catalog Co., New York, 1928.
56. Swietoslawski, W., *Microcalorimetry*, Reinhold, NewYork, 1964.
57. Calvet, E. and Prat, H., *Microcalorimétrie*, Masson et Cie, Paris, 1956.
58. Roth, W. A. and Becker, F., *Kalorimetrische Methoden zur Bestimmung chemischer Reaktionswärmen*, F. Vieweg, Braunschweig, 1956.
59. Rossini, F. D. (Editor), *Experimental Thermochemistry*, Vol. I, Interscience, New York, 1956.
60. Weissberger, A. (Editor), "*Calorimetry*" *in Technique of Organic Chemistry Vol. I. Physical Methods of Organic Chemistry*, Chap. X, Interscience, New York, 1959.
61. Skinner, H. A. (Editor), *Experimental Thermochemistry*, Vol. II, Interscience, London, 1962.

Numerical Data

Data Presentation and Related General Information

1. SCOPE OF COVERAGE

The materials studied in this volume consist of metallic elements and their alloys.* The elements are listed in the table of contents in alphabetical order according to chemical name while the alloys are in alphabetical order according to the major constituent element. The data presented are original experimental data on the specific heat of these materials as reported by various investigators. These data were extracted from the world's technical and scientific literature, United States Government Publications, Doctoral and Masters dissertations, data supplied by private companies, and special reports of major research centers throughout the world. The range of temperatures covered is from zero degree Kelvin to the melting point and beyond. For most high-temperature metals and alloys, no information is found in the liquid range.

2. PRESENTATION OF DATA

The data for all substances are presented in graphical and tabular form together with a specification table for each substance. The specification table gives the temperature range, the original reference number, the curve number, reported estimates of error, year of publication of the original document, specimen designation, and such other pertinent information as composition or purity of sample, test environment, mechanical, chemical, and thermal history of the test specimen, etc., to the extent provided in the original source document. The data for the specific heat of the materials are plotted on a log–log scale for comparative evaluation. When several sets of data are coincident, the graphical plotting of all of them would lead to confusion. For this reason, some of the sets of data points are omitted from the figures. They are, however, reported in the data tables and specification tables.

The numerical data are presented in double columns. The temperature T is in degrees Kelvin, and the specific heat C_p is in calories per gram per degree Kelvin. A unique curve number is assigned to each set of data. This corresponds exactly to the number which also appears in the specification table and on the figure.

The two general types of data that are obtainable from the literature are the true specific heat data obtained directly from the results of measurements using, for instance, the Nernst-type calorimeter and the derived true specific heat data, deduced from direct enthalpy measurements using the drop technique. In the latter type an empirical equation has been fitted by the authors to the enthalpy data by least squares technique and specific heat obtained by differentiation. The results are usually tabulated at rounded temperature intervals.

*Boron, which is a nonmetal, has been included in this volume because of its extensive use as an alloying element for most metallic alloys. However, boron has also been listed in Volume 5, as a nonmetal.

3. SYMBOLS AND ABBREVIATIONS USED IN THE FIGURES AND TABLES

Symbol	*Definition*	*Units*
T	Temperature	degree Kelvin, K
C_p	Constant pressure specific heat	cal g^{-1} K^{-1}
C_v	Constant volume specific heat	cal g^{-1} K^{-1}
M.P.	Melting point	degree Kelvin, K
T.P.	Transition point	degree Kelvin, K
s.c.	Superconducting	
N	Normal	
c	Cubic	
f.c.c.	Face-centered cubic	

CONVERSION FACTORS FOR UNITS OF SPECIFIC HEAT

MULTIPLY by appropriate factor to OBTAIN →	cal_{th} $g\text{-}mol^{-1}$ C^{-1}	cal_{th} g^{-1} C^{-1}	cal_{IT} $g\text{-}mol^{-1}$ C^{-1}	cal_{IT} g^{-1} C^{-1}	J $g\text{-}mol^{-1}$ K^{-1}	J g^{-1} K^{-1}	J $kg\text{-}mol^{-1}$ K^{-1}	J kg^{-1} K^{-1}	Btu_{th} lb^{-1} F^{-1}	Btu_{IT} lb^{-1} F^{-1}
cal_{th} $g\text{-}mol^{-1}$ C^{-1}	1	1/M	0.999331	0.999331/M	4.184	4.184/M	4.184 x 10^3	4.184/M x 10^3	1/M	0.999331/M
cal_{th} g^{-1} C^{-1}	M	1	0.999331M	0.999331	4.184M	4.184	4.184M x 10^3	4.184 x 10^3	1	0.999331
cal_{IT} $g\text{-}mol^{-1}$ C^{-1}	1.00067	1.00067/M	1	1/M	4.1868	4.1868/M	4.1868 x 10^3	(4.1868/M) x 10^3	1.00067/M	1/M
cal_{IT} g^{-1} C^{-1}	1.00067M	1.00067	M	1	4.1868M	4.1868	4.1868M x 10^3	4.1868 x 10^3	1.00067	1
J $g\text{-}mol^{-1}$ K^{-1}	0.239006	0.239006/M	0.238846	0.238846/M	1	1/M	1 x 10^3	1 x 10^3/M	0.239006/M	0.238846/M
J g^{-1} K^{-1}	0.239006M	0.239006	0.238846M	0.238846	M	1	M x 10^3	10^3	0.239006	0.238846
J $kg\text{-}mol^{-1}$ K^{-1}	2.39006 x 10^{-4}	(2.39006/M) x 10^{-4}	2.38846 x 10^{-4}	(2.38846/M) x 10^{-4}	10^{-3}	10^{-3}/M	1	1/M	(2.39006/M) x 10^{-4}	(2.38846/M) x 10^{-4}
J kg^{-1} K^{-1}	2.39006M x 10^{-4}	2.39006 x 10^{-4}	2.38846M x 10^{-4}	2.38846 x 10^{-4}	M x 10^{-3}	10^{-3}	M	1	2.39006 x 10^{-4}	2.38846 x 10^{-4}
Btu_{th} lb^{-1} F^{-1}	M	1	0.999331M	0.999331	4.184M	4.184	4.184M x 10^3	4.184 x 10^3	1	0.999331
Btu_{IT} lb^{-1} F^{-1}	1.00067M	1.00067	M	1	4.1868M	4.1868	4.1868M x 10^3	4.1868 x 10^3	1.00067	1

Classification of Materials

Classification			Limits of composition (weight percent)*			
			X_1	$X_1 + X_2$	X_2	X_3
1. Metallic elements			>99.5	—	<0.2	<0.2
2. Nonferrous alloys ($X_1 \neq Fe$)	A. Binary alloys		—	≥ 99.5	≥ 0.2	≤ 0.2
	B. Multiple alloys		—	≥ 99.5	>0.2	>0.2
			—	<99.5	≥ 0.2	≤ 0.2
			—	<99.5	>0.2	>0.2
			≤ 99.5	—	<0.2	<0.2
			X_1	X_2	X_3	Mn, P, S, or Si
3. Ferrous Alloys ($X_1 = Fe$)	A. Carbon steels	Group I	Fe	$C \leq 2.0$	≤ 0.2	≤ 0.6
		Group II	Fe	$C \leq 2.0$	≤ 0.2	>0.6
			Fe	$C \leq 2.0$	>0.2	≤ 0.6
			Fe	$C \leq 2.0$	>0.2	>0.6
	B. Cast irons	Group I	Fe	$C > 2.0$	≤ 0.2	≤ 0.6
		Group II	Fe	$C > 2.0$	≤ 0.2	>0.6
			Fe	$C > 2.0$	>0.2	≤ 0.6
			Fe	$C > 2.0$	>0.2	>0.6
	C. Alloy steels†	Group I	Fe	$\neq C$	≤ 0.2 and $C \leq 2.0$	≤ 0.6
		Group II	Fe	$\neq C$	≤ 0.2	>0.6
			Fe	$\neq C$	>0.2	≤ 0.6
			Fe	$\neq C$	>0.2	>0.6

*$X_1 \geq X_2 \geq X_3 \geq X_4 \ldots$

†In case Mn, P, S, or Si represents X_2, this particular element is dropped from the last column. Alloy cast irons are also included in Group II of this category.

b.c.c. Body-centered cubic
h Hexagonal
c.p.h. Close-packed hexagonal

4. CONVERSION FACTORS FOR UNITS OF SPECIFIC HEAT

The conversion factors given in the table on page 20a are based upon the following basic definitions:

$$1 \text{ lb} = 0.45359237 \text{ kg*}$$
$$1 \text{ cal}_{th} = 4.184 \text{ (exactly) J*}$$
$$1 \text{ cal}_{IT} = 4.1868 \text{ (exactly) J*}$$
$$1 \text{ Btu}_{th} \text{ lb}^{-1} \text{ F}^{-1} = 1 \text{ cal}_{th} \text{ g}^{-1} \text{ C}^{-1}\dagger$$
$$1 \text{ Btu}_{IT} \text{ lb}^{-1} \text{ F}^{-1} = 1 \text{ cal}_{IT} \text{ g}^{-1} \text{ C}^{-1}\dagger$$

*National Bureau of Standards, "New Values for the Physical Constants Recommended by NAS–NRC," *NBS Tech. News Bull.* **47**(10), 175–7, 1963.

†Mueller, E. F. and Rossini, F. D., "The Calory and the Joule in Thermodynamics and Thermochemistry," *Am. J. Phys.* **12**(1), 1–7, 1944.

The subscripts "th" and "IT" designate "thermochemical" and "International Steam Table," respectively.

5. CLASSIFICATION OF MATERIALS

The classification scheme as shown in the table for metallic elements and alloys contained in this volume is based upon the chemical composition of the material. This scheme is mainly for the convenience of material grouping and data organization and is not intended to be used as definitions for the various material groups.

6. CONVENTION FOR BIBLIOGRAPHIC CITATION

For the following types of documents the bibliographic information is cited in the sequences given below.

Journal Article:

a. Author(s)—The names and initials of all authors are given. The last name is written first, followed by initials.
b. Title of article—In this volume, the titles of the journal articles listed in the *References to Text* are given, but not of those listed in the *References to Data Sources.*
c. Journal title—The abbreviated title of the journal as in *Chemical Abstracts* is given.
d. Series, volume, and number—If the series is designated by a letter, no comma is used between the letter for series and the numeral for volume, and they are underlined together. In case series is also designated by a numeral, a comma is used between the numeral for series and the numeral for volume, and only the numeral representing volume is underlined. No comma is used between the numerals representing volume and number. The numeral for number is enclosed in parentheses.
e. Pages—The inclusive page numbers of the article.
f. Year—The year of publication.

Report:

a. Author(s)
b. Title of report—In this volume, the titles of the reports listed in the *References to Text* are given, but not of those listed in the *References to Data Sources.*
c. Name of the responsible organization.
d. Report, or bulletin, circular, technical note, etc.
e. Number
f. Part
g. Pages
h. Year
i. ASTIA'S AD number—This is given in square brackets whenever available.

Book:

a. Author(s)
b. Title
c. Volume
d. Edition
e. Publisher
f. Place of publication
g. Pages
h. Year

7. CRYSTAL STRUCTURES, TRANSITION TEMPERATURES, AND OTHER PERTINENT PHYSICAL CONSTANTS OF THE ELEMENTS

The table on the following pages contains information on the crystal structure, transition temperatures, and certain other pertinent physical constants of each element. This information is very useful in data analysis and synthesis. However, no attempt has been made to critically evaluate the temperatures/constants given in the table and they should not be considered recommended values. This table has an independent series of numbered references which immediately follow the table.

CRYSTAL STRUCTURES, TRANSITION TEMPERATURES, AND OTHER PERTINENT PHYSICAL CONSTANTS OF THE ELEMENTS

Name	Atomic Number	Atomic Weight[a]	Density,[b] $kg\,m^{-3}\cdot 10^{-3}$	Crystal Structure	Phase Transition Temp., K	Superconducting Transition Temp., K	Curie Temp., K	Néel Temp., K	Debye Temperature at 0 K, K	Debye Temperature at 298 K, K	Melting Point, K	Boiling Point, K	Critical Temp., K
Actinium	89	(227)	10.07[1c]	f.c.c.[2]					124[3]	100[4] (at~50 K)	1323[5]	3200 ±300[6]	
Aluminum	13	26.9815	2.702[5]	f.c.c.[7]		1.196[5] 1.17[8] 1.18[9]			423 ±5[3]	390[3]	933.2[3,10]	2723[29]	8650[11] 7740[109]
Americium	95	(243)	11.7[5]	Double c.p.h.[2]							1473[29]	2880[108]	
Antimony	51	121.75	6.684[29]	r.[2] (?) ? (?) ? (?)	367.8[13] (?–?) 690[13] (?–?)	2.6[8] (Sb II, high-pressure modification)			150[3]	200[14]	903.7[13] 903.65[23]	1907 ±10[3]	2989[15]
Argon	18	39.948	0.0017824[29] (at 273.2 K and 1 atm)	f.c.c.[16]						90[4] (at~45 K)	83.8[17]	87.29[13]	151[15]
Arsenic	33	74.9216	5.73[29] (gray, at 287.2 K) 4.7[29] (black) 2.0[29] (yellow)	r.[7] (gray) c.[5] (yellow)					236[3]	275[18]	1090[13] (35.8 atm) subl. 886[5]	1090[13] (35.8 atm)	
Astatine	85	(210)									573.2[19]	650[20]	
Barium	56	137.34	3.5[29]	b.c.c.[2] (α) ? (β)[13]	648[13,21] (α–β)				110.5 ±1.8[22]	116[23]	998.2[5]	1910[3]	3663[15] 3920[109]
Berkelium	97	(249)											
Beryllium	4	9.0122	1.85[29]	c.p.h.[2] (α) b.c.c.[2] (β)	1533[24] (α–β)	~6[108] ~8.4[108]			1160[25]	1031[3]	1550[26]	3142 ±100[3]	6153[15]
Bismuth	83	208.980	9.78[29]	r.[2]		3.9[8] (Bi II, at 25 kbar) 7.2[8] (Bi III, at 27 kbar)			119 ±2[3]	116 ±5[3]	544.525[3,111]	1824 ±8[3]	4620[27]
Boron	5	10.811	2.50[42]	Simple r.[2] (α) r.[2] (β)	1473[2] (α–β)				1315[53]	1362[3]	2573[5]	4050 ±100[30]	
Bromine	35	79.909	3.119[29]	orthorh.[16]							266.0[17]	331.93[29]	~~589~~[15] 584

[a] Atomic weights are based on $^{12}C = 12$ as adopted by the International Union of Pure and Applied Chemistry in 1961; those in parentheses are the mass numbers of the isotopes of longest known half-life.

[b] Density values are given at 293.2 K unless otherwise noted.

[c] Superscript numbers designate references listed at the end of the table.

Name	Atomic Number	Atomic Weight[a]	Density,[b] kg m⁻³·10⁻³	Crystal Structure	Phase Transition Temp., K	Superconducting Transition Temp., K	Curie Temp., K	Néel Temp., K	Debye Temperature at 0 K, K	Debye Temperature at 298 K, K	Melting Point, K	Boiling Point, K	Critical Temp., K
Cadmium	48	112.40	8.65[29]	c.p.h.[2] b.c.c.[4] (?)		0.56[5] 0.52[9]			252±48[3]	221[3] 170[4] (b.c.c., at~85 K)	594.18[3,10] Subl. 594.1[13] (at 0.11 mm Hg)	1038[3]	1903[15] 3560[109]
Calcium	20	40.08	1.55[29]	f.c.c.[7] (α) b.c.c.[7] (β)	737[62] (α–β)				234±5[3]	230[3]	1123[19] Subl. 1123[13] (at 0.35 mm Hg)	1765[3]	3267[15]
Californium	98	(251)											
Carbon (amorphous)	6	12.01115	1.8~2.1[29]								Subl. 3925–3970[5]	4473[5]	
Carbon (diamond)	6	12.01115	3.51[29]	d.[16]					2240±5[31]	1874[3]	>3823[5]	5100[5]	
Carbon (graphite)	6	12.01115	2.26[29] (α)	h.[2] (α) r.[2] (β)					402±11[3]	1550[3]	Subl. 3925–3970[5]	4473[5]	
Cerium	58	140.12	6.90[29]	f.c.c. (α)[32] Double c.p.h.?[8] (β) f.c.c. (γ)[32] b.c.c. (δ)[32]	103±5[33] (α–β) 263±5[33] (β–γ) 1003[32] (γ–δ)			13[32]	146[3]	138[34]	1077[26]	3972[3]	10400[109]
Cesium	55	132.905	1.873[29]	b.c.c.[2]					40±5[3]	43[23]	301.9[29] Subl. 301.9[13] (at 1.2 μHg)	939[35]	2060[113,114,115] 1900[109]
Chlorine	17	35.453	0.003214[29] (at 273.2 K)	t.[16]						115[4,36] (at~58 K)	172.2[26]	239.10[13]	417[15]
Chromium	24	51.996	7.16[42]	c.p.h.[17,d] (α) b.c.c.[7] (β)	~299[17] (α–β)[d]			311[37]	598±32[3]	424[3]	2118[38]	2918±35[3]	
Cobalt	27	58.9332	8.862[42]	c.p.h.[7] (α) f.c.c.[17] (β)	690[39] (α–β)		1400[40]		452±17[3]	386[3]	1765[3,10]	3229[3]	
Copper	29	63.54	8.933[29]	f.c.c.[2]					342±2[3]	310[3]	1356[3,10]	2811±20[41]	8500[11] 8280[109]
Curium	96	(247)	7[42]	Double c.p.h.[8]									
Dysprosium	66	162.50	8.556[42]	c.p.h.[2] (α) b.c.c.[2] (β)	Near m.p.[2] (α–β)			174[43] 83.5[43] (ferro-antiferromag.)	172±35[3]	158[44]	1773[12]	3011[44]	7640[109]

[d] Close-packed hexagonal crystalline modification of chromium may be formed by electrodeposition below 293 K under special conditions of deposition process. This c.p.h. form is unstable and will irreversibly transform into b.c.c. form on heating.

Name	Atomic Number	Atomic Weight[a]	Density,[b] kg m^{-3} · 10^{-3}	Crystal Structure	Phase Transition Temp., K	Superconducting Transition Temp., K	Curie Temp., K	Neel Temp., K	Debye Temperature at 0 K, K	Debye Temperature at 298 K, K	Melting Point, K	Boiling Point, K	Critical Temp., K
Einsteinium	99	(254)											
Erbium	68	167.26	9.06[42]	c.p.h.[2] (α) b.c.c.[2] (β)	1643[2] (α-β)		19[4]	80[4]	134±10[45]	163[44]	1770[26]	3000[3]	7250[109]
Europium	63	151.96	5.245[28]	b.c.c.[7]				~90[4]	127[3]		1099[5]	1971[46]	4600[109]
Fermium	100	(253)											
Fluorine	9	18.9984	0.001695[29] (at 273.2 K and 1 atm)	c.[108] (β-F_2)							53.58[5]	85.24[13]	144[15]
Francium	87	(223)							39[3]		300.2[19]	879[108]	
Gadolinium	64	157.25	7.87[42]	c.p.h.[2] (α) b.c.c.[2] (β)	1535[32] (α-β)		292[40]		170[3]	155±3[3]	1579[19]	3540[3]	8670[109]
Gallium	31	69.72	5.91[29]	orthorh.[4] (α) t.[4] (β)	275.6[13] (α-β) (at 8.86 x 10^6 mm Hg)	1.091[5] 7.2[38] (Ga II, high-pressure modification)			317[3]	240[14] 125[4] (tetra at ~63 K)	302.93[5] 275.6[13] (at 8.86 x 10^6 mm Hg)	2510[3]	7620[27]
Germanium	32	72.59	5.36[29]	d.[7]		5.5[47] (at ~118 kbar) 8.4[108]			378±22[3]	403[3]	1210.6[5]	3100[3]	5642[15]
Gold	79	196.967	19.3[42]	f.c.c.[7]					165±1[3]	178±8[3]	1336.2[3,10] 1336.15[23]	3240[3]	9500[11] 8060[109]
Hafnium	72	178.49	13.28[42]	c.p.h.[48] (α) b.c.c.[48] (β)	2023±20[48] (α-β)	0.16[9] 0.35[108]			256±5[3]	213[23]	2495[19]	4575±150[49]	
Helium	2	4.0026	0.0001785[29] (at 273.2 K and 1 atm)	c.p.h.[16]						30[4] (at~15 K)	3.45[29] 1.8±0.2[17] (at 30 atm)	4.216[13] 4.22[23]	5.3[15]
Holmium	67	164.930	8.80[29]	c.p.h.[2] (α) b.c.c.[2] (β)	Near m.p.[50] (α-β)		20[4]	132[4]	114±7[45]	161[44]	1734[19]	3228[51]	
Hydrogen	1	1.00797	0.00008987[29] (at 273.2 K and 1 atm)	c.p.h.[16]						116[36] (para., at~58 K) 105[36] (ortho., at~53 K)	13.8±0.1[17]	20.39[13] 20.37[23]	33.3[15]
Indium	49	114.82	7.3[29]	f.c.t.[7]		3.4035[5]			108.8±0.3[3]	129[14]	429.76[3,110]	2279±6[3]	4377[15] 7050[109]
Iodine	53	126.9044	4.93[29]	orthorh.[16]						105[4] (at~53 K)	386.8[29] subl. 298.16[13] (at 0.31 mm Hg)	457.50[29]	785[15]
Iridium	77	192.2	22.5[42]	f.c.c.[7]		0.14[5,9]			425±5[3]	228[3]	2716[3,10]	4820±30[3]	

Name	Atomic Number	Atomic Weight[a]	Density,[b] kg m^{-3} ·10^{-3}	Crystal Structure	Phase Transition Temp., K	Superconducting Transition Temp., K	Curie Temp., K	Néel Temp., K	Debye Temperature at 0 K, K	Debye Temperature at 298 K, K	Melting Point, K	Boiling Point, K	Critical Temp., K
Iron	26	55.847	7.87[28]	b.c.c.-ferromag.[7] (α) b.c.c.-paramag.[7] (β) f.c.c.[7] (γ) b.c.c.[7] (δ)	1183[2] (β–γ) 1673[13] (γ–δ)		1043[40]		457 ±12[3]	373[3]	1810[19]	3160[20]	~~10350~~[27] 6750[123] 9400[109]
Krypton	36	83.80	0.003708[29] (at 273.2 K and 1 atm)	f.c.c.[16]						60[4] (at ~30 K)	116.6[5]	119.93[13]	209.4[15]
Lanthanum	57	138.91	6.18[42]	Double c.p.h.[8] (α) f.c.c.[2] (β) b.c.c.[2] (γ)	583[32] (α–β) 1141[32] (β–γ)	4.9[8] (α) 6.3[8] (β)			142 ±3[52]	135 ±5[44]	1193[5]	3713 ±70[3]	10500[109]
Lawrencium	103	(257)											
Lead	82	207.19	11.34[29]	f.c.c.[2]		7.193[5]			102 ±5[3]	87 ±1[3]	600.576[3,111]	2022 ±10[41]	5400[27] 4760[109]
Lithium	3	6.939	0.534[29]	b.c.c.[7]	Martensitic transformation at low temp.[56]				352 ±17[3]	448[3]	453.7[19]	1599[13]	4150[11] 3720[109]
Lutetium	71	174.97	9.85[29]	c.p.h.[2] (α) b.c.c.[2] (β)	Near m.p.[50] (α–β)				210[54]	116[3]	1923[19]	4140[3]	
Magnesium	12	24.312	1.74[29]	c.p.h.[7]					396 ±54[3]	330[3]	923[55]	1385[3]	3530[109]
Manganese	25	54.9380	7.43(α)[28] 7.29(β)[28] 7.18(γ)[28]	~~c.[7] (α)~~ b.c.c.[93] (α) c.[7] (β) ~~f.c.t.[7] (γ)~~ f.c.c.[93] (γ) b.c.c.[7] (δ)	1000[13] (α–β) 1374[13] (β–γ) 1410[13] (γ–δ)			95[5]	418 ±32[3]	363[3]	1517 ±3[5]	2360[13]	6050[109]
Mendelevium	101	(256)											
Mercury	80	200.59	13.546[29] 14.19[29] (at 234.25 K)	r.[7] (α) b.c.t.-pressure induced structure (β)	Martensitic transformation at low temp.[56]	4.153[5] (α) 3.949[5] (β)			~ 75[58]	92 ±8[3]	234.28[3,10]	629.73[3,10]	1733[27] 1705[109]
Molybdenum	42	95.94	10.24[42]	b.c.c.[2]		0.92[5,9]			459 ±11[3]	377[3]	2883[13]	5785 ±175[3]	17000[11] 16800[109]
Neodymium	60	144.24	7.007[29]	Double c.p.h.[8] (α) b.c.c.[32] (β)	1135[32] (α–β)			8[4] (ordinary) 19[4] (special)	159[3]	148 ±8[3]	1292[19]	2956[60]	7900[109]
Neon	10	20.183	0.0009002[29] (at 273.2 K and 1 atm)	f.c.c.[16]						60[4] (at ~30 K)	24.48[5]	27.23[5] 27.06[23]	44.5[15]

Name	Atomic Number	Atomic Weight[a]	Density,[b] kg m^{-3} · 10^{-3}	Crystal Structure	Phase Transition Temp., K	Superconducting Transition Temp., K	Curie Temp., K	Néel Temp., K	Debye Temperature at 0 K, K	Debye Temperature at 298 K, K	Melting Point, K	Boiling Point, K	Critical Temp., K
Neptunium	93	(237)	20.46[42]	orthorh.[2] (α) t.[2] (β) b.c.c.[2] (γ)	551[2] (α–β) 813[2] (β–γ)				121[3]	163[3]	913.2[5]	4150[3]	
Nickel	28	58.71	8.90[42]	f.c.c.[7]			631[40]		427 ± 14[3]	345[3]	1726[3, 10] 1726 ± 4[61]	3055[63]	6294[15] 11750[109]
Niobium	41	92.906	8.57[42]	b.c.c.[7]		9.13[5] 9.09[8] 9.1[9]			241 ± 13[3]	260[64]	2741 ± 27[3] 2688[65]	4813[66]	19000[109]
Nitrogen	7	14.0067	0.0012506[29]	c.[16] (α) h.[107] (β)	35.62[13] (α–β)					70[4] (at ~35 K)	63.29[5]	77.34[13, 23]	126.2[15]
Nobelium	102	(254)											
Osmium	76	190.2	22.48[29]	c.p.h.[2]		0.655[5] 0.65[8]			500[67]	400[68]	3283 ± 10[69]	5300 ± 100[70]	
Oxygen	8	15.9994	0.001429[29] (at 273.2 K and 1 atm)	b.c. orthorh.[7] (α) r.[7] (β) c.[7] (γ)	23.876 ± 0.01[112] (α–β) 43.818 ± 0.01[112] (β–γ)					250[4] (at ~125 K) 500[36] (at ~250 K)	54.8[5]	90.19[13] 90.18[23]	154.8[15]
Palladium	46	106.4	12.02[28]	f.c.c.[2]					283 ± 16[3]	275[14]	1825[3, 10]	3200[3]	
Phosphorus	15	30.9738	1.82[29] (β) 2.22[29] (γ) 2.69[29] (δ)	h. ?[7] (α) b.c.c.[7] (β) c.[7] (γ) f.c. orthorh.[17] (δ)	196[71] (α–β) 298.16[13] (β–γ) 298.16[13] (β–δ)				193[3] (white) 325[3] (red)	576[3] (white) 800[3] (red)	317.3[5] (white) 1300[72] (black)	553[13]	993.8[15]
Platinum	78	195.09	21.45[29]	f.c.c.[2]					234 ± 1[3]	225 ± 5[3]	2042[3, 10]	4100[3]	8280[15]
Plutonium	94	(242)	19.737[29] (at 298.2 K)	Simple monocl.[2] (α) b.c. monocl.[2] (β) f.c. orthorh.[2] (γ) f.c.c.[2] (δ) b.c.t.[2] (δ′) b.c.c.[2] (ε)	396.7[73] (α–β) 475[73] (β–γ) 591.4[73] (γ–δ) 729[73] (δ–δ′) 757 ± 3[73] (δ′–ε)				171[74]	176[74]	912.7[5]	3727[75]	
Polonium	84	(210)	9.3[29] (α) 9.5[29] (β)	Simple c.[7] (α) r.[7] (β)	327 ± 1.5[76] (α–β)				81[3]		527.2[5]	1235[20]	2281[15]
Potassium	19	39.102	0.86[29]	b.c.c.[7]					89.4 ± 0.5[3]	100[3]	336.8[5]	1027[35]	2450[11] 2140[109]
Praseodymium	59	140.907	6.769[29]	Double c.p.h.[8] (α) b.c.c.[2] (β)	1071[32] (α–β)			25[77]	85 ± 1[45]	138[78]	1192 ± 2[79]	3616[80]	8900[109]

Name	Atomic Number	Atomic Weight[a]	Density,[b] kg m^{-3} · 10^{-3}	Crystal Structure	Phase Transition Temp., K	Superconducting Transition Temp., K	Curie Temp., K	Néel Temp., K	Debye Temperature at 0 K, K	Debye Temperature at 298 K, K	Melting Point, K	Boiling Point, K	Critical Temp., K
Promethium	61	(145)		h.[7] (α) b.c.c.[120] (β)	1185[120] (α–β)			6[120]			1353 ± 10[81]	2730[3]	
Protactinium	91	(231)	15.37[42]	b.c.t.[2]		1.4[9]			159[3]	262[3]	1503[5]	4680[3]	
Radium	88	(226)	5[29]						89[3]		973.2[5]	1900[3]	
Radon	86	(222)	0.00973[29] (at 273.2 K and 1 atm)	f.c.c.[7]						400[4] (at ~200 K)	202.2[5]	211[13]	377.16[15]
Rhenium	75	186.2	21.1[42]	c.p.h.[2]		1.698[26]			429 ± 22[3]	275[23]	3453[5]	6035 ± 135[3]	20000[11]
Rhodium	45	102.905	12.45[42]	f.c.c.[7]	possible transformation at 1373–1473 K[57]				480 ± 32[3]	350[3]	2233[3, 10, 82]	3960 ± 60[3]	
Rubidium	37	85.47	1.53[29]	b.c.c.[2]					54 ± 4[3]	59[23]	312.04[5]	959[35]	2100[113,115,116] 2030[109]
Ruthenium	44	101.07	12.2[29]	c.p.h.[7] (α) ? (β) ? (γ) ? (δ)	1308[13, 121] (α–β) 1473[13, 121] (β–γ) 1773[13, 121] (γ–δ)	0.49[5, 9]			600[67]	415[3]	2523 ± 10[69]	4325 ± 25[3]	
Samarium	62	150.35	7.54[29]	r.[32] (α) b.c.c.[32] (β)	1190[32] (α–β)		14[8]	106[8]	116[45]	184 ± 4[3]	1345.2[83]	2140[3]	5400[109]
Scandium	21	44.956	3.00[42]	c.p.h.[2] (α) b.c.c.[2] (β)	1607[2] (α–β)				470 ± 80[52]	476[3]	1812[5]	3537 ± 30[3]	
Selenium	34	78.96	4.50[29] (α) 4.80[29] (β)	monocl.[7] (α) h.[7] (β) amorphous[7]	304[84, 117] (vitrification) 398[13] (vit.–β) 423[13] (α–β)	7.3[85] (at ~118 kbar)			151.7 ± 0.4[86]	89[36] (at ~45 K) 150[4] (at ~75 K)	490.2[5]	1009[13] (Se_6) 958.0[13] ($Se_{4.37}$) 1027[13] (Se_2)	1757[15]
Silicon	14	28.086	2.33[42]	d.[7]		7.5[47] (at 118–128 kbar)			647 ± 11[3]	692[3]	1685 ± 2[87]	2753[28]	5159[15]
Silver	47	107.870	10.5[29]	f.c.c.[2]					228 ± 3[3]	221[3]	1234.0[3, 13]	2468 ± 15[41]	7460[11]
Sodium	11	22.9898	0.9712[29]	b.c.c.[2]	Martensitic transformation at low temp.[56]				157 ± 1[3]	155 ± 5[3]	371.0[13]	1154[35]	2800[11] 2400[109]
Strontium	38	87.62	2.60[28]	f.c.c.[88] (α) c.p.h.[7] (β) b.c.c.[7] (γ)	488[88] (α–β) 878[88] (β–γ)				147 ± 1[22]	148[23]	1042[5]	1645[3]	3059[15] 3810[109]
Sulfur	16	32.064	2.07[29] (α) 1.96[29] (β)	r.[7] (α) monocl.[7] (β)	368.6[13] (α–β)				200[3] (β)	527[89] (α) 250[89] (α, at 40 K)	386.0[5] (α) 392.2[5] (β) Subl. 368.6[13] (at 0.0047 mm Hg)	717.75[3, 10]	1313[15]
Tantalum	73	180.948	16.6[42]	b.c.c.[2]		4.483[5] 4.48[9]			247 ± 13[3]	225[14]	3269[5]	5760 ± 60[3]	22000[11]

Name	Atomic Number	Atomic Weight[a]	Density,[b] kg m^{-3} · 10^{-3}	Crystal Structure	Phase Transition Temp., K	Superconducting Transition Temp., K	Curie Temp., K	Neel Temp., K	Debye Temperature at 0 K, K	Debye Temperature at 298 K, K	Melting Point, K	Boiling Point, K	Critical Temp., K
Technetium	43	(99)	11.50[29]	c.p.h.[2]		8.22[5] 11.2[9]			351[3]	422[3]	2473 ± 50[5]	5300[3]	
Tellurium	52	127.60	6.24[29] (α) 6.00[5] (amorph.)	h.[7] (α) ? (β)[7] amorph.[5]	621[13] (α–β)	3.3[8] (Te II, at 56 kbar)			141 ± 12[3]		722.7[5]	1163 ± 1[3]	2329[15]
Terbium	65	158.924	8.25[29]	c.p.h.[2,32] (α) b.c.c.[2] (β)	Near m.p.[2] (α–β)		219[90]	230[90]	150[91]	158[44]	1629[19]	3810[3]	
Thallium	81	204.37	11.85[29]	c.p.h.[2] (α) b.c.c.[2] (β)	508.3[5] (α–β)	2.39[5] 2.38[8] 2.37[9]			88 ± 1[3]	96[14]	576.2[19]	1939[92]	3219[15]
Thorium	90	232.038	11.7[42]	f.c.c.[2] (α) b.c.c.[2] (β)	1673 ± 25[93] (α–β)	1.368[5] 1.37[9]			170[94]	100[14]	2023[19]	4500[20]	14550[109]
Thulium	69	168.934	9.32[29]	c.p.h.[2] (α) b.c.c.[2] (β)	Near m.p.[50] (α–β)		22[95] (ferro.–antiferro.)	53[96]	127 ± 1[45]	167[44]	1818[5]	2266[97]	6430[109]
Tin	50	118.69	5.750[29] (α) 7.31[29] (β)	f.c.c.[7] (α) b.c.t.[7] (β) r.[29] (?)	286.2 ± 3[98] (α–β)	3.722[5] (β)			236 ± 24[3] (gray) 196 ± 9[3] (white)	254[3] (gray) 170[14] (white)	505.06[3,10]	2766 ± 14[3]	8000[11] 9300[109]
Titanium	22	47.90	4.5[29]	c.p.h.[7] (α) b.c.c.[7] (β)	1155[13] (α–β)	0.39[5,9]			426 ± 5[3]	380[14]	1953[99]	3586[100]	
Tungsten	74	183.85	19.3[29]	b.c.c.[2]		0.011[122]			388 ± 17[3]	312 ± 3[3]	3653[3,10,13]	6000 ± 200[3]	23000[11]
Uranium	92	238.03	19.07[28]	orthorh.[7] (α) t.[7] (β) b.c.c.[7] (γ)	37 ± 2[118] (α_0–α) 938[13] (α–β) 1049[13] (β–γ)	0.68[5] (α) 1.80[5] (γ)			200[94]	300[3]	1405.6 ± 0.6[101]	3950 ± 250[102]	12500[27] 12000[109]
Vanadium	23	50.942	6.1[28]	b.c.c.[2]		5.3[5] 5.03[9]			326 ± 54[3]	390[14]	2192 ± 2[61]	3582 ± 42[3]	11200[109]
Xenon	54	131.30	0.005851[29] (at 273.2 K and 1 atm)	f.c.c.[16]							161.2[26]	165.1[13]	289.75[15]
Ytterbium	70	173.04	7.02[42]	f.c.c.[32] (α) b.c.c.[32] (β)	1071[2,5] (α–β)				118[103]		1097[12]	1970[3]	4420[109]
Yttrium	39	88.905	4.47[29]	c.p.h.[32] (α) b.c.c.[32] (β)	1753[119] (α–β)				268 ± 32[3]	214[104]	1798[119]	3670[105]	8950[109]
Zinc	30	65.37	7.140[29]	c.p.h.[2]		0.875[5] 0.85[9]			316 ± 20[3]	237 ± 3[3]	692.655[3,110]	1175[106]	2169[15] 2910[109]
Zirconium	40	91.22	6.57[59]	c.p.h.[7] (α) b.c.c.[7] (β)	1135[13] (α–β)	0.546[5] 0.55[9]			289 ± 24[3]	250[14]	2125[19]	4650[20]	12300[109]

REFERENCES

(Crystal Structures, Transition Temperatures, and Other Pertinent Physical Constants of the Elements)

1. Farr, J.D., Giorgi, A.L., and Bowman, M.G., USAEC Rept. LA-1545, 1-13, 1953.
2. Elliott, R.P., Constitution of Binary Alloys, 1st Suppl., McGraw-Hill, 1965.
3. Gschneider, K.A, Jr., Solid State Physics (Sietz, F. and Turnbull, D., Editors), 16, 275-426, 1964.
4. Gopal, E.S.R., Specific Heat at Low Temperatures, Plenum Press, 1966.
5. Weast, R.C. (Editor), Handbook of Chemistry and Physics, 47th Ed., The Chemical Rubber Co., 1966-67.
6. Foster, K.W. and Fauble, L.G., J. Phys. Chem., 64, 958-60, 1960.
7. The Institution of Metallurgists, Annual Yearbook, pp. 68-73, 1960-61.
8. Meaden, G.T., Electrical Resistance of Metals, Plenum Press, 1965.
9. Matthias, B.T., Geballe, T.H., and Compton, V.B., Rev. Mod. Phys., 35, 1-22, 1963.
10. Stimson, H.F., J. Res. NBS, 42, 209, 1949.
11. Grosse, A.V., Rev. Hautes Tempér. et Réfract., 3, 115-46, 1966.
12. Spedding, F.H. and Daane, A.H., J. Metals, 6 (5), 504-10, 1954.
13. Rossini, F.D., Wagman, D.D., Evans, W.H., Levine, S., and Jaffe, I., NBS Circ. 500, 537-822, 1952.
14. deLaunay, J., Solid State Physics, 2, 219-303, 1956.
15. Gates, D.S. and Thodos, G., AIChE J., 6 (1), 50-4, 1960.
16. Gray, D.E. (Coordinating Editor), American Institute of Physics Handbook, McGraw-Hill, 1957.
17. Sasaki, K. and Sekito, S., Trans. Electrochem. Soc., 59, 437-60, 1931.
18. Anderson, C.T., J. Am. Chem. Soc., 52, 2296-300, 1930.
19. Trombe, F., Bull. Soc. Chim. (France), 20, 1010-2, 1953.
20. Stull, D.R. and Sinke, G.C., Thermodynamic Properties of the Elements in Their Standard State, American Chemical Soc., 1956.
21. Rinck, E., Ann. Chim. (Paris), 18 (10), 455-531, 1932.
22. Roberts, L.M., Proc. Phys. Soc. (London), B70, 738-43, 1957.
23. Zemansky, M.W., Heat and Thermodynamics, 4th Ed., McGraw-Hill, 1957.
24. Martin, A.J. and Moore, A., J. Less-Common Metals, 1, 85, 1959.
25. Hill, R.W. and Smith, P.L., Phil. Mag., 44 (7), 636-44, 1953.
26. Moffatt, W.G., Pearsall, G.W., and Wulff, J., The Structure and Properties of Materials, Vol. I, pp. 205-7, 1964.
27. Grosse, A.V., Temple Univ. Research Institute Rept., 1-40, 1960.
28. Lyman, T. (Editor), Metals Handbook, Vol. 1, 8th Ed., American Soc. for Metals, 1961.
29. Lange, N.A. (Editor), Handbook of Chemistry, Revised 10th Edition, McGraw-Hill, 1967.
30. Paule, R.C., Dissertation Abstr., 22, 4200, 1962.
31. Burk, D.L. and Friedberg, S.A., Phys. Rev., 111 (5), 1275-82, 1958.
32. Spedding, F.H. and Daane, A.H. (Editors), The Rare Earths, John Wiley, 1961.
33. McHargue, C.J., Yakel, H.L., and Letter, C.K., ACTA Cryst., 10, 832-33, 1957.
34. Arajs, S. and Colvin, R.V., J. Less-Common Metals, 4, 159-68, 1962.
35. Bonilla, C.F., Sawhney, D.L., and Makansi, M.M., Trans. Am. Soc. Metals, 55, 877, 1962.
36. Rosenberg, H.M., Low Temperature Solid State Physics, Oxford at Clarendon Press, 1965.
37. Arajs, S., J. Less-Common Metals, 4, 46-51, 1962.
38. Edwards, A.R. and Johnstone, S.T.M., J. Inst. Metals, 84 (8), 313-7, 1956.
39. Lagneborg, R. and Kaplow, R., ACTA Metallurgica, 15 (1), 13-24, 1967.
40. Kittel, C., Introduction to Solid State Physics, 3rd Ed., John Wiley, 1967.
41. Kirshenbaum, A.D. and Cahill, J.A., J. Inorg. and Nucl. Chem., 25 (2), 232-34, 1963.
42. Touloukian, Y.S. (Ed.), Thermophysical Properties of High Temperature Solid Materials, MacMillan, Vol. 1, 1967.
43. Griffel, M., Skochdopole, R.E., and Spedding, F.H., J. Chem. Phys., 25 (1), 75-9, 1956.
44. Gschneidner, K.A., Jr., Rare Earth Alloys, Van Nostrand, 1961.
45. Dreyfus, B., Goodman, B.B., Lacaze, A., and Trolliet, G., Compt. Rend., 253, 1764-6, 1961.

46. Spedding, F.H., Hanak, J.J., and Daane, A.H., Trans. AIME, 212, 379, 1958.
47. Buckel, W. and Wittig, J., Phys. Lett. (Netherland), 17 (3), 187-8, 1965.
48. Deardorff, D.K. and Kata, H., Trans. AIME, 215, 876-7, 1959.
49. Panish, M.B. and Reif, L., J. Chem. Phys., 38 (1), 253-6, 1963.
50. Miller, A.E. and Daane, A.H., Trans. AIME, 230, 568-72, 1964.
51. Spedding, F.H. and Daane, A.H., USAEC Rept. IS-350, 22-4, 1961.
52. Montgomery, H. and Pells, G.P., Proc. Phys. Soc. (London), 78, 622-5, 1961.
53. Kaufman, L. and Clougherty, E.V., ManLabs, Inc., Semi-Annual Rept. No. 2, 1963.
54. Lounasmaa, O.V., Proc. 3rd Rare Earth Conf., 1963, Gordon and Breach, New York, 1964.
55. Baker, H., WADC TR 57-194, 1-24, 1957.
56. Reed, R.P. and Breedis, J.F., ASTM STP 387, pp. 60-132, 1966.
57. Hansen, M., Constitution of Binary Alloys, 2nd Edition, McGraw-Hill, p. 1268, 1958.
58. Smith, P.L., Conf. Phys. Basses Temp., Inst. Intern. du Froid, Paris, 281, 1956.
59. Powell, R.W. and Tye, R.P., J. Less-Common Metals, 3, 202-15, 1961.
60. Yamamoto, A.S., Lundin, C.E., and Nachman, J.F., Denver Res. Inst. Rept., NP-11023, 1961.
61. Oriena, R.A. and Jones, T.S., Rev. Sci. Instr., 25, 248-51, 1954.
62. Smith, J.F., Carlson, O.N., and Vest, R.W., J. Electrochem. Soc., 103, 409-13, 1956.
63. Edwards, J.W. and Marshal, A.L., J. Am. Chem. Soc., 62, 1382, 1940.
64. Morin, F.J. and Maita, J.P., Phys. Rev., 129 (3), 1115-20, 1963.
65. Pendleton, W.N., ASD-TDR-63-164, 1963.
66. Woerner, P.F. and Wakefield, G.F., Rev. Sci. Instr., 33 (12), 1456-7, 1962.
67. Walcott, N.M., Conf. Phys. Basses Temp., Inst. Intern. du Froid, Paris, 286, 1956.
68. White, G.K. and Woods, S.B., Phil. Trans. Roy. Soc. (London), A251 (995), 273-302, 1959.
69. Douglass, R.W. and Adkins, E.F., Trans. AIME, 221, 248-9, 1961.
70. Panish, M.B. and Reif, L., J. Chem. Phys., 37 (1), 128-31, 1962.
71. Bridgman, P.W., J. Am. Chem. Soc., 36 (7), 1344-63, 1914.
72. Slack, G.A., Phys. Rev., A139 (2), 507-15, 1965.
73. Sandenaw, T.A. and Gibney, R.B., J. Phys. Chem. Solids, 6 (1), 81-8, 1958.
74. Sandenaw, T.A., Olsen, C.E., and Gibney, R.B., Plutonium 1960, Proc. 2nd Intern. Conf. (Grison, E., Lord, W.B.H., and Fowler, R.D., Editors), 66-79, 1961.
75. Mulford, R.N.R., USAEC Rept. LA-2813, 1-11, 1963.
76. Goode, J.M., J. Chem. Phys., 26 (5), 1269-71, 1957.
77. Cable, J.W., Moon, R.M., Koehler, W.C., and Wollan, E.O., Phys. Rev. Letters, 12 (20), 553-5, 1964.
78. Murao, T., Progr. Theoret. Phys. (Kyoto), 20 (3), 277-86, 1958.
79. Grigor'ev, A.T., Sokolovskaya, E.M., Budennaya, L.D., Iyutina, I.A., and Maksimona, M.V., Zhur. Neorg. Khim., 1, 1052-63, 1956.
80. Daane, A.H., USAEC AECD-3209, 1950.
81. Weigel, F., Angew. Chem., 75, 451, 1963.
82. Nassau, K. and Broyer, A.M., J. Am. Ceram. Soc., 45 (10), 474-8, 1962.
83. McKeown, J.J., State Univ. of Iowa, Ph.D. Dissertation, 1-113, 1958.
84. Abdullaev, G.B., Mekhtiyeva, S.I., Abdinov, D.Sh., and Aliev, G.M., Phys. Letters, 23 (3), 215-6, 1966.
85. Wittig, J., Phys. Rev. Letters, 15 (4), 159, 1965.
86. Fukuroi, T. and Muto, Y., Tohoku Univ. Res. Inst. Sci. Rept., A8, 213-22, 1956.
87. Olette, M., Compt. Rend., 244, 1033-6, 1957.
88. Sheldon, E.A., and King, A.J., ACTA Cryst., 6, 100, 1953.
89. Eastman, E.D. and McGavock, W.C., J. Am. Chem. Soc., 59, 145-51, 1937.
90. Arajs, S. and Colvin, R.V., Phys. Rev., A136 (2), 439-41, 1964.
91. Roach, P.R. and Lounasmaa, O.V., Bull. Am. Phys. Soc., 7, 408, 1962.
92. Shchukarev, S.A., Semenov, G.A., and Rat'kovskii, I.A., Zh. Neorgan. Khim., 7, 469, 1962.
93. Pearson, W.B., A Handbook of Lattice Spacings and Structures of Metals and Alloys, Pergamon Press, 1958.

94. Smith, P.L. and Walcott, N.M., Conf. Phys. Basses Temp., Inst. Intern. du Froid, 283, 1956.

95. Davis, D.D. and Bozorth, R.M., Phys. Rev., 118 (6), 1543-5, 1960.

96. Aliev, N.G. and Volkenstein, N.V., Soviet Physics - JETP, 22 (5), 997-8, 1966.

97. Spedding, F.H., Barton, R.J., and Daane, A.H., J. Am. Chem. Soc., 79, 5160, 1957.

98. Raynor, G.V. and Smith, R.W., Proc. Roy. Soc. (London), A244, 101-9, 1958.

99. Savitskii, E.M. and Burhkanov, G.S., Zhur. Neorg. Khim., 2, 2609-16, 1957.

100. Argent, B.B. and Milne, J.G.C., Niobium, Tantalum, Molybdenum and Tungsten, Elsevier Publ. Co. (Quarrell, A.G., Editor), pp. 160-8, 1961.

101. Argonne National Laboratory, USAEC Rept. ANL-5717, 1-67, 1957.

102. Holden, A.N., Physical Metallurgy of Uranium, Addison-Wesley, 1958.

103. Lounasmaa, O.V., Phys. Rev., 129, 2460-4, 1963.

104. Jennings, L.D., Miller, R.E., and Spedding, F.H., J. Chem. Phys., 33 (6), 1849-52, 1960.

105. Ackerman, R.J. and Rauh, E.G., J. Chem. Phys., 36 (2), 448-52, 1962.

106. Rosenblatt, G.M. and Birchenall, C.E., J. Chem. Phys., 35 (3), 788-94, 1961.

107. Streib, W.E., Jordan, T.H., and Lipscomb, W.N., J. Chem. Phys., 37 (12), 2962-5, 1962.

108. Samsonov, G.V. (Editor), Handbook of the Physicochemical Properties of the Elements, Plenum Press, 1968.

109. Kopp, I.Z., Russ. J. Phys. Chem., 41 (6), 782-3, 1967.

110. Stimson, H.F., in Temperature, Its Measurement and Control in Science and Industry (Herzfeld, C.M., Ed.), Vol. 3, Part 1, Reinhold, New York, pp. 59-66, 1962.

111. McLaren, E.H., in Temperature, Its Measurement and Control in Science and Industry (Herzfeld, C.M., Ed.), Vol. 3, Part 1, Reinhold, New York, pp. 185-98, 1962.

112. Orlova, M.P., in Temperature, Its Measurement and Control in Science and Industry (Herzfeld, C.M., Ed.), Vol. 3, Part 1, Reinhold, New York, pp. 179-83, 1962.

113. Grosse, A.V., J. Inorg. Nucl. Chem., 28, 2125-9, 1966.

114. Hochman, J.M. and Bonilla, C.F., in Advances in Thermophysical Properties at Extreme Temperatures and Pressures (Gratch, S., Ed.), ASME 3rd Symposium on Thermophysical Properties, Purdue University, March 22-25, 1965, ASME, pp. 122-30, 1965.

115. Dillon, I.G., Illinois Institute of Technology, Ph.D. Thesis, June 1965.

116. Hochman, J.M., Silver, I.L., and Bonilla, C.F., USAEC Rept. CU-2660-13, 1964.

117. Abdullaev, G.B., Mekhtieva, S.I., Abdinov, D.Sh., Aliev, G.M., and Alieva, S.G., Phys. Status Solidi, 13(2), 315-23, 1966.

118. Fisher, E.S. and Dever, D., Phys. Rev., 2, 170(3), 607-13, 1968.

119. Beaudry, B.J., J. Less-Common Metals, 14(3), 370-2, 1968.

120. Williams, R.K. and McElroy, D.L., USAEC Rept. ORNL-TM 1424, 1-32, 1966.

121. Jaeger, F.M. and Rosenbaum, E., Proc. Nederland Akademie van Wetenschappen, 44, 144-52, 1941.

122. Gibson, J.W. and Hein, R.A., Phys. Letters, 12(25), 688-90, 1964.

123. Grosse, A.V., Research Institute of Temple Univ., Report on USAEC Contract No. AT (30-1) - 2082, 1-71, 1965

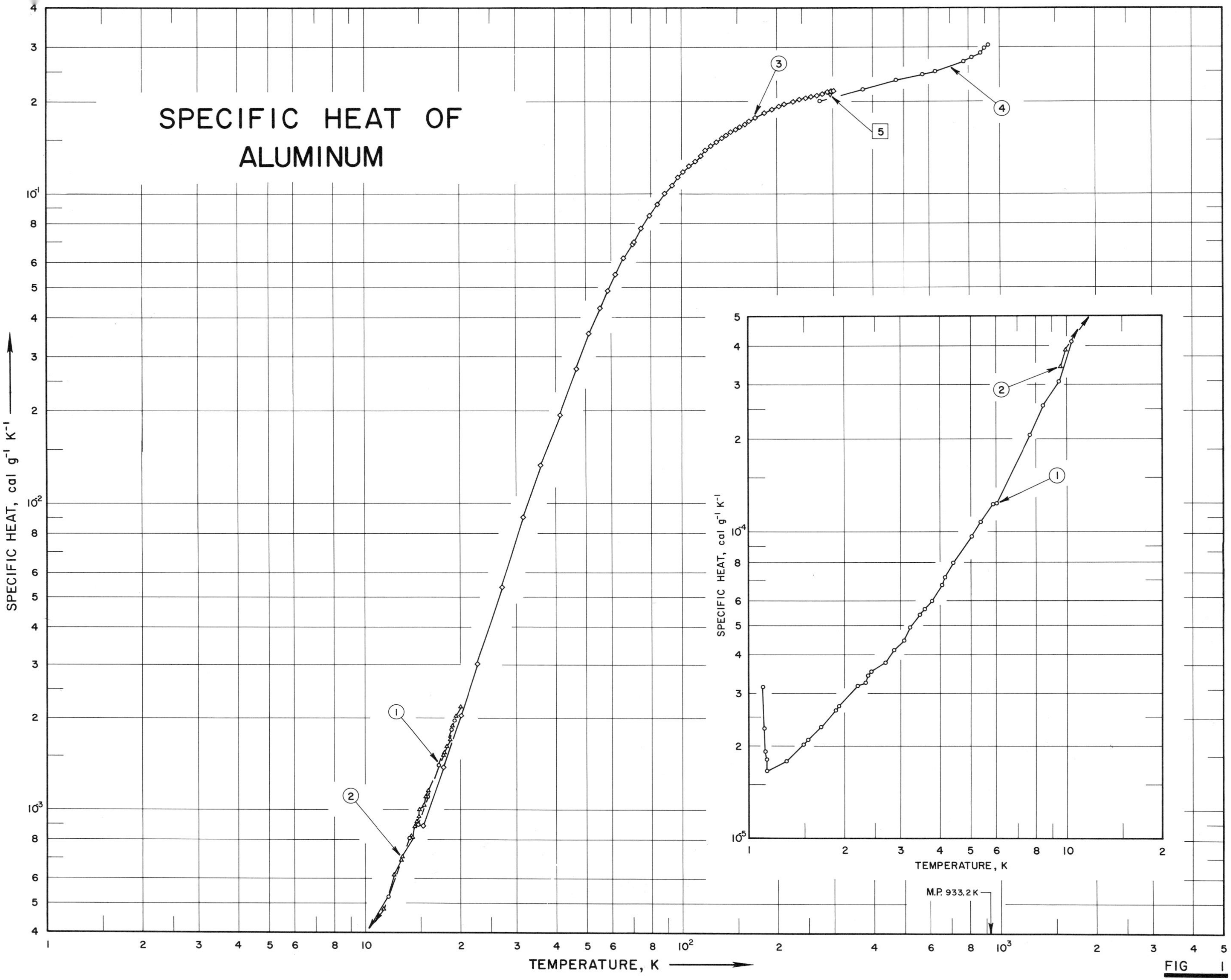

SPECIFIC HEAT OF ALUMINUM

FIG 1

SPECIFICATION TABLE NO. 1 SPECIFIC HEAT OF ALUMINUM

(Impurity <0.20% each; total impurities <0.50%)

[For Data Reported in Figure and Table No. 1]

Curve No.	Ref. No.	Year	Temp. Range, K	Reported Error, %	Name and Specimen Designation	Composition (weight percent), Specifications and Remarks
1	2	1937	1.1-19			99.7 Al; liquid helium atmosphere.
2	2	1937	15-20			99.7 Al; solid and liquid hydrogen atmosphere.
3	4	1941	15-302	0.1-3.0		99.944 Al; sample supplied by the Research Lab of the Aluminum Company of America; melted, and cooled for 2 days to produce single crystals.
4	3	1953	273-923	<5.0	Al - wire	99.9 Al, 0.05 SiO_2, and 0.03 B.
5	1	1961	295	±5.0		
6	5	1962	1.1-1.2	0.88		99.995 Al, 0.00025 Fe, <0.0001 Si, and 0.00005 Cu; 1.4×10^{-5} mm Hg vacuum; melted, etched in dilute aqua regia, annealed under vacuum at 600 C for 91 hrs and cooled gradually to room temperature during 24 hrs; etched again for 10 min; annealed for 165 hrs at 585 C and cooled gradually to room temperature during 114 hrs.
7	6	1962	1.17-1.18	±3.0	Al - I	99.99 Al, 0.009 Si, 0.001 Mg, <0.0008 Cu, <0.0006 Fe, and 0.0003 Mn; two large single crystals; zone-refined.
8	6	1962	1.18-1.19	±3.0	Al - II	99.99 Al, 0.01 Si, 0.001 Mn, 0.0005 Ti, <0.0005 Cu, <0.0005 Fe, and 0.0002 Mg; about 6 single crystals of equal size; zone-refined.
9	6	1962	1.18-1.19	±3.0	Al - III	99.99 Al, 0.009 Si, 0.004 Mg, 0.001 Mn, 0.0005 Ti, <0.0005 Cu, and <0.0005 Fe; polycrystalline.
10	102	1962	323-573	3.0-5.0		
11	5	1962	1.0-1.2	0.88	Superconducting	99.995 Al, 0.00025 Fe, <0.0001 Si, and 0.00005 Cu; 1.4×10^{-5} mm Hg vacuum; melted, etched in dilute aqua regia, annealed under vacuum at 600 C for 91 hrs and cooled gradually to room temperature during 24 hrs; etched again for 10 min; annealed for 165 hrs at 585 C and cooled gradually to room temperature during 114 hrs.
12	6	1962	1.1-1.2	±3.0	Al - III cooling	99.99 Al, 0.009 Si, 0.004 Mg, 0.001 Mn, 0.0005 Ti, <0.0005 Cu, and <0.0005 Fe; polycrystalline.
13	179	1924	373-873	1		NBS standard.
14	261	1934	55-296			99.985 Al; annealed; heated in high vacuum for 18 hrs at 460 C.
15	261	1934	56-291			99.985 Al; single crystal; hard drawn aluminum.
16	262	1937	398-673			
17	263	1939	373-873			
18	264	1951	90-373			99.5 Al.
19	265	1959	0.1-4.0	<0.5		99.998 Al, 0.002 Cu; polycrystalline with grain size 3-5 mm; cast in a vacuum; vacuum annealed at 450 C for 48 hrs; 300 gauss magnetic field.
20	265	1959	0.2-1.2	<0.5		Same as above; zero magnetic field; superconducting.

DATA TABLE NO. 1 SPECIFIC HEAT OF ALUMINUM

[Temperature, T, K; Specific Heat, C_p, Cal g^{-1} K^{-1}]

CURVE 1

T	C_p
1.110	3.152×10^{-5}
1.111	3.181*
1.122	3.109*
1.129	2.291
1.139	1.926
1.140	1.802
1.141	1.668
1.312	1.798
1.482	2.038
1.544	2.100
1.690	2.321
1.873	2.636
1.911	2.716
2.050	2.742*
2.201	3.175
2.269	3.263*
2.311	3.149*
2.330	3.240
2.353	3.468*
2.371	3.370
2.432	3.520
2.486	3.505*
2.551	3.781*
2.692	3.766
2.859	4.133
3.070	4.463
3.153	4.852*
3.213	4.918
3.268	4.951*
3.453	5.140
3.589	5.667
3.652	5.730*
3.786	5.997
3.901	6.179*
4.072	6.768
4.148	7.153
4.400	7.987
4.485	8.284*
5.022	9.674
5.093	9.826*
5.367	1.109×10^{-4}*
5.382	1.072
5.393	1.060*
5.871	1.236
6.003	1.245

CURVE 1 (contd)

T	C_p
7.664	2.061×10^{-4}
8.426	2.567
9.463	3.072
10.479	4.133
11.722	5.208
12.185	6.067*
13.726	8.095
14.713	8.918
15.817	1.100×10^{-3}
17.121	1.398
17.951	1.527*
18.875	1.821
19.280	1.958

CURVE 2

T	C_p
9.581	3.432×10^{-4}
9.766	3.618*
9.901	3.870
10.288	3.664*
11.503	4.785
12.424	6.075
13.072	6.887
13.219	7.072
14.217	8.106
14.655	9.188
14.727	8.814
14.785	9.451
14.899	1.006×10^{-3}
15.083	1.047*
15.307	1.035
15.458	9.896×10^{-4}*
15.600	1.106
15.742	1.090*
15.820	1.158
17.649	1.507
17.763	1.536
18.095	1.609
18.388	1.759*
18.598	1.694
18.850	1.875
19.286	1.995*
19.354	1.930*
19.442	2.026
20.003	2.194

CURVE 3

T	C_p
15.29	8.9×10^{-4}
17.68	1.4×10^{-3}
20.08	2.0
22.67	3.0
27.01	5.4
31.61	9.0
35.94	1.33×10^{-2}
40.68	1.94
45.98	2.73
51.11	3.59
55.18	4.29
58.59	4.88
61.90	5.46
65.79	6.13
70.16	6.90
70.52	6.96
74.66	7.69
79.35	8.45
84.00	9.21
88.52	9.90
93.63	1.069×10^{-1}
97.97	1.135
102.24	1.183
106.65	1.235
111.17	1.288
115.78	1.342
119.44	1.388*
119.74	1.389
124.85	1.434
129.77	1.481
134.95	1.524
138.91	1.557
143.55	1.595
148.50	1.628
153.54	1.665
158.60	1.697
163.66	1.736
167.26	1.762*
172.78	1.782
178.45	1.816*
183.79	1.841
189.05	1.865*
194.06	1.890
199.45	1.913*
204.92	1.932

CURVE 3 (contd)

T	C_p
209.38	1.943×10^{-1}*
210.52	1.959
214.35	1.958*
216.53	1.974*
220.40	1.977*
225.75	1.993
231.37	2.009*
236.95	2.031
242.44	2.044*
247.80	2.051
253.32	2.065*
257.99	2.077
263.44	2.091*
268.80	2.095
273.03	2.109*
278.57	2.121
284.01	2.129*
289.65	2.152
295.40	2.155*
295.94	2.150
301.60	2.169

CURVE 4

T	C_p
273.15	2.00×10^{-1}
373	2.19
473	2.35
573	2.46
673	2.57*
773	2.70
823	2.78
873	2.89
898	2.97
923	3.05

CURVE 5

T	C_p
295	2.1×10^{-1}

CURVE 6*

T	C_p
1.056_3	1.461×10^{-5}
1.068_0	1.481
1.079_1	1.496
1.090_6	1.513
1.107_4	1.541
1.118_0	1.565
1.126_0	1.571
1.140_6	1.586
1.148_5	1.598
1.156_7	1.615
1.162_3	1.620
1.166_5	1.629
1.170_0	1.630
1.173_5	1.640
1.177_3	1.638
1.180_8	1.650
1.182_5	1.652
1.184_3	1.655
1.186_1	1.650
1.189_0	1.653

CURVE 7*

T	C_p
1.170_7	3.64×10^{-5}
1.171_1	3.63
1.171_4	3.59
1.171_4	3.62
1.171_7	3.66
1.171_9	3.68
1.172_0	3.61
1.172_1	3.62
1.172_1	3.67
1.172_3	3.56
1.172_3	3.66
1.172_5	3.46
1.172_6	3.28
1.172_6	3.32
1.172_8	2.82
1.172_8	3.05
1.172_9	2.62
1.173_1	2.41
1.173_2	2.33
1.173_5	2.26
1.173_6	2.16
1.173_9	2.11

CURVE 7 (contd)

T	C_p
1.173_9	2.17×10^{-5}
1.174_3	2.11
1.174_4	2.10
1.174_6	2.14
1.174_8	2.09
1.175_0	2.12
1.175_1	2.08
1.175_3	2.08
1.175_6	2.02
1.175_7	2.03
1.175_9	1.98
1.176_0	1.77
1.176_1	1.71
1.176_2	1.67
1.176_4	1.60
1.176_4	1.58
1.176_6	1.59
1.176_8	1.58
1.176_8	1.57
1.177_3	1.57
1.177_8	1.54
1.178_5	1.58
1.178_9	1.58

CURVE 8*

T	C_p
1.182_3	3.56×10^{-5}
1.183_0	3.60
1.183_5	3.59
1.183_8	3.60
1.184_0	3.56
1.184_3	3.52
1.184_8	3.50
1.185_2	3.47
1.185_5	3.48
1.185_9	3.46
1.186_2	3.41
1.186_4	3.36
1.186_6	3.31
1.186_8	3.26
1.187_1	3.21
1.187_6	3.06
1.188_0	2.95
1.188_2	2.76
1.188_3	2.63

*Not shown on plot

DATA TABLE NO. 1 (continued)

T	C_p
CURVE 8 (cont.)*	
1.188_4	2.22×10^{-5}
1.188_5	2.04
1.188_6	1.58
1.189_0	1.57
1.189_2	1.58
1.189_5	1.57
1.190_0	1.58
CURVE 9*	
1.180_2	3.57×10^{-5}
1.180_4	3.54
1.180_7	3.54
1.181_0	3.52
1.181_2	3.49
1.181_4	3.49
1.181_4	3.48
1.181_7	3.44
1.181_8	3.37
1.182_0	3.30
1.182_2	3.17
1.182_4	3.04
1.182_5	2.99
1.182_5	2.95
1.182_6	2.83
1.182_6	2.82
1.182_7	2.80
1.182_7	2.82
1.182_8	2.64
1.182_8	2.62
1.183_0	2.47
1.183_1	2.37
1.183_2	2.44
1.183_2	2.38
1.183_3	2.36
1.183_3	2.26
1.183_4	2.16
1.183_5	2.10
1.183_6	2.01
1.183_6	1.69
1.183_6	1.73
1.183_7	1.72
1.183_8	1.62
1.183_8	1.61
1.184_2	1.58
1.184_3	1.47
CURVE 9 (cont.)*	
1.184_3	1.52×10^{-5}
1.184_5	1.47
1.184_5	1.51
1.184_6	1.51
1.184_7	1.47
1.184_7	1.51
1.184_9	1.47
1.184_9	1.51
1.185_1	1.47
1.185_2	1.50
1.185_2	1.48
1.185_3	1.54
1.185_6	1.50
1.185_8	1.49
CURVE 10*	
323.0	1.94×10^{-1}
373.0	2.07
423.0	2.17
473.0	2.26
523.0	2.34
573.0	2.41
CURVE 11*	
1.071_2	3.038×10^{-5}
1.077_9	3.063
1.084_6	3.080
1.093_5	3.109
1.102_1	3.204
1.107_0	3.229
1.112_5	3.241
1.118_1	3.297
1.137_0	3.385
1.148_7	3.433
1.154_6	3.477
1.157_3	3.495
1.160_2	3.497
1.163_0	3.516
1.165_2	3.529
1.167_2	3.533
1.168_8	3.555
1.170_5	3.561
1.172_2	3.580
1.173_8	3.577
1.176_0	3.590
CURVE 12*	
1.181_8	3.40×10^{-5}
1.182_0	3.22
1.182_2	3.11
1.182_6	2.95
1.182_8	2.91
1.182_9	2.82
1.183_0	2.43
1.183_0	1.97
1.183_0	2.91
1.183_1	1.81
1.183_1	2.16
1.183_2	1.69
1.183_2	1.64
1.183_2	1.70
1.183_3	1.64
1.183_3	1.62
1.183_4	1.57
1.183_4	1.49
1.183_5	1.54
1.183_7	1.60
1.183_8	1.53
1.183_8	1.60
1.184_2	1.57
1.184_3	1.41
1.184_3	1.56
1.184_6	1.50
1.184_8	1.57
1.185_0	1.46
1.185_3	1.46
1.185_9	1.50
CURVE 13*	
373.0	2.2832×10^{-1}
473.0	2.3795
573.0	2.4759
673.0	2.5760
773.0	2.6723
873.0	2.7687
CURVE 14*	
54.80	4.184×10^{-2}
58.42	4.777
61.84	5.456
69.99	6.879
CURVE 14 (cont.)*	
83.96	9.106×10^{-2}
97.68	1.110×10^{-1}
112.40	1.309
124.60	1.427
141.00	1.571
164.20	1.722
186.20	1.827
199.40	1.890
218.90	1.967
243.80	2.033
257.50	2.060
278.90	2.112
296.30	2.128
CURVE 15*	
56.18	4.362×10^{-2}
59.60	5.070
63.86	5.789
75.44	7.753
79.99	8.510
84.45	9.203
90.24	1.009×10^{-1}
92.85	1.041
100.60	1.150
112.50	1.295
126.50	1.440
141.70	1.574
163.40	1.719
185.00	1.821
200.40	1.892
219.60	1.964
241.10	2.021
257.40	2.052
272.80	2.088
290.60	2.123
CURVE 16*	
398.0	2.200×10^{-1}
473.0	2.310
573.0	2.405
673.0	2.427
CURVE 17*	
373.0	2.127×10^{-1}
473.0	2.210
573.0	2.290
673.0	2.380
773.0	2.468
873.0	2.593
CURVE 18*	
90.0	1.79×10^{-1}
293.0	1.79
293.0	2.24
373.0	2.24
CURVE 19*	
Series 1	
1.2182	1.492×10^{-5}
1.3017	1.601
1.4064	1.740
1.5300	1.906
1.6774	2.108
1.8286	2.321
1.9832	2.544
2.1669	2.817
2.3866	3.153
2.6017	3.496
2.8060	3.838
3.0409	4.247
3.3163	4.750
3.6431	5.398
4.0097	6.194
Series 2	
1.1879	1.456×10^{-5}
1.2565	1.546
1.3532	1.673
1.4668	1.822
1.6012	2.005
1.7632	2.228
1.9313	2.469
2.1047	2.721
2.3121	3.037
2.5138	3.356
CURVE 19 (cont.)*	
Series 2 (cont.)	
2.7040	3.669×10^{-5}
2.9229	4.042
3.1783	4.496
3.4773	5.059
3.8296	5.791
Series 3	
2.7818	3.800×10^{-5}
3.0774	4.154
3.3951	4.903
3.7374	5.597
Series 4	
0.1185	1.54×10^{-6}
0.1346	1.74
0.1535	2.00
0.1760	2.20
0.1976	2.49
0.2171	2.73
0.2418	2.99
0.2688	3.31
0.2945	3.59
0.3222	3.91
0.3562	4.32
0.3875	4.76
Series 5	
0.118	1.40×10^{-6}
0.1307	1.72
0.1704	2.18
0.1863	2.37
0.2002	2.55
0.2130	2.67
0.2326	2.90
0.2576	3.15
0.2794	3.42
0.3040	3.74
0.3320	4.00
0.3640	4.43
0.3689	4.50
0.4080	4.93

*Not shown on plot

DATA TABLE NO. 1 (continued)

T	C_p
CURVE 19 (cont.)*	
Series 5 (cont.)	
0.4643	5.56×10^{-6}
0.5334	6.42
0.5908	7.12
0.6524	7.87
0.3413	4.17
0.4164	5.08
0.4755	5.75
0.5422	6.56
0.6046	7.32
0.6668	8.18
0.7240	8.80
0.7925	9.691
0.8712	1.066×10^{-5}
0.9582	1.179
1.0499	1.301
Series 6	
0.3082	3.77×10^{-6}
0.3599	4.41
0.4045	4.91
0.4565	5.51
0.5139	6.45
0.5529	6.72
0.5946	7.30
0.6357	7.72
0.6945	8.44
0.7567	9.247
0.8303	1.020×10^{-5}
0.8697	1.072
0.9319	1.148
1.0159	1.256
1.1018	1.364
1.1924	1.464
CURVE 20*	
Series 1	
1.1290	3.14×10^{-5}
1.1361	3.17
1.1513	3.22
1.1526	3.23

T	C_p
CURVE 20 (cont.)*	
Series 2	
0.2402	2.03×10^{-7}
0.2695	3.56
0.2953	5.21
0.3182	8.11
0.3272	8.77
0.3503	1.213×10^{-6}
0.3889	1.860
0.4210	2.418
0.4731	3.610
0.5329	5.240
0.5932	7.255
0.6530	9.345
0.7099	1.156×10^{-5}
0.7732	1.425
0.8296	1.655
0.8471	1.748
0.9096	2.035
0.9961	2.452
1.0745	2.848
Series 3	
0.2017	1.080×10^{-7}
0.2055	9.740×10^{-8}
0.2058	9.660
0.2264	1.440×10^{-7}
0.2314	1.550
0.2323	1.800
Series 4	
0.1887	5.800×10^{-8}
0.2239	1.580×10^{-7}
0.2423	2.040
0.2600	2.770
Series 5	
0.1718	7.700×10^{-8}
0.2103	1.180×10^{-7}
0.2302	1.480
0.2386	1.780
0.2284	1.600
0.2514	2.49

T	C_p
CURVE 20 (cont.)*	
Series 5 (cont.)	
0.2693	3.53×10^{-7}
0.2891	4.840
0.3054	6.62
0.3126	7.21
0.9049	2.010×10^{-5}
0.9736	2.356
1.0381	2.659
1.1124	3.051
Series 6	
0.3191	8.020×10^{-7}
0.3840	1.750×10^{-6}
0.3852	1.780
0.4432	2.980
0.5210	5.030
0.5803	6.840
0.6382	8.800
0.6934	1.090×10^{-5}
0.7462	1.309
1.0333	2.646
1.1078	3.038

*Not shown on plot

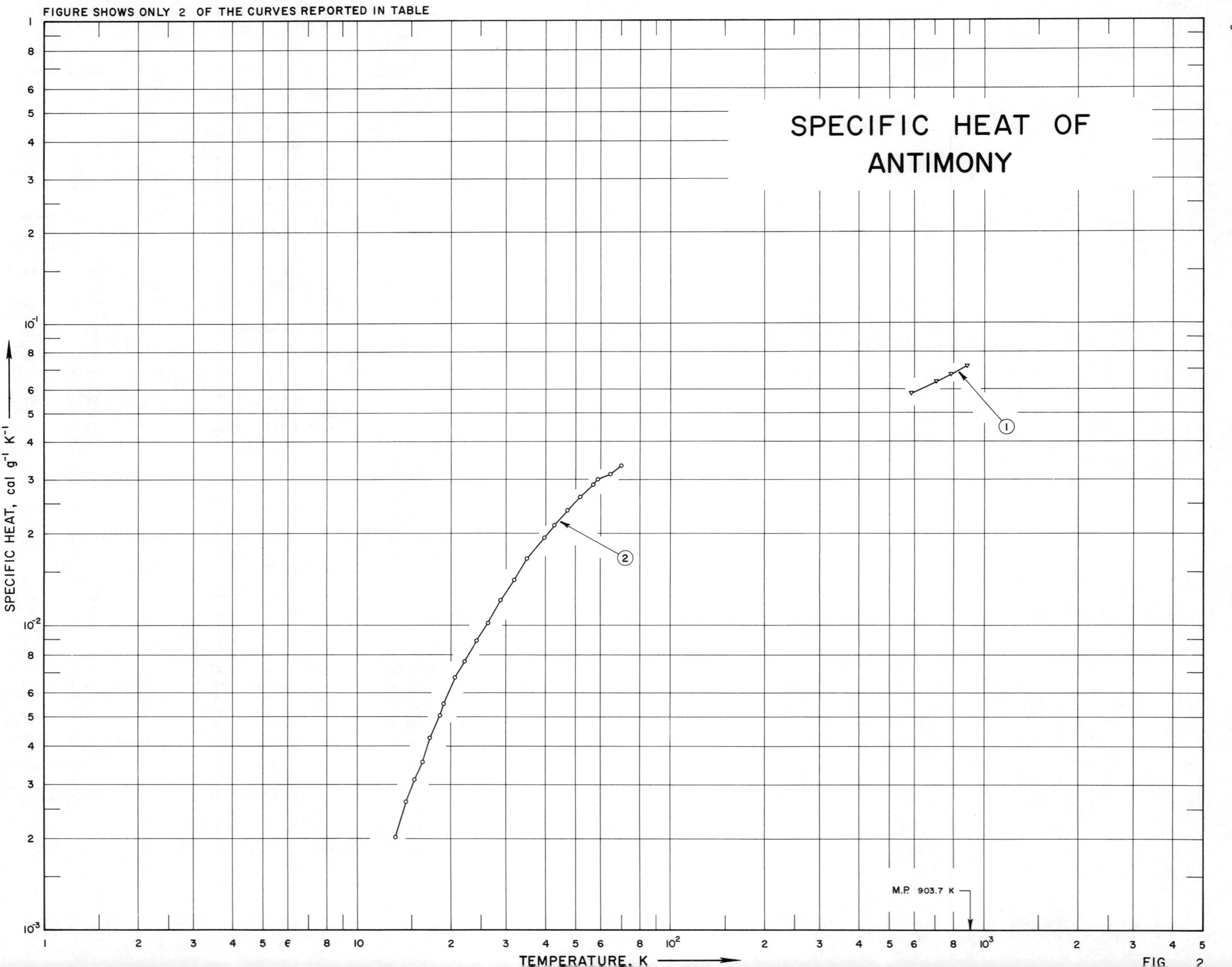
FIGURE SHOWS ONLY 2 OF THE CURVES REPORTED IN TABLE
SPECIFIC HEAT OF
ANTIMONY
1
2
M.P. 903.7 K
SPECIFIC HEAT, cal g⁻¹ K⁻¹
TEMPERATURE, K
FIG 2

SPECIFICATION TABLE NO. 2 SPECIFIC HEAT OF ANTIMONY

(Impurity <0.20% each; total impurities <0.50%)

[For Data Reported in Figure and Table No. 2]

Curve No.	Ref. No.	Year	Temp. Range, K	Reported Error, %	Name and Specimen Designation	Composition (weight percent), Specifications and Remarks
1	152	1961	587-885	≤0.2		99.99 Sb; purified by zone recrystallization.
2	152	1953	13-70			99.90 Sb, 0.03 As, 0.015 Fe, 0.015 S, and 0.01 Pb; sample supplied by the Johnson, Matthey Co.
3	267	1920	80-98			
4	268	1926	373-1273			99.735 Sb, 0.145 S, 0.06 Fe, 0.04 Pb, and 0.02 Cu.
5	180	1930	56-293	1		<0.2 total of As, Pb, and insoluble matter; density = 6.74 g cm^{-3} at 24.1 C.

DATA TABLE NO. 2 SPECIFIC HEAT OF ANTIMONY

[Temperature, T, K; Specific Heat, C_p, Cal $g^{-1}K^{-1}$]

T	C_p	T	C_p
CURVE 1		CURVE 4*	
587.0	5.81×10^{-2}	423.0	5.05×10^{-2}
706.0	6.36	523.0	5.20
786.0	6.73	623.0	5.37
885.0	7.19	723.0	5.52
		823.0	5.69
CURVE 2		948.0	6.56
		1023.0	6.56
13.21	2.02×10^{-3}	1123.0	6.56
13.44	2.04*	1223.0	6.56
14.44	2.64		
15.32	3.11	CURVE 5*	
16.19	3.56		
17.10	4.28	66.0	3.217×10^{-2}
17.26	4.29*	69.6	3.354
18.49	5.06	75.7	3.562
18.86	5.52	85.8	3.780
20.57	6.77	93.1	3.860
22.18	7.63	104.6	4.092
24.06	8.94	114.8	4.291
26.26	1.048×10^{-2}	127.7	4.395
28.78	1.211	139.7	4.461
31.73	1.418	157.9	4.598
34.92	1.653	169.4	4.674
39.55	1.941	181.9	4.703
42.64	2.137	193.0	4.755
47.09	2.374	202.6	4.792
51.70	2.628	214.5	4.828
56.58	2.894	227.0	4.840
58.95	3.020	240.6	4.881
64.33	3.136	252.9	4.865
69.76	3.339	264.5	4.879
		278.4	4.963
CURVE 3*		293.2	4.988
80.4	4.26×10^{-2}		
81.6	4.16		
83.7	4.14		
85.6	4.29		
86.2	4.29		
92.0	4.38		
93.3	4.39		
96.0	4.40		
98.1	4.49		

*Not shown on plot

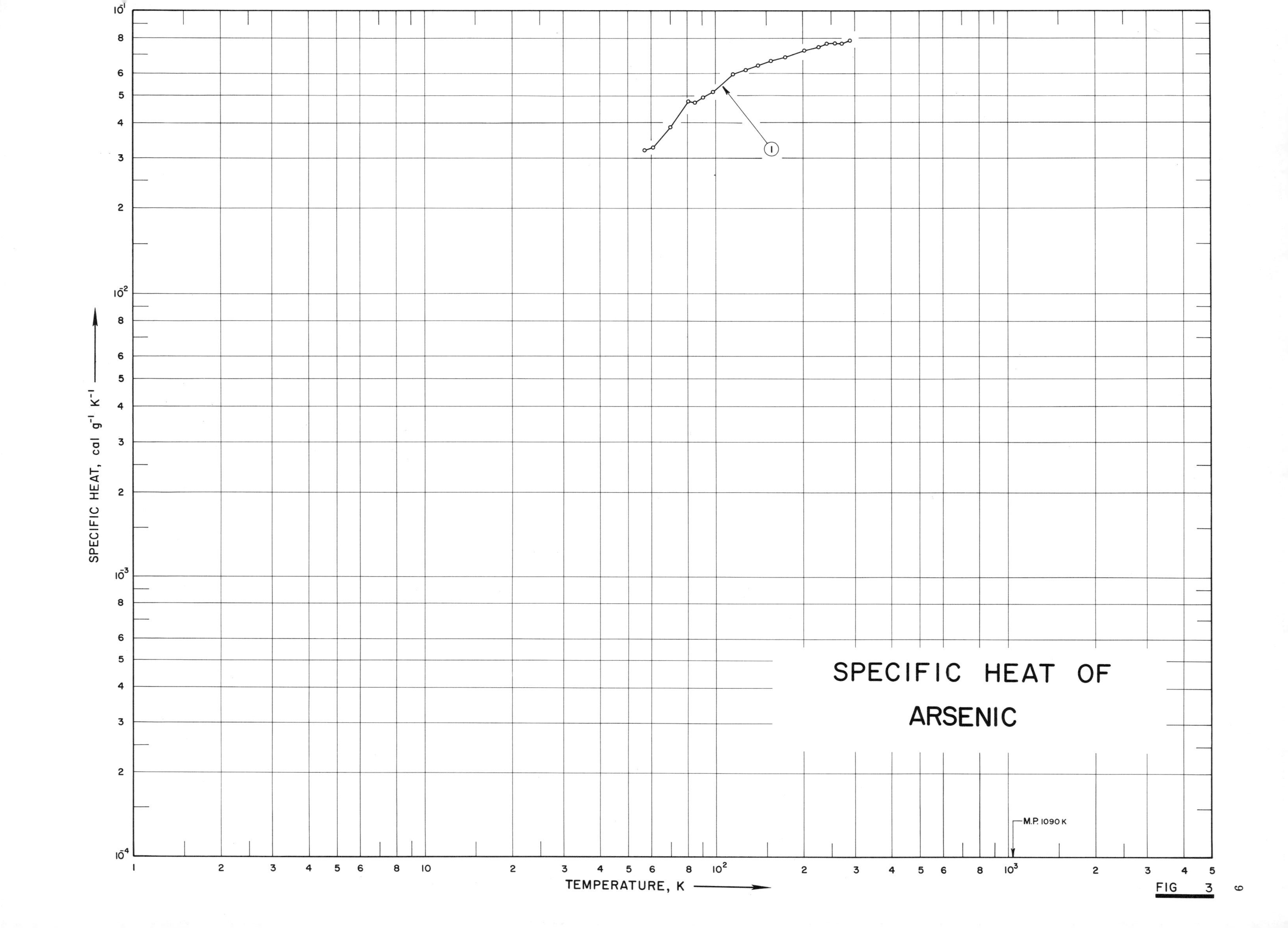

SPECIFIC HEAT OF
ARSENIC
1
M.P. 1090 K
SPECIFIC HEAT, cal g^{-1} K^{-1}
TEMPERATURE, K
FIG 3

SPECIFICATION TABLE NO. 3 SPECIFIC HEAT OF ARSENIC

(Impurity $<0.20\%$ each; total impurities $<0.50\%$)

[For Data Reported in Figure and Table No. 3]

Curve No.	Ref. No.	Year	Temp. Range, K	Reported Error, %	Name and Specimen Designation	Composition (weight percent), Specifications and Remarks
1	260	1930	57-291			>99.8 As and 0.13 Sb; density = 5.48 g cm^{-3}.
2	424	1967	0.6-4.5	1.0-2.0		99.9999 As; sample supplied by Cominco Products Inc.

DATA TABLE NO. 3 SPECIFIC HEAT OF ARSENIC

[Temperature, T, K; Specific Heat, C_p, Cal g^{-1} K^{-1}]

T	C_p
CURVE 1	
57.2	3.202×10^{-2}
61.1	3.279
70.2	3.854
80.9	4.781
85.2	4.717
91.0	4.940
98.6	5.180
115.5	5.978
128.2	6.171
140.6	6.403
155.5	6.657
174.4	6.853
202.7	7.268
212.7	7.385*
227.1	7.460
242.4	7.661
258.9	7.693
267.2	7.649*
272.8	7.676
285.7	7.801*
291.0	7.871
CURVE 2	
0.582	4.300×10^{-7}
0.595	4.447
0.603	4.479
0.615	4.578
0.615	4.610
0.620	4.635
0.624	4.683
0.628	4.731
0.630	4.693
0.631	4.737
0.635	4.747
0.637	4.788
0.644	4.855
0.646	4.878
0.658	4.983
0.660	4.986
0.661	5.002
0.662	4.989
0.663	4.750
0.673	5.091

T	C_p
CURVE 2 (cont.)*	
0.675	5.111×10^{-7}
0.675	5.117
0.676	5.130
0.683	5.187
0.683	5.187
0.688	5.229
0.688	5.229
0.698	5.327
0.704	5.378
0.719	5.602
0.720	5.592
0.723	5.634
0.730	5.691
0.733	5.723
0.737	5.745
0.739	5.742
0.742	5.790
0.743	5.812
0.746	5.838
0.747	5.822
0.752	5.873
0.764	6.007
0.770	6.058
0.778	6.182
0.779	6.189
0.781	6.214
0.789	6.291
0.790	6.300
0.794	6.355
0.801	6.412
0.806	6.476
0.812	6.540
0.818	6.635
0.821	6.667
0.823	6.699
0.825	6.699
0.827	6.699
0.829	6.699
0.834	6.795
0.844	6.891
0.844	6.891
0.845	6.922
0.847	6.954
0.861	7.050

T	C_p
CURVE 2 (cont.)*	
0.872	7.178×10^{-7}
0.875	7.273
0.876	7.210
0.878	7.273
0.879	7.337
0.882	7.337
0.889	7.433
0.893	7.465
0.895	7.497
0.897	7.497
0.899	7.624
0.910	7.742
0.917	7.816
0.927	7.943
0.936	8.039
0.940	8.039
0.941	8.135
0.950	8.230
0.957	8.326
0.958	8.294
0.959	8.326
0.959	8.390
0.964	8.422
0.976	8.581
0.976	8.613
0.993	8.805
0.998	8.996
0.999	8.900
1.003	8.996
1.011	9.124
1.013	9.124
1.026	9.347
1.029	9.347
1.029	9.379
1.030	9.379
1.031	9.411
1.034	9.443
1.035	9.443
1.036	9.475
1.057	9.730
1.060	9.794
1.061	9.762
1.106	1.053×10^{-6}
1.119	1.081

T	C_p
CURVE 2 (cont.)*	
1.128	1.094×10^{-6}
1.130	1.088
1.131	1.097
1.147	1.126
1.147	1.126
1.155	1.142
1.167	1.164
1.174	1.171
1.181	1.184
1.188	1.196
1.204	1.228
1.208	1.228
1.214	1.233
1.228	1.263
1.243	1.305
1.243	1.298
1.255	1.327
1.267	1.346
1.285	1.378
1.286	1.378
1.287	1.381
1.295	1.394
1.310	1.423
1.312	1.432
1.314	1.439
1.318	1.448
1.333	1.467
1.345	1.506
1.350	1.512
1.353	1.515
1.357	1.522
1.371	1.547
1.386	1.592
1.387	1.595
1.397	1.605
1.399	1.617
1.416	1.652
1.424	1.713
1.432	1.704
1.439	1.700
1.447	1.732
1.468	1.774
1.471	1.780
1.483	1.809

T	C_p
CURVE 2 (cont.)*	
1.485	1.822×10^{-6}
1.488	1.825
1.499	1.853
1.520	1.911
1.524	1.908
1.535	1.933
1.543	1.965
1.564	2.016
1.570	2.029
1.585	2.077
1.586	2.061
1.619	2.172
1.625	2.185
1.634	2.214
1.637	2.208
1.661	2.281
1.668	2.322
1.688	2.367
1.693	2.373
1.708	2.424
1.725	2.539
1.725	2.479
1.750	2.562
1.757	2.584
1.785	2.661
1.802	2.705
1.818	2.785
1.837	2.846
1.864	2.919
1.881	3.021
1.901	3.040
1.927	3.155
1.949	3.251
1.966	3.324
1.987	3.433
2.000	3.436
2.033	3.598
2.039	3.605
2.078	3.752
2.109	3.908
2.126	3.975
2.175	4.185
2.183	4.265
2.209	4.335

T	C_p
CURVE 2 (cont.)*	
2.253	4.616×10^{-6}
2.256	4.581
2.290	4.756
2.332	5.031
2.346	5.044
2.372	5.177
2.380	5.248
2.434	5.563
2.443	5.557
2.455	5.627
2.503	5.953
2.533	6.125
2.558	6.329
2.569	6.326
2.624	6.667
2.646	6.827
2.650	6.891
2.688	7.178
2.728	7.401
2.762	7.624
2.788	7.848
2.843	8.262
2.864	8.390
2.897	8.741
2.917	8.837
2.962	9.187
2.972	9.283
3.011	9.666
3.061	1.008×10^{-5}
3.095	1.030
3.119	1.065
3.168	1.104
3.234	1.161
3.241	1.174
3.294	1.225
3.330	1.270
3.381	1.321
3.394	1.330
3.432	1.375
3.459	1.404
3.542	1.496
3.551	1.506
3.590	1.544
3.591	1.557

*Not shown on plot

DATA TABLE NO. 3 (continued)

T	C_p
CURVE 2 (cont.)*	
3.714	1.719 x 10^{-5}
3.727	1.723
3.735	1.758
3.766	1.777
3.899	1.971
3.933	2.026
3.987	2.096
4.101	2.278
4.142	2.316
4.195	2.434
4.231	2.498
4.343	2.702
4.353	2.708
4.526	3.133

*Not shown on plot

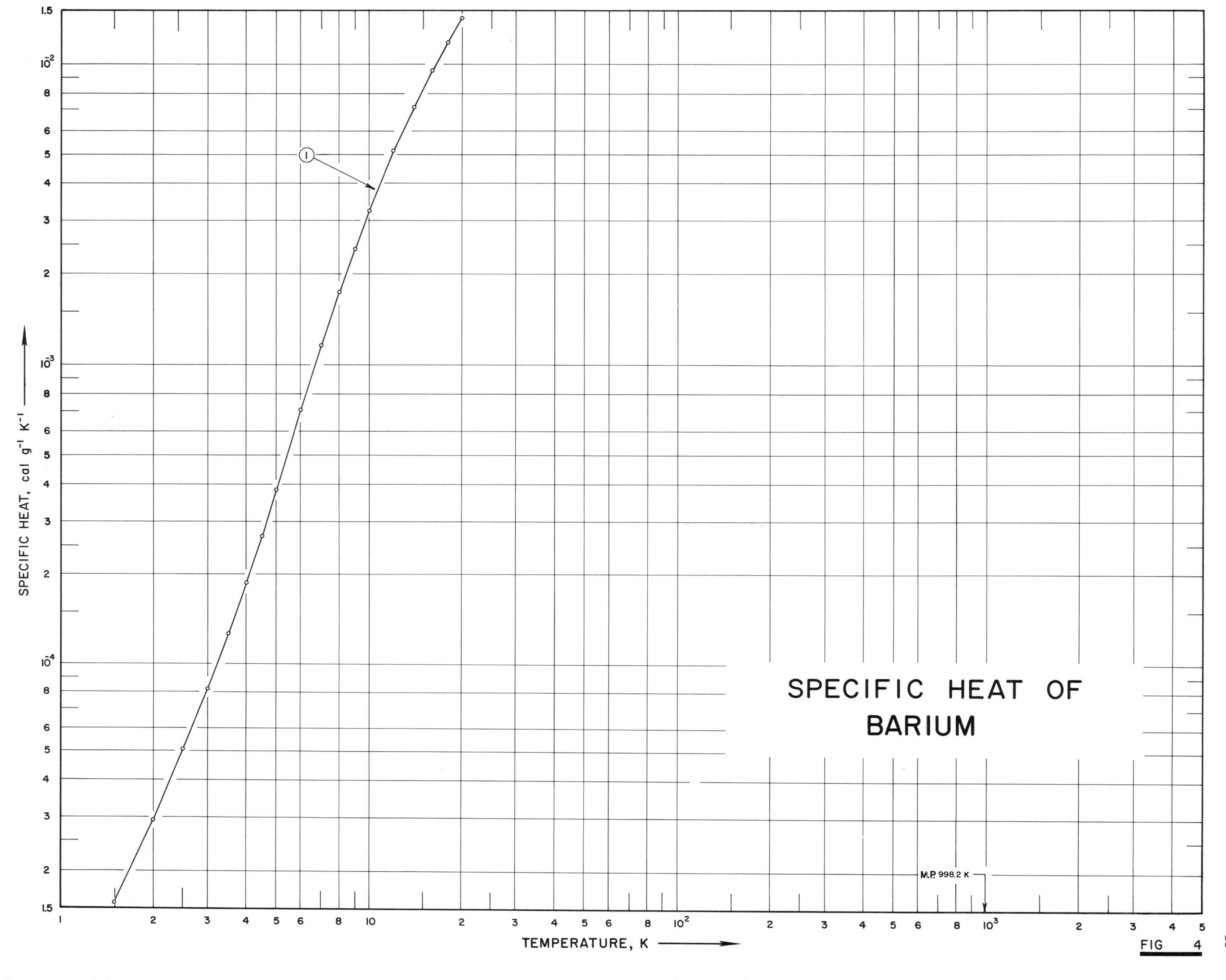
SPECIFIC HEAT OF
BARIUM
SPECIFIC HEAT, cal g⁻¹ K⁻¹
TEMPERATURE, K
M.P. 998.2 K
1
FIG 4

SPECIFICATION TABLE NO. 4 SPECIFIC HEAT OF BARIUM

(Impurity $<0.20\%$ each; total impurities $<0.50\%$)

[For Data Reported in Figure and Table No. 4]

Curve No.	Ref. No.	Year	Temp. Range, K	Reported Error, %	Name and Specimen Designation	Composition (weight percent), Specifications and Remarks
1	212	1957	1.5-20	2		0.2 Sr, 0.05 Fe and 0.02 each Al, Sn; sample supplied by Messrs New Metals and Chemicals Ltd.

DATA TABLE NO. 4 SPECIFIC HEAT OF BARIUM

[Temperature, T, K; Specific Heat, C_p, Cal $g^{-1}K^{-1}$]

T	C_p
CURVE 1	
1.5	1.566×10^{-5}
2.0	2.941
2.5	5.099
3.0	8.197
3.5	1.262×10^{-4}
4.0	1.862
4.5	2.680
5.0	3.829
6.0	7.100
7.0	1.157×10^{-3}
8.0	1.740
9.0	2.402
10.0	3.219
12.0	5.116
14.0	7.152
16.0	9.467
18.0	1.183×10^{-2}
20.0	1.420

* Not shown on plot

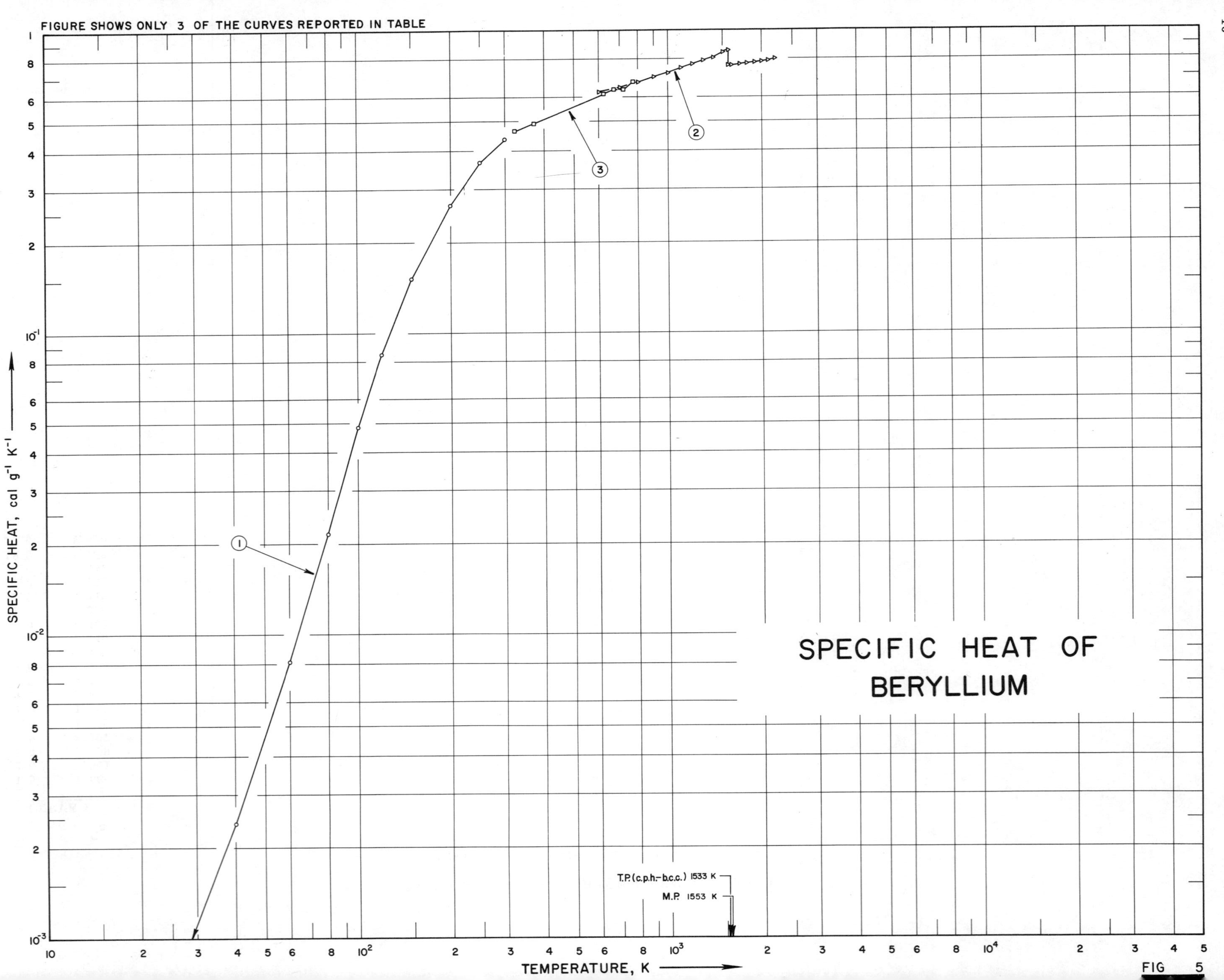

FIGURE SHOWS ONLY 3 OF THE CURVES REPORTED IN TABLE
SPECIFIC HEAT OF BERYLLIUM
T.P. (c.p.h.–b.c.c.) 1533 K
M.P. 1553 K
TEMPERATURE, K
SPECIFIC HEAT, cal g⁻¹ K⁻¹
1
2
3

FIG 5

SPECIFICATION TABLE NO. 5 SPECIFIC HEAT OF BERYLLIUM

(Impurity <0.20% each; total impurities <0.50%)

[For Data Reported in Figure and Table No. 5]

Curve No.	Ref. No.	Year	Temp. Range, K	Reported Error, %	Name and Specimen Designation	Composition (weight percent), Specifications and Remarks
1	78	1953	5-30			99.5 Be, 0.15 Cl, 0.10 O and 0.05 others; prepared from powder by high temperature extrusion process.
2	79	1961	600-2200			99.8 Be; pulverized and tightly filled into ampules; measured under 10-15 mm Hg argon atmosphere.
3	7	1959	323-773	1.8-2.0		99.8 Be.
4	269	1966	1.3-20		Sample I	Obtained by seven zone-refining passes from vacuum distilled starting material; sample supplied by Nuclear Metals Inc; Sample I included polycrystalline trailing end of the zone-refining; the portion of Sample I which was single crystal had a resistance ratio of $\frac{R_{300}}{R_{4.2}} = 380$.
5	269	1966	1.3-30		Sample II	Obtained by seven zone-refining passes from vacuum distilled starting material; sample supplied by Nuclear Metals Inc; Sample II was spark cut from center section of the rod; single crystal had resistance ratio of $\frac{R_{300}}{R_{4.2}} = 1100$.
6	201	1929	373-1173			
7	270	1929	98-463			Commercially pure Be, traces of Al, Mn, and Cr, also small traces of Fe, Mg, and Si, the total about 0.5%; sample supplied by Beryllium Company of America.
8	271	1931	286			
9	272	1934	273-1073	1.0-1.2		Pure crystallized Be lumps.
10	273	1939	303-1073			99.962 Be.
11	273	1939	303-473			99.781 Be.

DATA TABLE NO. 5 SPECIFIC HEAT OF BERYLLIUM

[Temperature, T, K; Specific Heat, C_p, Cal $g^{-1}K^{-1}$]

T	C_p
CURVE 1	
5	3.4×10^{-5}*
10	9.2*
20	3.83×10^{-4}*
40	2.40×10^{-3}
60	8.14
80	2.16×10^{-2}
100	4.85
120	8.46
150	1.52×10^{-1}
200	2.66
250	3.67
300	4.38
CURVE 2	
600	6.25×10^{-1}
700	6.49
800	6.73
900	6.97
1000	7.22
1100	7.46
1200	7.70
1300	7.94
1400	8.18
1500	8.42
(s) 1560	8.57
(l) 1560	7.64
1600	7.66
1700	7.72
1800	7.77
1900	7.83
2000	7.89
2100	7.94
2200	8.00
CURVE 3	
323	4.65×10^{-1}
373	4.90
623	6.15
673	6.35
723	6.55
773	6.72
CURVE 4*	
Series 1	
1.359	6.244×10^{-6}
1.360	6.191
1.361	6.206
1.665	7.574
2.047	9.633
2.494	1.169×10^{-5}
3.030	1.442
4.463	2.176
5.176	2.557
5.565	2.781
6.059	3.069
6.438	3.309
6.831	3.601
Series 2	
4.376	2.113×10^{-5}
4.668	2.278
4.998	2.457
5.342	2.655
8.748	5.051
8.770	5.112
9.149	5.348
9.205	5.404
9.537	5.689
9.603	5.727
9.953	6.054
10.447	6.494
11.159	7.249
12.026	8.378
13.247	9.982
13.968	1.104×10^{-4}
14.828	1.229
14.681	1.202
15.357	1.326
16.054	1.427
16.894	1.604
17.889	1.822
18.544	1.954
19.219	2.146
20.013	2.364
CURVE 4* (cont.)	
Series 3	
4.872	2.373×10^{-5}
5.200	2.562
5.596	2.804
6.090	3.100
6.631	3.448
7.120	3.803
7.641	4.156
8.306	4.709
9.105	5.332
9.329	5.605
Series 4	
4.469	2.166×10^{-5}
5.039	2.465
5.321	2.625
5.611	2.789
5.917	2.958
6.233	3.196
6.545	3.394
6.886	3.637
7.240	3.846
7.583	4.086
7.821	4.224
8.140	4.480
8.489	4.805
8.941	5.136
9.895	5.918
10.441	6.489
11.092	7.165
Series 5	
3.185	1.513×10^{-5}
3.383	1.612
3.583	1.701
3.819	1.823
4.042	1.940
4.272	2.059
CURVE 4* (cont.)	
Series 6	
3.132	1.473×10^{-5}
3.290	1.557
3.465	1.622
3.639	1.726
3.833	1.821
4.056	1.950
4.323	2.077
Series 7	
2.359	1.090×10^{-5}
2.490	1.154
2.639	1.235
2.821	1.327
3.054	1.445
CURVE 5*	
Series 1	
20.710	2.520×10^{-4}
21.849	2.877
23.140	3.329
24.550	3.923
26.241	4.668
28.280	5.859
Series 2	
19.805	2.279×10^{-4}
20.698	2.516
21.689	2.809
22.843	3.213
24.203	3.749
27.823	5.526
Series 3	
15.092	1.267×10^{-4}
15.804	1.392
16.642	1.552
CURVE 5* (cont.)	
17.670	1.768×10^{-4}
18.932	2.065
20.563	2.519
Series 4	
8.741	5.041×10^{-5}
9.192	5.434
9.908	6.074
10.236	6.387
10.571	6.752
10.963	7.152
11.401	7.643
11.905	8.226
12.521	9.022
13.261	9.996
14.197	1.134×10^{-4}
15.419	1.329
Series 5	
8.572	4.892×10^{-5}
8.821	5.111
9.357	5.587
9.641	5.816
9.994	6.147
10.374	6.516
Series 6	
9.089	5.341×10^{-5}
10.729	6.906
11.127	7.336
11.591	7.863
12.132	8.512
12.792	9.387
13.611	1.050×10^{-4}
14.643	1.199
Series 7	
9.776	5.914×10^{-5}
9.929	6.062
10.084	6.236

* Not shown on plot

DATA TABLE NO. 5 (continued)

T	C_p
CURVE 5* (cont.)	
10.241	6.396×10^{-5}
10.409	6.605
10.562	6.720
10.673	6.865
10.772	6.968
10.869	7.083
10.969	7.152
11.077	7.278
11.187	7.408
11.300	7.503
11.418	7.636
11.290	7.497
11.382	7.578
11.462	7.684
11.544	7.757
11.630	7.890
11.718	7.986
11.808	8.124
11.901	8.230
11.996	8.392
12.093	8.496
12.469	8.898
12.682	9.249
12.884	9.481
13.096	9.767
13.314	1.015×10^{-4}
13.314	1.010
13.501	1.041
13.691	1.063
13.891	1.099
14.104	1.126
14.564	1.183
14.813	1.491
15.081	1.261
15.363	1.304
15.672	1.350
16.001	1.414
17.181	1.659
17.763	1.781
18.302	1.902
18.980	2.058
19.805	2.281
20.673	2.521
21.671	2.827
Series 8	
18.247	1.901×10^{-4}
18.904	2.050
19.651	2.244
20.484	2.478
21.456	2.760
Series 9	
18.994	2.064×10^{-4}
19.965	2.337
20.860	2.599
Series 10	
19.241	2.147×10^{-4}
19.030	2.077
19.918	2.306
20.775	2.545
21.761	2.846
22.917	3.246
24.276	3.777
25.907	4.428
27.899	5.567
Series 11	
19.199	2.114×10^{-4}
20.415	2.443
21.348	2.709
22.444	3.079
23.715	3.560
25.220	4.209
27.056	5.045
Series 12	
4.363	2.106×10^{-5}
4.757	2.324
4.939	2.425
5.122	2.540
5.316	2.653
5.528	2.775
CURVE 5* (cont.)	
5.775	2.915×10^{-5}
6.062	3.092
6.388	3.312
6.770	3.589
7.232	3.904
Series 13	
4.617	2.249×10^{-5}
4.943	2.432
5.215	2.594
5.442	2.726
5.677	2.855
5.946	3.022
6.253	3.222
6.612	3.478
6.891	3.679
7.415	4.004
8.043	4.477
8.429	4.768
8.887	5.169
9.441	5.615
Series 14	
3.491	1.650×10^{-5}
3.646	1.739
3.825	1.831
4.039	1.941
4.294	2.077
Series 15	
3.082	1.443×10^{-5}
3.191	1.496
3.305	1.557
3.433	1.624
3.579	1.698
3.750	1.787
3.952	1.890
4.188	2.013
4.471	2.170
4.826	2.363
CURVE 5* (cont.)	
Series 16	
1.833	8.450×10^{-6}
1.947	8.938
2.083	9.581
2.217	1.020×10^{-5}
2.304	1.063
2.382	1.103
2.460	1.139
2.550	1.183
2.653	1.254
2.775	1.295
2.871	1.340
3.034	1.420
3.130	1.469
3.308	1.560
3.490	1.651
Series 17	
1.347	6.252×10^{-6}
1.388	6.401
1.460	6.742
1.440	6.630
1.419	6.531
1.505	7.024
1.523	6.975
1.661	7.610
1.894	8.713
1.566	7.167
1.736	7.953
1.855	8.527
2.018	9.276
2.064	9.471
2.169	9.949
2.268	1.048×10^{-5}
2.284	1.055
2.425	1.127
2.587	1.203
2.781	1.301
Series 18	
22.358	3.044×10^{-4}
23.569	3.463
CURVE 5* (cont.)	
26.796	4.865×10^{-4}
24.064	3.669
25.141	4.139
26.307	4.631
27.625	5.337
28.194	5.739
29.988	6.896
CURVE 6*	
373	4.776×10^{-1}
423	5.118
473	5.410
523	5.657
573	5.865
623	6.038
673	6.182
723	6.301
773	6.401
823	6.481
873	6.564
923	6.637
973	6.712
1023	6.792
1073	6.884
1123	6.984
1173	7.060
CURVE 7*	
97.6	3.89×10^{-2}
208.2	2.70×10^{-1}
282.7	3.99
377.0	5.19
463.2	5.93
CURVE 8*	
286.15	3.97×10^{-1}
CURVE 9*	
273	4.140×10^{-1}
373	4.825
473	5.402

* Not shown on plot

DATA TABLE NO. 5 (continued)

T	C_p
CURVE 9*(cont.)	
573	5.839 x 10^{-1}
673	6.149
773	6.364
873	6.585
973	6.924
1073	7.536
CURVE 10*	
303.15	5.08 x 10^{-1}
373.15	5.24
473.15	5.56
673.15	5.89
873.15	6.38
1073.15	6.79
CURVE 11*	
303.15	4.88 x 10^{-1}
373.15	4.98
473.15	5.13

* Not shown on plot

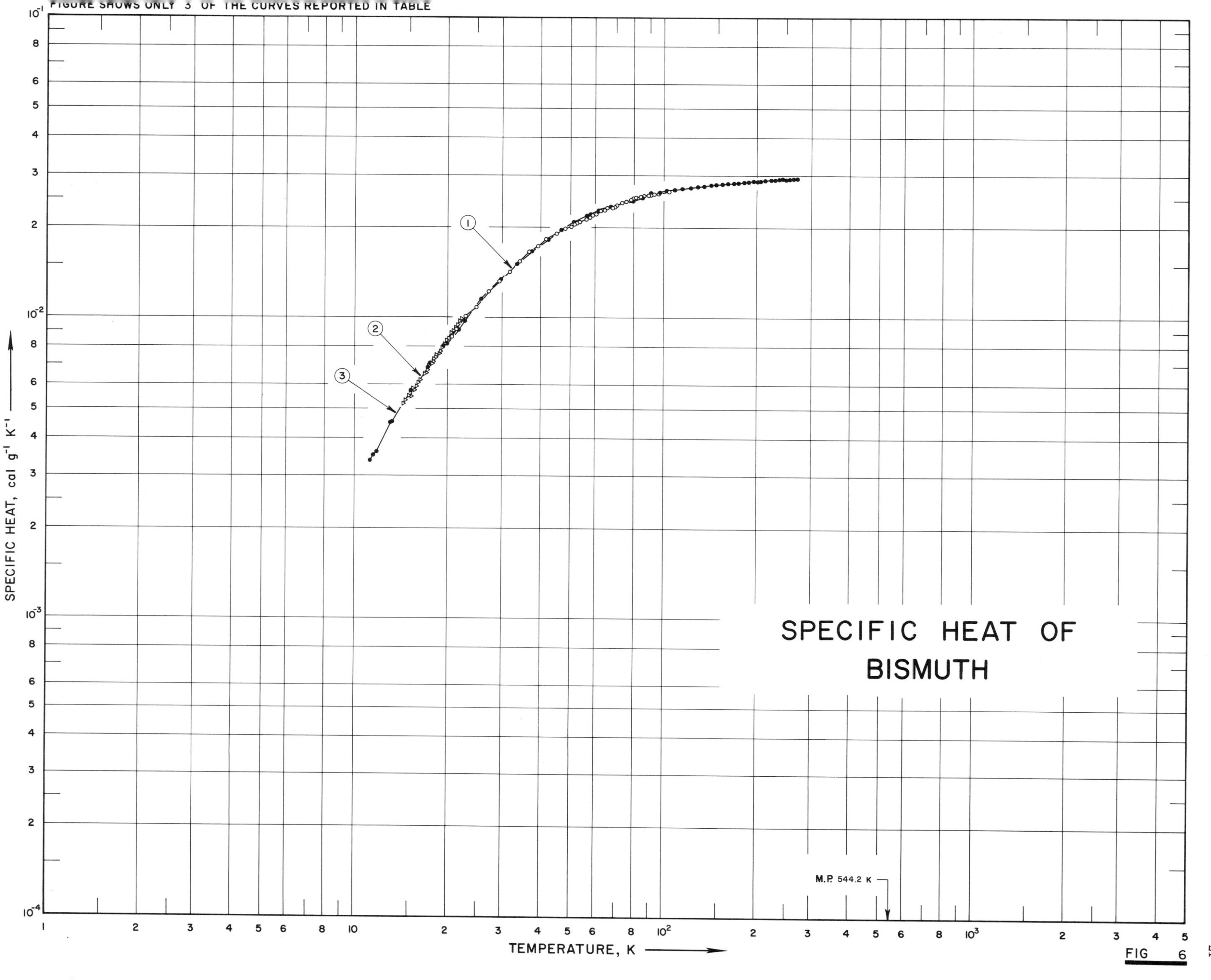
FIGURE SHOWS ONLY 3 OF THE CURVES REPORTED IN TABLE
SPECIFIC HEAT OF BISMUTH
SPECIFIC HEAT, cal g⁻¹ K⁻¹
TEMPERATURE, K
M.P. 544.2 K
1
2
3
FIG 6

SPECIFICATION TABLE NO. 6 SPECIFIC HEAT OF BISMUTH

(Impurity <0.20% each; total impurities <0.50%)

[For Data Reported in Figure and Table No. 6]

Curve No.	Ref. No.	Year	Temp. Range, K	Reported Error, %	Name and Specimen Designation	Composition (weight percent), Specifications and Remarks
1	90	1958	21-105			0.00005 Cu.
2	60	1949	15-21	0.6		99.998 Bi.
3	91	1964	11-271			99.9 Bi.
4	268	1928	348-873			99.85 Bi, 0.1 Cu, 0.015 Fe, and 0.01 Al.
5	274	1927	272-426			
6	275	1930	61-298			High purity sample supplied by American Smelting and Refining Co.; density = 9.86 g cm^{-3} at 20.6 C.
7	276	1930	3.6-19			Sample supplied by Kahlbaum and Co.
8	277	1932	306-644			0.1 Pb, 0.002 Ag and very minute traces of Ca, Fe, Mg, Na, Ni, Sn, and Th.
9	278	1938	193-393	0.1		<0.02 S, 0.0022 Cu, <0.002 Pb, and <0.00005 As; sample supplied by the Consolidated Mining and Smelting Co.
10	266	1954	0.9-5.0			99.99 Bi; polycrystalline; resistance ratio: $\frac{R_{4.2}}{R_{273}} = 0.112$.
11	280	1960	545-802			99.9999 Bi, and trace of Fe; pelleted sample supplied by Consolidated Mining and Smelting Co.

DATA TABLE NO. 6 SPECIFIC HEAT OF BISMUTH

[Temperature, T, K; Specific Heat, C_p, Cal g^{-1} K^{-1}]

CURVE 1

T	C_p
20.64	8.632 x 10^{-3}
21.48	9.144
22.81	1.004 x 10^{-2}
24.82	1.084
27.14	1.220
29.47	1.322
31.72	1.420
34.19	1.544
36.89	1.655
39.09	1.734
41.92	1.823
45.00	1.914
48.19	1.988
49.50	2.007
50.26	2.022
51.57	2.058
52.05	2.064
53.03	2.070*
53.51	2.095
56.46	2.140
57.86	2.169
58.96	2.203
60.24	2.229*
60.94	2.250
62.43	2.270
62.75	2.287*
64.59	2.287
64.77	2.299*
65.66	2.314
68.32	2.333
69.06	2.356
69.38	2.364*
70.40	2.374
71.20	2.396*
73.32	2.436
75.80	2.467
78.78	2.491
79.08	2.507
81.32	2.536
82.22	2.547*
84.22	2.544
86.21	2.558
89.21	2.571

CURVE 1 (cont.)

T	C_p
90.52	2.577 x 10^{-2}
93.14	2.592
96.31	2.605
104.62	2.654

CURVE 2

T	C_p
14.40	5.220 x 10^{-3}
14.60	5.373
15.07	5.541
15.09	5.474*
15.24	5.512
15.59	5.852
15.62	5.799
15.88	5.933
16.14	6.110
16.26	6.239
16.43	6.239*
16.86	6.522
17.04	6.593
17.27	6.727
17.38	6.938
17.53	6.904*
17.76	7.062
17.99	7.153
18.08	7.301
18.33	7.402
18.47	7.483
18.72	7.603
18.88	7.220
19.20	7.976
19.23	7.909*
19.62	8.167
19.93	8.402
20.00	8.431*
20.25	8.560
20.26	8.622*
20.35	8.656*
20.58	8.900
20.59	8.818*
20.65	8.852*
20.91	9.091
20.97	8.976*

CURVE 2 (cont.)

T	C_p
21.21	9.278 x 10^{-3}
21.24	9.239*
21.60	9.435
21.64	9.536*
21.91	9.775
21.96	9.699*
22.23	9.842*
22.25	9.952

CURVE 3

T	C_p
11.32	3.378 x 10^{-3}
11.59	3.513
11.80	3.619
11.82	3.639*
13.18	4.504
13.29	4.563
13.36	4.565*
15.26	5.766
15.32	5.761*
15.43	5.847*
15.51	5.809*
17.35	6.876
17.66	7.057
17.67	7.048*
18.87	7.660*
19.57	8.057
19.96	8.230
20.08	8.254*
21.70	9.148
22.75	9.775
25.89	1.156 x 10^{-2}
29.67	1.341
33.56	1.518
37.44	1.681
42.09	1.838
46.74	1.968
51.34	2.087
56.77	2.194
57.69	2.213
61.92	2.279
62.19	2.283*
67.09	2.350

CURVE 3 (cont.)

T	C_p
67.54	2.353 x 10^{-2}*
73.68	2.426*
79.43	2.485
85.11	2.533
91.54	2.601
97.51	2.633
103.25	2.656
109.92	2.680
115.88	2.709
123.47	2.729
129.74	2.752
135.65	2.767
142.59	2.790
148.66	2.797
155.63	2.807
161.81	2.822
169.60	2.835
175.81	2.840
182.90	2.856
188.99	2.864
196.21	2.874
200.53	2.865
202.30	2.873*
204.49	2.884*
206.26	2.875
209.07	2.880*
213.25	2.893
216.86	2.885*
222.41	2.904
225.86	2.909*
229.85	2.901
230.85	2.909*
236.96	2.919
242.43	2.923
242.65	2.917*
248.15	2.917
250.14	2.922*
255.63	2.931
255.96	2.933*
262.67	2.935
265.14	2.928*
270.59	2.942

CURVE 4*

T	C_p
348	2.96 x 10^{-2}
398	3.20
423	3.32
448	3.44
473	3.60
(s) 523	3.86
(l) 573	3.73
673	3.73
773	3.73
873	3.73

CURVE 5*

T	C_p
271.75	2.90 x 10^{-2}
289.65	2.98
319.15	2.64
350.15	2.68
377.15	2.50
382.15	2.60
387.15	2.50
392.15	2.80
426.15	2.90

CURVE 6*

T	C_p
60.8	2.216 x 10^{-2}
63.4	2.312
64.7	2.283
71.1	2.411
74.6	2.496
101.2	2.614
111.2	2.715
125.1	2.781
137.2	2.756
150.2	2.783
162.5	2.808
176.8	2.822
187.7	2.834
198.9	2.859
208.8	2.858
218.8	2.861
258.1	2.811
266.3	2.899

CURVE 6* (cont.)

T	C_p
272.8	2.911 x 10^{-2}
285.3	2.937
295.2	2.913
298.2	2.921

CURVE 7*

Series 1

T	C_p
14.26	5.55 x 10^{-3}
14.72	6.03
15.27	6.99
16.41	7.51
16.86	7.66
17.11	8.33
17.76	9.04

Series 2

T	C_p
5.53	4.23 x 10^{-4}
6.09	6.08
6.61	6.84
7.81	1.34 x 10^{-3}
8.16	1.60
10.27	2.80
10.70	4.25
14.31	5.69

Series 3

T	C_p
3.52	6.89 x 10^{-5}
4.17	1.53 x 10^{-4}
4.52	2.23
4.88	2.62

Series 4

T	C_p
2.77	5.84 x 10^{-5}
3.64	1.02 x 10^{-4}
4.38	1.72
5.35	3.81
5.84	5.31
6.45	6.94

*Not shown on plot

DATA TABLE NO. 6 (continued)

T	C_p
CURVE 7* (cont.)	
10.33	3.01×10^{-3}
12.33	4.39
14.49	7.66
Series 5	
14.04	7.48×10^{-3}
15.14	6.84
15.72	7.23
16.14	7.27
16.59	7.80
17.07	8.99
17.65	8.90
18.10	8.52
18.57	8.57
19.21	9.24
CURVE 8*	
305.9	2.970×10^{-2}
306.5	2.938
372.5	3.040
373.0	3.041
427.5	3.140
428.4	3.142
477.9	3.199
478.6	3.187
505.1	3.250
505.8	3.274
520.8	3.397
521.7	3.399
522.4	3.394
523.1	3.443
534.7	4.326
535.4	4.602
535.8	4.294
536.4	4.459
537.9	5.134
538.5	5.670
539.4	6.446
540.0	7.163
540.8	9.814
(s)541.4	1.242×10^{-1}
(l)545.6	3.453×10^{-2}
546.2	3.445
CURVE 8* (cont.)	
558.0	3.432×10^{-2}
558.6	3.441
576.8	3.420
577.6	3.447
600.2	3.407
601.2	3.386
643.1	3.336
644.1	3.347
CURVE 9*	
193.15	1.203×10^{-1}
213.15	1.212
233.15	1.221
253.15	1.230
273.15	1.239
293.15	1.247
313.15	1.256
333.15	1.265
353.15	1.274
373.15	1.283
393.15	1.292
CURVE 10*	
Series 1	
1.297	3.300×10^{-6}
1.299	3.410
1.299	3.258
1.300	3.222
1.301	3.247
1.302	3.215
1.549	5.245
1.562	5.464
1.562	5.442
1.688	7.059
1.712	7.132
1.913	9.950
1.921	1.006×10^{-5}
2.128	1.354
2.159	1.390
2.385	1.945
2.409	2.045
2.706	2.890
2.731	3.017
CURVE 10* (cont.)	
2.960	3.867×10^{-5}
3.009	4.083
3.249	5.362
3.282	5.566
3.453	6.963
3.497	7.238
3.534	7.517
3.576	7.897
3.640	8.365
3.679	8.813
3.716	8.965
3.751	9.331
3.782	9.542
3.817	9.959
3.850	1.016×10^{-4}
3.888	1.062
3.952	1.132
3.989	1.179
4.019	1.204
4.060	1.279
4.106	1.336
4.146	1.366
4.183	1.425
4.220	1.468
4.252	1.486
4.287	1.562
4.347	1.630
4.376	1.698
4.485	1.888
4.507	1.933
4.601	2.072
4.635	2.132
4.781	2.409
4.819	2.514
5.008	2.963
5.036	3.115
Series 2	
1.344	3.569×10^{-6}
1.358	3.540
1.363	3.612
1.371	3.606
1.374	3.624
1.374	3.617
1.410	4.031
CURVE 10* (cont.)	
1.434	4.126×10^{-6}
1.462	4.347
1.469	4.386
1.530	5.534
1.572	5.534
1.611	5.921
1.659	6.402
1.689	6.705
1.710	6.893
Series 3	
0.953	1.315×10^{-6}
0.956	1.379
0.982	1.455
0.994	1.490
1.008	1.567
1.023	1.544
1.026	1.696
1.028	1.570
1.029	1.654
1.034	1.562
1.036	1.618
1.053	1.681
1.056	1.752
1.062	1.751
1.079	1.799
1.090	1.911
1.095	1.943
1.111	2.000
1.129	2.067
1.138	2.149
1.167	2.315
1.190	2.437
1.193	2.423
1.215	2.564
1.221	2.561
1.255	2.787
1.291	3.049
1.295	3.076
1.319	3.250
1.346	3.472
1.357	3.517
1.396	3.867
1.437	4.219
1.470	4.487
CURVE 10* (cont.)	
1.505	4.760×10^{-6}
1.533	5.137
1.606	5.888
1.639	6.330
1.710	7.094
1.752	7.684
1.804	8.305
1.839	8.847
1.907	9.879
1.958	1.074×10^{-5}
1.987	1.111
2.027	1.182
CURVE 11*	
545.2	3.479×10^{-2}
544.8	3.512
546.6	3.493
545.4	3.508
545.4	3.484
545.9	3.503
546.6	3.488
558.5	3.450
558.5	3.455
577.4	3.412
606.8	3.311
606.3	3.364
653.4	3.311
698.1	3.283
698.1	3.249
755.3	3.235
755.2	3.268
801.7	3.201
801.8	3.216

*Not shown on plot

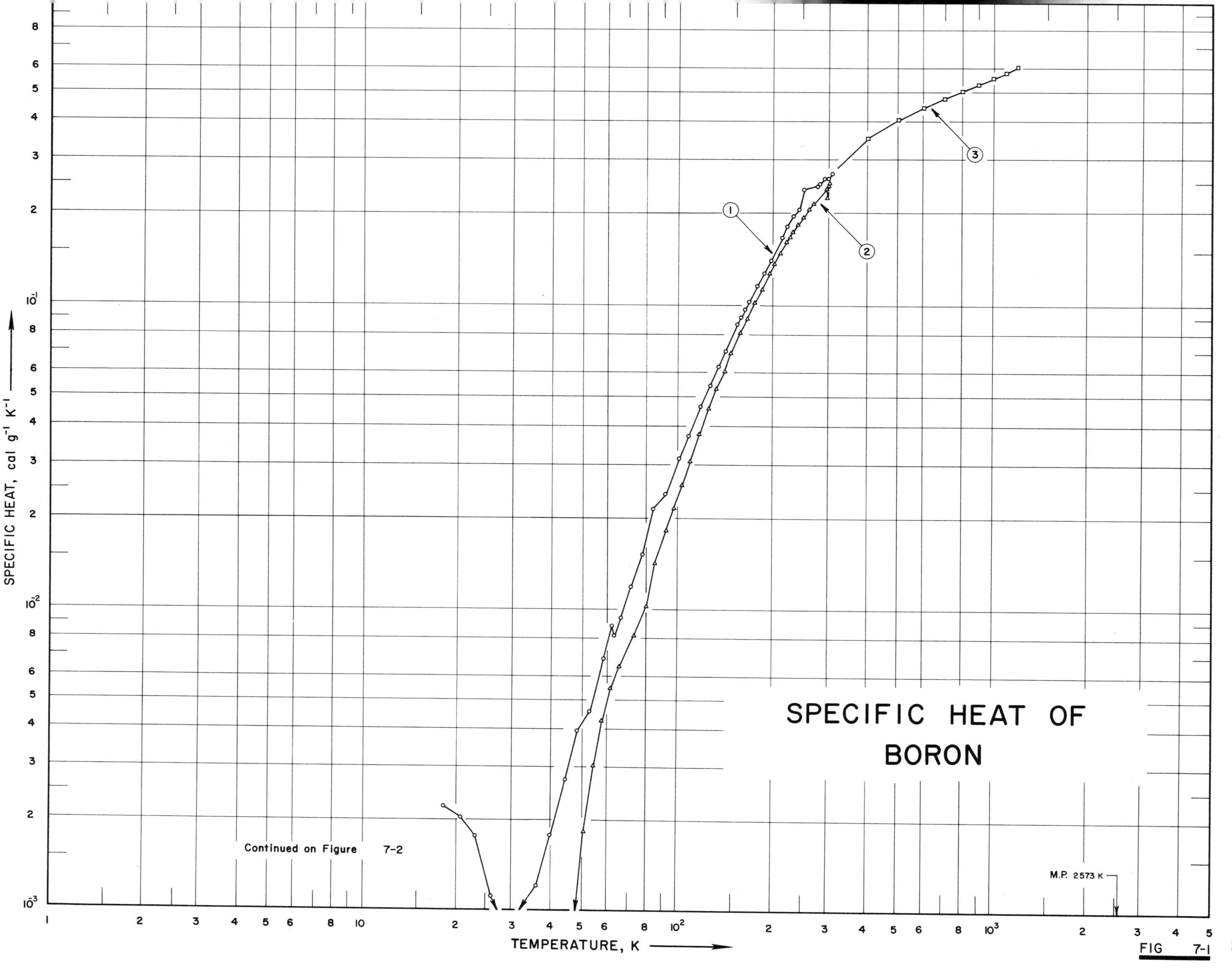
SPECIFIC HEAT OF
BORON
SPECIFIC HEAT, cal g^{-1} K^{-1}
TEMPERATURE, K
M.P. 2573 K
Continued on Figure 7-2
1
2
3
FIG 7-1

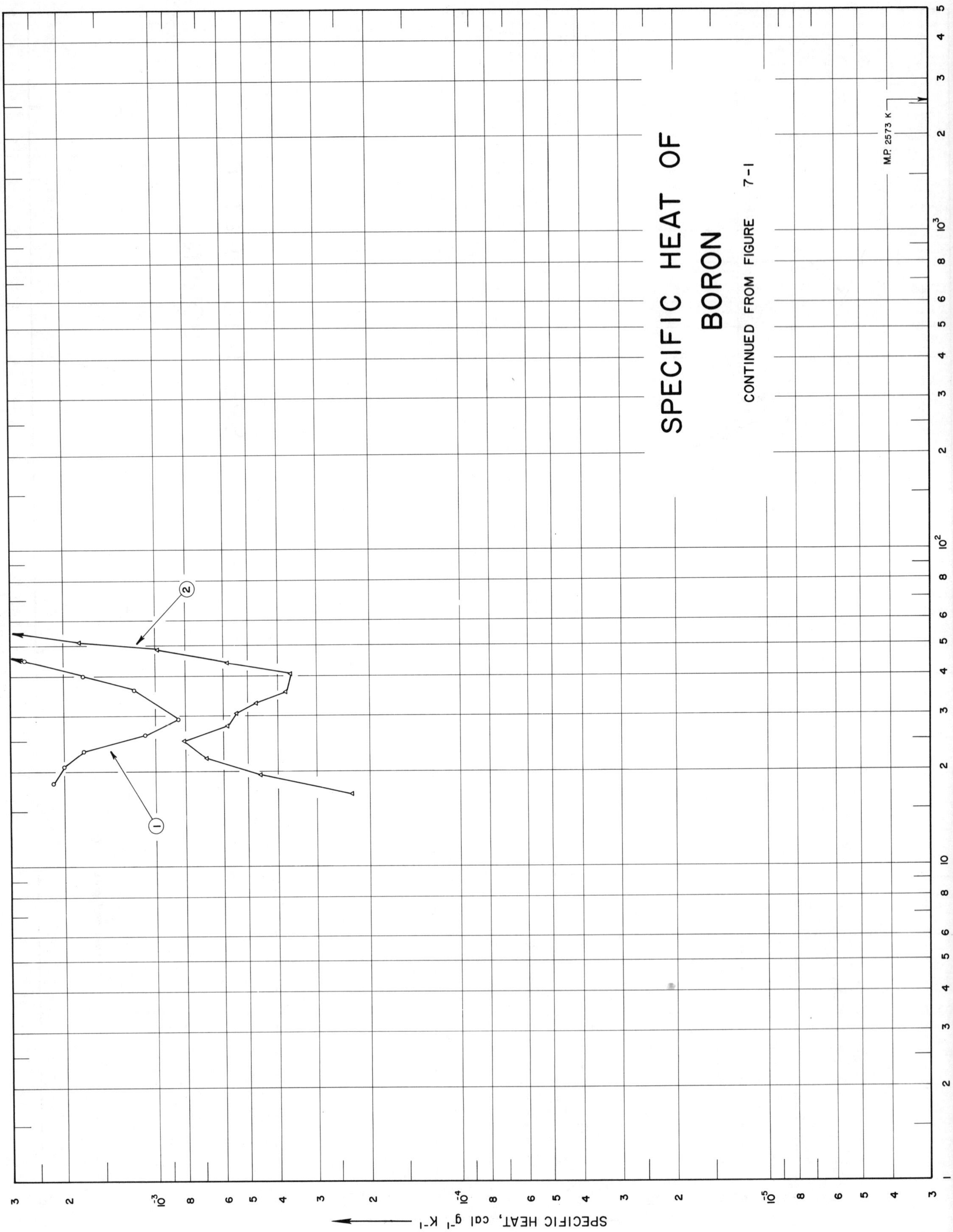

SPECIFIC HEAT OF BORON

CONTINUED FROM FIGURE 7-1

SPECIFICATION TABLE NO. 7 SPECIFIC HEAT OF BORON

(Impurity $< 0.20\%$ each; total impurities $< 0.50\%$)

[For Data Reported in Figure and Table No. 7]

Curve No.	Ref. No.	Year	Temp. Range, K	Reported Error, %	Name and Specimen Designation	Composition (weight percent), Specifications and Remarks
1	92	1951	18-308			Extremely pure; amorphous.
2	92	1951	17-304			Extremely pure; crystalline; heated under vacuum to 1700-1900 C.
3	162	1960	298-1200		Boron III	0.08 Si, 0.06 Na, 0.04 Fe and 0.02 Ni; amorphous; sample supplied by the Fairmount Chemical Company; sealed in gold ampules.

DATA TABLE NO. 7 SPECIFIC HEAT OF BORON

[Temperature, T, K; Specific Heat, C_p, Cal g^{-1} K^{-1}]

T	C_p
CURVE 1	
18.25	2.201×10^{-3}
20.55	2.044
23.04	1.757
25.89	1.110
29.08	8.602×10^{-4}
35.98	1.202×10^{-3}
39.70	1.776
44.43	2.710
48.52	3.940
52.97	4.551
58.71	6.873
62.25	8.806
63.10	8.121
66.10	9.379
71.20	1.193×10^{-2}
77.03	1.539
83.79	2.154
91.78	2.405
100.83	3.154
108.93	3.737
118.06	4.625
127.04	5.457
135.43	6.299
142.40	7.067
155.10	8.676
159.49	9.120
163.88	9.694
168.71	1.027×10^{-1}
178.62	1.156
187.96	1.281
197.89	1.405
215.29	1.680
223.73	1.833
233.59	1.973
243.69	2.083
252.69	2.415
277.62	2.477
279.62	2.471*
283.18	2.476*
283.85	2.530
288.97	2.547*
291.10	2.622
296.58	2.610*
300.26	2.629
CURVE 1 (cont.)	
303.26	2.629×10^{-1}*
308.29	2.725
CURVE 2	
16.90	2.312×10^{-4}
19.47	4.625
21.89	6.937
24.90	8.232
27.84	5.920
30.48	5.550
32.74	4.782
35.47	3.811
40.48	3.691
43.87	5.938
48.12	1.508×10^{-3}
50.96	1.831
54.51	3.006
57.77	4.246
61.46	5.485
65.23	6.484
72.71	8.158
79.58	1.023×10^{-2}
84.74	1.434
91.66	1.828
97.02	2.167
103.11	2.585
109.72	3.080
116.81	3.765
125.43	4.588
133.00	5.309
140.54	6.086
147.98	6.928
157.86	8.103
166.08	9.000
175.54	1.020×10^{-1}
185.96	1.141
195.77	1.280
202.71	1.375
211.43	1.490
220.70	1.617
227.43	1.690
232.75	1.752
241.07	1.850
CURVE 2 (cont.)	
251.28	1.954×10^{-1}
261.67	2.078
270.29	2.172
297.74	2.263
296.44	2.434
301.79	2.487
303.71	2.516
CURVE 3	
298	2.643×10^{-1}*
400	3.562
500	4.099
600	4.490
700	4.814
800	5.102
900	5.369
1000	5.623
1100	5.868
1200	6.108

*Not shown on plot

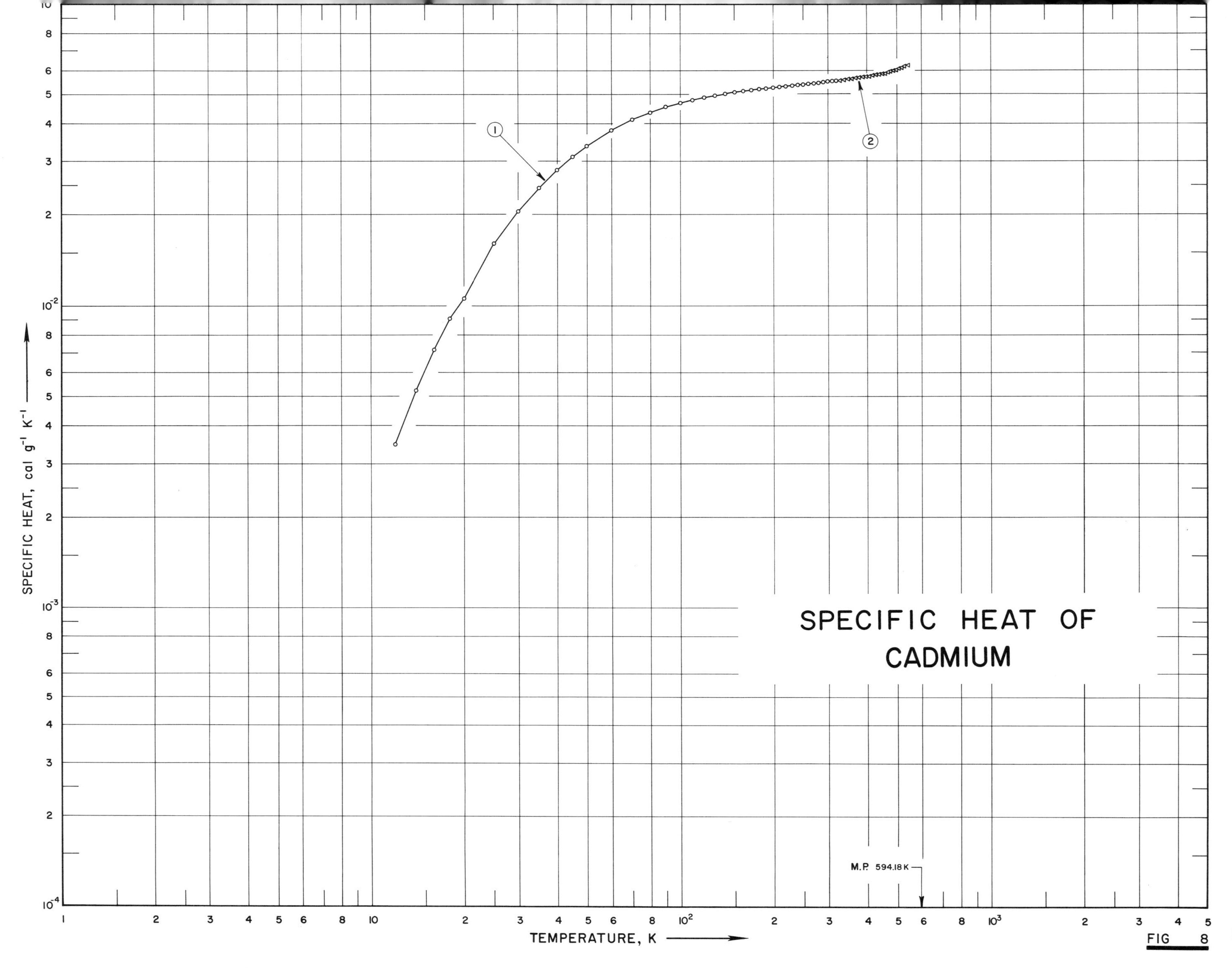
SPECIFIC HEAT OF
CADMIUM
SPECIFIC HEAT, cal g⁻¹ K⁻¹
TEMPERATURE, K
M.P. 594.18 K
1
2
FIG 8

SPECIFICATION TABLE NO. 8 SPECIFIC HEAT OF CADMIUM

(Impurity $< 0.20\%$ each; total impurities $< 0.50\%$)

[For Data Reported in Figure and Table No. 8]

Curve No.	Ref. No.	Year	Temp. Range, K	Reported Error, %	Name and Specimen Designation	Composition (weight percent), Specifications and Remarks
1	48	1953	12-320	0.10		99.9 Cd.
2	47	1957	298-543	0.4		99.9 Cd.
3	281	1956	513-688			
4	189	1923	70-298	1.0		Analyzed as being a very high purity sample.
5	282	1924	10-373			
6	268	1926	348-923			99.959 Cd, with 0.02 Al, and Fe, also 0.001 Cu.
7	283	1928	10-594			
8	182	1936	193-383	0.05-0.1		99.97 Cd, with 0.022 Zn, 0.01 Pb, 0.002 Cu, and 0.0001 Fe.
9	284	1956	1.5-20			99.99 Cd; sample supplied by Johnson, Matthey and Co.

DATA TABLE NO. 8 SPECIFIC HEAT OF CADMIUM

[Temperature, T, K; Specific Heat, C_p, Cal $g^{-1}K^{-1}$]

T	C_p
CURVE 1	
12	3.488 x 10^{-3}
14	5.267
16	7.153
18	9.075
20	1.103 x 10^{-2}
25	1.604
30	2.052
35	2.456
40	2.810
45	3.117
50	3.383
60	3.810
70	4.134
80	4.377
90	4.571
100	4.701
110	4.816
120	4.909
130	4.989
140	5.057
150	5.112
160	5.159
170	5.199
180	5.235
190	5.269
200	5.299
210	5.327
220	5.354
230	5.380
240	5.403
250	5.423
260	5.444
270	5.466
280	5.490
290	5.517
298.16	5.537*
300	5.542
310	5.565
320	5.587
CURVE 2	
298.16	5.527 x 10^{-2}*
300	5.532
CURVE 2 (cont.)	
310	5.551 x 10^{-2}*
320	5.569*
330	5.590
340	5.611
350	5.635
360	5.659
370	5.685
380	5.708
390	5.732
400	5.756
410	5.782
420	5.807
430	5.833
440	5.859
450	5.886
460	5.918
470	5.954
480	5.986
490	6.022
500	6.062
510	6.111
520	6.163
530	6.217
540	6.276*
543.16	6.295
CURVE 3*	
513	5.82 x 10^{-2}
550	5.82
(s)593	5.82
(l)598	5.82
600	5.82
650	5.82
688	5.82
CURVE 4*	
69.66	4.15 x 10^{-2}
70.00	4.16
72.40	4.22
74.97	4.28
77.56	4.34
80.00	4.39
CURVE 4* (cont.)	
80.09	4.41 x 10^{-2}
82.59	4.43
85.06	4.48
87.70	4.55
89.91	4.58
90.00	4.58
92.39	4.64
94.70	4.65
97.08	4.68
99.37	4.72
100.00	4.73
298.00	5.56
CURVE 5*	
10.0	9.79 x 10^{-4}
20.0	7.74 x 10^{-3}
30.0	1.83 x 10^{-2}
40.0	2.76
60.0	3.89
80.0	4.48
120.0	4.98
200.0	5.34
273.2	5.50
373.2	5.66
CURVE 6*	
348	5.51 x 10^{-2}
373	5.58
423	5.57
473	5.91
523	6.06
573	6.20
623	6.17
673	6.17
773	6.17
873	6.17
923	6.17
CURVE 7*	
10	1.913 x 10^{-3}
20	1.023 x 10^{-2}
CURVE 7* (cont.)	
30	2.011 x 10^{-2}
40	2.847
50	3.470
60	3.888
70	4.155
80	4.386
90	4.600
100	4.733
110	4.831
120	4.911
130	4.982
140	5.044
150	5.098
160	5.142
170	5.178
180	5.214
190	5.249
200	5.276
210	5.311
220	5.338
230	5.365
240	5.391
250	5.409
260	5.436
270	5.454
280	5.472
290	5.489
300	5.516
320	5.561
340	5.596
360	5.641
380	5.694
400	5.721
450	5.819
500	5.934
550	6.050
(s)594	6.174
(l)594	6.673
CURVE 8*	
193.15	5.272 x 10^{-2}
203.15	5.304
213.15	5.332
CURVE 8* (cont.)	
223.15	5.359 x 10^{-2}
233.15	5.382
243.15	5.406
253.15	5.430
263.15	5.454
273.15	5.476
283.15	5.497
293.15	5.521
303.15	5.543
313.15	5.564
323.15	5.588
333.15	5.609
343.15	5.633
353.15	5.657
363.15	5.681
373.15	5.703
383.15	5.727
CURVE 9*	
1.5	5.000 x 10^{-6}
2	7.607
3	2.153 x 10^{-5}
4	5.792
5	1.512 x 10^{-4}
6	3.461
8	1.014 x 10^{-3}
10	2.135
12	3.648
14	5.338
16	7.198
18	9.235
20	1.157 x 10^{-2}

*Not shown on plot

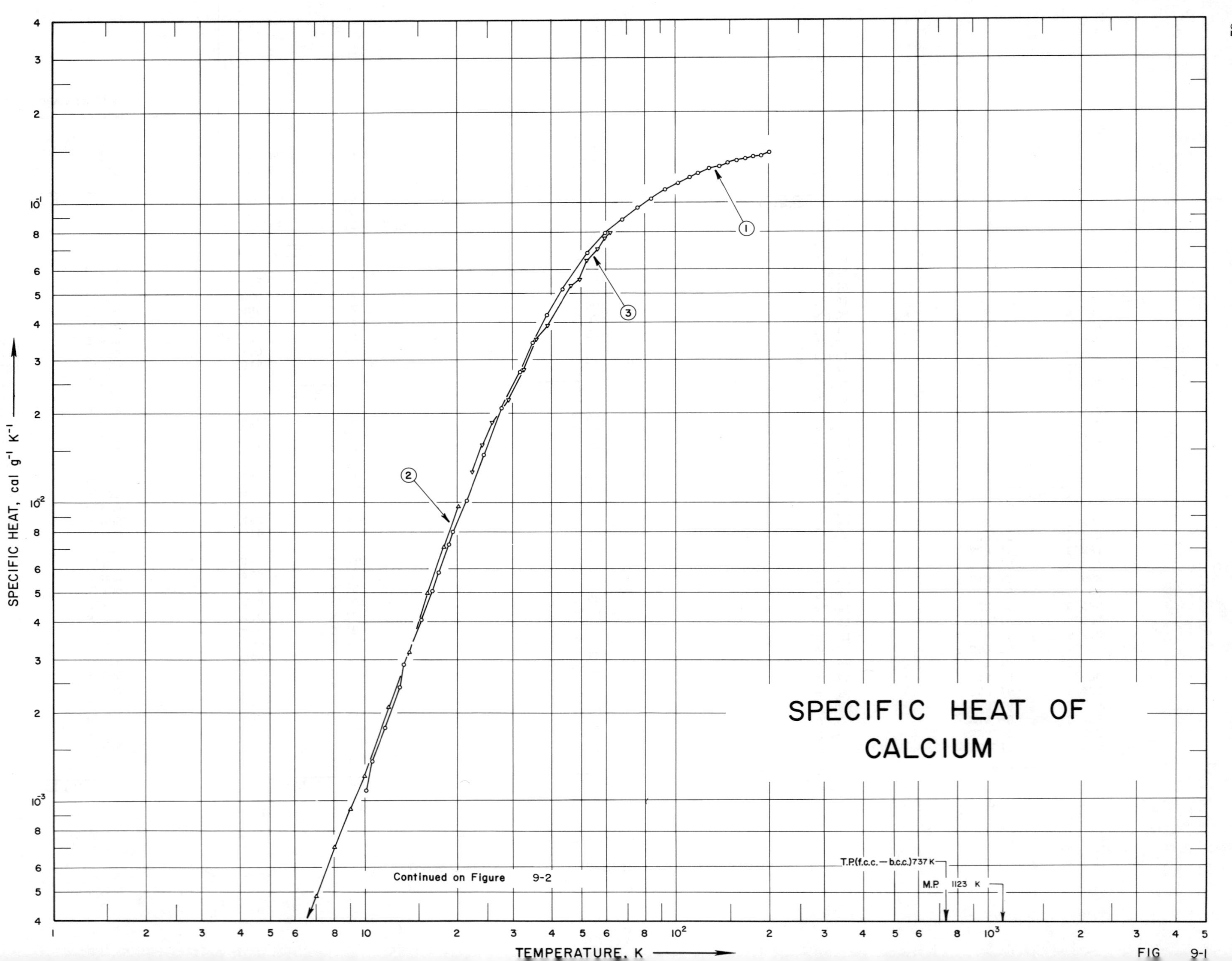
SPECIFIC HEAT OF
CALCIUM
SPECIFIC HEAT, cal g⁻¹ K⁻¹
TEMPERATURE, K
Continued on Figure 9-2
T.P.(f.c.c.—b.c.c.) 737 K
M.P. 1123 K
1
2
3
FIG 9-1

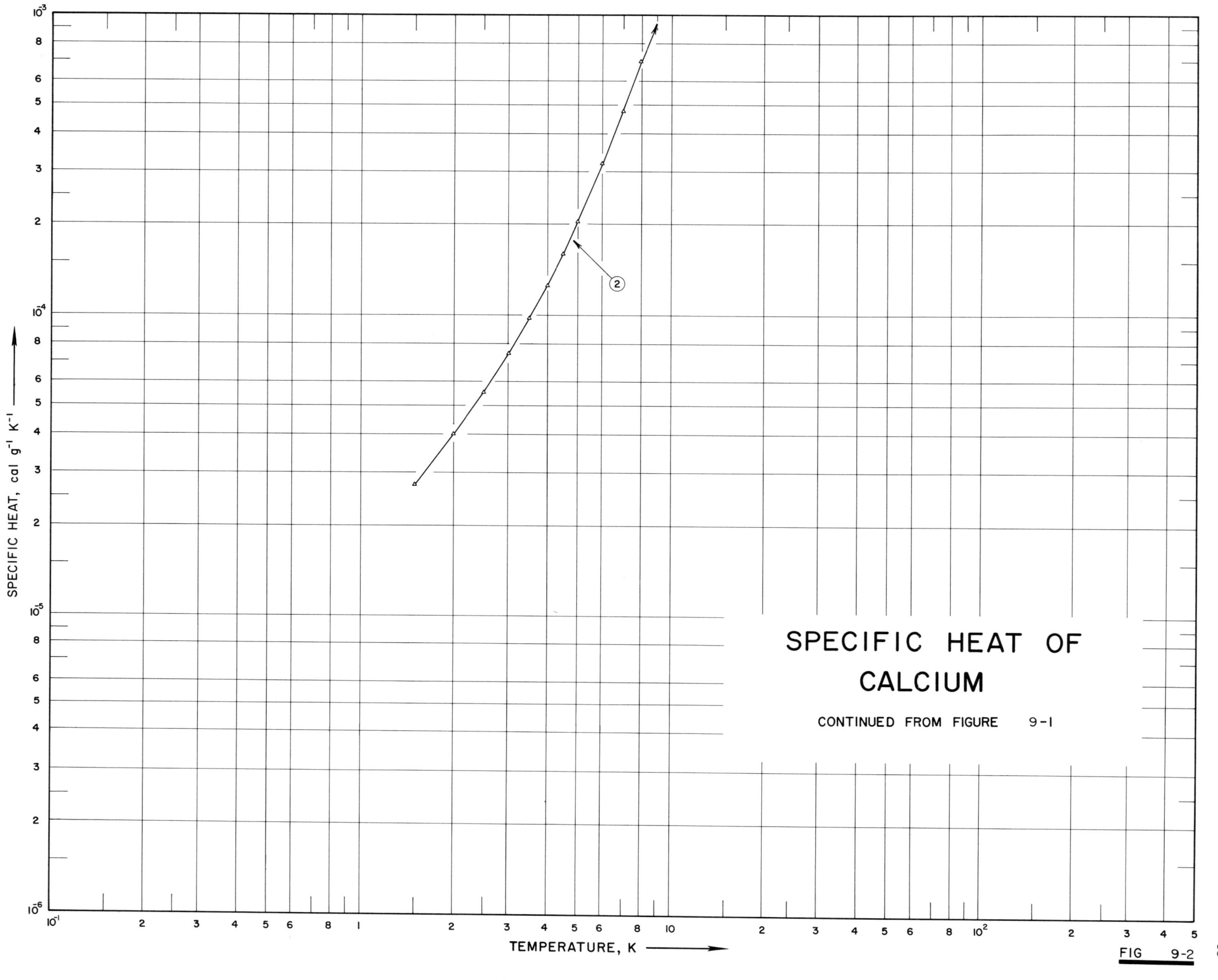
SPECIFIC HEAT OF
CALCIUM
CONTINUED FROM FIGURE 9-1
SPECIFIC HEAT, cal g^{-1} K^{-1}
TEMPERATURE, K
2

FIG 9-2

SPECIFICATION TABLE NO. 9 SPECIFIC HEAT OF CALCIUM

(Impurity $< 0.20\%$ each; total impurities $< 0.50\%$)

[For Data Reported in Figure and Table No. 9]

Curve No.	Ref. No.	Year	Temp. Range, K	Reported Error, %	Name and Specimen Designation	Composition (weight percent), Specifications and Remarks
1	186	1930	10-201	<3		Purest commercial metal; traces of Fe, N_2, Si and $CaCl_2$.
2	212	1957	1.5-20	2		0.1 Ba, 0.05 each Al, Fe, Mg, Mn, Si, Sn and Sr.
3	221	1916	22-62			

DATA TABLE NO. 9 SPECIFIC HEAT OF CALCIUM

[Temperature, T, K; Specific Heat, C_p, Cal $g^{-1}K^{-1}$]

T	C_p
CURVE 1	
10.16	1.10×10^{-3}
10.66	1.37
11.72	1.77
11.76	1.80*
13.02	2.42
13.49	2.89
15.34	4.04
16.58	5.06
17.43	5.81
18.73	7.24
19.36	7.98
21.40	1.01×10^{-2}
24.20	1.45
27.60	2.08
31.70	2.730
34.80	3.438
38.80	4.252
43.30	5.180
52.20	6.846
59.60	7.959
67.70	8.847
75.70	9.678
83.50	1.047×10^{-1}
92.70	1.115
102.20	1.173
111.50	1.226*
118.40	1.261
128.60	1.306*
138.80	1.334
147.20	1.367*
157.00	1.394
168.00	1.410*
178.00	1.442
189.40	1.448*
200.80	1.480
CURVE 2	
1.5	2.767×10^{-5}
2	4.031
2.5	5.576
3	7.514
3.5	9.839
CURVE 2 (cont.)	
4	1.270×10^{-4}
4.5	1.610
5	2.051
6	3.220
7	4.830
8	7.037
9	9.422
10	1.222×10^{-3}
12	2.075
14	3.280
16	4.967
18	7.096*
20	9.720
CURVE 3	
22.3	1.270×10^{-2}
24.0	1.564
25.9	1.856
29.2	2.221
32.6	2.794
35.8	3.533
38.8	3.910
46.3	5.319
49.3	5.599
52.0	6.452
56.5	7.023
59.7	7.640
62.0	7.989

* Not shown on plot

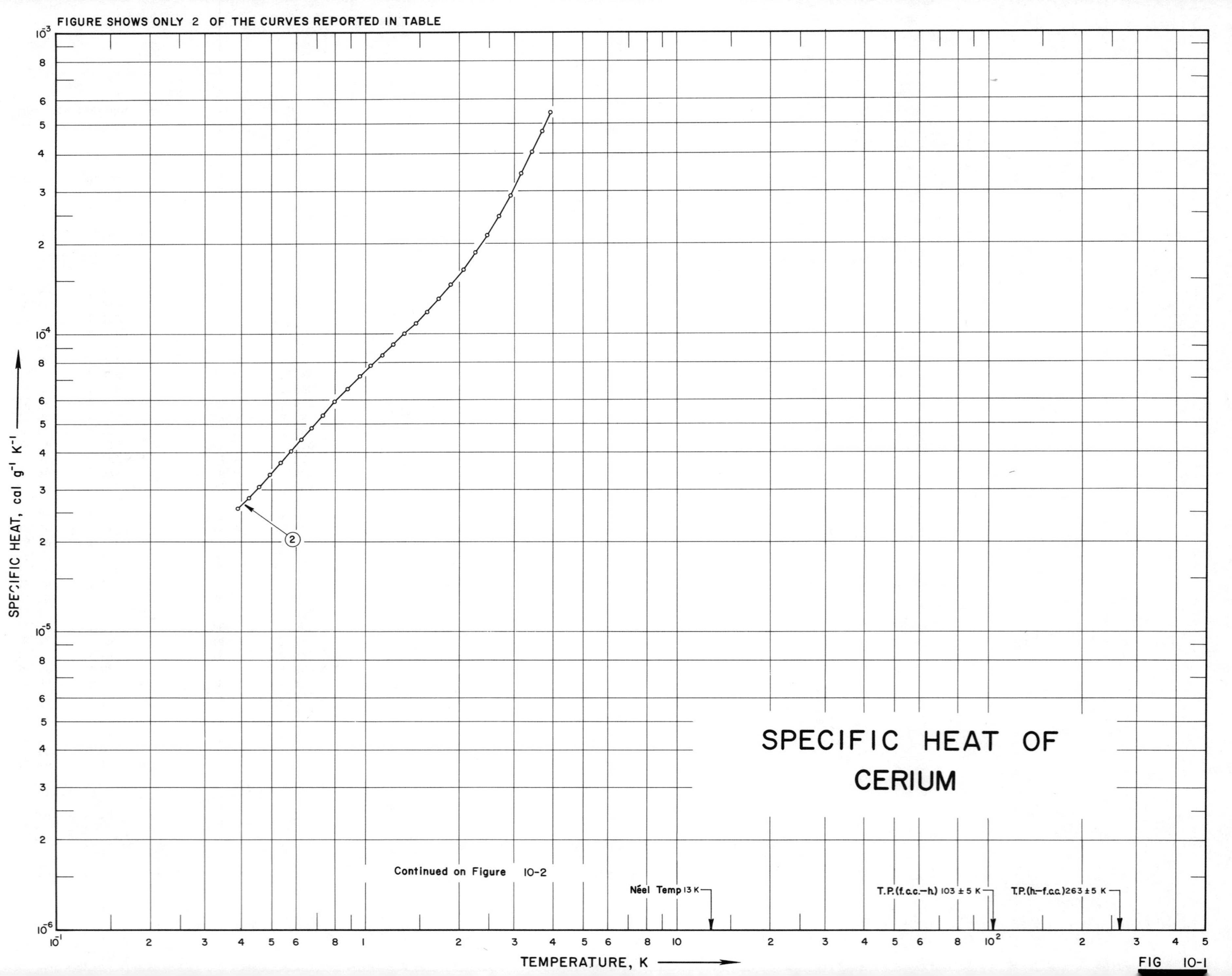
FIGURE SHOWS ONLY 2 OF THE CURVES REPORTED IN TABLE
SPECIFIC HEAT OF CERIUM
SPECIFIC HEAT, cal g⁻¹ K⁻¹
TEMPERATURE, K
2
Continued on Figure 10-2
Néel Temp 13 K
T.P. (f.c.c.–h.) 103 ± 5 K
T.P. (h.–f.c.c.) 263 ± 5 K

FIG 10-1

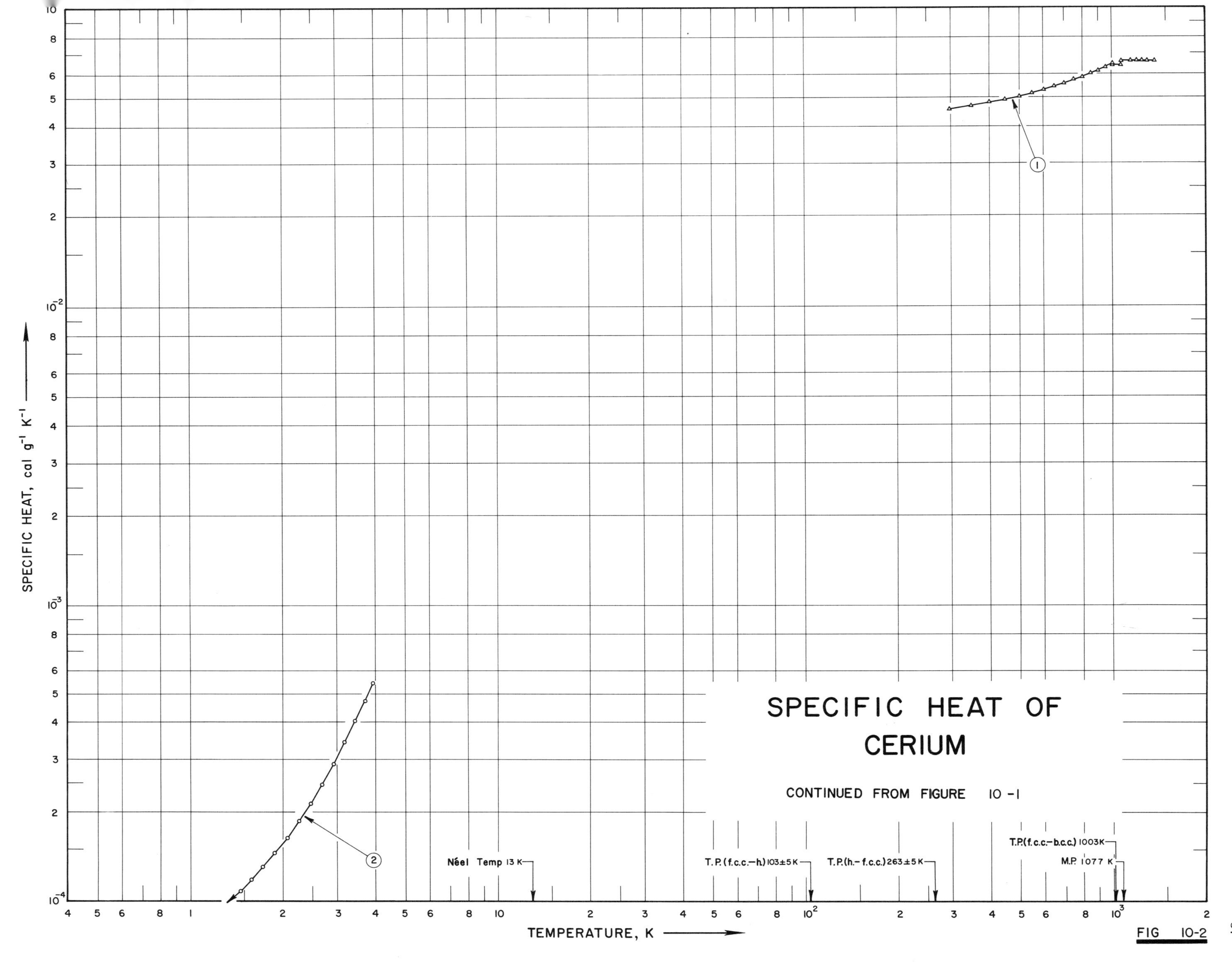
SPECIFIC HEAT OF
CERIUM
CONTINUED FROM FIGURE 10 -1
SPECIFIC HEAT, cal g⁻¹ K⁻¹
TEMPERATURE, K
Néel Temp 13 K
T. P. (f.c.c.–h.) 103±5K
T. P. (h.– f.c.c.) 263±5K
T.P.(f.c.c.–b.c.c.) 1003K
M.P. 1077 K
1
2

FIG 10-2

SPECIFICATION TABLE NO. 10 SPECIFIC HEAT OF CERIUM

(Impurity <0.20% each; total impurities < 0.50%)

[For Data Reported in Figure and Table No. 10]

Curve No.	Ref. No.	Year	Temp. Range, K	Reported Error, %	Name and Specimen Designation	Composition (weight percent), Specifications and Remarks
1	73	1960	298-1373	≤0.14		>99.92 Ce, ≤0.05 Cu, ≤0.02 Si; and ≤0.01 La.
2	86	1964	0.4-4	<1.5	Expt. I Run I	0.08 Er, 0.074 O_2, 0.072 C, 0.034 F, 0.022 H_2, 0.018 Ag, 0.018 Fu, 0.018 Th, 0.015 Fe, 0.01 Ta, 0.0044 N_2, 0.003 Ni, 0.0025 Y, 0.002 Dy, 0.002 Gd, 0.0015 Mo, 0.001 Ca, 0.001 Nd, 0.0005 K, 0.0004 La; 60% α Ce and 40% β Ce; cycled between 77 and 293 K each warming and cooling period of 1 hr; distilled, remelted and cast into tantalum crucible; cooled to room temperature in 1 1/2 hrs.
3	86	1964	0.4-3.8	<1.5	Expt. I Run II and III	Same as above.
4	86	1964	1.2-4	<1.5	Expt. II	Same as above except 35% α Ce, and 65% β Ce.
5	285	1958	273-1373			0.05 Ca, 0.01 La, and 0.02 Si.

DATA TABLE NO. 10 SPECIFIC HEAT OF CERIUM

[Temperature, T, K; Specific Heat, C_p, Cal $g^{-1} K^{-1}$]

CURVE 1

T	C_p
298.15	4.60 x 10^{-2}
300	4.60*
350	4.71
400	4.82
450	4.94
500	5.07
550	5.19
600	5.32
650	5.46
700	5.60
750	5.74
800	5.89
850	6.04
900	6.19
950	6.35
1000	6.52
1003.15	6.53*
1003.15	6.46*
1050	6.46
(s)1077.15	6.46*
(l)1077.15	6.67
1100	6.67*
1150	6.67
1200	6.67
1250	6.67
1300	6.67
1350	6.67*
1373.15	6.67

CURVE 2

T	C_p
0.389_5	2.594* x 10^{-5}*
0.422_4	2.816*
0.457_4	3.075*
0.494_7	3.362*
0.534_9	3.693*
0.577_1	4.037*
0.621_6	4.411*
0.670_7	4.825*
0.729_3	5.330*
0.798_4	5.895*
0.875_4	6.523*
0.957_9	7.173*
1.044_1	7.821*
1.135_7	8.484*
1.235_9	9 201*

CURVE 2 (cont.)

T	C_p
1.343_8	1.002 x 10^{-4}
1.457_4	1.083
1.580_4	1.185
1.722_4	1.305
1.886_0	1.453
2.067_6	1.639
2.259_5	1.863
2.465_9	2.130
2.690_7	2.460
2.931_5	2.888
3.185_1	3.420
3.441_8	4.029
3.702_1	4.698
3.948_0	5.443

CURVE 3

T	C_p
0.375_1	2.501* x 10^{-5}*
0.405_6	2.704*
0.439_5	2.942*
0.475_7	3.215*
0.514_3	3.522*
0.555_3	3.862*
0.598_9	4.225*
0.646_0	4.626*
0.699_9	5.080*
0.763_5	5.617*
0.836_6	6.221*
0.916_2	6.860*
1.000_5	7.488*
1.090_2	8.158*
1.186_6	8.856*
1.291_1	9.622*
1.402_4	1.046 x 10^{-4}
1.520_1	1.138
1.652_3	1.246
1.805_1	1.383
1.977_2	1.548
2.165_2	1.755
2.362_5	1.992
2.574_4	2.289
2.803_2	2.676
3.047_1	3.139
3.311_9	3.710
3.591_2	4.428
3.862_6	5.245

CURVE 4

T	C_p
1.187_1	1.103 x 10^{-4}
1.265_4	1.177
1.351_1	1.263
1.442_2	1.352
1.542_3	1.455
1.659_0	1.584
1.794_9	1.745
1.947_7	1.941
2.111_2	2.178
2.289_7	2.460
2.485_8	2.816
2.697_8	3.275
2.924_0	3.823
3.154_2	4.496
3.387_9	5.267
3.608_3	6.095
3.800_0	6.915
3.970_5	7.720

CURVE 5*

T	C_p
273.2	4.55 x 10^{-2}
323.2	4.65
373.2	4.76
423.2	4.88
473.2	5.00
523.2	5.12
573.2	5.25
623.2	5.39
673.2	5.52
723.2	5.67
773.2	5.81
823.2	5.96
873.2	6.12
923.2	6.27
973.2	6.43
1003.2	6.53
1003.2	6.46
1048.2	6.46
1077.2	6.46
1077.2	6.67
1123.2	6.67
1173.2	6.67
1223.2	6.67
1273.2	6.67
1323.2	6.67
1373.2	6.67

*Not shown on plot

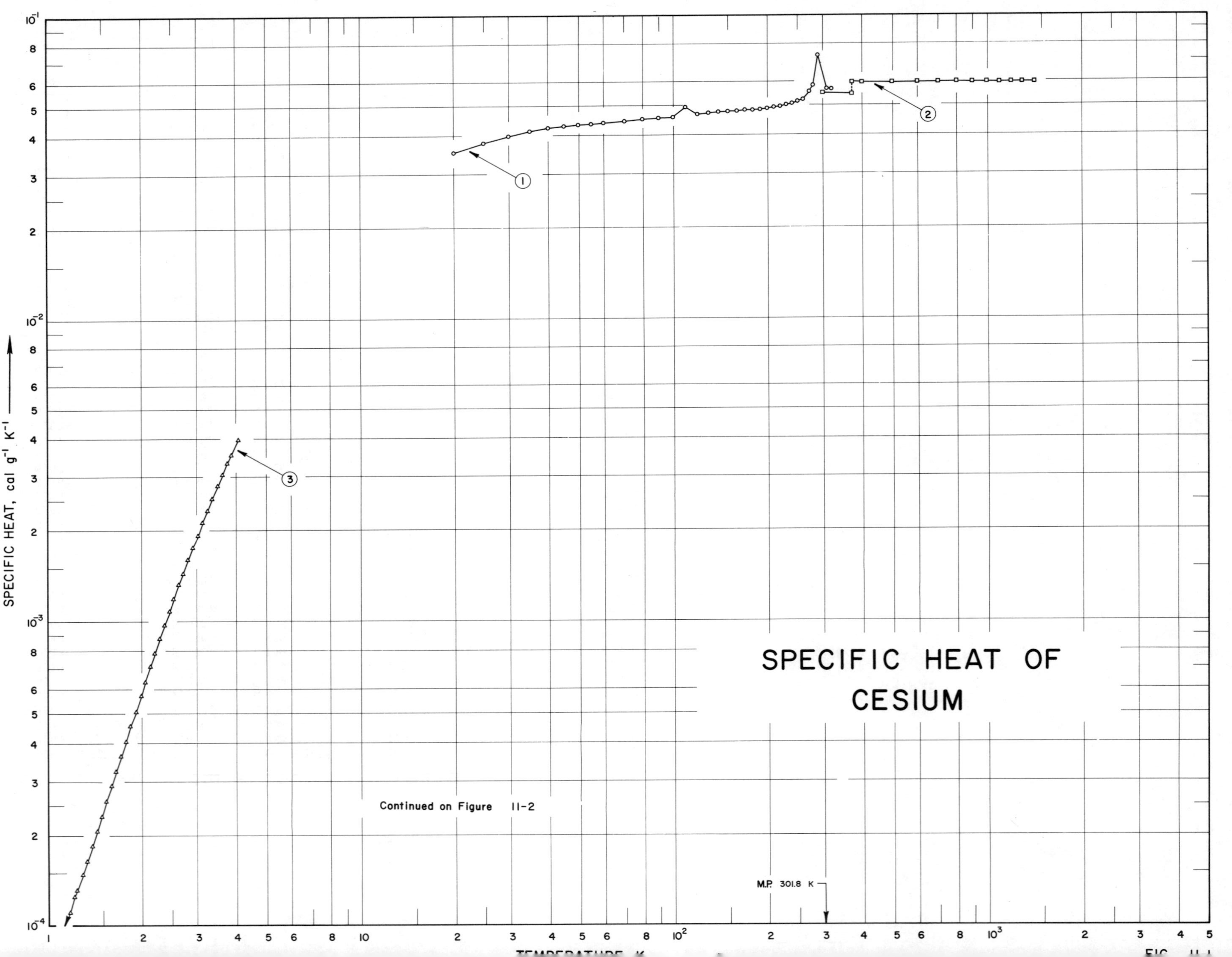
SPECIFIC HEAT OF CESIUM
SPECIFIC HEAT, cal g⁻¹ K⁻¹
M.P. 301.8 K
Continued on Figure 11-2
1
2
3
10⁻¹
10⁻²
10⁻³
10⁻⁴
1
10
10²
10³

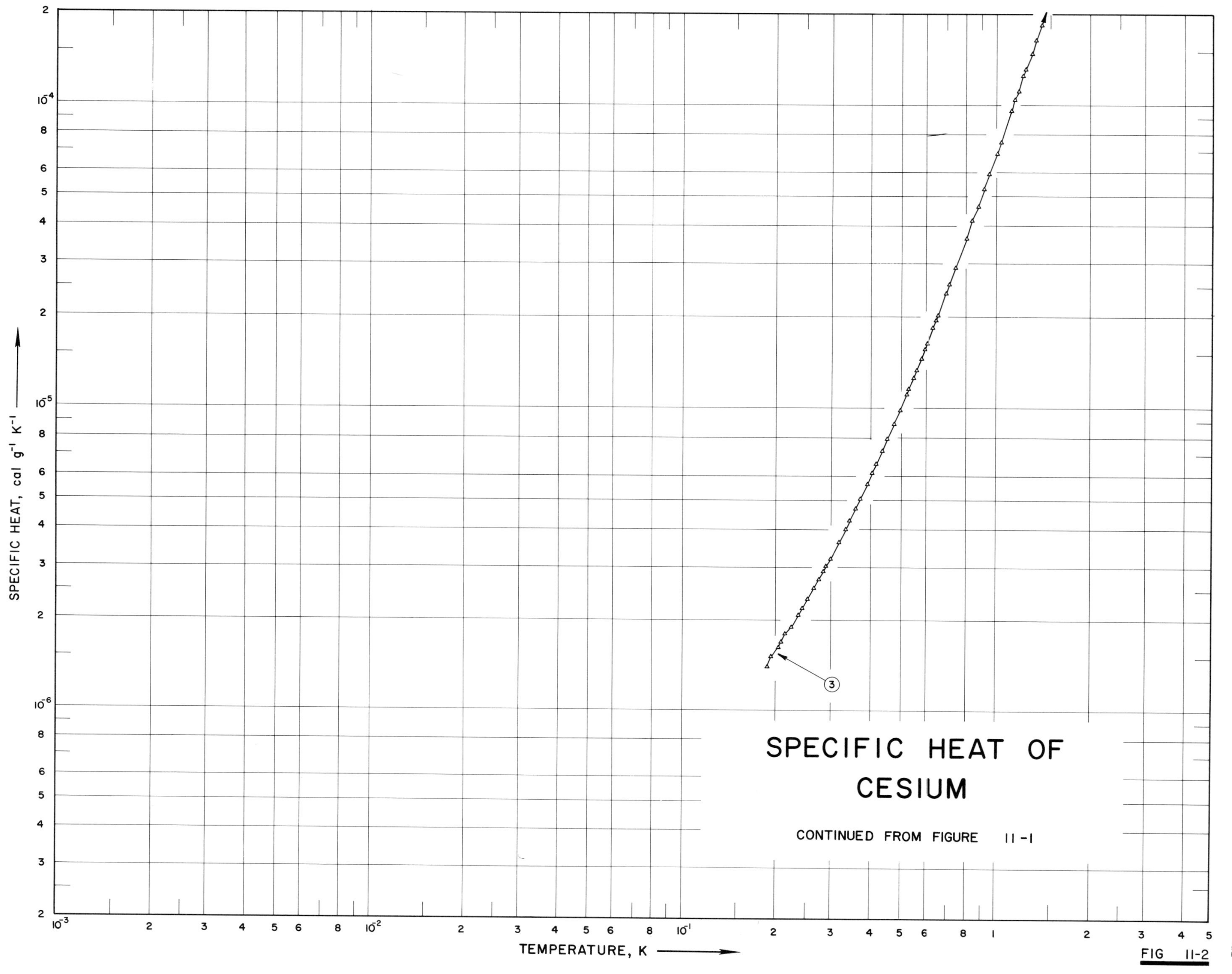

SPECIFIC HEAT OF CESIUM
CONTINUED FROM FIGURE 11-1
3
SPECIFIC HEAT, cal g⁻¹ K⁻¹
TEMPERATURE, K

FIG 11-2

SPECIFICATION TABLE NO. 11 SPECIFIC HEAT OF CESIUM

(Impurity < 0.20% each; total impurities < 0.50%)

[For Data Reported in Figure and Table No. 11]

Curve No.	Ref. No.	Year	Temp. Range, K	Reported Error, %	Name and Specimen Designation	Composition (weight percent), Specifications and Remarks
1	176	1955	20-320			0.3 O_2.
2	286	1964	273-1423			
3	356	1964	0.18-4			99.8 Cs stated purity; measured under helium atmosphere.

DATA TABLE NO. 11 SPECIFIC HEAT OF CESIUM

[Temperature, T, K; Specific Heat, C_p, Cal $g^{-1}K^{-1}$]

T	C_p
CURVE 1	
20	3.55×10^{-2}
25	3.82
30	4.03
35	4.18
40	4.27
45	4.33
50	4.36
55	4.39
60	4.43
70	4.49
80	4.54
90	4.60
100	4.63
110	4.98
120	4.72
130	4.76
140	4.80
150	4.82
160	4.85
170	4.86
180	4.88
190	4.91
200	4.94
210	4.97
220	5.00
230	5.06
240	5.12
250	5.23
260	5.28
273.15	5.61
280	5.87
290	7.37
310	5.73
320	5.73
CURVE 2	
273.15	5.59×10^{-2}*
300	5.59
301.52	5.59*
301.52	5.54*
373.15	5.54
373.15	6.0
400	6.0
CURVE 2 (cont.)	
500	6.0×10^{-2}
600	6.0
700	6.0
800	6.0
900	6.0
1000	6.0
1100	6.0
1200	6.0
1300	6.0
1400	6.0*
1423	6.0
CURVE 3	
Series 1	
0.1874	1.447×10^{-6}
0.1923	1.511
0.2073	1.699
0.2141	1.804
0.2351	2.084
0.2416	2.205
0.2629	2.568
0.2883	3.041
0.3166	3.652
0.3434	4.298
0.3719	5.095
0.4056	6.170
0.4370	7.285
0.4755	8.927
0.5214	1.117×10^{-5}
0.5612	1.344
0.6059	1.647
0.6550	2.030
0.7130	2.561
Series 2	
0.2080	1.708×10^{-6}*
0.2225	1.908
0.2361	2.117*
0.2519	2.361
0.2568	2.440*
0.2739	2.741
CURVE 3 (cont.)	
0.2832	2.920×10^{-6}
0.2984	3.212
0.3143	3.575*
0.3155	3.589*
0.3339	4.041
0.3598	4.728
0.3916	5.679
0.4190	6.645
0.4539	7.972
0.4963	9.873
0.5480	1.265×10^{-5}
0.5952	1.570
0.6461	1.955
0.6951	2.394
0.7468	2.911
0.8072	3.633
0.8801	4.643
0.9524	5.927
1.0020	6.909
1.0340	7.549
1.0990	9.166*
1.1180	9.594
1.1460	1.042×10^{-4}
1.2000	1.205*
1.2170	1.256
Series 3	
0.5293	1.160×10^{-5}
0.5800	1.465
0.6321	1.845
0.6884	2.325
0.7503	2.955*
0.8136	3.712*
0.8457	4.154
0.9181	5.296
1.2031	1.180×10^{-4}*
1.2932	1.475
1.3883	1.849
1.4897	2.316
1.6006	2.906
1.7185	3.647
1.8435	4.553
1.9803	5.706
CURVE 3 (cont.)	
2.1302	7.148×10^{-4}
2.2875	8.848
2.4529	1.089×10^{-3}
2.6285	1.323
2.8182	1.605
3.0238	1.937
3.2512	2.336
3.5012	2.816
3.7667	3.347
4.0437	3.920*
Series 4	
1.1779	1.107×10^{-4}
1.2459	1.315
1.3371	1.642
1.4374	2.063
1.5444	2.591
1.6575	3.241
1.7799	4.064
1.9111	5.073
2.0522	6.326
2.2053	7.898
2.3645	9.733
2.5355	1.192×10^{-3}
2.7183	1.448
2.9183	1.759
3.1364	2.131
3.3691	2.555
3.6290	3.061
3.8761	3.557
4.0880	3.987

*Not shown on plot

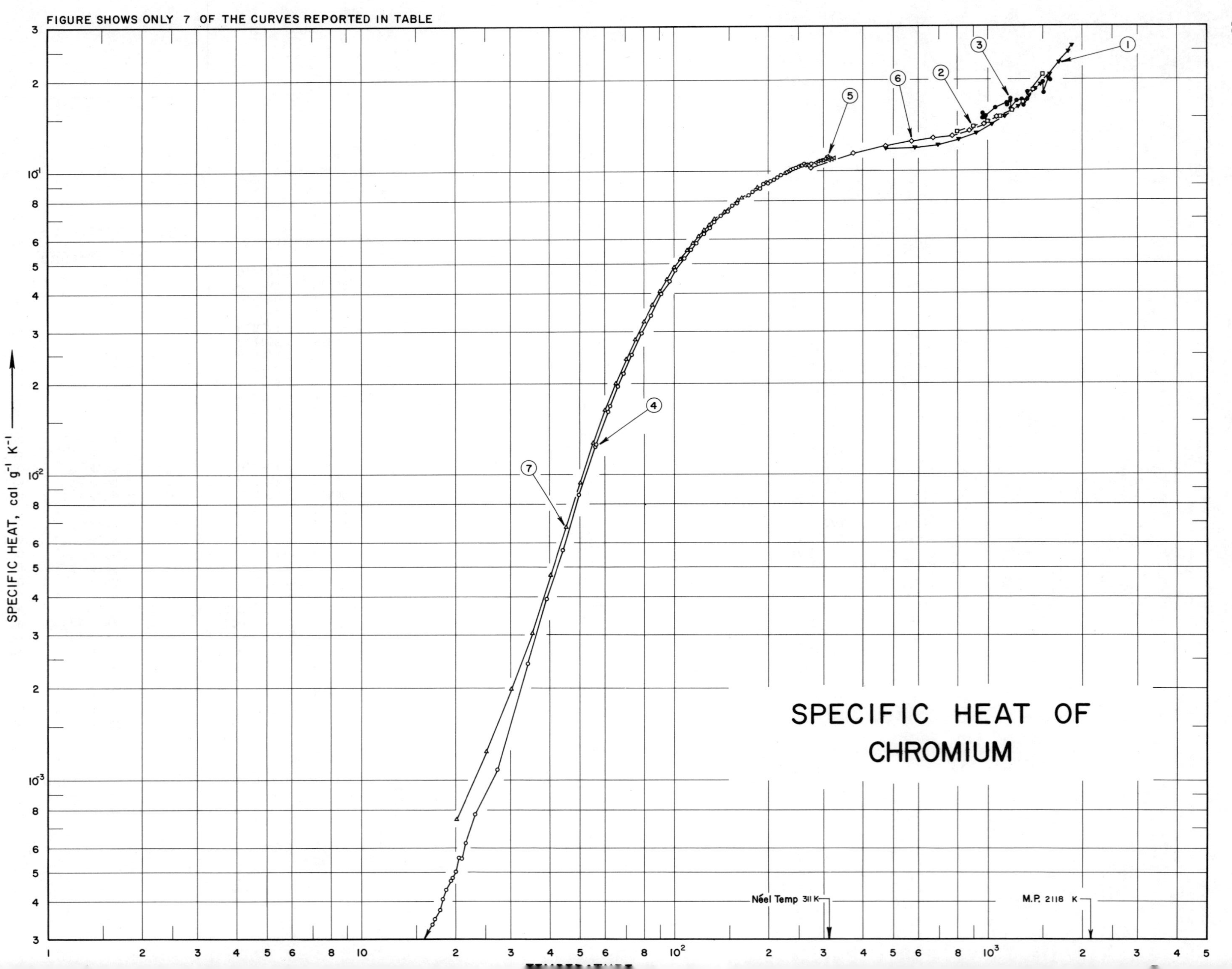

FIGURE SHOWS ONLY 7 OF THE CURVES REPORTED IN TABLE
SPECIFIC HEAT OF CHROMIUM
SPECIFIC HEAT, cal g⁻¹ K⁻¹
Néel Temp 311 K
M.P. 2118 K
1
2
3
4
5
6
7

SPECIFICATION TABLE NO. 12 SPECIFIC HEAT OF CHROMIUM

(Impurity < 0.20% each; total impurities < 0.50%)

[For Data Reported in Figure and Table No. 12]

Curve No.	Ref. No.	Year	Temp. Range, K	Reported Error, %	Name and Specimen Designation	Composition (weight percent), Specifications and Remarks
1	10	1958	297-1922			Chemically pure; ductile; specimen sealed in capsule; density = 448 lb ft^{-3} at 75 F.
2	20	1959	800-1500	± 0.3		100 Cr; specimen under argon atmosphere.
3	13	1958	964-1598	0.7-2.0		99.96 Cr, 0.04 O_2.
4	22	1962	14-274			99.9 Cr.
5	23	1960	268-324	0.13	Ductile Chromium	99.998 Cr; specimen produced by Aeronautic's Research Laboratory, Melbourne.
6	24	1950	273-1073	1.0		0.1 impurities, mostly chromous oxide with absorbed H_2 and some Ca and Na; electrolytic flakes.
7	25	1954	20-200			99.9 Cr.
8	125	1965	1273-2103	1.42		99.99 Cr; form of crystals made by vapor decomposition of the iodide.
9	287	1937	56-291			Electrolytic Cr; after treatment specimen contained 0.65 O_2, assumed to be in the form of Cr_2O_3; evacuated and heated to 1100 C to remove H_2; corrected for Cr_2O_3 impurities.
10	288	1952	1.8-4.0			99.9 Cr.

DATA TABLE NO. 12 SPECIFIC HEAT OF CHROMIUM

[Temperature, T, K; Specific Heat, C_p, Cal g^{-1} K^{-1}]

T	C_p
CURVE 1	
478	1.18 x 10^{-1}
589	1.19
700	1.22
811	1.27
922	1.33
1033	1.42
1144	1.51
1255	1.63
1366	1.76
1478	1.91
1589	2.08
1700	2.27
1811	2.47
1866	2.57
1922	2.69
CURVE 2	
800	1.35 x 10^{-1}
900	1.40
1000	1.46
1100	1.52
1200	1.59
1300	1.68
1400	1.85
1500	2.08
CURVE 3	
964	1.49 x 10^{-1}
969	1.55
977	1.49
989	1.52
1067	1.62
1067	1.66*
1068	1.55*
1159	1.67
1160	1.68*
1160	1.65
1176	1.73
1177	1.70
1177	1.66*
1180	1.59
1183	1.63*
1250	1.62*
CURVE 3 (contd)	
1250	1.70 x 10^{-1}
1252	1.71*
1295	1.72
1302	1.65
1348	1.72
1350	1.82
1351	1.79*
1351	1.82*
1424	1.86
1431	1.79*
1507	1.95
1510	1.80
1513	2.04*
1514	1.84*
1588	2.01
1590	2.15*
1598	1.97
CURVE 4	
14.10	2.29 x 10^{-4}*
14.21	2.27*
14.22	2.27*
14.47	2.42*
15.15	2.67*
16.24	3.12*
16.37	3.21*
16.90	3.35
17.09	3.48
17.77	3.75
18.28	4.04
18.64	4.19
19.47	4.69
19.52	4.77
20.01	5.00
20.59	5.58
20.77	5.54
21.61	6.21
23.18	7.71
27.35	1.079 x 10^{-3}
34.05	2.408
39.28	3.914
44.14	5.695
49.75	8.522
CURVE 4 (contd)	
56.05	1.235 x 10^{-2}
56.45	1.255
61.80	1.615
62.60	1.686
66.33	1.958
69.04	2.164
73.22	2.491
78.94	2.956
84.50	3.383
91.66	3.989
97.21	4.395
102.30	4.748
108.74	5.202
113.84	5.550
118.72	5.850
125.16	6.250
130.40	6.543
135.38	6.835
142.44	7.143
148.21	7.397
153.92	7.710
159.48	7.877
173.22	8.351
178.97	8.568
188.46	8.787
194.88	9.006
199.22	9.143
200.98	9.128
203.24	9.285*
204.96	9.256
209.27	9.378*
209.29	9.353
215.04	9.520
215.14	9.524*
215.43	9.553*
220.44	9.678
220.48	9.683*
229.44	9.851
230.17	9.862
235.43	9.964
235.76	1.001 x 10^{-1}
240.99	1.013
241.01	1.008*
246.33	1.018
CURVE 4 (contd)	
251.41	1.026 x 10^{-1}
255.68	1.032
257.50	1.033
261.24	1.050
263.10	1.039
267.11	1.047
272.15	1.049
274.43	1.052*
CURVE 5	
267.861	1.045 x 10^{-1}*
269.490	1.047*
271.111	1.047*
272.727	1.052*
274.340	1.054*
275.947	1.057*
277.551	1.058*
279.153	1.059*
280.750	1.063*
282.343	1.065*
283.932	1.067*
284.729	1.066
286.756	1.073*
288.781	1.070*
290.947	1.074
293.244	1.077*
295.526	1.080
297.794	1.084*
298.054	1.083*
299.443	1.085*
300.048	1.088*
300.830	1.086*
301.234	1.086
302.213	1.092*
302.384	1.095*
303.527	1.092*
303.591	1.094*
304.691	1.097*
304.966	1.094*
305.850	1.098
306.240	1.100*
307.002	1.103*
308.647	1.104*
CURVE 5 (contd)	
309.281	1.109 x 10^{-1}*
309.782	1.112*
310.409	1.117*
310.912	1.117
311.535	1.119*
311.750	1.111*
312.042	1.105
312.351	1.103*
312.670	1.099*
312.955	1.095*
313.191	1.098*
313.562	1.095*
313.760	1.091*
314.348	1.091*
314.918	1.092*
315.506	1.097*
316.073	1.093
316.662	1.096*
317.227	1.095*
317.816	1.098*
318.378	1.097*
318.969	1.097*
319.530	1.094*
320.680	1.098
321.827	1.099*
322.972	1.101*
324.115	1.101
CURVE 6	
273.15	1.038 x 10^{-1}
373	1.146
473	1.216
573	1.259
673	1.278
773	1.308
873	1.359
973	1.424
1073	1.505
CURVE 7	
20	7.441 x 10^{-4}
25	1.242 x 10^{-3}
30	1.986
35	3.042
40	4.716
45	6.766
50	9.373
55	1.272 x 10^{-2}
60	1.645
65	2.017
70	2.415
75	2.811
80	3.228
85	3.655
90	4.053
95	4.443
100	4.836
105	5.191
110	5.512
115	5.828
120	6.126
125	6.412
130	6.679
135	6.945
140	7.187*
145	7.399
150	7.424*
155	7.822*
160	8.014
165	8.193
170	8.373*
175	8.541*
180	8.683*
185	8.819
190	8.939*
195	9.025*
200	9.081*

* Not shown on plot

DATA TABLE NO. 12 (continued)

T	C_p
CURVE 8*	
1273	1.544 x 10^{-1}
1373	1.682
1473	1.820
1573	1.958
1673	2.097
1773	2.235
1873	2.373
1973	2.511
2073	2.650
2103	2.691
CURVE 9*	
56.1	1.219 x 10^{-2}
59.5	1.446
61.1	1.559
63.5	1.636
66.9	2.003
68.0	2.090
72.4	2.415
74.2	2.609
81.6	3.165
84.8	3.415
93.2	4.064
96.1	4.301
105.8	5.043
123.1	5.820
141.6	7.143
162.9	8.037
181.6	8.668
200.1	9.260
222.1	9.696
244.1	1.008 x 10^{-1}
274.0	1.046
281.5	1.053
291.1	1.066
CURVE 10*	
1.833	1.432 x 10^{-5}
2.020	1.559
2.210	0.752
2.343	1.750
2.451	1.828
2.572	2.173
2.658	2.192

T	C_p
CURVE 10* (cont.)	
2.731	2.250 x 10^{-5}
2.808	2.307
3.885	2.326
2.967	2.403
2.781	2.403
2.911	2.442
3.013	2.346
3.134	2.596
3.232	2.500
3.313	2.692
3.392	2.999
3.462	2.999
3.517	3.288
3.578	3.096
3.637	3.307
3.709	3.326
3.772	3.653
3.823	3.230
3.949	3.634
3.995	3.788

*Not shown on plot

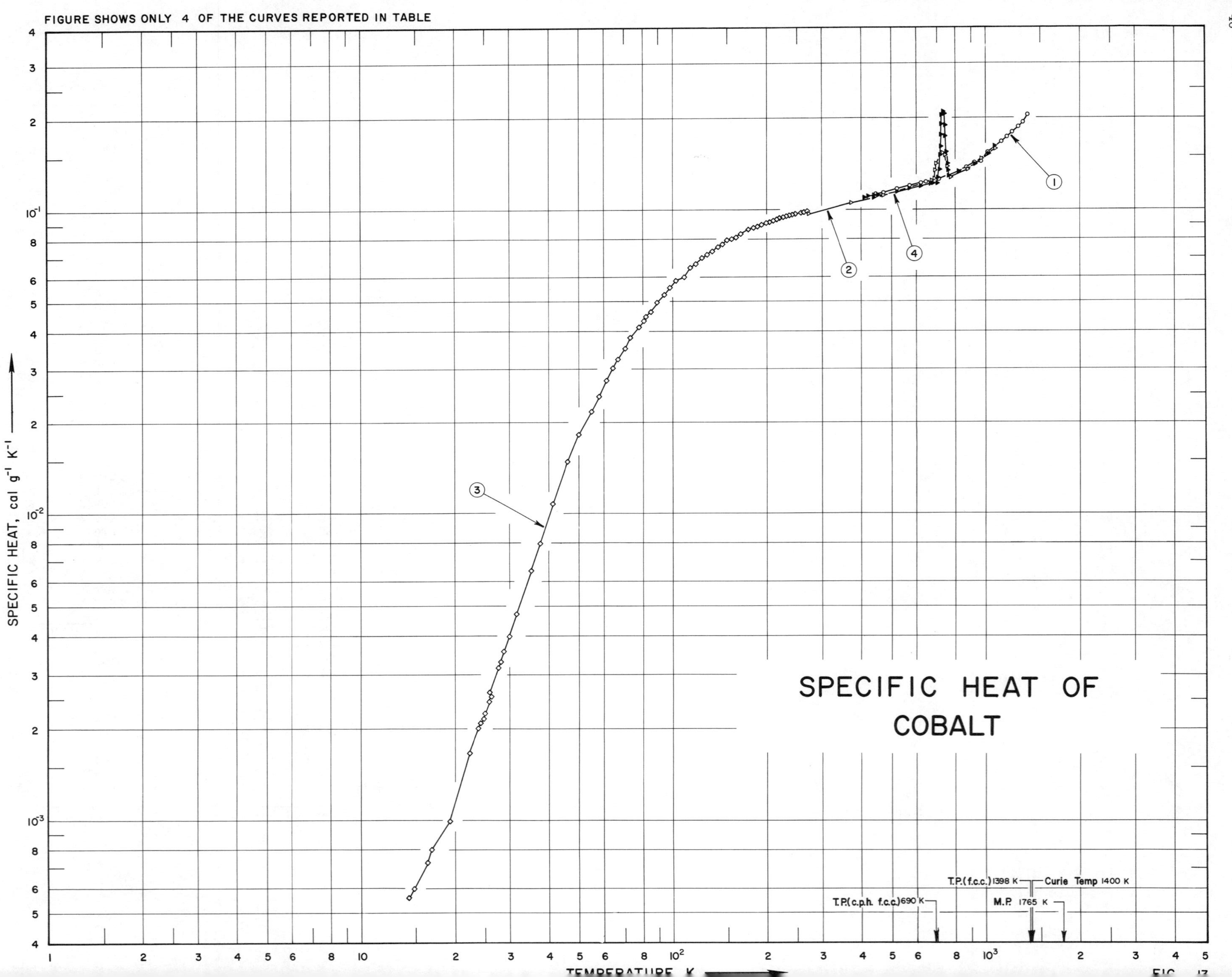
FIGURE SHOWS ONLY 4 OF THE CURVES REPORTED IN TABLE
SPECIFIC HEAT OF
COBALT
SPECIFIC HEAT, cal g⁻¹ K⁻¹
T.P.(c.p.h. f.c.c.) 690 K
T.P.(f.c.c.) 1398 K
Curie Temp 1400 K
M.P. 1765 K
1
2
3
4

SPECIFICATION TABLE NO. 13 SPECIFIC HEAT OF COBALT

(Impurity <0.20% each; total impurities < 0.50%)

[For Data Reported in Figure and Table No. 13]

Curve No.	Ref. No.	Year	Temp. Range, K	Reported Error, %	Name and Specimen Designation	Composition (weight percent), Specifications and Remarks
1	93	1940	448-1673			99.95 Co, 0.01-0.06 Fe; deposited electrolytically; density = 551 lb ft^{-3}.
2	24	1950	273-1073		Electrolytic	< 0.01 impurities.
3	94	1952	14-270			96.95 Co, 2.50 Fe, 0.26 Cu, 0.20 Ni, 0.08 Si, 0.01 Mn; data corrected for impurities; density = 551 lb ft^{-3}.
4	31	1959	413-1073	1.0		Sealed in argon.
5	164	1932	273-1473			
6	223	1959	0.3-1	~5		99.9 Co, with the principal impurity Fe; hexagonal closed packed; sample supplied by African Metals Corporation.

DATA TABLE NO. 13 SPECIFIC HEAT OF COBALT

[Temperature, T, K; Specific Heat, C_p, Cal g^{-1} K^{-1}]

CURVE 1

T	C_p
448	1.13 x 10^{-1}
473	1.14
523	1.17
573	1.20
623	1.23
648	1.24
713	1.26
873	1.37
923	1.43
973	1.49
1023	1.55
1073	1.61*
1123	1.68
1173	1.74
1223	1.80
1273	1.87
1323	1.94
1373	2.05

CURVE 2

T	C_p
273.15	9.83 x 10^{-2}
373	1.06 x 10^{-1}
473	1.13
573	1.20*
673	1.25
683	1.26*
691	1.27
693	1.36
698	1.42
733	1.54
743	1.52
753	1.39
773	1.28
873	1.36
973	1.47
1073	1.60

CURVE 3

T	C_p
14.27	5.60 x 10^{-4}
14.85	5.89
16.36	7.26
16.86	8.06
19.24	9.94
22.36	1.67 x 10^{-3}
23.77	2.01
24.10	2.09
24.75	2.16
24.90	2.24
25.72	2.46
25.73	2.63
26.18	2.56
27.67	3.16
27.72	3.17*
28.23	3.31
28.34	3.33*
28.62	3.58
29.92	3.99
31.60	4.72
35.16	6.50
37.59	7.99
41.24	1.07 x 10^{-2}
46.96	1.48
50.81	1.82
54.93	2.17
57.67	2.44
61.54	2.75
64.32	3.02
66.98	3.236
70.53	3.528
73.86	3.801
78.28	4.142
81.35	4.327
82.59	4.459
85.23	4.629
89.51	4.960
94.25	5.277
98.84	5.576
103.84	5.885
109.21	5.998
114.56	6.458
119.50	6.697
124.51	6.949
129.17	7.130
134.20	7.320
140.02	7.554
145.34	7.739
150.48	7.909
155.69	8.077
160.67	8.186 x 10^{-2}
166.12	8.355
177.64	8.611
183.51	8.729
187.85	8.825
194.82	8.942
200.40	9.044
200.58	9.056*
205.86	9.126
205.94	9.127*
211.33	9.221
216.64	9.319
221.80	9.402
227.08	9.475
232.84	9.543
237.98	9.611
243.50	9.682
249.31	9.743
254.31	9.789*
259.52	9.830
265.25	9.877
269.83	9.911

CURVE 4

T	C_p
413	1.10 x 10^{-1}
423	1.11
443	1.10
443	1.12
453	1.13
473	1.13*
523	1.15
573	1.17
623	1.20
673	1.22
683	1.22
703	1.22
708	1.27
713	1.29*
718	1.36
719	1.52
723	1.62
724	1.76
725	1.91
725	2.05
733	2.09 x 10^{-1}
738	2.07
743	1.90
745	1.75
758	1.56
763	1.42
764	1.35
765	1.33*
768	1.29
773	1.29*
823	1.34
873	1.38*
923	1.42
973	1.46*
1023	1.54
1073	1.62

CURVE 5*

T	C_p
273.15	1.055 x 10^{-1}
373.15	1.094
473.15	1.160
573.15	1.228
673.15	1.318
773.15	1.416
873.15	1.534
973.15	1.653
1073.15	1.781
1173.15	1.945
1273.15	2.040
1373.15	2.120
1473.15	2.195

CURVE 6*

Series 1

T	C_p
0.612	4.50 x 10^{-5}
0.624	5.11
0.625	4.70
0.637	4.99
0.646	5.11
0.641	4.91

Series 2

T	C_p
0.565	5.68 x 10^{-5}
0.578	5.60
0.586	5.56
0.593	5.52
0.599	5.43
0.604	5.31
0.625	5.48
0.640	5.07
0.655	4.91
0.670	4.79
0.625	5.15
0.640	4.70
0.669	5.35

Series 3

T	C_p
0.327	1.03 x 10^{-4}
0.360	9.69 x 10^{-5}

Series 4

T	C_p
0.336	9.53 x 10^{-5}
0.349	9.69
0.370	1.37 x 10^{-4}
0.402	7.38 x 10^{-5}
0.446	7.42
0.489	5.56
0.530	6.69
0.583	5.03
0.637	5.35
0.696	4.14
0.756	3.97
0.815	3.6
0.898	4.02
0.999	4.30

Series 5

T	C_p
0.548	5.31 x 10^{-5}
0.552	5.52
0.565	5.35

Series 6

T	C_p
0.485	5.76 x 10^{-5}
0.498	6.08
0.510	5.48

*Not shown on Plot

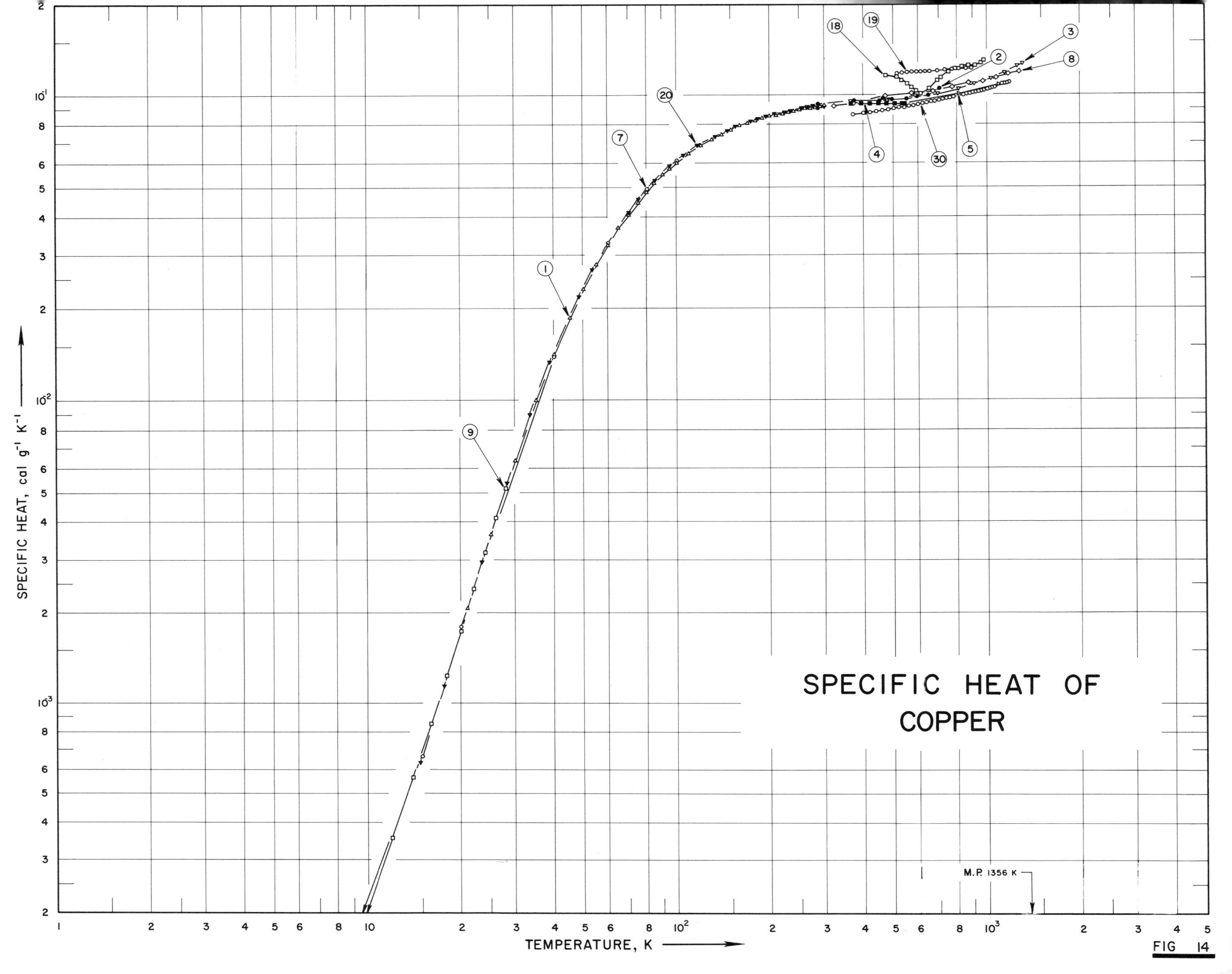
SPECIFIC HEAT OF COPPER
SPECIFIC HEAT, cal g⁻¹ K⁻¹
TEMPERATURE, K
M.P. 1356 K
1
2
3
4
5
6
7
8
9
18
19
20
30

FIG 14

SPECIFICATION TABLE NO. 14 SPECIFIC HEAT OF COPPER

(Impurity <0.20% each; total impurities <0.50%)

[For Data Reported in Figure and Table No. 14]

Curve No.	Ref. No.	Year	Temp. Range, K	Reported Error, %	Name and Specimen Designation	Composition (weight percent), Specifications and Remarks
1	8	1960	20-300	<2.0		99.999 Cu; sample-supplied by American Smelting and Refining Co.; heavily cold worked below room temperature with hydraulic press until strength increased 50%.
2	9	1959	288-701			
3	11	1958	366-1310			Electrolytic tough-pitch copper; cold drawn; under helium atmosphere; density = 40 lb ft^{-3} at 75 F.
4	11	1961	366-544		Calibration Specimen	
5	12	1962	533-1089	≤5.0		100.0 electrolytic copper.
6	1	1961	295	±5.0		
7	14	1959	5-298		OFHC	99.96 Cu, 0.001-0.01 Ag, 0.001-0.01 Sb, <0.001 each, Bi, Ca, Fe, Mg, Ni, Pb, and Si; oxygen free high conductivity copper.
8	15	1959	323-1273	1.0		99.9 Cu.
9	16	1961	0.4-30	1.0	High purity copper	99.999 Cu, <0.0001 each Se and S; sample supplied by the American Smelting & Refining Co.
10	17	1960	421			
11	8	1960	20-300	<2.0	I (a)	Commercially pure cold rolled copper; sample supplied by the American Smelting & Refining Co.
12	8	1960	20-280	<2.0	I (b)	Same as above.
13	8	1960	20-290	<2.0	II	Lighter commercially pure cold-rolled copper; sample supplied by the American Smelting & Refining Co.
14	8	1960	20-300	<2.0		99.999 Cu; annealed; melted by induction heating under high vacuum and cooled slowly for 4 hrs.
15	18	1956	363-873	±0.5	Electrolytic copper	99.99 Cu, major metallic impurities are Ag and Fe.
16	16	1961	0.4-30	1.0	0.05 Fe dilute copper alloy	99.949 Cu, <0.0001 each Se, and S; melted and cooled to room temperature.
17	16	1961	0.4-30	1.0	0.10 Fe dilute copper alloy	99.899 Cu, <0.0001 each Se, and S; melted at 1300 C; annealed for 72 hrs at 870 C to homogenize; cooled rapidly to room temperature.
18	19	1962	473-973		Copper powder compact first heating	99.8 Cu; reduced for 3 hrs. at 250 C in dry and purified hydrogen stream; heated under 10^{-4} mm Hg vacuum at 400 C until degassing from powder is completed.
19	19	1962	513-893		Copper powder compact second heating	Same as above.
20	4	1941	15-300			99.96 Cu; single crystals; melted and solidified 5 days in a nitrogen atmosphere; density = 558.91 lb ft^{-3}.

SPECIFICATION TABLE NO. 14 (continued)

Curve No.	Ref. No.	Year	Temp. Range, K	Reported Error, %	Name and Specimen Designation	Composition (weight percent), Specifications and Remarks
21	21	1956	811-1311		QQC 576	Electrolytic tough pitch; density = 551.4 lb ft^{-3}.
22	37	1962	284-300	0.10		99.999 Cu; annealed.
23	37	1962	284-303	0.10		99.999 Cu; cold worked.
24	38	1961	727-1783			
25	101	1958	337-946			Specimen's surface plated with platinum black.
26	105	1948	589-794			99.92 Cu; capsule.
27	55	1930	95-215	1.5	Recrystallized copper	Cold deformed; recrystallized for 10 hrs at 1000 C under nitrogen atmosphere.
28	55	1930	84-183	1.5	Compressed copper	Deformed by hydraulic press sidewise.
29	154	1956	273-1338		QQC 502	Electrolytic; tough pitch.
30	83	1954	373-1183			
31	268	1926	373-1723			Electrolytic copper.
32	289	1927	373-1073			0.1 Ni; vacuum melted.
33	170	1931	291-973		Electrolytic copper	
34	290	1936	573-1173		Electrolytic copper	Sample melted and allowed to solidify.
35	291	1933	203-389	0.05		>99.5 Cu; cold rolled.
36	261	1934	54-294			Hard drawn wire sample from Bell Telephone.
37	261	1934	53-293			Sample annealed 16 hrs at 400 C in high vacuum.
38	167	1936	1.2-20			99.9 Cu.
39	262	1937	373-764			
40	292	1937	29-194	0.05		Commercially pure sample; cold rolled.
41	263	1939	373-1273			
42	293	1941	82-273			99.92 Cu, with 0.02 Fe, 0.01 C, and 0.003 S impurities.
43	294	1952	2.1-4			99.60 Cu, with principal impurity Pb, and traces of Fe, Ni, and Ag.
44	295	1955	90-300		Sample I	99.999 Cu, single crystal; before deformation copper was annealed at 400 C.
45	295	1955	90-300		Sample II	Polycrystalline; before deformation copper was annealed at 400 C.
46	296	1955	1.1-4.8	.05		99.999^{+} Cu, with a trace of Ag; sample was supplied by the American Smelting and Refining Company; annealed 3 hrs at 1000 C under vacuum of 1 x 10^{-6} mm Hg, and allowed to cool in vacuum to room temperature at 200 C per hr.

SPECIFICATION TABLE NO. 14 (continued)

Curve No.	Ref. No.	Year	Temp. Range, K	Reported Error, %	Name and Specimen Designation	Composition (weight percent), Specifications and Remarks
47	223	1959	0.4-0.5	5		
48	297	1959	2.7-5		OFHC copper	Annealed at 300 C for 1 hr.
49	298	1963	77-1357			
50	85	1963	0.1-1.1			99.999 Cu, specimen supplied by American Smelting and Refining Co.; sample was annealed in vacuum several hours at 600 C.
51	299	1964	0.1-1.0	1-3		
52	300	1966	1-9	0.1		Single crystal; resistance ratio $\frac{R_{300}}{R_{4.2}}$ = 3000-6000; etched.
53	300	1966	1-24	0.1		99.999 Cu, polycrystalline sample supplied by American Smelting and Refining Co.; annealed 3 1/2 hrs at 650 C in a vacuum by induction heating and heavily etched; the pressure before heating was 5 x 10^{-9} mm Hg, pressure during heating increased to a maximum 6 x 10^{-5} and decreased to 2 x 10^{-8} mm Hg.
54	300	1966	1-27	0.1	Standard 19th Annual Calorimetry Conference Sample	Chill cast sample; heavily etched.
55	300	1966	1-24	0.1		Cold worked sample; prepared from annealed sample; heavily etched.

DATA TABLE NO. 14 SPECIFIC HEAT OF COPPER

[Temperature, T, K; Specific Heat, C_p, Cal g^{-1} K^{-1}]

CURVE 1

T	C_p
21	2.06×10^{-3}
25	3.60
30	6.37
35	1.00×10^{-2}
40	1.41
45	1.857
50	2.317
55	2.786
60	3.244
65	3.676
70	4.087
75	4.484
80	4.849
85	5.179
90	5.505
95	5.773
100	6.018
110	6.467
120	6.859
130	7.195
140	7.466
150	7.705
160	7.918
170	8.097
180	8.250
190	8.390
200	8.513
210	8.617
220	8.709
230	8.793
240	8.870
250	8.947
260	9.019
270	9.086
273.15	9.103
280	9.139
290	9.183
298.15	9.219*
300	9.227

CURVE 2

T	C_p
288	9.08×10^{-2}
288	9.30
375	9.58
453	9.52*
453	9.55
465	9.72
495	9.64
551	9.72
598	9.89
648	9.96
701	1.06×10^{-1}

CURVE 3

T	C_p
366	9.50×10^{-2}
477	9.60
589	9.90
700	1.01×10^{-1}
811	1.05
922	1.08
1033	1.13
1144	1.18
1255	1.24
1311	1.27

CURVE 4

T	C_p
366	9.34×10^{-2}
394	9.34
422	9.34
450	9.34
478	9.34
505	9.34
533	9.34
544	9.34

CURVE 5

T	C_p
533	9.40×10^{-2}*
811	1.01×10^{-1}
1089	1.08

CURVE 6*

T	C_p
295	9.70×10^{-2}

CURVE 7

T	C_p
5	4.72×10^{-5}*
10	2.20×10^{-4}
15	6.61
20	1.79×10^{-3}
40	1.39×10^{-2}
60	3.266
80	4.926
100	6.109
120	6.892*
140	7.452*
160	7.932*
180	8.295*
200	8.595*
220	8.798*
240	8.945*
260	9.037*
280	9.115*
290	9.152*
298.15	9.180*

CURVE 8

T	C_p
323	9.20×10^{-2}
373	9.40
473	9.90
573	1.01×10^{-1}
673	1.03
773	1.06
873	1.09
973	1.12
1073	1.14
1173	1.17
1273	1.19

CURVE 9

T	C_p
0.4	1.05×10^{-6}
0.5	1.32*
0.7	1.88*
0.9	2.46*
1.1	3.09*
1.2	3.49*
1.3	3.77*
1.5	4.50*
1.5	4.53*
2.0	6.58*
2.5	9.35*
3.0	1.28×10^{-5}*
3.0	1.26*
3.5	1.69*
4.0	2.20*
4.0	2.19*
5.0	3.55*
6.0	5.46*
7.0	8.04*
8.0	1.141×10^{-4}*
9.0	1.569*
10.0	2.103
12.0	3.547
14.0	5.614
16.0	8.470
18.0	1.232×10^{-3}
20.0	1.737
22.0	2.384
24.0	3.176
26.0	4.100
28.0	5.149
30.0	6.325*

CURVE 10*

T	C_p
921	9.15×10^{-2}

CURVE 11*

T	C_p
21	2.08×10^{-3}
25	3.62
30	6.39
35	1.003×10^{-2}
40	1.410
45	1.854
50	2.313
55	2.784
60	3.236
65	3.667
70	4.079
75	4.471
80	4.838
85	5.176
90	5.478
95	5.755
100	5.998
110	6.465
120	6.846
130	7.181
140	7.472
150	7.710
160	7.916
170	8.094
180	8.251
190	8.380
200	8.500
210	8.612
220	8.711
230	8.801
240	8.884
250	8.950
260	9.008
270	9.071
273.15	9.090
280	9.125
290	9.172
298.15	9.207
300	9.215

CURVE 12*

T	C_p
21	2.08×10^{-3}
25	3.64
30	6.40
35	1.001×10^{-2}
40	1.412
45	1.860
50	2.359
55	2.784
60	3.244
65	3.686
70	4.089
75	4.460
80	4.833
85	5.173
90	5.364
95	5.763
100	6.013
110	6.470
120	6.859
130	7.195
140	7.474
150	7.715
160	7.924
170	8.100
180	8.255
190	8.385
200	8.505
210	8.610
220	8.705
230	8.791
240	8.864
250	8.941
260	9.015
270	9.084
273.15	9.103
280	9.145

*Not shown on plot

DATA TABLE NO. 14 (continued)

T	C_p
CURVE 13*	
22	2.41 x 10^{-3}
25	3.64
30	6.42
35	1.003 x 10^{-2}
40	1.418
45	1.865
50	2.325
55	2.792
60	3.248
65	3.684
70	4.093
75	4.482
80	4.861
85	5.195
90	5.502
95	5.788
100	6.040
110	6.481
120	6.873
130	7.206
140	7.496
150	7.726
160	7.935
170	8.115
180	8.272
190	8.415
200	8.530
210	8.635
220	8.722
230	8.802
240	8.881
250	8.955
260	9.029
270	9.093
273.15	9.112
280	9.152
290	9.197
CURVE 14*	
20	1.78 x 10^{-3}
21	2.08
25	3.62
30	6.36
35	9.98
40	1.405 x 10^{-2}
45	1.856
50	2.309
CURVE 14 (contd)*	
55	2.776 x 10^{-2}
60	3.234
65	3.673
70	4.075
75	4.471
80	4.838
85	5.173
90	5.489
95	5.768
100	6.021
110	6.478
120	6.852
130	7.184
140	7.474
150	7.716
160	7.915
170	8.086
180	8.242
190	8.384
200	8.497
210	8.602
220	8.698
230	8.780
240	8.853
250	8.927
260	8.999
270	9.067
273.15	9.086
280	9.119
290	9.166
298.15	9.196
300	9.200
CURVE 15*	
363	9.42 x 10^{-2}
383	9.53
393	9.57
403	9.61
423	9.65
433	9.68
453	9.72
473	9.76
483	9.77
503	9.84
513	9.87
523	9.90
543	9.94
CURVE 15 (contd)*	
553	9.96 x 10^{-2}
563	9.99
583	1.003 x 10^{-1}
593	1.004
603	1.006
623	1.008
633	1.011
643	1.015
663	1.018
673	1.017
683	1.019
703	1.022
713	1.025
723	1.027
753	1.035
793	1.041
833	1.051
873	1.061
883	1.067
CURVE 16*	
0.4	1.94 x 10^{-6}
0.5	2.31
0.7	3.10
0.9	3.95
1.1	4.78
1.2	5.30
1.3	5.73
1.5	6.77
1.5	6.70
2.0	9.25
2.5	1.24 x 10^{-5}
3.0	1.59
3.0	1.61
3.5	2.02
4.0	2.54
4.0	2.58
5.0	3.96
6.0	5.89
7.0	8.59
8.0	1.185 x 10^{-4}
9.0	1.616
10.0	2.132
12.0	3.577
14.0	5.639
16.0	8.552
18.0	1.239 x 10^{-3}
CURVE 16 (contd)*	
20.0	1.734 x 10^{-3}
22.0	2.364
24.0	3.162
26.0	4.138
28.0	5.216
30.0	6.368
CURVE 17*	
0.4	3.64 x 10^{-6}
0.5	4.08
0.7	5.00
0.9	6.04
1.1	7.14
1.2	7.95
1.3	8.29
1.5	9.49
1.5	9.38
2.0	1.22 x 10^{-5}
2.5	1.55
3.0	1.93
3.0	1.96
4.0	2.90
4.0	2.93
5.0	4.32
6.0	6.22
7.0	8.86
8.0	1.212 x 10^{-4}
9.0	1.640
10.0	2.169
12.0	3.615
14.0	5.686
16.0	8.552
18.0	1.247 x 10^{-3}
20.0	1.758
22.0	2.386
24.0	3.204
26.0	4.169
28.0	5.277
30.0	6.426
CURVE 18	
473	1.16 x 10^{-1}
493	1.17*
513	1.15
533	1.12
553	1.09
CURVE 18 (contd)	
573	1.06 x 10^{-1}
593	1.03
613	1.02
633	1.03*
653	1.05
673	1.08*
693	1.12
713	1.14
733	1.16*
753	1.19
773	1.21*
793	1.22
833	1.24
873	1.24
913	1.26
933	1.26*
958	1.28
973	1.30
CURVE 19	
513	1.17 x 10^{-1}
533	1.18
553	1.18
573	1.19
593	1.19
613	1.19
633	1.20
653	1.20
673	1.20*
693	1.21
713	1.21*
733	1.22
753	1.22*
773	1.22
813	1.24
853	1.25
893	1.26
CURVE 20	
14.70	6.30 x 10^{-4}
14.82	6.30*
17.63	1.133 x 10^{-3}
19.75	1.684*
19.87	1.731
23.35	2.943
28.21	5.351
CURVE 20 (contd)	
33.52	8.971 x 10^{-3}
38.86	1.322 x 10^{-2}
44.21	1.789*
48.17	2.175
53.34	2.671
59.08	3.184*
65.12	3.691
70.12	4.100
75.36	4.526
80.60	4.907*
85.62	5.234
90.73	5.529*
95.78	5.829
101.24	6.127*
106.72	6.388
112.25	6.626*
117.86	6.810
123.40	7.016*
128.99	7.167
134.54	7.326
140.20	7.480*
146.02	7.622
151.01	7.749*
156.91	7.869
162.77	7.973*
168.28	8.116*
174.05	8.163
179.36	8.247*
184.72	8.344
190.18	8.409*
195.81	8.492
201.39	8.552*
207.07	8.629
213.03	8.709*
218.90	8.719*
224.23	8.780
229.66	8.818*
235.25	8.870
240.71	8.928*
245.98	8.977*
251.58	9.015*
256.60	9.024
261.34	9.060*
266.61	9.095
272.18	9.112
277.69	9.160*
283.59	9.177*
289.51	9.171*

* Not shown on Plot

DATA TABLE NO. 14 (continued)

T	C_p
CURVE 20 (cont.)	
294.76	9.213×10^{-1}*
300.15	9.191
CURVE 21*	
811	1.042×10^{-1}
978	1.080
1144	1.118
1311	1.156
CURVE 22*	
283.27	8.856×10^{-2}
283.81	8.869
283.89	8.853
284.36	8.845
284.46	8.871
285.05	8.854
285.34	8.804
285.62	8.876
285.84	8.874
286.83	8.877
287.32	8.866
288.34	8.871
288.81	8.862
289.80	8.879
290.29	8.874
291.78	8.895
293.27	8.894
293.42	8.898
294.75	8.885
294.75	8.897
296.24	8.900
296.29	8.909
297.73	8.950
297.78	8.920
299.22	8.906
299.22	8.917
299.72	8.940
300.70	8.924
302.20	8.946
303.69	8.927
CURVE 23*	
283.86	8.877×10^{-2}
285.34	8.882
286.83	8.895
CURVE 23 (cont.)*	
288.31	8.880×10^{-2}
290.29	8.929
291.78	8.912
293.27	8.923
295.25	8.708
296.74	8.932
298.72	8.941
300.21	8.953
301.70	8.950
303.19	8.950
CURVE 24*	
690	9.638×10^{-2}
700	9.662
800	9.886
900	1.011×10^{-1}
1000	1.033
1100	1.056
1210	1.078
CURVE 25*	
337	9.60×10^{-2}
356	9.40
400	9.60
465	9.70
565	9.99
604	9.90
676	1.04×10^{-1}
781	1.03
875	1.07
946	1.07
CURVE 26*	
589	9.60×10^{-2}
596	9.90
627	1.01×10^{-1}
794	1.19
CURVE 27*	
94.61	5.732×10^{-2}
113.21	6.627
122.11	6.967
129.92	7.214
134.62	7.369
CURVE 27 (cont.)*	
136.68	7.428×10^{-2}
154.92	7.850
174.97	8.180
204.10	8.575
215.44	8.742
CURVE 28*	
83.83	5.063×10^{-2}
87.43	5.241
94.72	5.710
119.55	6.853
139.05	7.491
145.43	7.646
183.11	8.313
CURVE 29*	
273	9.40×10^{-2}
366	9.50
477	9.60
589	9.90
700	1.01×10^{-1}
811	1.05
9.22	1.08
1033	1.13
1144	1.18
1255	1.24
1311	1.27
1338	1.29
CURVE 30	
373	8.640×10^{-2}
403	8.735
423	8.795
443	8.860
463	8.920
483	8.981
503	9.045
523	9.105
543	9.158
563	9.220
583	9.285
603	9.336
623	9.407
643	9.463
663	9.529
683	9.592
CURVE 30 (cont.)	
703	9.650×10^{-2}
723	9.715
743	9.771
763	9.828
783	9.893*
803	9.942*
823	1.001×10^{-1}*
843	1.008
863	1.014
883	1.019
903	1.024
923	1.030
943	1.036
963	1.042
983	1.048
1003	1.053
1023	1.053
1043	1.065
1063	1.071*
1083	1.077*
1103	1.082*
1123	1.088
1143	1.094
1163	1.099
1183	1.105
CURVE 31*	
373	9.70×10^{-2}
473	9.98
573	1.034×10^{-1}
673	1.063
773	1.124
873	1.131
973	1.170
1073	1.213
1173	1.250
1273	1.310
1298	1.336
1523	1.220
1623	1.220
1723	1.220
CURVE 32*	
373	9.470×10^{-2}
473	9.780
573	1.004×10^{-1}
CURVE 32* (cont.)	
673	1.028×10^{-1}
773	1.050
873	1.073
973	1.096
1073	1.120
CURVE 33*	
291.15	9.08×10^{-2}
373.15	9.38
473.15	9.68
573.15	9.94
673.15	1.02×10^{-1}
773.15	1.04
873.15	1.06
973.15	1.09
CURVE 34*	
573.15	9.880×10^{-2}
673.15	1.009×10^{-1}
773.15	1.030
873.15	1.051
973.15	1.072
1073.15	1.092
1173.15	1.113
CURVE 35*	
Series 1	
298.66	9.180×10^{-2}
309.74	9.230
320.74	9.280
331.68	9.327
342.56	9.367
353.40	9.407
364.18	9.443
374.91	9.479
378.33	9.490
379.36	9.496
389.07	9.521
Series 2	
203.16	8.481×10^{-2}
212.46	8.600
223.05	8.700
CURVE 35* (cont.)	
234.26	8.787×10^{-2}
245.86	8.876
256.66	8.952
267.36	9.019
278.70	9.081
289.24	9.142
299.00	9.191
CURVE 36*	
54.14	2.729×10^{-2}
56.72	2.916
60.18	3.308
69.63	4.122
83.23	5.033
101.50	6.116
126.30	7.096
153.50	7.918
190.30	8.363
213.40	8.669
240.40	8.870
261.00	8.999
279.30	9.084
294.30	9.174
CURVE 37*	
53.29	2.720×10^{-2}
57.72	3.072
60.56	3.348
72.29	4.300
84.19	5.052
99.20	5.995
123.10	6.994
152.80	7.794
188.20	8.311
209.60	8.626
241.70	8.883
267.50	9.048
288.10	9.116
291.90	9.136
293.00	9.131

* Not shown on Plot

DATA TABLE NO. 14 (continued)

T	C_p
CURVE 38*	
Series 1	
1.171	3.577 x 10^{-6}
1.172	3.738
1.477	4.584
1.484	4.684
1.658	5.598
1.669	5.411
2.010	6.893
2.032	7.572
2.270	8.640
2.298	9.578
2.573	9.964
2.451	9.238
2.503	9.778
2.732	1.132 x 10^{-5}
3.818	1.229
3.085	1.383
3.164	1.449
3.448	1.720
3.573	1.903
3.784	2.022
3.908	2.251
3.969	2.345
4.093	2.502
4.112	2.534
4.260	2.729
4.631	3.259
5.138	4.026
5.845	5.607
6.548	6.950
6.831	7.880
9.148	1.637 x 10^{-4}
10.321	2.260
11.222	2.960
12.114	3.947
13.405	5.236
14.239	5.341
15.121	7.509
16.059	9.262
17.298	1.131 x 10^{-3}
18.124	1.306
18.910	1.495
19.611	1.687
CURVE 38* (cont.)	
Series 2	
9.978	2.065 x 10^{-4}
11.444	3.118
12.711	4.290
13.760	5.494
14.256	5.936
14.689	6.953
CURVE 39*	
373	9.08 x 10^{-2}
410	8.91
473	9.35
498	9.84
596	9.53
636	9.87
673	1.37 x 10^{-1}
764	1.20
CURVE 40*	
28.64	5.448 x 10^{-3}
35.93	1.061 x 10^{-2}
42.58	1.669
50.13	2.355
59.24	3.199
67.21	3.871
74.64	4.468
87.45	5.350
87.88	5.374
92.79	5.651
93.18	5.670
97.41	5.885
103.08	6.162
108.51	6.406
113.73	6.624
119.38	6.836
125.42	7.042
131.29	7.228
132.97	7.271
137.48	7.388
144.24	7.577
151.08	7.732
CURVE 40* (cont.)	
157.79	7.868 x 10^{-2}
164.71	7.993
171.83	8.122
178.84	8.227
186.33	8.330
194.29	8.430
CURVE 41*	
373	9.46 x 10^{-2}
473	9.72
573	9.99
673	1.03 x 10^{-1}
773	1.06
873	1.09
973	1.13
1073	1.16
1173	1.21
1273	1.27
CURVE 42*	
82.1	5.02 x 10^{-2}
88.9	5.40
101.0	6.03
104.9	6.14
112.1	6.52
125.2	7.03
130.1	7.24
145.3	7.74
151.0	7.88
163.2	8.06
173.25	8.29
180.1	8.42
191.8	8.53
212.0	8.80
227.9	8.97
235.0	8.99
242.3	9.05
253.4	9.13
265.2	9.18
273.2	9.18
CURVE 43*	
2.068	1.0308 x 10^{-2}
2.174	9.1440 x 10^{-3}
2.402	1.0151 x 10^{-2}
2.429	1.0355
2.610	1.2197
2.711	1.2071
2.829	1.3928
2.894	1.3535
3.061	1.3251
3.402	1.7941
3.538	2.1246
3.639	2.1561
CURVE 44*	
90	4.15 x 10^{-2}
300	9.20
CURVE 45*	
90	4.12 x 10^{-2}
300	9.55
CURVE 46*	
Series 1	
1.144	3.29508 x 10^{-6}
1.185	3.37031
1.213	3.44554
1.248	3.59600
1.382	4.09252
1.579	4.73197
1.775	5.57455
1.974	6.37199
2.208	7.89916
2.375	8.76431
2.570	9.70468
2.772	1.12469 x 10^{-5}
2.974	1.23754
3.181	1.38047
3.376	1.53846
3.579	1.72653
CURVE 46* (cont.)	
3.781	1.97479 x 10^{-5}
3.982	2.21929
4.178	2.34718
4.374	2.55782
4.601	2.77599
4.790	3.04306
Series 2	
1.174	3.37031 x 10^{-6}
1.214	3.45682
1.247	3.60728
1.285	3.73894
1.465	4.38592
1.737	5.37895
1.897	6.07107
2.089	6.84970
2.278	8.12485
2.468	9.21569
2.662	1.03065 x 10^{-5}
2.915	1.19240
3.083	1.29772
3.279	1.44818
3.480	1.64002
3.693	1.87699
3.899	2.49388
4.091	2.32461
4.296	2.50140
4.487	2.67067
4.693	2.91141
4.864	3.18223
Series 3	
1.169	3.20856 x 10^{-6}
1.208	3.52077
1.401	4.14894
1.582	4.72069
1.800	5.63850
2.000	6.48107
2.373	8.76431
2.577	9.77991
2.772	1.10964 x 10^{-5}
2.977	1.30148

*Not shown on plot

DATA TABLE NO. 14 (continued)

T	C_p
CURVE 47*	
0.362	2.26 x 10^{-6}
0.366	2.19
0.370	2.44
0.376	2.35
0.377	2.51
0.377	2.46
0.393	2.76
0.406	2.39
0.408	2.81
0.509	2.85
0.529	3.67
0.544	3.42
0.548	3.53
CURVE 48*	
2.700	1.07712 x 10^{-5}
2.974	1.32956
3.151	1.43233
3.385	1.68555
3.615	1.90903
3.846	2.12779
4.089	2.40006
4.221	2.51967
4.356	2.74001
4.511	2.92414
4.656	3.18067
4.786	3.39629
CURVE 49*	
77	4.690 x 10^{-2}
100	6.043
120	6.862
140	7.476
160	7.916
180	8.247
200	8.514
220	8.703
240	8.861
260	8.971
280	9.112
300	9.207
350	9.364
CURVE 49* (cont.)	
400	9.506 x 10^{-2}
450	9.632
500	9.742
550	9.852
600	9.978
650	1.007 x 10^{-1}
700	1.020
750	1.031
800	1.042
900	1.067
1000	1.091
1100	1.114
1200	1.138
1300	1.162
1357	1.176
CURVE 50*	
Series 1	
0.0966	2.458 x 10^{-7}
0.0973	2.498
0.1020	2.650
0.1065	2.791
0.1074	2.819
0.1077	2.840
1.1078	2.836
0.1166	3.110
0.1171	3.121
0.1187	3.171
0.1190	3.177
0.1266	3.395
0.1276	3.420
0.1321	3.542
0.1451	3.856
0.1507	3.965
0.1586	4.224
0.1735	4.604
0.1911	5.071
0.2118	5.623
0.2368	6.301
0.2661	7.064
CURVE 50* (cont.)	
Series 2	
0.1389	3.667 x 10^{-7}
0.1546	4.149
0.1672	4.446
0.1709	4.536
0.1852	4.931
0.1885	4.984
0.2072	5.499
0.2274	6.033
0.2392	6.353
0.2516	6.677
0.2637	6.996
0.2763	7.320
0.2898	7.696
0.3085	8.174
0.3203	8.490
0.3426	9.080
0.3523	9.340
0.3780	1.002 x 10^{-6}
0.4085	1.086
0.4171	1.109
0.4540	1.209
0.5035	1.363
Series 3	
0.1004	2.613 x 10^{-7}
0.1302	3.472
0.1388	3.690
0.1500	3.938
0.2991	7.921
0.3370	8.937
0.3804	1.008 x 10^{-6}
0.8444	2.338
0.8902	2.460
1.0129	2.858
Series 4	
0.4678	1.246 x 10^{-6}
0.5264	1.408
0.5652	1.519
0.5956	1.599
CURVE 50* (cont.)	
0.6187	1.666 x 10^{-6}
0.6825	1.849
0.7528	2.058
1.0065	2.856
1.0736	3.065
1.0992	3.178
CURVE 51*	
Series 1	
0.1318	3.411 x 10^{-7}
0.1457	3.830
0.1651	4.336
0.1876	4.919
0.2179	5.710
0.2406	6.324
0.2626	6.910
0.2861	7.513
0.3178	8.339
0.3544	9.319
0.3893	1.026 x 10^{-6}
0.4261	1.126
0.4653	1.233
Series 2	
0.1313	3.383 x 10^{-7}
0.1485	3.890
0.1658	4.343
0.1855	4.861
0.2047	5.366
0.2292	6.011
0.2503	6.568
0.2762	7.254
0.3018	7.929
0.3316	8.695
0.3653	9.605
0.4044	1.067 x 10^{-6}
0.4433	1.173
0.4816	1.278
0.5287	1.407
CURVE 51* (cont.)	
Series 3	
0.1273	3.286 x 10^{-7}
0.1403	3.649
0.1754	4.597
0.1923	5.036
0.2096	5.499
0.2276	5.968
0.2470	6.480
0.2711	7.113
0.2943	7.737
0.3208	8.414
0.3516	9.228
0.3832	1.010 x 10^{-6}
0.4126	1.089
0.4505	1.192
0.4964	1.317
0.5470	1.459
0.6124	1.641
0.7008	1.895
Series 4	
0.5401	1.437 x 10^{-6}
0.5959	1.593
0.6531	1.756
0.7143	1.933
0.7964	2.174
0.9056	2.504
1.0221	2.877
Series 5	
0.5851	1.565 x 10^{-6}
0.6373	1.716
0.6972	1.885
0.7741	2.108
0.8748	2.408
0.9857	2.756
Series 6	
0.8171	2.236 x 10^{-6}
0.8882	2.451

*Not shown on plot

DATA TABLE NO. 14 (continued)

T	C_p
CURVE 51* (cont.)	
0.9847	2.748×10^{-6}
1.0872	3.082
CURVE 52*	
1.259	3.710×10^{-6}
1.267	3.720
1.272	3.742
1.289	3.804
1.313	3.887
1.321	3.909
1.335	3.956
1.371	4.075
1.388	4.140
1.410	4.213
1.423	4.258
1.446	4.316
1.447	4.373
1.481	4.463
1.500	4.538
1.503	4.564
1.519	4.623
1.545	4.712
1.570	4.822
1.573	4.826
1.591	4.897
1.599	4.931
1.624	5.024
1.658	5.157
1.682	5.257
1.724	5.435
1.767	5.606
1.799	5.757
1.805	5.790
1.820	5.878
1.855	5.999
1.887	6.142
1.914	6.270
1.961	6.484
1.962	6.488
1.982	6.571
2.063	6.947
2.104	7.164
2.159	7.434
CURVE 52* (cont.)	
2.232	7.836×10^{-6}
2.276	8.066
2.367	8.602
2.421	8.905
2.505	9.400
2.547	9.661
2.609	1.004×10^{-5}
2.723	1.078
2.781	1.118
2.838	1.156
2.944	1.231
3.065	1.320
3.111	1.358
3.204	1.431
3.233	1.455
3.366	1.568
3.400	1.599
3.556	1.741
3.5640	1.749
3.719	1.900
3.880	2.069
4.055	2.261
4.258	2.501
4.341	2.608
4.495	2.810
4.605	2.966
4.849	3.321
5.049	3.641
5.238	3.957
5.441	4.316
5.669	4.754
5.931	5.295
6.242	5.993
6.604	6.925
7.035	8.122
7.306	4.925
7.706	1.026×10^{-4}
8.044	1.148
8.430	1.302
8.887	1.504
9.443	1.775
CURVE 53*	
1.276	3.776×10^{-6}
1.283	3.777
CURVE 53* (cont.)	
1.286	3.944×10^{-6}
1.337	3.961
1.368	4.017
1.379	4.108
1.396	4.161
1.472	4.437
1.500	4.536
1.511	4.595
1.515	4.568
1.577	4.844
1.621	5.015
1.670	5.212
1.679	5.249
1.722	5.433
1.741	5.500
1.767	5.613
1.796	5.741
1.862	6.025
1.879	6.105
1.911	6.246
1.936	6.370
1.965	6.490
2.025	6.776
2.059	6.944
2.077	7.063
2.118	7.248
2.254	7.942
2.282	8.094
2.339	8.426
2.436	8.993
2.450	9.054
2.552	9.681
2.622	1.012×10^{-5}
2.693	1.057
2.758	1.123
2.868	1.176
2.894	1.196
2.902	1.200
3.043	1.303
3.068	1.323
3.172	1.405
3.274	1.488
3.319	1.527
3.451	1.643
3.498	1.684
3.580	1.763
CURVE 53* (cont.)	
3.710	1.888×10^{-5}
3.715	1.893
3.866	2.047
3.972	2.164
4.039	2.238
4.238	2.470
4.470	2.771
4.493	2.807
4.676	3.058
4.851	3.314
4.919	3.424
5.034	3.603
5.220	3.918
5.323	4.103
5.408	4.253
5.630	4.670
5.893	5.203
5.924	5.153
5.938	4.690
6.193	5.873
6.236	5.978
6.537	6.726
6.576	6.830
6.977	7.935
6.997	7.997
7.513	9.600
7.987	1.126×10^{-4}
8.404	1.290
8.581	1.363
8.726	1.427
8.869	1.494
8.947	1.526
9.148	1.622
9.377	1.734
9.403	1.753
9.532	1.817
9.817	1.970
9.998	2.075
10.085	2.131
10.305	2.261
10.550	2.418
10.922	2.673
11.078	2.783
11.559	3.152
11.642	3.215
11.691	3.263
CURVE 53* (cont.)	
12.038	3.549×10^{-4}
12.095	3.601
12.339	3.821
12.662	4.128
12.737	4.210
13.022	4.492
13.422	4.921
13.551	5.069
13.872	5.422
14.391	6.119
14.592	6.347
14.994	6.891
15.643	7.870
15.675	7.927
16.412	9.217
16.581	9.507
17.199	1.069×10^{-3}
17.597	1.151
17.938	1.222
18.637	1.380
18.854	1.436
19.415	1.575
20.276	1.817
20.536	1.883
21.234	2.118
21.324	2.145
22.355	2.505
22.741	2.643
24.267	3.273
CURVE 54*	
1.312	3.867×10^{-6}
1.322	3.900
1.368	4.055
1.392	4.142
1.416	4.209
1.446	4.320
1.447	4.323
1.466	4.408
1.483	4.485
1.502	4.543
1.522	4.620
1.558	4.759
1.564	4.782
1.588	4.875

*Not shown on plot

DATA TABLE NO. 14 (continued)

T	C_p
CURVE 54* (cont.)	
1.638	5.084×10^{-6}
1.668	5.193
1.719	5.415
1.731	5.463
1.749	5.540
1.769	5.621
1.784	5.690
1.806	5.795
1.838	5.927
1.851	5.981
1.871	6.075
1.902	6.207
1.950	6.423
1.985	6.585
2.044	6.862
2.084	7.052
2.117	7.221
2.156	7.411
2.183	7.555
2.209	7.692
2.226	7.796
2.256	7.955
2.308	8.231
2.344	8.452
2.368	8.585
2.403	8.799
2.445	9.037
2.513	9.436
2.564	9.745
2.647	1.027×10^{-5}
2.708	1.067
2.716	1.072
2.753	1.096
2.810	1.136
2.811	1.137
2.848	1.162
2.901	1.200
2.940	1.227
2.999	1.270
3.046	1.304
3.121	1.363
3.176	1.407
3.265	1.479
3.329	1.533
CURVE 54* (cont.)	
3.368	1.569×10^{-5}
3.434	1.626
3.484	1.672
3.509	1.694
3.594	1.774
3.722	1.899
3.729	1.907
3.879	2.016
4.002	2.197
4.053	2.252
4.257	2.494
4.381	2.654
4.441	2.734
4.498	2.809
4.657	3.032
4.971	3.504
5.082	3.685
5.120	3.750
5.290	4.040
5.359	4.165
5.417	4.267
5.509	4.437
5.545	4.507
5.657	4.722
5.757	4.920
5.927	5.277
6.044	5.530
6.238	5.978
6.375	6.313
6.601	6.890
6.763	7.332
6.906	7.728
7.036	8.094
7.148	8.434
7.233	8.695
7.411	9.252
7.703	1.023×10^{-4}
8.040	1.145
8.375	1.275
8.422	1.296
8.429	1.299
8.661	1.396
8.783	1.452
8.987	1.544
CURVE 54* (cont.)	
9.129	1.613×10^{-4}
9.304	1.698
9.365	1.728
9.524	1.812
9.569	1.835
9.798	1.958
9.833	1.979
9.983	2.067
10.122	2.148
10.304	2.258
10.465	2.361
10.536	2.410
10.613	2.459
10.855	2.623
11.030	2.747
11.301	2.950
11.503	3.105
11.821	3.365
12.054	3.565
12.435	3.911
12.717	4.184
13.182	4.662
13.536	5.047
14.134	5.751
14.591	6.341
15.397	7.499
15.775	8.071
16.378	9.105
16.777	9.843
17.414	1.110×10^{-3}
17.897	1.211
18.703	1.395
19.310	1.547
20.348	1.836
20.555	1.894
20.748	1.954
21.136	2.081
21.187	2.092
21.820	2.305
21.925	2.341
22.744	2.639
23.082	2.770
23.665	3.002
24.581	3.396
CURVE 54* (cont.)	
24.691	3.445×10^{-3}
25.849	3.988
26.254	4.190
27.179	4.673
CURVE 55*	
1.294	3.824×10^{-6}
1.299	3.853
1.310	3.896
1.327	3.942
1.363	4.073
1.367	4.074
1.381	4.133
1.413	4.232
1.447	4.361
1.500	4.576
1.502	4.579
1.549	4.760
1.610	4.998
1.619	5.039
1.696	5.352
1.733	5.510
1.781	5.722
1.849	6.013
1.851	6.044
1.887	6.181
1.959	6.505
1.965	6.543
2.024	6.823
2.062	6.995
2.128	7.332
2.171	7.551
2.311	8.314
2.478	9.297
2.632	1.024×10^{-5}
2.761	1.110
2.906	1.211
3.079	1.340
3.294	1.516
3.484	1.684
3.639	1.831
3.820	2.005
4.034	2.248
CURVE 55* (cont.)	
4.291	2.557×10^{-5}
4.602	2.977
4.885	3.392
4.993	3.568
5.170	3.860
5.430	4.321
5.714	4.869
6.025	5.525
6.353	6.323
6.739	7.311
7.213	8.687
7.798	1.064×10^{-4}
8.387	1.285
8.881	1.501
9.457	1.782
10.089	2.135
10.654	2.494
11.351	2.998
12.351	3.855
13.351	4.858
14.280	5.953
17.502	1.133×10^{-3}
18.784	1.421
20.322	1.835
20.584	1.913
21.266	2.138
22.018	2.394
22.835	2.689
23.750	3.056

*Not shown on plot

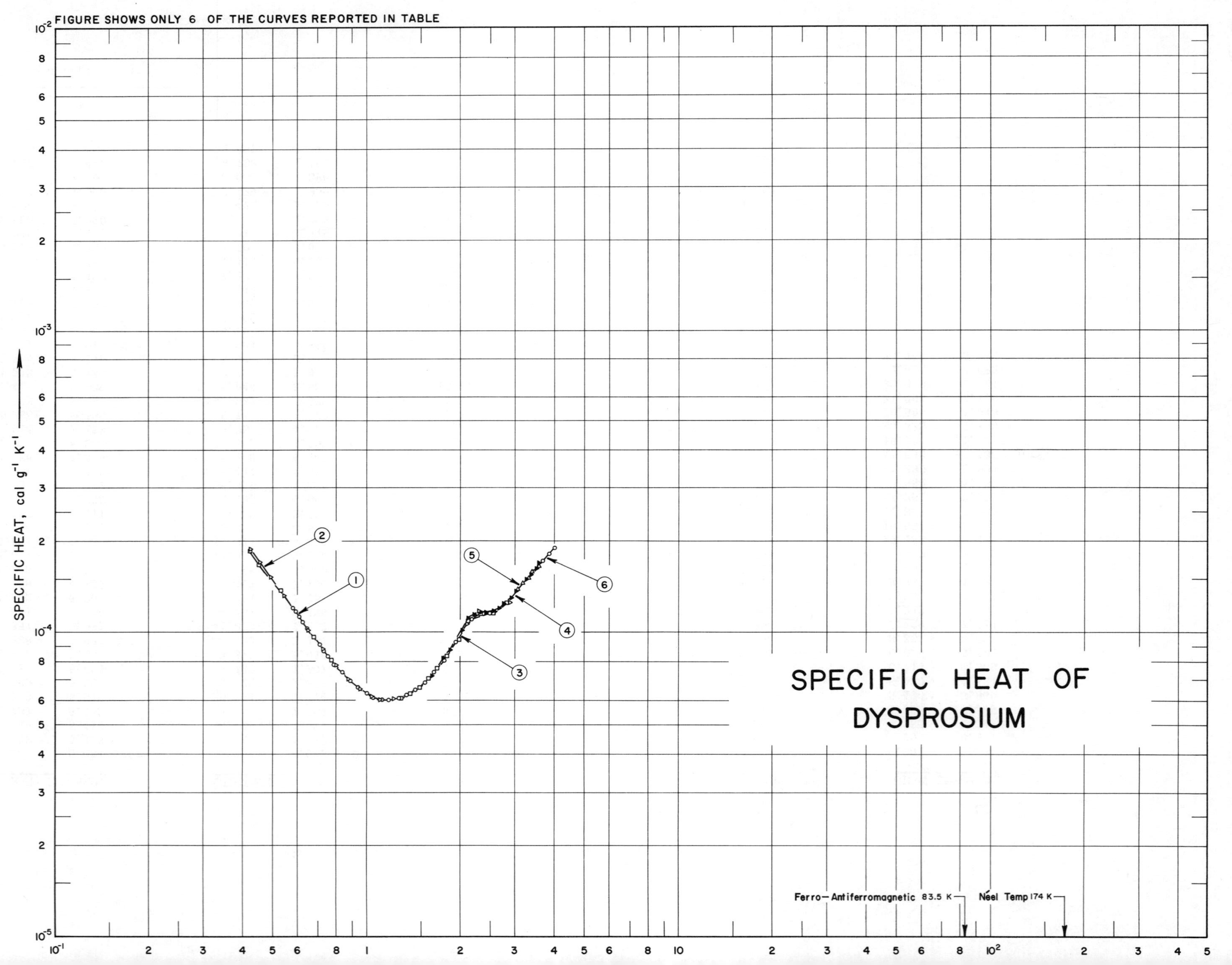
FIGURE SHOWS ONLY 6 OF THE CURVES REPORTED IN TABLE
SPECIFIC HEAT OF
DYSPROSIUM
SPECIFIC HEAT, cal g⁻¹ K⁻¹
Ferro—Antiferromagnetic 83.5 K
Néel Temp 174 K
1
2
3
4
5
6

SPECIFICATION TABLE NO. 15 SPECIFIC HEAT OF DYSPROSIUM

(Impurity $<$ 0.20% each; total impurities $<$0.50%)

[For Data Reported in Figure and Table No. 15]

Curve No.	Ref. No.	Year	Temp. Range, K	Reported Error, %	Name and Specimen Designation	Composition (weight percent), Specifications and Remarks
1	88	1962	0.6-2	<2.0	Run I-A	99.86 Dy, 0.08 O_2, 0.03 H_2, 0.03 Ta; vacuum distilled; remelted in vacuum and cast into tantalum crucible.
2	88	1962	0.4-3	<2.0	Run I-B	Same as above.
3	88	1962	0.4-3	<2.0	Run I-C	Same as above.
4	88	1962	1-4	<2.0	Run II-A	Same as above.
5	88	1962	3-3.96	<2.0	Run II-B	Same as above.
6	88	1962	2-4	<2.0	Run III	Same as above.
7	301	1966	298-2000	<2.0		0.04 N_2, 0.03 Ca, and Ni, 0.02 Fe, 0.007 Si, 0.006 Mg, 0.003 Al, and 0.002 O_2; prepared by metallothermic reduction of the fluoride with calcium and purified by distillation.

DATA TABLE NO. 15 SPECIFIC HEAT OF DYSPROSIUM

[Temperature, T, K; Specific Heat, C_p, Cal $g^{-1} K^{-1}$]

T	C_p
CURVE 1	
0.582_2	1.204×10^{-4}
0.594_2	1.170
0.610_7	1.123
0.627_9	1.081
0.645_9	1.028
0.710_4	9.081×10^{-5}
0.731_7	8.706
0.753_8	8.387
0.788_5	7.810
0.837_1	7.410
0.888_8	6.978
0.943_5	6.632
1.008_0	6.354
1.060_0	6.167
1.120_5	6.069
1.192_7	6.032
1.275_4	6.119
1.356_8	6.261
1.449_1	6.510
1.550_5	6.894
1.657_4	7.422
1.783_2	8.145
1.946_5	9.334
2.188_2	1.111×10^{-4}
2.492_4	1.162
CURVE 2	
0.424_1	1.884×10^{-4}
0.455_9	1.710
0.494_9	1.527
0.542_9	1.335*
0.592_8	1.174*
0.650_4	1.025
0.722_0	8.865×10^{-5}
0.796_7	7.832
0.872_9	7.086
0.956_2	6.558
1.045_1	6.208*
1.137_5	6.044*
1.231_0	6.054*
1.322_5	6.201*
1.439_5	6.489*
1.549_2	6.905*
1.289_3	6.132*
CURVE 2 (contd)	
1.398_2	6.366×10^{-5}*
1.510_1	6.767*
1.639_5	7.363*
1.794_1	8.260*
1.951_3	9.410*
2.114_9	1.079×10^{-4}
2.316_4	1.173*
2.566_9	1.173*
2.842_9	1.261
CURVE 3	
0.426_6	1.870×10^{-4}
0.458_2	1.699*
0.493_1	1.533*
0.532_0	1.374*
0.575_4	1.226*
0.624_2	1.087*
0.679_2	9.640×10^{-5}*
0.707_5	9.137*
0.773_8	8.131*
0.847_6	7.314*
0.928_5	6.705*
1.015_9	6.307*
1.107_5	6.095*
1.200_9	6.039*
1.294_1	6.120
1.385_2	6.316
1.481_0	6.630
1.585_8	7.080
1.695_4	7.651
1.817_4	8.409
1.962_4	9.485
2.130_8	1.090×10^{-4}
2.331_7	1.151
2.567_0	1.169
2.820_1	1.252
CURVE 4	
1.365_6	6.288×10^{-5}*
1.421_7	6.455*
1.485_6	6.642*
1.555_5	6.939*
1.629_2	7.276
CURVE 4 (contd)	
1.705_5	7.697×10^{-5}*
1.784_6	8.207
1.866_3	8.729*
1.951_8	9.346*
2.041_6	1.011×10^{-4}
2.135_3	1.090
2.232_2	1.151
2.333_1	1.182
2.440_4	1.175
2.553_3	1.181
2.670_1	1.209
2.787_6	1.250
2.906_5	1.302
3.024_2	1.372
3.267_3	1.502
3.389_7	1.566
3.512_3	1.626
3.641_0	1.699
CURVE 5	
3.065_8	1.392×10^{-4}
3.188_5	1.457
3.313_1	1.521
3.441_7	1.589
3.575_0	1.655
CURVE 6	
2.294_7	1.154×10^{-4}*
2.420_7	1.158*
2.552_2	1.159*
2.687_2	1.204*
2.822_7	1.260*
2.957_7	1.320*
3.094_5	1.390*
3.235_2	1.465*
3.380_1	1.541*
3.530_8	1.629*
3.689_9	1.722
3.856_6	1.821
4.034_1	1.917
CURVE 7*	
298.15	4.14×10^{-5}
300	4.14
400	4.08
500	4.07
600	4.12
700	4.20
800	4.34
900	4.52
1000	4.74
1100	5.02
1200	5.35
1300	5.72
1400	6.14
1500	6.603
1600	7.114
1657	7.428
1657	4.123
1682	4.123
1682	7.342
1700	7.342
1800	7.342
1900	7.342
2000	7.342

*Not shown on Plot

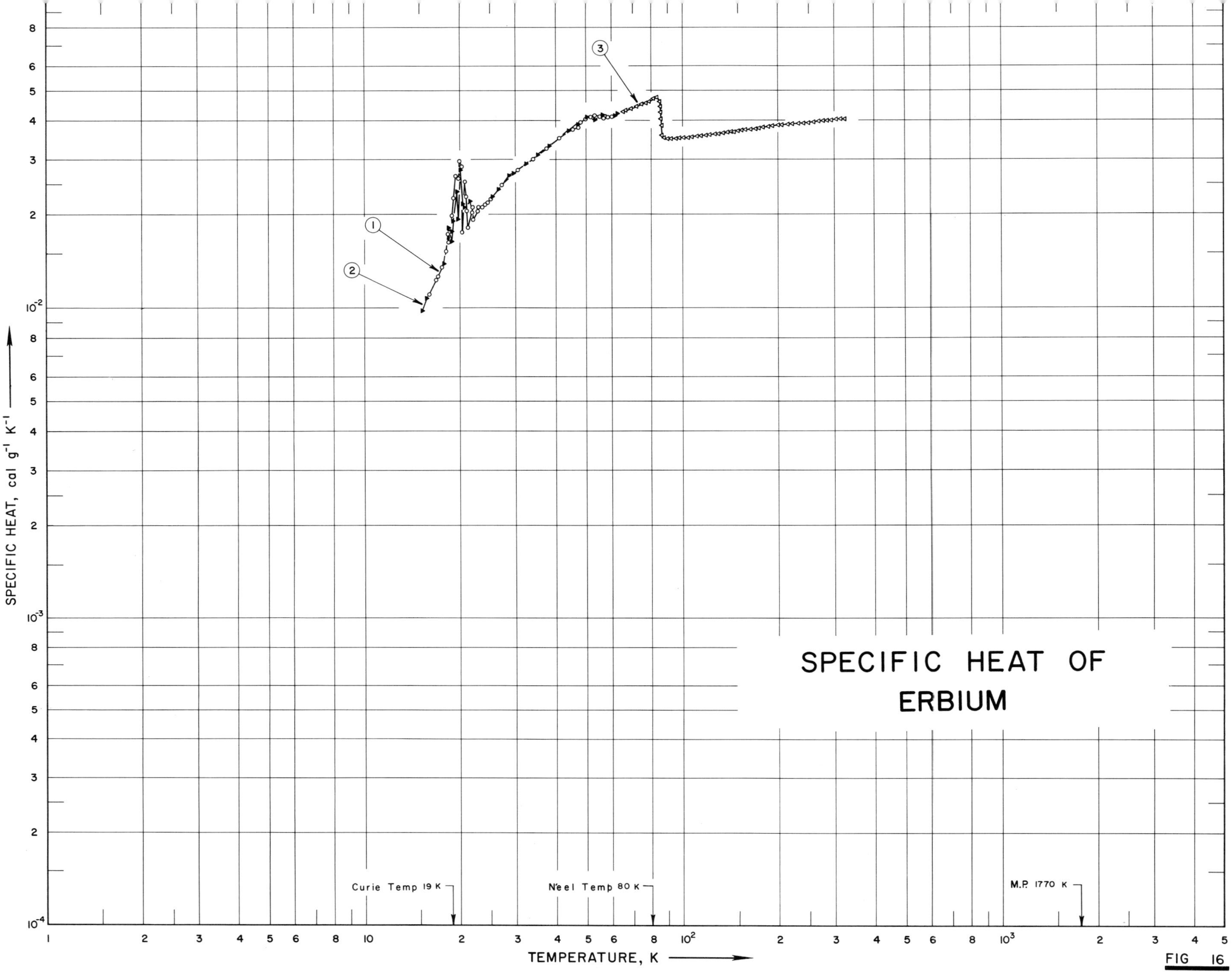

SPECIFIC HEAT OF ERBIUM

FIG 16

SPECIFICATION TABLE NO. 16 SPECIFIC HEAT OF ERBIUM

(Impurity < 0.20% each; total impurities < 0.50%)

[For Data Reported in Figure and Table No. 16]

Curve No.	Ref. No.	Year	Temp. Range, K	Reported Error, %	Name and Specimen Designation	Composition (weight percent), Specifications and Remarks
1	95	1955	16-60			< 0.10 each Ca, Mg, Si, and Y.
2	95	1955	15-64			Same as above.
3	95	1955	60-325			Same as above.
4	301	1966	298-1900	<2		0.04 F_2, 0.02 Fe and 0.02 Mg, 0.013 N_2, 0.01 Cr, 0.005 Si, 0.004 C, 0.003 O_2, and 0.002 Ca; prepared by metallothermic reduction of the fluoride with calcium and purified by distillation.

DATA TABLE NO. 16 SPECIFIC HEAT OF ERBIUM

[Temperature, T, K; Specific Heat, C_p, Cal g^{-1} K^{-1}]

CURVE 1

T	C_p
16.01	1.111 x 10^{-2}
16.79	1.233
17.03	1.266
17.60	1.356
18.09	1.525
18.34	1.731
18.46	1.629
18.88	1.981
19.04	2.250
19.16	2.227*
19.44	2.662
19.77	2.609
19.92	2.976
19.93	2.571*
20.34	2.852
20.39	1.754
20.46	2.237*
20.70	2.548
20.94	2.273
21.15	2.052
21.22	2.281*
21.23	1.818
21.87	2.029
21.99	2.119
22.02	1.934
22.27	2.069*
22.60	2.037*
22.94	2.042
22.98	2.104
23.48	2.093*
23.65	2.115
24.00	2.134*
24.04	2.157
24.51	2.182
24.98	2.230*
25.06	2.242
27.16	2.492
30.42	2.775
34.01	3.013
37.61	3.267
41.34	3.518
45.22	3.767
47.36	3.808
48.01	3.959
49.71	4.002*
49.84	4.043

CURVE 1 (contd)

T	C_p
51.50	4.115 x 10^{-2}*
51.61	4.128
53.25	4.156*
53.36	4.168
54.99	4.129*
55.09	4.135
56.73	4.095*
56.83	4.097
58.47	4.101*
58.57	4.104
60.29	4.139

CURVE 2

T	C_p
15.21	9.847 x 10^{-3}
15.79	1.074 x 10^{-2}
16.93	1.253*
17.75	1.395
18.48	1.806
18.84	1.641
18.90	1.774
19.23	1.902
19.65	2.365
19.67	1.937
20.01	2.785
20.54	2.155
20.71	2.104
20.75	2.036*
21.59	2.207*
22.01	2.146*
22.13	2.026*
22.53	2.035*
23.00	2.065*
23.76	2.121*
24.19	2.119*
24.56	2.189*
24.57	2.191*
25.49	2.277
26.29	2.372*
26.50	2.400
26.98	2.472
28.62	2.672
29.69	2.729
32.49	2.901
35.38	3.110
38.33	3.317

CURVE 2 (contd)

T	C_p
41.34	3.520 x 10^{-2}*
44.28	3.711
47.19	3.889
50.07	4.058
50.27	4.067*
52.19	4.149*
53.35	4.027
53.63	4.157*
54.46	4.146*
54.72	4.055*
54.81	4.048*
56.08	4.081
56.15	4.057*
56.82	4.097*
56.98	4.107*
57.44	4.075*
57.48	4.080*
58.78	4.103*
58.80	4.101*
59.13	4.115*
60.08	4.135*
60.10	4.132*
60.11	4.138*
61.40	4.165*
61.41	4.170*
61.42	4.167
62.72	4.205

CURVE 3

T	C_p
59.62	4.116 x 10^{-2}*
60.82	4.152*
62.83	4.204*
63.57	4.226*
65.96	4.292
66.66	4.310
69.01	4.370
69.68	4.372*
72.01	4.441
72.38	4.449*
74.84	4.528
74.93	4.512*
77.06	4.567
77.79	4.588*
78.82	4.618
80.26	4.653*

CURVE 3 (contd)

T	C_p
80.59	4.666 x 10^{-2}*
81.44	4.701
82.47	4.739*
83.27	4.777
83.33	4.770*
83.47	4.788*
83.49	4.784*
83.58	4.787*
83.74	4.787*
83.90	4.787*
84.05	4.787*
84.21	4.788*
84.37	4.787*
84.50	4.770*
84.52	4.787*
84.68	4.786*
84.84	4.715*
85.00	4.615
85.16	4.454
85.33	4.250
85.51	4.036
85.60	4.032*
86.30	3.858
86.82	3.593
88.11	3.540
89.41	3.523*
89.63	3.523*
90.87	3.519
92.65	3.514
93.07	3.514*
94.59	3.514*
96.49	3.515
96.51	3.514*
98.90	3.517
102.27	3.540
105.61	3.533
108.93	3.553
112.23	3.569
115.50	3.584
119.23	3.596
123.43	3.616
127.59	3.635
131.71	3.646
135.81	3.661
139.87	3.675
142.55	3.682

CURVE 3 (contd)

T	C_p
143.91	3.691 x 10^{-2}*
146.53	3.697
147.92	3.704*
150.49	3.713
152.27	3.720*
154.43	3.723
156.23	3.722*
158.34	3.731
160.18	3.734*
162.23	3.744*
164.10	3.739*
166.11	3.755
168.02	3.743*
169.97	3.770*
171.91	3.769
173.81	3.784*
175.77	3.788
177.62	3.803*
179.61	3.808
183.43	3.820*
187.67	3.836
192.44	3.850
197.20	3.859*
201.94	3.868
206.66	3.881
211.35	3.883*
216.04	3.893
220.70	3.899*
222.87	3.909
232.49	3.911
237.08	3.921*
241.66	3.929
246.23	3.933*
250.78	3.944
255.31	3.950*
259.84	3.950
264.34	3.962*
268.83	3.975
273.30	3.973*
277.77	3.990
282.03	3.991*
286.27	3.998
290.69	4.007*
295.09	4.015
297.12	4.014*
299.49	4.020*

CURVE 3 (contd)

T	C_p
301.86	4.018 x 10^{-2}*
306.59	4.024
311.29	4.031*
315.98	4.040
320.66	4.040*
325.32	4.048

CURVE 4*

T	C_p
298.15	4.02 x 10^{-2}
300	4.02
400	4.06
500	4.11
600	4.17
700	4.25
800	4.35
900	4.46
1000	4.59
1100	4.73
1200	4.89
1300	5.06
1400	5.25
1500	5.46
1600	5.69
1700	5.93
1795	6.18
1795	5.53
1800	5.53
1900	5.53

*Not shown on plot

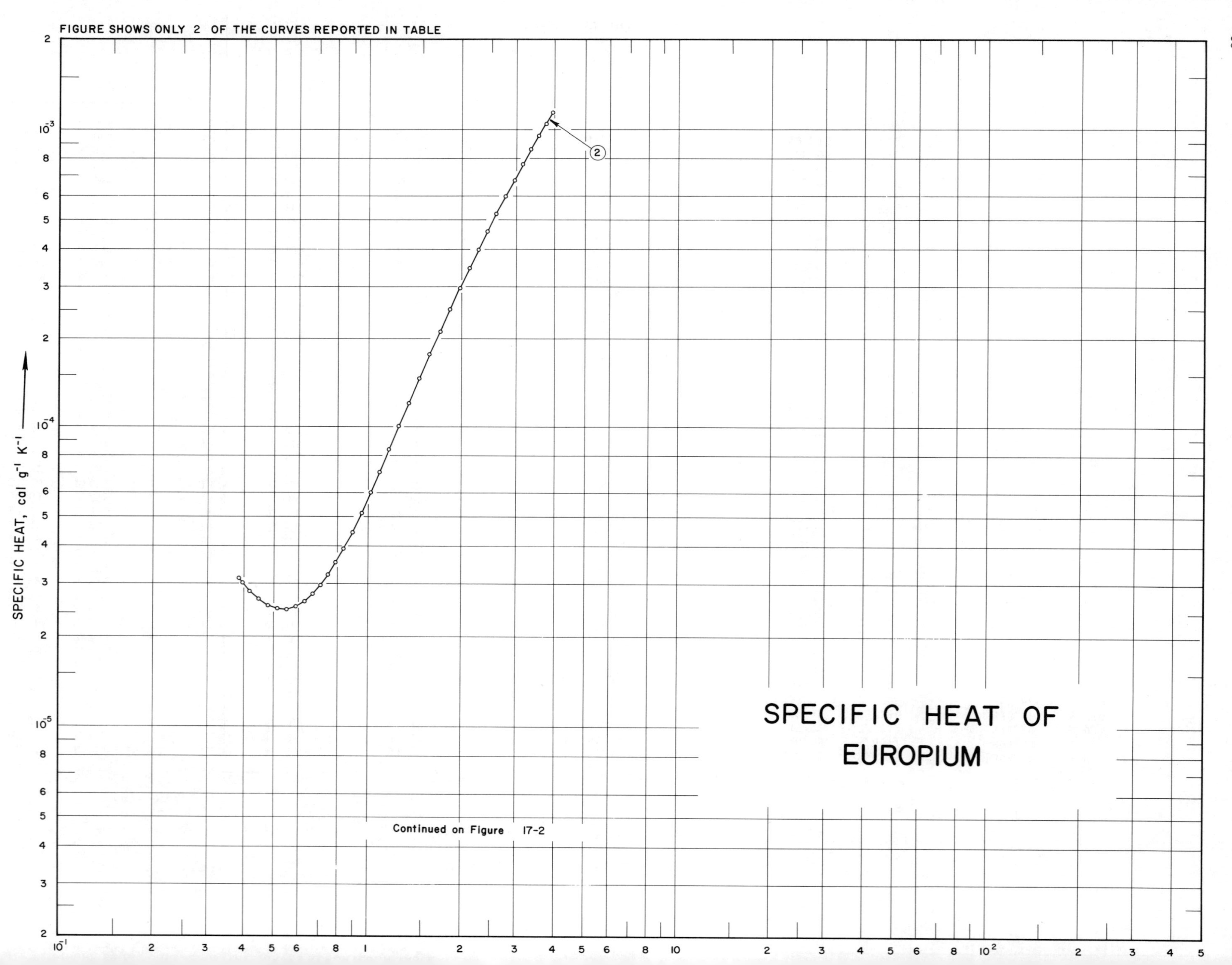

FIGURE SHOWS ONLY 2 OF THE CURVES REPORTED IN TABLE
SPECIFIC HEAT OF EUROPIUM
SPECIFIC HEAT, cal g⁻¹ K⁻¹
Continued on Figure 17-2
2

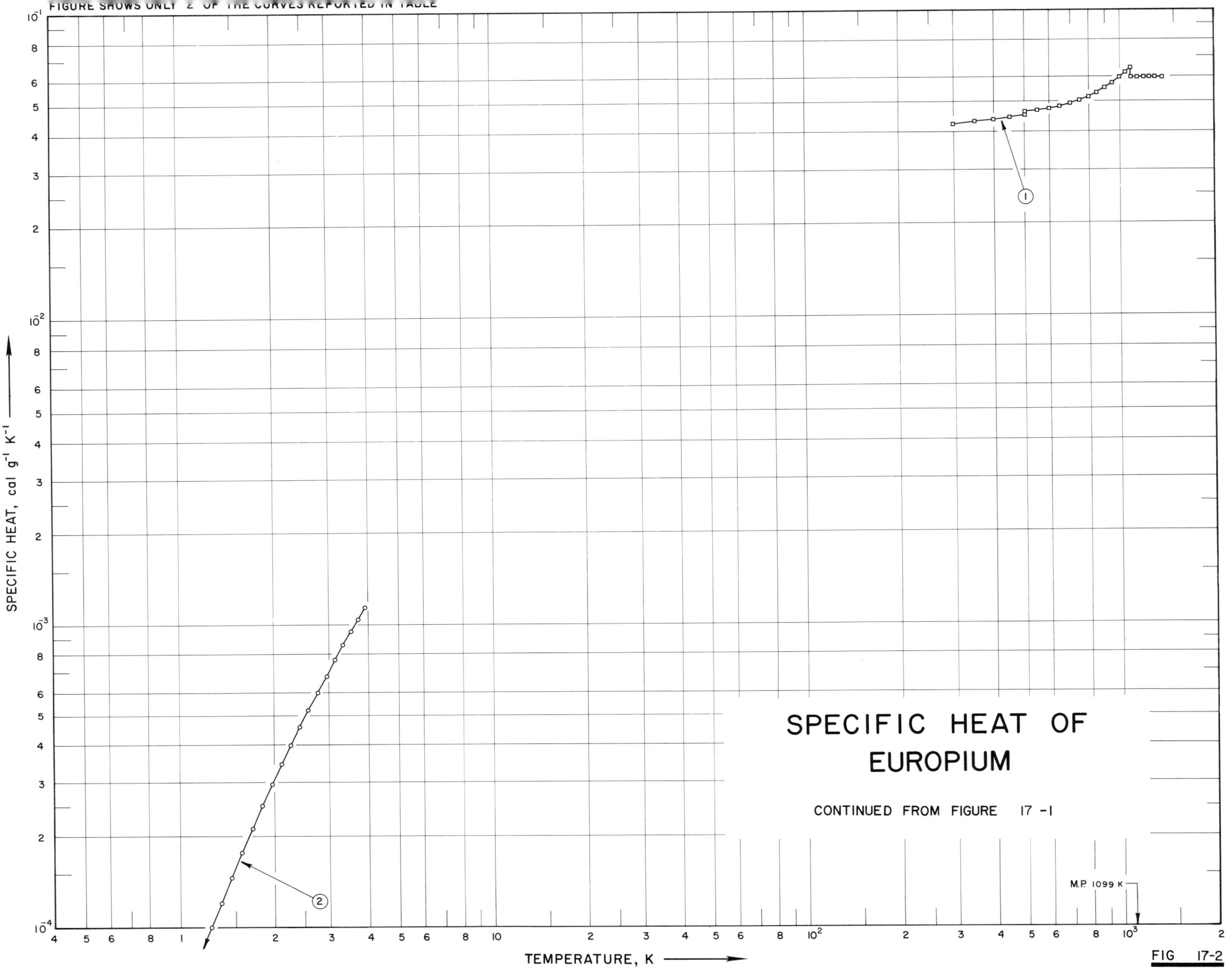
FIGURE SHOWS ONLY 2 OF THE CURVES REPORTED IN TABLE
SPECIFIC HEAT OF EUROPIUM
CONTINUED FROM FIGURE 17 -1
SPECIFIC HEAT, cal g⁻¹ K⁻¹
TEMPERATURE, K
M.P. 1099 K
1
2

FIG 17-2

SPECIFICATION TABLE NO. 17 SPECIFIC HEAT OF EUROPIUM

(Impurity < 0.20% each; total impurities < 0.50%)

[For Data Reported in Figure and Table No. 17]

Curve No.	Ref. No.	Year	Temp. Range, K	Reported Error, %	Name and Specimen Designation	Composition (weight percent), Specifications and Remarks
1	36	1961	298-1373			>99.971 Eu, <0.02 Sm, 0.019 O_2, 0.009 Gd, 0.0045 C, 0.003 N_2, and 0.0012 H_2; prepared by reduction of sintered Europium oxide with Lanthanum metal; cast into 1/2-inch rods from which 3/4-inch long samples were prepared; sealed under helium in tantalum crucibles.
2	86	1964	0.4-4	<1.5	Run I & II	0.079 C, 0.06 Mg, 0.026 H_2, 0.023 N_2, <0.01 Ta, 0.008 Ag, 0.006 Ce, 0.003 F, 0.003 Sm, and 0.001 each Al, La, Mn, Nd; polished in glove box.
3	86	1964	0.3-4	<1.5	Run III & IV	Same as above.
4	205	1966	3-25	0.5		

DATA TABLE NO. 17 SPECIFIC HEAT OF EUROPIUM

[Temperature, T, K; Specific Heat, C_p, Cal $g^{-1}K^{-1}$]

T	C_p
CURVE 1	
298.15	4.26 x 10^{-2}
300	4.26*
350	4.33
400	4.40
450	4.47
500	4.53
503.15	4.54*
503.15	4.66*
550	4.70
600	4.76
650	4.83
700	4.93
750	5.05
800	5.19
850	5.36
900	5.55
950	5.74
1000	5.97
1050	6.22
(s)1090.15	6.43
(l)1090.15	5.99
1100	5.99*
1150	5.99
1200	5.99
1250	5.99
1300	5.99
1350	5.99*
1373.15	5.99
CURVE 2	
0.383_5	3.135 x 10^{-5}*
0.394_7	3.020*
0.417_8	2.833*
0.446_7	2.661*
0.478_8	2.539*
0.512_7	2.474*
0.549_0	2.461*
0.587_8	2.506*
0.626_7	2.609*
0.665_5	2.757*
0.705_0	2.954*
0.745_6	3.202*
0.789_1	3.520*
0.836_7	3.926*
CURVE 2 (cont.)	
0.890_6	4.461* x 10^{-5}
0.951_8	5.146*
1.017_5	6.003*
1.089_1	7.073*
1.168_3	8.399*
1.255_4	1.004 x 10^{-4}
1.352_7	1.208
1.460_1	1.459
1.577_0	1.757
1.702_8	2.111
1.835_5	2.513
1.973_5	2.960
2.115_9	3.451
2.263_6	3.993
2.419_2	4.577
2.586_7	5.247
2.764_7	5.986
2.954_3	6.799
3.147_6	7.675
3.342_4	8.588
3.538_3	9.511
3.734_0	1.049 x 10^{-3}
3.929_4	1.147
CURVE 3	
0.360_9	3.396 x 10^{-5}*
0.380_9	3.149*
0.404_8	2.929*
0.432_9	2.730*
0.463_5	2.589*
0.495_1	2.501*
0.530_0	2.458*
0.568_1	2.476*
0.607_2	2.553*
0.446_0	2.674*
0.684_9	2.852*
0.726_0	3.076*
0.768_8	3.363*
0.814_8	3.732*
0.865_4	4.196*
0.922_8	4.810*
0.986_4	5.593*
1.055_1	6.551*
1.130_0	7.737*
CURVE 3 (cont.)	
1.213_6	9.226*
1.306_0	1.108 x 10^{-4}
1.407_4	1.335
1.516_8	1.601
1.640_4	1.933
1.769_3	2.307
1.903_4	2.727
2.043_7	3.194
2.189_9	3.710
2.342_5	4.284
2.503_8	4.913
2.677_9	5.621
2.863_8	6.411
3.057_6	7.290
3.255_9	8.179
3.446_2	9.070
3.629_0	9.909
3.823_7	1.093 x 10^{-3}
4.029_2	1.196
CURVE 4*	
Series 1	
3.0185	7.06823 x 10^{-4}
3.3602	8.63597
3.6959	1.02793 x 10^{-3}
4.0253	1.19763
4.3527	1.37772
4.6905	1.57278
5.0537	1.79416
5.4756	2.06579
6.0171	2.44268
6.7256	2.97650
7.4764	3.60234
8.1672	4.22518
8.8367	4.88127
9.5328	5.62654
10.2930	6.50888
11.1380	7.57085
12.0420	8.82504
12.9660	1.02021 x 10^{-2}
13.8450	1.15729
14.4920	1.27468
14.9530	1.36291
CURVE 4 (cont.)*	
15.3840	1.46076 x 10^{-2}
15.7350	1.55151
15.9670	1.62588
16.1160	1.66559
16.2390	1.53985
16.3760	1.31643
16.5290	1.25435
16.7250	1.24663
17.0170	1.26176
17.4460	1.29343
18.1340	1.36307
19.1520	1.47399
20.5710	1.63675
22.4580	1.85970
24.7570	2.15025
Series 2	
3.1337	7.58503 x 10^{-4}
3.4585	9.10708
3.7854	1.07300 x 10^{-3}
4.1222	1.25010
4.4452	1.43003
4.7913	1.63218
5.2535	1.92005
5.8776	2.34279
6.5306	2.81138
7.1054	3.28580
7.7068	3.80559
8.4097	4.46105
9.2276	5.29408
10.1410	6.32769
11.0250	7.42589
11.7890	8.45320
12.5440	9.54195
13.2960	1.07000
13.9750	1.18187
14.5900	1.29358
15.1520	1.40845
15.6000	1.51638
15.8910	1.60051
16.1160	1.64731
16.3540	1.36717
16.6270	1.24253
16.9830	1.25624
CURVE 4 (cont.)*	
17.5520	1.30540 x 10^{-2}
18.4720	1.39931
19.7960	1.54679
21.5960	1.75556
23.9090	2.03759
Series 3	
4.5273	1.48124 x 10^{-3}
4.9243	1.73050
5.6156	2.14410
6.5253	2.82619

* Not shown on plot

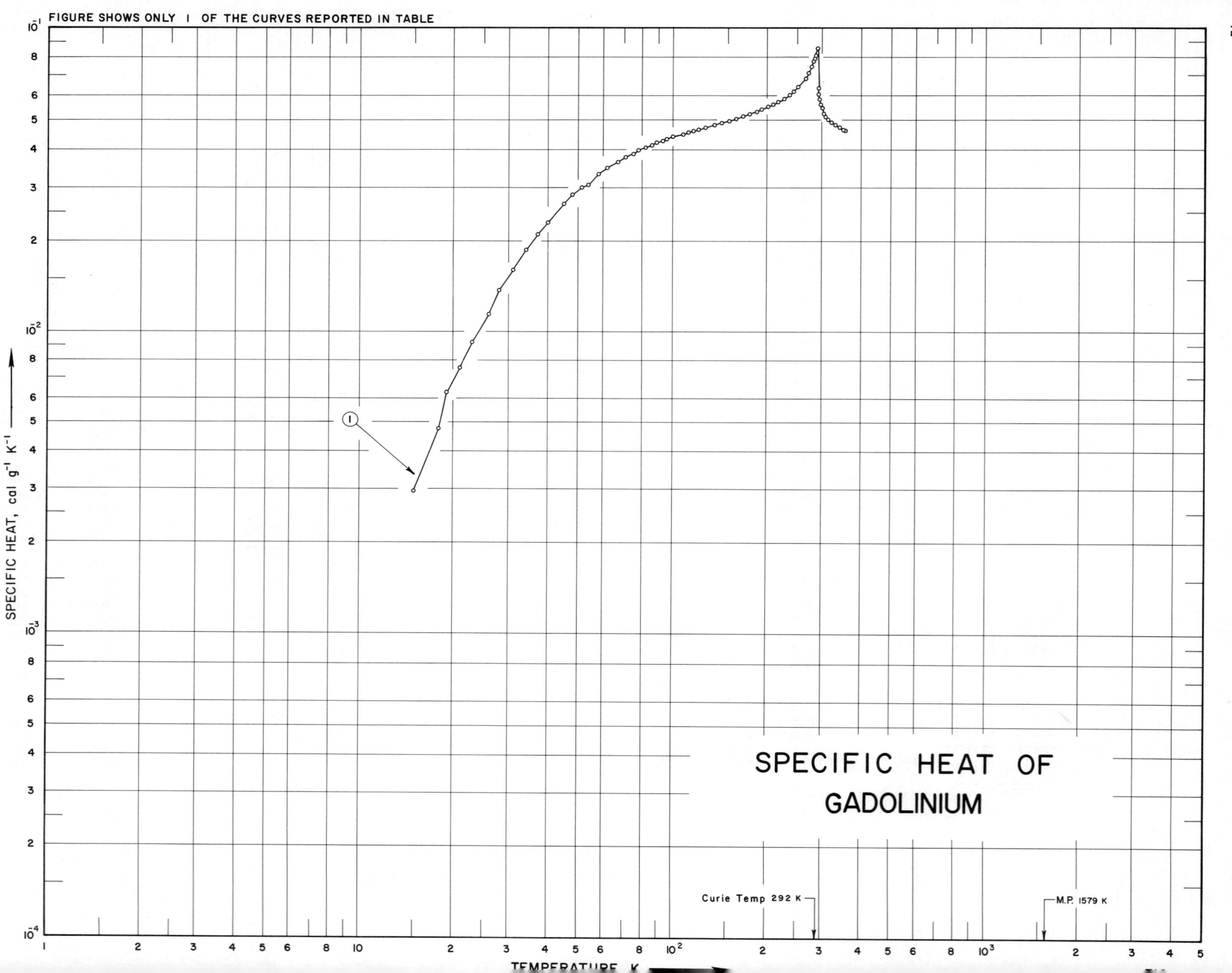

FIGURE SHOWS ONLY 1 OF THE CURVES REPORTED IN TABLE
SPECIFIC HEAT OF
GADOLINIUM
SPECIFIC HEAT, cal g^{-1} K^{-1}
TEMPERATURE, K
Curie Temp 292 K
M.P. 1579 K
1

SPECIFICATION TABLE NO. 18 SPECIFIC HEAT OF GADOLINIUM

(Impurity < 0.20% each; total impurities < 0.50%)

[For Data Reported in Figure and Table No. 18]

Curve No.	Ref. No.	Year	Temp. Range, K	Reported Error, %	Name and Specimen Designation	Composition (weight percent), Specifications and Remarks
1	217	1954	16-358	0.1-1		< 0.3 Ta, 0.1 Sn, < 0.1 Y, < 0.04 Ca, and < 0.01 each Fe, Mg, Si.
2	301	1966	300-1700	< 2.0		Sample I: 0.14 O_2, 0.1 Y, 0.05 Fe, 0.02 Mg, 0.02 Si, 0.013 C, 0.008 F, 0.005 Ca, and 0.005 N_2, Sample II: 0.041 O_2, 0.01 C, 0.0082 F_2, 0.0048 N_2, 0.0045 Fe, 0.0025 Ca, 0.002 Y, and 0.0003 Si; prepared by metallothermic reduction of the fluoride with calcium and purified by distillation; the data is based on the combined values of Sample I and Sample II.

DATA TABLE NO. 18 SPECIFIC HEAT OF GADOLINIUM

[Temperature, T, K; Specific Heat, C_p, Cal $g^{-1}K^{-1}$]

T	C_p
CURVE 1	
15.10	2.976 x 10^{-3}
18.00	4.806
19.96	6.303
21.51	7.578
23.45	9.261
26.05	1.156 x 10^{-2}
28.66	1.388
31.30	1.617
34.28	1.876
37.31	2.120
40.19	2.329
45.10	2.661
48.43	2.860
51.54	3.029
54.48	3.173
58.57	3.358
62.89	3.532*
62.97	3.532
67.16	3.692
71.19	3.801
75.10	3.910
78.90	4.006
82.59	4.099
86.21	4.180
89.75	4.247
93.24	4.305
96.67	4.356
100.52	4.416
104.78	4.477*
108.97	4.532
113.10	4.585
117.18	4.639
121.21	4.683
125.19	4.725*
128.69	4.764
132.59	4.805*
136.46	4.844
140.28	4.880*
144.08	4.924
147.84	4.962
151.57	4.991
155.69	5.033*

T	C_p
CURVE 1 (cont.)	
160.18	5.082 x 10^{-2}
164.62	5.129*
169.01	5.177
173.36	5.219*
177.67	5.263
181.94	5.305*
186.18	5.350
190.36	5.396*
194.51	5.442
198.62	5.489*
202.69	5.540
206.72	5.597*
210.87	5.644
215.42	5.706*
219.91	5.767
224.35	5.841*
228.74	5.905
233.08	5.978*
237.36	6.062
241.57	6.144*
245.74	6.220
249.83	6.338*
253.87	6.442
257.84	6.564*
266.31	6.851
273.60	7.170
274.43	7.192*
276.44	7.313*
278.42	7.417*
279.62	7.508
280.54	7.558*
282.68	7.724*
282.95	7.839
284.61	7.893*
285.41	7.980
286.27	8.053*
287.05	8.152
287.63	8.205*
288.65	8.381*
288.70	8.363*
289.74	8.547*
290.21	8.649

T	C_p
CURVE 1 (cont.)	
290.66	8.720 x 10^{-2}*
291.47	8.827*
291.75	8.528*
292.30	8.147*
293.23	6.519*
293.54	6.405
294.39	6.110
295.67	5.874
295.85	5.871*
297.93	5.658
297.95	5.660*
300.26	5.505
300.43	5.498*
302.64	5.390*
303.90	5.333*
305.02	5.287
307.49	5.212*
308.78	5.183*
309.97	5.164
312.49	5.099*
313.75	5.080*
315.03	5.063
317.59	4.989*
318.80	4.993*
323.90	4.927
329.05	4.886*
334.24	4.844
337.33	4.769*
339.47	4.788*
342.43	4.751*
344.74	4.730
347.54	4.702*
352.70	4.667
357.82	4.644
CURVE 2*	
300	4.37 x 10^{-2}
400	4.39
500	4.45
600	4.52
700	4.60

T	C_p
CURVE 2 (cont.)*	
800	4.70 x 10^{-2}
900	4.81
1000	4.94
1100	5.09
1200	5.25
1300	5.42
1400	5.61
1500	5.81
1533	5.88
1533	4.30
1585	5.65
1600	5.65
1700	5.65

*Not shown on plot

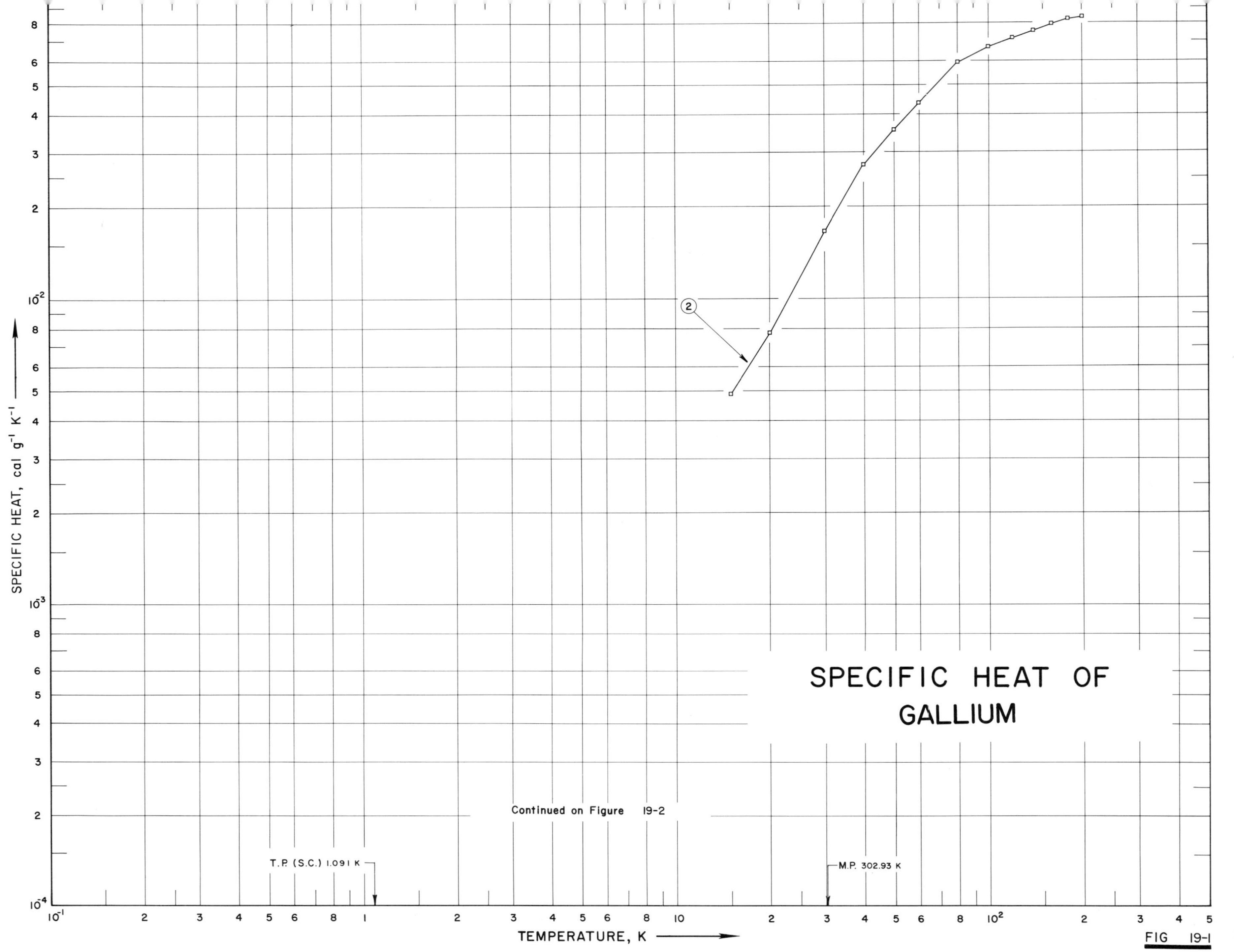
SPECIFIC HEAT OF
GALLIUM
SPECIFIC HEAT, cal g⁻¹ K⁻¹
TEMPERATURE, K
T.P. (S.C.) 1.091 K
M.P. 302.93 K
Continued on Figure 19-2
2

FIG 19-1

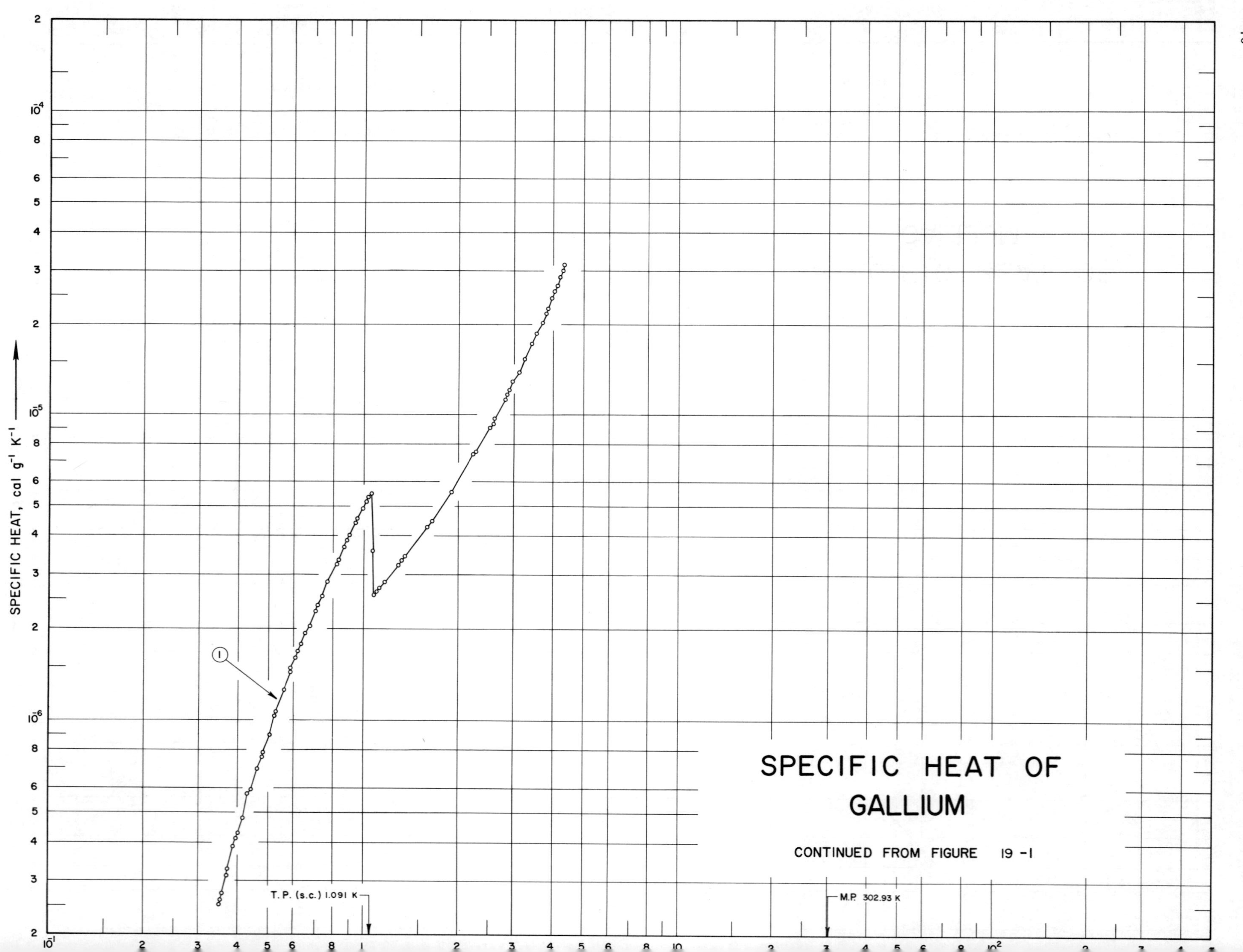

SPECIFIC HEAT OF
GALLIUM
CONTINUED FROM FIGURE 19 -1
T. P. (s.c.) 1.091 K
M.P. 302.93 K
1
SPECIFIC HEAT, cal g⁻¹ K⁻¹

SPECIFICATION TABLE NO. 19 SPECIFIC HEAT OF GALLIUM

(Impurity $<0.20\%$ each; total impurities $<0.50\%$)

[For Data Reported in Figure and Table No. 19]

Curve No.	Ref. No.	Year	Temp. Range, K	Reported Error, %	Name and Specimen Designation	Composition (weight percent), Specifications and Remarks
1	218	1958	0.4-4			99.999 Ga; sample supplied by the Eagle-Picher Co.
2	207	1928	15-200			

DATA TABLE NO. 19 SPECIFIC HEAT OF GALLIUM

[Temperature, T, K; Specific Heat, C_p, Cal g^{-1} K^{-1}]

T	C_p
CURVE 1	
0.353_2	2.516×10^{-7}
0.353_5	2.550*
0.356_7	2.609
0.360_1	2.746
0.371_2	3.133
0.375_0	3.281
0.392_6	3.798
0.399_7	4.124
0.403_0	4.186*
0.405_7	4.309
0.420_4	4.810
0.436_3	5.735
0.438_7	5.687*
0.445_0	5.917
0.465_7	6.925
0.469_0	6.993*
0.482_1	7.576
0.485_5	7.816
0.488_6	7.953*
0.510_3	8.982
0.513_8	9.119*
0.537_0	1.035×10^{-6}
0.540_2	1.063
0.543_4	1.063*
0.572_3	2.173*
0.575_3	1.251
0.602_8	1.440
0.606_4	1.474
0.626_8	1.597
0.635_7	1.656
0.655_9	1.765
0.674_5	1.920
0.690_2	2.029
0.696_5	2.077*
0.721_5	2.266
0.736_6	2.389
0.756_5	2.540
0.763_1	2.598*
0.788_2	2.814
0.841_1	3.212
0.852_2	3.342

T	C_p
CURVE 1 (cont.)	
0.861_6	3.442×10^{-6}*
0.886_7	3.658
0.901_5	3.853
0.925_9	4.004
0.961_5	4.412
0.976_9	4.535
1.012_0	4.909
1.028_0	5.060*
1.041_0	5.170
1.052_0	5.265*
1.056_0	5.338
1.061_0	5.451*
1.070_0	5.482
1.074_0	5.553
1.091_0	3.565
1.103_0	2.564
1.124_0	2.629
1.125_0	2.612*
1.150_0	2.705
1.177_0	2.770*
1.199_0	2.845
1.327_0	3.205
1.359_0	3.342
1.385_0	3.432
1.630_0	4.271
1.683_0	4.491
1.940_0	5.574
1.951_0	5.639*
2.264_0	7.439
2.287_0	7.439*
2.308_0	7.542
2.552_0	9.084
2.577_0	9.256*
2.610_0	9.393
2.638_0	9.701
2.854_0	1.138×10^{-5}
2.892_0	1.179
2.948_0	1.214
3.049_0	1.310
3.154_0	1.399
3.290_0	1.543

T	C_p
CURVE 1 (cont.)	
3.458_0	1.721×10^{-5}
3.577_0	1.868
3.716_0	2.050
3.815_0	2.197
3.867_0	2.273
3.984_0	2.451
4.060_0	2.585
4.143_0	2.708
4.227_0	2.890
4.309_0	3.037
4.378_0	3.164
CURVE 2	
15	4.87×10^{-3}
20	7.74
30	1.66×10^{-2}
40	2.73
50	3.56
60	4.39
80	5.95
100	6.70
120	7.13
140	7.54
160	7.99
180	8.26
200	8.40

*Not shown on plot

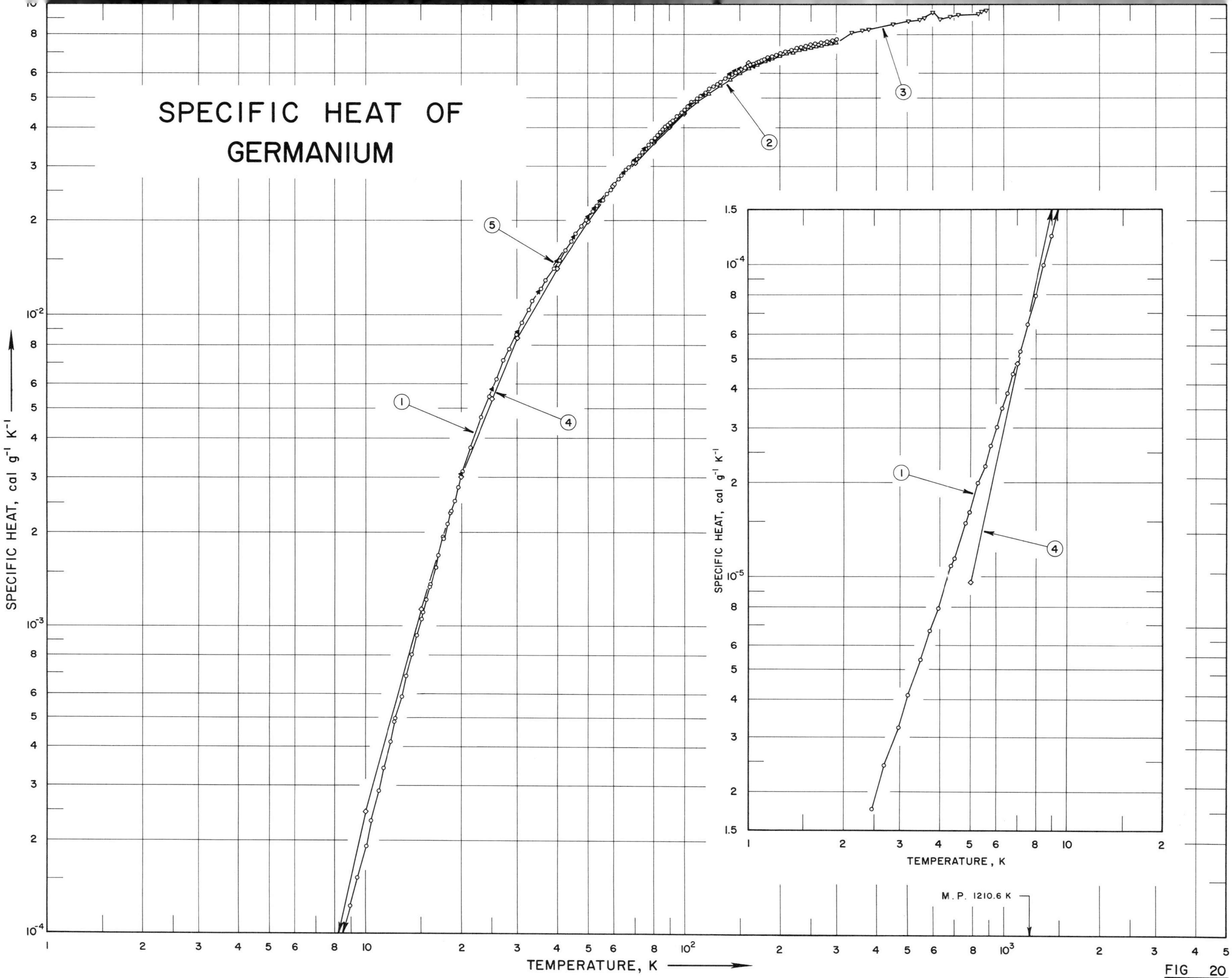

SPECIFIC HEAT OF GERMANIUM

FIG 20

SPECIFICATION TABLE NO. 20 SPECIFIC HEAT OF GERMANIUM

(Impurity <0.20% each; total impurities <0.50%)

[For Data Reported in Figure and Table No. 20]

Curve No.	Ref. No.	Year	Temp. Range, K	Reported Error, %	Name and Specimen Designation	Composition (weight percent), Specifications and Remarks
1	40	1959	2-300	±2.0		Sample supplied by Societe des Mines et Founderies de Zinc de la Vielle Montagne, Belgium; broken into pieces of 3 mm size; evacuated to a pressure of 10^{-6} mm Hg and then sealed with a small amount of helium exchange gas inside.
2	41	1959	80-300	≤7.0		Single crystals; p-type.
3	42	1964	296-884	≤3.0		Resistivity 0.01 ohm-cm; n-type.
4	43	1952	5-160	±1.0		Spectroscopically pure.
5	44	1952	20-200			Pure sample (2.8 x 10^{14} impurity centers cm^{-3} (298 K).
6	44	1952	20-200			Intermediate purity (9.5 x 10^{17} impurity centers cm^{-3} (302 K); n-type.
7	44	1952	20-200			0.00223 Al (2.2 x 10^{20} impurity centers cm^{-3} (297 K); p-type.
8	302	1934	10-200			
9	303	1963	12-273	≤2.0		Single crystals.
10	304	1966	298-1500	±0.3		99.99^{+} Ge; specimen under vacuum.

DATA TABLE NO. 20 SPECIFIC HEAT OF GERMANIUM

[Temperature, T. K; Specific Heat, C_p, Cal g^{-1} K^{-1}]

CURVE 1

T	C_p
2.461	1.767 x 10^{-6}*
2.675	2.432*
2.971	3.219*
3.175	4.143*
3.481	5.397*
3.713	6.716*
3.961	7.938*
4.364	1.093 x 10^{-5}*
4.471	1.152*
4.814	1.483*
4.963	1.612*
5.283	1.992*
5.503	2.253*
5.774	2.638*
6.024	3.014*
6.284	3.461*
6.506	3.859*
6.784	4.472*
7.131	5.296*
7.510	6.430*
7.974	7.930*
8.465	9.945*
8.948	1.233 x 10^{-4}
9.434	1.521
10.006	1.916
10.442	2.313
11.015	2.897
11.458	3.415
12.011	4.186
12.366	4.837
12.471	4.902
12.902	5.767*
13.008	5.875
13.478	6.813
14.002	8.029
14.536	9.321
15.002	1.053 x 10^{-3}
15.192	1.117
15.579	1.216
15.945	1.344
15.961	1.354
16.578	1.552
16.979	1.696
17.579	1.931
17.588	1.927 x 10^{-3}
18.051	2.113
18.493	2.307
18.628	2.343
19.041	2.515
19.632	2.792
20.233	3.107
21.449	3.731
23.186	4.672
24.548	5.441
25.86	6.204
27.22	7.016
28.46	7.756
29.87	8.632
31.16	9.420
32.68	1.034 x 10^{-2}
33.91	1.109
35.72	1.219
36.93	1.291
39.00	1.413
40.46	1.497
42.43	1.609
44.29	1.708
45.89	1.807
47.87	1.919
49.37	2.001
51.90	2.145
53.59	2.234
55.63	2.346
57.15	2.431
59.02	2.534
60.39	2.610
62.35	2.719
63.76	2.795
65.81	2.908
67.30	2.988
69.22	3.084
70.73	3.158
72.54	3.255
74.07	3.335
76.00	3.431
77.45	3.507
79.38	3.605
81.09	3.693
82.75	3.778 x 10^{-2}
84.36	3.858
85.97	3.938
87.50	4.010
89.06	4.095
90.70	4.153
92.22	4.211
93.98	4.297*
95.46	4.359
97.17	4.430*
98.76	4.499
100.27	4.558
103.23	4.672
106.87	4.824
110.39	4.953
113.82	5.072
117.17	5.190
120.44	5.302
123.84	5.410
127.36	5.518
130.81	5.618
134.18	5.716*
135.93	5.755
139.22	5.847
142.42	5.934
145.75	6.010
149.23	6.098
152.66	6.164
156.05	6.238
159.39	6.304
162.69	6.377
166.06	6.430
169.50	6.490
172.90	6.547
176.26	6.606
178.49	6.645
183.82	6.721
189.31	6.792
192.91	6.841*
196.48	6.886
200.02	6.934
203.54	6.977*
207.02	7.012
210.49	7.051*
213.92	7.087 x 10^{-2}*
217.34	7.125
222.82	7.189*
226.53	7.207
230.21	7.240*
233.87	7.281
237.50	7.299*
241.11	7.335
244.70	7.368*
246.40	7.368*
249.98	7.402
253.51	7.424*
257.02	7.459
262.81	7.489*
268.85	7.522
272.84	7.545*
276.81	7.567*
280.72	7.589
280.90	7.596*
284.80	7.610*
286.55	7.627*
288.66	7.639*
292.38	7.654
292.48	7.658*
296.23	7.687*
296.24	7.575*
300.01	7.726
300.05	7.713*

CURVE 2

T	C_p
80	3.65 x 10^{-2}
90	4.09
100	4.50
110	4.88
120	5.18
130	5.48
140	5.74
150	6.01
160	6.23
170	6.43
180	6.60
190	6.74
200	6.86
210	6.97 x 10^{-2}
220	7.05
230	7.13
240	7.20
250	7.27
260	7.33
270	7.40
280	7.45
290	7.49
300	7.53

CURVE 3

T	C_p
296.1	7.65 x 10^{-2}*
294.6	7.50*
335.8	8.12
363.2	8.28
380.0	8.32
452.6	8.62
501.9	8.84
548.1	8.93
568.7	9.02
599.9	9.45
631.9	8.98
681.4	9.12
720.9	9.29
836.7	9.38
852.7	9.49
883.9	9.56

CURVE 4

T	C_p
5	9.64 x 10^{-6}
7	4.82 x 10^{-5}
10	2.48 x 10^{-4}
15	1.13 x 10^{-3}
20	3.06
25	5.39
30	8.44
40	1.90 x 10^{-2}
50	2.00
60	2.59
70	3.09
80	3.57
90	4.02 x 10^{-2}
100	4.44
120	5.19*
140	5.87*
160	6.45

CURVE 5

T	C_p
20	3.086 x 10^{-3}
25	5.761
30	8.796
35	1.182 x 10^{-2}
40	1.480
45	1.765
50	2.062
52.5	2.199
55	2.312
60	2.591*
65	2.865
70	3.138
75	3.405
80	3.667*
85	3.911*
90	4.131*
95	4.350*
100	4.558*
105	4.773
110	4.957*
115	5.141
120	5.313*
125	5.480*
130	5.629*
135	5.771*
140	5.914
145	6.027
150	6.134
155	6.230*
160	6.307*
165	6.389
170	6.466*
175	6.538*
180	6.610*
185	6.680
190	6.741*
195	6.810*
200	6.877*

*Not shown on plot

DATA TABLE NO. 20 (continued)

CURVE 6*

T	C_p
20	3.091 x 10^{-3}
25	5.765
30	8.796
35	1.183 x 10^{-2}
40	1.480
45	1.765
50	2.062
52.5	2.199
55	2.319
60	2.597
65	2.865
70	3.144
75	3.412
80	3.680
85	3.917
90	4.149
95	4.363
100	4.582
105	4.784
110	4.975
115	5.154
120	5.331
125	5.498
130	5.634
135	5.778
140	5.914
145	6.027
150	6.134
155	6.230
160	6.307
165	6.389
170	6.466
175	6.538
180	6.610
185	6.680
190	6.741
195	6.811
200	6.876

CURVE 7*

T	C_p
20	3.091 x 10^{-3}
25	5.767
30	8.799
35	1.183 x 10^{-2}
40	1.481
45	1.766
50	2.062
52.5	2.200

CURVE 7 (contd)*

T	C_p
55	2.324 x 10^{-2}
60	2.616
65	2.889
70	3.174
75	3.443
80	3.697
85	3.941
90	4.185
95	4.399
100	4.614
105	4.815
110	5.005
115	5.196
120	5.369
125	5.517
130	5.654
135	5.785
140	5.915
145	6.028
150	6.134
155	6.230
160	6.308
165	6.391
170	6.468
175	6.539
180	6.611
185	6.690
190	6.742
195	6.812
200	6.878

CURVE 8*

T	C_p
10	1.2 x 10^{-3}
20	8.7
30	2.2 x 10^{-2}
40	3.3
50	4.1
60	4.7
70	5.0
75	7.0
80	6.8
90	5.1
100	5.4
110	5.6
140	6.5
170	7.0
180	7.2
200	7.4

CURVE 9*

T	C_p
12	4.64 x 10^{-4}
15	1.14 x 10^{-3}
20	3.00
25	5.77
30	8.81
35	1.180 x 10^{-2}
40	1.479
45	1.769
50	2.050
60	2.595
70	3.120
80	3.629
90	4.112
100	4.550
110	4.937
120	5.282
130	5.587
140	5.877
150	6.114
160	6.321
170	6.501
180	6.667
190	6.824
200	6.971
210	7.107
220	7.219
230	7.317
240	7.401
250	7.472
260	7.530
270	7.584
273.2	7.601

CURVE 10*

T	C_p
298.15	7.69 x 10^{-2}
400	8.06
500	8.20
600	8.31
700	8.40
800	8.53
900	8.72
1000	8.95
1100	9.20
1200	9.45
(s)1210	9.46
(l)1210	9.09
1300	9.09
1400	9.09
1500	9.09

*Not shown on Plot

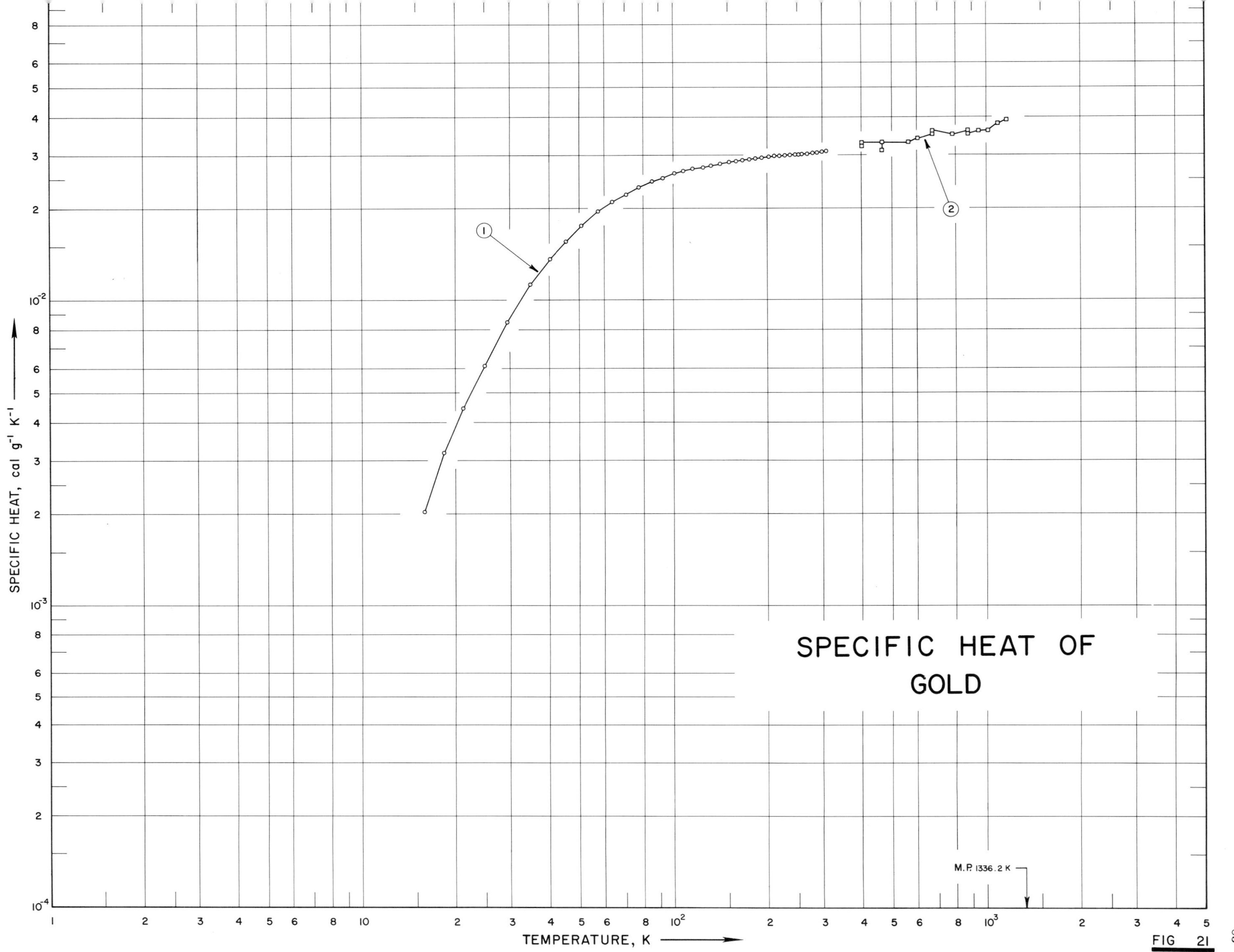
SPECIFIC HEAT OF
GOLD
SPECIFIC HEAT, cal g⁻¹ K⁻¹
TEMPERATURE, K
M.P. 1336.2 K
1
2
FIG 21

SPECIFICATION TABLE NO. 21 SPECIFIC HEAT OF GOLD

(Impurity $<0.20\%$ each; total impurities $<0.50\%$)

[For Data Reported in Figure and Table No. 21]

Curve No.	Ref. No.	Year	Temp. Range, K	Reported Error, %	Name and Specimen Designation	Composition (weight percent), Specifications and Remarks
1	96	1952	16-309			99.99 Au; single crystal.
2	101	1958	400-1164			Pure; specimen's surface plated with platinum black.
3	268	1926	373-1498			
4	207	1928	15-213			
5	290	1932	473-1236			Specimen was in a perfectly pure state.
6	305	1955	1-5	0.5		99.99^{+} Au, with traces of Cu, Fe, Mg, Si, Sn and Ag; sample supplied by J. Bishop and Co.; vacuum melted from gold sponge; annealed under vacuum of 1×10^{-6} mm Hg for 4 hrs at 950 C and cooled in vacuo at the rate of 200 C per hr.
7	306	1963	13-273	<0.5		99.99 Au.
8	184	1966	3-30	±0.5		Semiquantitative spectrographic analysis: <0.00005 Ag, 0.00003 Mg, 0.00002 Si, <0.00002 Cu, and <0.00001 Fe; large crystals; annealed condition.

DATA TABLE NO. 21 SPECIFIC HEAT OF GOLD

CURVE 1

T	C_p
15.81	2.04 x 10^{-3}
18.35	3.19
21.16	4.43
24.70	6.14
29.19	8.48
34.76	1.129 x 10^{-2}
40.12	1.363
45.17	1.558
50.98	1.760
57.41	1.951
63.85	2.102
70.66	2.234
78.27	2.358
85.67	2.452
93.19	2.527
100.82	2.602
108.65	2.666
116.89	2.708
125.05	2.744
133.54	2.783
142.25	2.817
150.88	2.852
159.58	2.877
167.82	2.896
176.12	2.918
184.40	2.934
192.70	2.944
202.79	2.966
205.46	2.974
211.14	2.980
211.30	2.966*
219.92	2.982
220.50	2.986*
228.60	2.992
230.01	2.995*
237.38	3.000
239.82	3.006*
246.12	3.016
249.95	3.021*
253.81	3.027
259.68	3.032
263.63	3.039*
269.96	3.045
272.47	3.045*
280.20	3.054
288.41	3.065
298.46	3.088
309.02	3.091

CURVE 2

T	C_p
400	3.20 x 10^{-2}
400	3.20*
400	3.30
465	3.30
465	3.10
465	3.30*
565	3.30
565	3.30*
604	3.40
604	3.40*
676	3.50
676	3.60
781	3.50
875	3.60
875	3.50
946	3.60
1017	3.60
1090	3.80
1164	3.90

CURVE 3*

T	C_p
373	3.38 x 10^{-2}
473	3.42
573	3.41
673	3.46
773	3.60
873	3.68
973	3.82
1073	3.95
1173	4.10
1273	4.27
1373	3.49
1473	3.49
1498	3.49

CURVE 4*

T	C_p
14.96	1.812 x 10^{-3}
15.29	2.016
15.43	2.021
15.73	2.188
17.65	2.838
18.02	2.965
20.33	3.970
21.70	4.579
24.90	6.356
32.50	1.017 x 10^{-2}

CURVE 4 (cont.)*

T	C_p
35.70	1.178 x 10^{-2}
44.00	1.544
51.10	1.817
63.60	2.115
72.70	2.262
82.50	2.409
94.50	2.528
105.00	2.631
147.50	2.879
176.50	2.960
212.50	3.004

CURVE 5*

T	C_p
473	3.175 x 10^{-2}
673	3.264
873	3.390
1073	3.554
1173	3.755
1236	3.826

CURVE 6*

Series 1

T	C_p
1.188	1.954 x 10^{-6}
1.248	2.160
1.317	2.378
1.418	2.767
1.575	3.446
1.816	4.720
1.984	5.764
2.185	7.548
2.370	8.713
2.597	1.173 x 10^{-5}
2.772	1.385
2.971	1.635
3.179	1.965
3.381	2.313
3.581	2.721
3.779	3.193
3.982	3.775
4.169	4.240
4.371	4.627
4.563	5.154
4.759	5.763
4.926	6.344

CURVE 6 (cont.)*

Series 2

T	C_p
1.203	2.087 x 10^{-6}
1.257	2.184
1.292	2.257
1.189	1.772
1.563	3.410
1.695	4.077
1.883	5.230
2.097	6.686
2.295	8.640
2.489	1.041 x 10^{-5}
2.673	1.255
2.880	1.527
3.084	1.801
3.289	2.124
3.469	2.511
3.694	2.999
3.893	3.530
4.098	4.049
4.282	4.417
4.461	4.819
4.670	5.377
4.843	6.099

CURVE 7*

T	C_p
12.71	1.122 x 10^{-3}
12.84	1.153
13.00	1.176
13.19	1.255
14.06	1.533
14.08	1.526
14.30	1.594
15.65	2.008
16.04	2.160
16.41	2.297
16.52	2.338
17.32	2.623
18.00	2.962
18.49	3.112
18.65	3.302
19.42	3.515
20.85	4.240
24.49	6.030
31.20	9.584
35.44	1.154 x 10^{-2}
39.99	1.364

CURVE 7 (cont.)*

T	C_p
45.28	1.576 x 10^{-2}
50.52	1.756
56.31	1.922
57.45	1.953
61.57	2.050
61.62	2.056
66.90	2.159
66.90	2.167
72.32	2.261
78.26	2.354
83.77	2.436
89.26	2.508
94.72	2.574
100.91	2.616
106.54	2.661
112.16	2.704
117.78	2.706
124.42	2.771
130.28	2.790
136.11	2.819
141.90	2.843
149.49	2.874
155.68	2.883
161.82	2.901
167.89	2.914
176.08	2.922
182.48	2.941
188.84	2.942
195.06	2.955
201.75	2.961
202.52	2.960
208.03	2.979
208.82	2.976
214.50	2.984
216.78	2.988
220.47	2.996
222.84	3.001
229.52	3.009
231.86	3.012
236.11	3.022
238.47	3.021
247.44	3.035
244.80	3.030
248.54	3.030
250.81	3.035
258.54	3.049
260.28	3.053
265.37	3.045

CURVE 7 (cont.)*

T	C_p
266.86	3.050 x 10^{-2}
271.75	3.065
272.92	3.059

CURVE 8

Series 1

T	C_p
3.12	1.890 x 10^{-5}
3.32	2.257
3.50	2.595
3.67	2.948
3.86	3.374
4.08	3.930
4.30	4.528

Series 2

T	C_p
3.28	2.173 x 10^{-5}
3.49	2.560
3.66	2.917
3.84	3.319
4.04	3.833
4.30	4.472
4.57	5.360
4.82	6.231
5.11	7.354
5.42	8.685
5.73	1.014 x 10^{-4}
6.03	1.75
6.30	1.337
6.56	1.488
6.87	1.715
7.18	1.951
7.48	2.196
7.83	2.516
8.23	2.918
8.58	3.294
8.90	3.675
9.25	4.122
9.66	4.691
10.07	5.364
10.44	6.001
10.82	6.704
11.31	7.653
11.80	8.691
12.23	9.696
12.67	1.085 x 10^{-3}

*Not shown on plot

DATA TABLE NO. 21 (continued)

T	C_p
CURVE 8 (cont.)*	
Series 2 (cont.)	
13.11	1.198×10^{-3}
13.66	1.356
14.23	1.535
14.68	1.679
15.10	1.822
15.49	1.955
15.84	2.088
16.21	2.224
16.57	2.365
16.91	2.499
17.24	2.641
17.55	2.769
17.88	2.914
18.21	3.047
18.59	3.228
19.00	3.415
19.47	3.634
19.96	3.845
20.40	4.045
20.81	4.244
21.22	4.437
21.22	4.437
21.62	4.627
22.01	4.841
22.47	5.087
23.01	5.344
23.57	5.631
24.15	5.935
24.74	6.232
25.35	6.544
25.95	6.844
26.53	7.166
27.19	7.533
27.87	7.872
28.54	8.202
29.19	8.538
29.76	8.798
Series 3	
15.10	1.813×10^{-3}
15.56	1.985
16.13	2.196
16.66	2.403
17.14	2.599×10^{-3}
17.58	2.782
18.00	2.969
18.41	3.144
18.83	3.327
19.23	3.511
19.71	3.740
20.27	3.994
20.84	4.257
21.44	4.552
22.02	4.842
22.58	5.133
23.16	5.414
23.74	5.717
24.33	6.021
24.98	6.347
25.68	6.717
26.36	7.073
27.06	7.450
27.76	7.822
28.46	8.159
29.16	8.511
29.87	8.840
Series 4	
2.92	1.573×10^{-5}
3.11	1.891
3.27	2.164
3.44	2.471
3.61	2.813
3.82	3.279
3.98	3.656
4.14	4.029
4.31	4.550
4.49	5.096
4.65	5.610
2.82	1.462
3.14	1.940
3.37	2.330
3.56	2.712
3.75	3.108
3.96	3.611
4.18	4.194
4.43	4.915
4.72	5.851
4.99	6.858×10^{-5}
5.24	7.872
5.48	8.960
5.72	1.009×10^{-4}
5.95	1.126
6.20	1.264
6.46	1.434
6.70	1.592
6.96	1.774
7.31	2.046
7.74	2.420
8.16	2.832
8.51	3.217
8.86	3.617
9.26	4.119
9.68	4.706
10.07	5.335
10.49	6.057
10.89	6.783
11.38	7.760
11.93	8.989
12.44	1.020
12.94	1.152
13.49	1.308
14.07	1.479
14.59	1.649
15.07	1.811
15.49	1.958

*Not shown on plot

FIGURE SHOWS ONLY 1 OF THE CURVES REPORTED IN TABLE

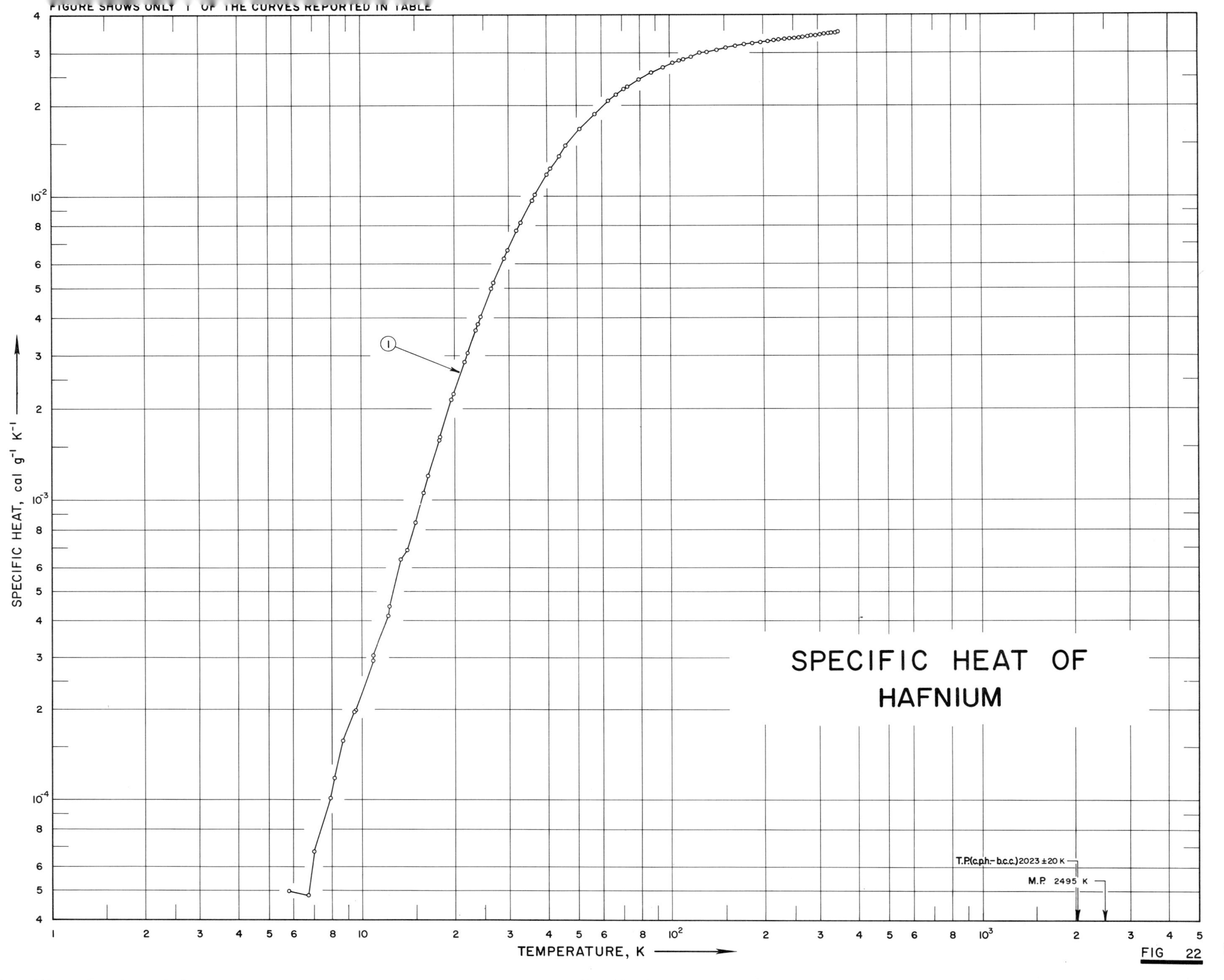

FIG 22

SPECIFICATION TABLE NO. 22 SPECIFIC HEAT OF HAFNIUM

(Impurity $< 0.20\%$ each; total impurities $< 0.50\%$)

[For Data Reported in Figure and Table No. 22]

Curve No.	Ref. No.	Year	Temp. Range, K	Reported Error, %	Name and Specimen Designation	Composition (weight percent), Specifications and Remarks
1	97	1964	6-348			99.95 Hf, 0.05 Zr, 0.02 Fe, 0.015 C, 0.0065 Si, 0.0043 Mo, 0.0018 O_2, 0.0007 Ni, <0 0005 N_2, 0.0003 Cu, 0.0001 W, and <0.0001 H_2.
2	307	1963	298-1346	0.4		2.8 Zr, 0.020 Fe, 0.010 Ni, and 0.008 O_2; corrected for impurities.
3	308	1957	1-20			99 9 Hf, and 0.01 other impurities; prepared by Van Arkel iodide process.
4	302	1934	13-210			

DATA TABLE NO. 22 SPECIFIC HEAT OF HAFNIUM

[Temperature, T, K; Specific Heat, C_p, Cal $g^{-1} K^{-1}$]

CURVE 1

T	C_p
5.82	4.99×10^{-5}
5.85	4.99*
6.74	4.82
7.00	6.78
7.92	1.014×10^{-4}
8.17	1.182
8.70	1.586
9.45	1.961
9.56	1.972
10.84	2.936
10.84	3.059
12.17	4.151
12.35	4.476
13.47	6.432
14.01	6.936
14.89	8.477
15.85	1.062×10^{-3}
16.46	1.204
17.84	1.583
17.95	1.615
19.65	2.150
19.89	2.246
21.56	2.867
22.06	3.064
23.45	3.646
23.88	3.829
24.36	4.032
26.43	4.995
26.80	5.218
29.05	6.278
29.86	6.688
31.92	7.738
32.83	8.222
35.77	9.720
36.54	1.012×10^{-2}
39.98	1.181
41.18	1.240
43.94	1.364
46.23	1.471
51.38	1.683
57.07	1.889
63.65	2.094
67.00	2.182
71.27	2.282
73.17	2.322

CURVE 1 (cont.)

T	C_p
79.59	2.456×10^{-2}
79.93	2.462*
87.25	2.594
87.56	2.600*
95.14	2.695
103.26	2.784
108.62	2.846
111.40	2.867
111.41	2.863*
117.22	2.916
125.15	3.002
133.80	3.032
142.71	3.079
152.94	3.127
164.49	3.172
175.81	3.208
186.93	3.237
197.89	3.264
200.73	3.270*
207.97	3.285
211.62	3.294*
217.16	3.307
226.25	3.325
235.23	3.342
244.12	3.360
252.94	3.363
261.68	3.379
269.30	3.394
278.03	3.414
286.69	3.425
295.93	3.438
305.74	3.452
315.48	3.460
325.15	3.485
333.76	3.493
341.34	3.499
348.55	3.508

CURVE 2

T	C_p
298.15	3.445×10^{-2}
300	3.445
400	3.552
500	3.653
600	3.754

CURVE 2 (cont.)*

T	C_p
700	3.854×10^{-2}
800	3.955
900	4.062
1000	4.162
1100	4.263
1200	4.364
1300	4.471
1350	4.521

CURVE 3*

T	C_p
1.25	4.555×10^{-6}
1.29	4.891
1.75	6.891
1.77	7.227
1.83	7.339
2.15	9.300
2.20	9.300
2.25	9.973
2.54	1.126×10^{-5}
2.56	1.171
2.90	1.434
2.93	1.412
3.25	1.664
3.28	1.692
3.48	1.798
3.58	1.938
3.63	1.961
3.90	2.247
3.92	2.247
4.62	3.059
4.96	3.642
5.30	4.292
5.65	4.964
5.95	5.575
6.22	6.465
6.50	7.171
6.80	8.012
7.12	8.796
7.40	1.003×10^{-4}
7.68	1.104
7.97	1.233
8.26	1.361
8.56	1.490
8.85	1.647

CURVE 3 (cont.)*

T	C_p
9.16	1.843×10^{-4}
9.47	2.006
9.75	2.252
10.04	2.465
10.35	2.695
10.70	2.997
11.06	3.345
11.45	3.647
11.83	4.140
12.28	4.622
12.72	5.289
13.25	5.978
13.75	6.835
14.30	7.844
14.91	9.132
15.57	1.053×10^{-3}
16.22	1.216
16.95	1.389
17.70	1.630
18.48	1.871
19.25	2.157
20.15	2.471

CURVE 4*

T	C_p
13	2.0×10^{-4}
20	7.3
30	3.4×10^{-3}
40	9.0
50	2.2
60	4.0
70	5.6
75	6.2
80	5.0
90	3.0
100	3.0
110	3.1
150	3.2
200	3.4
210	3.4

*Not shown on Plot

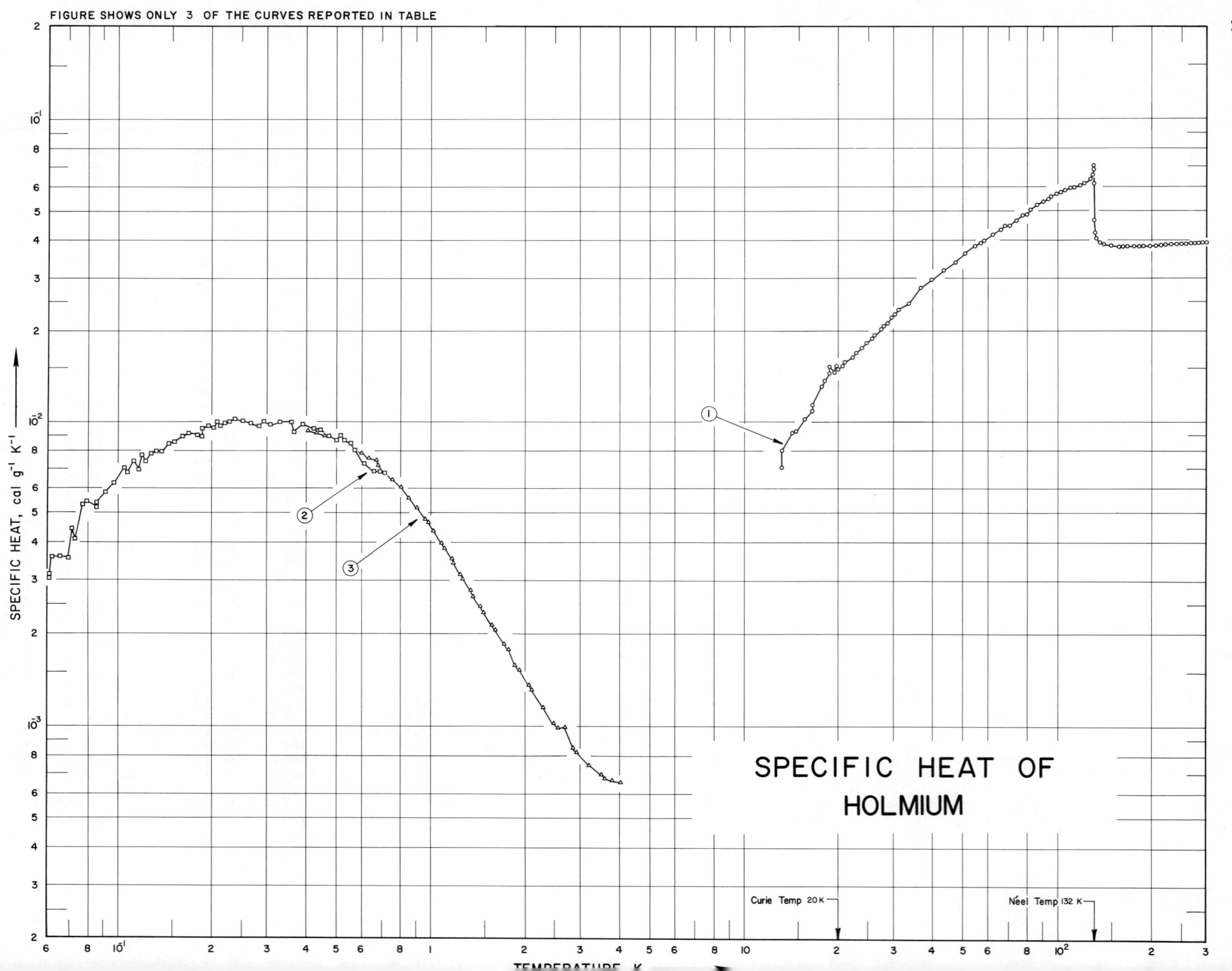
FIGURE SHOWS ONLY 3 OF THE CURVES REPORTED IN TABLE
SPECIFIC HEAT OF HOLMIUM
SPECIFIC HEAT, cal g⁻¹ K⁻¹
Curie Temp 20 K
Néel Temp 132 K
1
2
3

SPECIFICATION TABLE NO. 23 SPECIFIC HEAT OF HOLMIUM

(Impurity $< 0.20\%$ each; total impurities $< 0.50\%$)

[For Data Reported in Figure and Table No. 23]

Curve No.	Ref. No.	Year	Temp. Range, K	Reported Error, %	Name and Specimen Designation	Composition (weight percent), Specifications and Remarks
1	219	1957	15-300	0.3-2		0.2 Si, 0.15 Ca, <0.2 Fe, <0.1 Ta, <0.05 Dy, <0.02 Y, <0.01 Mg, Fr, and Tm; sample prepared by ion exchange separation and reduction of anhydrous fluoride with Ca; sublimed at 7500 C and 10^{-5} mm Hg vacuum.
2	220	1964	0.1-0.7			Hexagonal closed packed.
3	258	1962	0.3-4.0	0.2-3		0.21 O_2, 0.07 Na, 0.07 C, and 0.005 H_2; sample supplied by Research Chemicals Inc.
4	301	1966	298-1800	<2		0.03 Ca, 0.012 N_2, 0.005 Al, Cr, Fe, and Mg, and 0.002 O_2; prepared by metallothermic reduction of the fluoride with calcium and purified by distillation.
5	380	1966	3-24	0.6-2.0		0.21 O_2, 0.007 C, 0.007 N_2, and 0.005 H_2.

DATA TABLE NO. 23 SPECIFIC HEAT OF HOLMIUM

[Temperature, T, K; Specific Heat, C_p, Cal g^{-1} K^{-1}]

T	C_p
CURVE 1	
Series 1	
14.43	9.119 x 10^{-3}
15.66	1.017 x 10^{-2}
17.73	1.303
19.93	1.533
22.25	1.624
24.80	1.824
27.50	2.029
30.41	2.262
33.57	2.466
36.71	2.784
39.92	2.956
43.51	3.180
47.11	3.395
50.92	3.605
54.95	3.812
58.71	3.993
62.25	4.163
66.03	4.330
70.03	4.497
74.03	4.663
78.04	4.844
82.24	5.035
86.34	5.219
90.39	5.381
93.35	5.473
95.28	5.527
97.41	5.598*
99.98	5.693
103.22	5.761
106.82	5.836
110.59	5.907
114.48	5.974
118.55	6.040
122.96	6.142
127.48	6.367
131.37	6.174
134.62	4.067
137.82	3.939
141.04	3.880
144.81	3.874*
149.11	3.848
CURVE 1 (cont.)	
Series 1 (cont.)	
153.39	3.836 x 10^{-2}*
157.65	3.831
162.94	3.830
168.21	3.830
172.41	3.822*
176.58	3.825
180.74	3.828*
184.88	3.829
189.00	3.832
193.36	3.835*
197.96	3.840
202.54	3.843*
206.85	3.846
210.89	3.851*
215.16	3.854
219.67	3.858*
222.70	3.863
227.19	3.865*
231.64	3.871
236.08	3.872*
240.51	3.879
244.92	3.886*
249.32	3.898
253.71	3.897*
258.08	3.892
262.45	3.898*
266.79	3.900
271.12	3.908*
275.42	3.910
279.72	3.913*
284.00	3.923
288.27	3.933*
292.52	3.932
296.76	3.927*
300.97	3.935
Series 2	
13.26	7.058 x 10^{-3}
14.78	9.277
16.53	1.148 x 10^{-2}
CURVE 1 (cont.)	
Series 2 (cont.)	
17.53	1.285 x 10^{-2}*
18.13	1.367
18.81	1.512
19.57	1.586*
20.31	1.497*
20.87	1.572*
21.42	1.579
21.94	1.623*
22.45	1.669*
22.95	1.699
23.44	1.732*
23.92	1.759
24.39	1.796*
24.85	1.815*
25.31	1.865*
25.75	1.892
26.18	1.941
26.66	1.957*
27.18	2.006*
27.69	2.046*
28.18	2.075
28.66	2.130
29.68	2.205
31.34	2.364
Series 3	
18.85	1.443 x 10^{-2}
19.48	1.459
20.11	1.487
20.72	1.533
21.32	1.567*
21.89	1.615*
22.50	1.652*
23.82	1.755*
25.82	1.899*
28.00	2.067*
30.63	2.277*
33.50	2.502*
36.85	2.737*
40.39	2.990*
CURVE 1 (cont.)	
Series 3 (cont.)	
43.90	3.203 x 10^{-2}*
47.43	3.412*
50.71	3.593*
54.17	3.771*
58.31	3.972*
65.35	4.303*
69.34	4.469*
73.33	4.638*
77.32	4.809*
80.32	4.935
84.02	5.119*
87.92	5.286*
91.84	5.430*
97.64	5.607*
101.87	5.742*
Series 4	
105.80	5.621 x 10^{-2}*
110.23	5.769*
114.68	5.975*
118.89	6.069*
121.85	6.117*
123.81	6.147*
125.75	6.205*
126.95	6.304*
127.46	6.302*
127.98	6.310*
128.50	6.434*
129.04	6.424*
129.56	6.468*
130.02	6.540
130.49	6.649*
130.94	6.825
131.39	7.037
131.84	6.331*
132.42	4.691
133.13	4.246
133.87	4.126*
134.85	4.060*
136.09	3.996*
CURVE 1 (cont.)	
Series 4 (cont.)	
137.46	3.955 x 10^{-2}*
139.51	3.920*
142.38	3.874*
145.51	3.862*
149.28	3.867*
154.12	3.834*
159.16	3.832*
163.73	3.831*
168.07	3.833*
172.39	3.825*
176.68	3.828*
180.96	3.831*
185.21	3.834*
189.45	3.837*
193.67	3.833*
197.88	3.840*
202.06	3.840*
206.23	3.843*
210.59	3.851*
215.15	3.853*
219.69	3.850*
224.20	3.867*
228.69	3.867*
232.82	3.877*
237.28	3.883*
241.72	3.880*
246.16	3.885*
250.98	3.899*
255.78	3.889*
260.17	3.892*
264.55	3.897*
268.91	3.905*
273.25	3.911*
277.75	3.950*
282.44	3.909*
287.11	3.919*
291.77	3.930*
296.42	3.934*
301.03	3.949*
CURVE 1 (cont.)	
Series 5	
13.35	8.034 x 10^{-3}
14.55	9.246*
15.61	1.025 x 10^{-2}*
16.69	1.093
17.64	1.282*
18.40	1.424*
19.10	1.561*
19.71	1.498*
20.26	1.499*
20.92	1.550*
Series 6*	
18.91	1.438 x 10^{-2}
19.47	1.451
20.10	1.454
20.82	1.563
Series 7*	
126.71	6.246 x 10^{-2}
127.25	6.292
127.84	6.324
128.48	6.386
Series 8*	
126.76	6.241 x 10^{-2}
127.23	6.277
127.87	6.307
128.44	6.382
129.01	6.392
129.58	6.475
130.13	6.566
130.69	6.730
131.24	6.921
131.79	6.394
132.42	4.669
133.14	4.252
133.88	4.085
134.63	4.071

*Not shown on plot

DATA TABLE NO. 23 (continued)

CURVE 1 (cont.)

Series 8 (cont.)*

T	C_p
135.64	4.017×10^{-2}
137.05	3.975
138.87	3.925
141.51	3.896
144.67	3.870

Series 9

T	C_p
57.21	3.911×10^{-2}
62.19	4.160*
66.83	4.362*
71.22	4.545*
75.40	4.723*
79.40	4.900*
83.23	5.081*
86.93	5.245*
90.51	5.384*
94.00	5.490*
97.40	5.602*
100.73	5.739*
104.01	5.766*
107.23	5.829*
110.57	5.889*
114.02	5.955*
117.43	6.024*
120.79	6.089*

Series 10

T	C_p
58.90	4.002×10^{-2}*
63.92	4.232*
68.61	4.444
73.05	4.622*
77.61	4.819*
81.65	5.012*
85.90	5.203*
90.37	5.385*
94.68	5.513*

Series 11*

T	C_p
98.68	5.641×10^{-2}
101.35	5.708
103.80	5.767

CURVE 1 (cont.)

Series 11 (cont.)*

T	C_p
106.39	5.815×10^{-2}
109.13	5.866
111.84	5.913
114.68	5.967
117.82	6.027
121.09	6.097
124.31	6.174

Series 12*

T	C_p
94.40	5.505×10^{-2}
97.22	5.598
106.69	5.820

CURVE 2

Series 1

T	C_p
0.120	7.75×10^{-3}
0.133	8.00
0.152	8.64
0.168	9.12
0.180	9.03
0.186	8.91
0.195	9.64
0.209	9.96
0.220	9.88
0.228	9.98
0.238	1.02×10^{-2}
0.252	1.00
0.294	10.00×10^{-3}
0.308	9.74
0.330	9.97
0.360	9.97
0.390	9.74
0.424	9.45
0.448	9.38
0.472	8.93
0.501	8.67
0.534	8.67
0.575	8.03
0.617	7.25
0.670	6.88
0.716	6.75

CURVE 2 (cont.)

Series 2

T	C_p
0.061	3.13×10^{-3}
0.063	3.61
0.071	4.42
0.078	5.33
0.086	5.48
0.091	5.84
0.097	6.29
0.105	7.07
0.114	7.41
0.123	7.41
0.133	8.01*
0.146	8.43
0.162	8.72
0.182	9.14*

Series 3

T	C_p
0.061	3.03×10^{-3}
0.065	3.61
0.070	3.59
0.073	4.13
0.078	5.48
0.086	5.26
0.091	5.83*
0.099	6.30*
0.108	6.80
0.117	7.00
0.128	7.90
0.138	7.94
0.152	8.48*
0.163	8.77*
0.175	8.88*
0.186	9.43
0.193	9.56*
0.202	9.52
0.213	9.62
0.223	9.91*
0.239	9.74*
0.252	1.03×10^{-2}*
0.266	9.87×10^{-3}
0.284	9.64
0.308	9.56*
0.338	9.75*
0.366	9.27

CURVE 2 (cont.)

Series 3 (cont.)

T	C_p
0.399	9.55×10^{-3}*
0.443	9.39*
0.489	8.80*
0.517	8.97
0.558	8.48
0.690	6.83

CURVE 3

Series 1

T	C_p
0.3817	9.5309×10^{-3}*
0.4066	9.3426
0.4345	9.1749
0.4662	8.9991
0.4985	8.7513*
0.5311	8.5166*
0.5654	8.1238*
0.6008	7.8224
0.6375	7.5413
0.6753	7.4326
0.7152	6.7602*
0.7583	6.4008
0.8040	6.0154
0.8529	5.5951
0.9057	5.1821
0.9625	4.7691
1.0246	4.3691
1.0939	3.9214
1.1705	3.5040
1.2558	3.1388
1.3515	2.7867
1.4599	2.4592
1.5809	2.1462
1.7230	1.8578
1.8876	1.5984
2.0796	1.3636
2.3047	1.1579
2.5697	9.8396×10^{-4}
2.8795	8.4484
3.2344	7.4485
3.6295	6.7819
4.0497	6.5645

CURVE 3 (cont.)

Series 2

T	C_p
0.4471	9.0600×10^{-3}*
0.4735	8.9600*
0.5032	8.7571*
0.5361	8.5238*
0.5702	8.2586*
0.6054	7.9384*
0.6422	7.5746*
0.6808	7.1993
0.7218	6.7834*
0.7652	6.3660*
0.8117	5.9299*
0.8617	5.4980*
0.9159	5.0618*
0.9746	4.6633
1.0389	4.2344*
1.1107	3.8068
1.1901	3.4113
1.2788	3.0432
1.3787	2.6881
1.4921	2.3592
1.6217	2.0592
1.7713	1.7767
1.9369	1.5317
2.1117	1.3260
2.2925	1.1680*
2.4878	1.0274
2.7089	9.1005×10^{-4}
2.9559	8.2021
3.2287	7.4630*
3.5246	6.9558
3.8353	6.6660
4.1533	6.6515*

CURVE 4*

T	C_p
298.15	4.14×10^{-2}
300	4.14
400	4.06
500	4.03
600	4.04
700	4.12
800	4.24
900	4.41
1000	4.63

CURVE 4 (cont.)*

T	C_p
1100	4.90×10^{-2}
1200	5.22
1300	5.60
1400	6.02
1500	6.494
1600	7.015
1700	7.591
1701	7.597
1701	4.062
1743	4.062
1743	6.375
1800	6.375

CURVE 5*

Series 1

T	C_p
3.0545	7.908×10^{-4}
3.2704	7.375
3.5281	6.966
3.7979	6.708
4.0739	6.643

Series 2

T	C_p
3.1423	7.657×10^{-4}
3.4022	7.141
3.6878	6.801
3.9587	6.643
4.2846	6.733
4.7093	7.183
5.1655	8.086
5.6606	9.535
6.2162	1.175×10^{-3}
6.7863	1.461
7.3642	1.809
7.9833	2.240
8.6381	2.754
9.3994	3.406
10.3020	4.251
11.3500	5.327
12.5400	6.637
13.8350	8.135
15.2450	9.847
16.7720	1.153×10^{-2}
18.5120	1.249

*Not shown on plot

DATA TABLE NO. 23 (continued)

T	C_p
CURVE 5 (cont.)*	
Series 2 (cont.)	
20.5100	1.438×10^{-2}
22.4830	1.599
24.2620	1.747
Series 3	
4.4014	6.836×10^{-4}
4.8322	7.428
5.3828	8.732
5.9430	1.066×10^{-3}
6.5040	1.320
7.0924	1.646
7.6923	2.039
8.3191	2.507
9.0147	3.076
9.8307	3.823
10.7620	4.733
11.8550	5.882
13.1020	7.296
14.4470	8.889
15.8370	1.056×10^{-2}
17.2760	1.185
19.0010	1.294
20.9680	1.482
22.8740	1.635
24.6060	1.784
Series 4	
11.772	5.805×10^{-3}
12.854	7.017
14.055	8.435
15.403	1.008
16.894	1.176
18.536	1.255
Series 5	
14.017	8.385×10^{-3}
14.661	9.140
15.253	9.898
15.804	1.055×10^{-2}
16.320	1.116
16.810	1.164
17.283	1.194
CURVE 5 (cont.)*	
17.749	1.195×10^{-2}
18.211	1.215
18.662	1.254
19.095	1.299
19.514	1.343
Series 6	
16.542	1.134×10^{-2}
16.855	1.183
17.158	1.208
17.457	1.217
17.756	1.201

*Not shown on plot

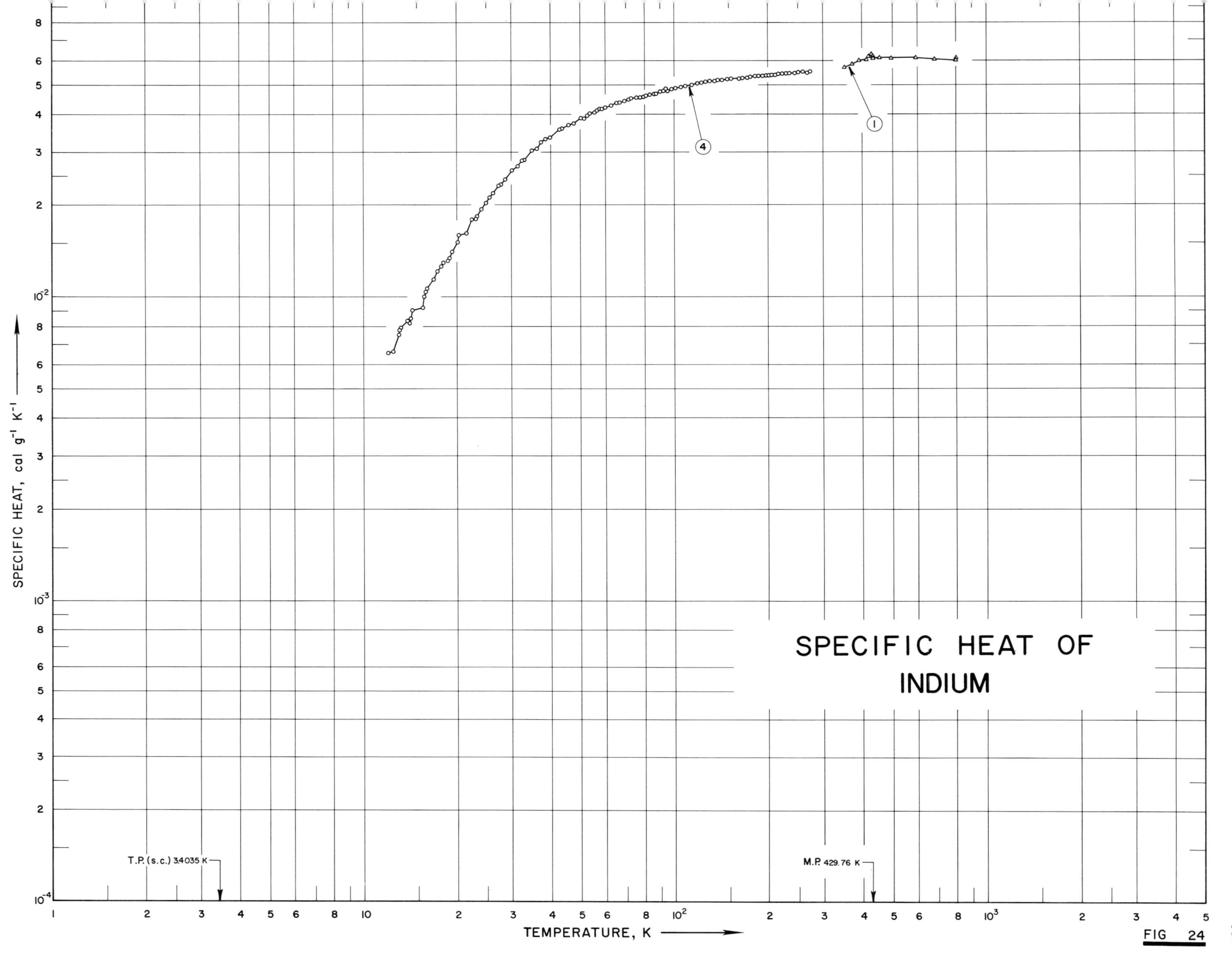

SPECIFIC HEAT OF
INDIUM
FIG 24
SPECIFIC HEAT, cal g⁻¹ K⁻¹
TEMPERATURE, K
T.P. (s.c.) 3.4035 K
M.P. 429.76 K
1
4

SPECIFICATION TABLE NO. 24 SPECIFIC HEAT OF INDIUM

(Impurity < 0.20% each; total impurities < 0.50%)

[For Data Reported in Figure and Table No. 24]

Curve No.	Ref. No.	Year	Temp. Range, K	Reported Error, %	Name and Specimen Designation	Composition (weight percent), Specifications and Remarks
1	77	1961	353-800			99.999 In.
2	85	1965	0.1-4.1	1.0		99.999 In; sample supplied by the American Smelting and Refining Co.; H = O magnetic field; vacuum cast; single crystal.
3	85	1965	0.08-4.1	1.0		99.999 In; sample supplied by the American Smelting and Refining Co.; H = 1000 Oe magnetic field; vacuum cast; single crystal.
4	155	1952	12-273			99.8 In; heated slowly under vacuum to 185 C; cooled slowly to room temperature.

DATA TABLE NO. 24 SPECIFIC HEAT OF INDIUM

[Temperature, T, K; Specific Heat, C_p, Cal $g^{-1}K^{-1}$]

CURVE 1

	T	C_p
	353.71	5.719 x 10^{-2}
	353.71	5.716*
	373.95	5.837
	391.75	6.000
	415.81	6.044
	415.81	6.082*
	415.89	6.154*
	421.28	6.203
	421.41	6.189*
	421.64	6.203*
	423.42	6.264*
	423.48	6.239*
	425.43	6.279*
	425.55	6.235*
	427.77	6.272*
(s)	428.90	6.351*
(l)	429.76	6.384*
	429.89	6.353
	430.07	6.325*
	430.73	6.260*
	430.85	6.279*
	431.03	6.217*
	434.41	6.105
	434.43	6.186
	438.30	6.159*
	439.88	6.104*
	439.97	6.146*
	440.79	6.144*
	440.90	6.167*
	459.07	6.151
	459.11	6.126*
	459.25	6.128*
	499.28	6.135
	499.48	6.076*
	598.01	6.113
	598.15	6.121*
	682.21	6.087
	682.25	6.096*
	800.64	6.036
	800.76	6.108

CURVE 2*

T	C_p
0.105_1	5.37 x 10^{-9}
0.106_4	6.45
0.113_1	6.20
0.114_9	6.95
0.116_8	7.62
0.117_5	7.14
0.132_9	9.41
0.140_7	1.049 x 10^{-8}
0.141_2	1.057
0.142_2	1.097
0.151_5	1.228
0.152_4	1.240
0.160_3	1.403
0.160_8	1.386
0.163_9	1.492
0.164_3	1.476
0.167_9	1.574
0.168_9	1.628
0.184_1	1.942
0.192_8	2.210
0.196_5	2.350
0.211_3	2.848
0.213_0	2.964
0.232_9	3.722
0.238_5	3.957
0.250_6	4.525
0.258_1	4.942
0.261_0	5.060
0.270_8	5.681
0.280_8	6.228
0.283_4	6.415
0.287_6	6.711
0.306_6	8.035
0.317_8	8.922
0.322_9	9.488
0.324_5	9.496
0.340_2	1.097 x 10^{-7}
0.349_4	1.180
0.363_5	1.331
0.368_6	1.359

CURVE 2 (cont.)*

T	C_p
0.379_2	1.496 x 10^{-7}
0.379_6	1.507
0.400_6	1.740
0.417_5	1.962
0.438_8	2.277
0.454_3	2.569
0.476_6	2.997
0.493_2	3.341
0.509_1	3.612
0.519_1	3.855
0.522_3	3.965
0.560_4	4.867
0.563_3	5.010
0.618_3	6.744
0.664_7	8.705
0.682_1	9.502
0.761_8	1.396 x 10^{-6}
0.845_0	1.989
0.891_5	2.404
0.950_1	3.025
0.975_0	3.247
1.022_1	3.826
1.206	6.705
1.222	7.117
1.274	8.041
1.381	1.058 x 10^{-5}
1.444	1.229
1.503	1.358
1.572	1.613
1.630	1.789
1.767	2.296
1.866	2.725
1.919	2.937
2.044	3.582
2.092	3.795
2.230	4.636
2.297	5.010
2.407	5.805
2.506	6.478
2.591	7.096

CURVE 2 (cont.)*

T	C_p
2.715	8.218 x 10^{-5}
2.811	9.176
2.945	1.047 x 10^{-4}
3.049	1.168
3.191	1.334
3.296	1.465
3.565	1.635
3.710	1.834
3.862	2.088
3.985	2.277
4.158	2.612

CURVE 3*

T	C_p
0.085_{60}	5.749 x 10^{-7}
0.086_{59}	5.714
0.087_{74}	5.731
0.088_{97}	5.747
0.092_{31}	5.735
0.095_{32}	5.837
0.095_{75}	5.743
0.098_{84}	5.735
0.099_{95}	5.687
0.105_5	5.847
0.105_5	5.691
0.116_6	5.712
0.117_0	5.770
0.121_4	5.776
0.128_8	5.833
0.129_6	5.964
0.134_4	6.103
0.142_2	6.216
0.142_5	6.207
0.157_0	6.621
0.165_8	6.813
0.170_5	6.927
0.172_5	6.988
0.182_4	7.294
0.190_4	7.550
0.190_6	7.567

CURVE 3 (cont.)*

T	C_p
0.200_8	7.897 x 10^{-7}
0.211_8	8.268
0.221_0	8.620
0.236_2	9.149
0.243_9	9.413
0.263_0	1.016 x 10^{-6}
0.270_3	1.046
0.290_8	1.123
0.299_0	1.161
0.300_3	1.165
0.320_7	1.252
0.329_8	1.290
0.330_4	1.293
0.353_5	1.394
0.362_0	1.433
0.365_2	1.447
0.399_7	1.609
0.401_1	1.616
0.435_7	1.786
0.439_6	1.808
0.472_6	1.984
0.477_3	2.010
0.523_9	2.273
0.576_3	2.598
0.656_3	3.162
0.721_6	3.672
0.795_5	4.315
0.872_1	5.052
1.216	9.691
1.244	1.020 x 10^{-5}
1.297	1.120
1.351	1.224
1.409	1.353
1.474	1.500
1.533	1.647
1.608	1.854
1.661	2.012
1.746	2.265
1.800	2.448
1.894	2.789

CURVE 3 (cont.)*

T	C_p
1.952	3.018 x 10^{-5}
2.056	3.443
2.120	3.728
2.225	4.248
2.303	4.648
2.406	5.235
2.496	5.797
2.599	6.459
2.842	8.335
2.935	9.167
3.079	1.050 x 10^{-4}
3.189	1.168
3.289	1.273
3.449	1.473
3.536	1.583
3.720	1.824
3.818	2.005
4.009	2.311
4.145	2.579

CURVE 4

T	C_p
12.09	6.578 x 10^{-3}
12.51	6.623
13.00	7.506
13.04	7.796
13.17	7.796
13.94	8.368
14.16	8.210
14.25	8.511
14.47	9.066
15.62	9.232
15.64	9.990
15.92	1.049 x 10^{-2}
16.12	1.060
16.22	1.052*
16.82	1.149
17.33	1.206
17.81	1.264
18.16	1.245

*Not shown on plot

DATA TABLE NO. 24 (continued)

T	C_p	T	C_p	T	C_p
CURVE 4 (contd)		CURVE 4 (contd)		CURVE 4 (contd)	
18.76	1.265×10^{-2}	69.23	4.426×10^{-2}	166.98	5.251×10^{-2}
18.89	1.349	69.55	4.484*	172.02	5.293
19.27	1.414	71.39	4.480	176.83	5.319
20.11	1.515	72.30	4.510	182.13	5.326
20.24	1.518*	75.61	4.527	187.69	5.320
20.26	1.596	77.90	4.568	192.44	5.327
21.51	1.661	79.34	4.592	197.91	5.354
22.44	1.791	80.48	4.612	201.47	5.376
23.16	1.799	80.95	4.620*	202.96	5.362*
23.28	1.835	81.10	4.616	204.31	5.407*
24.00	1.948	83.40	4.652	207.17	5.388
24.87	2.041	84.26	4.656*	212.14	5.395
25.59	2.126	84.26	4.672*	217.94	5.422
25.69	2.128	86.10	4.672	223.94	5.435
26.32	2.199	87.69	4.684	229.75	5.432
27.37	2.323	87.74	4.710*	235.79	5.443
27.83	2.345	88.26	4.693*	241.68	5.464
28.70	2.433	90.83	4.750	250.28	5.504
30.00	2.606	91.12	4.751*	254.32	5.513*
31.40	2.693	92.21	4.791	258.30	5.509
32.40	2.804	93.11	4.775*	262.08	5.501*
32.86	2.827	94.52	4.850	267.44	5.462
34.91	3.040	95.28	4.770	269.21	5.537*
36.11	3.078	96.69	4.814*	272.72	5.513
37.31	3.243	98.25	4.925		
39.58	3.316	101.71	4.856		
39.97	3.350	102.42	4.856*		
42.92	3.558	106.50	4.918		
43.62	3.563	108.96	4.927		
45.90	3.690	114.16	4.994		
47.67	3.730	114.75	4.998*		
50.20	3.897	118.70	5.061		
51.57	3.878	119.80	5.035*		
52.96	3.955	122.79	5.061		
53.56	4.008	124.88	5.061*		
55.38	4.065	126.52	5.103		
56.32	4.102	129.97	5.110		
58.42	4.172	130.51	5.114*		
58.52	4.144	135.61	5.132		
60.39	4.219	137.41	5.179		
60.69	4.250*	143.88	5.185		
62.84	4.278	148.67	5.218		
65.14	4.356	153.32	5.240		
66.68	4.355	162.43	5.239		
66.98	4.365				

*Not shown on plot

FIGURE SHOWS ONLY 1 OF THE CURVES REPORTED IN TABLE

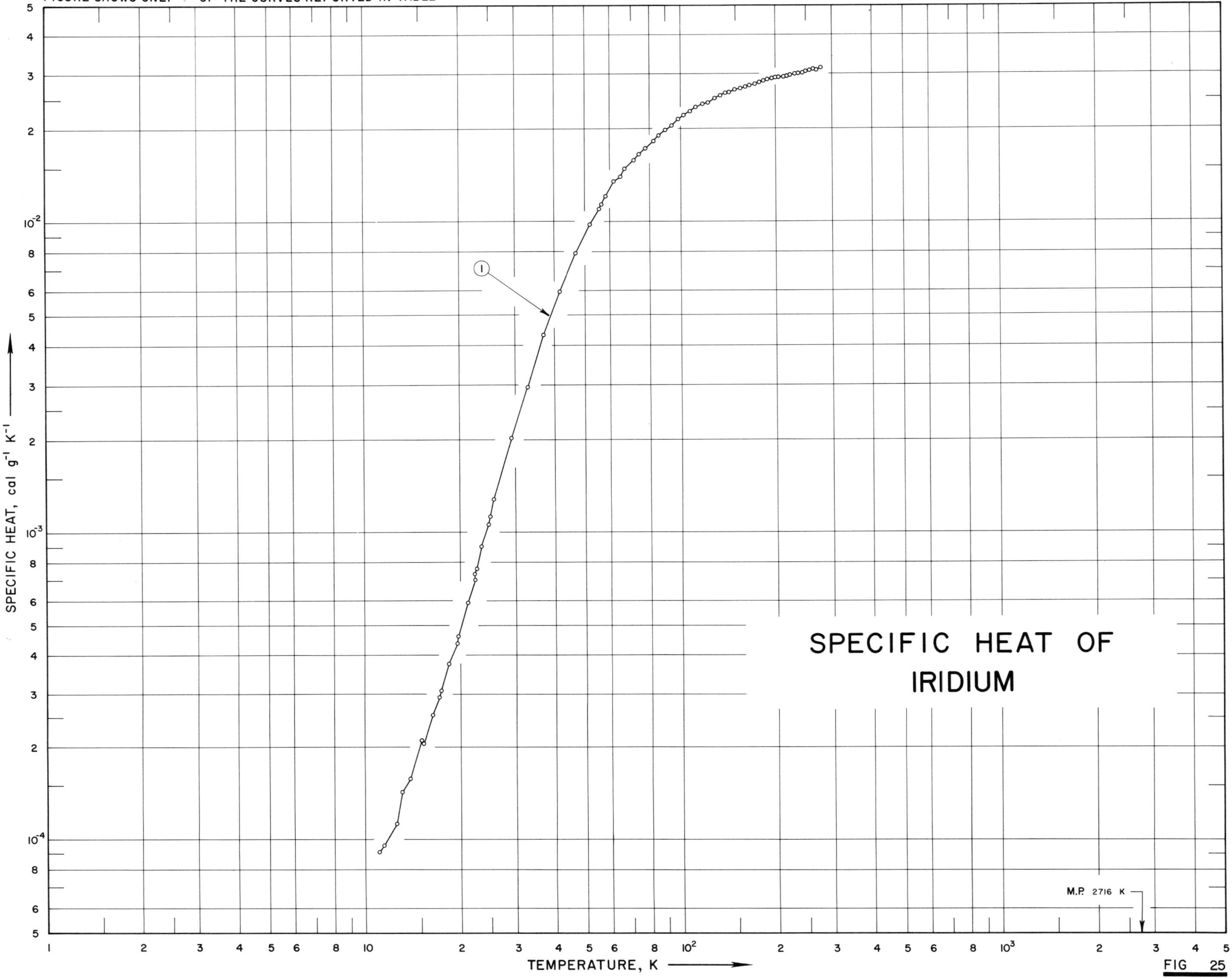

SPECIFIC HEAT OF IRIDIUM

FIG 25

SPECIFICATION TABLE NO. 25 SPECIFIC HEAT OF IRIDIUM

(Impurity $< 0.20\%$ each; total impurities $< 0.50\%$)

[For Data Reported in Figure and Table No. 25]

Curve No.	Ref. No.	Year	Temp. Range, K	Reported Error, %	Name and Specimen Designation	Composition (weight percent), Specifications and Remarks
1	107	1955	11-276			99.96 Ir, 0.0 X Pt type metals, traces Ag, Cu, Fe; cast.
2	215	1931	273-1973			Sample in purest form.
3	309	1933	406-781			

DATA TABLE NO. 25 SPECIFIC HEAT OF IRIDIUM

[Temperature, T, K; Specific Heat, C_p, Cal g^{-1} K^{-1}]

CURVE 1

T	C_p
11.13	9.10×10^{-5}
11.45	9.52
12.58	1.13×10^{-4}
13.06	1.43
13.86	1.58
15.04	2.10
15.24	2.06
16.37	2.56
17.10	2.92
17.38	3.08
18.46	3.76
19.58	4.39
19.67	4.66
21.04	5.91
22.22	7.31
22.25	7.01
22.57	7.66
23.42	9.00
24.55	1.068×10^{-3}
24.92	1.132
25.69	1.286
29.08	2.020
32.82	2.935
36.97	4.331
41.57	5.968
46.80	7.929
51.99	9.813
55.49	1.100×10^{-2}
56.81	1.149
58.75	1.215
61.77	1.353
61.94	1.303*
64.90	1.406
66.99	1.494
67.92	1.491*
71.16	1.580
74.38	1.666
77.54	1.741
82.15	1.822
85.91	1.904
89.98	1.988
94.01	2.050
98.74	2.153
103.55	2.218

CURVE 1 (cont.)

T	C_p
108.42	2.272×10^{-2}
113.33	2.353
118.19	2.404
123.04	2.443
129.05	2.515
134.09	2.569
139.19	2.618
143.62	2.632
149.52	2.687
155.10	2.702
160.95	2.742
166.22	2.760
172.32	2.795
177.71	2.824
184.22	2.853
189.67	2.890
196.51	2.911
199.45	2.904*
199.87	2.922
202.04	2.924*
205.45	2.929
207.60	2.915*
208.20	2.943*
212.15	2.939
213.83	2.944*
218.19	2.957
219.17	2.948*
223.96	2.980
226.03	2.981*
230.72	3.002
232.93	3.017*
236.71	3.014
240.27	3.027*
243.47	3.027
247.35	3.050*
249.53	3.056
253.43	3.055*
256.62	3.087
260.68	3.059*
262.76	3.119
266.73	3.096*
269.48	3.081
273.45	3.131*
276.02	3.126

CURVE 2*

T	C_p
273.15	3.07×10^{-2}
373.15	3.15
473.15	3.22
573.15	3.29
673.15	3.37
773.15	3.44
873.15	3.52
973.15	3.59
1073.15	3.66
1173.15	3.74
1273.15	3.81
1373.15	3.89
1473.15	3.96
1573.15	4.03
1673.15	4.11
1773.15	4.18
1873.15	4.26
1973.15	4.33

CURVE 3*

T	C_p
406.28	3.34×10^{-2}
431.15	3.36
481.08	3.43
531.11	3.50
579.91	3.56
630.57	3.63
681.20	3.69
731.39	3.76
781.33	3.84

* Not shown on plot

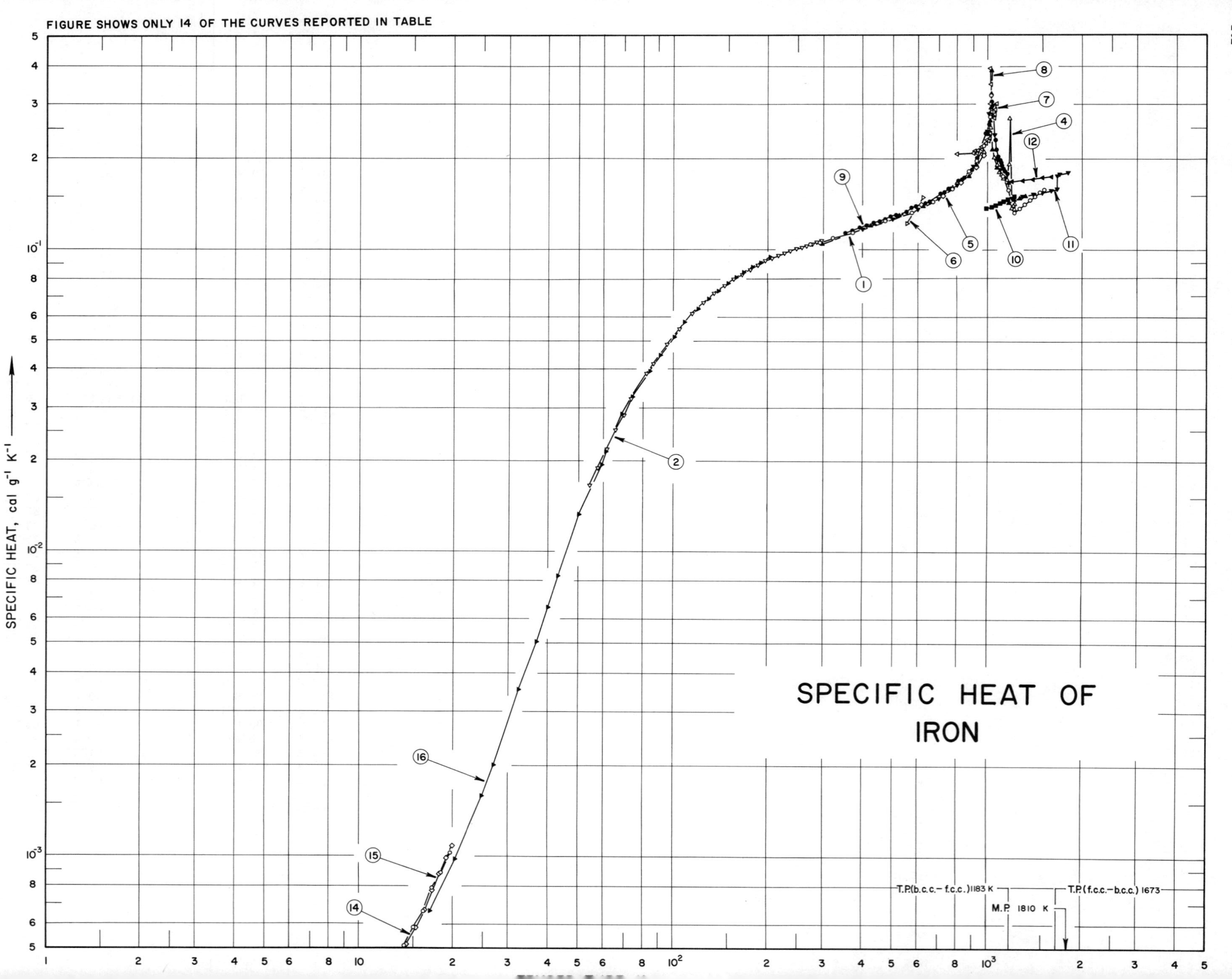
FIGURE SHOWS ONLY 14 OF THE CURVES REPORTED IN TABLE
SPECIFIC HEAT OF IRON
SPECIFIC HEAT, cal g⁻¹ K⁻¹
T.P.(b.c.c.– f.c.c.) 1183 K
T.P.(f.c.c.–b.c.c.) 1673
M.P. 1810 K

SPECIFICATION TABLE NO. 26 SPECIFIC HEAT OF IRON

(Impurity < 0.20% each; total impurities < 0.50%)

[For Data Reported in Figure and Table No. 26]

Curve No.	Ref. No.	Year	Temp. Range, K	Reported Error, %	Name and Specimen Designation	Composition (weight percent), Specifications and Remarks
1	26	1949	273-1523	±2.0		99.99 Fe; annealed; density = 491.59 lb ft^{-3}.
2	27	1943	54-295	0.3		99.94 Fe, 0.031 N, <0.008 Si, <0.005 C, 0.004 Cu, <0.004 S, 0.003 Mn, 0.002 P, <0.001 O; annealed at 1190 C in dry hydrogen until O_2 content was below 0.001 %.
3	28	1963	1073-1673		α and γ iron	Impurities, 0.004 P, 0.004 Si (insol.), 0.003 Al (sol.), 0.0026 S, 0.002 Si (sol.), 0.001 C, 0.001 N_2, <0.001 Al (insol.), 0.0006 O_2, and 0.00001 H_2.
4	29	1940	373-1223			99.99 Fe.
5	30	1955	618-973	3.0-5.0	Armco iron	99.75 Fe, nominal composition.
6	17	1961	553, 623	<2.4		
7	31	1959	800-1071		Armco iron	99.75 Fe, nominal composition; measured in an argon atmosphere.
8	32	1960	298-1323	<2.0	High purity iron	Impurities 0.10 N_2, 0.03 C, 0.01 O_2, <0.0005 Ni, 0.0001 Cu, <0.0001 Ag, <0.0001 Mg, <0.0001 Na, and <0.0001 Si; sample supplied by the Johnson, Matthey and Co. Ltd; furnace under vacuum.
9	33	1957	353-1173	≤0.9	Electrolytic iron	Impurities, 0.016 C, 0.009 S, <0.005 Mn, <0.005 Si, <0.002 P, and traces of Al, Cu, and Ni.
10	33	1957	1013-1218	≤0.9	γ - iron	Same as above.
11	34	1962	298-1809		solid iron	99.945 Fe, 0.031 Ni, 0.008 Si, 0.005 C, 0.004 S, 0.004 other metals, 0.002 P, and 0.001 other non metals.
12	34	1962	1184-1665		α - iron	Same as above.
13	104	1946	348-1198	2.0		> 99.9 Fe.
14	82	1939	1.5-20			0.1 Mn, < 0.1 Ni, 0.04 Cu, 0.01 C, 0.005 Si, and 0.003 P; melted under vacuum.
15	82	1939	1.7-20			Same as above.
16	55	1930	17-206	1.5	Electrolytic iron	
17	83	1954	343-1208		Electrolytic iron	
18	28	1963	1198-1623		γ - iron	0.004 P, 0.004 Si (insol.), 0.003 Al (sol.), 0.0026 S, 0.002 Si (sol.), 0.001 C, 0.001 N, <0.001 Al (insol.), 0.0006 O, 0.00001 H.
19	104	1946	978-1193	2.0		> 99.9 Fe.
20	310	1963	20-1663		γ - iron	Pure iron.
21	310	1963	20-1663		α - iron	Pure iron.
22	311	1925	73-198			99.88 Fe, with small amounts of C, Si, Mn, and S.
23	312	1926	373-1523			100.0 Fe; specific heat corrected for carbon content.

SPECIFICATION TABLE NO. 26 (continued)

Curve No.	Ref. No.	Year	Temp. Range, K	Reported Error, %	Name and Specimen Designation	Composition (weight percent), Specifications and Remarks
24	268	1926	373-1903			99.70 Fe, 0.110 Cu, 0.073 Mn, 0.042 Si, 0.040 C, 0.029 P, and 0.006 S.
25	289	1927	373-1273			Chemically pure specimen; vacuum melted.
26	313	1929	1123-1833		Electrolytic iron	
27	164	1932	273-1873			
28	314	1935	30-220			
29	315	1935	298.65			Annealed.
30	316	1938	378-1773		Electrolytic	Heated several times at 900 C to expel H_2 gas.
31	201	1939	1.2-20			Very pure sample, 0.01 Si; pressed and sintered in hydrogen at 1350 C for 9 hrs., sintered again above 1400 C for 16 hrs; density = 7.25 g cm^{-3}.
32	317	1945	1.5-20			0.1 Mn, 0.1 Ni, 0.04 Cu, 0.01 C, 0.005 Si, and 0.003 P.
33	318	1954	1181-1193			99.99 Fe; annealed.
34	319	1958	293-1030			
35	320	1959	1.8-5.3			Pure iron.
36	1	1961	295	± 5		Specimen probably in pure state.
37	318	1954	1175-1196			99.99 Fe; annealed.
38	318	1954	1174-1193			Same as above.

DATA TABLE NO. 26 SPECIFIC HEAT OF IRON

[Temperature, T, K; Specific Heat, C_p, Cal $g^{-1} K^{-1}$]

CURVE 1

T	C_p
273.15	1.04 x 10^{-1}
293	1.05
323	1.08
373	1.14
473	1.24
573	1.32
673	1.44
773	1.58
873	1.80
923	1.95
973	2.17
1023	2.70
1033	3.18
1073	1.99*
1123	1.74
1173	1.57
1223	1.32
1273	1.36
1323	1.40
1373	1.45
1423	1.49
1473	1.53
1523	1.56

CURVE 2

T	C_p
54.6	1.655 x 10^{-2}
57.8	1.893
61.4	2.172
65.6	2.519
69.4	2.835
73.7	3.189
82.2	3.857
86.2	4.165
95.1	4.824
105.0	5.483
115.3	6.111
125.5	6.670
135.5	7.150
145.9	7.587
155.8	7.956
166.0	8.299
176.3	8.591
186.0	8.849
196.5	9.118

CURVE 2 (cont.)

T	C_p
206.2	9.318 x 10^{-2}
216.0	9.501
225.9	9.687
236.6	9.863
246.3	1.003 x 10^{-1}
256.2	1.017
266.1	1.032
276.1	1.046
285.8	1.056
295.1	1.066

CURVE 3*

T	C_p
1073	2.125 x 10^{-1}
1098	2.022
1123	1.936
1148	1.868
1173	1.814
1673	1.783

CURVE 4

T	C_p
373	1.14* x 10^{-1}
473	1.24*
573	1.32*
673	1.43*
773	1.60
873	1.80*
973	2.09
1033	2.49
1053	2.00
1073	1.89
1093	1.81
1113	1.76
1133	1.72
1153	1.69
1173	1.90
1183	2.68
1193	1.37
1203	1.39
1213	1.41
1223	1.37

CURVE 5

T	C_p
618	1.40 x 10^{-1}
723	1.50
823	1.66
923	1.87
973	2.04

CURVE 6

T	C_p
553	1.22 x 10^{-1}
623	1.49

CURVE 7

T	C_p
800	2.05 x 10^{-1}
905	2.07
912	2.08
912	2.10*
921	2.07
921	2.11*
927	2.07*
927	2.11*
936	2.08*
936	2.12
944	2.11*
944	2.13*
951	2.14
951	2.16
960	2.14*
960	2.18
967	2.14*
967	2.18
976	2.22*
977	2.18*
987	2.22*
990	2.27*
996	2.22
996	2.28
1006	2.27
1006	2.32
1014	2.31*
1021	2.38
1022	2.98
1023	3.90
1026	3.46

CURVE 7 (cont.)

T	C_p
1034	2.82* x 10^{-1}
1050	2.73*
1051	2.68*
1057	2.76
1065	2.86
1071	2.99

CURVE 8

T	C_p
298	1.04 x 10^{-1}
323	1.08*
373	1.14*
423	1.20
473	1.25*
523	1.30
573	1.35
623	1.40
673	1.44*
723	1.50*
773	1.57*
823	1.64*
873	1.74
923	1.89
973	2.15*
1003	2.38
1023	2.65
1033	2.91
1038	3.21
1042	3.85
1043	2.13
1053	1.99*
1073	1.87
1123	1.73*
1173	1.66
1173	1.44
1223	1.46
1273	1.48
1323	1.50

CURVE 9

T	C_p
353	1.127 x 10^{-1}
373	1.150
393	1.175
413	1.197
433	1.219
453	1.238*
473	1.254
493	1.272
513	1.295*
533	1.313
553	1.337
573	1.352
593	1.377*
613	1.393
633	1.413
653	1.435*
673	1.457*
693	1.479
713	1.512
733	1.530
753	1.571*
773	1.586*
783	1.597*
793	1.616*
803	1.636
813	1.669*
823	1.679
833	1.696
843	1.713*
853	1.734*
863	1.750*
873	1.780*
883	1.815
893	1.835*
903	1.869*
913	1.910*
923	1.949
933	1.993*
943	2.032*
953	2.084*
963	2.145*
973	2.212*
983	2.280
993	2.371*
1003	2.458

CURVE 9 (cont.)

T	C_p
1013	2.565* x 10^{-1}
1023	2.691*
1033	2.935
1038	3.015*
1043	2.924*
1053	2.353
1063	2.270
1073	2.116*
1083	2.059
1093	2.003
1103	1.961
1113	1.930
1123	1.895
1133	1.874
1143	1.847
1153	1.827
1163	1.809
1173	1.795

CURVE 10

T	C_p
1013	1.376* x 10^{-1}
1023	1.382
1033	1.387
1038	1.390*
1043	1.393*
1053	1.398*
1063	1.403
1073	1.409*
1083	1.415*
1093	1.420*
1103	1.426
1113	1.431*
1123	1.437*
1133	1.444
1143	1.449*
1153	1.455*
1163	1.459
1173	1.467*
1198	1.486*
1208	1.488*
1218	1.496

*Not shown on plot

DATA TABLE NO. 26 (continued)

CURVE 11

T	C_p
298.15	1.074 x 10^{-1}*
400	1.164
500	1.257
600	1.357
700	1.475
800	1.642
900	1.877
1000	2.428
1020	2.790
1033	3.187*
1040	2.972
1060	2.381
1080	1.969
1100	1.832
1120	1.746*
1140	1.710
1160	1.690*
1184	1.671*
1184	1.456*
1200	1.461
1300	1.488
1400	1.515
1500	1.542
1600	1.569
1665	1.586
1665	1.748
1700	1.758
1800	1.791*
1809	1.794

CURVE 12

T	C_p
1184	1.67 x 10^{-1}*
1200	1.67
1300	1.68
1400	1.69
1500	1.71
1600	1.73
1665	1.75*

CURVE 13*

T	C_p
348	1.12 x 10^{-1}
398	1.17
448	1.22
498	1.26
548	1.30
598	1.35
648	1.40
698	1.46
748	1.55
798	1.65
848	1.75
898	1.85
948	1.98
998	2.32
1048	2.18
1098	1.80
1148	1.70
1198	2.26

CURVE 14

T	C_p
1.50	3.35 x 10^{-5}*
1.95	3.96*
2.92	6.09*
3.63	8.04*
4.45	1.04 x 10^{-4}*
5.34	1.31*
6.23	1.51*
8.48	2.33
9.94	2.83*
11.17	3.51*
12.40	4.10*
13.38	4.80*
14.27	5.16
15.22	5.86
16.23	6.79
17.24	7.83
18.15	8.79
19.15	9.80
20.12	1.08 x 10^{-3}

CURVE 15

T	C_p
1.69	3.78 x 10^{-5}*
2.03	4.58*
2.65	5.73*
3.02	6.43*
3.55	8.18*
3.82	8.51*
8.59	2.35 x 10^{-4}*
10.02	2.99*
11.06	3.42*
12.08	3.99*
13.08	4.57*
14.02	5.12
15.01	5.89
16.13	6.61
17.36	7.88
18.45	8.88
19.55	1.06 x 10^{-3}

CURVE 16

T	C_p
16.90	6.683 x 10^{-4}*
20.25	9.789*
24.66	1.596 x 10^{-3}
26.80	2.014
29.35	2.564
32.40	3.549
36.85	5.042
40.00	6.577
43.00	8.296
50.82	1.324 x 10^{-2}
59.05	1.945
69.80	2.809
74.95	3.228
84.00	3.907
91.25	4.453
100.55	5.118
109.90	5.730
120.50	6.362
129.95	6.881
139.02	7.295

CURVE 16 (cont.)

T	C_p
148.52	7.721 x 10^{-2}
158.10	8.058
167.44	8.380
178.80	8.711
189.64	8.980
197.23	9.173*
205.59	9.392

CURVE 17*

T	C_p
343	1.122 x 10^{-1}
363	1.148
383	1.184
403	1.215
423	1.240
443	1.268
463	1.282
483	1.308
503	1.337
523	1.362
543	1.383
563	1.407
583	1.424
603	1.443
623	1.462
643	1.483
663	1.499
683	1.524
703	1.542
723	1.572
743	1.596
763	1.623
783	1.646
803	1.678
823	1.704
843	1.745
863	1.784
883	1.820
903	1.873
923	1.938

CURVE 17 (cont.)

T	C_p
943	2.014 x 10^{-1}
963	2.088
983	2.191
1003	2.320
1018	2.448
1028	2.607
1038	2.814
1048	2.275
1058	2.085
1068	2.067
1078	2.018
1088	1.995
1098	2.044
1108	1.958
1118	1.860
1128	1.911
1138	1.901
1148	1.855
1158	1.847
1168	1.847
1178	1.845
1198	1.595
1208	1.568

CURVE 18*

T	C_p
1198	1.481 x 10^{-1}
1223	1.490
1248	1.499
1273	1.508
1323	1.526
1373	1.543
1423	1.561
1473	1.581
1523	1.599
1573	1.617
1623	1.635

CURVE 19*

T	C_p
978	2.1 x 10^{-1}
988	2.2
998	2.3
1008	2.4
1018	2.5
1028	2.7
1038	2.2
1048	2.1
1058	2.0
1068	1.7
1178	5.4
1188	2.8
1193	2.5

CURVE 20*

T	C_p
20	1.284 x 10^{-3}
30	3.853
40	8.986
50	1.754 x 10^{-2}
60	2.653
70	3.552
80	4.450
90	5.263
100	6.035
120	7.318
140	8.346
160	9.116
180	9.759
200	1.027 x 10^{-1}
220	1.066
240	1.100
260	1.130
280	1.156
300	1.181
350	1.233
400	1.275
450	1.309
500	1.344
550	1.370
600	1.395
656	1.421

*Not shown on plot

DATA TABLE NO. 26 (continued)

T	C_p
CURVE 20 (cont.)*	
700	1.442×10^{-1}
750	1.459
800	1.481
850	1.498
900	1.498
950	1.532
1000	1.545
1020	1.549
1033	1.554
1040	1.558
1060	1.562
1080	1.566
1100	1.571
1120	1.575
1140	1.583
1160	1.588
1180	1.596
1183	1.596
1200	1.605
1250	1.618
1300	1.676
1350	1.639
1400	1.652
1450	1.665
1500	1.678
1550	1.690
1600	1.703
1650	1.711
1663	1.716
CURVE 21*	
20	8.558×10^{-4}
30	2.567×10^{-3}
40	5.564
50	1.283×10^{-2}
60	2.054
70	2.868
80	3.681
90	4.450
100	5.179
120	6.377
140	7.361
160	8.088
180	8.731
CURVE 21 (cont.)*	
200	9.202×10^{-2}
220	9.587
240	9.928
260	1.023×10^{-1}
280	1.049
300	1.074
350	1.130
400	1.177
450	1.224
500	1.275
550	1.322
600	1.374
650	1.434
700	1.494
750	1.558
800	1.643
850	1.754
900	1.887
950	2.088
1000	2.440
1020	2.782
1033	3.226
1040	2.430
1060	2.067
1080	1.990
1100	1.939
1120	1.896
1140	1.866
1160	1.840
1180	1.819
1183	1.819
1200	1.810
1250	1.802
1300	1.798
1350	1.806
1400	1.810
1450	1.819
1500	1.832
1550	1.840
1600	1.853
1650	1.862
1663	1.866
CURVE 22*	
72.86	3.26×10^{-2}
75.64	3.55
77.85	3.67
80.17	3.96
82.43	4.33
84.55	4.40
86.53	4.55
88.51	4.80
90.37	4.91
196.44	9.63
197.58	9.72
198.04	9.76
CURVE 23*	
373	1.150×10^{-1}
473	1.218
573	1.309
673	1.431
773	1.579
873	1.774
973	2.102
1023	2.430
1073	2.710
1098	2.760
1123	1.690
1173	1.700
1198	1.680
1223	1.690
1273	1.660
1373	1.660
1473	1.660
1498	1.660
1523	1.660
CURVE 24*	
373	1.13×10^{-1}
473	1.21
573	1.32
673	1.44
773	1.59
873	1.78
973	2.15
1023	2.51
CURVE 24 (cont.)*	
1073	2.73×10^{-1}
1123	2.62
1173	1.70
1263	1.73
1273	1.76
1473	1.80
1573	1.82
1673	1.86
1773	1.88
1843	2.22
1873	2.22
1903	2.22
CURVE 25*	
373	1.150×10^{-1}
473	1.278
573	1.400
673	1.510
773	1.626
873	1.885
973	2.300
1033	3.200
1073	2.095
1173	1.860
1181	1.600
1273	1.620
CURVE 26*	
1123	1.85×10^{-1}
1173	1.85
1223	1.61
1273	1.63
1323	1.65
1373	1.67
1423	1.69
1473	1.71
1523	1.74
1573	1.76
1623	1.79
1723	1.85
1773	1.85
1813	1.94
1833	1.94
CURVE 27*	
273.15	1.054×10^{-1}
373.15	1.168
473.15	1.282
573.15	1.396
673.15	1.509
773.15	1.623
873.15	1.737
973.15	1.850
993.15	1.873
998.15	1.879
1008.15	2.830
1018.15	3.080
1028.15	3.760
1038.15	3.440
1048.15	2.676
1058.15	1.592
1068.15	1.592
1192.15	1.448
1473.15	1.448
1573.15	1.449
1677.65	2.142
1678.15	2.142
1801.50	1.501
1873.50	1.501
CURVE 28*	
30	3.22×10^{-3}
40	6.63
50	1.31×10^{-2}
60	2.06
70	2.90
80	3.76
90	4.49
100	5.21
110	5.91
120	6.48
130	6.95
140	7.41
150	7.77
160	8.06
180	8.63
200	9.19
220	9.63
CURVE 29*	
298.65	1.107×10^{-1}
298.65	1.110
CURVE 30*	
378	1.12×10^{-1}
383	1.13
393	1.13
403	1.14
413	1.18
418	1.24
423	1.32
428	1.39
433	1.36
438	1.25
443	1.15
448	1.11
453	1.11
458	1.12
463	1.18
468	1.22
473	1.23
478	1.25
483	1.28
493	1.28
503	1.32
513	1.32
523	1.34
533	1.35
573	1.36
623	1.43
673	1.46
723	1.51
773	1.60
798	1.64
828	1.69
848	1.71
873	1.75
883	1.77
893	1.79
903	1.80
913	1.83
923	1.86
933	1.92
943	1.93

* Not shown on plot

DATA TABLE NO. 26 (continued)

T	C_p
CURVE 30 (cont.)*	
953	1.98×10^{-1}
963	2.04
973	2.13
983	2.18
993	2.27
1003	2.42
1013	2.59
1023	2.81
1028	2.87
1033	2.88
1038	2.80
1043	2.70
1048	2.60
1053	2.50
1058	2.38
1063	2.25
1068	2.15
1073	2.13
1083	2.05
1093	1.99
1103	1.95
1113	1.92
1123	1.90
1133	1.88
1143	1.87
1153	1.85
1163	1.83
1173	1.82
1248	1.42
1273	1.44
1298	1.46
1323	1.48
1348	1.49
1373	1.51
1398	1.53
1423	1.56
1448	1.59
1473	1.62
1498	1.65
1523	1.69
1548	1.73
1573	1.76
1598	1.77
1623	1.78
1648	1.79
1673	1.86
1698	1.99×10^{-1}
1723	2.00
1748	2.00
1773	2.00
CURVE 31*	
Series 1	
1.256	3.076×10^{-5}
1.731	3.784
2.144	4.468
2.526	5.676
2.597	5.753
2.702	5.825
2.775	5.970
2.935	6.323
2.982	6.324
3.034	6.588
3.089	6.645
3.148	6.702
3.197	6.808
Series 2	
1.125	2.772×10^{-5}
1.138	2.595
1.296	2.792
1.296	3.225
1.300	3.194
2.236	4.913
2.261	5.024
2.290	5.035
2.328	5.234
2.464	5.596
2.618	5.929
2.737	5.993
3.072	6.840
3.348	7.454
3.443	7.599
3.516	7.848
3.604	8.110
3.665	8.280
3.730	8.394
3.787	8.547
3.845	8.715
CURVE 31 (cont.)*	
3.897	8.844×10^{-5}
3.947	8.901
3.992	9.078
4.038	9.225
4.093	9.250
Series 3	
9.48	2.865×10^{-4}
9.92	3.026
10.51	3.261
11.33	3.715
11.94	4.170
15.24	6.337
15.39	6.378
15.68	6.462
16.28	6.620
16.95	7.569
18.19	8.974
19.10	1.016×10^{-3}
19.92	1.077
Series 4	
12.59	4.434×10^{-4}
13.25	5.363
14.41	5.776
14.83	5.925
15.42	6.344
15.99	6.894
16.52	7.309
17.01	7.800
17.38	7.918
17.82	8.661
18.43	9.415
19.09	9.680
19.62	1.015×10^{-3}
20.27	1.061
Series 5	
10.06	2.906×10^{-4}
10.36	3.273
12.85	4.720
12.89	4.974
13.20	5.152
CURVE 31 (cont.)*	
13.72	5.436×10^{-4}
14.20	5.669
14.73	5.973
15.11	6.183
15.56	6.407
16.03	6.804
CURVE 32*	
Series 1	
1.50	3.349×10^{-5}
1.95	3.958
2.92	6.089
3.63	8.041
4.45	1.039×10^{-4}
5.34	1.314
6.23	1.511
8.40	2.328
9.94	2.829
11.17	3.510
12.40	4.101
13.38	4.799
14.27	5.158
15.22	5.856
16.23	6.787
17.24	7.826
18.15	8.793
19.15	9.796
20.12	1.083×10^{-3}
Series 2	
1.69	3.778×10^{-5}
2.03	4.584
2.65	5.730
3.02	6.429
3.55	8.184
3.82	8.506
8.59	2.346×10^{-4}
10.02	2.991
11.06	3.420
12.08	4.029
13.08	4.566
14.02	5.122
15.01	5.892
CURVE 32 (cont.)*	
16.13	6.608×10^{-4}
17.36	7.880
18.45	8.882
19.55	1.062×10^{-3}
CURVE 33*	
1181.2	1.56×10^{-1}
1182.2	1.54
1184.3	1.58
1185.8	1.68
1187.2	2.86
1187.9	2.63
1188.7	1.83
1190.5	1.49
1192.0	1.42
1192.9	1.38
CURVE 34*	
293	1.080×10^{-1}
333	1.115
373	1.153
393	1.173
413	1.190
433	1.210
453	1.228
473	1.248
493	1.268
513	1.286
533	1.305
553	1.325
573	1.345
593	1.367
613	1.390
633	1.415
653	1.440
673	1.465
693	1.493
713	1.521
733	1.549
753	1.577
773	1.606
793	1.635
813	1.665
833	1.705
CURVE 34 (cont.)*	
853	1.750×10^{-1}
873	1.800
893	1.862
913	1.944
933	2.040
953	2.155
973	2.280
993	2.430
1003	2.520
1013	2.650
1023	2.850
1030.6	3.040
CURVE 35*	
1.823	4.000×10^{-5}
1.924	4.216
2.069	4.540
2.303	4.985
2.546	5.547
2.717	5.913
2.866	6.256
2.985	6.514
3.075	6.783
3.121	6.804
3.142	6.946
3.216	7.041
3.299	7.291
3.437	7.662
3.548	7.898
3.627	8.120
3.836	8.482
3.855	8.529
4.072	9.064
4.38	9.843
4.829	1.102×10^{-4}
5.266	1.231
CURVE 36*	
295.15	1.1×10^{-1}
CURVE 37*	
1175.0	1.56×10^{-1}
1183.6	1.54

* Not shown on plot

DATA TABLE NO. 26 (continued)

T	C_p
CURVE 37 (cont.)*	
1185.6	1.56 x 10^{-1}
1187.0	1.78
1187.6	2.06
1187.9	1.94
1187.9	1.52
1189.6	1.44
1191.0	1.36
1192.4	1.34
1193.5	1.34
1196.2	1.32
CURVE 38*	
1174.3	1.52 x 10^{-1}
1176.4	1.54
1177.3	1.56
1179.4	1.56
1181.5	1.54
1182.3	1.50
1183.7	1.52
1186.4	1.54
1187.8	1.82
1188.1	2.02
1189.2	1.62
1190.2	1.47
1191.7	1.39
1193.0	1.37

* Not shown on plot

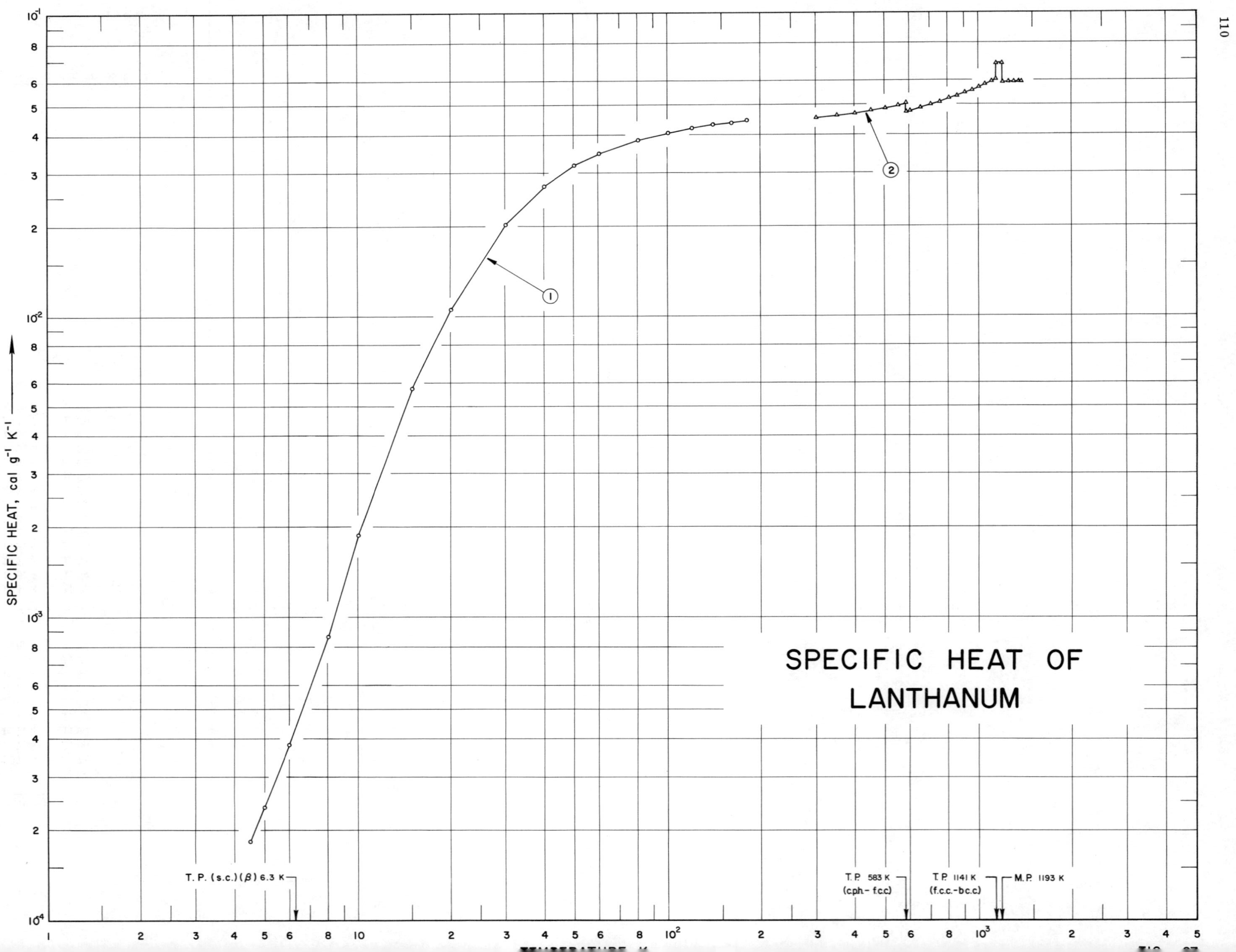
SPECIFIC HEAT OF LANTHANUM
SPECIFIC HEAT, cal g⁻¹ K⁻¹
T. P. (s.c.)(β) 6.3 K
T.P. 583 K (c.p.h- f.c.c)
T.P. 1141 K (f.c.c.-b.c.c)
M.P. 1193 K
1
2

SPECIFICATION TABLE NO. 27 SPECIFIC HEAT OF LANTHANUM

(Impurity < 0.20% each; total impurities < 0.50%)

[For Data Reported in Figure and Table No. 27]

Curve No.	Ref. No.	Year	Temp. Range, K	Reported Error, %	Name and Specimen Designation	Composition (weight percent), Specifications and Remarks
1	35	1951	5-180			Spectroscopically pure, < 0.05 Ca, < 0.02 rare earth, < 0.01 Be, < 0.002 Fe; hexagonal close packed and cubic closed packed crystal structure.
2	36	1962	298-1373			> 99.76 La, 0.2 Nd, < 0.10 Ta, < 0.05 Ca, 0.0455 O_2, < 0.03 Ce, < 0.03 Pr, < 0.02 Mg, 0.0152 C, < 0.01 Cr, 0.0033 H_2, 0.0013 N_2; prepared by metallothermic reduction of anhydrous lanthanum fluoride and calcium metal; cast into 1/2 in. rods; sealed under helium in tantalum crucible.
3	422	1967	3-25	0.4-2.0		0.03 F, 0.023 Na, 0.02 C, 0.01 H_2, 0.005 each Cu and K, 0.003 Na, 0.002 Fe, 0.001 Bi and 0.001 Al.

DATA TABLE NO. 27 SPECIFIC HEAT OF LANTHANUM

[Temperature, T, K; Specific Heat, C_p, Cal $g^{-1} K^{-1}$]

T	C_p
CURVE 1	
4.5	1.83×10^{-4}
5	2.38
6	3.82
8	8.64
10	1.87×10^{-3}
15	5.72
20	1.06×10^{-2}
30	2.02
40	2.70
50	3.18
60	3.48
80	3.84
100	4.06
120	4.21
140	4.30
160	4.38
180	4.46
CURVE 2	
298.15	4.50×10^{-2}*
300	4.51
350	4.60
400	4.69
450	4.79
500	4.88
550	4.97
583.15	5.04
583.15	4.74
600	4.78
650	4.88
700	4.99
750	5.10
800	5.21
850	5.32
900	5.44
950	5.56
1000	5.69
1050	5.81
1100	5.94
1141.15	6.05
1141.15	6.80
1150	6.80*
(s) 1193.15	6.80
(l) 1193.15	5.91
CURVE 2 (contd)	
1200	5.91×10^{-2}*
1250	5.91
1300	5.91
1350	5.91
1373.15	5.91
CURVE 3*	
Series I	
3.1058	9.159×10^{-5}
3.3601	1.161×10^{-4}
3.5660	1.388
3.7408	1.601
3.8938	1.8075
4.0306	2.0045
4.1541	2.1953
4.2671	2.3815
4.3715	2.5609
4.4682	2.7316
4.5587	2.8952
4.6448	3.0423
4.7268	3.1580
4.8071	3.1660
4.8919	2.8708
4.9845	2.6069
5.0991	2.7189
5.2310	2.8538
5.3523	3.0262
5.4667	3.1994
5.5750	3.3799
5.6783	3.5380
5.7764	3.6862
5.8714	3.7818
5.9650	3.8293
6.0564	3.9336
6.1445	4.1137
6.2290	4.2937
6.3098	4.4570
6.4855	4.8649
6.9215	5.9962
7.4575	7.6566
7.9665	9.4942
8.5701	1.1984×10^{-3}
9.1686	1.4866
CURVE 3 (cont.)*	
9.8187	1.8481×10^{-3}
10.743	2.4288
11.938	3.2920
13.029	4.1803
14.019	5.0477
15.168	6.1304
16.522	7.4604
18.095	9.0761
19.776	1.0845×10^{-2}
21.604	1.2787
23.415	1.4737
25.001	1.6543
Series II	
3.2151	1.017×10^{-4}
3.4837	1.294
3.7855	1.661
4.1277	2.1542
4.4800	2.7531
4.6930	3.1186
4.7745	3.1858
4.8574	3.0217
4.9475	2.6667
5.2522	2.8880
5.6310	3.4725
5.8000	3.7204
5.9092	3.8001
6.1582	4.1504
6.5730	5.0896
7.1277	6.6139
7.8876	9.1896
8.6433	1.2380×10^{-3}
9.4025	1.6155
10.344	2.1647
11.316	2.8259
12.301	3.5804
13.366	4.4661
14.483	5.4714
15.748	6.6879
17.206	8.1573
18.808	9.8193
20.533	1.1645×10^{-2}
22.329	1.3562
24.039	1.5408

*Not shown on plot

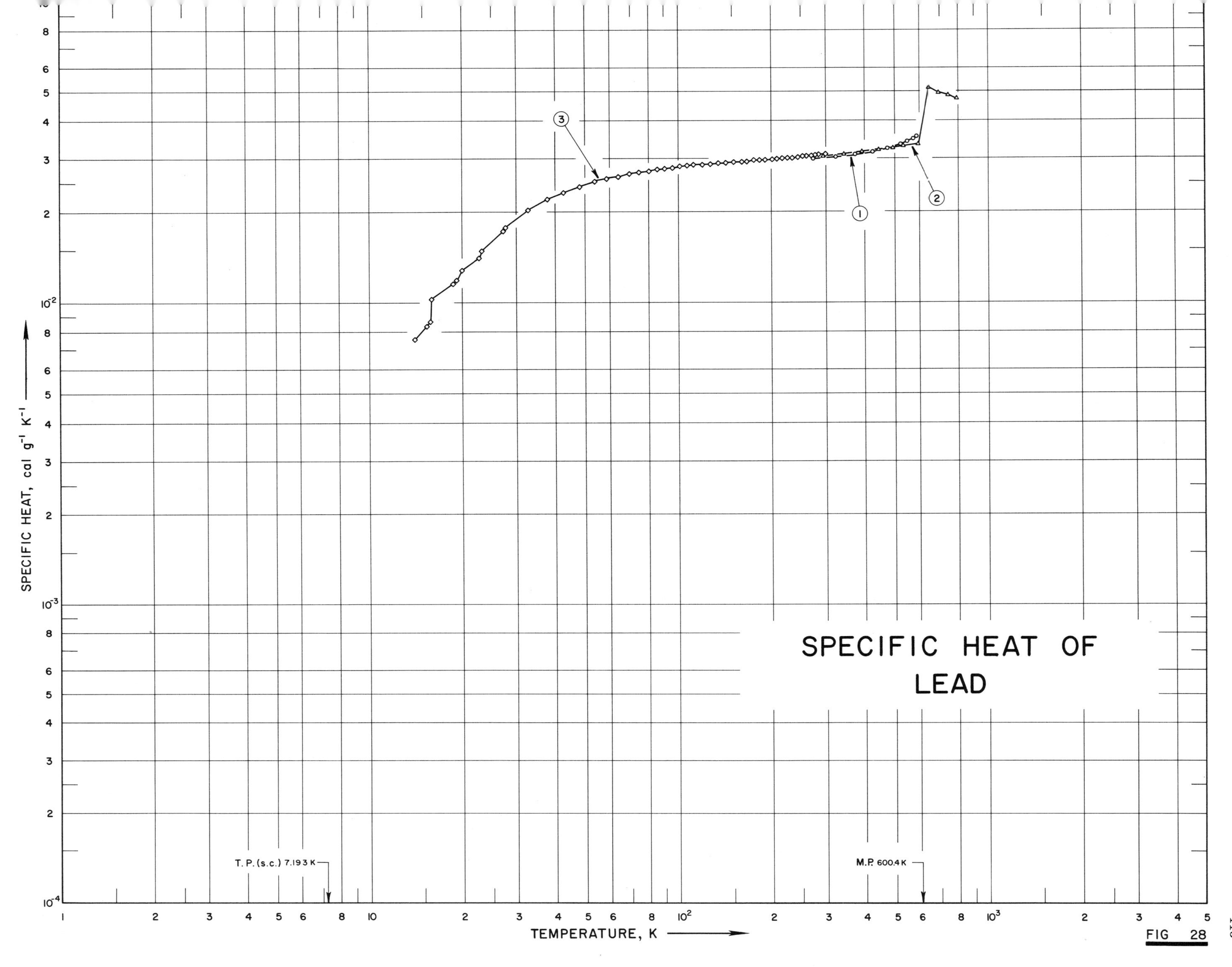
SPECIFIC HEAT OF LEAD
SPECIFIC HEAT, cal g⁻¹ K⁻¹
TEMPERATURE, K
T. P. (s.c.) 7.193 K
M.P. 600.4 K
1
2
3

FIG 28

SPECIFICATION TABLE NO. 28 SPECIFIC HEAT OF LEAD

(Impurity < 0.20% each; total impurities < 0.50%)

[For Data Reported in Figure and Table No. 28]

Curve No.	Ref. No.	Year	Temp. Range, K	Reported Error, %	Name and Specimen Designation	Composition (weight percent), Specifications and Remarks
1	3	1953	273-588	< 5.0	Lead wire	99.9 Pb
2	39	1961	293-793		No. 6	
3	103	1941	15-300			99.999 Pb; melted and crystallized in a vacuum furnace; cooled over a period of 4 days
4	84	1965	0.4-4	± 2.0		99.9999 Pb; single crystals, normal state.
5	84	1965	0.4-4	± 2.0		99.999 Pb; single crystals, superconducting state.
6	268	1926	348-1023			99.977 Pb, 0.02 Fe and 0.003 Cu.
7	289	1927	323-773			
8	274	1927	14-78			
9	214	1927	627-732	1.0		Merck's C.P. grade; granular form; fused in an atmosphere of hydrogen.
10	276	1930	2-16			
11	182	1936	203-493			99.997 Pb, 0 0009 Cu, 0.0007 Sb, 0 0006 Ag, and 0.0002 Bi; smoothed.
12	193	1947	273-601			
13	322	1952	1-77			99.99 Pb; normal state.
14	323	1954	298-1200			≥ 99.9 Pb, 0.001 - 0.1 Bi, < 0.05 Na, and < 0.01 Al, Ca, Cr, Cu, Fe, Mg, Si, Ag, and Sn.
15	324	1966	1-40			99.999 Pb.

DATA TABLE NO. 28 SPECIFIC HEAT OF LEAD

[Temperature, T, K; Specific Heat, C_p, Cal $g^{-1}K^{-1}$]

T	C_p
CURVE 1	
273.15	2.99 x 10^{-2}
323	3.04
373	3.09
423	3.15
473	3.23
523	3.34
548	3.41
573	3.48
588	3.55
CURVE 2	
293	3.05 x 10^{-2}
343	3.10
393	3.16
443	3.20
493	3.25
533	3.29
593	3.34
643	5.12
693	4.96
743	4.84
793	4.75
CURVE 3	
14.16	7.53 x 10^{-3}
15.47	8.374
15.74	8.688
16.95	1.038 x 10^{-2}
18.77	1.164
19.06	1.191
20.11	1.287
22.74	1.467
23.08	1.492
23.62	1.579*
27.06	1.737
27.78	1.780
32.85	2.034
37.79	2.200
42.87	2.322
48.23	2.435
53.42	2.527
58.85	2.576
64.54	2.629
CURVE 3 (cont.)	
69.52	2.671 x 10^{-2}
74.84	2.704
80.39	2.730
86.13	2.760
90.96	2.770
96.00	2.795
101.77	2.834
107.54	2.846
113.39	2.863
120.37	2.864
127.72	2.875
135.26	2.893
143.40	2.914
151.63	2.918
160.03	2.929
167.97	2.938
176.40	2.962
184.48	2.967
192.61	2.963
200.75	2.975
209.04	2.987
217.61	3.007
226.00	3.008
234.99	3.029
244.07	3.032
252.58	3.054
260.98	3.061
269.44	3.067
277.89	3.083
284.66	3.095
292.60	3.078*
299.91	3.098
CURVE 4*	
0.431	1.66 x 10^{-6}
0.485	1.93
0.545	2.19
0.616	2.58
0.741	3.35
0.908	4.57
1.004	5.43
1.168	7.19
1.282	8.60
1.414	1.05 x 10^{-5}
CURVE 4 (cont.)*	
1.546	1.27 x 10^{-5}
1.680	1.50
1.716	1.59
1.819	1.82
1.907	2.03
1.983	2.24
2.046	2.41
2.238	3.01
2.440	3.78
2.658	4.76
2.889	5.98
3.129	7.55
3.376	9.49
3.700	1.26 x 10^{-4}
4.072	1.72
4.41	2.25
CURVE 5*	
0.363	1.04 x 10^{-7}
0.379	1.13
0.395	1.34
0.434	1.69
0.472	2.22
0.498	2.56
0.526	3.01
0.588	4.18
0.609	4.64
0.667	6.07
0.700	6.96
0.768	9.39
0.827	1.15 x 10^{-6}
0.900	1.50
0.964	1.81
1.067	2.53
1.098	2.70
1.175	3.37
1.272	4.25
1.357	5.11
1.408	5.76
1.578	8.15
1 723	1.06 x 10^{-5}
1.753	1.12
1.958	1.59
2.032	1.76
CURVE 5 (cont.)*	
2.180	2.25 x 10^{-5}
2.316	2.91
2.514	3.56
2.529	3.54
2.662	4.28
2.818	4.97
2.851	5.40
3.004	6.45
3.158	7.58
3.347	8.80
3.599	1.13 x 10^{-4}
3.889	1.45
4.117	1.81
4.45	2.41
CURVE 6*	
348	2.68 x 10^{-2}
373	2.79
423	3.01
473	3.20
523	3.42
573	3.66
648	3.26
723	3.26
823	3.26
923	3.26
1023	3.26
CURVE 7*	
323	3.12 x 10^{-2}
373	3.20
423	3.29
473	3.38
523	3.46
573	3.56
(s)600	3.62
(l)600	3.88
633	3.75
723	3.66
773	3.70
CURVE 8*	
14.45	7.7 x 10^{-3}
18.08	1.14 x 10^{-2}
20.37	1.32
56.61	2.55
68.63	2.69
78.00	2.77
CURVE 9*	
627	3.39 x 10^{-2}
638	3.35
648	3.36
651	3.38
691	3.35
692	3.35
732	3.35
CURVE 10* Series I	
9.96	3.38 x 10^{-3}
12.15	5.45
14.84	7.53
15.73	8.88
Series II	
2.20	2.78 x 10^{-4}
3.55	1.94
4.49	3.40
4.77	4.77
5.58	7.34
Series III	
2.98	1.39 x 10^{-4}
3.56	1.69
4.06	2.41
4.61	4.32
4.73	4.12
5.48	6.42
6.16	8.98
7.61	1.75 x 10^{-3}
8.46	2.24
9.44	3.02
CURVE 11*	
203	2.963 x 10^{-2}
213	2.972
223	2.982
233	2.989
243	2.998
253	3.008
263	3.017
273	3.027
283	3.037
293	3.048
303	3.058
313	3.068
323	3.079
333	3.089
343	3.101
353	3.111
363	3.120
373	3.132
383	3.142
393	3.154
CURVE 12*	
273	3.00 x 10^{-2}
373	3.19
473	3.43
600.75	3.68
CURVE 13* Series I	
64.21	2.66 x 10^{-2}
65.18	2.68
66.82	2.69
71.79	2.69
72.58	2.71
76.80	2.77
Series II	
14.37	8.01 x 10^{-3}
14.51	7.96
14.58	8.16
14.66	8.21
15.99	9.46

* Not shown on plot

DATA TABLE NO. 28 (continued)

T	C_p	
CURVE 13 (cont.)*		
16.16	9.60	x 10^{-3}
16.33	9.56	
16.83	9.85	
16.98	9.70	
17.15	9.94	
17.86	1.08	x 10^{-2}
18.02	1.09	
18.15	1.11	
19.45	1.14	
19.65	1.16	
19.80	1.18	
20.79	1.26	
20.92	1.25	
Series III		
1.40	1.23	x 10^{-5}
1.61	1.31	
1.86	2.84	
1.89	2.21	
2.09	2.85	
3.30	1.04	x 10^{-4}
3.41	1.11	
3.54	1.20	
4.12	1.91	
CURVE 14*		
298.16	3.05	x 10^{-2}
300	3.06	
325	3.08	
350	3.11	
375	3.14	
400	3.17	
425	3.20	
450	3.22	
475	3.25	
500	3.28	
525	3.31	
550	3.34	
575	3.36	
600	3.39	
(s) 600.6	3.39	
(l) 600.6	3.53	
650	3.51	
700	3.50	
750	3.48	

T	C_p	
CURVE 14 (cont.)*		
800	3.46	x 10^{-2}
850	3.45	
900	3.43	
950	3.41	
1000	3.39	
1050	3.37	
1100	3.35	
1150	3.34	
1200	3.32	
CURVE 15*		
2	2.9	x 10^{-5}
3	7.4	
4	1.7	x 10^{-4}
5	3.73	
6	7.65	
7	1.27	x 10^{-3}
8	1.86	
10	3.27	
15	7.98	
20	1.30	x 10^{-2}
25	1.63	
30	1.93	
40	2.32	

* Not shown on plot

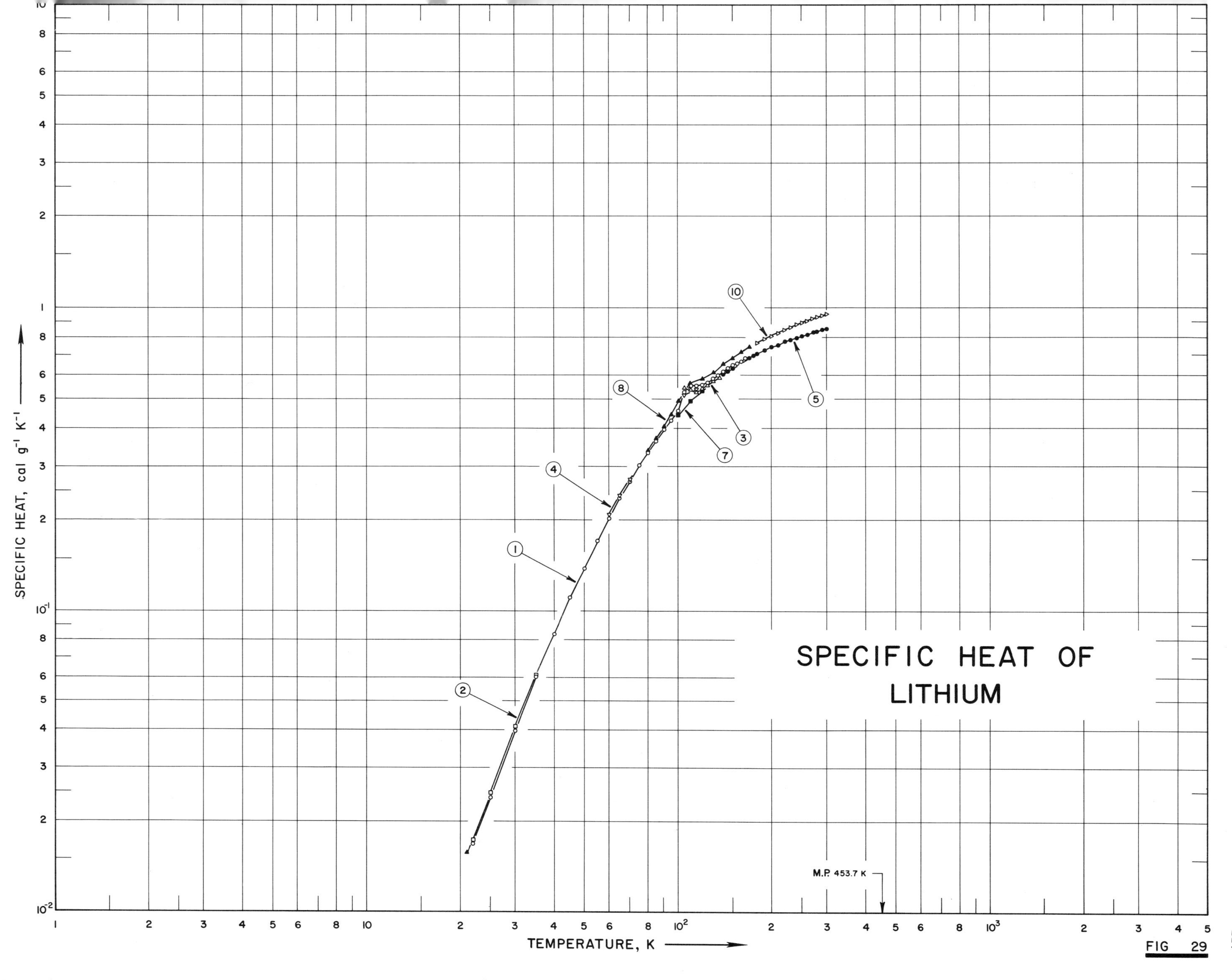
SPECIFIC HEAT OF LITHIUM
SPECIFIC HEAT, cal g⁻¹ K⁻¹
TEMPERATURE, K
M.P. 453.7 K
1
2
3
4
5
7
8
10

FIG 29

SPECIFICATION TABLE NO. 29 SPECIFIC HEAT OF LITHIUM

(Impurity < 0.20% each; total impurities < 0.50%)

[For Data Reported in Figure and Table No. 29]

Curve No.	Ref. No.	Year	Temp. Range, K	Reported Error, %	Name and Specimen Designation	Composition (weight percent), Specifications and Remarks
1	50	1960	22-165	≤ 2.0		99.95 Li, 0.02 N, 0.01 Ca, 0.01 K, 0.005 Na, and 0.001 Fe; cooled to 4 K.
2	50	1960	22-165	≤ 2.0		Same as above; cooled to 20 K.
3	50	1960	95-135	≤ 2.0		Same as above; not annealed.
4	50	1960	60-165	≤ 2.0		Same as above; annealed at 300 K.
5	50	1960	90-300	≤ 2.0		Same as above; cooled to 85 K.
6	50	1960	22-160	≤ 2.0		Same as above; cooled to 20 K.
7	50	1960	90-300	≤ 2.0		Same as above; cooled to 85 K.
8	51	1959	21-170	≤ 2.0		99.3 ^{6}Li, 0.70 ^{7}Li; cooled to 20 K.
9	51	1959	90-170	≤ 2.0		Same as above; cooled to 85 K.
10	51	1959	180-300	≤ 2.0		99.3 ^{6}Li, 0.70 ^{7}Li.
11	314	1935	15-300			
12	325	1950	459-773	± 10		
13	326	1950	473-773	± 10		Impurities: 0.1 Ca, 0.1 Si, < 0.1 Hg, < 0.1 P, and < 0.01 each Al, B, Cr, Cu, Fe, K, Na, and Ni; specimen supplied by The Maywood Chemical Co.
14	327	1951	773-1273	1.0		
15	328	1955	298-1200	0.3-0.5		Impurities: sample 1, 0.028 O_2, 0.003 N_2, 0.0036 Fe, 0.0006 Ni, 0.029 Ca, and 0.016 Na, sample 2, 0.003 Na, 0.001 Ca, 0.0006 Fe and 0.0003 Ni.

DATA TABLE NO. 29 SPECIFIC HEAT OF LITHIUM

[Temperature, T, K; Specific Heat, C_p, Cal $g^{-1} K^{-1}$]

T	C_p
CURVE 1	
22	1.69×10^{-2}
25	2.39
30	3.96
35	6.04
40	8.39
45	1.11×10^{-1}
50	1.39
55	1.71
60	2.03
65	2.36
70	2.68
75	3.01
80	3.32
85	3.64
90	3.95
95	4.26
100	4.60
105	5.28
110	5.52
115	5.51
120	5.58
125	5.69
130	5.83
135	5.98
140	6.15
145	6.31
150	6.45
155	6.56
160	6.68
165	6.81
CURVE 2	
22	1.73×10^{-2}
25	2.46
30	4.05
35	6.12
40	8.47*
45	1.12×10^{-1}*
50	1.40*
55	1.71*
60	2.04*
65	2.37*
70	2.69*
75	3.02*
80	3.34*
CURVE 2 (contd)	
85	3.65×10^{-1}*
90	3.95*
95	4.25*
100	4.60*
105	5.25*
110	5.57*
115	5.53*
120	5.58*
125	5.68*
130	5.80*
135	5.95*
140	6.11*
145	6.26*
150	6.39*
155	6.52*
160	6.66*
165	6.80*
CURVE 3	
95	4.26×10^{-1}*
100	4.60*
105	5.46*
110	5.39
115	5.30
120	5.41
125	5.56
130	5.72
137	5.88
CURVE 4	
60	2.08×10^{-1}
65	2.40
70	2.72
75	3.03*
80	3.35*
85	3.66*
90	3.95*
95	4.24*
100	4.57*
105	5.14*
110	5.34*
115	5.38*
120	5.48*
125	5.62*
CURVE 4 (contd)	
130	5.78×10^{-1}*
135	5.95*
140	6.11*
145	6.25*
150	6.40*
155	6.53*
160	6.66*
165	6.79*
CURVE 5	
90	3.97×10^{-1}*
95	4.23*
100	4.47*
105	4.70*
110	4.92*
115	5.13*
120	5.33*
125	5.52*
130	5.70*
135	5.88*
140	6.04
145	6.19
150	6.33
155	6.46*
160	6.60*
165	6.72*
170	6.85
175	6.95
180	7.06
190	7.25
200	7.42
210	7.57
220	7.71
230	7.83
240	7.95
250	8.07
260	8.18
270	8.29*
273.15	8.32
280	8.39
290	8.49
298.15	8.56*
300	8.58
CURVE 6*	
22	1.74×10^{-2}
25	2.49
30	4.14
35	6.17
40	8.56
45	1.12×10^{-1}
50	1.41
55	1.72
60	2.04
65	2.37
70	2.69
75	3.02
80	3.34
85	3.65
90	3.96
95	4.25
100	4.60
110	5.56
120	5.54
130	5.80
140	6.10
150	6.39
160	6.64
CURVE 7	
90	3.96×10^{-1}*
95	4.22*
100	4.46
110	4.92
120	5.34
130	5.72*
140	6.03*
150	6.34*
160	6.61*
170	6.84*
180	7.05*
190	7.25*
200	7.42*
210	7.56*
220	7.71*
230	7.84*
240	7.96*
250	8.07*
260	8.18*
270	8.28*
CURVE 7 (contd)	
273.15	8.32×10^{-1}*
280	8.38*
290	8.48*
298.15	8.55*
300	8.57*
CURVE 8	
21	1.57×10^{-2}
25	2.46*
30	4.07*
35	6.06*
40	8.43*
45	1.11×10^{-1}*
50	1.39*
55	1.71*
60	2.04*
65	2.38*
70	2.71*
75	3.05*
80	3.39
85	3.73
90	4.08
95	4.43
100	4.91
110	5.63
120	5.82
130	6.15
140	6.52
150	6.86
160	7.18
170	7.46
CURVE 9*	
90	4.07×10^{-1}
95	4.37
100	4.64
110	5.17
120	5.63
130	6.08
140	6.47
150	6.82
160	7.13
170	7.43
CURVE 10	
180	7.68×10^{-1}
190	7.90
200	8.09
210	8.28
220	8.47
230	8.65
240	8.82
250	8.97
260	9.10
270	9.22
273.15	9.26
280	9.33
290	9.44
298.15	9.52
300	9.54
CURVE 11*	
15	6.48×10^{-3}
20	1.37×10^{-2}
25	2.44
30	3.93
35	5.95
40	8.26
45	1.11×10^{-1}
50	1.44
60	2.06
70	2.71
80	3.34
90	3.88
100	4.39
110	4.84
120	5.24
130	5.58
140	5.88
150	6.14
160	6.38
180	6.74
200	7.09
220	7.42
240	7.61
260	7.84
280	8.01
300	8.16

* Not shown on plot

DATA TABLE NO. 29 (continued)

T	C_p
CURVE 12*	
459	1.4×10^{0}
573	1.3
673	1.2
773	1.0
CURVE 13*	
473	8.04×10^{-1}
573	1.03×10^{0}
673	1.11
773	1.14
CURVE 14*	
773	9.62×10^{-1}
800	9.62
900	9.62
1000	9.62
1100	9.62
1200	9.62
1273	9.62
CURVE 15*	
298.16	8.491×10^{-1}
300	8.501
320	8.645
340	8.828
360	9.040
380	9.270
400	9.510
420	9.749
440	9.976
(s)453.7	1.011×10^{0}
(l)453.7	1.047
460	1.046
480	0.142
500	1.038
520	1.034
540	1.029
560	1.026
580	1.021
600	1.017
620	1.014
640	1.009
660	1.005
680	1.001
CURVE 15 (cont.)*	
700	9.984×10^{-1}
750	9.975
800	9.967
850	9.958
900	9.950
950	9.941
1000	9.932
1050	9.924
1100	9.915
1150	9.906
1200	9.898

* Not shown on plot

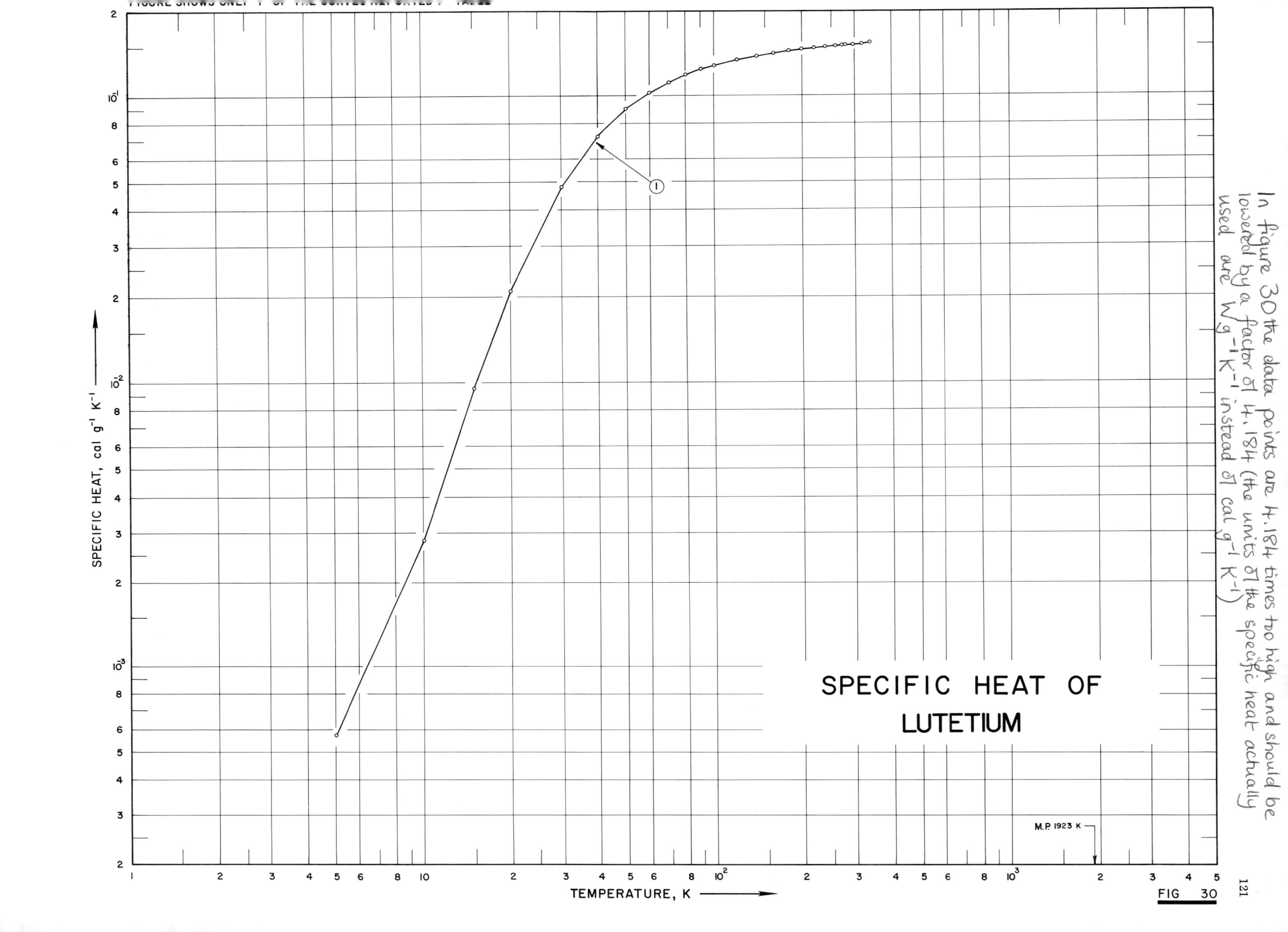

In figure 30 the data points are 4.184 times too high and should be lowered by a factor of 4.184 (the units of the specific heat actually used are $W g^{-1} K^{-1}$ instead of $cal\ g^{-1} K^{-1}$)

SPECIFICATION TABLE NO. 30 SPECIFIC HEAT OF LUTETIUM

(Impurity < 0 20% each; total impurities < 0. 50%)

[For Data Reported in Figure and Table No. 30]

Curve No.	Ref. No.	Year	Temp. Range, K	Reported Error, %	Name and Specimen Designation	Composition (weight percent), Specifications and Remarks
1	158	1960	5-340	0. 1-2. 0		96. 16 Lu, < 0. 20 total of Ca, Cr, Cu, Fe, Mg, Sc, Si, Tm, Y and Yb, 0. 08 N_2, and 0. 01 C; after heat capacity measurements, chemical analysis showed 1. 62 Ta, 1. 97 Lu OF; (corrected for impurities).
2	301	1966	298-1936	< 2. 0		0. 035 Fe, 0. 027 Ca, 0. 025 Si, 0. 013 N_2, 0. 005 Al, 0. 005 Mg, 0. 004 Ni, 0. 003 O_2, and 0. 002 Cu; prepared by metallothermic reduction of the fluoride with calcium and purified by distillation.

DATA TABLE NO. 30 SPECIFIC HEAT OF LUTETIUM

[Temperature, T, K; Specific Heat, C_p Cal $g^{-1}K^{-1}$]

T	C_p
CURVE 1	
5	5.715×10^{-4}
10	2.800×10^{-3}
15	9.544
20	2.092×10^{-2}
30	4.841
40	7.264
50	9.053
60	1.034×10^{-1}
70	1.128
80	1.199
90	1.253
100	1.291
120	1.351
140	1.395
160	1.427
180	1.453
200	1.473
220	1.488
240	1.500
260	1.515
273.15	1.522
280	1.525
298.15	1.535*
300	1.536
320	1.545
340	1.553
CURVE 2*	
298.15	3.64×10^{-2}
300	3.64
400	3.63
500	3.65
600	3.69
700	3.75
800	3.83
900	3.95
1000	4.08
1100	4.24
1200	4.42
1300	4.62
1400	4.85
1500	5.10
1600	5.38
1700	5.68
CURVE 2 (cont.)*	
1800	6.00×10^{-2}
1900	6.35
1936	6.48

* Not shown on plot

123 5a-30a Replace the values of curve 1 by the following:

T	C_p	T	C_p
5	1.37×10^{-4}	140	3.333×10^{-2}
10	6.69	160	3.411
15	2.28×10^{-3}	180	3.472
20	5.00	200	3.521
30	1.16×10^{-2}	220	3.556
40	1.736	240	3.586
50	2.164	260	3.620
60	2.472	273.15	3.638
70	2.696	280	3.646
80	2.866	298.15	3.668
90	2.996	300	3.672
100	3.086	320	3.694
120	3.229	340	3.711

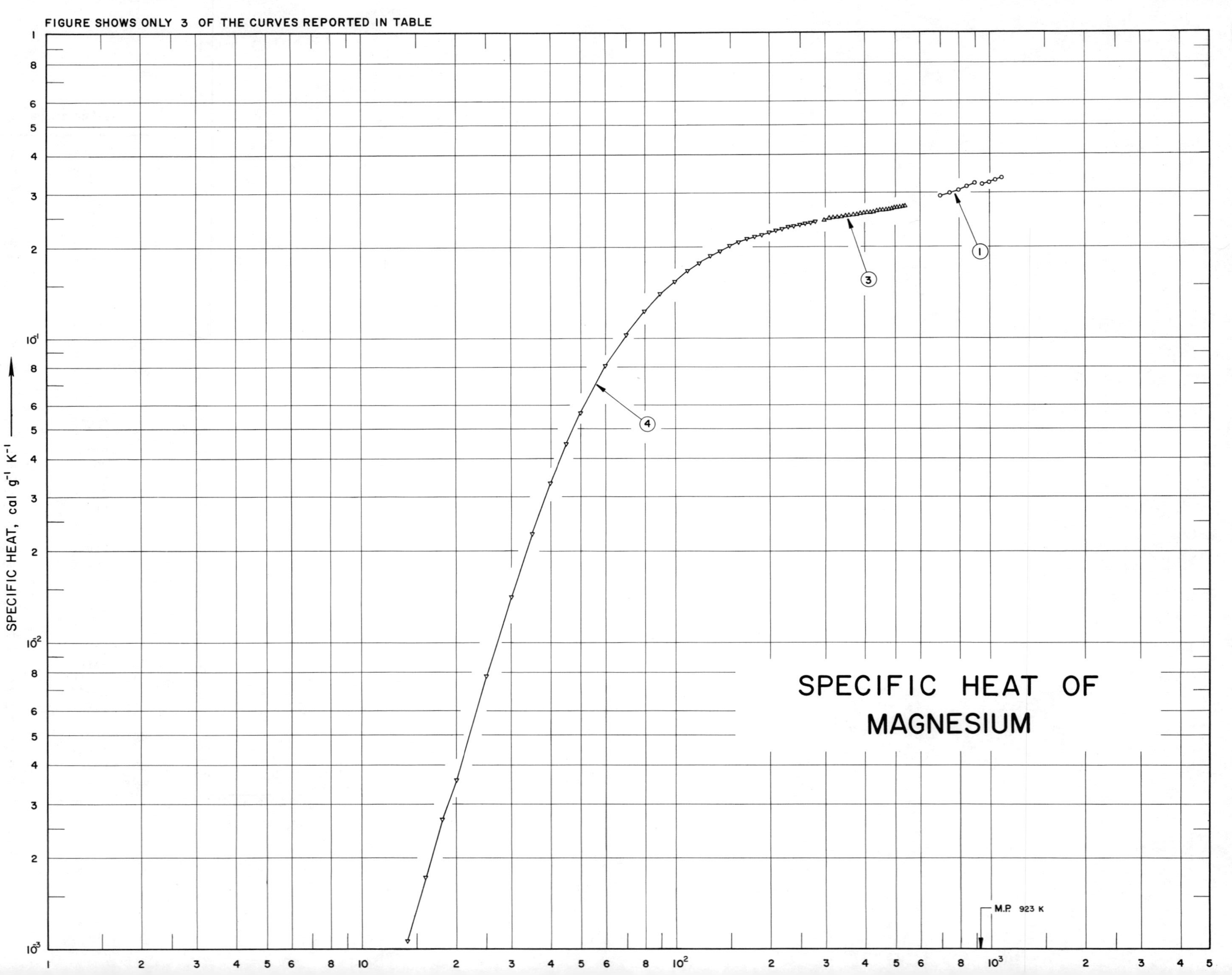
FIGURE SHOWS ONLY 3 OF THE CURVES REPORTED IN TABLE
SPECIFIC HEAT, cal g⁻¹ K⁻¹
SPECIFIC HEAT OF MAGNESIUM
M.P. 923 K
1
3
4

SPECIFICATION TABLE NO. 31 SPECIFIC HEAT OF MAGNESIUM

(Impurity < 0.20% each; total impurities < 0.50%)

[For Data Reported in Figure and Table No. 31]

Curve No.	Ref. No.	Year	Temp. Range, K	Reported Error, %	Name and Specimen Designation	Composition (weight percent), Specifications and Remarks
1	45	1955	700-1100			99.95-99.98 Mg.
2	47	1957	298-543			
3	48	1953	12-320	0.1		99.99 Mg.
4	49	1960	190-300			Pure; polycrystalline; degassed at 500 C in high vacuum.
5	170	1918	75-289	1.0		Small amounts of Al, Fe, and SiO_2.
6	179	1924	373-873	1.0		Small amounts of Al, Fe, and SiO_2; commercial product; cast in sticks.
7	186	1930	11-228			Impurities: 0.035 Mn, 0.02 Si, 0.018 Fe, and 0.006 Al.
8	187	1931	291-773			
9	208	1935	273-873			99.93 Mg.
10	294	1952	2-4			99.96 Mg; sample supplied by the Dow Chemical Co.
11	329	1957	3-13	3.0	Sample 1	0.043 Mn.
12	329	1957	3-13	3.0	Sample 2	0.013 Fe.
13	49	1960	80-290			Sample 1: 99.956 Mg, 0.044 Si, sample 2: 99.918 Mg, 0.082 Si, sample 3: 99.78 Mg, and 0.22 Si; data is average of the 3 samples.

DATA TABLE NO. 31 SPECIFIC HEAT OF MAGNESIUM

[Temperature, T, K; Specific Heat, C_p, Cal g^{-1} K^{-1}]

CURVE 1

T	C_p
700	2.92×10^{-1}
750	2.98
800	3.05
850	3.13
(s)900	3.21
(l) 950	3.19
1000	3.24
1050	3.29
1100	3.35

CURVE 2*

T	C_p
700	2.92×10^{-1}
750	2.98
800	3.05
850	3.13
(s)900	3.21
923	M.P.
(l)950	3.19
1000	3.24
1050	3.29
1100	3.35

CURVE 3

T	C_p
298.16	2.444×10^{-1}*
300	2.447
310	2.461
320	2.473
330	2.485
340	2.498
350	2.511
360	2.523
370	2.535
380	2.546
390	2.557
400	2.567
410	2.577
420	2.587
430	2.597
440	2.607
450	2.617
460	2.627
470	2.638
480	2.649
490	2.660×10^{-1}
500	2.671
510	2.681
520	2.690
530	2.699
540	2.708*
543.16	2.711*

CURVE 4

T	C_p
12	6.58×10^{-4}*
14	1.07×10^{-3}
16	1.73
18	2.67
20	3.54
25	7.73
30	1.40×10^{-2}
35	2.26
40	3.30
45	4.426
50	5.623
60	8.033
70	1.027×10^{-1}
80	1.226
90	1.400
100	1.544
110	1.667
120	1.772
130	1.862
140	1.941
150	2.006
160	2.062
170	2.111
180	2.154
190	2.193
200	2.229
210	2.257
220	2.283
230	2.308
240	2.331
250	2.352
260	2.372
270	2.390
280	2.407
290	2.425*
298.16	2.439×10^{-1}*
300	2.442*
310	2.456*
320	2.467*

CURVE 5*

T	C_p
190	2.193×10^{-1}
200	2.229
210	2.257
220	2.283
230	2.308
240	2.331
250	2.352
260	2.372
280	2.407
300	2.442

CURVE 6*

T	C_p
373	2.57×10^{-1}
473	2.68
573	2.79
673	2.89
773	3.00
873	3.11

CURVE 7*

T	C_p
11.31	5.3×10^{-4}
11.43	5.3
14.14	1.1×10^{-3}
14.28	1.2
16.94	2.0
17.24	2.1
19.50	3.1
19.64	3.0
21.7	4.61
24.3	6.95
27.1	1.01×10^{-2}
30.2	1.45
34.2	2.19
37.6	2.86
41.7	3.78
45.2	4.614
48.7	5.469×10^{-2}
51.8	6.242
55.1	7.089
63.7	9.104
86.7	1.335×10^{-1}
93.2	1.458
99.2	1.529
106.7	1.595
115.4	1.727
124.2	1.820
136.2	1.908
145.4	1.986
163.4	1.209
172.6	2.135
182.0	2.150
191.2	2.191
217.2	2.245
228.4	2.270

CURVE 8*

T	C_p
291	2.42×10^{-1}
373	2.55
473	2.67
573	2.76
673	2.87
773	2.99

CURVE 9*

T	C_p
273	2.413×10^{-1}
373	2.518
473	2.624
573	2.729
673	2.834
773	2.939
873	3.045

CURVE 10*

T	C_p
2.040	3.22×10^{-5}
2.207	3.57
2.364	3.81
2.538	4.19
2.672	4.56
2.778	4.85×10^{-5}
2.896	5.14
2.957	5.10
3.054	5.43
3.150	5.47
3.223	5.80
3.301	6.21
3.351	6.25
3.410	6.41
3.469	6.78
3.510	7.16
3.571	7.24
3.631	7.65

CURVE 11*

T	C_p
3.14	5.2×10^{-5}
4.55	7.6
5.60	1.33×10^{-4}
6.44	1.64
7.22	2.18
7.89	2.60
8.47	3.11
9.10	3.76
9.71	4.28
10.30	4.26
10.83	5.31
11.35	6.33
11.84	6.87
12.27	8.40
12.62	8.12

CURVE 12*

T	C_p
3.14	5.3×10^{-5}
4.55	7.8
5.60	1.27×10^{-4}
6.44	1.62
7.22	2.08
7.89	2.70
8.47	3.19
9.10	3.52
9.71	4.24
10.30	4.71
10.83	5.37×10^{-4}
11.35	6.29
11.84	6.92
12.27	7.36
12.62	7.82

CURVE 13*

T	C_p
80	1.229×10^{-1}
90	1.377
100	1.506
110	1.625
120	1.736
130	1.830
140	1.906
150	1.971
160	2.021
170	2.064
180	2.101
190	2.140
200	2.176
210	2.209
220	2.240
230	2.267
240	2.290
250	2.309
260	2.326
270	2.341
280	2.354
290	2.364

* Not shown on plot

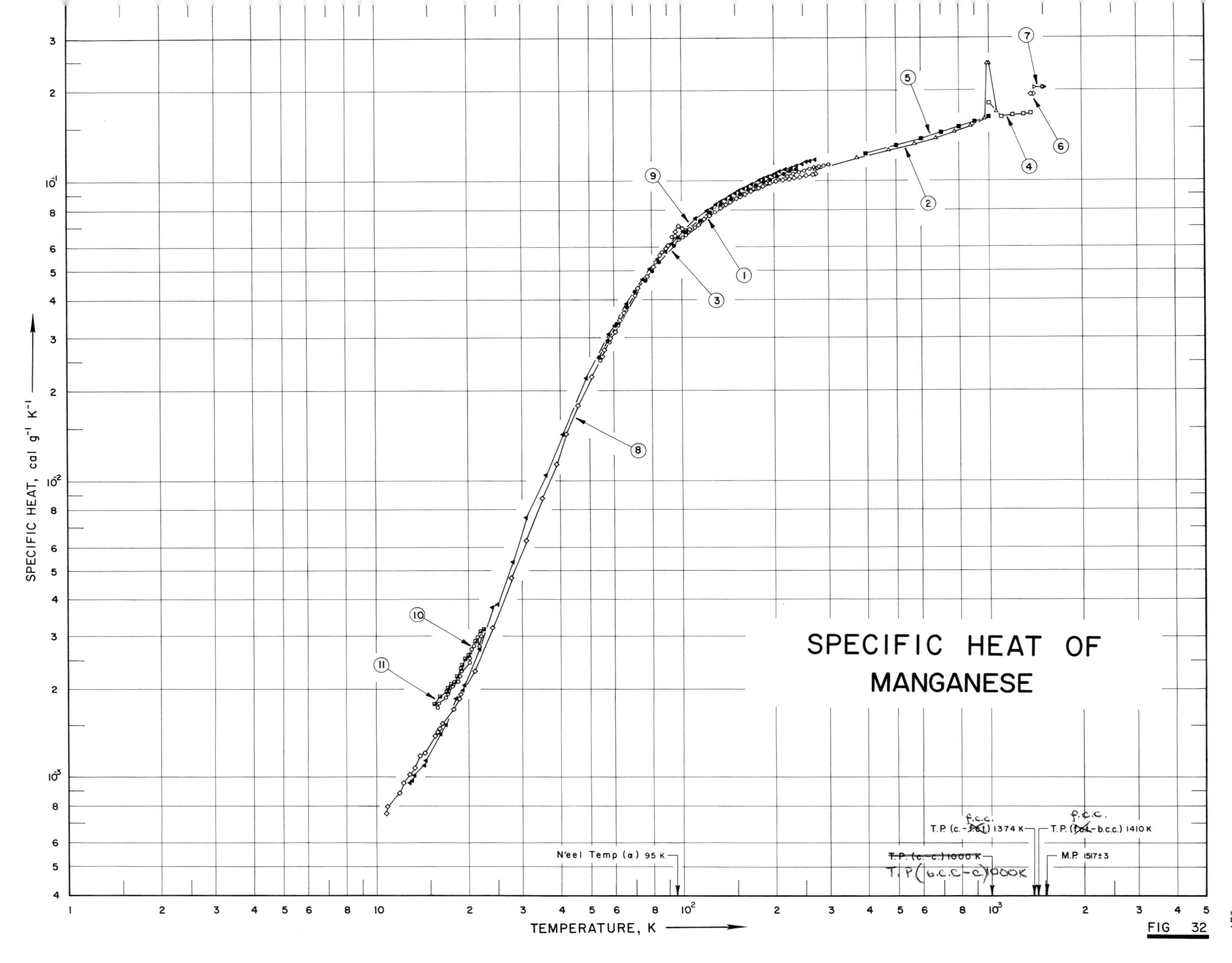
SPECIFIC HEAT OF
MANGANESE
SPECIFIC HEAT, cal g⁻¹ K⁻¹
TEMPERATURE, K
N'eel Temp (α) 95 K
T.P. (c.-c.) 1000 K
T.P. (b.c.c.-c.) 1000 K
T.P. (c.-f.c.c.) 1374 K
T.P. (f.c.c.-b.c.c.) 1410 K
M.P. 1517±3
FIG 32

SPECIFICATION TABLE NO. 32 SPECIFIC HEAT OF MANGANESE

(Impurity < 0.20% each; total impurities < 0.50%)

[For Data Reported in Figure and Table No. 32]

Curve No.	Ref. No.	Year	Temp. Range, K	Reported Error, %	Name and Specimen Designation	Composition (weight percent), Specifications and Remarks
1	330	1945	54-301		α - manganese	99.9 Mn; prepared by heating gradually to 550 C in 10^{-4} mm Hg vacuum.
2	24	1950	273-1073		Electrolytic	< 0.01 impurities, mainly Mg.
3	330	1945	54-237		γ - manganese	99.9 Mn; prepared by electrolysis of manganous sulfate in the presence of sulfites
4	57	1946	1000-1374		β - manganese	99.9+ Mn, 0.03 - 0.07 Si samples in capsule of silica glass.
5	57	1946	298-1000		α - manganese	Same as above.
6	57	1946	1374-1410		γ - manganese	Same as above.
7	57	1946	1410-1517		δ - manganese	Same as above.
8	58	1964	11-273		α - manganese	99.95, 0.01 each C, Cu, and Fe; 0.001-0.01 O, 0.001 Si, 0.001 N and 2.00 H_2 per 199 g; density = 464 lb ft^3
9	58	1964	13-273		γ - manganese	Same as above; γ- manganese stabilized by adding 0.06 Cu.
10	59	1940	16-22		Powdered manganese	
11	60	1949	16-18		Powdered manganese	
12	221	1939	54-290			98.94 Mn, 1.06 Mn O_2, 0.24 O_2; (corrected for impurities).
13	331	1945	300-1450		Electrolytic	99.9 Mn; sample A, degassed by heating to 850 C under 10^{-5} mm vacuum and cooled slowly.
14	332	1955	12-20		β - manganese	Traces of Mg, and Ca; standardized Mn from Johnson Matthey and Co., Ltd., Lab. No 4135; annealed at 1120 C for 16 hrs under argon atmosphere followed by rapid quenching.
15	332	1955	12-20		α - manganese	Same composition as above; obtained from β - Mn by heating to 1100 C followed by slow cooling to room temperature and afterwards held at 600 C for 3 hrs followed by slow cooling to room temperature
16	223	1959	0.64-3.1	5.0		> 99.9 Mn; α - phase body centered cubic crystal structures; sample supplied by the Bureau of Mines, Battlesville, Oklahoma.

DATA TABLE NO. 32 SPECIFIC HEAT OF MANGANESE

[Temperature, T, K; Specific Heat, C_p, Cal $g^{-1}K^{-1}$]

T	C_p
CURVE 1	
54.2	2.512 x 10^{-2}
58.2	2.898
62.1	3.288
70.3	4.120
77.8	4.884*
82.2	5.330
86.5	5.786
90.2	6.168*
93.3	6.486
95.9	6.526*
98.4	6.383
101.2	6.441
104.3	6.597
111.5	6.991
129.3	7.894
149.8	8.704*
172.6	9.412
190.4	9.829
222.5	1.044 x 10^{-1}
232.0	1.059
241.2	1.072
251.1	1.087
260.9	1.102
271.0	1.114
280.8	1.125
291.0	1.137
301.3	1.147
CURVE 2	
273.15	1.111 x 10^{-1}*
373	1.200
473	1.274
573	1.336
673	1.398
773	1.458
873	1.530
973	1.615
1073	1.701
CURVE 3	
53.7	2.572 x 10^{-2}
57.4	2.918
61.7	3.330
CURVE 3 (cont.)	
66.4	3.772 x 10^{-2}
71.8	4.271
76.1	4.651
80.3	4.997
84.8	5.350
94.7	6.080
104.5	6.725
115.2	7.349
124.3	7.815
135.0	8.311
145.8	8.729
155.5	9.064
165.6	9.386
175.8	9.692
185.6	9.936
196.1	1.017 x 10^{-1}
205.9	1.040
216.5	1.065
226.3	1.080
236.5	1.099
CURVE 4	
1000	1.64 x 10^{-1}
1100	1.65
1200	1.66
1300	1.67
1374	1.68
CURVE 5	
298.15	1.144 x 10^{-1}
300	1.146
400	1.241
500	1.318
600	1.388
700	1.454
800	1.519
900	1.583
1000	1.646
1374	1.950
1400	1.950
1410	1.950
CURVE 7	
1410	2.06 x 10^{-1}
1500	2.06
1517	2.06
CURVE 8	
10.74	7.55 x 10^{-4}
10.87	7.99
11.95	8.88
12.22	9.52
12.28	9.65
12.92	1.03 x 10^{-3}
13.40	1.07
13.94	1.19
13.98	1.12
14.50	1.22
15.62	1.39
15.86	1.43
16.20	1.46
16.53	1.53
17.92	1.70
18.70	1.867
18.81	1.915
18.94	1.929
21.13	2.372
24.02	3.231
27.84	4.709
31.27	6.336
35.29	8.759
39.20	1.148 x 10^{-2}
42.99	1.449
46.90	1.794
51.58	2.230*
55.14	2.585
56.95	2.738
58.96	2.963
61.03	3.140
63.54	3.429
65.69	3.615
67.62	3.808*
71.43	4.188*
72.99	4.350
76.37	4.678*
77.55	4.776
81.02	5.133
CURVE 8 (cont.)	
81.88	5.206 x 10^{-2}*
85.50	5.601
86.18	5.588*
87.00	5.706*
87.03	5.613*
88.52	5.868*
89.10	5.910
90.75	6.081
91.35	6.174*
91.69	6.203*
94.69	6.451
94.93	6.514*
95.58	6.560
96.02	6.600*
96.20	6.675*
97.82	6.818*
97.92	6.815*
98.06	7.024
99.99	7.035*
101.96	6.929
104.01	6.888*
104.84	6.793*
105.46	6.840*
106.09	6.867
108.25	6.886*
109.99	6.973*
110.09	6.933*
110.40	6.968*
114.67	7.184
119.96	7.424
124.58	7.657
129.60	7.803*
135.33	8.120
140.25	8.335
145.77	8.524
151.14	8.711
152.03	8.759*
156.18	8.848
157.23	8.890*
161.74	9.046
162.38	9.101*
166.73	9.172*
168.84	9.277
174.18	9.387*
179.30	9.479
CURVE 8 (cont.)	
185.25	9.601 x 10^{-2}
190.33	9.747*
196.16	9.860
199.55	9.905*
201.22	9.956*
204.73	1.004 x 10^{-1}
206.48	1.008*
208.52	1.012*
211.21	1.016*
213.97	1.021
216.61	1.021*
226.18	1.045
228.97	1.049*
231.34	1.045*
234.41	1.059
237.20	1.063*
239.62	1.064*
244.95	1.070
246.58	1.082*
250.76	1.074*
252.37	1.089*
256.15	1.091
259.81	1.085*
263.67	1.100
269.58	1.112
273.00	1.107
CURVE 9	
12.90	9.57 x 10^{-4}
13.20	9.72
13.25	9.74*
13.46	1.03 x 10^{-3}
14.44	1.11
14.51	1.15
14.55	1.14*
14.67	1.16*
16.18	1.41
16.32	1.44
16.73	1.51
16.78	1.51*
18.39	1.85
18.47	1.86*
19.11	1.99
19.46	2.08
CURVE 9 (cont.)	
21.95	2.74 x 10^{-3}
22.06	2.71
24.80	3.78
25.04	3.87
28.13	5.36
31.97	7.52
36.35	1.056 x 10^{-2}
41.01	1.430
45.17	1.797
49.55	2.206
58.67	3.082
61.71	3.295
66.48	3.855
66.77	3.862*
71.34	4.285*
75.67	4.680
79.86	5.064
84.72	5.466
89.13	5.779
93.51	6.123
93.83	6.121*
98.37	6.494
102.90	6.747
103.11	6.787*
107.69	7.100
112.22	7.406
121.43	7.916
126.14	8.076
130.74	8.309
135.94	8.526
140.76	8.688
145.48	8.942
150.76	9.094
155.61	9.265
160.39	9.428
165.60	9.563
170.42	9.771
181.84	1.001 x 10^{-1}
186.87	1.016
191.84	1.032
198.81	1.045
198.98	1.046*
204.18	1.057
209.44	1.071
210.32	1.072*

* Not shown on plot

DATA TABLE NO. 32 (continued)

T	C_p
CURVE 9 (cont.)	
214.48	1.083 x 10^{-1}
215.77	1.080*
218.97	1.085
223.05	1.095
224.66	1.101*
228.79	1.109
230.14	1.106*
234.47	1.117
238.01	1.127
239.94	1.130*
243.95	1.139
247.75	1.142
249.60	1.147*
253.66	1.152*
255.00	1.160
259.29	1.161*
262.73	1.167
267.38	1.175*
268.63	1.177*
273.38	1.186
CURVE 10	
15.90	1.726 x 10^{-3}
15.91	1.709*
16.03	1.798
16.67	1.920*
16.85	1.877
17.16	1.929
17.32	1.995*
17.25	1.888*
17.41	1.980*
17.57	2.000*
17.73	2.055*
17.79	2.050*
17.80	2.181*
18.27	2.162*
18.51	2.131
18.75	2.228
18.99	2.324
19.23	2.304*
19.31	2.303*
19.35	2.335*
19.87	2.417*
19.96	2.510*
19.97	2.401*
20.14	2.474

T	C_p
CURVE 10 (cont.)	
20.28	2.550 x 10^{-3}
20.41	2.588*
20.59	2.721
20.74	2.743*
20.91	2.788
20.97	2.727*
21.50	2.996
21.85	2.783
22.10	3.018
CURVE 11	
15.52	1.771 x 10^{-3}
16.24	1.891
16.34	1.867*
16.81	1.951*
17.07	1.966
17.29	2.029
17.64	2.080
17.65	2.080
18.08	2.122
18.43	2.228
18.62	2.235*
18.92	2.243
19.25	2.415
19.53	2.545
19.83	2.474*
20.14	2.596
20.55	2.725*
20.80	2.781*
21.14	2.880
21.39	2.947
21.32	2.894*
22.02	3.012*
22.16	3.122
22.48	3.167
CURVE 12*	
53.7	2.483 x 10^{-2}
56.8	2.791
60.3	3.159
64.0	3.554
68.0	3.961
74.8	4.631
80.4	5.178
88.2	5.698

T	C_p
CURVE 12 (cont.)*	
96.9	6.371 x 10^{-2}
105.8	6.836
115.0	7.273
124.1	7.698
133.2	8.092
143.0	8.465
153.5	8.844
163.1	9.137
172.9	9.403
183.2	9.668
193.1	9.910
203.3	1.008 x 10^{-1}
213.0	1.028
222.5	1.046
231.5	1.056
244.1	1.077
254.4	1.093
265.8	1.107
275.0	1.116
278.1	1.119
282.0	1.127
285.9	1.125
289.8	1.126
CURVE 13*	
300	1.146 x 10^{-1}
400	1.241
500	1.318
600	1.388
700	1.441
800	1.509
900	1.583
α 1000	1.646
β 1000	1.636
1100	1.648
1200	1.660
β 1374	1.681
γ 1374	1.948
1400	1.948
γ 1410	1.948
δ 1410	2.057
1420	2.057
1430	2.057
1440	2.057
1450	2.057

T	C_p
CURVE 14*	
12	1.13 x 10^{-2}
13	1.16
14	1.26
15	1.35
16	1.49
17	1.63
18	1.78
19	1.97
20	2.11
CURVE 15*	
12	3.46 x 10^{-3}
13	4.00
14	4.55
15	5.10
16	5.82
17	6.55
18	7.28
19	8.01
20	8.91
CURVE 16*	
0.635	1.350 x 10^{-4}
0.640	1.163
0.647	1.016
0.649	1.267
0.670	1.503
0.675	1.184
0.689	1.447
0.696	1.390
0.703	1.489
0.715	1.329
0.715	1.250
0.730	1.315
0.733	1.444
0.737	1.440
0.737	1.514
0.742	1.325
0.767	1.544
0.768	1.530
0.770	1.494
0.776	1.289
0.776	1.358
0.803	1.479
0.812	1.343

T	C_p
CURVE 16 (cont.)*	
0.828	1.269 x 10^{-4}
0.833	1.487
0.843	1.503
0.862	1.370
0.871	1.581
0.885	1.617
0.901	1.645
0.975	1.620
1.132	2.057
1.240	2.091
1.364	2.453
1.381	2.457
1.419	2.518
1.453	2.487
1.548	2.895
1.636	2.943
1.749	3.160
1.851	3.367
2.081	3.948
2.111	4.152
2.171	4.183
2.570	5.136
3.086	5.938

* Not shown on plot

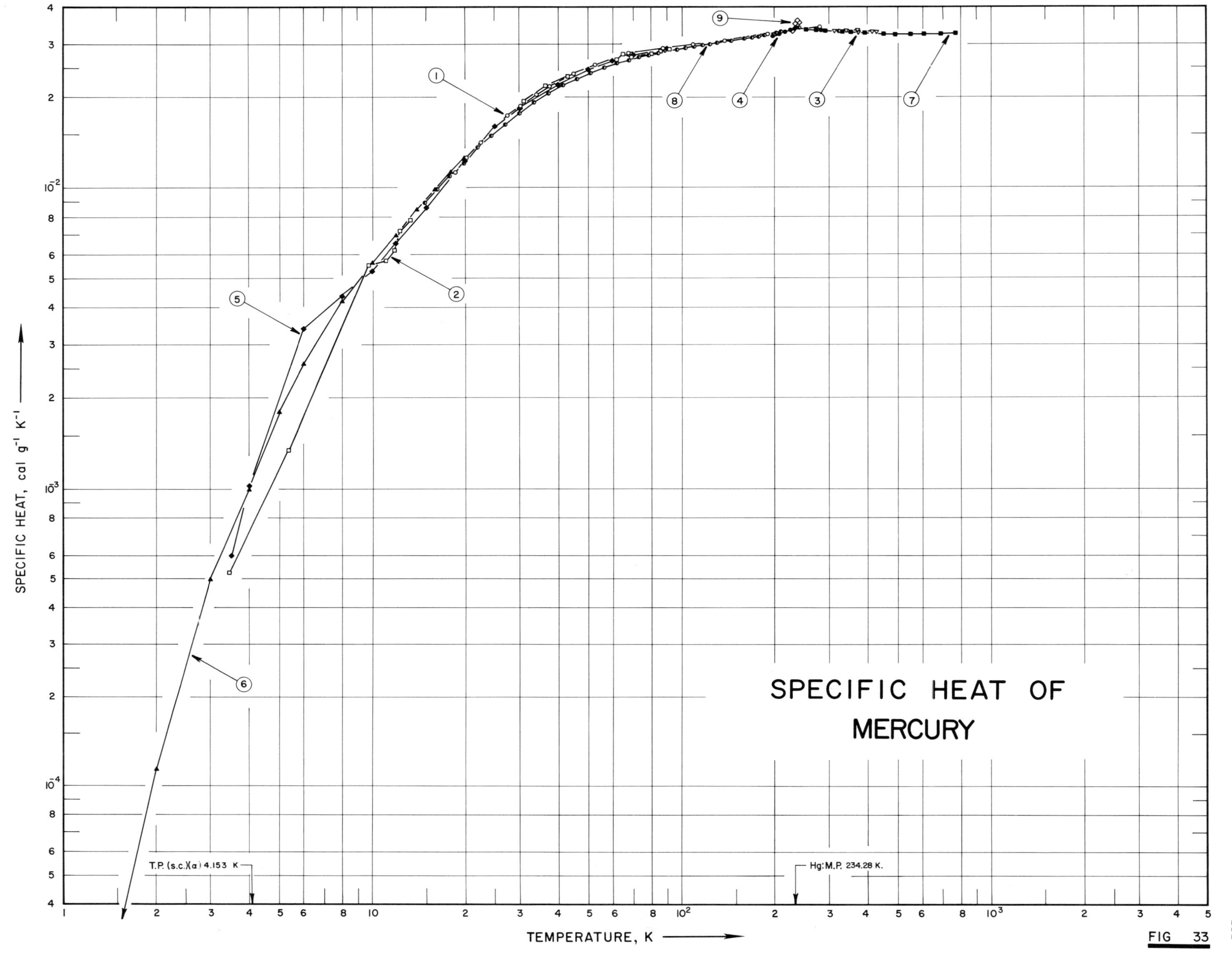

SPECIFIC HEAT OF
MERCURY
SPECIFIC HEAT, cal g⁻¹ K⁻¹
TEMPERATURE, K
T.P. (s.c.)(α) 4.153 K
Hg:M.P. 234.28 K.
FIG 33

SPECIFICATION TABLE NO. 33 SPECIFIC HEAT OF MERCURY

(Impurity < 0.20% each; total impurities < 0.50%)

[For Data Reported in Figure and Table No. 33]

Curve No.	Ref. No.	Year	Temp. Range, K	Reported Error, %	Name and Specimen Designation	Composition (weight percent), Specifications and Remarks
1	333	1922	19-282			Vacuum distilled.
2	334	1923	3-92			
3	214	1927	314-416	1.0		Electrolytically purified; twice distilled under low pressure of current air.
4	335	1930	198-285	1.0	P sample	Trace of Sn; purified.
5	336	1948	4-90			Very pure.
6	284	1956	1-20			99.999 Hg; sample supplied from Johnson, Matthey and Co.
7	337	1951	234-773			< 0.01 each Cu and Ni, and 0.00001 non volatile impurity, mostly silver; sample was distilled 3 times.
8	338	1953	15-318	0.1-3.0		0.0006 max. total impurity; specimen from Mallinckrodt.
9	423	1911	61-243			

DATA TABLE NO. 33 SPECIFIC HEAT OF MERCURY

[Temperature, T, K; Specific Heat, C_p, Cal $g^{-1}K^{-1}$]

T	C_p
CURVE 1	
18.7	1.123 x 10^{-2}
20.3	1.259
22.6	1.404
27.5	1.724
30.3	1.866
34.2	2.035
37.7	2.167
45.0	2.381
52.9	2.550
67.9	2.790
87.5	2.891
109.4	2.983
138.0	3.073
191.7	3.241
281.8	3.421
CURVE 2	
Series I	
9.78	5.519 x 10^{-3}
10.17	5.738
10.89	6.202
11.09	6.401*
12.35	7.194
12.55	7.388*
13.35	7.827
Series II	
3.45	5.24 x 10^{-4}
5.37	1.35 x 10^{-3}
9.78	5.53*
10.17	5.73*
10.89	6.18*
11.09	6.38*
12.35	7.18*
12.55	7.38*
13.35	7.83*
18.7	1.12 x 10^{-2}*
20.3	1.26*
22.6	1.41*
27.5	1.73*
30.3	1.86*
31.1	1.94
34.2	2.03*
CURVE 2 (cont.)	
36.6	2.17 x 10^{-2}
37.7	2.17*
43.0	2.34
45.0	2.38*
52.9	2.55*
62.0	2.66
65.0	2.68
67.9	2.79*
69.0	2.71
80.5	2.78
86.0	2.83
87.5	2.89*
92.0	2.89
CURVE 3	
314	3.32 x 10^{-2}
343	3.33
343	3.29*
373	3.33
374	3.27
374	3.28*
374	3.31*
375	3.29*
376	3.34*
376	3.28*
376	3.32*
376	3.32*
376	3.29*
376	3.32*
416	3.30
416	3.32*
430	3.29
CURVE 4	
197.6	3.20 x 10^{-2}
199.4	3.24*
201.1	3.24*
207.5	3.24*
208.1	3.25
208.6	3.29*
209.9	3.27*
212.9	3.24*
214.5	3.25*
CURVE 4 (cont.)	
217.6	3.29 x 10^{-2}
218.9	3.33*
222.9	3.32*
227.4	3.35
229.2	3.39*
229.7	3.37*
231.0	3.37*
231.0	3.36*
232.4	3.36*
232.7	3.38*
(s) 233.8	3.37
(l) 236.5	3.39
285.2	3.34
CURVE 5	
3.5	6.0 x 10^{-4}
4	1.0 x 10^{-3}
6	3.4
8	4.4
10	5.28
12	6.53
15	8.58
20	1.23 x 10^{-2}
25	1.59
30	1.84
40	2.20
50	2.47
60	2.64
70	2.75
80	2.83*
90	2.89
CURVE 6	
1.2	1.41 x 10^{-5}*
1.5	3.05*
2.0	1.15 x 10^{-4}
3.0	4.99
4.0	9.97
5.0	1.80 x 10^{-3}
6.0	2.59
8.0	4.17
10.0	5.61
12.0	6.98
CURVE 6 (cont.)	
14.0	8.48 x 10^{-3}
16.0	9.87
18.0	1.13 x 10^{-2}
20.2	1.26
CURVE 7	
234	3.379 x 10^{-2}*
253	3.353
273	3.338
293	3.324
298	3.321*
313	3.311
333	3.299
353	3.288
373	3.278*
393	3.269
413	3.261*
433	3.254*
453	3.248
473	3.243*
793	3.239
513	3.236*
533	3.234*
553	3.232
573	3.232*
593	3.233*
610	3.236
613	3.234*
633	3.237*
653	3.240*
673	3.244*
693	3.248
713	3.254*
733	3.260*
753	3.266*
773	3.259
CURVE 8	
Series I*	
197.57	3.243 x 10^{-2}
203.71	3.264
209.92	3.289
216.42	3.317
CURVE 8 (cont.)	
222.94	3.346 x 10^{-2}
229.26	3.370
249.35	3.375
255.80	3.378
Series II*	
288.00	3.344 x 10^{-2}
Series III	
325.89	3.306 x 10^{-2}
Series IV*	
242.17	3.387 x 10^{-2}
248.18	3.383
254.32	3.372
261.79	3.363
268.04	3.360
274.22	3.355
280.50	3.353
286.50	3.346
293.03	3.338
298.49	3.332
299.48	3.332
304.99	3.329
311.35	3.321
317.77	3.316
Series V*	
222.46	3.340 x 10^{-2}
229.09	3.372
239.02	3.397
Series VI	
14.90	8.868 x 10^{-3}
16.28	9.855
17.90	1.089 x 10^{-2}
19.84	1.215
22.02	1.353
24.31	1.477
27.04	1.613
CURVE 8 (cont.)	
30.20	1.765 x 10^{-2}
33.54	1.916
37.35	2.054
41.60	2.182
46.20	2.294
51.16	2.397
56.53	2.495
62.27	2.582
67.80	2.649
73.21	2.707
78.68	2.749
84.48	2.793
90.64	2.834*
97.08	2.873
103.84	2.915
110.84	2.943
117.69	2.973
124.52	2.998
131.52	3.025
138.81	3.049*
146.32	3.072
161.89	3.123
170.04	3.141
178.00	3.174
186.27	3.201
194.44	3.230*
202.89	3.260
211.39	3.294
219.79	3.328*
228.13	3.365
241.86	3.391
CURVE 9	
61.5	2.670 x 10^{-2}*
62	2.665*
65	2.685*
66	2.688*
69	2.715*
79	2.770*
82	2.808*
86	2.830*
90	2.848*
94	2.875*
201	3.210*

* Not shown on plot

DATA TABLE NO. 33 (continued)

T	C_p
CURVE 9 (cont.)	
206	3.215 x 10^{-2}*
209	3.260*
214	3.290*
228	3.310*
230	3.305
233	3.475*
233	3.350*
236	3.490
238	3.545*
238	3.590
243	3.545

*Not shown on plot

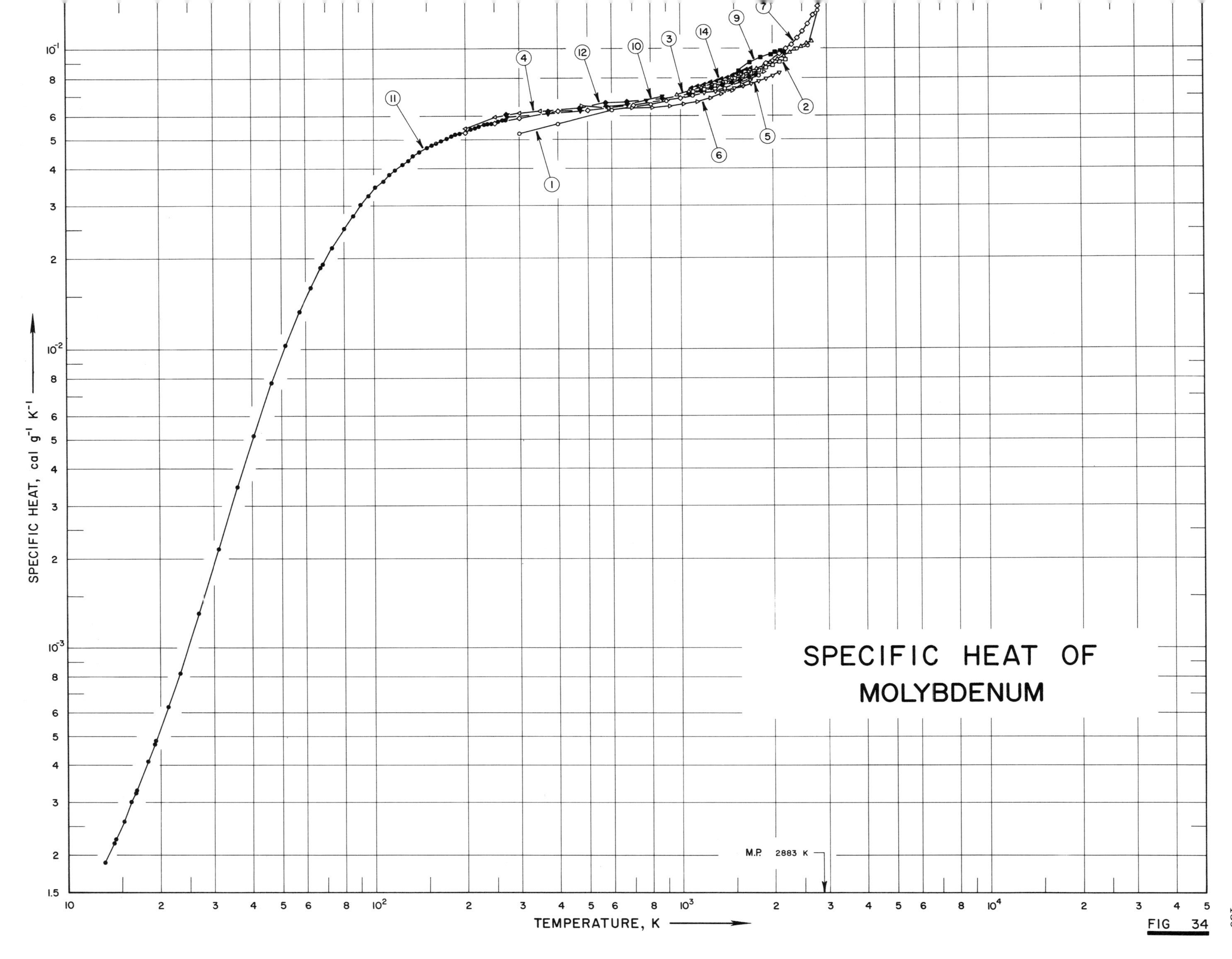

SPECIFIC HEAT OF
MOLYBDENUM
SPECIFIC HEAT, cal g⁻¹ K⁻¹
TEMPERATURE, K
M.P. 2883 K
1
2
3
4
5
6
7
9
10
11
12
14

FIG 34

SPECIFICATION TABLE NO. 34 SPECIFIC HEAT OF MOLYBDENUM

(Impurity < 0.20% each; total impurities < 0.50%)

[For Data Reported in Figure and Table No. 34]

Curve No.	Ref. No.	Year	Temp. Range, K	Reported Error, %	Name and Specimen Designation	Composition (weight percent), Specifications and Remarks
1	17	1960	300-2800	1.7		
2	61	1961	1100-2200	5.0		99.98 Mo, and 0.02 MoO; sample supplied by Moscow Hard Alloys Plant.
3	62	1962	973-2673	± 1.2		Polished surface; under vacuum and argon atmosphere.
4	63	1961	200-350			
5	64	1963	1200-2100	≤ 3.0	wire sample	99.93 Mo, 0.01 each C and Fe, 0.005 O_2, 0.001 H_2, 0.001 N_2, and 0.02 distilled residue; sample supplied by Metallwerk Plansee, Reutte Austria; outgassed and mounted in evacuated (1 x 10^{-6} mm Hg) glass envelope.
6	10	1958	478-1866			Sample supplied by the Climax Molybdenum Co.; measured under an atmosphere of 95% argon and 5% hydrogen; density = 640 lb ft^{-3} at 75 F.
7	65	1961	200-2860	4.0		
8	66	1951	473-1273			
9	67	1960	1550-2180	± 10.0		
10	68	1962	273-2673			
11	69	1959	13-271			99.9896 Mo, 0.004-0.006 Fe, 0.002 O, 0.002 Si, 0.0001 H, and 0.0001-0.0003 Co, Cu, Mg, Ni, and W; sample rods formed by powder metallurgy.
12	70	1934	273-1873			Pure; heated at 1400 C for 4, 8, and 12 hrs.
13	71	1960	1089-1700		Mo-9-8	99.90 Mo, <0.005 Fe, and < 0.003 C.
14	71	1960	1089-1700		Mo-11-5	Same as above.
15	71	1960	1089-1700		Mo-11-10	Same as above
16	72	1965	1250-1600	0.09		99.9 Mo; sample supplied by the Fansteel Metallurgical Corp.; annealed at 1425 K.
17	101	1958	400-946			Specimen's surface plated with platinum black.
18	213	1926	16-275			
19	339	1929	233-523	2.4		
20	340	1952	298-2650			99.9 Mo; sample supplied by the Fansteel Metallurgical Corp.
21	175	1953	1-10			Same as above.
22	341	1964	0.4-4		Mo-1	Very pure molybdenum single crystals; zone-refined; normal state.
23	341	1964	0.4-4		Mo-2	Ultra pure molybdenum single crystals; zone refined; normal state.
24	341	1964	0.4-0.9		Mo-2	Ultra pure molybdenum single crystals; zone refined; superconducting state.
25	341	1964	0.4-0.9		Mo-1	Ultra pure molybdenum single crystals; zone refined; superconducting state.

DATA TABLE NO. 34 SPECIFIC HEAT OF MOLYBDENUM

[Temperature, T, K; Specific Heat, C_p, Cal $g^{-1} K^{-1}$]

CURVE 1

T	C_p
300	5.25 x 10^{-2}
400	5.65
600	6.30
800	6.55
900	6.72
1100	7.00
1400	7.40
1800	8.15
2000	9.14
2400	1.00 x 10^{-1}
2600	1.03
2800	1.35

CURVE 2

T	C_p
1100	7.21 x 10^{-2}
1200	7.39
1300	7.57
1400	7.76
1500	7.94
1600	8.13
1700	8.31
1800	8.49
1900	8.68
2000	8.86
2100	9.04
2200	9.23

CURVE 3

T	C_p
973	7.074 x 10^{-2}
1073	7.251
1173	7.432
1273	7.617
1373	7.807
1473	8.002
1573	8.201
1673	8.405
1773	8.614
1873	8.827
1973	9.045
2073	9.267
2173	9.494
2273	9.726
2373	9.962 x 10^{-2}
2473	1.020 x 10^{-1}
2573	1.045
2673	1.070

CURVE 4

T	C_p
200	5.47 x 10^{-2}
250	5.95
300	6.13
350	6.22

CURVE 5

T	C_p
1200	7.12 x 10^{-2}
1300	7.21
1400	7.31
1500	7.41
1600	7.53
1700	7.66
1800	7.81
1900	7.96
2000	8.14
2100	8.34

CURVE 6

T	C_p
478	6.5 x 10^{-2}
589	6.4
700	6.4
811	6.4
922	6.5
1033	6.6
1144	6.7
1255	6.9
1366	7.1
1478	7.3
1589	7.6
1700	7.9
1811	8.2*
1866	8.4

CURVE 7

T	C_p
200	5.30 x 10^{-2}
250	5.70
300	5.93
400	6.21
500	6.29
600	6.39
700	6.50
800	6.60
900	6.75
1000	6.88
1100	7.03
1200	7.21
1300	7.40
1400	7.64
1500	7.86
1600	8.12
1700	8.38
1800	8.68
1900	8.96
2000	9.28
2100	9.65
2200	1.00 x 10^{-1}
2300	1.04
2400	1.09
2500	1.15
2600	1.22
2700	1.30
2800	1.39
2860	1.45

CURVE 8*

T	C_p
473	6.75 x 10^{-2}
673	6.75
873	6.75
1073	6.75
1273	6.75

CURVE 9

T	C_p
1550	8.4 x 10^{-2}
1690	9.0
1830	9.4
1970	9.6
2040	9.7 x 10^{-2}
2110	9.8
2180	9.8

CURVE 10

T	C_p
273.15	5.970 x 10^{-2}
373	6.114
473	6.262
573	6.416
673	6.573
773	6.736
873	6.903
973	7.074*
1073	7.251*
1173	7.432*
1273	7.617*
1373	7.807*
1473	8.002*
1573	8.201*
1673	8.405*
1773	8.614*
1873	8.827*
1973	9.045*
2073	9.267*
2173	9.494*
2273	9.726*
2373	9.962*
2473	1.020 x 10^{-1}*
2573	1.045*
2673	1.070*

CURVE 11

T	C_p
13.23	1.90 x 10^{-4}
14.14	2.18
14.38	2.25
15.27	2.60
16.10	3.00
16.61	3.24
16.71	3.28
18.34	4.11
19.24	4.70
19.39	4.77*
19.46	4.82 x 10^{-4}
21.30	6.28
23.16	8.22
26.81	1.307 x 10^{-3}
31.13	2.152
35.98	3.472
40.77	5.186
46.62	7.724
51.85	1.032 x 10^{-2}
57.56	1.332
62.60	1.601
67.52	1.856
68.42	1.906
73.46	2.171
73.53	2.169*
80.61	2.519
86.23	2.788
91.57	3.046
96.69	3.250
102.73	3.463
108.36	3.647
113.27	3.823
118.39	3.965
125.51	4.164
131.14	4.296
136.73	4.435
142.22	4.563
149.98	4.707
155.80	4.828
161.44	4.898
166.97	4.988
174.57	5.071
180.88	5.142
186.74	5.213
192.37	5.275
200.32	5.351*
201.57	5.371*
202.37	5.374*
202.63	5.412*
207.25	5.424*
208.03	5.418
212.53	5.470*
212.74	5.480*
215.86	5.498
218.05	5.531*
221.86	5.560* x 10^{-2}
225.64	5.589*
227.59	5.590*
231.42	5.638
235.29	5.656*
236.95	5.682
241.36	5.701*
244.84	5.700
247.05	5.720*
250.64	5.760*
256.22	5.785*
256.31	5.786
262.57	5.812*
264.80	5.817
268.30	5.837*
270.89	5.845

CURVE 12

T	C_p
273.15	6.105 x 10^{-2}
373	6.229
473	6.355
573	6.484
673	6.614
773	6.746*
873	6.881
973	7.018*
1073	7.157
1173	7.297
1273	7.440
1373	7.585
1473	7.732
1573	7.881
1673	8.033
1773	8.186
1873	8.342*

CURVE 13*

T	C_p
1088.9	7.28 x 10^{-2}
1144.4	7.32
1200.0	7.38
1255.5	7.46
1311.1	7.56
1366.6	7.66

*Not shown on Plot

DATA TABLE NO. 34 (continued)

CURVE 13 (cont.)

T	C_p
1422.2	7.57 x 10^{-2}
1477.8	7.88
1533.3	7.98
1588.9	8.07
1644.4	8.14
1700.0	8.21

CURVE 14

T	C_p
1088.9	7.42 x 10^{-2}
1144.4	7.52
1200.0	7.64
1255.5	7.75
1311.1	7.86
1366.6	7.97
1422.2	8.08
1477.8	8.19
1533.3	8.30
1588.9	8.41
1644.4	8.51
1700	8.63

CURVE 15*

T	C_p
1088.9	7.52 x 10^{-2}
1144.4	7.73
1200.0	7.88
1255.5	8.02
1311.1	8.13
1366.6	8.21
1422.2	8.26
1477.8	8.29
1533.3	8.31
1588.9	8.29
1644.4	8.27
1700.0	8.23

CURVE 16*

T	C_p
1250	7.459 x 10^{-2}
1271	7.506
1300	7.576
1322	7.626
1350	7.700
1370	7.758
1400	7.830
1416	7.875

CURVE 16 (cont.)

T	C_p
1450	7.964 x 10^{-2}
1461	7.994
1489	8.074
1500	8.106
1518	8.156
1547	8.243
1550	8.256
1575	8.338
1600	8.410

CURVE 17*

T	C_p
400	6.08 x 10^{-2}
465	6.18
565	6.45
604	6.49
676	6.81
781	6.78
875	6.65
946	7.09

CURVE 18*

T	C_p
15.97	4.2 x 10^{-1}
17.97	4.6
20.66	7.6
27.54	1.79 x 10^{-3}
31.52	2.75
34.04	3.45
38.1	4.78
56.0	1.28 x 10^{-2}
64.3	1.70
69.5	1.98
79.4	2.49
91.6	3.00
98.6	3.27
105.6	3.52
112.8	3.78
120.4	3.99
144.6	4.57
200.2	5.31
206.1	5.36
209.1	5.39
238.6	5.62
274.7	5.76
238.4	5.89

CURVE 19*

T	C_p
233	5.64 x 10^{-2}
248	5.71
273	5.89
298	5.97
323	6.09
348	6.13
373	6.12
398	6.14
423	6.16
448	6.20
473	6.24
498	6.25
523	6.32

CURVE 20*

T	C_p
298.16	6.150 x 10^{-2}
400	6.254
500	6.369
600	6.483
700	6.608
800	6.744
900	6.890
1000	7.036
1100	7.192
1200	7.359
1300	7.526
1400	7.703
1500	7.890
1600	8.088
1700	8.286
1800	8.495
1900	8.714
2000	8.933
2100	9.162
2200	9.402
2300	9.652
2400	9.902
2500	1.016 x 10^{-1}
2600	1.043
2650	1.057

CURVE 21*

T	C_p
1.339	7.21 x 10^{-6}
1.365	7.73
1.392	7.55
1.420	7.73
2.412	1.47 x 10^{-5}
2.523	1.51
2.741	1.59
2.878	1.54
2.980	1.69
3.478	2.08
4.495	2.91
4.603	2.70
4.769	3.07
4.873	3.05
9.179	9.17
9.429	9.54
9.910	1.12 x 10^{-4}
10.445	1.30

CURVE 22*

T	C_p
0.3622	1.61 x 10^{-6}
0.3714	1.64
0.3604	1.71
0.4034	1.73
0.4179	1.81
0.4346	1.90
0.4442	1.95
0.4475	1.95
0.4679	2.01
0.4926	2.12
0.5098	2.16
0.5188	2.21
0.5280	2.27
0.5598	2.43
0.5794	2.47
0.5925	2.52
0.6020	2.59
0.6186	2.69
0.6272	2.79
0.6443	2.86
0.6607	2.96
0.6793	3.06
0.6859	3.09
0.7025	3.14
0.7208	3.21
0.7546	3.44
0.7624	3.46

CURVE 22 (cont.)*

T	C_p
0.7947	3.66 x 10^{-6}
0.8163	3.71
0.8313	3.76
0.8470	3.89
0.8656	4.01
0.8830	3.96
0.9050	4.11
0.9269	4.24
0.9405	4.26
0.9538	4.33
1.085	4.86
1.155	5.18
1.276	5.78
1.407	6.35
1.453	6.60
1.617	7.50
1.717	8.05
1.811	8.54
2.000	9.47
2.176	1.04 x 10^{-5}
2.241	1.09
2.314	1.12
2.346	1.14
2.410	1.19
2.624	1.30
2.695	1.35
2.782	1.42
2.785	1.41
2.834	1.44
2.921	1.51
2.948	1.52
3.108	1.63
3.387	1.84
3.428	1.88
3.460	1.88

CURVE 23*

T	C_p
0.3587	1.63 x 10^{-6}
0.3926	1.81
0.4129	1.88
1.4151	1.88
1.4324	1.91
0.4537	2.07
0.4846	2.23
0.4929	2.18
0.5091	2.29

CURVE 23 (cont.)*

T	C_p
0.5265	2.32 x 10^{-6}
0.5457	2.49
0.5610	2.52
0.5773	2.64
0.5941	2.69
0.6129	2.77
0.6223	2.86
0.6461	2.94
0.6581	3.09
0.6715	3.01
0.6815	3.09
0.6948	3.16
0.7045	3.31
0.7221	3.31
0.7344	3.39
0.7483	3.44
0.7527	3.44
0.7573	3.59
0.7767	3.61
0.7897	3.69
0.8101	3.71
0.8210	3.86
0.8327	3.79
0.8533	3.99
0.8620	3.96
0.8754	4.14
0.8964	3.99
0.9026	4.11
0.9116	4.41
0.9170	4.09
0.9284	4.26
0.9358	4.36
0.9475	4.36
0.9687	4.46
1.100	5.08
1.174	5.43
1.331	6.15
1.565	7.42
1.661	8.27
1.736	8.37
1.846	8.79
1.958	9.42
2.085	9.96
2.140	1.04 x 10^{-5}
2.288	1.12
2.344	1.16
2.393	1.17

* Not shown on plot

DATA TABLE NO. 34 (continued)

T	C_p
CURVE 23 (cont.)*	
2.417	1.17 x 10^{-5}
2.468	1.23
2.528	1.26
2.570	1.29
2.641	1.32
2.807	1.42
2.908	1.46
3.020	1.53
3.072	1.59
3.089	1.62
3.180	1.65
3.242	1.72
3.281	1.73
3.389	1.77
3.443	1.84
3.448	1.81
3.497	1.85
3.562	1.89
CURVE 24*	
0.3787	1.47 x 10^{-6}
0.4008	1.65
0.4173	1.81
0.4340	1.97
0.4629	2.24
0.4715	2.38
0.4850	2.52
0.5187	2.94
0.5286	3.04
0.5386	3.29
0.5490	3.31
0.5591	3.46
0.5689	3.54
0.5771	3.69
0.5844	3.76
0.6189	4.19
0.6299	4.33
0.6393	4.61
0.6517	4.86
0.6603	4.83
0.6778	5.21
0.6875	5.43
0.7009	5.63
0.7105	5.80
0.7244	5.98
0.7502	6.50
CURVE 24 (cont.)*	
0.7600	6.63 x 10^{-6}
0.7755	6.93
0.7918	7.20
0.8078	7.45
0.8123	7.62
0.8212	7.90
0.8470	8.10
0.8568	8.30
0.8659	8.57
0.8756	8.69
0.8820	9.07
0.8857	8.87
0.8917	9.22
CURVE 25*	
0.3771	1.33 x 10^{-6}
0.4062	1.57
0.4130	1.62
0.4261	1.74
0.4289	1.81
0.4446	1.95
0.4626	2.15
0.4811	2.32
0.4997	2.54
0.5150	2.77
0.5350	3.01
0.5519	3.19
0.5763	3.54
0.5832	3.61
0.5996	3.91
0.6137	4.16
0.6270	4.24
0.6420	4.48
0.6558	4.68
0.6683	4.96
0.6890	5.26
0.6985	5.38
0.7083	5.58
0.7174	5.73
0.7242	5.83
0.7365	5.98
0.7614	6.30
0.7742	6.65
0.7845	6.90
0.7938	7.20
0.8031	4.72
CURVE 25 (cont.)*	
0.8125	7.47 x 10^{-6}
0.8223	7.65
0.8323	8.05
0.8426	8.07
0.8514	8.32
0.8641	8.47
0.8708	8.60
0.8890	9.07
0.8946	9.19
0.9061	9.27
0.9065	9.69

* Not shown on plot

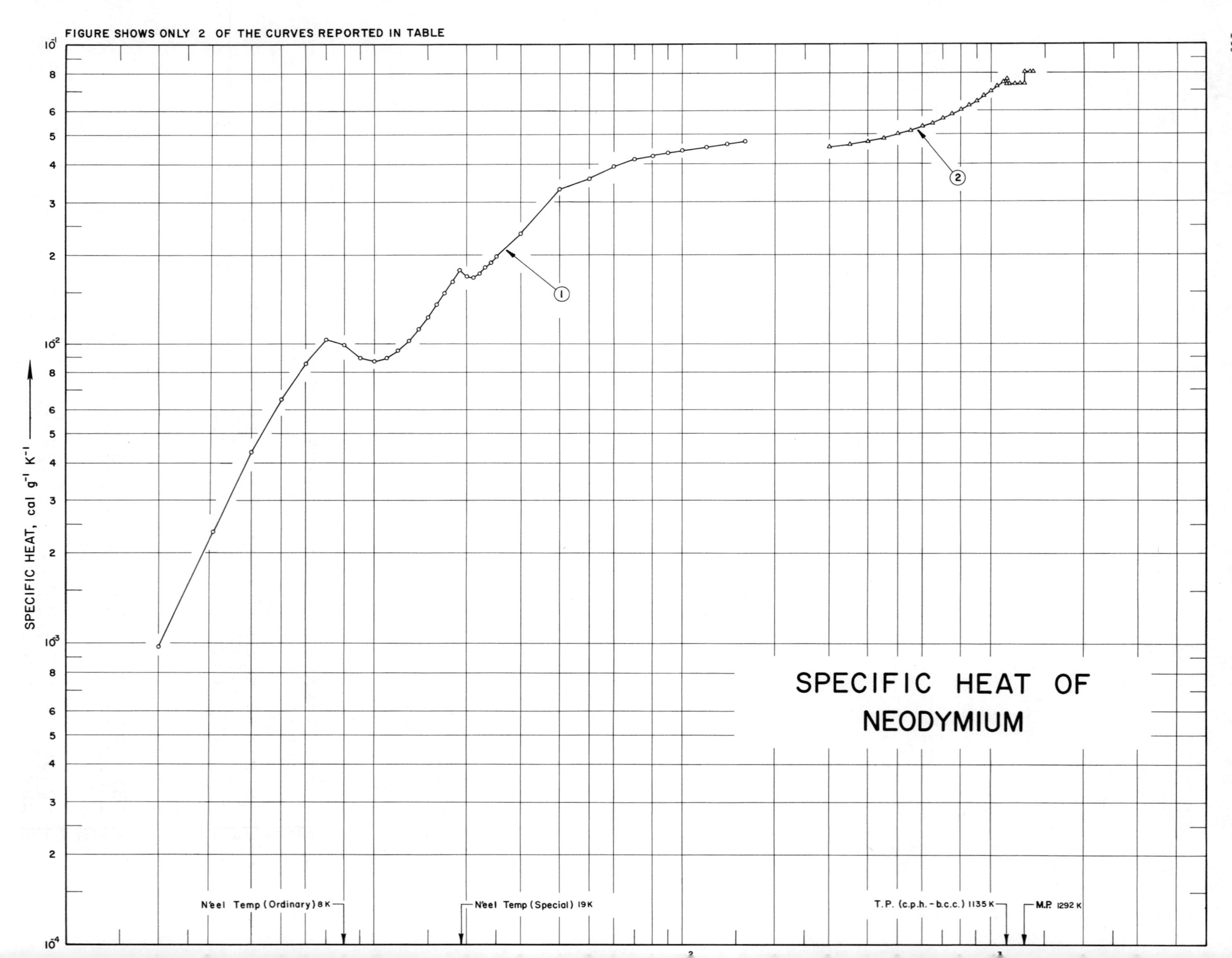
FIGURE SHOWS ONLY 2 OF THE CURVES REPORTED IN TABLE
SPECIFIC HEAT OF
NEODYMIUM
SPECIFIC HEAT, cal g⁻¹ K⁻¹
Néel Temp (Ordinary) 8 K
Néel Temp (Special) 19 K
T.P. (c.p.h.-b.c.c.) 1135 K
M.P. 1292 K
1
2
10⁻¹
10⁻²
10⁻³
10⁻⁴

SPECIFICATION TABLE NO. 35 SPECIFIC HEAT OF NEODYMIUM

(Impurity < 0.20% each; total impurities < 0.50%)

[For Data Reported in Figure and Table No. 35]

Curve No.	Ref. No.	Year	Temp. Range, K	Reported Error, %	Name and Specimen Designation	Composition (weight percent), Specifications and Remarks
1	35	1951	2-160			0.052 Fe, 0.025 Mg, and <0.05 rare earth metals.
2	73	1960	298-1373	0.14		>99.78 Nd, 0.1 Ca, ≤ 0.04 Pr, ≤0.03 Sm, ≤ 0.02 Si, ≤ 0.02 Ta, and ≤ 0.01 Fe.
3	87	1964	0.4-4	<1.5	Expt. I	0.13 O_2, 0.12 Ta, 0.07 N_2, 0.065 Fe, 0.05 Na, 0.025 C, 0.015 Al, 0.0045 B, 0.0045 Ba, 0.004 F, 0.002 Gd, 0.002 K, 0.0015 Ni, 0.0015 Y, 0.0012 H_2, and traces (total 0.0028) Cu, Cr, Er, La, Li, Lu, Mg, Sc, Sr, and Zn; vacuum distilled; remelted in vacuum and cast into tantalum crucible; machined in argon atmosphere.
4	87	1964	0.4-4	<1.5	Expt. II	Same as above.
5	321	1936	373-973			99.5 Nd, traces of Fe, Si, and Al.
6	285	1958	273-1373			Impurities: 0.2 Ta, 0.1 Ca, 0.04 Pr, 0.03 Sm, 0.02 Si, and 0.01 Fe.

DATA TABLE NO. 35 SPECIFIC HEAT OF NEODYMIUM

[Temperature, T, K; Specific Heat, C_p, Cal g^{-1} K^{-1}]

CURVE 1

T	C_p
2	9.71×10^{-4}
3	2.36×10^{-3}
4	4.37
5	6.52
6	8.60
7	1.04×10^{-2}
8	9.91×10^{-3}
9	8.94
10	8.74
11	8.94
12	9.50
13	1.03×10^{-2}
14	1.13
15	1.24
16	1.37
17	1.50
18	1.64
19	1.78
20	1.70
21	1.68
22	1.74
23	1.82
24	1.89
25	1.98
30	2.35
40	3.03
50	3.58
60	3.92
70	4.15
80	4.26
90	4.35
100	4.41
120	4.54
140	4.64
160	4.74

CURVE 2

T	C_p
298.15	4.56×10^{-2}*
300	4.56
350	4.66
400	4.76
450	4.89
500	5.02
550	5.16
600	5.32
650	5.48×10^{-2}
700	5.66
750	5.84
800	6.04
850	6.25
900	6.48
950	6.71
1000	6.95
1050	7.21
1100	7.47
1135.15	7.67
1135.15	7.38
1150	7.38
1200	7.38
1250	7.38
(s) 1297.15	7.38
(l) 1297.15	8.08*
1300	8.08
1350	8.08
1373.15	8.08

CURVE 3*

T	C_p
0.396_6	9.405×10^{-5}
0.427_3	9.000
0.463_2	8.785
0.503_9	8.350
0.547_6	8.764
0.596_4	8.930
0.647_0	9.556
0.697_9	1.055×10^{-4}
0.749_9	1.186
0.803_0	1.350
0.857_3	1.549
0.912_9	1.785
0.873_1	2.051
1.033_1	2.381
1.092_4	2.749
1.161_1	3.123
1.245_9	3.624
1.346_2	4.275
1.457_4	5.117
1.578_0	6.141
1.710_1	7.349
1.857_8	8.729
2.018_2	1.041×10^{-3}
2.184_8	1.231×10^{-3}
2.303_6	1.605
2.371_3	1.700
2.477_1	1.848
2.687_4	2.156
2.919_3	2.522
3.163_1	2.925
3.424_5	3.397
3.723_6	3.983

CURVE 4*

T	C_p
0.398_2	9.607×10^{-5}
0.424_0	9.180
0.450_9	8.930
0.480_1	8.817
0.513_3	8.829
0.550_3	8.974
0.591_3	9.460
0.637_0	1.031×10^{-4}
0.685_7	1.151
0.735_3	1.303
0.785_7	1.483
0.837_8	1.695
0.891_8	1.950
0.947_0	2.237
1.004_8	2.575
1.064_9	2.943
1.128_4	3.382
1.203_3	3.939
1.296_5	4.681
1.406_0	5.644
1.522_7	6.747
1.645_1	7.970
1.773_1	9.337
1.910_4	1.096×10^{-3}
2.062_9	1.281
2.223_1	1.487
2.408_8	1.742
2.630_1	2.070
2.871_2	2.436
3.142_1	2.888
3.447_6	3.428
3.772_9	4.000

CURVE 5*

T	C_p
373	5.22×10^{-2}
473	6.18
573	6.86
623	7.085
673	7.245
723	7.335
823	6.669
873	6.905
923	7.169
973	7.446

CURVE 6*

T	C_p
273.2	4.52×10^{-2}
323.2	4.60
373.2	4.71
423.2	4.83
473.2	4.96
523.2	5.09
573.2	5.23
623.2	5.40
673.2	5.57
723.2	5.75
773.2	5.94
823.2	6.14
873.2	6.36
923.2	6.59
973.2	6.83
1023.2	7.078
1073.2	7.335
1135.2	7.681
1135.2	7.384
1173.2	7.384
1223.2	7.384
1273.2	7.384
1297.2	7.384
1297.2	8.084
1323.2	8.084
1373.2	8.084

* Not shown on plot

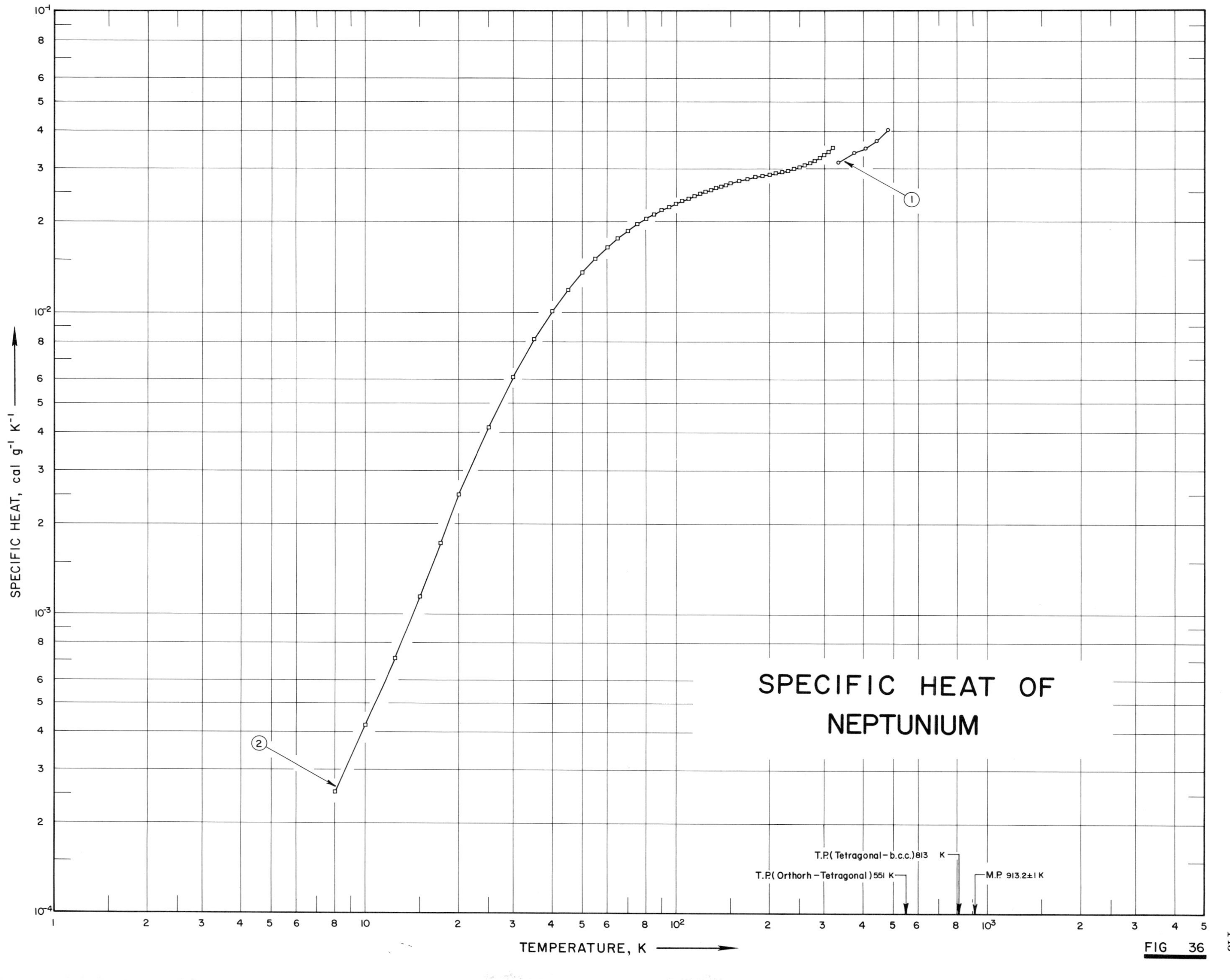
SPECIFIC HEAT OF
NEPTUNIUM
SPECIFIC HEAT, cal g⁻¹ K⁻¹
TEMPERATURE, K
T.P.(Tetragonal – b.c.c.) 813 K
T.P.(Orthorh – Tetragonal) 551 K
M.P. 913.2±1 K
1
2

FIG 36

SPECIFICATION TABLE NO. 36 SPECIFIC HEAT OF NEPTUNIUM

(Impurity $<$ 0.20% each; total impurities $<$ 0.50%)

[For Data Reported in Figure and Table No. 36]

Curve No.	Ref. No.	Year	Temp. Range, K	Reported Error, %	Name and Specimen Designation	Composition (weight percent), Specifications and Remarks
1	425	1959	333–480			0.34 Ca and 0.22 U.
2	426	1965	8–320			$<$ 0.5 each Ce, La, Nd, Pr, Sm, Sc, Na, Ti and Yb, $<$ 0.4 each B, Li, and Zn, $<$0.05 each Cr, Mn, Fe, Co, Ni, Y and Zr, 0.01–0.1 Th and 0.03 Pu^{238}.

DATA TABLE NO. 36 SPECIFIC HEAT OF NEPTUNIUM

[Temperature, T,K; Specific Heat, C_p, Cal $g^{-1} K^{-1}$]

T	C_p
CURVE 1	
333	3.14×10^{-2}
375	3.38
407	3.49
442	3.70
480	4.02
CURVE 2	
8	2.5×10^{-4}
10	4.22
12.5	7.13
15	1.15×10^{-3}
17.5	1.75
20	2.51
25	4.18
30	6.12
35	8.19
40	1.02×10^{-2}
45	1.20
50	1.36
55	1.51
60	1.65
65	1.76
70	1.87
75	1.96
80	2.04
85	2.11
90	2.18
95	2.24
100	2.29
105	2.34
110	2.39
115	2.43
120	2.47
125	2.51
130	2.55
135	2.58
140	2.62
145	2.65
150	2.68
160	2.73
170	2.77
180	2.80
190	2.84
200	2.87
CURVE 2 (cont.)	
210	2.89×10^{-2}
220	2.92
230	2.95
240	2.98
250	3.02
260	3.07
270	3.13
273.15	3.14
280	3.19
290	3.26
298.15	3.32
300	3.33
310	3.42
320	3.51

*Not shown on plot

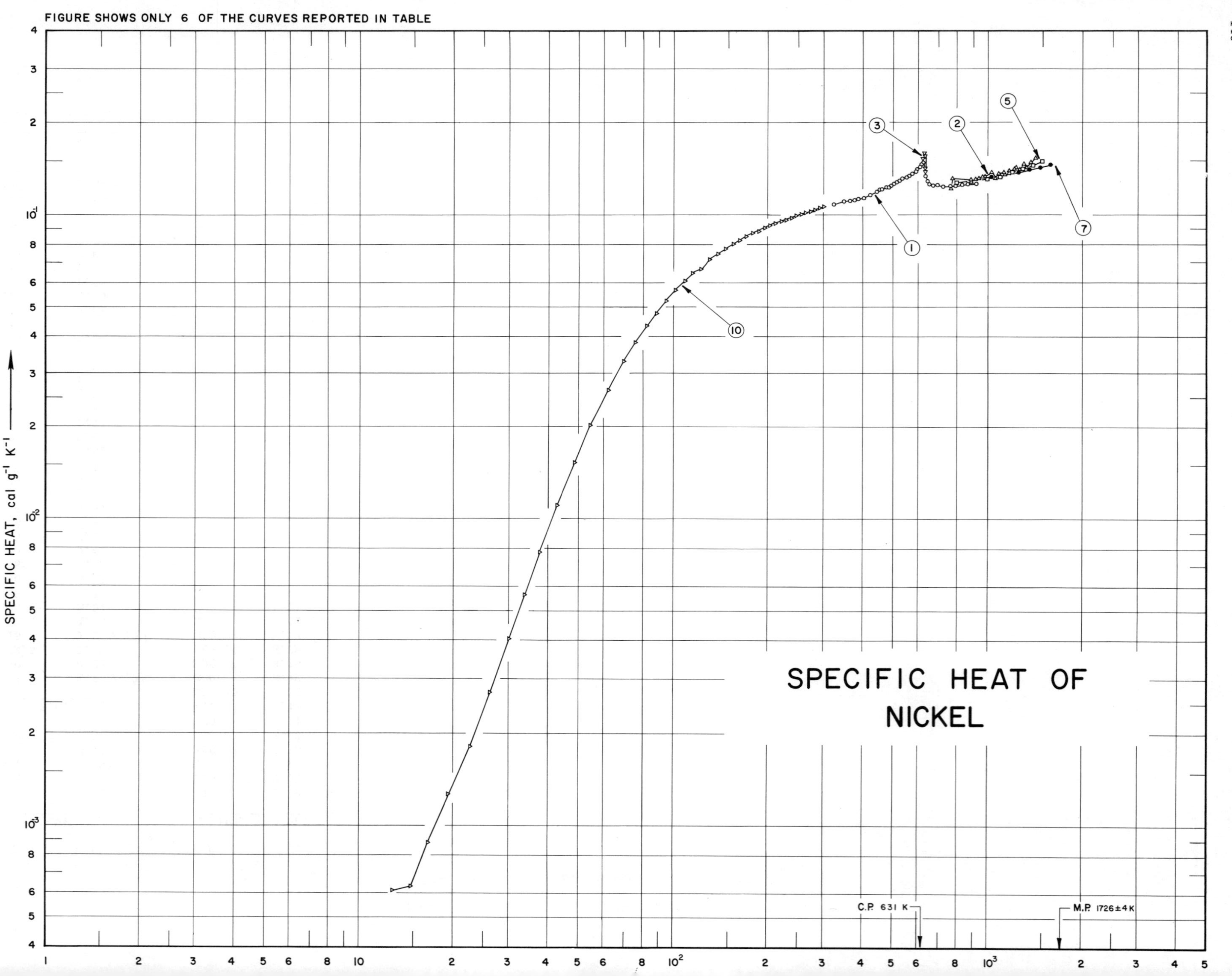
FIGURE SHOWS ONLY 6 OF THE CURVES REPORTED IN TABLE
SPECIFIC HEAT OF NICKEL
SPECIFIC HEAT, cal g⁻¹ K⁻¹
C.P. 631 K
M.P. 1726±4 K
1
2
3
5
7
10

SPECIFICATION TABLE NO. 37 SPECIFIC HEAT OF NICKEL

(Impurity < 0.20% each; total impurities < 0.50%)

[For Data Reported in Figure and Table No. 37]

Curve No.	Ref. No.	Year	Temp. Range, K	Reported Error, %	Name and Specimen Designation	Composition (weight percent), Specifications and Remarks
1	52	1936	325-922			0.10 Fe, 0.07 Mg, 0.003 S, Si, and trace Mn; aged for 3 hrs at 680 C and cooled slowly.
2	53	1955	770-1437	1.0		99.97 Ni, 0.0008 As, 0.0006 Fe, <0.0004 Si, <0.0003 Cr, <0.0003 Cu, 0.0001 Mn, re-melted several times; heated several times to 1100 C and cooled slowly.
3	53	1955	423-1423	1.0		Same as above.
4	1	1961	295	± 5.0		
5	20	1959	800-1500	± 0.3		Sealed in argon.
6	18	1956	323-883	± 0.5	Electrolytic nickel	99.95 Ni, <0.05 Co, and the rest Al, Cu, Fe and Si.
7	38	1961	466-1584			
8	54	1954	673-1123			99.9 Ni.
9	55	1930	18-189	1.5	Electrolytic nickel	Cold deformed; recrystallized for 10 hrs. at 1000 C under nitrogen atmosphere.
10	56	1952	13-303			0.014 C, 0.0018 Co, 0.0009 Cu, and very slight trace Al, B, Ca, and Fe; annealed for 2 hrs at 900 C in H_2; heated for 5 hrs. at 1000-1100 C in 2 x 10^{-5} mm Hg and 5 hrs. in 8 x 10^{-6} mm Hg vacuum; cooled in vacuum to 800 C in 1 hr. and then to 100 C in 17 hrs.
11	101	1958	337-1164			Specimen's surface plated with platinum black.
12	55	1930	15-204	1.5	Electrolytic nickel	Cold deformed; recrystallized for 10 hrs. at 1000 C under nitrogen atmosphere.
13	268	1926	373-1903			99.920 Ni, 0.06 Fe, 0.013 Cu, and 0.007 Si.
14	342	1928	93-735			Pure.
15	164	1932	273-873			
16	343	1934	86-726	≤ 2.0		0.03 impurities.
17	315	1935	297.9		Electrolytic nickel	Forged annealed.
18	344	1935	1-19			99.81 Ni, 0.083 Fe, 0.04 C, 0.037 Mg, and 0.017 Cu.
19	345	1936	10-26			
20	182	1936	203-383	0.1		99.69 Ni, 0.13 Fe, 0.10 Si, 0.03 Cu, and 0.03 C.
21	183	1936	373-773	0.2-1.0		Same as above.
22	346	1936	373-1273			Pure.
23	347	1938	333-873	≤ 2.0	Sample II	0.031 Fe, 0.025 C, 0.007 Cu, 0.0004 S and negligible amount of Co; Mond pellets; melted under vacuum of $<10^{-3}$ mm Hg and hammered into suitable shape.
24	347	1938	333-873	≤ 2.0	Sample III	0.04 Fe, trace of C and O_2, and negligible amount of Co; pressed powder; prepared by sintering a block of pressed powder in vacuo at 900 C; cooled at rate of one degree C per minute to room temperature.

SPECIFICATION TABLE NO. 37 (continued)

Curve No.	Ref. No.	Year	Temp. Range, K	Reported Error, %	Name and Specimen Designation	Composition (weight percent), Specifications and Remarks
25	347	1938	333-410	≤ 2.0	Sample IV	Cathode nickel.
26	177	1938	291-813			99.91 Ni.
27	348	1939	813-1280			99.51 Ni.
28	293	1941	82-273			Impurities: 0.02 Fe, 0.01 C, 0.002 Si, 0.001 P, 0.001 Mn, and 0.0003 S.
29	319	1958	293-643			
30	349	1962	2-4	≤ 2.0		99.979 Ni, 0.01 Cu, 0.01 Fe, and 0.001 S.
31	350	1962	309-670		Mond nickel	99.85 Ni, 0.14 Fe, and trace of Co.

DATA TABLE NO. 37 SPECIFIC HEAT OF NICKEL

[Temperature, T, K; Specific Heat, C_p, Cal $g^{-1} K^{-1}$]

T	C_p
CURVE 1	
325.2	1.084×10^{-1}
351.2	1.101
366.2	1.108
379.8	1.118
389.6	1.131
406.4	1.146
425.0	1.166
447.2	1.196
458.2	1.211
467.6	1.219
480.2	1.237
486.0	1.237
494.6	1.250
506.4	1.267
517.6	1.278
527.6	1.295
539.8	1.310
555.4	1.331
566.4	1.345
578.2	1.365
593.0	1.389
598.6	1.404
611.0	1.440
618.4	1.460
624.4	1.487
636.0	1.341
647.1	1.290
656.4	1.261
675.5	1.250
695.4	1.252
722.8	1.243
767.4	1.249
798.0	1.249
835.0	1.251
869.2	1.256
922.8	1.266
CURVE 2	
769.6	1.233×10^{-1}
777.0	1.319
781.2	1.262*
877.0	1.265*
885.4	1.304
891.0	1.290
915.8	1.311
CURVE 2 (cont.)	
939.6	1.319×10^{-1}
966.6	1.336
984.6	1.336
1030.8	1.379
1063.2	1.317
1080.4	1.365
1128.6	1.378
1141.0	1.366
1176.4	1.368
1178.8	1.395
1222.4	1.406
1243.0	1.429
1272.4	1.401
1306.6	1.473
1340.2	1.434
1377.7	1.492
1435.4	1.542*
1437.0	1.537*
CURVE 3	
453	$1.198^* \times 10^{-1}$
463	1.208*
473	1.220*
483	1.232*
493	1.245*
503	1.259*
513	1.276*
523	1.292*
533	1.308*
543	1.323*
553	1.338*
563	1.353*
573	1.368*
583	1.384*
593	1.401*
603	1.421*
613	1.448*
618	1.467*
623	1.496*
625	1.512*
627	1.534*
629	1.567
630	1.598
631	1.558
633	1.420
CURVE 3 (cont.)	
635	1.378×10^{-1}
638	1.347*
643	1.318*
648	1.303*
653	1.291*
663	1.275*
673	1.267*
693	1.262*
713	1.258*
733	1.256*
753	1.256*
773	1.260*
823	1.267*
873	1.294*
923	1.311*
973	1.328*
1023	1.346*
1073	1.363*
1123	1.380*
1173	1.397*
1223	1.415*
1273	1.436*
1323	1.460*
1373	1.490*
1423	1.525*
CURVE 4*	
295	1.2×10^{-1}
CURVE 5	
800	1.27×10^{-1}
900	1.29*
1000	1.31
1100	1.34
1200	1.38
1300	1.42
1400	1.46
1500	1.50
CURVE 6*	
323	1.083×10^{-1}
363	1.113
383	1.139
403	1.161
423	1.181
443	1.198
463	1.218
483	1.238
503	1.265
523	1.292
543	1.318
563	1.349
578	1.372
588	1.392
598	1.411
608	1.437
618	1.470
624	1.496
626	1.536
628	1.524
628	1.469
638	1.332
648	1.290
658	1.275
668	1.270
683	1.255
703	1.249
723	1.250
743	1.250
763	1.251
783	1.251
803	1.255
823	1.259
843	1.263
863	1.267
883	1.271
CURVE 7	
466.0	1.201×10^{-1}
477.6	1.204
533.1	1.217
588.7	1.230
644.3	1.243
699.8	1.255
755.4	1.268
CURVE 7 (cont.)	
810.9	1.281×10^{-1}
866.5	1.294
922.0	1.307
977.6	1.319
1033.1	1.332
1088.7	1.345
1144.3	1.358
1199.8	1.371
1255.3	1.383
1310.9	1.396
1366.5	1.409
1422.0	1.422
1477.6	1.435
1533.2	1.447
1584.0	1.459
CURVE 8*	
673	1.273×10^{-1}
723	1.267
773	1.273
823	1.288
873	1.305
923	1.326
973	1.348
1023	1.367
1073	1.379
1123	1.385
CURVE 9*	
17.70	9.499×10^{-4}
22.30	1.562×10^{-3}
24.60	2.138
27.20	2.826
30.75	4.268
32.25	4.802
84.90	4.493×10^{-2}
85.00	4.512
87.40	4.624
139.32	7.498
144.40	7.687
171.76	8.525
175.23	8.608
188.51	8.912
CURVE 10	
12.95	6.13×10^{-4}
14.68	6.30
16.71	8.86
19.46	1.28×10^{-3}
22.78	1.82
26.37	2.71
30.14	4.02
33.83	5.66
37.86	7.80
42.93	1.102×10^{-2}
48.93	1.537
54.89	2.043
62.12	2.644
69.74	3.292
75.91	3.810
82.22	4.318
88.52	4.771
95.13	5.226
102.07	5.682
109.01	6.100
116.23	6.470
123.62	6.655
131.77	7.165
139.64	7.488
147.50	7.778
155.38	8.046
163.33	8.276
171.37	8.501
179.48	8.710
187.68	8.896
196.03	9.068
204.43	9.240
212.96	9.393
221.33	9.516
229.59	9.635
238.59	9.761
238.75	9.767*
247.95	9.913
247.96	9.917*
256.73	1.004×10^{-1}
265.59	1.016
274.81	1.028
284.34	1.040
293.99	1.056
302.98	1.073

*Not shown on plot

DATA TABLE NO. 37 (continued)

CURVE 11*

T	C_p
337	1.10 x 10^{-1}
356	1.13
400	1.17
400	1.10
465	1.25
465	1.25
565	1.40
565	1.43
604	1.33
633	1.30
633	1.30
676	1.29
781	1.27
781	1.23
875	1.27
875	1.30
946	1.32
946	1.35
1017	1.31
1017	1.36
1090	1.31
1090	1.39
1164	1.34

CURVE 12*

T	C_p
15.05	7.913 x 10^{-4}
18.06	1.066 x 10^{-3}
22.11	1.652
25.20	2.436
28.00	3.286
31.30	4.381
34.55	6.013
37.70	7.898
40.93	9.601
47.10	1.438 x 10^{-2}
5.70	2.129
67.13	3.109
74.73	3.727
82.30	4.319
92.95	5.030
104.00	5.783
114.33	6.340
123.96	6.813
133.38	7.276
141.71	7.609
149.96	7.913

CURVE 12 (cont.)*

T	C_p
159.91	8.222 x 10^{-2}
168.74	8.474
178.49	8.692
185.57	8.866
194.81	9.079
204.05	9.252

CURVE 13*

T	C_p
373	1.11 x 10^{-1}
473	1.16
573	1.22
763	1.31
773	1.34
873	1.23
973	1.24
1073	1.26
1173	1.30
1273	1.32
1373	1.34
1473	1.36
1573	1.39
1673	1.42
1773	1.81
1823	1.81
1903	1.81

CURVE 14*

T	C_p
98	5.5 x 10^{-2}
195	8.9
202	9.02
227	9.52
257	1.010 x 10^{-1}
283	1.062
294	1.075
323	1.092
325	1.093
389	1.140
447	1.198
530	1.315
555	1.350
576	1.391
612	1.470
623	1.541
625	1.560
626	1.577

CURVE 14 (cont.)*

T	C_p
627	1.577 x 10^{-1}
628	1.450
628	1.445
629	1.425
630	1.371
630	1.360
631	1.330
633	1.290
633	1.278
634	1.126
638	1.252
735	1.265
648	1.265
654	1.272
704	1.278
733	1.288

CURVE 15*

T	C_p
273	1.080 x 10^{-1}
373	1.133
473	1.237
573	1.320
673	1.245
773	1.255
873	1.260

CURVE 16*
Series I

T	C_p
295.4	1.077 x 10^{-1}
295.5	1.078
296.1	1.078
296.9	1.070
385.0	1.148
452.2	1.230
479.9	1.258
510.2	1.290
546.3	1.332
568.6	1.380
578.6	1.402
599.0	1.465
601.3	1.465
618.7	1.537
625.3	1.557
627.4	1.587
631.4	1.432

CURVE 16 (cont.)*

T	C_p
634.0	1.417 x 10^{-1}
634.7	1.442
635.0	1.392
638.1	1.384
640.6	1.360
643.2	1.316
646.2	1.311
650.9	1.330
661.2	1.292
683.9	1.288
684.1	1.286
726.2	1.307

Series II

T	C_p
86.2	5.0 x 10^{-2}
202.2	9.4
216.3	9.9
255.8	1.03 x 10^{-1}
293.4	1.077

CURVE 17*

T	C_p
297.9	1.07 x 10^{-1}

CURVE 18*
Series I

T	C_p
1.11	2.964 x 10^{-5}
1.14	3.117
1.23	3.475
1.31	3.900
1.41	4.156
1.48	4.446
1.52	4.718
1.67	5.059
1.77	5.314
1.92	5.842
2.08	6.524
2.24	6.677
2.39	7.324
2.49	7.682
2.56	7.835
2.71	8.312
2.81	8.772
2.82	8.584
3.00	9.249

CURVE 18 (cont.)*

T	C_p
3.15	9.862 x 10^{-5}
3.26	1.015 x 10^{-4}
3.37	1.049
3.46	1.092
3.57	1.119

Series II

T	C_p
1.10	3.168 x 10^{-5}
1.10	3.100
1.10	3.134
1.13	2.913
1.14	3.185
1.66	5.144
3.49	1.048 x 10^{-4}
3.63	1.094
3.73	1.110
3.83	1.175
3.89	1.223
3.97	1.252
4.04	1.260
4.05	1.277
4.12	1.310
4.20	1.337
4.27	1.352
4.31	1.366

Series III

T	C_p
3.11	9.487 x 10^{-5}
3.26	1.003 x 10^{-4}
3.37	1.032
3.51	1.058
3.71	1.105
3.78	1.121
4.03	1.252
4.16	1.306
4.27	1.332
4.42	1.381
4.58	1.436
4.70	1.487
4.86	1.562
5.01	1.671
5.28	1.805
5.45	1.891
5.62	1.925
5.75	1.942
5.86	1.942

CURVE 18 (cont.)*
Series IV

T	C_p
5.72	1.771 x 10^{-4}
5.90	1.891
6.30	2.112
6.48	2.180
6.54	2.180
7.51	2.793
7.69	2.878
7.93	2.981
8.05	3.083
8.21	2.998
8.39	3.168
8.70	3.185
8.91	3.372
9.22	3.475
14.34	6.660
14.84	7.392
15.41	7.443
15.78	8.074
16.46	8.533
17.07	9.453
17.59	9.521
18.08	1.017 x 10^{-3}
18.49	1.073
19.02	1.129

CURVE 19*

T	C_p
10.0	3.9 x 10^{-4}
10.2	3.9
10.6	4.3
10.8	4.3
10.8	4.3
11.6	4.9
11.7	4.9
11.7	4.9
12.0	5.1
12.5	5.5
12.9	5.8
13.2	6.0
13.5	6.3
13.6	6.3
14.6	7.2
15.0	7.5
15.1	7.5
16.7	9.0
16.8	9.0

* Not shown on plot

DATA TABLE NO. 37 (continued)

CURVE 19 (cont.)*

T	C_p
17.0	9.4 x 10^{-4}
17.0	7.7
18.9	1.2 x 10^{-3}
19.0	1.2
21.2	1.5
21.5	1.6
21.6	1.6
22.1	1.70
23.4	1.99
23.8	2.06
24.5	2.25
25.6	2.55
25.9	2.64

CURVE 20*

T	C_p
203	9.252 x 10^{-2}
213	9.427
223	9.582
233	9.732
243	9.876
253	1.001 x 10^{-1}
260	1.015
273	1.027
283	1.039
293	1.051
303	1.062
313	1.072
323	1.083
333	1.093
343	1.104
353	1.114
363	1.124
373	1.134
383	1.145

CURVE 21*

T	C_p
373	1.13 x 10^{-1}
423	1.18
473	1.24
523	1.30
573	1.37
623	1.50
675	1.27
723	1.26
773	1.26

CURVE 22*

T	C_p
373	1.128 x 10^{-1}
423	1.189
473	1.252
523	1.318
573	1.385
618	1.448
633	1.261
673	1.267
773	1.281
873	1.295
973	1.310
1073	1.324
1173	1.339
1273	1.354

CURVE 23*

T	C_p
333	1.055 x 10^{-1}
343	1.079
353	1.096
363	1.111
373	1.125
383	1.138
393	1.149
403	1.160
413	1.170
423	1.180
433	1.188
443	1.197
453	1.205
463	1.214
473	1.225
483	1.236
493	1.249
503	1.262
513	1.279
523	1.294
533	1.310
543	1.324
553	1.340
563	1.355
573	1.371
583	1.388
593	1.407
603	1.428
613	1.454
618	1.470

CURVE 23 (cont.)*

T	C_p
623	1.492 x 10^{-1}
625	1.504
627	1.520
629	1.543
630	1.562
631	1.586
632	1.460
633	1.420
635	1.384
637	1.364
643	1.330
648	1.312
653	1.302
663	1.282
673	1.269
683	1.261
693	1.256
703	1.252
713	1.251
723	1.251
733	1.253
743	1.257
753	1.260
763	1.265
773	1.270
783	1.276
793	1.281
803	1.287
813	1.293
823	1.299
833	1.305
843	1.311
853	1.317
863	1.322
873	1.324

CURVE 24*

T	C_p
343	1.066 x 10^{-1}
353	1.088
363	1.106
373	1.121
383	1.136
393	1.146
403	1.158
413	1.170
423	1.181

CURVE 24 (cont.)*

T	C_p
433	1.191 x 10^{-1}
443	1.199
453	1.207
463	1.215
473	1.225
483	1.236
493	1.248
503	1.260
513	1.275
523	1.289
533	1.303
543	1.318
553	1.334
563	1.348
573	1.363
583	1.379
593	1.405
603	1.426
613	1.454
618	1.471
623	1.495
625	1.508
627	1.524
629	1.547
630	1.563
630.8	1.576
631	1.470
632	1.395
633	1.373
635	1.352
637	1.338
643	1.308
648	1.290
653	1.280
663	1.271
673	1.264
683	1.260
693	1.257
703	1.255
713	1.253
723	1.252
733	1.252
743	1.252
753	1.254
763	1.256
773	1.259
783	1.264

CURVE 24 (cont.)*

T	C_p
793	1.269 x 10^{-1}
803	1.275
813	1.281
823	1.286
833	1.293
843	1.301
853	1.307
863	1.315
873	1.324

CURVE 25*

T	C_p
333	1.060 x 10^{-1}
343	1.080
353	1.096
363	1.112
373	1.128
383	1.140
393	1.154
403	1.165
413	1.177
423	1.187
433	1.196
443	1.204
453	1.213
463	1.222
473	1.231
483	1.243
493	1.256
503	1.269
513	1.283
523	1.296
533	1.310
543	1.325
553	1.340
563	1.354
573	1.369
583	1.385
593	1.400
603	1.419
613	1.441
618	1.454
623	1.470
627	1.487
629	1.498
631	1.511
633	1.534

CURVE 25 (cont.)*

T	C_p
634	1.550 x 10^{-1}
635	1.578
636	1.600
637	1.420
638	1.390
643	1.335
648	1.310
653	1.294
663	1.275
673	1.265
683	1.260

CURVE 26*

Series I

T	C_p
394.0	1.154 x 10^{-1}
465.4	1.238
535.5	1.314
607.9	1.449
619.5	1.487
627.5	1.539
631.2	1.556
636.8	1.375
641.8	1.341
649.3	1.332
663.4	1.313

Series II

T	C_p
334.1	1.096 x 10^{-1}
364.9	1.129
429.6	1.202
498.6	1.289
576.4	1.385
634.5	1.404
638.4	1.383
644.2	1.342
651.3	1.337

Series III

T	C_p
484.8	1.268 x 10^{-1}
659.7	1.322
668.4	1.315
683.8	1.305
699.0	1.305
730.0	1.297

*Not shown on plot

DATA TABLE NO. 37 (continued)

T	C_p
CURVE 26 (cont.)*	
730.4	1.301 x 10^{-1}
760.9	1.303
813.3	1.307
Series IV	
287.1	1.057 x 10^{-1}
287.3	1.052
291.4	1.060
292.9	1.057
308.2	1.074
CURVE 27*	
697.6	1.359 x 10^{-1}
800.7	1.339
813.4	1.350
891.1	1.378
976.5	1.393
982.2	1.398
1044.5	1.406
1116.1	1.438
1191.9	1.462
1175.4	1.446
1232.6	1.471
1279.6	1.484
CURVE 28*	
81.75	4.29 x 10^{-2}
83.55	4.46
92.45	5.08
98.30	5.47
107.16	6.00
119.19	6.61
130.68	7.10
145.91	7.73
157.49	8.18
165.62	8.47
176.65	8.75
191.30	9.10
213.67	9.54
223.41	9.66
231.29	9.78
240.20	9.83
253.10	9.90
273.20	9.95

T	C_p
CURVE 29*	
293	1.06 x 10^{-1}
333	1.089
373	1.115
393	1.133
413	1.154
433	1.177
453	1.201
473	1.225
493	1.250
513	1.274
533	1.301
553	1.328
573	1.358
593	1.395
613	1.416
633	1.443
643	1.483
CURVE 30*	
1.776	5.278 x 10^{-5}
1.843	5.490
1.894	5.662
1.943	5.878
1.993	6.011
2.057	6.227
2.139	6.498
2.235	6.781
2.346	7.145
2.438	7.471
2.508	7.747
2.603	8.036
2.728	8.404
2.889	8.944
3.068	9.634
3.201	1.008 x 10^{-4}
3.305	1.043
3.436	1.091
3.607	1.156
3.752	1.211
3.864	1.256
3.990	1.295
4.114	1.356

T	C_p
CURVE 31*	
308.7	1.070 x 10^{-1}
310.2	1.071
350.9	1.112
402.9	1.169
437.8	1.208
461.2	1.232
496.8	1.282
499.0	1.283
500.8	1.281
502.8	1.283
504.7	1.286
504.8	1.289
506.7	1.290
508.7	1.285
518.6	1.292
541.1	1.328
557.3	1.358
577.8	1.395
603.4	1.455
622.9	1.523
624.8	1.534
626.7	1.552
628.6	1.564
630.5	1.504
632.3	1.425
633.3	1.403
670.2	1.270

* Not shown on plot

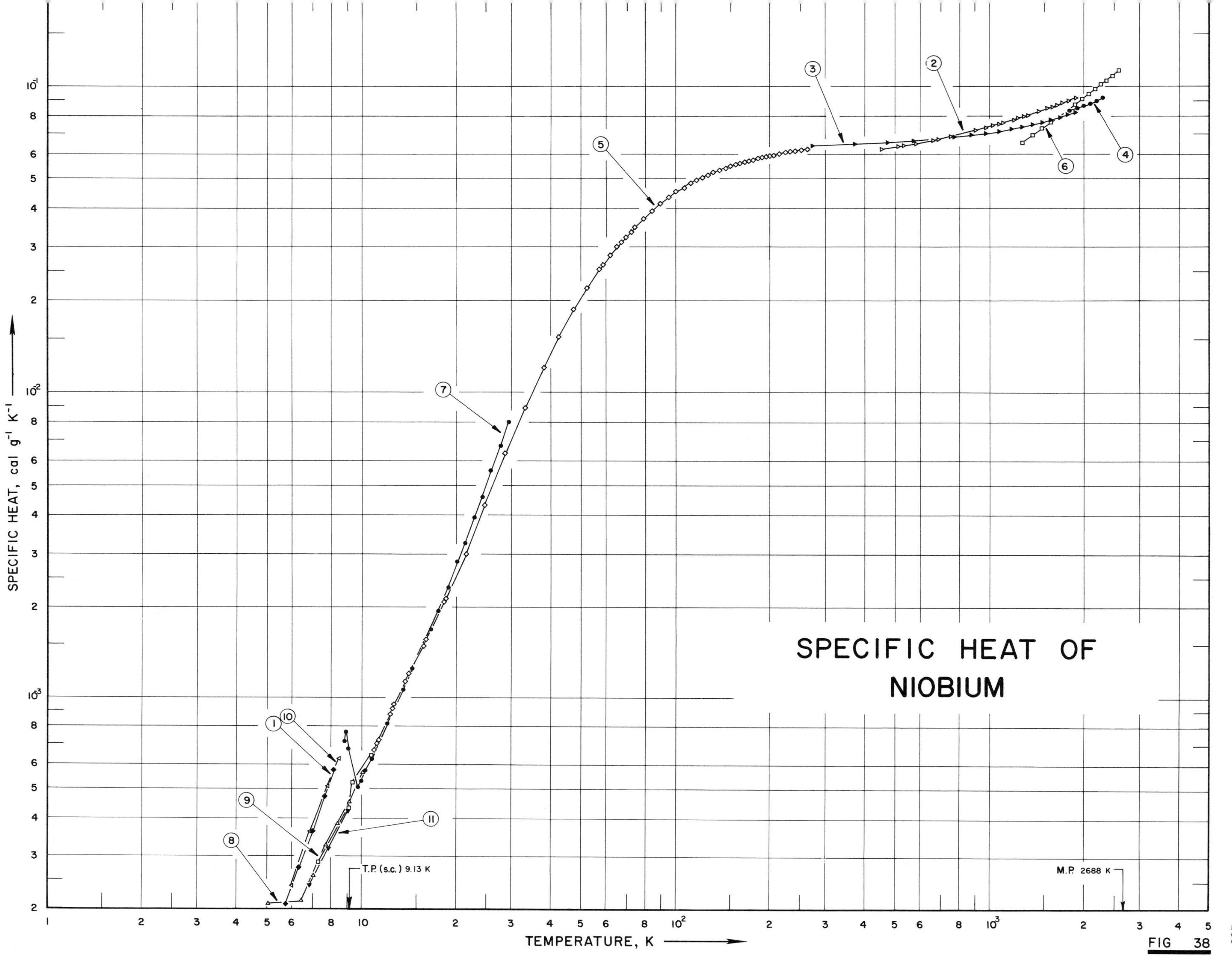
SPECIFIC HEAT OF
NIOBIUM
SPECIFIC HEAT, cal g⁻¹ K⁻¹
TEMPERATURE, K
T.P. (s.c.) 9.13 K
M.P. 2688 K

FIG 38

SPECIFICATION TABLE NO. 38 SPECIFIC HEAT OF NIOBIUM

(Impurity < 0.20% each; total impurities < 0.50%)

[For Data Reported in Figure and Table No. 38]

Curve No.	Ref. No.	Year	Temp. Range, K	Reported Error, %	Name and Specimen Designation	Composition (weight percent), Specifications and Remarks
1	74	1958	2-11	< 5.4	Nb-1, Run I	99.8 Nb, major impurity Ta; annealed; strain free, superconducting state; zero magnetic field.
2	75	1958	454-1882	≤ 2.9		Sealed in helium.
3	70	1934	273-1873			
4	64	1963	1400-2350	≤ 4.0	Wire sample	99.8 Nb, 0.08 Ta, 0.05 N_2, 0.05 O_2, 0.02 C, 0.02 Fe, 0.02 Ta, 0.02 Zr, 0.01 Ni, and 0.01 W; sample supplied by the Fansteel Metallurgical Corp.; outgassed and sealed in < 1 x 10^{-6} mm Hg glass envelope.
5	76	1960	11-271			97.708 Nb, 0.122 Ta, 0.08 O_2, 0.03 Si, 0.023 Ti, 0.02 Ni, 0.01 C, and 0.007 Fe.
6	125	1965	1273-2593	± 1.26		99.83 Nb; powder metallurgy product of 20-mil sheet.
7	74	1958	9-29	< 5.4	Nb-1, Run II	99.8 Nb, major impurity Ta; annealed; strain free; superconducting state; zero magnetic field.
8	74	1958	5-11	< 5.4	Nb-1, Run III	99.8 Nb, major impurity Ta; annealed; strain free; normal state; 2640 gauss magnetic field.
9	74	1958	5-11	< 5.4	Nb-1, Run IV	99.8 Nb, major impurity Ta; annealed; strain free; normal state; 3000 gauss magnetic field.
10	74	1958	1.4-11	< 5.4	Nb-II, Run I	99.8 Nb, major impurity Ta; annealed, strain free; superconducting; zero magnetic field.
11	74	1958	1.3-9	< 5.4	Nb-II, Run II	99.8 Nb, major impurity Ta; annealed; strain free; normal state; 4130 gauss magnetic field.
12	351	1953	2.6-10.5			99.8 Nb; annealed; strain free; superconducting; zero magnetic field.
13	352	1964	0.5-9			< 0.075 Ta, 0.0075 Si, 0.0054 N_2, 0.0034 O_2, and 0.0025 Fe; single crystal.
14	351	1953	3.2-8.6			99.8 Nb; annealed; strain free; normal state.

DATA TABLE NO. 38 SPECIFIC HEAT OF NIOBIUM

[Temperature, T, K; Specific Heat, C_p, Cal $g^{-1} K^{-1}$]

T	C_p
CURVE 1	
2.039	4.18 x 10^{-6}*
2.040	4.20*
2.297	7.21*
2.301	7.18*
2.425	8.10*
2.429	8.14*
2.481	9.35*
2.485	9.39*
2.725	1.53 x 10^{-5}*
2.729	1.54*
2.821	1.85*
2.825	1.86*
3.045	2.37*
3.048	2.39*
3.690	5.26*
3.691	5.29*
4.009	6.75*
4.009	6.80*
4.307	8.55*
4.310	8.48*
4.736	1.155 x 10^{-4}*
4.740	1.151*
5.207	1.521*
5.212	1.521*
5.786	2.086*
5.788	2.099*
6.383	2.763
7.012	3.645
7.637	4.751
8.202	5.782
10.081	5.520*
10.875	6.521*
CURVE 2	
454	6.28 x 10^{-2}
511	6.40
533	6.45
583	6.55
660	6.71
688	6.77*
753	6.92
793	7.00*
CURVE 2 (cont.)	
811	7.03*
903	7.23*
974	7.38
1031	7.50
1074	7.59*
1089	7.62*
1106	7.66
1189	7.83
1238	7.93
1283	8.03
1336	8.14
1366	8.20*
1444	8.36
1530	8.55
1589	8.67
1621	8.74*
1644	8.79*
1672	8.85*
1710	8.93*
1776	9.07
1832	9.18*
1882	9.29
CURVE 3	
273.15	6.430 x 10^{-2}
373	6.510
473	6.594
573	6.683
673	6.777*
773	6.875
873	6.978
973	7.086
1073	7.198
1173	7.316
1273	7.437
1373	7.564
1473	7.695
1573	7.852
1673	7.973
1773	8.118
1873	8.267
CURVE 4	
1400	7.85 x 10^{-2}*
1500	7.96*
1600	8.10*
1700	8.24*
1800	8.40
1900	8.56
2000	8.72
2100	8.88
2200	9.07
2300	9.27
CURVE 5	
10.98	6.67 x 10^{-4}
11.22	7.00
11.34	7.21
11.47	7.32*
12.29	8.72
12.56	9.15
12.68	9.47*
12.70	9.36*
13.77	1.13 x 10^{-3}
13.96	1.16*
14.08	1.20
14.11	1.18*
15.72	1.48
16.02	1.55
16.18	1.58*
16.20	1.57*
18.41	2.09
18.53	2.14
18.66	2.16*
18.69	2.15*
21.60	3.01
24.78	4.32
28.89	6.35
33.44	8.94
38.21	1.206 x 10^{-2}
42.98	1.539
47.77	1.876
52.52	2.202
57.46	2.543
59.01	2.631
CURVE 5 (cont.)	
62.38	2.833 x 10^{-2}
65.37	3.008
67.55	3.123
69.73	3.242
72.60	3.357
74.20	3.473
79.23	3.710
84.62	3.936
89.99	4.150
95.58	4.346
101.43	4.533
107.34	4.690
112.93	4.844
118.00	4.945
123.15	5.041
128.57	5.141
133.91	5.260
139.54	5.341
145.20	5.430
150.61	5.500
156.29	5.559
161.87	5.637
167.60	5.695
173.20	5.731
178.78	5.767
184.50	5.818
189.92	5.881
195.20	5.906
198.91	5.921*
200.81	5.952
201.16	5.945*
204.13	5.969
206.09	5.976*
206.51	5.996*
209.84	6.019*
211.68	6.013*
212.25	6.041*
215.09	6.037
216.50	6.049*
217.49	6.058*
220.80	6.069*
221.21	6.074*
222.91	6.089*
225.98	6.120
CURVE 5 (cont.)	
228.03	6.118 x 10^{-2}*
231.91	6.135*
233.76	6.154
237.37	6.164*
238.98	6.170*
243.73	6.174
244.35	6.198*
248.53	6.208*
249.20	6.214*
253.60	6.235
254.94	6.233*
259.03	6.243*
260.72	6.243*
265.30	6.264
265.80	6.267*
270.51	6.259*
270.64	6.286*
CURVE 6	
1273	6.575 x 10^{-2}
1373	6.938
1473	7.301
1573	7.665
1673	8.028*
1773	8.391*
1873	8.754
1973	9.117
2073	9.481
2173	9.844
2273	1.021 x 10^{-1}
2373	1.057
2473	1.093
2573	1.129*
2593	1.137
CURVE 7	
8.804	7.118 x 10^{-4}
8.932	7.729
9.030	6.780
9.712	5.027
9.951	5.298
10.300	5.732

*Not shown on plot

DATA TABLE NO. 38 (continued)

CURVE 7 (cont.)

T	C_p
10.757	6.283×10^{-4}
11.406	7.195*
12.034	8.148
12.740	9.194*
13.538	1.060×10^{-3}
14.594	1.258
15.648	1.487*
16.654	1.700
17.680	1.947
18.903	2.323
20.138	2.843
21.574	3.253
22.979	3.944
24.323	4.598
25.859	5.574
27.724	6.708
29.450	8.066

CURVE 8

T	C_p
5.078	2.091×10^{-4}
5.125	1.913*
6.440	2.184
7.061	2.578
7.734	3.262
8.469	3.851
9.138	4.518
10.121	5.652
10.944	6.646*

CURVE 9

T	C_p
5.267	1.843×10^{-4}*
7.313	2.862
9.001	4.329
9.853	5.234
10.744	6.348

CURVE 10

T	C_p
1.467	7.2×10^{-7}*
1 469	7.2*
1.729	1.40×10^{-6}*
1.732	1.36*

CURVE 10 (cont.)

T	C_p
1.863	2.54×10^{-6}*
1.865	2.56*
2.081	4.31*
2.082	4.32*
2.314	7.22*
2.318	7.18*
2.519	1.02×10^{-5}*
2.525	1.03*
2.788	1.61*
2.794	1.63*
3.146	2.63*
3.149	2.66*
3.602	4.28*
3.603	4.32*
3.928	6.14*
3.929	6.07*
4.344	8.59*
4.348	8.52*
4.804	1.216×10^{-4}*
4.809	1.212*
5.300	1.631*
5.305	1.637*
6.015	2.381*
6.017	2.390
6.923	3.618
6.923	3.628*
7.857	5.130
8.468	6.312
8.763	7.062*
9.927	5.285*
10.015	5.360*
11.132	7.00*

CURVE 11

T	C_p
1.353	2.93×10^{-5}*
1.353	2.94*
1.466	2.73*
1.467	2.70*
1.620	3.48*
1.623	3.43*
1.770	3.65*
1.774	3.65*
1.961	4.17*

CURVE 11 (cont.)

T	C_p
1.967	4.12×10^{-5}*
2.217	4.80*
2.229	4.65*
2.402	5.31*
2.418	5.19*
2.596	5.51*
2.613	5.53*
2.940	6.46*
2.954	6.63*
3.250	7.09*
3.258	7.23*
3.579	8.30*
3.583	8.36*
3.933	9.33*
4.436	9.34*
4.439	1.08×10^{-4}*
4.819	1.322*
4.877	1.310*
5.392	1.513*
5.392	1.515*
6.059	1.912*
6.887	2.415
7.921	3.170
9.110	4.243

CURVE 12*

T	C_p
2.563	1.257×10^{-5}
3.110	2.598
4.282	9.170
4.675	1.147×10^{-4}
5.228	1.808
7.420	4.762
10.475	5.854

CURVE 13*

T	C_p
0.5	1.01×10^{-5}
1.0	2.03
1.5	3.09
2.0	4.20
2.5	5.39
3.0	6.67
3.5	8.19
4.0	9.89

CURVE 13 (cont.)*

T	C_p
4.5	1.181×10^{-4}
5.0	1.398
5.5	1.641
6.0	1.914
6.5	2.219
7.0	2.561
7.5	2.939
8.0	3.357
8.5	3.818
9.0	4.325

CURVE 14*

T	C_p
3.233	7.481×10^{-5}
3.996	1.110×10^{-4}
6.144	2.056
8.63	4.032

*Not shown on plot

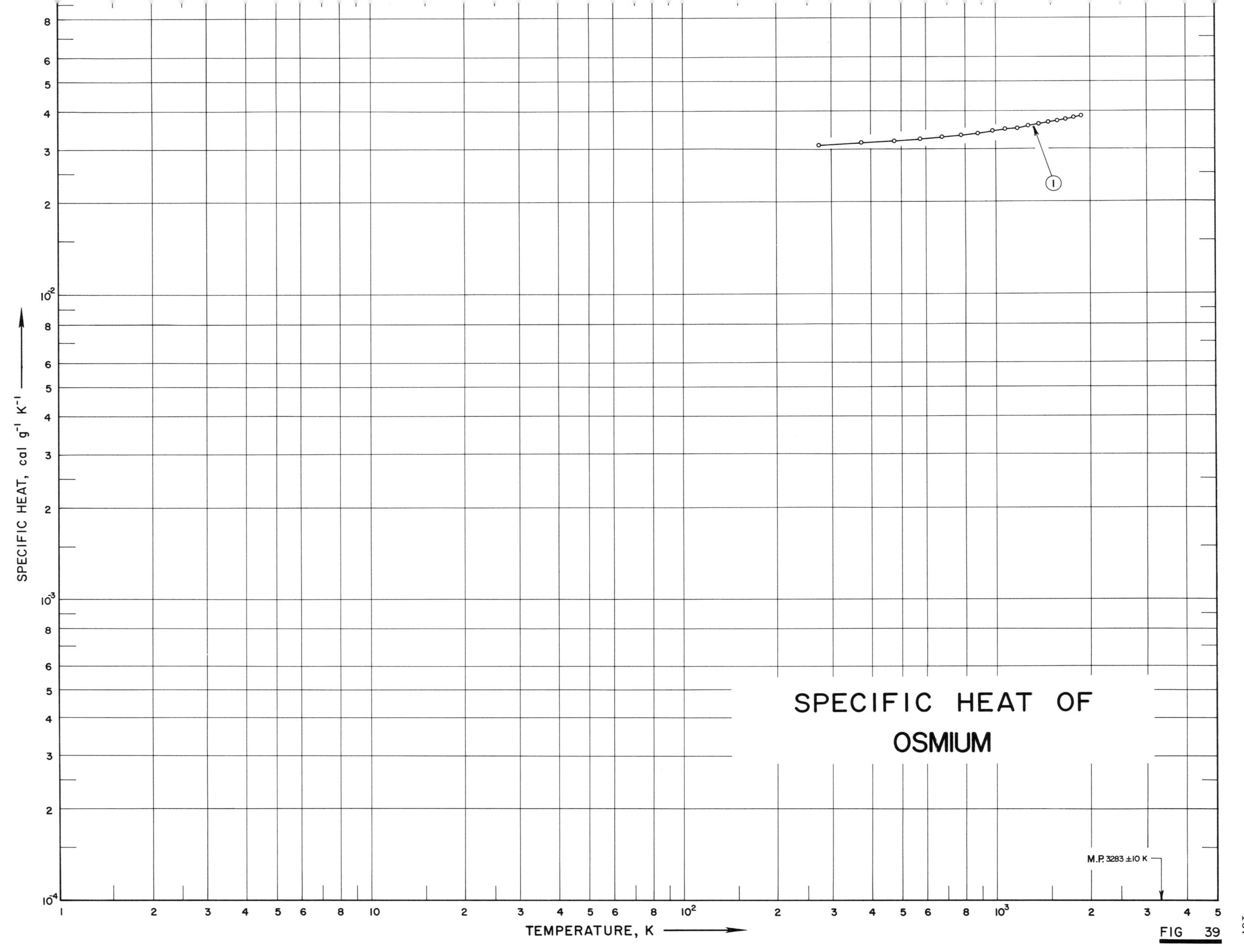
SPECIFIC HEAT OF
OSMIUM
M.P. 3283 ±10 K
1
TEMPERATURE, K
SPECIFIC HEAT, cal g^{-1} K^{-1}
FIG 39

SPECIFICATION TABLE NO. 39 SPECIFIC HEAT OF OSMIUM

(Impurity $< 0.20\%$ each; total impurities $< 0.50\%$)

[For Data Reported in Figure and Table No. 39]

Curve No.	Ref. No.	Year	Temp. Range, K	Reported Error, %	Name and Specimen Designation	Composition (weight percent), Specifications and Remarks
1	163	1931	273-1873			

DATA TABLE NO. 39 SPECIFIC HEAT OF OSMIUM

[Temperature, T, K; Specific Heat, C_p, Cal g^{-1} K^{-1}]

T	C_p
CURVE 1	
273.15	3.099 x 10^{-2}
373.15	3.146
473.15	3.193
573.15	3.240
673.15	3.287
773.15	3.335
873.15	3.382
973.15	3.429
1073.15	3.476
1173.15	3.503
1273.15	3.571
1373.15	3.618
1473.15	3.665
1573.15	3.712
1673.15	3.759
1773.15	3.807
1873.15	3.854

*Not shown on plot

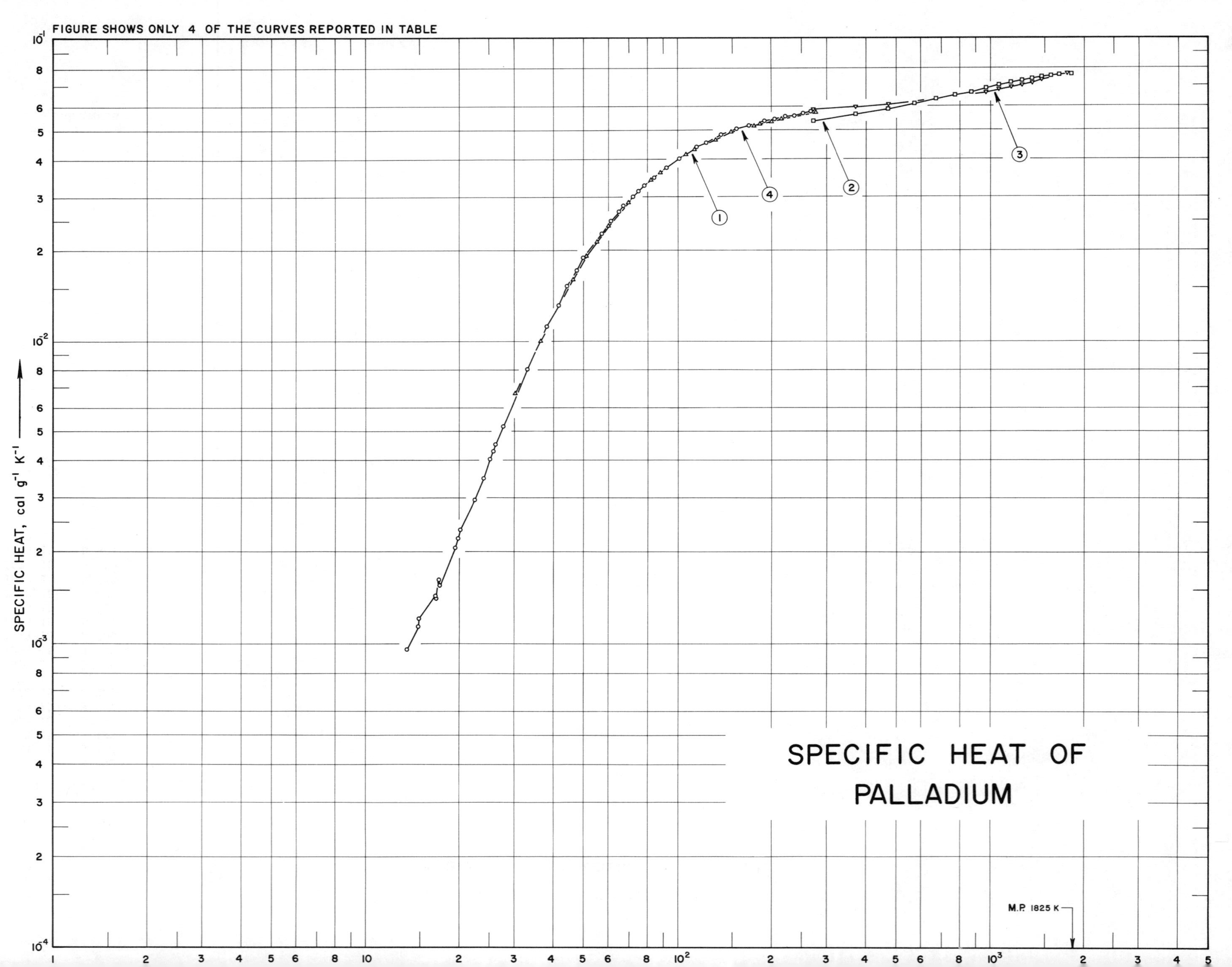
FIGURE SHOWS ONLY 4 OF THE CURVES REPORTED IN TABLE
SPECIFIC HEAT OF
PALLADIUM
SPECIFIC HEAT, cal g⁻¹ K⁻¹
M.P. 1825 K
1
2
3
4
10⁻¹
10⁻²
10⁻³
10⁻⁴
10
10²
10³

SPECIFICATION TABLE NO. 40 SPECIFIC HEAT OF PALLADIUM

(Impurity < 0.20% each; total impurities < 0.50%)

[For Data Reported in Figure and Table No. 40]

Curve No.	Ref. No.	Year	Temp. Range, K	Reported Error, %	Name and Specimen Designation	Composition (weight percent), Specifications and Remarks
1	156	1963	30-278	0.05		High purity.
2	164	1932	273-1811			
3	165	1936	273-1773			
4	166	1947	14-268			Heated slowly to 120 C.
5	336	1948	2-22			

DATA TABLE NO. 40 SPECIFIC HEAT OF PALLADIUM

[Temperature, T, K; Specific Heat, C_p, Cal g^{-1} K^{-1}]

T	C_p
CURVE 1	
30.43	6.708×10^{-3}
36.55	1.010×10^{-2}
41.87	1.319*
46.58	1.601
51.20	1.918
55.73	2.143
60.25	2.408
64.78	2.644*
69.90	2.887
74.45	3.091*
78.75	3.272*
82.38	3.432
88.60	3.628
94.53	3.843*
100.64	4.037*
107.42	4.172
114.42	4.331
125.57	4.563*
132.85	4.666
140.34	4.798*
148.82	4.909
157.88	5.021*
166.75	5.101*
175.49	5.193
184.09	5.261
192.55	5.344*
200.85	5.372
207.12	5.403*
216.55	5.466
225.88	5.506*
235.12	5.560*
244.27	5.614*
253.40	5.626*
262.36	5.694*
271.28	5.726*
277.59	5.720
CURVE 2	
273.15	5.377×10^{-2}
373.15	5.640
473.15	5.887
573.15	6.118
CURVE 2 (cont.)	
673.15	6.334×10^{-2}
773.15	6.534
873.15	6.717
973.15	6.885
1073.15	7.037
1173.15	7.173
1273.15	7.294
1373.15	7.398
1473.15	7.487
1573.15	7.559
1673.15	7.616
1773.15	7.657*
1793.15	7.664*
1803.15	7.641*
1810.15	7.640
CURVE 3	
273.15	5.838×10^{-2}
373.15	5.959
473.15	6.080
573.15	6.202*
673.15	6.324*
773.15	6.447*
873.15	6.570*
973.15	6.694
1073.15	6.819
1173.15	6.944
1273.15	7.069
1373.15	7.195
1473.15	7.321
1573.15	7.448*
1673.15	7.576*
1773.15	7.704
CURVE 4	
13.70	9.597×10^{-4}
14.84	1.144×10^{-3}
14.96	1.207
16.82	1.440
16.97	1.414
17.33	1.626
CURVE 4 (cont.)	
17.45	1.561×10^{-3}
17.49	1.568*
19.50	2.054
19.62	2.051*
19.94	2.218
20.37	2.360
22.51	2.946
22.69	3.002*
24.03	3.483
25.32	4.028
25.92	4.297
26.23	4.501
27.83	5.191
33.20	8.024
38.57	1.124×10^{-2}
41.88	1.319
44.59	1.523
47.90	1.711
50.07	1.898
57.50	2.261
57.92	2.290*
61.55	2.490
62.31	2.523*
65.08	2.675
67.38	2.804
72.14	3.009
75.10	3.140
76.66	3.215*
78.82	3.281
80.70	3.310*
81.42	3.405*
83.61	3.463*
84.43	3.474
85.34	3.526*
88.67	3.642*
89.77	3.679*
92.38	3.766
93.71	3.747*
95.95	3.871*
97.35	3.951*
101.63	4.040
105.82	4.157*
110.82	4.287*
CURVE 4 (cont.)	
115.53	4.401×10^{-2}
120.16	4.491*
124.73	4.572
129.18	4.661*
134.49	4.754*
138.77	4.820
143.67	4.886*
149.94	4.969*
155.31	5.025
159.99	5.091*
164.95	5.134*
170.26	5.191
179.53	5.279*
185.76	5.321*
191.05	5.380
196.69	5.382*
201.18	5.378*
201.87	5.404*
205.34	5.455
206.95	5.455*
210.52	5.518*
215.78	5.527*
222.09	5.543
226.04	5.581*
230.15	5.595*
237.31	5.598
242.16	5.628*
249.17	5.664*
253.02	5.699
258.67	5.717*
263.68	5.731*
268.44	5.748
CURVE 5*	
2	6.19×10^{-5}
4	1.24×10^{-4}
6	2.16
8	3.37
10	5.06
12	7.12
14	9.93
16	1.33×10^{-3}
CURVE 5 (cont.)*	
18	1.74×10^{-3}
20	2.25
22	2.87

*Not shown on plot

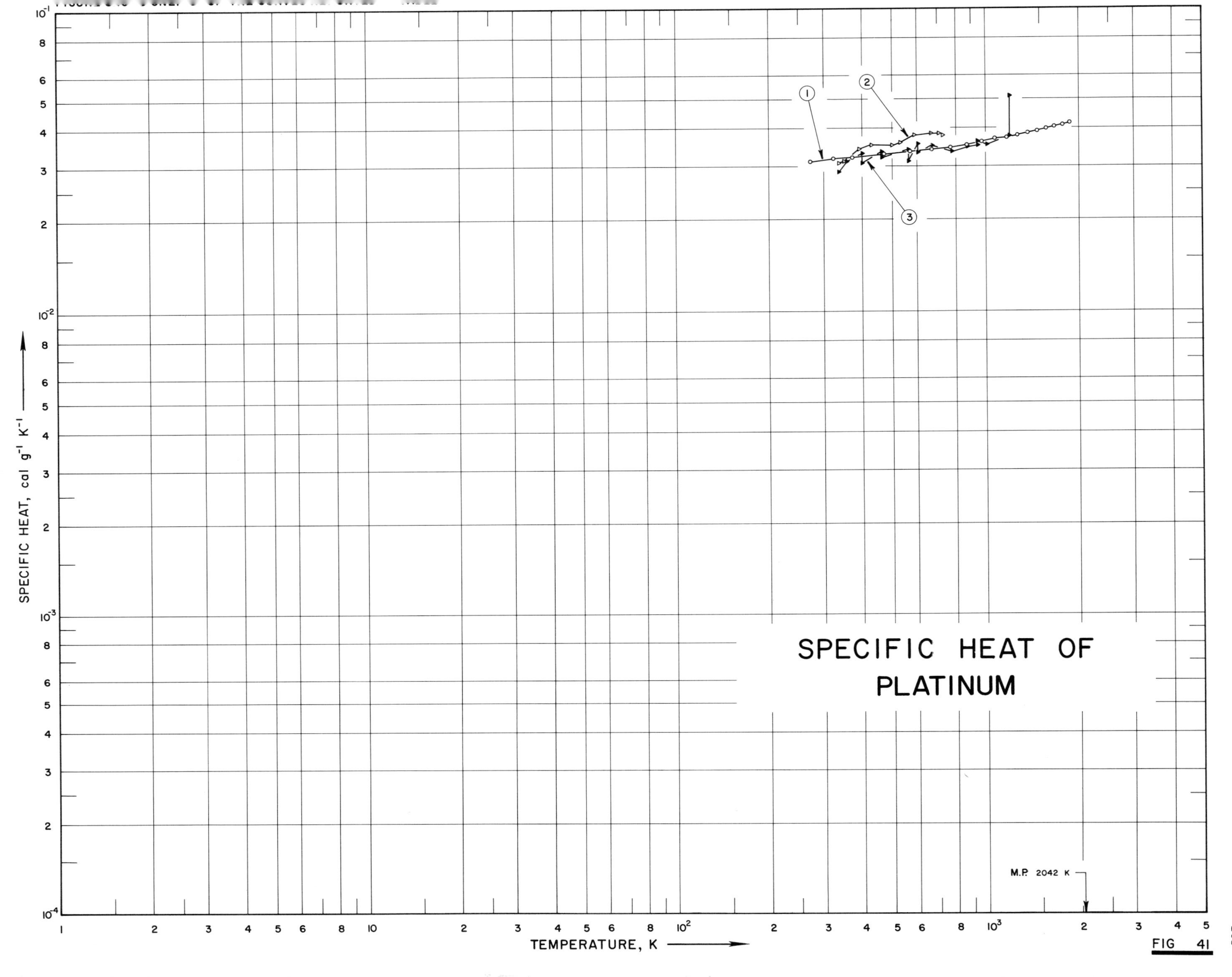
SPECIFIC HEAT OF PLATINUM
SPECIFIC HEAT, cal g⁻¹ K⁻¹
TEMPERATURE, K
M.P. 2042 K
1
2
3

FIG 41

SPECIFICATION TABLE NO. 41 SPECIFIC HEAT OF PLATINUM

(Impurity < 0.20% each; total impurities < 0.50%)

[For Data Reported in Figure and Table No. 41]

Curve No.	Ref. No.	Year	Temp. Range, K	Reported Error, %	Name and Specimen Designation	Composition (weight percent), Specifications and Remarks
1	99	1939	273-1873			
2	100	1955	338-727			Thermocouple grade; tested at pressure of 5 x 10^{-4} mm Hg.
3	101	1958	337-1164			Specimen's surface plated with platinum black.
4	167	1936	1.1-20			99.95 Pt.
5	168	1957	11-274			99.94 Pt, 0.03 Rh, 0.01 Pd, trace of Ag, Ca, Cu, Fe, and Mg.
6	169	1962	298-2043	0.3		99.99 Pt; measured in argon atmosphere.
7	213	1926	18-208			
8	261	1933	473-1373			

DATA TABLE NO. 41 SPECIFIC HEAT OF PLATINUM

[Temperature, T, K; Specific Heat, C_p, Cal $g^{-1} K^{-1}$]

T	C_p
CURVE 1	
273.15	3.136 x 10^{-2}
323	3.180
373	3.221
473	3.297
573	3.363
673	3.419
773	3.466
773	3.479*
873	3.543
973	3.606
1073	3.670
1173	3.733
1273	3.797
1373	3.860
1473	3.923
1573	3.987
1673	4.050
1773	4.114
1873	4.177
CURVE 2	
338	3.08 x 10^{-2}
350	3.13
393	3.43
430	3.54
500	3.53
532	3.59
590	3.82
669	3.85
710	3.85
727	3.79
CURVE 3	
337	2.87 x 10^{-2}
356	3.11
400	3.31
400	3.27*
465	3.36
465	3.33*
465	3.20
565	3.40
565	3.32*
CURVE 3 (cont.)	
565	3.13 x 10^{-2}
604	3.56
604	3.33
604	3.29*
676	3.51
676	3.47*
676	3.41*
781	3.45*
781	3.36
875	3.53*
875	3.48
946	3.52
946	3.63
1017	3.54
1190	3.77
1194	5.13
CURVE 4	
Series 1	
1.135	9.278 x 10^{-6}
1.143	1.008 x 10^{-5}
1.402	1.855
1.410	1.855
2.024	1.749
2.024	1.714
2.034	2.421
2.042	2.493
2.473	3.375
2.537	2.969
2.550	1.258
3.020	1.185
3.035	2.467
3.391	3.658
3.603	3.807
4.448	5.357
5.038	6.561
6.021	9.006
Series 2	
1.133	9.519 x 10^{-6}
1.149	9.939
CURVE 4 (cont.)	
Series 2 (cont.)	
1.187	9.098 x 10^{-6}
1.691	1.435 x 10^{-5}
1.718	1.498
1.947	1.748
1.982	1.750
2.363	2.270
Series 3	
1.341	2.251 x 10^{-5}
1.359	2.228
1.380	2.328
1.568	2.734
2.400	1.166
2.442	1.174
2.515	1.136
2.746	1.388
3.001	3.071
3.484	3.719
3.828	4.150
4.094	4.672
4.334	5.099
4.409	5.285
4.544	5.536
4.539	5.490
4.731	5.895
4.845	6.151
5.227	6.981
Series 4	
9.361	2.274 x 10^{-4}
9.728	2.591
10.501	2.983
11.240	3.514
12.432	4.789
12.903	5.351
14.515	7.397
15.469	8.535
16.862	1.081 x 10^{-3}
18.017	1.302
19.171	1.536
20.298	1.844
CURVE 5	
10.63	3.51 x 10^{-4}
10.81	3.66
11.18	3.97
11.23	4.07
12.27	4.93
12.24	5.05
12.88	5.628
12.95	5.915
13.29	6.074
13.89	7.094
14.14	7.407
14.38	7.535
14.57	7.755
15.33	9.283
15.61	1.007 x 10^{-3}
16.19	1.067
16.24	1.070
16.62	1.141
17.54	1.312
17.60	1.330
18.01	1.411
18.59	1.542
18.84	1.604
19.65	1.767
19.78	1.804
21.00	2.114
21.07	2.156
21.63	2.271
23.66	2.959
24.04	3.060
26.88	4.059
27.52	4.299
30.69	5.628
31.58	5.884
34.86	7.217
35.89	7.617
39.42	9.103
40.78	9.575
43.47	1.071 x 10^{-2}
45.02	1.126
47.13	1.210
49.70	1.308
50.83	1.345
53.58	1.432
CURVE 5 (cont.)	
54.22	1.457 x 10^{-2}
57.08	1.546
57.62	1.568
58.19	1.565
59.86	1.634
60.94	1.657
61.46	1.675
62.33	1.703
62.69	1.709
62.91	1.715
65.26	1.778
65.82	1.789
66.71	1.811
71.13	1.915
65.65	2.009
80.24	2.099
85.00	2.170
87.34	2.203
89.56	2.251
92.22	2.284
94.21	2.305
97.18	2.350
98.75	2.387
101.97	2.401
102.05	2.402
103.45	2.436
107.04	2.458
108.05	2.487
109.21	2.489
111.97	2.526
112.83	2.544
113.66	2.553
117.24	2.559
117.49	2.569
118.43	2.597
122.36	2.629
122.45	2.630
122.78	2.627
127.36	2.653
127.74	2.658
132.17	2.706
132.77	2.701
133.00	2.698
136.60	2.721
CURVE 5 (cont.)	
138.04	2.742 x 10^{-2}
138.83	2.741
143.43	2.780
144.32	2.767
148.65	2.796
150.34	2.804
154.23	2.839
156.11	2.823
158.59	2.840
162.36	2.861
165.34	2.874
168.15	2.891
170.74	2.901
174.35	2.907
176.72	2.932
180.05	2.930
182.67	2.931
186.34	2.943
188.25	2.963
191.84	2.965
194.34	2.972
198.17	2.971
200.01	2.974
201.39	2.969
203.25	2.993
203.78	2.999
206.16	3.004
206.87	3.012
209.59	3.001
211.78	3.018
213.33	3.030
215.32	3.015
217.33	3.019
217.85	3.023
218.87	3.040
221.78	3.044
223.09	3.031
225.39	3.050
227.56	3.051
229.54	3.051
230.99	3.068
235.13	3.070
235.45	3.073
237.31	3.090

*Not shown on plot

DATA TABLE NO. 41 (continued)

T	C_p
CURVE 5 (cont.)	
242.20	3.086×10^{-2}
242.37	3.092
242.82	3.082
247.95	3.080
248.21	3.077
249.58	3.103
254.85	3.110
254.95	3.120
255.16	3.117
260.65	3.117
260.69	3.123
261.67	3.127
267.04	3.127
267.18	3.123
267.83	3.101
273.53	3.161
273.71	3.146
CURVE 6	
298.15	3.17×10^{-2}
300	3.17
400	3.23
500	3.30
600	3.37
700	3.43
800	3.50
900	3.56
1000	3.63
1100	3.70
1200	3.76
1300	3.83
1400	3.89
1500	3.96
1600	4.02
1700	4.09
1800	4.16
1900	4.22
2000	4.29
2043	4.32

T	C_p
CURVE 7	
Series 1	
76.46	2.037×10^{-2}
81.40	2.112
91.70	2.317
98.85	2.389
201.50	3.040
206.10	3.050
Series 2	
17.50	1.225×10^{-3}
19.46	1.651
21.78	2.286
24.34	3.096
26.75	3.937
30.26	5.249
35.42	7.279
42.40	1.012×10^{-2}
50.50	1.312
57.60	1.543
96.10	2.353
101.90	2.409
108.00	2.507
114.30	2.563
121.00	2.609
198.20	3.045
203.10	3.050
208.30	3.070
CURVE 8	
473.15	3.26×10^{-2}
573.15	3.32
673.15	3.37
773.15	3.41
873.15	3.44
973.15	3.47
1073.15	3.48
1173.51	3.49
1273.15	3.50
1373.15	3.49

*Not shown on plot

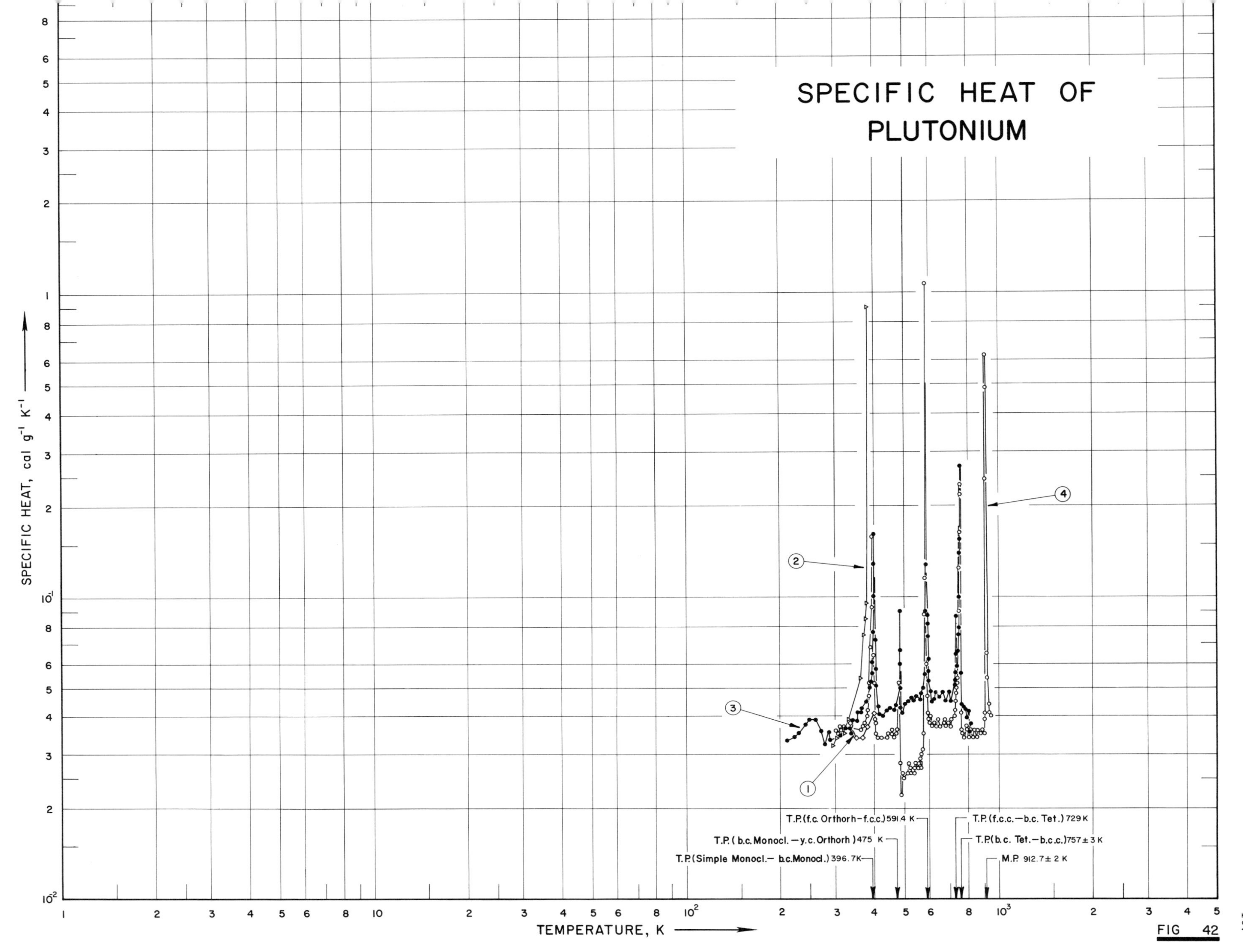
SPECIFIC HEAT OF PLUTONIUM
T.P. (f.c. Orthorh—f.c.c.) 591.4 K
T.P. (b.c. Monocl.—y.c. Orthorh) 475 K
T.P. (Simple Monocl.— b.c. Monocl.) 396.7 K
T.P. (f.c.c.—b.c. Tet.) 729 K
T.P. (b.c. Tet.—b.c.c.) 757 ± 3 K
M.P. 912.7 ± 2 K
1
2
3
4
SPECIFIC HEAT, cal g⁻¹ K⁻¹
TEMPERATURE, K
FIG 42

SPECIFICATION TABLE NO. 42 SPECIFIC HEAT OF PLUTONIUM

(Impurity $< 0.20\%$ each; total impurities $< 0.50\%$)

[For Data Reported in Figure and Table No. 42]

Curve No.	Ref. No.	Year	Temp. Range, K	Reported Error, %	Name and Specimen Designation	Composition (weight percent), Specifications and Remarks
1	121	1962	338-400			
2	122	1958	295-385	± 5.0		
3	123	1958	211-819	5.0		
4	124	1964	303-944	5.0		99.95 Pu.

DATA TABLE NO. 42 SPECIFIC HEAT OF PLUTONIUM

[Temperature, T, K; Specific Heat, C_p, Cal g^{-1} K^{-1}]

	T	C_p
	CURVE 1	
	338	3.8×10^{-2}
	352	3.4
	369	3.4
(α)	383	3.7
(β)	400	4.1
	CURVE 2	
	295	3.2×10^{-2}
	304	3.4
	321	3.5
	331	3.9
	363	5.4
	372	7.5
	376	8.5
	379	9.6
	385	8.99×10^{-1}
	CURVE 3	
	211.2	3.34×10^{-2}
	214.6	3.34*
	223	3.42
	229	3.52
	241	3.75
	248	3.89
	252	3.87*
	260	3.88
	271.6	3.57
	272.2	3.55*
	278.2	3.24
	286	3.55
	288	3.35
	393	3.38*
	302	3.41*
	311	3.45
	317	3.49*
	325	3.66
	335	3.66
	338	3.50
	343	3.87
	352	3.86
	354	4.13
	358	4.15*
	CURVE 3 (contd)	
	363	4.13×10^{-2}
	369	4.25
	377	4.49
	384	4.49*
	386	4.98
	391	5.28
	394	5.60
	395	6.09
	397	7.67
	399	1.017×10^{-1}
	399	1.292
	400	1.619
	403	9.90×10^{-2}*
	405	7.22
	406	5.79
	408	5.09
	413	4.31
	418	4.05
	425	3.99
	431	4.11*
	440	4.19
	444	4.20*
	449	4.25
	457	4.16*
	463	4.19
	465	4.33*
	470	4.34
	473	4.23*
	477	4.28*
	481	4.34*
	483	4.97
	484	9.03
	485	6.65
	485	5.99
	486	4.26
	491	4.08
	499	4.23*
	503	4.40
	509	4.30*
	515	4.48
	516	4.44*
	523	4.50*
	527	4.63
	533	4.52
	CURVE 3 (contd)	
	540	4.69×10^{-2}
	545	4.59*
	552	4.61*
	557	4.68*
	561	4.53
	564	4.78
	570	4.79*
	576	4.99
	578	5.05*
	580	6.07
	585	8.96
	588	1.281×10^{-1}
	591	8.73×10^{-2}
	593	8.20
	595	7.39
	597	6.23
	599	5.68
	600	5.28
	605	4.85
	607	4.71*
	616	4.46
	619	4.55*
	623	4.55
	629	4.81
	634	4.65*
	641	4.69*
	647	4.64
	649	4.69*
	652	4.71*
	656	4.64*
	662	4.83
	670	4.57*
	680	4.56
	685	4.78*
	693	4.86
	699	4.86*
	706	4.50
	715	4.70*
	722	4.97*
	723	5.02*
	725	5.10
	727	5.39
	729	5.66
	730	6.41
	CURVE 3 (contd)	
	732	8.68×10^{-2}
	734	6.29*
	737	5.86
	741	6.58
	743	7.50
	746	7.94
	746	1.006×10^{-1}
	750	1.416
	753	1.552
	754	1.578*
	756	2.681
	759	5.56×10^{-2}
	761	4.38
	764	4.34*
	771	4.31
	778	4.40*
	786	4.22
	791	3.94
	800	4.17
	806	3.54
	811	3.69*
	819	3.76
	CURVE 4	
	303	3.6×10^{-2}
	305	3.5
	307	3.5*
	309	3.7
	312	3.6
	316	3.6*
	318	3.5*
	320	3.7
	322	3.6*
	326	3.5*
	327	3.7*
	329	3.7*
	331	3.6*
	333	3.6*
	335	3.7*
	337	3.7*
	339	3.5*
	341	3.8*
	342	3.7*
	CURVE 4 (contd)	
	344	3.7×10^{-2}*
	346	3.8*
	350	3.8*
	352	3.8*
	354	3.9*
	356	3.9*
	357	3.8*
	359	3.8
	361	3.8*
	363	3.6
	365	3.8*
	367	3.7
	369	3.8*
	371	3.7*
	372	3.8
	374	3.8*
	376	3.9*
	378	4.0*
	380	4.0
	382	4.2
	383	4.3*
	386	4.7
	387	5.2
	389	6.8
	391	9.3
	395	1.59×10^{-1}
	397	9.4×10^{-2}*
	399	6.4
	401	5.2
	402	3.9
	404	3.8
	406	3.5
	408	3.5*
	410	3.4
	412	3.4*
	414	3.4*
	416	3.4*
	417	3.4*
	419	3.4*
	421	3.4*
	423	3.4*
	425	3.4
	427	3.4*
	429	3.4*
	CURVE 4 (contd)	
	431	3.4×10^{-2}*
	432	3.4*
	438	3.4*
	440	3.4
	442	3.5
	444	3.5*
	446	3.5*
	447	3.5*
	449	3.5*
	451	3.5*
	453	3.5*
	455	3.5
	457	3.6
	459	3.5*
	464	3.4
	466	3.5*
	468	3.5*
	470	3.5
	472	3.5*
	474	3.6
	476	3.6*
	477	4.4*
	481	5.2
	483	4.3*
	485	3.8
	489	3.2
	491	3.6
	493	3.6*
	494	3.5
	496	3.5*
	498	3.5*
	498	3.5*
	502	3.5*
	504	3.5*
	506	3.5*
	507	3.6*
	509	3.6*
	511	3.6
	513	3.8
	515	3.7*
	517	3.6*
	518	3.5*
	521	3.7
	522	3.6
	524	3.6*

* Not shown on plot

DATA TABLE NO. 42 (continued)

T	C_p	T	C_p	T	C_p	T	C_p	T	C_p
CURVE 4 (contd)		CURVE 4 (contd)		CURVE 4 (contd)		CURVE 4 (contd)		CURVE 4 (contd)	
526	3.6×10^{-2}*	614	3.8×10^{-2}*	702	3.7×10^{-2}*	790	3.7×10^{-2}	888	3.5×10^{-2}*
528	3.6*	616	3.8*	704	3.7	792	3.6	890	3.5*
530	3.6*	618	3.8*	706	3.9	794	3.5	892	3.5*
532	3.6*	620	3.7*	708	3.8*	796	3.5*	894	3.5*
534	3.6*	622	3.8*	710	3.6*	798	3.6*	895	3.7*
536	3.6*	624	3.8	712	3.7*	800	3.5*	897	3.5
537	3.6	626	3.9*	714	3.7*	804	3.4*	901	3.9
539	3.7	627	3.9*	715	3.9*	805	3.4	903	4.1
541	3.7*	629	3.8*	717	3.8*	807	3.5*	905	3.7*
543	3.8	631	3.8*	719	3.8*	809	3.6*	907	3.6*
545	3.7*	633	3.7	723	3.9*	811	3.5*	909	2.44×10^{-1}
547	3.7*	635	3.8*	725	4.0	813	3.5*	910	7.75
549	3.7*	637	3.8*	727	4.2	815	3.6*	912	4.86
550	3.7	639	3.9*	729	4.2*	817	3.6	914	1.35*
552	3.8	641	3.9	730	4.5	819	3.5*	916	6.5×10^{-2}
554	3.7*	642	3.8*	732	4.8	820	3.5*	918	4.3
556	3.7*	644	3.8*	734	4.9*	822	3.5*	920	5.4
558	3.7*	646	3.8*	736	5.0	824	3.5	922	4.4*
560	3.8*	648	3.8*	738	5.2	826	3.4	924	4.2*
562	3.9	650	3.7	740	5.4	823	3.5*	925	3.9
564	3.7	652	3.9*	742	5.3*	830	3.4*	927	4.0*
566	3.8*	654	3.8*	744	6.6*	832	3.4*	931	4.4
567	4.0	656	3.8*	745	7.8*	834	3.6	933	4.2*
569	4.0*	657	3.9*	747	9.0	835	3.5*	935	4.1
571	4.0*	659	3.8*	749	1.26×10^{-1}	837	3.5*	937	4.2*
573	4.1	661	3.8*	751	1.64	839	3.6*	939	4.0*
575	4.5	663	3.8*	753	2.17	841	3.4*	940	4.1*
577	4.5*	665	3.8*	755	2.34	843	3.4*	942	4.1*
579	5.4*	667	3.8*	757	1.75*	845	3.4*	944	4.0
581	8.8	669	3.8*	759	7.5×10^{-2}*	847	3.4*		
582	1.16×10^{-1}	671	3.8*	760	4.1	849	3.6*		
584	9.4×10^{-2}*	672	3.8	762	3.6	850	3.6*		
586	1.077×10^{0}	674	3.9	764	3.5	852	3.4		
588	6.0×10^{-2}	676	3.7	766	3.5*	854	3.6		
590	5.5*	678	3.8*	768	3.5*	856	3.5*		
592	4.7	680	3.8*	770	3.5*	858	3.5*		
594	4.1	682	3.8*	772	3.4	860	3.5*		
596	3.9	684	3.7*	774	3.5*	862	3.5*		
597	3.9*	685	3.9*	775	3.4*	864	3.4*		
599	3.8	687	3.8*	777	3.3*	865	3.5*		
601	3.8*	689	3.8*	779	3.6*	867	3.5*		
603	3.8*	691	3.8*	781	3.5*	869	3.6*		
607	4.0*	693	3.9	783	3.5*	879	3.5		
609	4.0	697	3.8*	785	3.4*	882	3.5*		
611	3.7	699	3.8*	787	3.4*	884	3.5*		
612	3.7*	700	3.8*	789	3.4*	886	3.6		

*Not shown on plot

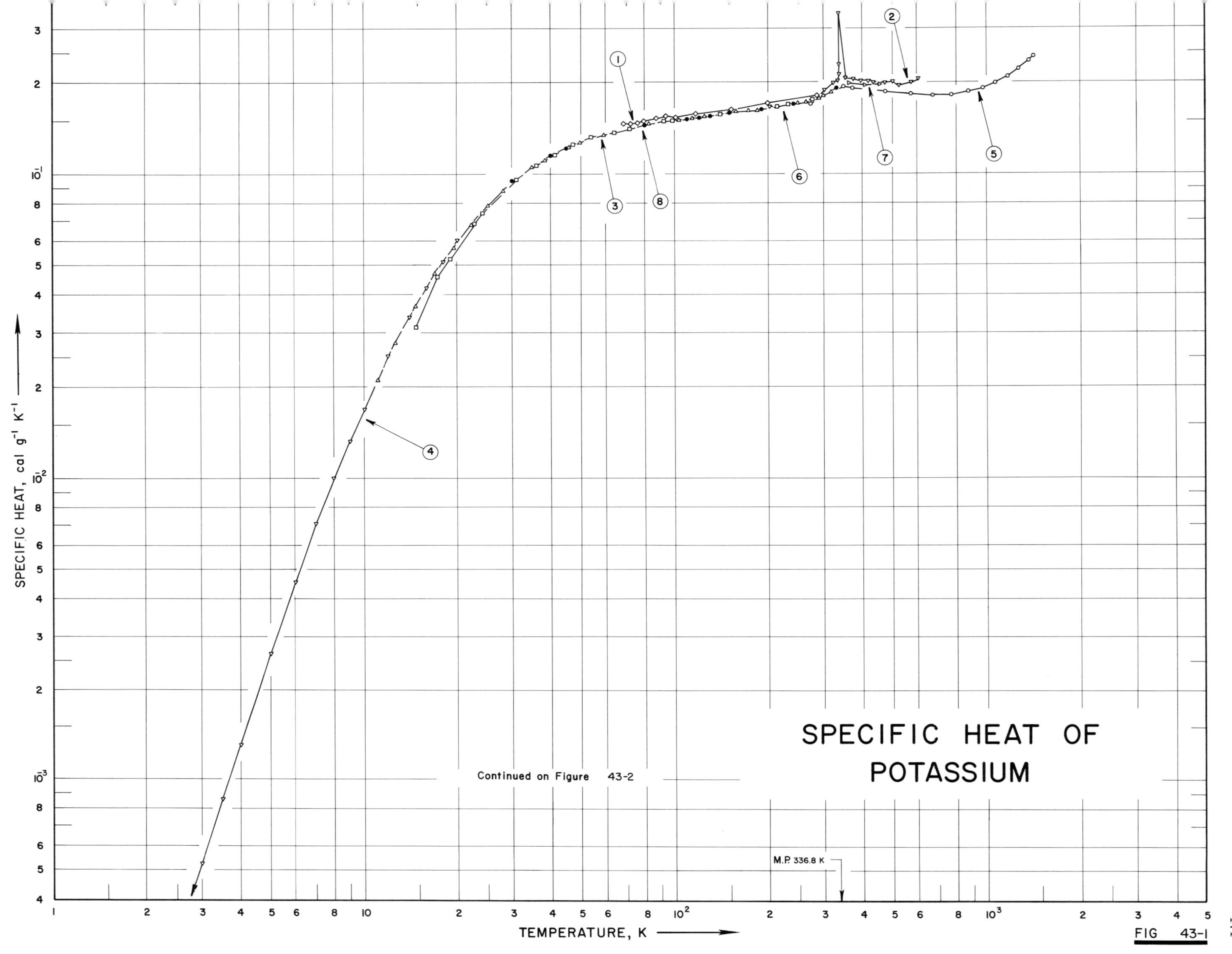
SPECIFIC HEAT OF
POTASSIUM
Continued on Figure 43-2
M.P. 336.8 K
SPECIFIC HEAT, cal g^{-1} K^{-1}
TEMPERATURE, K
FIG 43-1

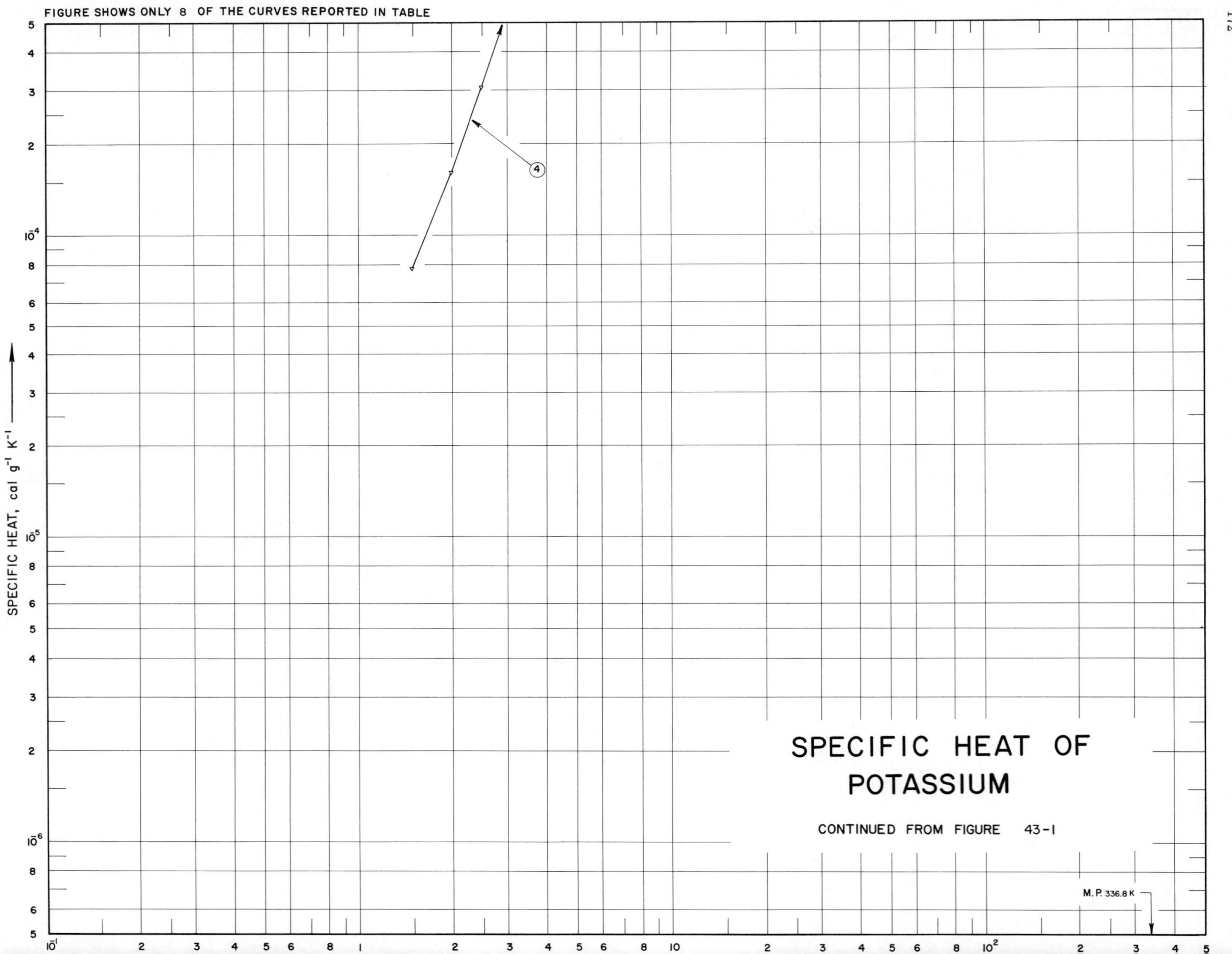
FIGURE SHOWS ONLY 8 OF THE CURVES REPORTED IN TABLE
SPECIFIC HEAT, cal g^{-1} K^{-1}
4
SPECIFIC HEAT OF
POTASSIUM
CONTINUED FROM FIGURE 43-I
M. P. 336.8 K

SPECIFICATION TABLE NO. 43 SPECIFIC HEAT OF POTASSIUM

(Impurity < 0.20% each; total impurities < 0.50%)

[For Data Reported in Figure and Table No. 43]

Curve No.	Ref. No.	Year	Temp. Range, K	Reported Error, %	Name and Specimen Designation	Composition (weight percent), Specifications and Remarks
1	170	1918	69-287	< 1		Kahlbaum's purity; melted under vacuum.
2	171	1939	203-609			Kahlbaum's purest potassium; distilled.
3	259	1957	11-323	0.1		Spectroscopic traces of Al, B and Si; triple distilled under high vacuum.
4	173	1957	1.5-2.0			99.995 K
5	174	1963	273-1423			0.0035-0.004 O_2; under argon atmosphere.
6	213	1926	15-277			
7	214	1927	363-454	1.0		Melted in vacuum and filtered into distillation bulbs.
8	176	1955	30-330			< 0.01 Na; specimen from Pure Metal Research Committee of the United Kingdom and prepared by Imperial Chemical Industry.
9	172	1954	11-322	0.15-0.5		99.96 ± 0.02 K, 0.01-0.1 Rb, 0.003-0.3 Ca, 0.001-0.1 each Cr, Fe, and Na, and 0.0003-0.002 Cu; 99^+ K sample supplied by the Mine Safety Appliance Co.; triple distilled.
10	353	1952	273-1073	2.0		Impurities: 0.01-0.1 Na, 0.001-0.01 Ca, and < 0.001 each Al, Cr, Fe, Mg, Rb, Mn, and Si; sample supplied by Baker Chemical Co.; prepared by triple distillation.
11	356	1964	0.3-4	0.3-1.5		99.99 K (stated purity), 0.2 Na (analyzed); measured under argon atmosphere.

DATA TABLE NO. 43 SPECIFIC HEAT OF POTASSIUM

[Temperature, T, K; Specific Heat, C_p, Cal g^{-1} K^{-1}]

T	C_p
CURVE 1	
68.6	1.47 x 10^{-1}
72.3	1.47
76.0	1.48
79.8	1.50
87.0	1.52
94.1	1.55
101.8	1.55
116.7	1.59
119.3	1.59*
152.2	1.64
199.5	1.71
203.5	1.72*
285.1	1.80*
286.7	1.82
CURVE 2	
Series 1	
334.2	2.032 x 10^{-1}
334.3	2.036*
334.5	2.034*
334.8	2.056*
334.8	2.074*
335.5	2.082*
335.7	2.300
335.9	2.323*
336.0	3.388
336.1	2.258 x 10^{0}*
336.2	9.995*
336.9	2.060 x 10^{-1}*
337.3	2.045*
337.7	2.127
337.8	2.065*
338.1	2.054*
338.2	2.033*
338.2	2.076*
340.1	2.053*
354.7	2.073
355.8	2.064*
376.7	2.057
377.8	2.033*
CURVE 2 (cont.)	
Series 1 (cont.)	
396.6	2.009 x 10^{-1}*
397.4	2.039
417.8	2.017*
419.0	2.039
436.4	2.025*
437.3	2.000
455.1	2.034*
455.9	2.006*
476.7	2.000
477.4	2.003*
500.6	2.013
501.6	2.031*
525.1	1.966
526.0	1.973*
575.0	1.992
575.8	2.001*
575.9	1.987*
605.9	2.055
609.5	2.075*
Series 2	
203.4	1.670 x 10^{-1}*
204.1	1.737*
275.7	1.772*
276.8	1.850*
287.0	1.834*
288.0	1.867*
301.6	1.923*
302.4	1.891
305.1	1.862*
305.9	1.933*
308.3	1.914*
309.5	1.958*
310.7	1.931*
311.5	1.944*
311.6	1.971*
312.2	1.947*
312.5	1.951*
313.3	1.956*
CURVE 2 (cont.)	
Series 2 (cont.)	
317.2	1.977 x 10^{-1}*
318.3	1.970*
322.8	2.000
323.7	1.993*
324.5	1.955*
327.7	1.985*
328.7	2.003*
328.9	1.995*
329.5	2.039*
329.9	2.028*
330.4	2.026*
330.6	2.054*
330.8	2.073*
330.8	2.048*
330.9	2.054*
331.2	2.053*
331.5	1.993*
331.5	1.996*
331.8	2.021*
332.0	1.998*
332.4	2.038*
332.7	2.045*
333.1	2.025*
333.5	2.064*
CURVE 3	
Series 1	
295.09	1.802 x 10^{-1}*
300.32	1.815
305.68	1.832*
311.50	1.843*
317.61	1.870
Series 2	
279.91	1.764 x 10^{-1}*
284.92	1.773*
290.08	1.784
CURVE 3 (cont.)	
Series 2 (cont.)	
295.34	1.799 x 10^{-1}*
300.63	1.808*
306.08	1.830*
311.69	1.846*
317.33	1.866*
322.80	1.894*
Series 3	
202.17	1.651 x 10^{-1}*
206.63	1.656*
211.31	1.663*
216.26	1.668*
221.34	1.674*
226.44	1.678*
231.69	1.689*
237.10	1.695*
242.54	1.704*
248.00	1.709
253.48	1.718*
258.92	1.725*
264.37	1.733
269.82	1.744*
275.37	1.753*
281.02	1.763*
286.62	1.778*
Series 4*	
205.79	1.655 x 10^{-1}
210.48	1.662
215.42	1.667
220.52	1.672
225.65	1.679
230.91	1.687
236.34	1.694
241.80	1.701
247.30	1.709
252.85	1.717
258.36	1.725
CURVE 3 (cont.)	
Series 4 (cont.)*	
263.85	1.733 x 10^{-1}
269.31	1.743
274.87	1.753
280.54	1.767
286.21	1.777
Series 5	
103.73	1.513 x 10^{-1}
109.03	1.526*
114.43	1.534
119.82	1.544*
125.10	1.552
130.43	1.560*
135.80	1.568*
141.23	1.576*
146.71	1.585*
152.12	1.592*
157.46	1.603
162.88	1.609*
168.36	1.613*
173.85	1.618
179.36	1.625*
184.82	1.631
190.42	1.636*
196.14	1.643*
Series 6	
11.12	2.148 x 10^{-2}
12.62	2.818
14.64	3.698
16.99	4.734
19.51	5.780
22.16	6.867
25.03	7.908
28.10	8.882
31.36	9.764*
34.76	1.056 x 10^{-1}
38.29	1.121
CURVE 3 (cont.)*	
Series 6 (cont.)	
41.98	1.181 x 10^{-1}*
45.82	1.227
49.80	1.273
54.14	1.313*
59.05	1.350
64.32	1.385*
69.67	1.411*
76.32	1.433*
81.26	1.458
87.32	1.479*
93.62	1.493*
100.05	1.506*
105.35	1.519*
112.46	1.531*
118.43	1.540*
124.27	1.551*
129.99	1.560*
136.81	1.568*
141.74	1.577*
147.57	1.587*
153.52	1.593*
159.61	1.600*
165.63	1.609*
171.77	1.615*
178.02	1.632*
184.20	1.629*
190.37	1.636*
196.38	1.643*
202.20	1.650*
207.39	1.656*
212.07	1.662*
217.01	1.667*
222.09	1.675*
227.18	1.680*
CURVE 4	
1.5	7.74896 x 10^{-5}
2.0	1.61117 x 10^{-4}
2.5	3.06890

*Not shown on plot

DATA TABLE NO. 43 (continued)

T	C_p
CURVE 4 (cont.)	
3.0	5.29385 x 10^{-4}
3.5	8.64406
4.0	1.31195 x 10^{-3}
5.0	2.63414
6.0	4.57777
7.0	7.05846
8.0	9.94834
9.0	1.32986 x 10^{-2}
10.0	1.68789
12.0	2.53184
14.0	3.37579
16.0	4.21973
18.0	5.11483
20.0	6.03550
CURVE 5	
273.15	1.70_5 x 10^{-1}
298.15	1.81_0*
323.15	1.91_5*
348.15	1.94_4
373.15	1.92_7
473.15	1.87_2
573.15	1.84_0
673.15	1.82_9
773.15	1.83_9
873.15	1.87_2
973.15	1.92_8
1073.15	2.00_4
1173.15	2.10_3
1273.15	2.22_3
1373.15	2.36_6
1423.15	2.44_6
CURVE 6	
14.71	3.15 x 10^{-2}
17.30	4.58
19.00	5.22
22.83	6.88
24.21	7.42
31.05	9.59
36.00	1.07 x 10^{-1}
41.20	1.164
47.20	1.25
54.00	1.33
CURVE 6 (cont.)	
64.10	1.38 x 10^{-1}
71.80	1.41
92.50	1.49
98.30	1.50
140.40	1.58
201.10	1.66*
207.80	1.66*
214.50	1.66
220.00	1.71*
225.50	1.69*
232.10	1.70
276.50	1.76
CURVE 7	
363	1.99 x 10^{-1}
409	1.96
454	1.97
CURVE 8	
30	9.46 x 10^{-2}
35	1.06 x 10^{-1}*
40	1.16
45	1.22
50	1.27*
55	1.32*
60	1.10*
70	1.41*
80	1.46
90	1.49*
100	1.51*
110	1.52
120	1.54
130	1.55
140	1.58*
150	1.59
160	1.61*
170	1.62*
180	1.63*
190	1.64
200	1.65*
210	1.66*
220	1.68*
230	1.69*
CURVE 8 (cont.)	
240	1.70 x 10^{-1}
250	1.71*
260	1.73*
270	1.74*
273.15	1.75*
280	1.76*
290	1.79*
298.15	1.81*
300	1.81*
310	1.84*
320	1.88*
330	1.94
CURVE 9*	
Series 1	
295.09	1.801 x 10^{-1}
300.32	1.813
305.68	1.830
311.50	1.841
317.61	1.869
Series 2	
279.94	1.763 x 10^{-1}
284.92	1.771
290.08	1.783
295.34	1.798
300.63	1.806
306.08	1.828
311.69	1.844
317.33	1.864
322.80	1.892
Series 3	
202.17	1.650 x 10^{-1}
206.63	1.655
211.31	1.661
216.26	1.666
221.34	1.672
226.44	1.677
231.69	1.688
237.10	1.693
CURVE 9 (cont.)	
Series 3 (cont.)	
242.54	1.702 x 10^{-1}
248.00	1.707
253.48	1.716
258.92	1.724
264.37	1.731
269.82	1.743
275.37	1.751
281.02	1.762
286.62	1.777
Series 4	
205.79	1.653 x 10^{-1}
210.48	1.660
215.42	1.666
220.52	1.670
225.65	1.678
230.91	1.685
236.34	1.693
241.80	1.700
247.30	1.708
252.85	1.716
258.36	1.724
263.85	1.731
269.31	1.742
274.87	1.752
280.54	1.766
286.21	1.775
Series 5	
103.73	1.512 x 10^{-1}
109.03	1.524
114.43	1.533
119.82	1.54[illegible]
125.10	1.551
130.43	1.558
135.80	1.567
141.23	1.575
146.71	1.583
152.12	1.591
157.46	1.601
162.88	1.607
CURVE 9 (cont.)	
Series 5 (cont.)	
168.36	1.612 x 10^{-1}
173.85	1.616
179.36	1.623
184.82	1.629
190.42	1.635
196.14	1.641
Series 6	
11.12	2.13 x 10^{-2}
12.62	2.80
14.64	3.68
16.99	4.72
19.51	5.76
22.16	6.85
25.03	7.89
28.10	8.87
31.36	9.75
34.76	1.055 x 10^{-1}
38.29	1.120
41.98	1.179
45.82	1.226
49.80	1.272
54.14	1.312
59.06	1.348
64.32	1.384
69.67	1.410
75.32	1.432
81.26	1.457
87.32	1.477
93.62	1.491
100.05	1.505
106.35	1.517
112.46	1.530
118.43	1.539
124.27	1.549
129.99	1.558
135.81	1.567
141.74	1.576
147.57	1.585
153.52	1.592
159.61	1.599
165.63	1.607
CURVE 9 (cont.)	
Series 6 (cont.)	
171.77	1.614 x 10^{-1}
178.02	1.631
184.20	1.628
190.37	1.635
196.38	1.641
202.20	1.649
207.39	1.655
212.07	1.661
217.01	1.666
222.09	1.674
227.15	1.678
CURVE 10*	
273.15	1.698 x 10^{-1}
298.15	1.822
323	1.946
348	1.957
373	1.941
473	1.888
573	1.850
673	1.826
773	1.819
873	1.826
973	1.847
1073	1.884
CURVE 11*	
Series 1	
0.260_4	3.577 x 10^{-6}
0.278_1	3.855
0.295_3	4.148
0.250_1	3.418
0.269_8	3.708
0.289_4	4.069
0.306_7	4.342
0.327_0	4.699
0.347_8	5.111
0.373_4	5.611
0.399_4	6.131
0.427_4	6.736
0.457_8	7.384

*Not shown on plot

DATA TABLE NO. 43 (continued)

T	C_p
CURVE 11 (cont.)*	
Series II	
0.265_0	3.649×10^{-6}
0.288_5	4.021
0.337_9	4.867
0.364_4	5.414
0.393_5	5.949
0.423_1	6.241
0.451_5	7.194
0.483_5	7.958
0.496_9	8.270
0.543_5	9.480
0.594_4	1.092×10^{-5}
0.641_4	1.239
0.690_1	1.408
Series III	
0.480_5	7.897×10^{-6}
0.525_9	8.991
0.566_1	1.014×10^{-5}
0.612_2	1.150
0.661_4	1.310
0.715_5	1.502
0.769_7	1.710
0.829_6	1.982
0.892_2	2.301
Series IV	
0.723_6	1.535×10^{-5}
0.778_5	1.759
0.833_2	2.023
0.890_2	2.296
0.871_0	2.211
0.933_4	2.496
1.013_0	2.995
1.101_0	3.597
1.180_0	4.202
1.218_0	4.517
1.238_0	4.704
Series V	
1.160_6	3.985×10^{-5}
1.239_2	4.636
1.324_3	5.439
1.416_2	6.412
CURVE 11 (cont.)*	
1.527_4	7.793×10^{-5}
1.645_8	9.474
1.758_2	1.125×10^{-4}
1.874_1	1.339
1.997_8	1.598
2.130_0	1.917
2.272_5	2.311
2.420_5	2.778
2.573_8	3.342
2.735_7	4.027
2.914_0	4.889
3.110_3	5.996
3.314_0	7.329
3.530_8	8.949
3.766_9	1.097×10^{-3}
4.030_9	1.364
Series VI	
1.142_1	3.837×10^{-5}
1.203_3	4.349
1.283_4	5.054
1.391_8	6.155
1.489_7	7.317
1.588_4	8.631
1.696_8	1.027×10^{-4}
1.816_2	1.232
1.935_8	1.467
2.058_5	1.740
2.193_2	2.085
2.341_5	2.524
2.500_7	3.072
2.660_0	3.697
2.827_9	4.479
3.008_4	5.417
3.207_8	6.638
3.414_3	8.075
3.632_5	9.817
3.881_9	1.211×10^{-3}
4.101_2	1.444

*Not shown on plot

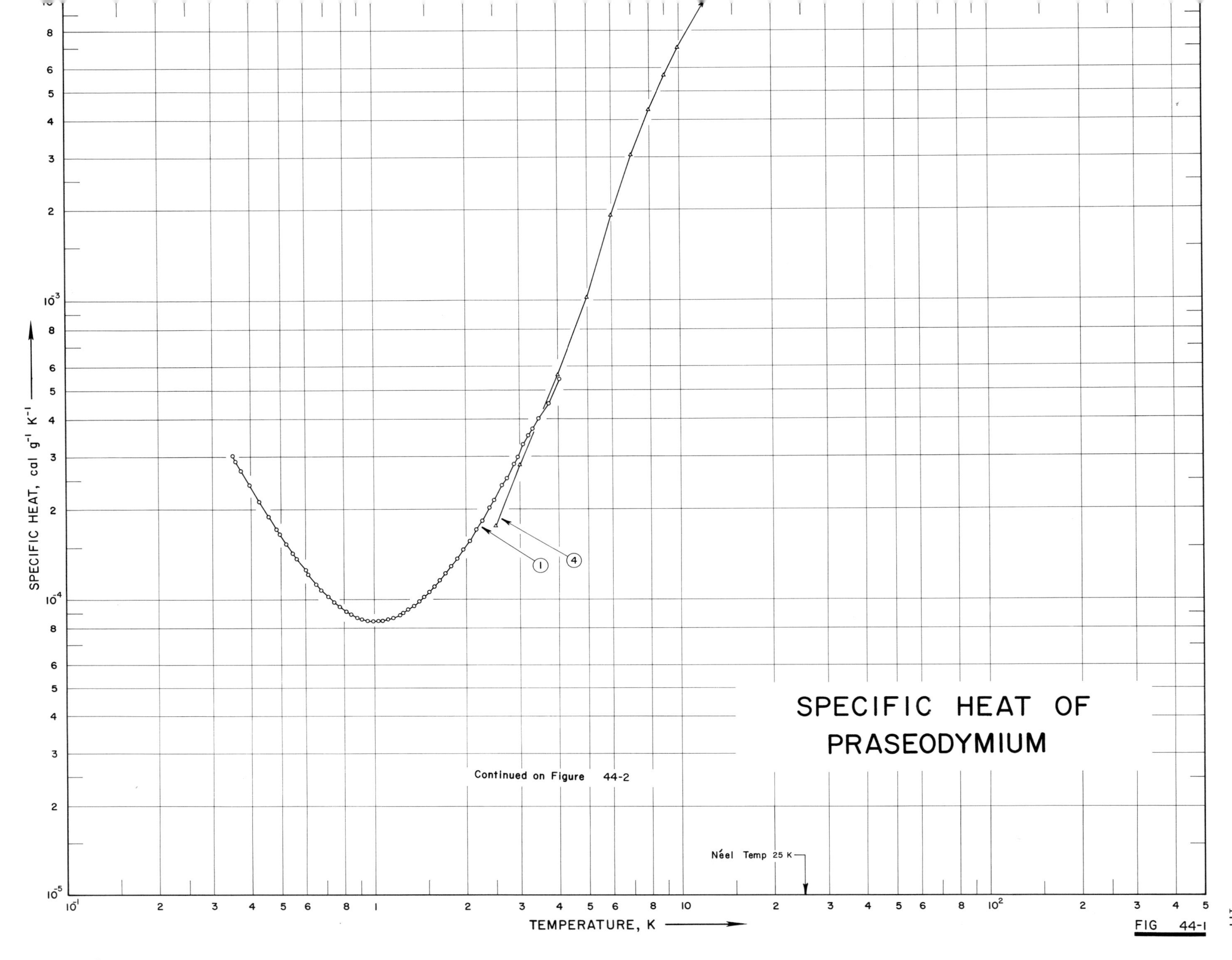
SPECIFIC HEAT, cal g^{-1} K^{-1}
1
4
SPECIFIC HEAT OF
PRASEODYMIUM
Continued on Figure 44-2
Néel Temp 25 K
TEMPERATURE, K
FIG 44-1

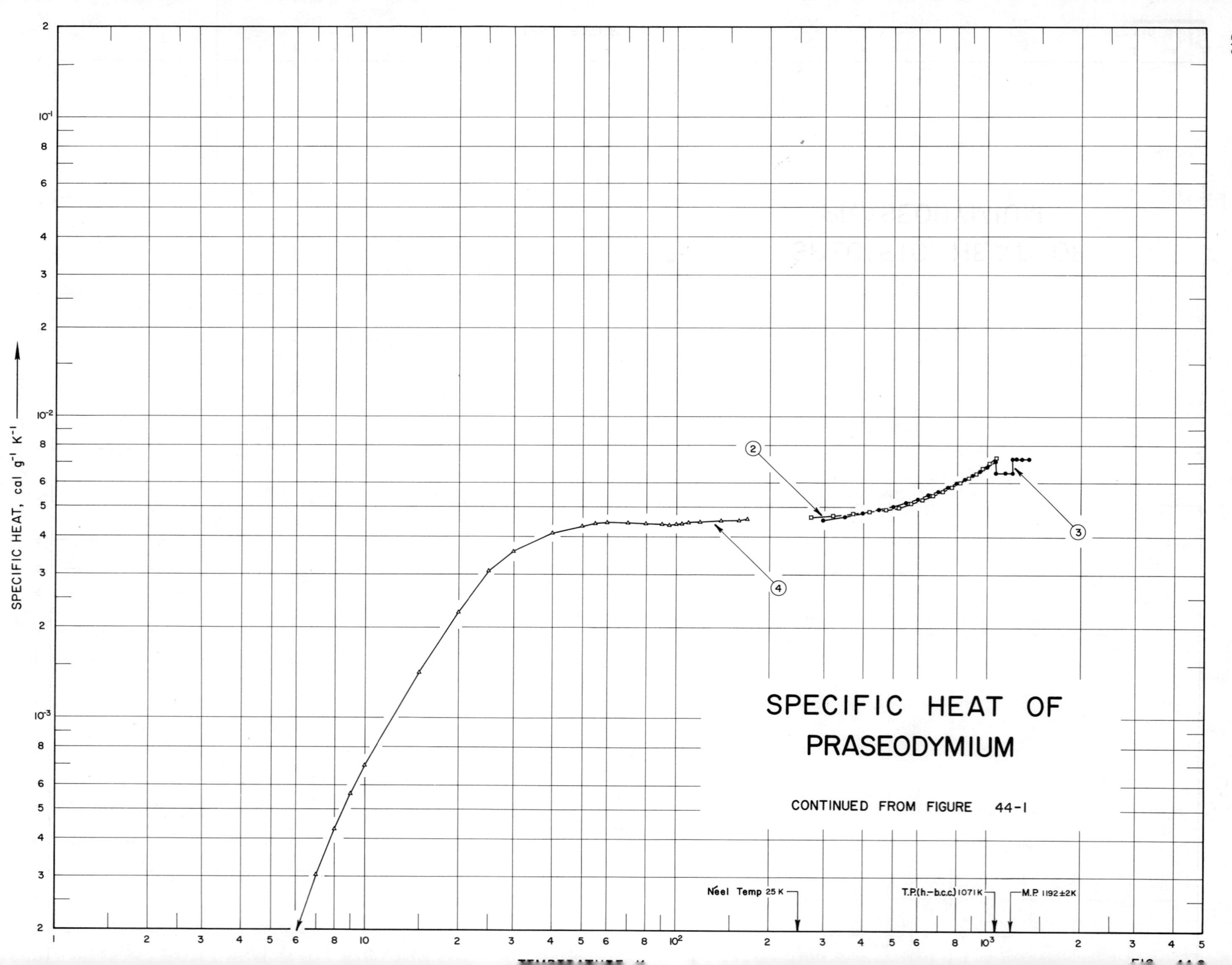
SPECIFIC HEAT OF
PRASEODYMIUM
CONTINUED FROM FIGURE 44-1
SPECIFIC HEAT, cal g⁻¹ K⁻¹
Néel Temp 25 K
T.P.(h.–b.c.c.) 1071 K
M.P. 1192±2K
2
3
4

SPECIFICATION TABLE NO. 44 SPECIFIC HEAT OF PRASEODYMIUM

(Impurity < 0.20% each; total impurities < 0.50%)

[For Data Reported in Figure and Table No. 44]

Curve No.	Ref. No.	Year	Temp. Range, K	Reported Error, %	Name and Specimen Designation	Composition (weight percent), Specifications and Remarks
1	87	1964	0.4-4	< 1.5		0.04 Ni, 0.029 F, 0.015 C, 0.011 O_2, 0.008 H_2, 0.004 N_2, 0.003 Na, 0.002 Ta, 0.0015 Fe, and trace amounts (total 0.036) of Ag, Al, B, Ca, Cu, Er, Gd, K, La, Li, Lu, Mn, Mo, Sr, V, and Y; vacuum distilled; remelted in vacuum and cast into tantalum crucible; machined in argon atmosphere.
2	285	1958	273-1071			Impurities: 0.1 each Ce, La, Nd, and Ta, 0.05 Ca, 0.02 Si, and 0.01 Fe.
3	36	1962	273-1373			< 0.3 Si, < 0.1 Ca, < 0.1 La, < 0.1 Nd, < 0.1 Ta, < 0.05 Ca, < 0.01 Fe and < 0.01 Mg.
4	35	1954	2.5-170			< 0.25 Mg, < 0.07 Fe, and 0.05 Ca.

DATA TABLE NO. 44 SPECIFIC HEAT OF PRASEODYMIUM

[Temperature, T, K; Specific Heat, C_p, Cal g^{-1} K^{-1}]

T	C_p
CURVE 1	
Series 1	
0.3585	2.899 x 10^{-4}
0.3972	2.409
0.4276	2.122
0.4578	1.895
0.4973	1.660
0.5483	1.431
0.6007	1.256
0.6537	1.126
0.7123	1.022
0.7765	9.424 x 10^{-5}
0.8452	8.881
0.9173	8.561
0.9912	8.420
1.0708	8.456
1.1577	8.652
Series 2	
1.2499	9.000 x 10^{-5}
1.3519	9.494
1.4623	1.020 x 10^{-4}
1.5822	1.107
1.7183	1.224
1.8748	1.378
2.0543	1.578
2.2607	1.839
2.4879	2.149
2.7241	2.538
2.9615	2.989
3.2024	3.511
3.4596	3.991
3.7416	4.500
Series 3	
0.3507	3.018 x 10^{-4}
0.3731	2.692
0.3981	2.398*
0.4249	2.144*
0.4534	1.925*
0.4853	1.725
CURVE 1 (cont.)	
Series 3 (cont.)	
0.5209	1.544 x 10^{-4}
0.5627	1.376
0.6153	1.215
0.6784	1.078
0.7442	9.777 x 10^{-5}
0.8113	9.109
0.8824	8.679
0.9562	8.471
1.0310	8.430
1.1125	8.547
1.2024	8.808
1.2994	9.234
1.4057	9.814
1.5209	1.061 x 10^{-4}
1.6482	1.163
1.7940	1.298
1.9613	1.479
2.1593	1.713
2.3922	2.018
2.6447	2.407
2.8832	2.838
3.0967	3.291
3.3012	3.710
3.5252	4.078*
3.7848	4.627*
4.0565	5.416
CURVE 2	
273.2	4.68 x 10^{-2}
323.2	4.72
373.2	4.78
423.2	4.85
473.2	4.95
523.2	5.07
573.2	5.19
623.2	5.34
673.2	5.49
723.2	5.67
773.2	5.86
CURVE 2 (cont.)	
823.2	6.07 x 10^{-2}
873.2	6.29
923.2	6.54
973.2	6.79
1023.2	7.07
1071.2	7.35
CURVE 3	
298.15	4.577 x 10^{-2}
300	4.584*
350	4.691
400	4.811
450	4.939
500	5.067
550	5.209
600	5.358
650	5.514
700	5.677
750	5.854
800	6.032
850	6.216
900	6.415
950	6.621
1000	6.827
1050	7.047*
1071.5	7.146
1071.5	6.521
1100	6.521*
1150	6.521
1200	6.521*
1205.15	6.521
1208.15	7.288
1250	7.288
1300	7.288
1350	7.288*
1373.15	7.288
CURVE 4	
2.5	1.77 x 10^{-4}
3	2.84
4	5.68
5	1.03 x 10^{-3}
6	1.92
7	3.05
8	4.33
9	5.68
10	6.99
15	1.43 x 10^{-2}
20	2.26
25	3.09
30	3.50
40	4.16
50	4.39
55	4.44
60	4.49
70	4.47
80	4.45
90	4.43
95	4.40
100	4.41
105	4.45
110	4.48
120	4.50
140	4.54
160	4.57
170	4.61

*Not shown on plot

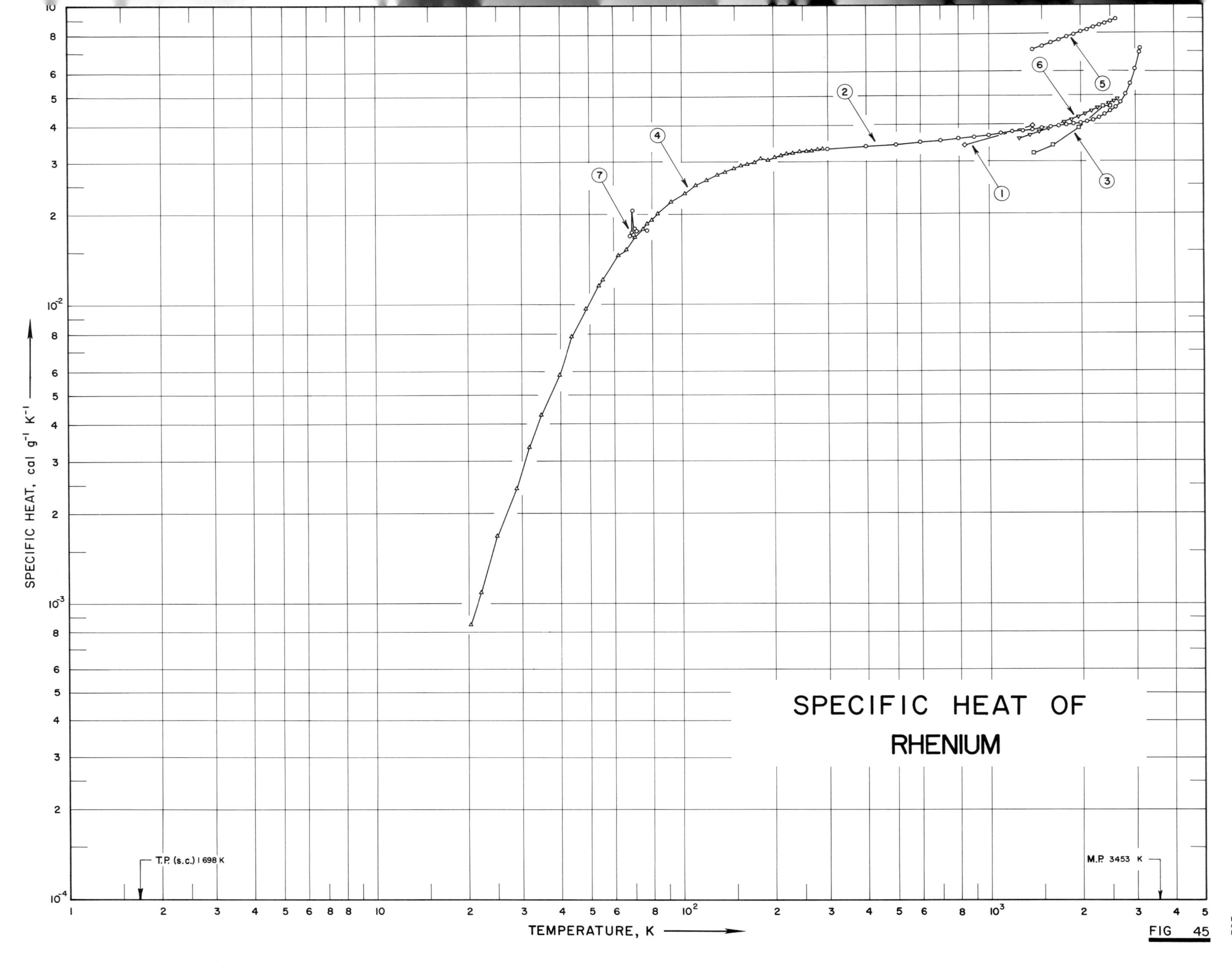

SPECIFIC HEAT OF
RHENIUM
SPECIFIC HEAT, cal g⁻¹ K⁻¹
TEMPERATURE, K
T.P. (s.c.) 1.698 K
M.P. 3453 K
FIG 45

SPECIFICATION TABLE NO. 45 SPECIFIC HEAT OF RHENIUM

(Impurity < 0.20% each; total impurities < 0.50%)

[For Data Reported in Figure and Table No. 45]

Curve No.	Ref. No.	Year	Temp. Range, K	Reported Error, %	Name and Specimen Designation	Composition (weight percent), Specifications and Remarks
1	17	1960	860, 1400	≤2.4		
2	65	1961	300-3120	4.0		
3	67	1962	1410-2720	± 10.0		
4	98	1953	20-300			99.9 Re; prepared by reducing ammonium perrhenate at 500-600 C in H_2; sintered at 1000 C; cooled in H_2
5	106	1956	1400-2600			99.942 Re, 0.015 Al, 0.014 Sn, 0.01 Ca, Si, 0.005 Mg, 0.0004 Mo, 0.0005 Cu, and 0.0 X Au; swaged; drawn; annealed for 2 hrs. at 1750 C.
6	125	1965	1273-2643	± 0.35		99.98 Re; powder metallurgy product of 20-mil sheet.
7	175	1953	68-77			99.8 Re; powder form; under helium atmosphere.

DATA TABLE NO. 45 SPECIFIC HEAT OF RHENIUM

[Temperature, T, K; Specific Heat, C_p, Cal g^{-1} K^{-1}]

T	C_p
CURVE 1	
860	3.40×10^{-2}
1400	3.96
CURVE 2	
300	3.32×10^{-2}
400	3.38
500	3.43
600	3.49
700	3.53
800	3.58
900	3.63
1000	3.68
1100	3.72
1200	3.77
1300	3.81
1400	3.84
1500	3.88
1600	3.91
1700	3.95
1800	3.98
1900	4.00
2000	4.03
2100	4.09
2200	4.12
2300	4.21
2400	4.31
2500	4.42
2600	4.54
2700	4.75
2800	5.02
2900	5.48
3000	6.10
3100	6.96
3120	7.19
CURVE 3	
1410	3.20×10^{-2}
1630	3.40
1980	3.9
2010	4.0*
2250	4.2*
2370	4.6
CURVE 3 (cont.)	
2500	4.6×10^{-2}
2610	4.6*
2720	4.8*
CURVE 4	
20.39	8.480×10^{-4}
22.00	1.090×10^{-3}
24.89	1.691
28.80	2.437
31.74	3.349
34.66	4.294
39.98	5.824
43.96	7.820
48.89	9.677
53.82	1.155×10^{-2}
55.42	1.207
62.26	1.465
66.27	1.527
70.82	1.683
75.39	1.788
77.39	1.860
80.05	1.912
84.00	2.005
92.88	2.210
103.04	2.360
112.19	2.502
121.82	2.621
131.01	2.714
139.85	2.792
148.41	2.857
156.76	2.919
164.92	2.958
173.88	2.997
183.64	3.096
193.20	3.064
202.65	3.114
211.97	3.161
222.30	3.200
233.62	3.213
244.81	3.265
255.86	3.267
266.79	3.286
277.59	3.319
CURVE 4 (cont.)	
288.30	3.303×10^{-2}
300.01	3.301*
CURVE 5	
1400	7.10×10^{-2}
1500	7.29
1600	7.48
1700	7.64
1800	7.82
1900	7.98
2000	8.13
2100	8.28
2200	8.43
2300	8.58
2400	8.70
2500	8.83
2600	8.98
CURVE 6	
1273	3.582×10^{-2}
1373	3.674
1473	3.767
1573	3.859
1673	3.952*
1773	4.044
1873	4.137
1973	4.229
2073	4.322
2173	4.414
2273	4.507
2373	4.599*
2473	4.692
2573	4.784
2643	4.877
CURVE 7	
68.00	1.692×10^{-2}
68.60	1.751*
69.00	1.735
69.40	2.057
69.70	1.713
CURVE 7 (cont.)	
70.00	1.665×10^{-2}*
70.20	1.810*
70.90	1.799
71.90	1.756
76.90	1.874*
77.10	1.767

*Not shown on plot

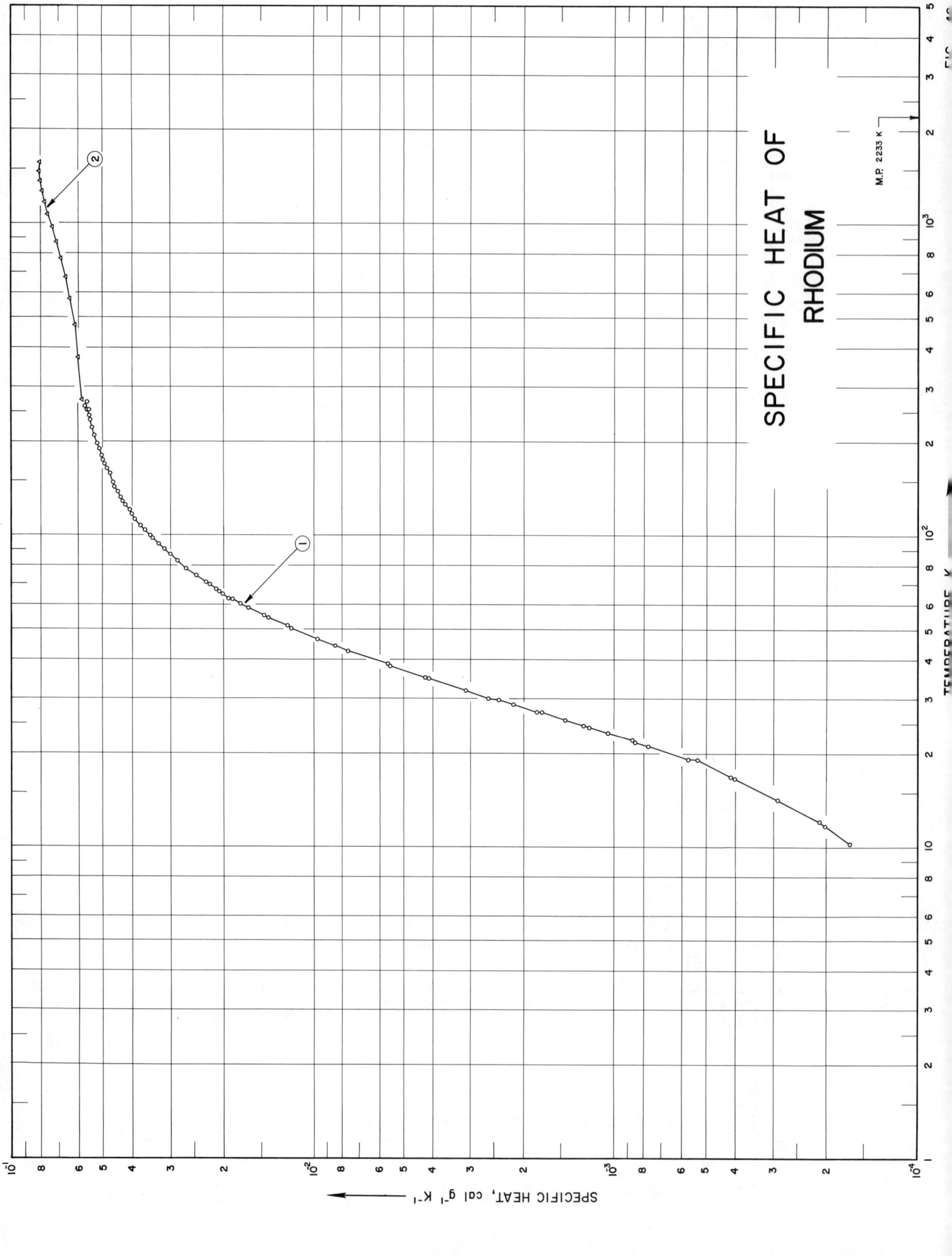

SPECIFIC HEAT OF RHODIUM

M.P. 2233 K

TEMPERATURE, K

SPECIFIC HEAT, cal g^{-1} K^{-1}

SPECIFICATION TABLE NO. 46 SPECIFIC HEAT OF RHODIUM

(Impurity $<0.20\%$ each; total impurities $<0.50\%$)

[For Data Reported in Figure and Table No. 46]

Curve No.	Ref. No.	Year	Temp. Range, K	Reported Error, %	Name and Specimen Designation	Composition (weight percent), Specifications and Remarks
1	107	1955	10-269			99.9 Rh, 0.0 X Pt type metals; traces of Ag, Cu, and Fe; cast.
2	164	1932	273-1573			

DATA TABLE NO. 46 SPECIFIC HEAT OF RHODIUM

[Temperature, T, K; Specific Heat, C_p, Cal g^{-1} K^{-1}]

T	C_p
CURVE 1	
10.28	1.681×10^{-4}
11.78	2.031
12.08	2.118
14.26	2.935
14.28	2.964*
16.66	4.033
16.82	4.179
19.09	5.384
19.14	5.743
19.19	5.763*
21.22	7.706
21.75	8.590
22.05	8.756
23.38	1.060×10^{-3}
24.29	1.219
24.53	1.276
25.64	1.460
27.12	1.748
27.16	1.811
28.75	2.177
29.62	2.437
30.12	2.619
31.94	3.130
32.05	3.176*
34.87	4.137
35.22	4.236
38.25	5.586
38.52	5.679
42.64	7.686
44.18	8.493
46.44	9.766
50.43	1.199×10^{-2}
51.34	1.235
54.60	1.427
55.60	1.472
58.58	1.658
60.52	1.757
62.47	1.873
62.92	1.942
65.12	2.016
65.29	2.015*
66.21	2.083
67.37	2.167
69.89	2.244
71.14	2.298
CURVE 1 (cont.)	
74.26	2.476×10^{-2}
78.66	2.683
82.79	2.864
87.10	3.022
90.49	3.163
93.62	3.294
95.53	3.353*
97.97	3.450
100.15	3.505
101.24	3.536*
104.67	3.666
105.01	3.690*
107.67	3.779
109.95	3.797
110.84	3.834*
113.81	3.953
115.33	3.974*
117.53	4.024
119.48	4.064*
121.57	4.100
124.35	4.183*
125.69	4.247
128.92	4.332
131.55	4.356*
133.40	4.393
136.29	4.464*
138.57	4.492
141.64	4.546
143.25	4.606*
146.50	4.634*
148.68	4.664
153.93	4.767*
158.71	4.774
159.57	4.827*
164.75	4.886
170.12	4.962
170.59	4.967
175.28	5.012
176.99	5.035*
181.40	5.076
182.92	5.090*
187.16	5.131*
189.95	5.179
192.22	5.178*
196.07	5.202*
CURVE 1 (cont.)	
197.39	5.235×10^{-2}
198.59	5.258*
199.32	5.221*
202.90	5.267*
204.61	5.244*
204.97	5.278*
208.09	5.304*
210.29	5.380
211.25	5.322*
213.62	5.329*
216.47	5.363*
216.75	5.390*
219.29	5.437*
222.29	5.435
222.48	5.462*
225.10	5.463*
228.22	5.498*
228.99	5.465*
230.97	5.559*
234.88	5.495*
234.98	5.511
241.18	5.550*
241.67	5.543
247.65	5.575*
247.75	5.573*
254.59	5.652*
254.74	5.585
260.87	5.713
261.76	5.656*
267.28	5.648*
268.66	5.661
CURVE 2	
273.15	5.893×10^{-2}
373.15	6.026
473.15	6.203
573.15	6.415
673.15	6.650
773.15	6.899
873.15	7.150
973.15	7.393
1073.15	7.618
1173.15	7.814
1273.15	7.969
CURVE 2 (cont.)	
1373.15	8.074×10^{-2}
1473.15	8.119
1573.15	8.092

*Not shown on plot

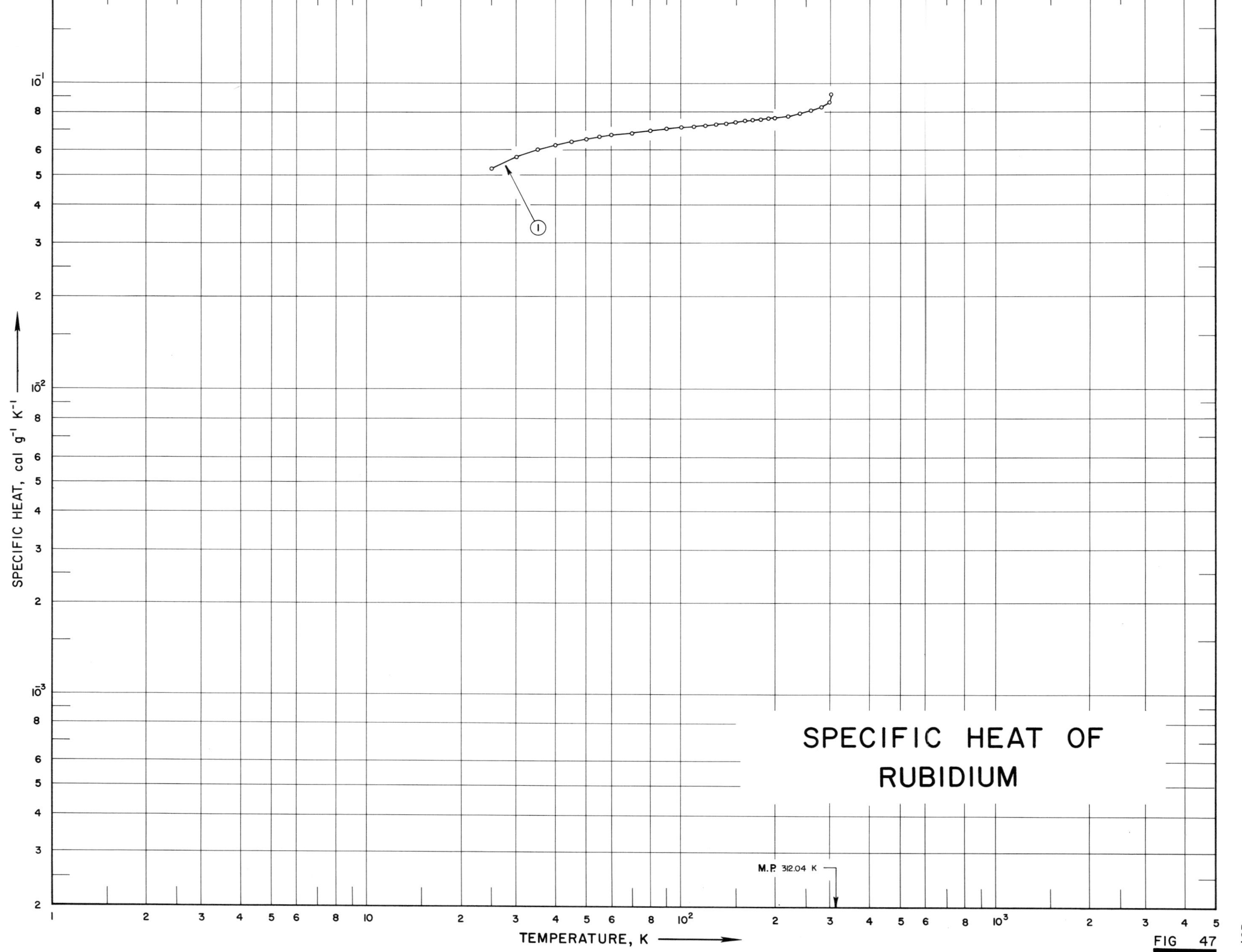
SPECIFIC HEAT OF
RUBIDIUM
FIG 47
M.P. 312.04 K
TEMPERATURE, K
SPECIFIC HEAT, cal g^{-1} K^{-1}
1

SPECIFICATION TABLE NO. 47 SPECIFIC HEAT OF RUBIDIUM

(Impurity $<0.20\%$ each; total impurities $<0.50\%$)

[For Data Reported in Figure and Table No. 47]

Curve No.	Ref. No.	Year	Temp. Range, K	Reported Error, %	Name and Specimen Designation	Composition (weight percent), Specifications and Remarks
1	176	1955	25-300			Order of magnitude of impurities 0.4.

DATA TABLE NO. 47 SPECIFIC HEAT OF RUBIDIUM

[Temperature, T, K; Specific Heat, C_p, Cal g^{-1} K^{-1}]

T	C_p
CURVE 1	
25	5.25 x 10^{-2}
30	5.73
35	6.03
40	6.26
45	6.40
50	6.53
55	6.65
60	6.74
70	6 87
80	6.98
90	7.09
100	7.14
110	7.18
120	7.24
130	7.30
140	7.37
150	7.43
160	7.50
170	7.53
180	7.58
190	7.64
200	7.69
210	7.73*
220	7.78
230	7.86*
240	7.93
250	8.03*
260	8.12
270	8.23*
273.15	8.26*
280	8.35
290	8.49*
298.15	8.63
300	9.15

*Not shown on plot

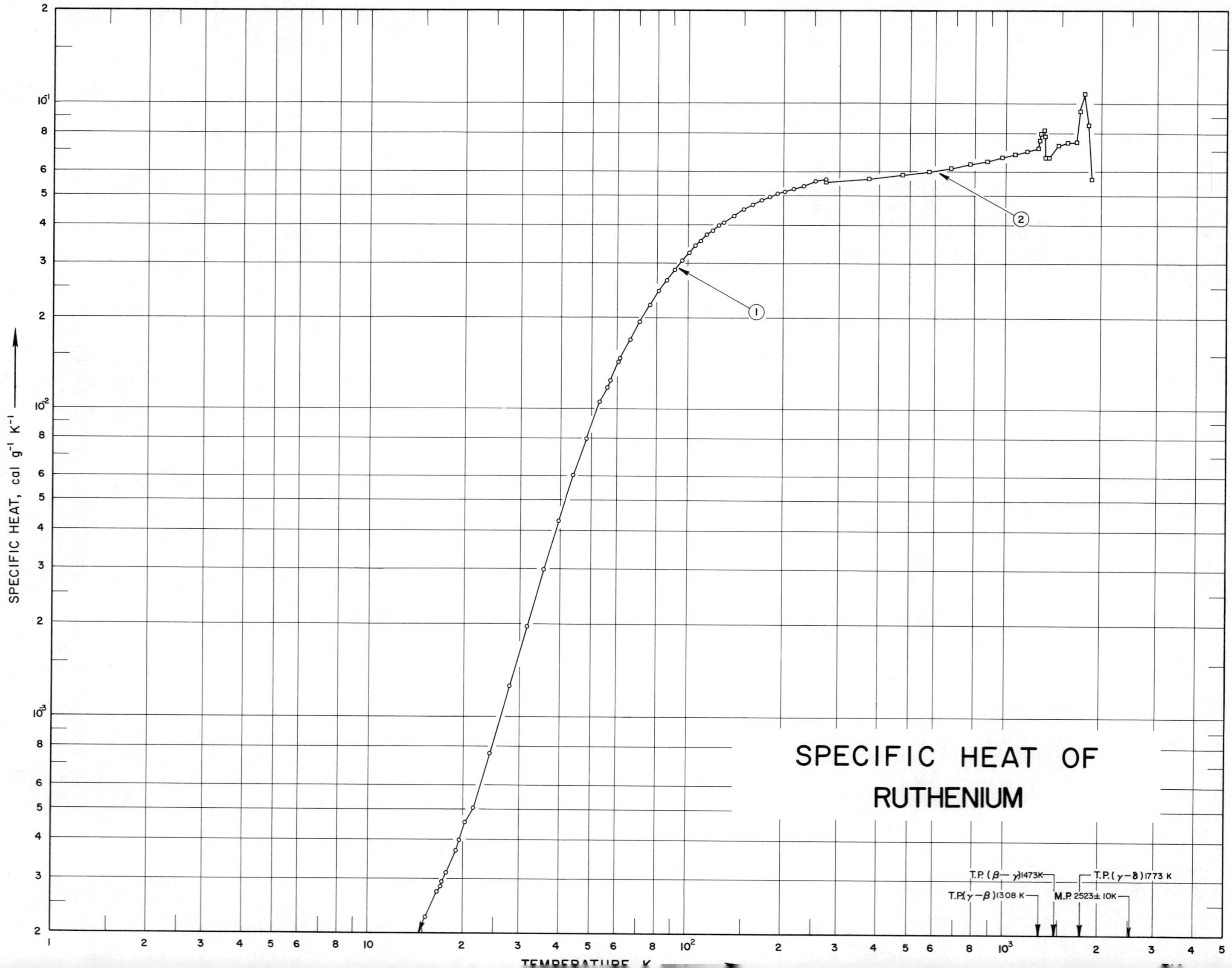

SPECIFIC HEAT OF RUTHENIUM

SPECIFIC HEAT, cal g⁻¹ K⁻¹

TEMPERATURE, K

T.P. (β — γ) 1473 K
T.P. (γ — δ) 1773 K
T.P. (γ — β) 1308 K
M.P. 2523 ± 10 K

1
2

SPECIFICATION TABLE NO. 48 SPECIFIC HEAT OF RUTHENIUM

(Impurity $<0.20\%$ each; total impurities $<0.50\%$)

[For Data Reported in Figure and Table No. 48]

Curve No.	Ref. No.	Year	Temp. Range, K	Reported Error, %	Name and Specimen Designation	Composition (weight percent), Specifications and Remarks
1	108	1959	11-272			Treated with H NO_3; washed with water and acetone; dried at 1400 C for 2 days.
2	215	1931	273-1873			Perfectly pure state.

DATA TABLE NO. 48 SPECIFIC HEAT OF RUTHENIUM

[Temperature, T, K; Specific Heat, C_p, Cal $g^{-1}K^{-1}$]

CURVE 1

T	C_p
11.39	1.217 x 10^{-4}*
11.46	1.276*
11.61	1.276*
11.90	1.346*
12.68	1.514*
12.81	1.494*
13.03	1.613*
13.05	1.583*
14.54	1.959*
14.84	2.117*
14.85	2.127*
15.39	2.246
16.67	2.701
17.06	2.820
17.30	2.919
17.80	3.127
19.03	3.681
19.09	3.661*
19.51	3.958
20.42	4.541
21.59	5.026
24.30	7.579
27.89	1.256 x 10^{-3}
31.65	1.951
35.68	2.989
39.62	4.296
44.02	6.006
48.41	7.965
53.08	1.054 x 10^{-2}
56.15	1.177
57.31	1.241
60.80	1.421
61.55	1.456
66.11	1.694
66.23	1.695*
70.82	1.946
71.00	1.950*
76.07	2.195
81.16	2.442
86.08	2.635
91.13	2.851
96.01	3.061
101.23	3.232

CURVE 1 (cont.)

T	C_p
106.01	3.411 x 10^{-2}
110.71	3.536
115.60	3.703
120.35	3.821
125.39	3.967
130.05	4.070*
135.44	4.183*
140.38	4.290
145.60	4.401*
150.59	4.491*
155.81	4.556*
160.83	4.661
166.15	4.727*
171.20	4.809
176.20	4.865*
181.59	4.922*
186.67	4.998*
191.71	5.037
195.01	5.066*
195.37	3.080*
196.41	5.086*
197.02	5.077*
198.16	5.082*
199.16	5.099*
200.41	5.115*
202.08	5.115
202.94	5.141*
203.80	5.129*
205.45	5.154*
207.67	5.161*
207.94	5.167*
208.70	5.175*
210.56	5.217*
212.78	5.219*
212.80	5.233*
213.68	5.223*
215.84	5.242
217.74	5.269*
217.97	5.274*
218.69	5.280*
220.90	5.288*
222.98	5.290*
223.77	5.298*

CURVE 1 (cont.)

T	C_p
226.34	5.335 x 10^{-2}*
228.68	5.343*
228.87	5.362*
231.50	5.360*
233.96	5.388*
234.06	5.396*
237.17	5.414*
239 14	5.406*
239.55	5.414*
242.42	5.433*
244.60	5.449*
245.30	5.441*
248.15	5.471*
249.79	5.490*
251.24	5.488*
253.50	5.521
255.92	5.528*
257.25	5.531*
259.60	5.552*
261.54	5.566*
262.94	5.562*
265.12	5.582*
265.22	5.593*
268.90	5.594*
270.80	5.621*
272.48	5.620

CURVE 2

T	C_p
273	5.51 x 10^{-2}
373	5.67
473	5.83
573	5.99
673	6.15
773	6.31
873	6.48
973	6.64
1073	6.80
1173	6.96
1273	7.12
1283	7.57
1293	7.92
1303	8.12*

CURVE 2 (cont.)

T	C_p
1313	8.17 x 10^{-2}
1323	8.06*
1333	7.81
1343	6.64
1373	6.65
1473	7.30
1573	7.45
1673	7.45
1723	9.40
1773	1.075 x 10^{-1}
1823	8.50 x 10^{-2}
1873	5.66

*Not shown on plot

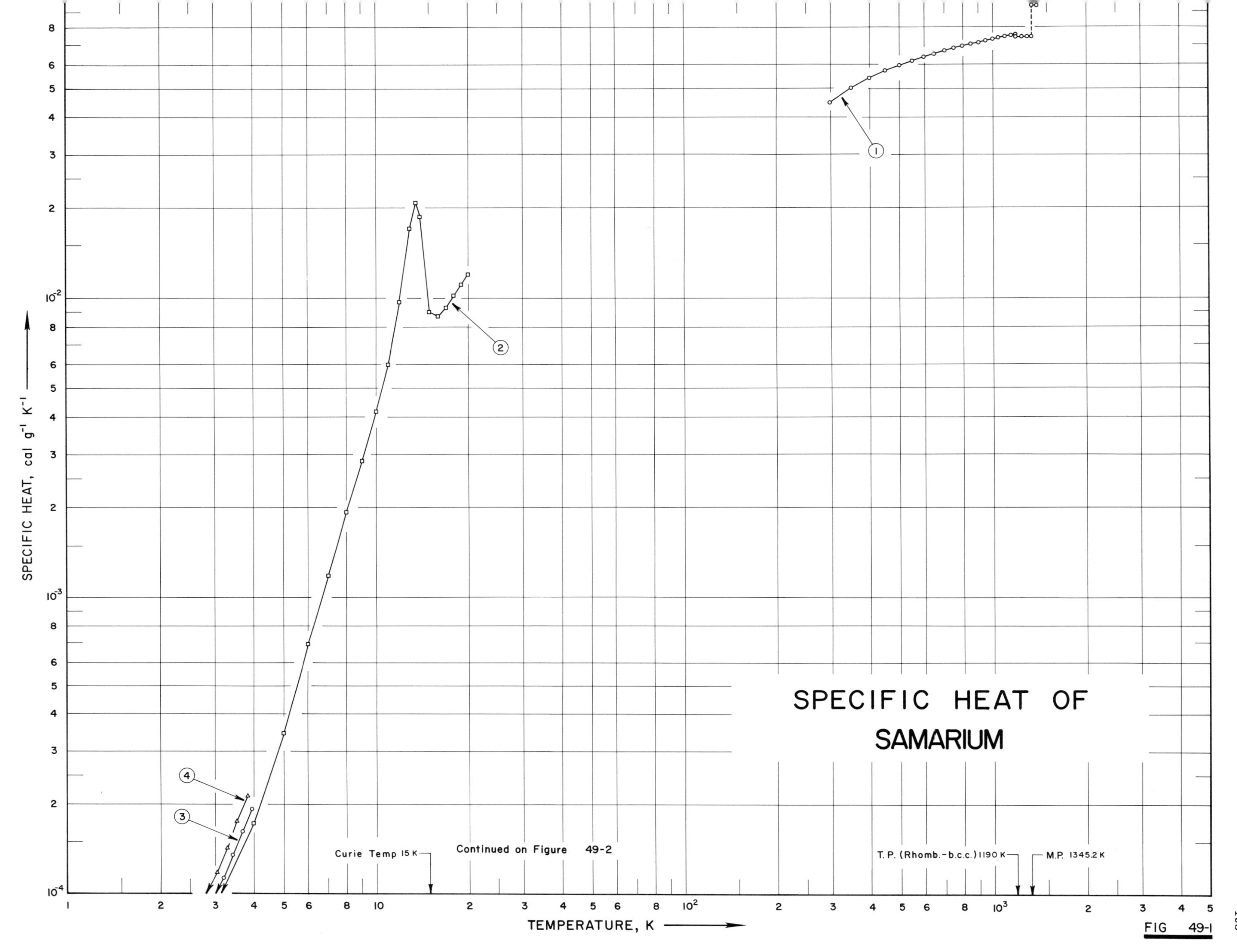
SPECIFIC HEAT OF
SAMARIUM
SPECIFIC HEAT, cal g⁻¹ K⁻¹
TEMPERATURE, K
Curie Temp 15 K
Continued on Figure 49-2
T. P. (Rhomb.-b.c.c.) 1190 K
M.P. 1345.2 K
1
2
3
4

FIG 49-1

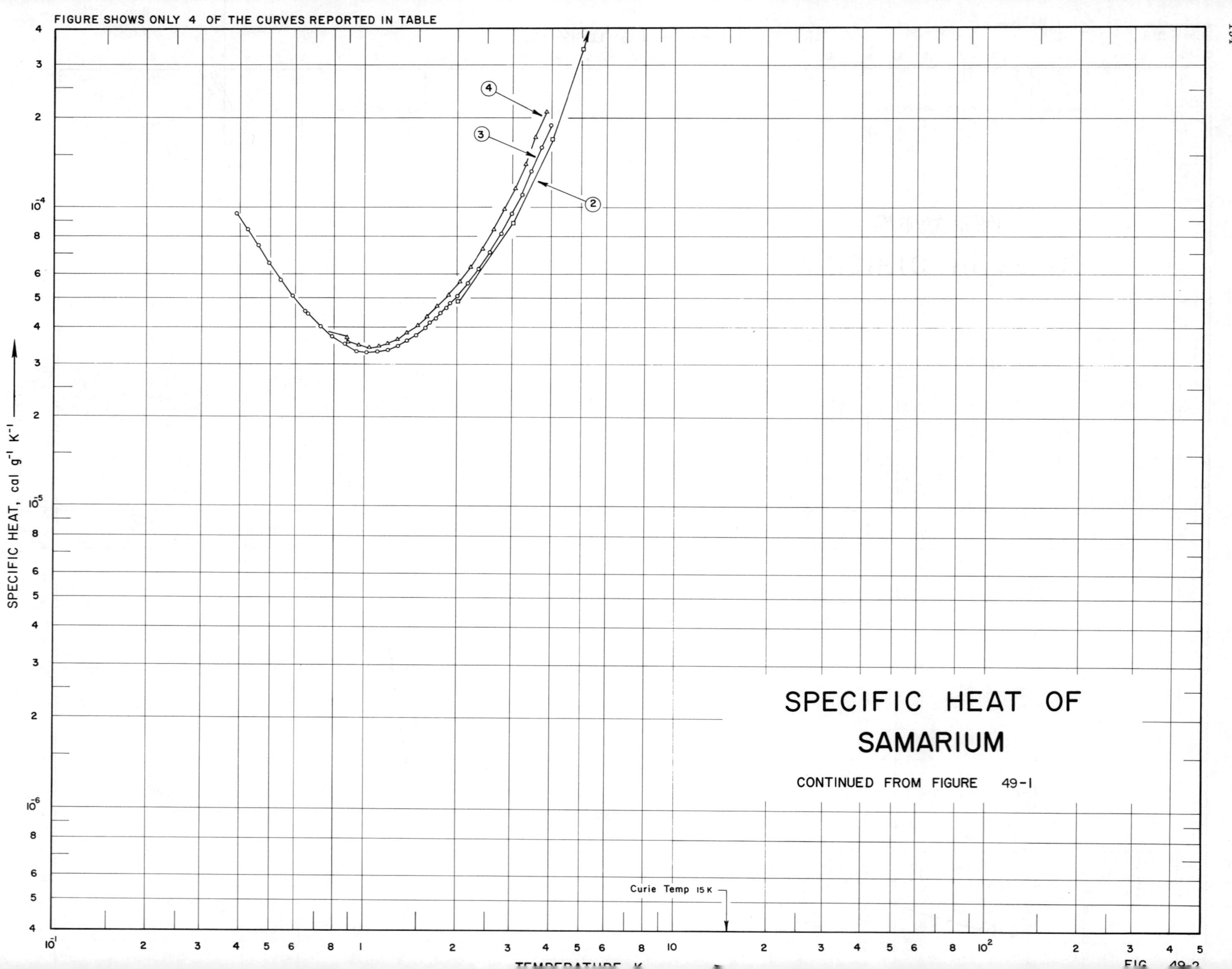
FIGURE SHOWS ONLY 4 OF THE CURVES REPORTED IN TABLE
SPECIFIC HEAT, cal g⁻¹ K⁻¹
4
3
2
SPECIFIC HEAT OF
SAMARIUM
CONTINUED FROM FIGURE 49-1
Curie Temp 15 K

SPECIFICATION TABLE NO. 49 SPECIFIC HEAT OF SAMARIUM

(Impurity $< 0.20\%$ each; total impurities $< 0.50\%$)

[For Data Reported in Figure and Table No. 49]

Curve No.	Ref. No.	Year	Temp. Range, K	Reported Error, %	Name and Specimen Designation	Composition (weight percent), Specifications and Remarks
1	73	1960	298-1398	0.14		≥ 99.71 Sm, ≤ 0.20 Eu, ≤ 0.05 Ca, 0.01 Fe, ≤ 0.01 La, ≤ 0.01 Mg, and ≤ 0.01 Si.
2	109	1957	2-20	± 2.0		Sample supplied by the Johnson, Matthey and Co.; evaporated on to thin tantalum strip.
3	151	1962	0.4-4	< 2.0		0.04 H_2, 0.02 C, and 0.008 O_2; vacuum distilled.
4	151	1962	0.4-4	< 2.0		Same as above.
5	285	1958	273-1398			0.2 Eu, 0.05 Ca, 0.01 Fe, 0.01 La, 0.01 Mg, and 0.01 Si.
6	356	1959	20-360			

DATA TABLE NO. 49 SPECIFIC HEAT OF SAMARIUM

[Temperature, T, K; Specific Heat, C_p, Cal g^{-1} K^{-1}]

	T	C_p
	CURVE 1	
	298.15	4.496 x 10^{-2}
	300	4.523*
	350	5.028
	400	5.421
	450	5.733
	500	5.986
	550	6.199
	600	6.385
	650	6.545
	700	6.684
	750	6.817
	800	6.930
	850	7.037
	900	7.137
	950	7.223
	1000	7.310
	1050	7.389
	1100	7.463
	1150	7.537
	1190.15	7.589
	1190.15	7.463
	1200	7.463*
	1250	7.463
	1300	7.463
(s)	1345.15	7.463
(l)	1345.15	9.471
	1350	9.471*
	1398.15	9.471
	CURVE 2	
	2	4.99 x 10^{-5}
	3	9.05
	4	1.73 x 10^{-4}
	5	3.46
	6	6.92
	7	1.19 x 10^{-3}
	8	1.93
	9	2.86
	10	4.19
	11	5.99
	12	9.71
	13	1.70 x 10^{-2}
	13.6	2.08

T	C_p
CURVE 2 (cont.)	
14	1.86 x 10^{-2}
15	8.98 x 10^{-3}
16	8.71
17	9.31
18	1.02 x 10^{-2}
19	1.11
20	1.20
CURVE 3	
Series 1A	
0.3890	9.597 x 10^{-5}
0.4208	8.454
0.4550	7.462
0.4938	6.561
0.5380	5.764
0.5879	5.101
0.6443	4.534
0.7071	4.098*
0.6577	4.437
0.7217	4.030
0.7913	3.736
0.8657	3.532
0.9437	3.396
1.0234	3.373
1.1033	3.383
1.1899	3.445
1.2824	3.550
1.3717	3.691
1.4669	3.863
1.5743	4.089
1.7009	4.391
1.8416	4.777
1.9892	5.208
2.1518	5.742
2.3316	6.413
2.5314	7.295
2.7484	8.389
2.9741	9.765
3.2004	1.141 x 10^{-4}
3.4400	1.351
3.6982	1.625

T	C_p
CURVE 3 (cont.)	
Series 1B*	
0.3951	9.369 x 10^{-5}
0.4259	8.269
0.4608	7.297
0.5004	6.413
0.5454	5.650
0.5964	5.004
0.6536	4.500
0.7171	4.057
0.7864	3.753
0.8604	3.543
0.9377	3.424
1.0169	3.375
1.0966	3.380
1.1755	3.432
1.2530	3.508
1.3283	3.618
1.4074	3.750
1.4975	3.929
1.5998	4.151
1.7224	4.457
1.8575	4.825
1.9977	5.244
2.1583	5.790
2.3357	6.468
2.5321	7.374
2.7456	8.460
2.9693	9.815
3.1947	1.147 x 10^{-4}
3.4318	1.359
3.6877	1.645
Series 2	
1.6332	4.271 x 10^{-5}
1.7542	4.567
1.8852	4.917
2.0240	5.314*
2.1834	5.832*
2.3003	6.502*
2.5580	7.363*
2.7717	8.506*

T	C_p
CURVE 3 (cont.)	
Series 2 (cont.)	
2.9935	9.889 x 10^{-5}*
3.2166	1.155 x 10^{-4}*
3.4519	1.367*
3.7081	1.630*
3.9703	1.939
CURVE 4	
0.4460	7.538 x 10^{-5}*
0.4769	6.796*
0.5151	6.063*
0.5626	5.368*
0.6161	4.782*
0.6756	4.319*
0.7408	3.974*
0.8810	3.729
0.8848	3.580
0.9608	3.512
1.0376	3.504
1.1140	3.547
1.1891	3.624
1.2764	3.745
1.3743	3.931
1.4776	4.162
1.5934	4.454
1.7172	4.813
1.8558	5.243
2.0142	5.805
2.1911	6.506
2.3872	7.435
2.5992	8.610
2.8188	1.005 x 10^{-4}
3.0486	1.188
3.2988	1.431
3.5642	1.756
3.8294	2.144

T	C_p
CURVE 5*	
273.2	4.17 x 10^{-2}
323.2	4.78
373.2	5.21
423.2	5.57
473.2	5.85
523.2	6.09
573.2	6.29
623.2	6.46
673.2	6.61
723.2	6.74
773.2	6.87
823.2	6.98
873.2	7.08
923.2	7.17
973.2	7.26
1023.2	7.34
1073.2	7.42
1123.2	7.50
1173.2	7.56
1190.2	7.59
1190.2	7.46
1223.2	7.46
1273.2	7.46
1323.2	7.46
1345.2	7.46
1345.2	9.34
1373.2	9.34
1398.0	9.34
CURVE 6*	
20.0	1.149 x 10^{-2}
25.0	1.598
30.0	2.067
35.0	2.507
40.0	2.906
45.0	3.267
50.0	3.588
55.0	3.887
60.0	4.165
65.0	4.418
70.0	4.661
75.0	4.888

T	C_p
CURVE 6 (cont.)*	
80.0	5.120 x 10^{-2}
85.0	5.354
90.0	5.578
95.0	5.790
100.0	6.038
102.0	6.177
104.0	6.411
104.2	6.449
104.4	6.489
104.6	6.543
104.8	6.626
105.0	6.683
105.2	6.704
105.4	6.635
105.6	6.355
105.8	5.953
106.0	5.427
106.5	4.691
107.0	4.519
107.5	4.446
108.0	4.397
110.0	4.311
115.0	4.230
120.0	4.192
130.0	4.170
140.0	4.168
150.0	4.179
160.0	4.200
170.0	4.224
180.0	4.246
190.0	4.270
200.0	4.303
210.0	4.335
220.0	4.368
230.0	4.405
240.0	4.440
250.0	4.486
260.0	4.529
270.0	4.562
273.15	4.575
280.0	4.607
290.0	4.655
298.15	4.694

* Not shown on plot

DATA TABLE NO. 49 (continued)

T	C_p
CURVE 6 (cont.)*	
300.0	4.702×10^{-2}
310.0	4.750
320.0	4.796
330.0	4.841
340.0	4.885
350.0	4.928
360.0	4.971

*Not shown on plot

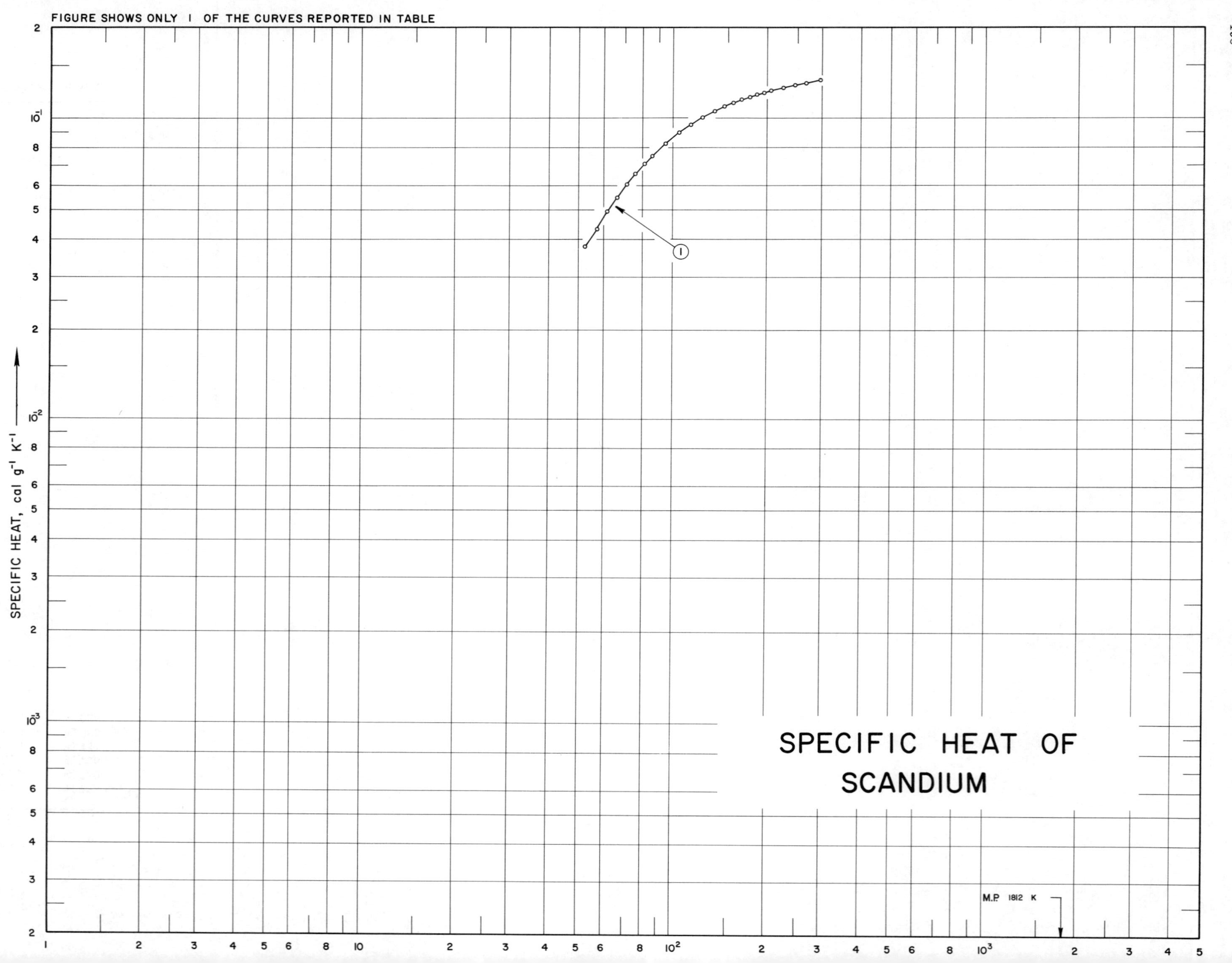
FIGURE SHOWS ONLY 1 OF THE CURVES REPORTED IN TABLE
SPECIFIC HEAT, cal g⁻¹ K⁻¹
SPECIFIC HEAT OF SCANDIUM
M.P. 1812 K
1

SPECIFICATION TABLE NO. 50 SPECIFIC HEAT OF SCANDIUM

(Impurity < 0.20% each; total impurities < 0.50%)

[For Data Reported in Figure and Table No. 50]

Curve No.	Ref. No.	Year	Temp. Range, K	Reported Error, %	Name and Specimen Designation	Composition (weight percent), Specifications and Remarks
1	157	1962	53-296	0.3		0.06 Cu, 0.02 Pb, 0.01 Al, 0.01 Fe, 0.01 Ti, and 0.01 Y; crystalline.
2	301	1966	298-1812			0.092 Cu, 0.06 Fe, 0.043 Ca, 0.026 Si, 0.024 N_2, 0.019 Cu, 0.015 Ni, 0.014 Al and 0.009 Mg; prepared by metallothermic reduction of the fluoride, with calcium and purified by distillation.

DATA TABLE NO. 50 SPECIFIC HEAT OF SCANDIUM

[Temperature, T, K; Specific Heat, C_p, Cal g^{-1} K^{-1}]

T	C_p	T	C_p
CURVE 1		CURVE 2 (cont.)*	
52.54	3.819 x 10^{-2}	1000	1.66 x 10^{-1}
57.08	4.377	1100	1.72
61.75	4.989	1200	1.79
66.46	5.567	1300	1.87
71.12	6.114	1400	1.95
75.91	6.628	1500	2.03
81.17	7.162	1600	2.12
86.01	7.607	1608	2.13
94.92	8.354	1608	2.351
105.17	9.093	1700	2.351
114.51	9.649	1800	2.351
124.56	1.018 x 10^{-1}	1812	2.351
136.38	1.072		
145.83	1.109		
155.97	1.143		
166.05	1.171		
176.19	1.195		
186.20	1.218		
196.01	1.235		
206.06	1.254		
216.20	1.269*		
225.84	1.283		
235.94	1.294*		
245.57	1.305		
256.30	1.319*		
266.25	1.327		
276.46	1.338*		
286.57	1.347*		
296.36	1.358		
CURVE 2*			
298.15	1.37 x 10^{-1}		
300	1.37		
400	1.39		
500	1.43		
600	1.46		
700	1.50		
800	1.55		
900	1.60		

*Not shown on plot

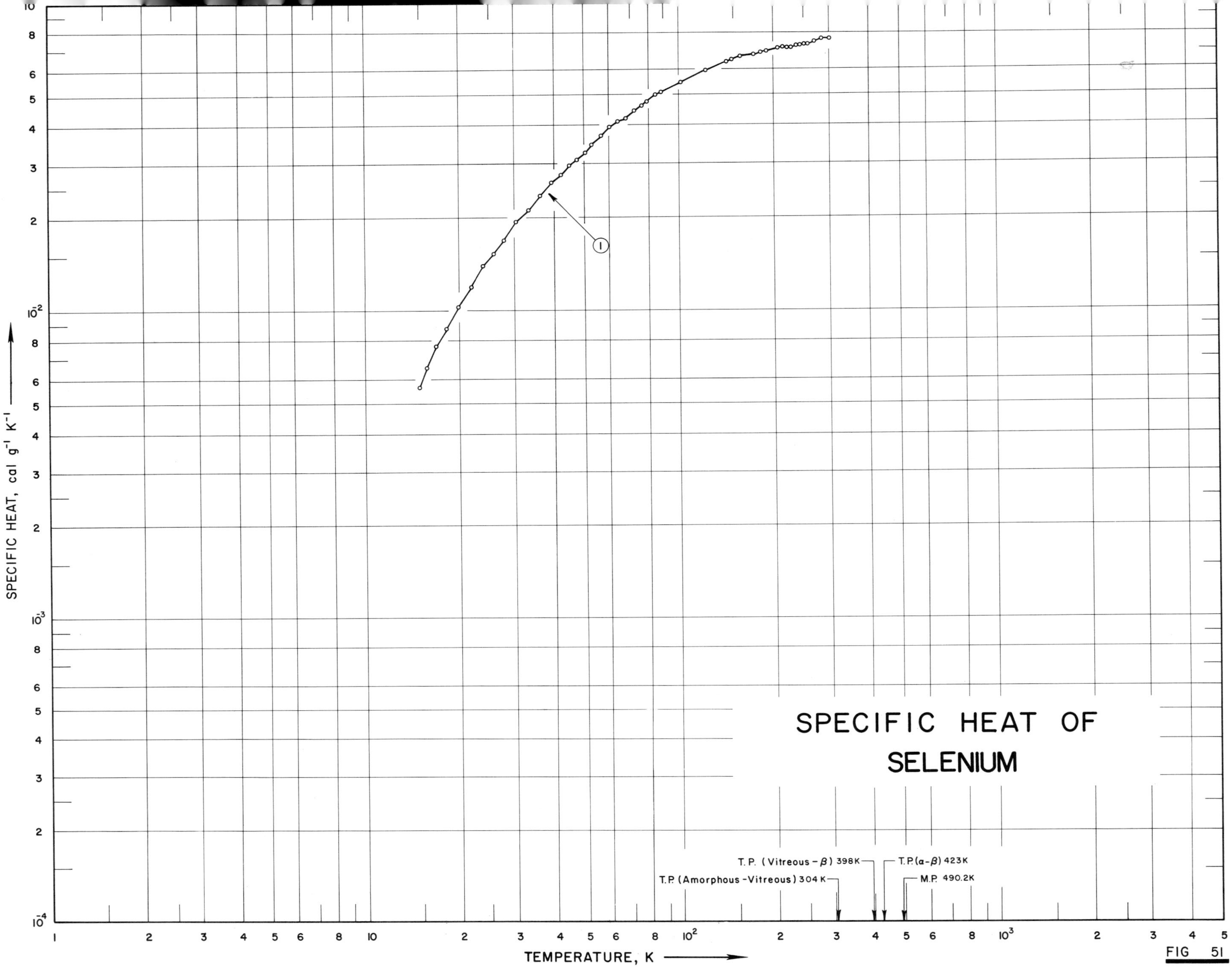

SPECIFIC HEAT OF SELENIUM
SPECIFIC HEAT, cal g^{-1} K^{-1}
TEMPERATURE, K
T.P. (Amorphous-Vitreous) 304 K
T.P. (Vitreous-β) 398K
T.P.(α-β) 423K
M.P. 490.2K
1

FIG 51

SPECIFICATION TABLE NO. 51 SPECIFIC HEAT OF SELENIUM

(Impurity < 0.20% each; total impurities < 0.50%)

[For Data Reported in Figure and Table No. 51]

Curve No.	Ref. No.	Year	Temp. Range, K	Reported Error, %	Name and Specimen Designation	Composition (weight percent), Specifications and Remarks
1	110	1953	15-300	<2.0		99.999 Se, 0.00009 Fe, 0.00004 Cu, 0.00001 Pb, and 0.00001 Te; kept at 130 C under vacuum for one week.
2	357	1932	98-278			Purified from Mallinckrodt grade Se.
3	358	1937	50-299			0.2 Te; glass.
4	358	1937	54-297			0.2 Te; crystals.

DATA TABLE NO. 51 SPECIFIC HEAT OF SELENIUM

[Temperature, T, K; Specific Heat, C_p, Cal $g^{-1} K^{-1}$]

CURVE 1

T	C_p
14.97	5.686 x 10^{-3}
15.73	6.586
16.87	7.687
18.17	8.726
19.87	1.030 x 10^{-2}
21.76	1.197
23.82	1.399
25.72	1.537
27.81	1.697
30.38	1.952
33.21	2.137
36.06	2.370
39.38	2.609
42.11	2.786
44.86	2.985
47.62	3.116
50.21	3.278
52.82	3.474
56.74	3.734
60.63	3.973
64.34	4.140
67.99	4.236
72.30	4.481
76.00	4.664
79.46	4.801
83.63	5.041
87.97	5.129
100.64	5.521
121.34	6.047
141.53	6.431
147.21	6.513
156.19	6.716
172.77	6.797
180.50	6.931
189.05	6.985
205.28	7.120
213.10	7.187
220.36	7.164
227.73	7.144
235.10	7.257
242.04	7.285
249.10	7.324
256.09	7.321
268.19	7.480

CURVE 1 (cont.)

T	C_p
275.27	7.572 x 10^{-2}*
281.82	7.622*
288.25	7.638*
294.85	7.663
300.30	7.643

CURVE 2*

T	C_p
98.3	7.50 x 10^{-2}
100.8	7.84
103.6	7.90
112.5	7.73
141.1	8.36
141.5	8.26
153.8	8.18
207.9	8.51
276.9	8.64
277.1	8.98
278.3	9.00

CURVE 3*

T	C_p
49.9	3.693 x 10^{-2}
52.5	3.766
55.5	3.883
56.2	3.889
58.1	3.996
59.5	4.037
63.3	4.214
68.2	4.449
71.9	4.602
78.3	4.810
88.9	5.155
107.7	5.710
118.6	5.978
132.0	6.259
150.7	6.554
162.7	6.715
177.4	6.874
188.4	6.985
205.0	7.100
216.2	7.224
227.3	7.199
249.0	7.649
299.1	7.760

CURVE 4*

T	C_p
54.4	3.528 x 10^{-2}
56.1	3.627
58.7	3.744
60.1	3.822
63.4	3.844
64.6	4.067
68.3	4.341
72.9	4.744
77.6	4.778
77.7	4.777
82.8	4.965
92.4	5.255
99.5	5.499
110.0	5.783
119.3	5.983
133.9	6.299
147.0	6.543
155.4	6.631
174.5	6.958
191.5	7.044
202.7	7.134
223.2	7.221
240.9	7.315
261.2	7.409
272.7	7.423
296.5	7.538

*Not shown on plot

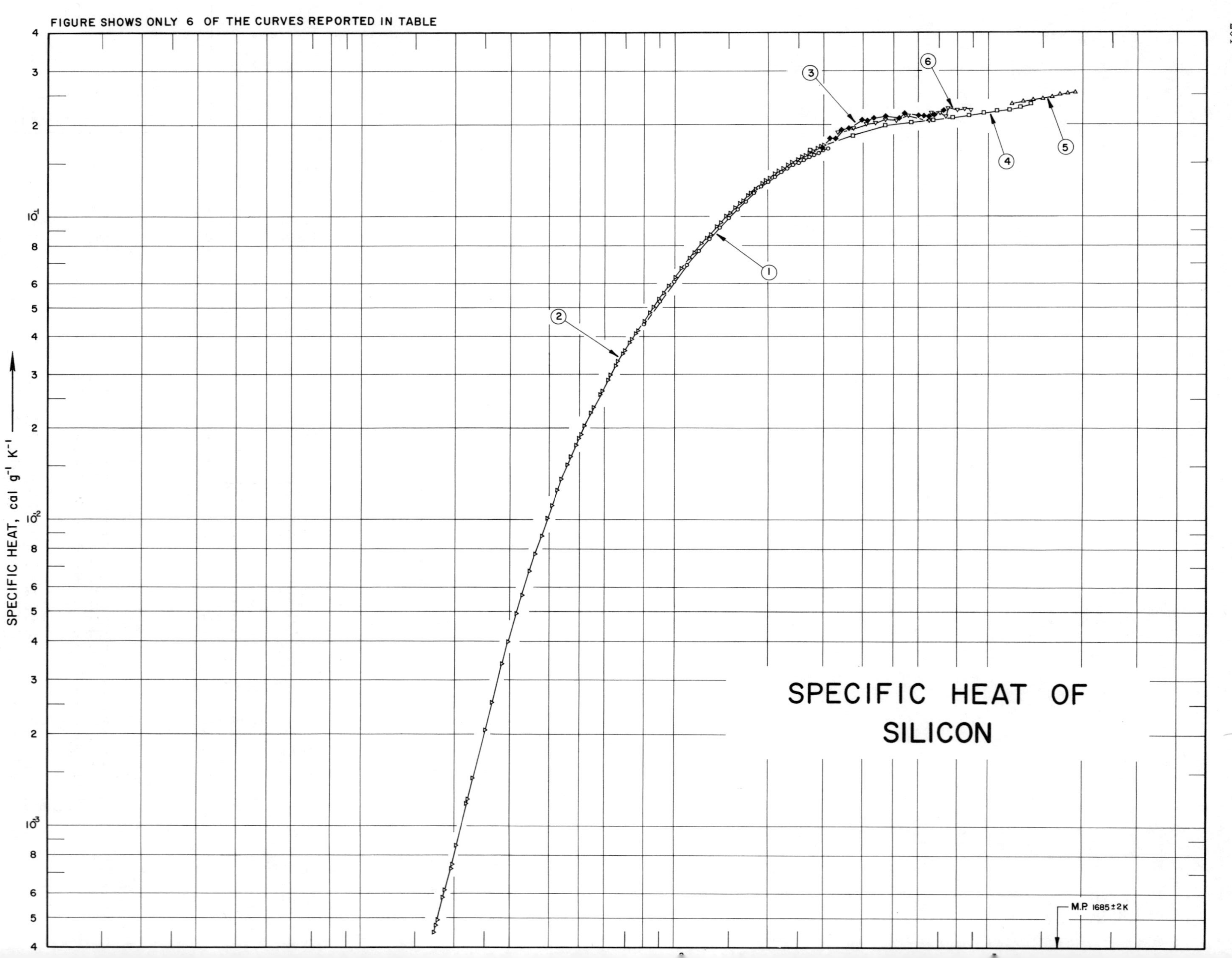

FIGURE SHOWS ONLY 6 OF THE CURVES REPORTED IN TABLE
SPECIFIC HEAT OF SILICON
SPECIFIC HEAT, cal g⁻¹ K⁻¹
M.P. 1685±2K
1
2
3
4
5
6
4
3
2
10⁻¹
8
6
5
4
3
2
10⁻²
8
6
5
4
3
2
10⁻³
8
6
5
4

SPECIFICATION TABLE NO. 52 SPECIFIC HEAT OF SILICON

(Impurity < 0.20% each; total impurities < 0.50%)

[For Data Reported in Figure and Table No. 52]

Curve No.	Ref. No.	Year	Temp. Range, K	Reported Error, %	Name and Specimen Designation	Composition (weight percent), Specifications and Remarks
1	41	1959	80-310	<7.0		Single crystals.
2	40	1959	8-300	<0.5		Hyper pure grade; single crystal slabs; sample supplied by du Pont; broken into 3 mm size; evacuated to a pressure of 10^{-6} mm Hg; sealed with a small amount of helium gas.
3	42	1964	295-723	<3.0	Si-4-690-2	Si, n-type; 0.003 ohm-cm resistivity.
4	111	1963	273-1373	0.5	4A	Si, p-type; single crystal; 1070 ohm-cm resistivity at 300 K; orientation (1, 1, 1).
5	112	1960	1200-1900			Highest purity.
6	42	1964	297-889	<3.0	Si-4-690-1	Si, n-type; 0.003 ohm-cm resistivity.
7	177	1930	61-296			99.7 Si; sample supplied by the Electro Metallurgical Co. of New York.
8	178	1952	1.7-100			Impurity concentration 1.5 x 10^{-3} B
9	359	1965	60-300			>99.999 Si.

DATA TABLE NO. 52 SPECIFIC HEAT OF SILICON

[Temperature, T, K; Specific Heat, C_p, Cal g^{-1} K^{-1}]

T	C_p
CURVE 1	
80	4.41 x 10^{-2}
90	5.27
100	6.12
110	6.94
120	7.73
130	8.47
140	9.22
150	9.93
160	1.06 x 10^{-1}
170	1.13
180	1.20
190	1.26
200	1.31
210	1.36
220	1.41
230	1.44
240	1.48
250	1.51
260	1.54
270	1.57
280	1.60
290	1.63
300	1.66
310	1.68
CURVE 2	
7.720	2.932 x 10^{-5}*
8.122	3.448*
8.465	3.880*
9.001	4.735*
9.471	5.486*
9.941	6.458*
10.455	7.593*
10.894	8.683*
11.501	1.028 x 10^{-4}*
11.908	1.166*
12.435	1.307*
12.592	1.398*
12.649	1.478*
12.927	1.487*
12.995	1.525*
13.032	1.590*
13.618	1.939*
CURVE 2 (contd)	
14.590	2.357 x 10^{-4}*
14.941	2.691*
15.302	2.860*
15.420	2.925*
15.527	2.979*
16.256	3.574*
16.505	3.738*
17.221	4.493
17.529	4.734
17.694	4.913
18.489	5.842
18.686	6.184
19.543	7.255
19.551	7.326*
19.606	7.455
20.253	8.601
21.967	1.192 x 10^{-3}
22.101	1.234
22.959	1.443
23.058	1.464*
25.17	2.069
26.47	2.545
28.48	3.395
29.80	3.991
31.62	4.938
32.90	5.675
34.67	6.775
36.15	7.743
37.75	8.864
39.43	1.011 x 10^{-2}
40.88	1.118
42.59	1.253
43.96	1.362
45.75	1.513
46.92	1.610
48.63	1.765
49.80	1.866
50.50	1.915
51.96	2.044
54.38	2.248
55.43	2.341
58.09	2.588
59.01	2.666
61.66	2.896
CURVE 2 (contd)	
62.71	2.989 x 10^{-2}
65.10	3.204
66.27	3.314
68.67	3.507
69.74	3.588
72.16	3.806
73.32	3.920
75.57	4.101
76.82	4.201
80.40	4.518
83.95	4.838
86.29	5.030
89.73	5.347
93.20	5.621
96.71	5.917
100.07	6.194*
101.46	6.308
103.46	6.476*
106.87	6.771
110.15	7.038*
113.53	7.312
117.02	7.608
120.39	7.882*
123.85	8.163
124.65	8.227*
128.05	8.501
131.42	8.768
134.75	9.032*
138.00	9.267
141.16	9.530
144.42	9.769*
147.78	1.002 x 10^{-1}
151.05	1.025
154.26	1.048*
157.51	1.072
159.72	1.087*
160.81	1.095*
162.97	1.109
166.35	1.131
169.87	1.155*
173.16	1.175
176.39	1.195
179.72	1.219*
183.00	1.235
184.61	1.243
CURVE 2 (contd)	
186.40	1.255 x 10^{-1}*
188.08	1.264*
189.94	1.275*
191.49	1.284
193.25	1.293*
194.86	1.301*
196.69	1.312
198.18	1.320*
200.08	1.331*
201.45	1.339*
203.42	1.348
204.95	1.354*
206.71	1.362*
208.67	1.375*
210.03	1.380
212.34	1.396*
213.38	1.397*
216.88	1.413
220.56	1.430*
224.19	1.446
227.78	1.462*
231.26	1.477
231.33	1.476*
234.77	1.490*
238.24	1.506
241.68	1.518*
245.09	1.534*
248.64	1.544
252.36	1.559*
256.03	1.573
259.68	1.588*
263.29	1.598
266.88	1.609*
270.43	1.623
273.77	1.631*
277.38	1.645
280.85	1.654*
283.36	1.656*
284.30	1.660*
287.71	1.672
287.68	1.677*
291.08	1.686*
292.90	1.685*
294.69	1.697*
295.52	1.698
CURVE 2 (contd)	
296.63	1.698 x 10^{-1}*
296.73	1.696*
298.41	1.703*
300.33	1.705*
300.45	1.707
CURVE 3	
295.2	1.68 x 10^{-1}*
315.7	1.81
328.8	1.80
343.1	1.92
364.8	1.94
397.5	2.07
416.9	2.05
434.3	2.10
471.7	2.12
521.4	2.10
543.3	2.18
545.0	2.17*
602.9	2.15
629.1	2.14
656.9	2.12
677.8	2.15
723.9	2.22
CURVE 4	
273	1.650 x 10^{-1}
373	1.840
473	1.970
573	2.025
673	2.065
773	2.105
873	2.145
973	2.180
1073	2.215
1173	2.250
1273	2.290
1373	2.345
CURVE 5	
1200	2.349 x 10^{-1}
1300	2.376
1400	2.402
1500	2.428
1600	2.454
1690	2.476*
1700	2.503
1800	2.525
1900	2.546
CURVE 6	
297.4	1.72 x 10^{-1}*
335.4	1.89
375.5	1.95
412.1	2.01
440.2	2.02
471.9	2.07
512.2	2.06
522.7	2.08*
559.4	2.14
592.0	2.12*
645.0	2.12*
652.6	2.07*
654.1	2.05
667.1	2.17
684.0	2.12*
709.6	2.18
739.0	2.12
746.9	2.26
800.0	2.23
849.0	2.25
888.6	2.23
CURVE 7*	
61.2	2.886 x 10^{-2}
65.1	3.466
75.0	4.128
88.4	5.171
98.8	6.032
116.7	7.694
129.8	8.694
133.6	8.964
145.5	9.822

* Not shown on plot

DATA TABLE NO. 52 (continued)

T	C_p
CURVE 7 (cont.)*	
157.1	1.062 x 10^{-1}
158.8	1.075
161.4	1.101
164.3	1.109
165.8	1.134
179.5	1.212
186.9	1.259
192.5	1.272
196.0	1.309
199.2	1.329
205.3	1.355
213.6	1.388
222.8	1.427
237.5	1.477
241.4	1.508
246.6	1.530
253.3	1.577
258.2	1.560
265.9	1.741
283.3	1.666
287.0	1.674
290.4	1.667
294.5	1.662
296.3	1.656
CURVE 8*	
1.7	5.763 x 10^{-7}
1.8	6.469
1.9	7.236
2.0	8.071
2.2	9.952
2.4	1.214 x 10^{-6}
2.6	1.466
2.8	1.755
3.0	2.082
3.2	2.452
3.4	2.866
3.6	3.328
3.8	3.839
4.0	4.404
4.2	5.025
4.4	5.704
4.6	6.444
4.8	7.249
5.0	8.120
12	1.105 x 10^{-4}
13	1.403

T	C_p
CURVE 8 (cont.)*	
14	1.786 x 10^{-4}
15	2.636
16	3.401
17	4.252
18	5.442
19	6.888
20	8.674
25	2.228 x 10^{-3}
30	4.337
35	7.313
40	1.105 x 10^{-2}
45	1.514
50	1.956
55	2.449
60	2.968
65	3.469
70	3.963
75	4.413
80	4.813
85	5.238
90	5.476
95	5.765
100	6.029
CURVE 9*	
60	1.16 x 10^{-1}
70	1.53
80	1.89
90	2.24
100	2.60
110	2.94
120	3.29
130	3.61
140	3.92
150	4.24
160	4.52
170	4.79
180	5.06
190	5.31
200	5.54
210	5.77
220	5.98
230	6.16
240	6.34
250	6.50
260	6.66
270	6.80

T	C_p
CURVE 9 (cont.)*	
273.15	6.84 x 10^{-1}
280	6.93
290	7.05
298.15	7.14
300	7.16

*Not shown on plot

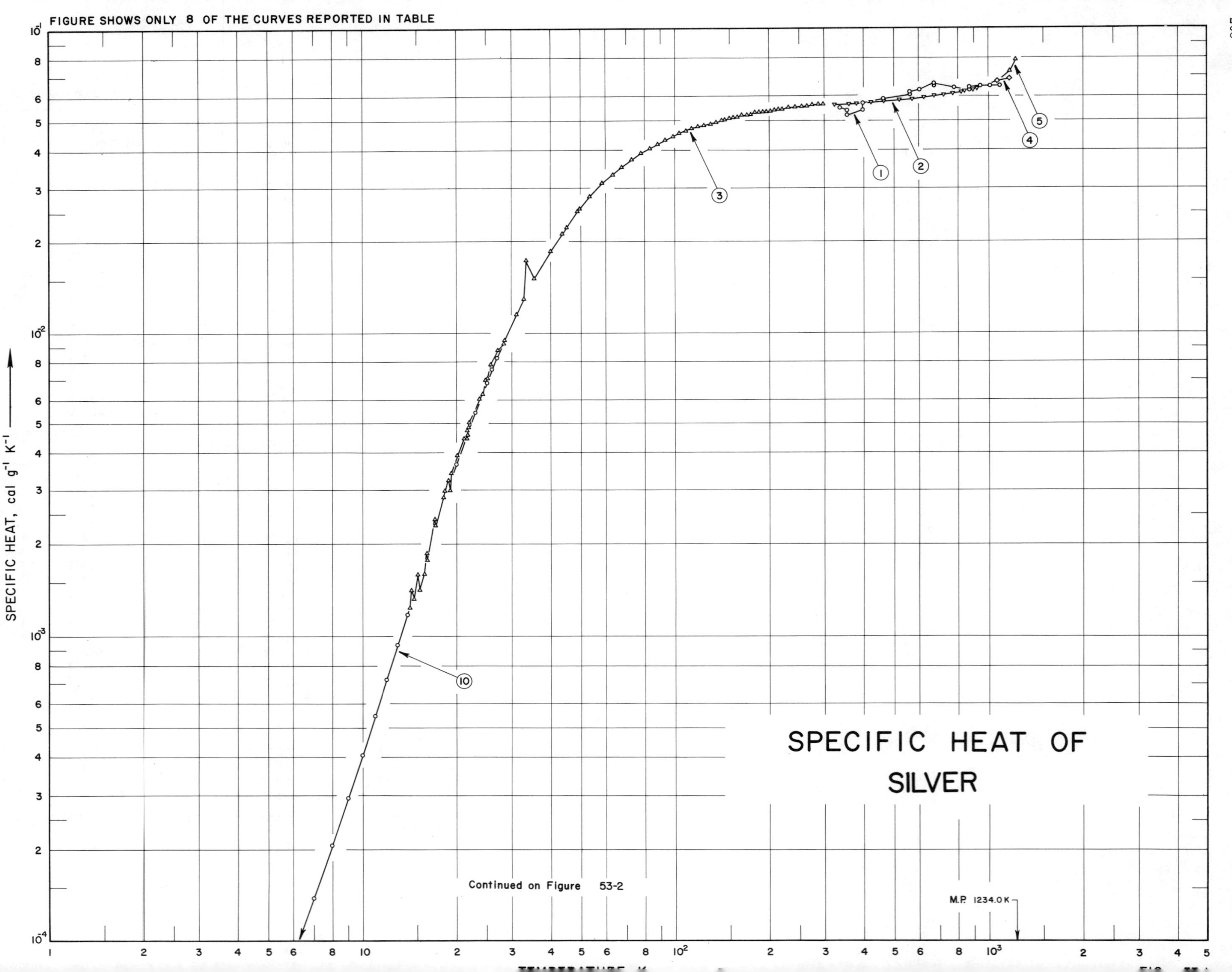

FIGURE SHOWS ONLY 8 OF THE CURVES REPORTED IN TABLE
SPECIFIC HEAT, cal g⁻¹ K⁻¹
SPECIFIC HEAT OF SILVER
Continued on Figure 53-2
M.P. 1234.0 K

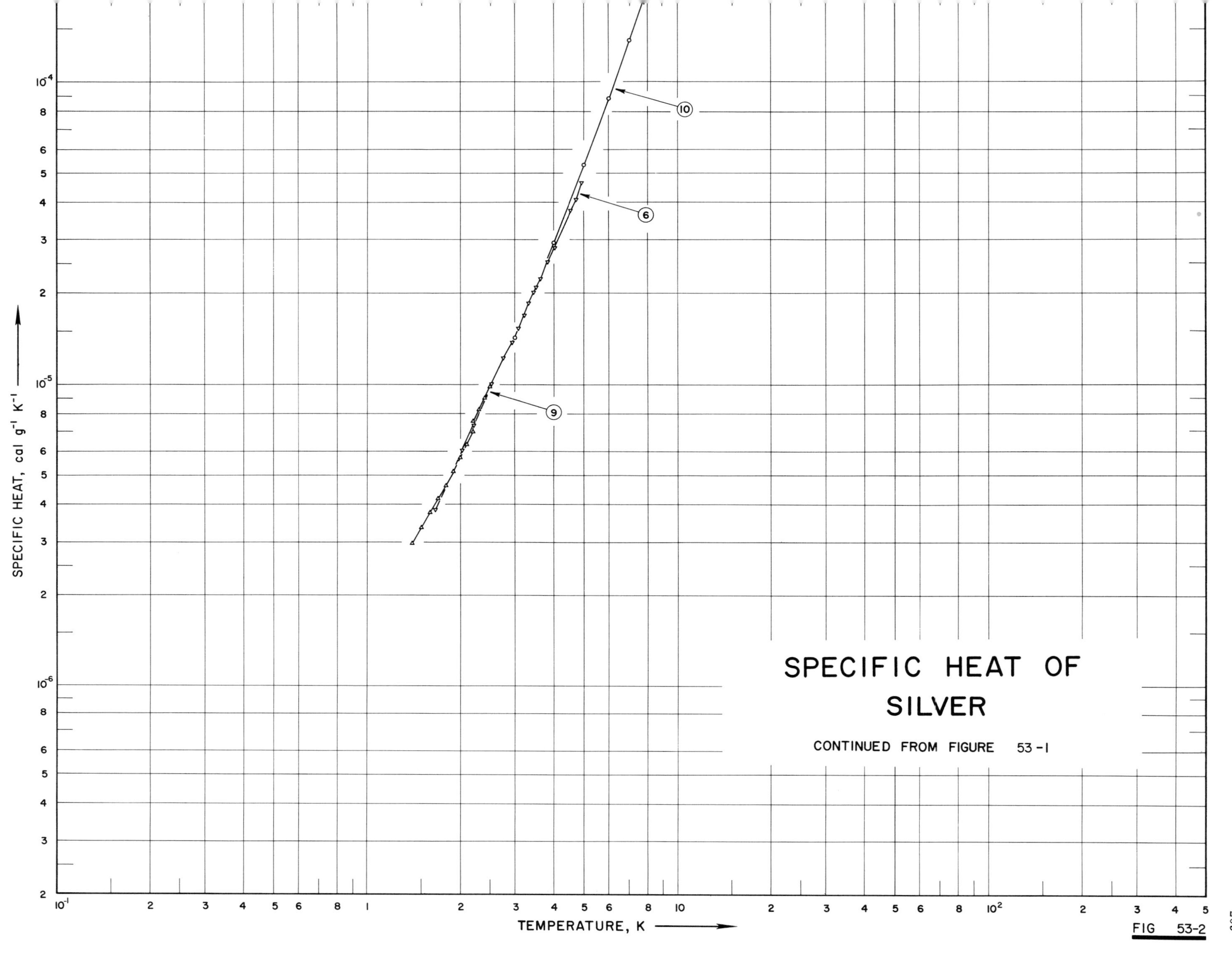

SPECIFIC HEAT OF
SILVER
CONTINUED FROM FIGURE 53-1
SPECIFIC HEAT, cal g⁻¹ K⁻¹
TEMPERATURE, K
10
6
9
FIG 53-2

SPECIFICATION TABLE NO. 53 SPECIFIC HEAT OF SILVER

(Impurity < 0.20% each; total impurities < 0.50%)

[For Data Reported in Figure and Table No. 53]

Curve No.	Ref. No.	Year	Temp. Range, K	Reported Error, %	Name and Specimen Designation	Composition (weight percent), Specifications and Remarks
1	101	1958	337-1090			Specimen's surface plated with platinum black.
2	52	1936	325-925			Inquartation silver; heated under reduced pressure in argon.
3	103	1941	15-298			99.99 Ag; melted and crystallized in nitrogen and cooled over a period of 5 days.
4	179	1924	373-1173	< 1.0		Melting sample.
5	180	1932	273-1073	≤ 0.2		Pure oxygen free silver.
6	181	1934	2-5			
7	182	1936	193-393	± .03		99.98+ Ag, 0.0095 Cu, and 0.0018 Fe.
8	183	1936	373-773	≤ 1.0		Same as above.
9	178	1952	1.4-2.5			99.999 Ag; single crystal.
10	184	1966	3-30	≤ ± 5.0		99.999 Ag, 0.0005 - 0.00005 Fe, < 0.00005 each, Ca and Mg, < 0.00002 Cu, and 0.0001 - 0.00001 Si; large crystals; cast by Consolidated Mining, Smelting and Refining Co., Ltd.; annealed condition.
11	268	1926	373-1573			Electrolytic silver.
12	360	1932	1-20			99.95 Ag; sample supplied by Zilverfabriek.
13	296	1955	1-5	0.5		99.98 Ag, 0.01 Cu, and 0.01 total, Fe, Pb, Mg, and Mn; sample supplied by Handy and Harman Co.; annealed under vacuum of 1×10^{-6} mm Hg for 4 hrs. at 700 C; cooled under vacuum at rate of 200 C per hr.

DATA TABLE NO. 53 SPECIFIC HEAT OF SILVER

[Temperature, T, K; Specific Heat, C_p, Cal $g^{-1}K^{-1}$]

T	C_p
CURVE 1	
337	5.497×10^{-2}
356	5.405
356	5.201
400	5.405
400	5.405*
400	5.701
465	5.905
465	5.905*
565	6.100
565	6.202
604	6.304
604	6.304*
676	6.499
676	6.601
781	6.397
781	6.499*
875	6.304
875	6.397
946	6.499
946	6.601*
1017	6.499
1017	6.499*
1090	6.499
1090	6.499*
CURVE 2	
324.75	5.63×10^{-2}
361.35	5.67
381.85	5.69
425.15	5.73
467.95	5.78
522.15	5.83
572.25	5.89
626.85	5.96
672.65	6.00
722.45	6.07
774.25	6.13
822.65	6.21
843.05	6.25
872.95	6.27*
896.45	6.31
925.35	6.35
CURVE 3	
Series 1	
63.78	3.340×10^{-2}
67.94	3.529
73.02	3.736
78.13	3.922
83.28	4.069
88.50	4.202
93.63	4.325
98.88	4.453
104.06	4.556
109.09	4.651
114.24	4.717
119.53	4.797
124.95	4.829
130.53	4.895
136.27	4.962
141.77	5.014
145.44	5.048
149.50	5.104
154.40	5.121
159.45	5.158
164.86	5.217
170.37	5.227
175.58	5.257
180.96	5.327
186.45	5.324
192.04	5.355
197.96	5.374
203.43	5.395
209.10	5.422
215.06	5.453
220.86	5.446
226.50	5.465*
232.30	5.515
238.25	5.511*
244.03	5.539
249.65	5.552*
255.40	5.567
260.36	5.586*
265.83	5.593
271.20	5.616*
276.83	5.635
CURVE 3 (cont.)	
Series 1 (cont.)	
282.54	5.650×10^{-2}*
288.11	5.665
291.37	5.594*
292.26	5.612*
297.81	5.652
Series 2	
15.34	1.418×10^{-3}
21.60	4.459
24.48	6.276
28.50	9.243
33.00	1.300×10^{-2}
33.63	1.746
43.83	2.131
49.00	2.514
53.74	2.829
58.85	3.112
Series 3	
16.19	1.771×10^{-3}
19.04	2.994
21.80	4.598
Series 4	
15.03	1.595×10^{-3}
17.06	2.401
19.30	3.430
22.00	5.025
24.98	6.981
Series 5	
15.07	1.567×10^{-3}*
17.07	2.299
18.93	3.226
Series 6	
14.43	1.409×10^{-3}
17.09	2.410*
20.11	3.903
CURVE 3 (cont.)	
Series 7	
16.99	2.271×10^{-3}*
18.99	3.282*
21.28	4.468
23.78	6.054
27.29	8.788
31.32	1.157×10^{-2}
35.56	1.519
40.07	1.869
45.16	2.238
49.80	2.557
Series 8	
14.22	1.242×10^{-3}
15.85	1.595
18.23	2.837
21.69	4.737
25.27	7.036
28.61	9.493
Series 9	
14.64	1.326×10^{-3}
16.15	1.863
18.49	2.976
21.90	4.848
25.93	7.843
CURVE 4	
373.15	5.72×10^{-2}*
473.15	5.86*
573.15	6.02*
673.15	6.15*
773.15	6.30*
873.15	6.44*
973.15	6.57*
1073.15	6.72
1173.15	6.85
CURVE 5	
Series 1*	
273.15	5.540×10^{-2}
373.15	5.682
473.15	5.822
573.15	5.958
673.15	6.091
773.15	6.220
873.15	6.347
973.15	6.469
1073.15	6.590
Series 2	
623.15	5.966×10^{-2}*
673.15	6.041*
773.15	6.220*
873.15	6.377*
973.15	6.491*
1073.15	6.678*
1173.15	7.276
1223.15	7.916
CURVE 6	
1.671	3.846×10^{-6}
2.037	6.060
2.211	7.338
2.535	1.009×10^{-5}
2.766	1.229
2.950	1.388
3.079	1.543
3.218	1.694
3.344	1.864
3.452	2.003
3.534	2.089
3.635	2.236
3.836	2.544
4.020	2.824
4.537	3.759
4.733	4.094
4.921	4.647
CURVE 7*	
193.15	5.3155×10^{-2}
CURVE 7 (cont.)*	
203.15	5.3561×10^{-2}
213.15	5.3896
223.15	5.4230
233.15	5.4517
243.15	5.4780
253.15	5.5043
263.15	5.5282
273.15	5.5473
283.15	5.5688
293.15	5.5880
303.15	5.6047
313.15	5.6238
323.15	5.6405
333.15	5.6573
343.15	5.6716
353.15	5.6883
363.15	5.7027
373.15	5.7170
383.15	5.7290
393.15	5.7409
CURVE 8*	
373.15	5.7099×10^{-2}
423.15	5.7672
473.15	5.8294
523.15	5.8867
573.15	5.9512
623.15	6.0086
673.15	6.0708
723.15	6.1186
773.15	6.1903
CURVE 9	
Series 1	
1.4	2.985×10^{-6}
1.5	3.354
1.6	3.756
1.7	4.192
1.8	4.665
1.9	5.176
2.0	5.728
2.1	6.323
2.2	6.994

*Not shown on plot

DATA TABLE NO. 53 (continued)

T	C_p
CURVE 9 (cont.)	
Series 2	
2.2	7.582 x 10^{-6}
2.3	8.293
2.4	9.053
2.5	9.864
CURVE 10	
3	1.428 x 10^{-5}
4	2.940
5	5.339
6	8.870
7	1.383 x 10^{-4}
8	2.052
9	2.929
10	4.058
11	5.468
12	7.208
13	9.334
14	1.186 x 10^{-3}
15	1.484*
16	1.834*
17	2.225*
18	2.662*
19	3.140*
20	3.661
21	4.226*
22	4.823*
23	5.453
24	6.121*
25	6.814
26	7.527
27	8.264
28	9.017*
29	9.789*
30	1.057 x 10^{-2}*

T	C_p
CURVE 11*	
373	5.80 x 10^{-2}
473	6.01
573	6.16
673	6.51
773	6.60
873	6.80
973	7.01
1073	7.17
1173	7.41
1273	6.92
1373	6.92
1473	6.92
1573	6.92
CURVE 12*	
Series I	
9.705	3.993 x 10^{-4}
10.786	5.591
11.180	6.177
11.742	7.178
12.345	8.579
13.562	1.118 x 10^{-3}
15.110	1.530
16.086	1.909
16.732	2.243
17.476	2.379
17.981	2.819
18.497	2.981
19.059	3.221
19.422	3.309
Series II	
9.831	4.202 x 10^{-4}
10.940	5.882
11.862	7.401

T	C_p
CURVE 12 (cont.)*	
12.748	9.127 x 10^{-4}
13.759	1.201 x 10^{-3}
16.062	1.878
14.387	1.290
15.085	1.569
15.817	1.768
16.459	2.045
17.160	2.351
17.812	2.734
18.586	3.112
19.439	3.425
20.314	3.999
Series III	
1.394	2.522 x 10^{-6}
1.614	3.627
1.637	3.956
2.086	6.191
2.308	7.760
2.514	9.799
3.029	1.527 x 10^{-5}
3.493	2.052
3.944	2.799
4.089	2.942
4.820	4.324
5.358	5.676
5.984	8.320
8.480	2.640 x 10^{-4}
8.988	3.153
10.275	4.802
11.330	6.688

T	C_p
CURVE 12 (cont.)*	
Series IV	
1.353	2.268 x 10^{-6}
1.402	2.506
1.492	3.155
2.474	8.696
2.837	1.284 x 10^{-5}
3.977	2.774
4.122	3.078
4.738	4.557
5.365	5.815
CURVE 13*	
Series I	
1.158	2.003 x 10^{-6}
1.222	2.375
1.281	2.530
1.323	2.630
1.356	2.767
1.546	3.534
1.777	4.436
1.988	5.349
2.209	7.283
2.215	7.263
2.378	8.499
2.584	1.012 x 10^{-5}
2.778	1.188
2.975	1.389
3.215	1.671
3.378	1.868
3.618	2.269
3.825	2.714
3.985	2.962
4.264	3.468
4.367	3.620
4.567	4.013
4.766	4.542
4.926	4.961

T	C_p
CURVE 13 (cont.)*	
Series II	
1.212	2.346 x 10^{-6}
1.319	2.739
1.380	2.945
1.431	3.111
1.676	4.092
1.966	5.417
2.087	6.153
2.294	7.733
2.507	9.421
2.667	1.079 x 10^{-5}
2.881	1.283
3.084	1.507
3.279	1.750
3.492	2.038
3.711	2.444
3.902	2.803
4.088	3.162
4.199	3.332
4.487	3.809
4.690	4.321
4.864	4.788

*Not shown on plot

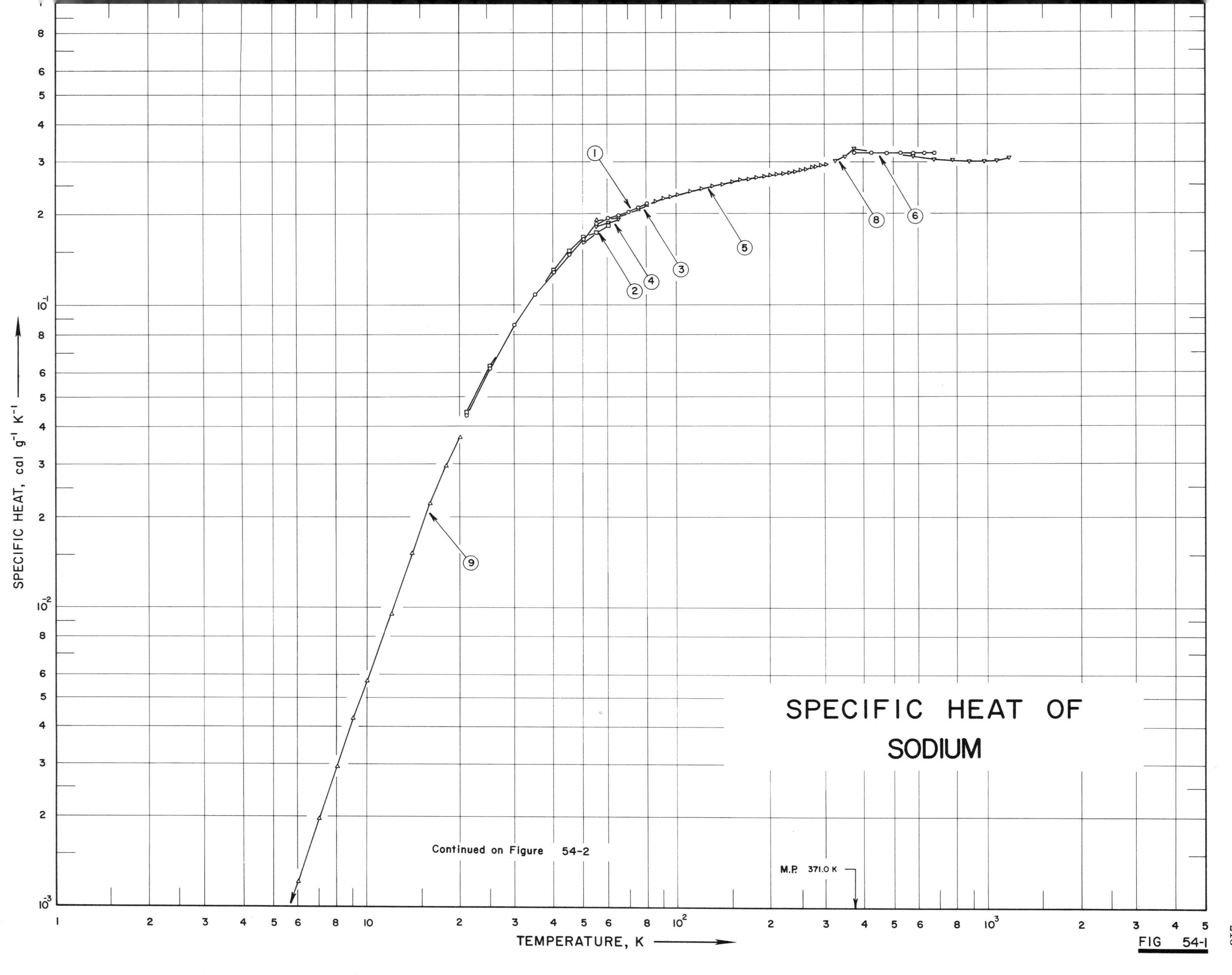

SPECIFIC HEAT OF
SODIUM
SPECIFIC HEAT, cal g⁻¹ K⁻¹
TEMPERATURE, K
M.P. 371.0 K
Continued on Figure 54-2
1
2
3
4
5
6
8
9
FIG 54-1

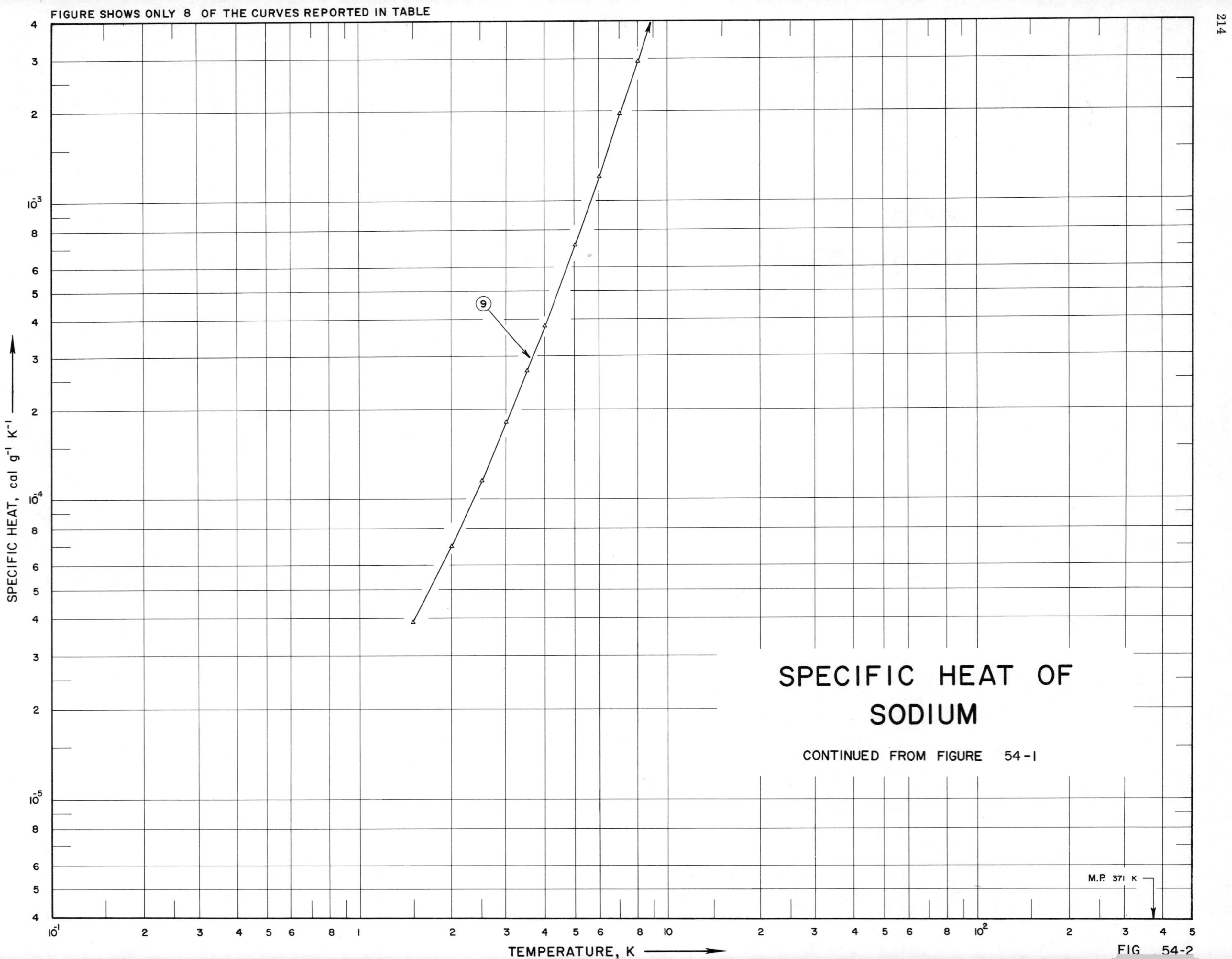
FIGURE SHOWS ONLY 8 OF THE CURVES REPORTED IN TABLE
9
SPECIFIC HEAT OF SODIUM
CONTINUED FROM FIGURE 54-1
M.P. 371 K
SPECIFIC HEAT, cal g⁻¹ K⁻¹
TEMPERATURE, K
FIG 54-2

SPECIFICATION TABLE NO. 54 SPECIFIC HEAT OF SODIUM

(Impurity $< 0.20\%$ each; total impurities $< 0.50\%$)

[For Data Reported in Figure and Table No. 54]

Curve No.	Ref. No.	Year	Temp. Range, K	Reported Error, %	Name and Specimen Designation	Composition (weight percent), Specifications and Remarks
1	113	1960	21-80	≤ 2.0		High purity; sealed in helium; cooled to 2 K.
2	113	1960	21-60	≤ 2.0		Same as above; cooled to 20 K.
3	113	1960	21-80	≤ 2.0		Same as above; cooled to 20 K; annealed at 300 K.
4	113	1960	40-65	≤ 2.0		Same as above; cooled to 35 K.
5	113	1960	40-300	≤ 2.0		Same as above; cooled to 36 K.
6	7	1959	373-673	≤ 2.0		99.99 Na; technical grade.
7	170	1918	65-294	< 1.0		Kahlbaum's purity; melted under vacuum.
8	185	1950	273-1173	1.0		0.001 - 0.01 K, and 0.0001 - 0.001 each Ca, and Li; under helium atm.
9	173	1957	1.5-20			99.995 Na.
10	267	1920	87-124			Kahlbaum's purity.
11	213	1926	17-118			Kahlbaum's purity.
12	214	1927	394-451	1.0		Commercial electrolytic metal; purified by melting in vacuum and filtering through capillary tubing.
13	336	1948	2-25			Pure metal, hydrogen free.
14	361	1954	55-320	± 0.1		
15	362	1955	1.5-20			Redistilled Na of very high purity; self annealed for several days at room temperature.
16	223	1959	0.4-2	5.0		> 99.99 Na; under pure argon atm; annealed.

DATA TABLE NO. 54 SPECIFIC HEAT OF SODIUM

[Temperature, T, K; Specific Heat, C_p, Cal g^{-1} K^{-1}]

T	C_p
CURVE 1	
21	4.328 x 10^{-2}
25	6.172
30	8.599
35	1.087 x 10^{-1}
40	1.287
45	1.466
50	1.651
55	1.855
60	1.940
65	1.994
70	2.047
75	2.107
80	2.170
CURVE 2	
21	4.454 x 10^{-2}
25	6.290
30	8.673*
35	1.095 x 10^{-1}*
40	1.300
45	1.506
50	1.686
55	1.747
60	1.844
CURVE 3	
21	4.367 x 10^{-2}*
25	6.203*
30	8.599*
35	1.087 x 10^{-1}*
40	1.287*
45	1.466*
50	1.639*
55	1.914
60	1.916*
65	1.965
70	2.026*
75	2.090
80	2.153
CURVE 4	
40	1.288 x 10^{-1}*
45	1.463*
50	1.673*
55	1.831
60	1.870
65	1.935
CURVE 5	
40	1.291 x 10^{-1}*
45	1.458*
50	1.608
55	1.739*
60	1.846*
65	1.935*
70	2.013*
75	2.082*
80	2.147*
85	2.208
90	2.256
95	2.296
100	2.330
110	2.391
120	2.448
130	2.495
140	2.535
150	2.569
160	2.601
170	2.628
180	2.653
190	2.679
200	2.702
210	2.722
220	2.743
230	2.767
240	2.789
250	2.813
260	2.836
270	2.860*
273.15	2.866
280	2.886
CURVE 5 (cont.)	
290	2.913 x 10^{-1}
298.15	2.935
300	2.940
CURVE 6	
373.15	3.20 x 10^{-1}
423.15	3.20
473.15	3.20
523.15	3.20
573.15	3.20
623.15	3.20
673.15	3.20
CURVE 7*	
64.6	1.966 x 10^{-1}
67.9	2.027
71.1	2.075
74.2	2.092
84.6	2.210
94.8	2.305
156.8	2.619
159.0	2.606
181.7	2.675
183.8	2.684
234.7	2.797
292.1	2.949
293.5	2.953
CURVE 8	
273.15	2.866 x 10^{-1}*
298.15	2.921*
323.15	3.001
348.15	3.124
370.95	3.259*
370.95	3.309
373.15	3.306*
473.15	3.201*
573.15	3.117
673.15	3.056
773.15	3.016
CURVE 8 (cont.)	
873.15	2.999 x 10^{-1}
973.15	3.004
1073.15	3.031
1173.15	3.080
CURVE 9	
1.5	3.871 x 10^{-5}
2.0	6.916
2.5	1.153 x 10^{-4}
3.0	1.805
3.5	2.684
4.0	3.828
5.0	7.134
6.0	1.214 x 10^{-3}
7.0	1.949
8.0	2.914
9.0	4.219
10.0	5.698
12.0	9.482
14.0	1.509 x 10^{-2}
16.0	2.210
18.0	2.953
20.0	3.697
CURVE 10*	
87.0	2.29 x 10^{-1}
87.9	2.31
89.7	2.30
91.5	2.35
92.2	2.38
105.2	2.43
107.2	2.45
108.0	2.45
121.9	2.47
124.0	2.47
CURVE 11*	
16.95	2.58 x 10_{2}
20.04	3.72
23.25	5.25
CURVE 11 (cont.)*	
26.17	6.74 x 10^{-2}
29.40	8.40
34.56	1.07 x 10^{-1}
41.40	1.34
49.50	1.65
58.10	1.86
117.60	2.45
CURVE 12*	
394	3.275 x 10^{-1}
451	3.197
CURVE 13*	
2	2.39 x 10^{-4}
3	3.48
4	5.87
5	1.17 x 10^{-3}
6	2.57
7	5.39
8	4.05
9	4.35
10	6.09
12	1.09 x 10^{-2}
14	1.67
16	2.35
18	3.09
20	3.91
25	6.31
CURVE 14*	
55	1.74 x 10^{-1}
60	1.84
70	2.01
80	2.15
90	2.26
100	2.34
110	2.40
120	2.44
130	2.48
140	2.52
150	2.55
CURVE 14 (cont.)*	
160	2.58 x 10^{-1}
170	2.61
180	2.64
190	2.67
200	2.69
210	2.71
220	2.74
230	2.75
240	2.77
250	2.80
260	2.82
270	2.84
273.15	2.85
280	2.87
290	2.90
298.15	2.92
300	2.93
310	2.96
320	2.99
CURVE 15*	
1.5	4.57 x 10^{-3}
1.8	6.31
2.0	7.83
2.5	1.26 x 10^{-2}
3.0	1.94
3.5	2.91
4.0	4.28
4.5	6.31
5.0	8.48
6.0	1.50 x 10^{-1}
7.0	2.26 x 10^{-2}
8.0	3.39
9.0	4.61
10.0	6.18
12.0	1.09 x 10^{-1}
14.0	1.67
16.0	2.33
18.0	3.09
20.0	3.91

*Not shown on plot

DATA TABLE NO. 54 (continued)

T	C_p
CURVE 16*	
Series I	
0.493	2.911×10^{-6}
0.517	3.152
0.579	3.561
0.649	4.093
0.721	4.604
0.883	5.929
Series II	
0.403	2.564×10^{-6}
0.408	2.428
0 411	2 411
0.415	2.503
0.422	2.577
0.456	2.553
0.492	2.788
0.515	3.244
1.009	7.196
1.264	1.010×10^{-5}
1.520	1.353
1.648	1.585
1.837	2.015
1.935	2.228
1.989	2.426
Series III	
0.417	2.593×10^{-6}
0.454	2.797
0.536	3.626
0.588	3.458
0.633	4.048
0.706	4.629
0.758	4.917
0.799	5.818
0.828	5.508
0.866	6.231
0.879	6.186
0.892	6.823
0.908	6.430
0.918	6.016
0.936	6.414
0.956	6.579
0.976	6.976
1.001	7.167

*Not shown on plot

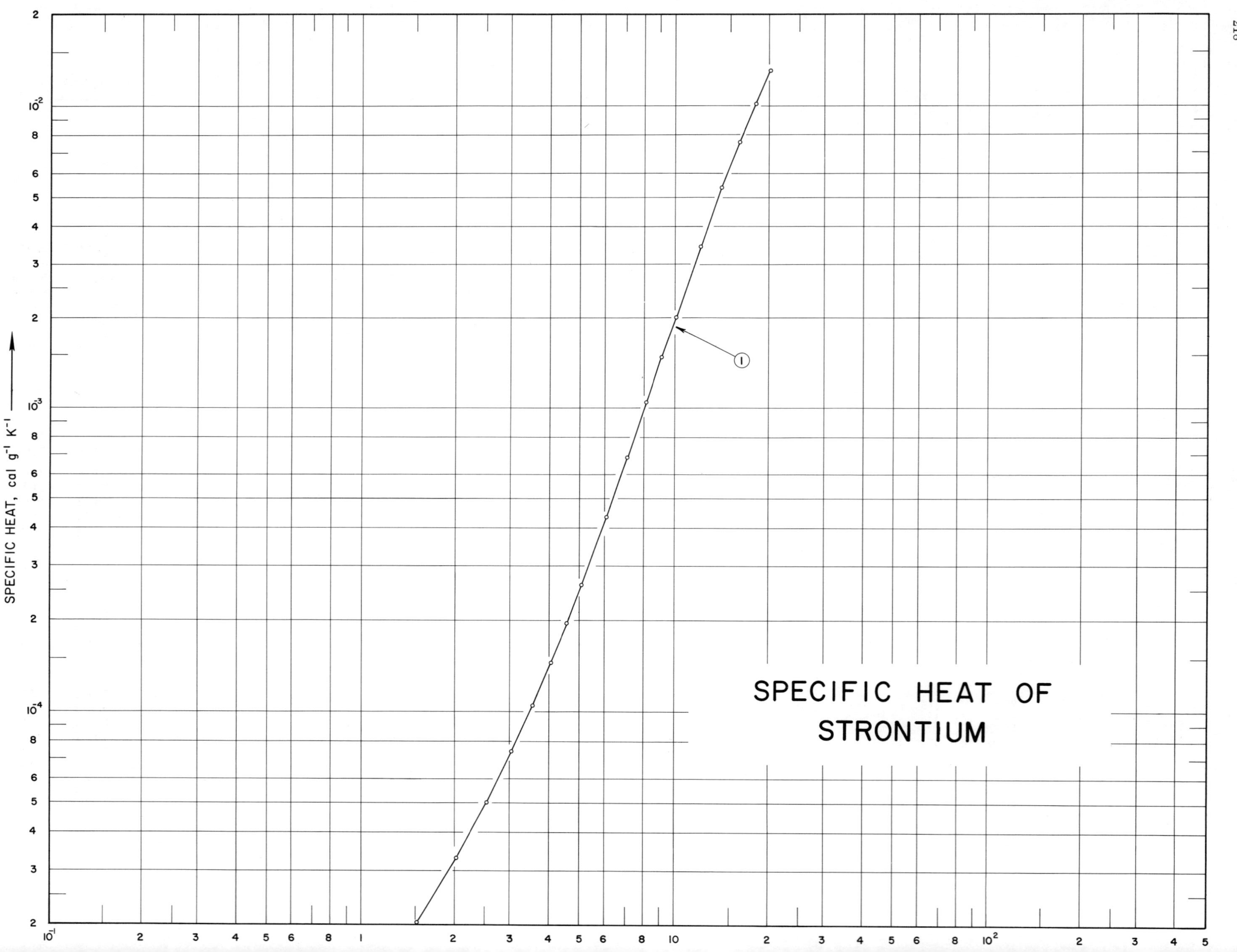
SPECIFIC HEAT OF
STRONTIUM
SPECIFIC HEAT, cal g⁻¹ K⁻¹
1

SPECIFICATION TABLE NO. 55 SPECIFIC HEAT OF STRONTIUM

(Impurity $<0.20\%$ each; total impurities $<0.50\%$)

[For Data Reported in Figure and Table No. 55]

Curve No.	Ref. No.	Year	Temp. Range, K	Reported Error, %	Name and Specimen Designation	Composition (weight percent), Specifications and Remarks
1	7	1957	1.5-20	2.0		0.5 each Ba, Fe, and Mn, 0.2 Ca, and 0.05 Si.

DATA TABLE NO. 55 SPECIFIC HEAT OF STRONTIUM

[Temperature, T,K; Specific Heat, C_p, Cal $g^{-1}K^{-1}$]

T	C_p
CURVE 1	
1.5	2.05×10^{-5}
2.0	3.33
2.5	5.07
3.0	7.47
3.5	1.06×10^{-4}
4.0	1.46
4.5	1.96
5.0	2.62
6.0	4.39
7.0	6.90
8.0	1.05×10^{-3}
9.0	1.48
10.0	2.02
12.0	3.46
14.0	5.43
16.0	7.66
18.0	1.03
20.0	1.31

* Not shown on plot

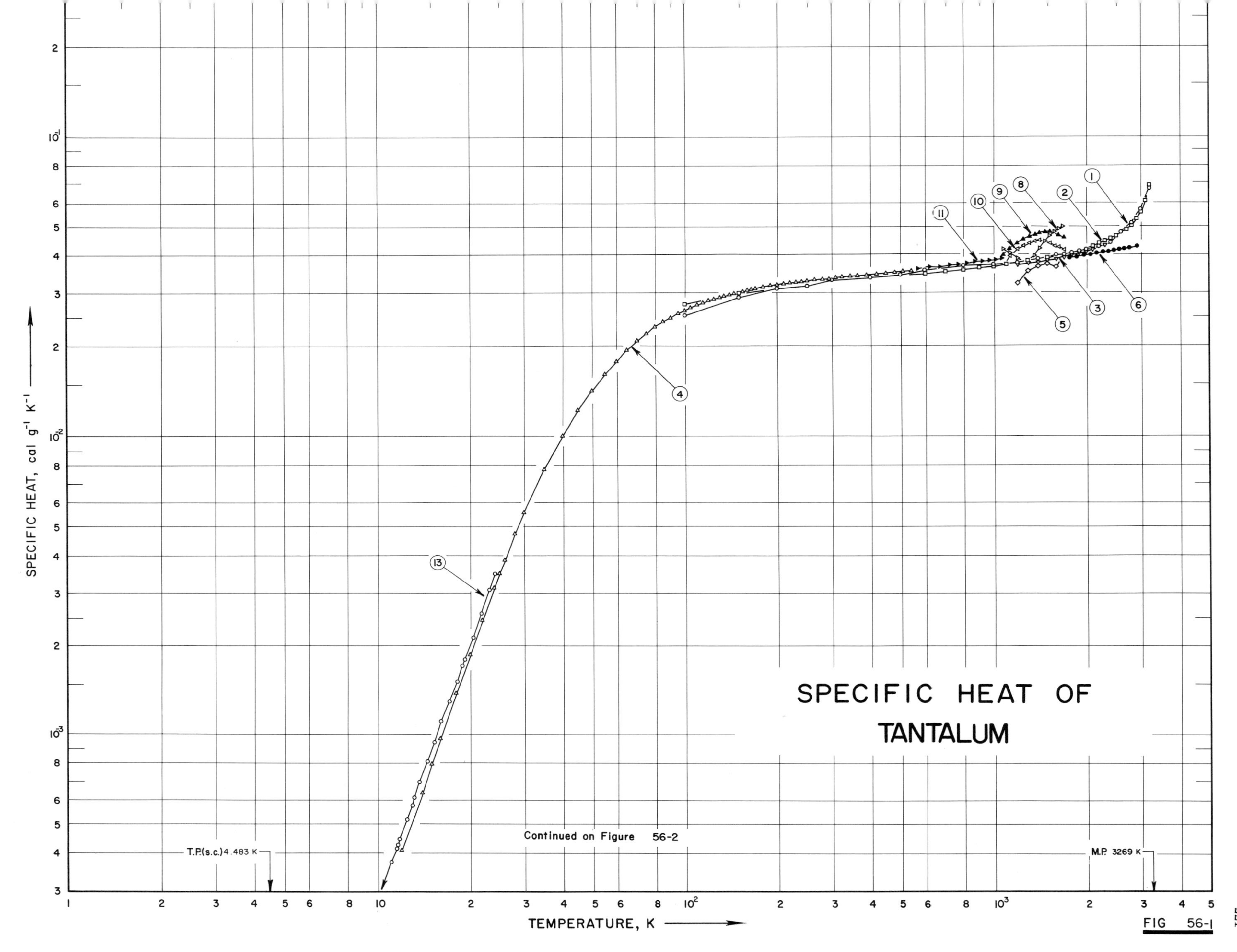

SPECIFIC HEAT OF
TANTALUM
SPECIFIC HEAT, cal g⁻¹ K⁻¹
TEMPERATURE, K
T.P.(s.c.) 4.483 K
M.P. 3269 K
Continued on Figure 56-2
FIG 56-1

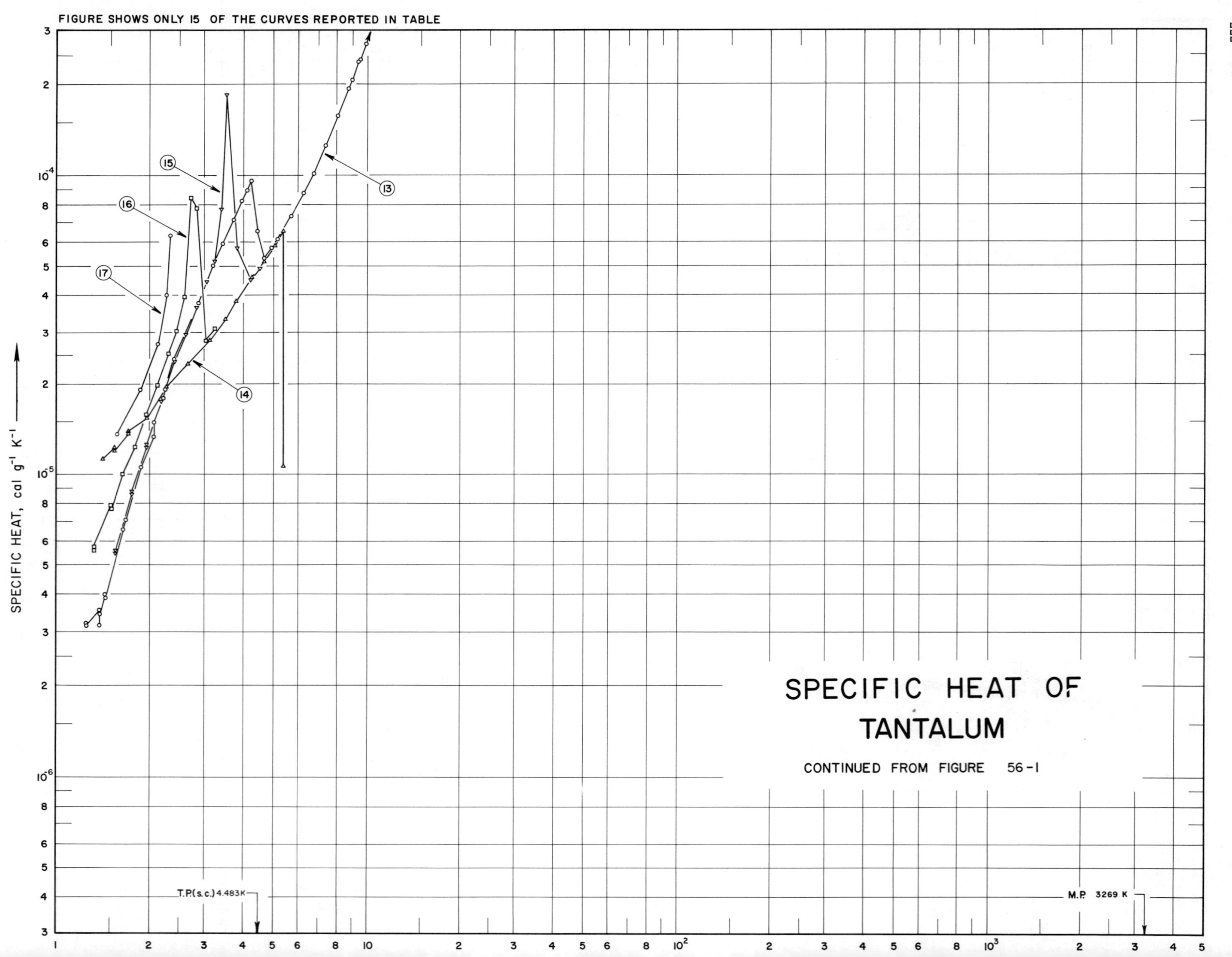
FIGURE SHOWS ONLY 15 OF THE CURVES REPORTED IN TABLE
SPECIFIC HEAT, cal g⁻¹ K⁻¹
SPECIFIC HEAT OF
TANTALUM
CONTINUED FROM FIGURE 56-1
T.P.(s.c.) 4.483K
M.P. 3269 K
13
14
15
16
17

SPECIFICATION TABLE NO. 56 SPECIFIC HEAT OF TANTALUM

(Impurity < 0.20% each; total impurities < 0.50%)

[For Data Reported in Figure and Table No. 56]

Curve No.	Ref. No.	Year	Temp. Range, K	Reported Error, %	Name and Specimen Designation	Composition (weight percent), Specifications and Remarks
1	17	1960	100-3200			
2	65	1961	100-3195	4.0		
3	64	1963	1200-2400	≤ 2.5	Wire	99.90 Ta, 0.05 Nb, 0.02 W, 0.015 C, 0.015 O_2, 0.01 each Fe, Mo, Si, Ti, and Zr, and 0.005 N_2; sample supplied by the Fansteel Metallurgical Corp.; outgassed and sealed in < 1 x 10^{-6} mm Hg glass envelope.
4	363	1957	12-539			99.9+ Ta, 0.04 Nb, 0.013 Fe, 0.01 W, < 0.01 each C, Si, and < 0.005 Mo; sample supplied by the Fansteel Metallurgical Corp.; annealed.
5	115	1962	1200-1700	< 10		
6	116	1961	1273-2939	1.0		99.9 Ta; sealed in vacuum; degassed for 2 hrs. at 2150 C.
7	117	1963	1200-2900	1.0		Sealed under vacuum of 2 x 10^{-6} mm Hg.
8	71	1960	1089-1700		Tan 9-4	> 99.854 Ta, 0.05 Nb, < 0.02 each Ti, W, 0.014 O_2, 0.005 C, 0.004 Fe, and 0.003 Mo.
9	71	1960	1089-1700		Tan 10-22	Same as above.
10	71	1960	1089-1700		Tan 10-29	Same as above.
11	201	1929	573-1173			
12	70	1934	273-1873	≤ 0.2		Sample supplied by the Fansteel Metallurgical Corp.; stabilized by heating 3 to 6 hrs. in vacuum at 1400 C and slowly cooling.
13	204	1958	1.3-24			99.9 Ta; sample supplied by the Fansteel Metallurgical Corp.; annealed in vacuum of 1 x 10^{-5} to 1 x 10^{-6} mm Hg for 24 hrs. at 1800 - 2400 C; zero magnetic field, superconducting.
14	204	1958	1.4-5.4			Same as above; 1930 gauss external magnetic field.
15	204	1958	1.6-4.5			Same as above; 237 gauss external magnetic field.
16	204	1958	1.3-3.2			Same as above; 454 gauss external magnetic field.
17	204	1958	1.5-2.3			Same as above; 557 gauss external magnetic field.
18	364	1940	53-295		Sample A	Pure wrought Ta; sheared in lengths of 0.5 cm; washed consecutively with HCL solution, NaOH, distilled water and alcohol; dried near 100 C.
19	364	1940	80-295		Sample B	0.067:1 ratio of hydrogen to tantalum; made from sample A by heating to 710 C in vacuum (10^{-5} mm Hg at room temperature to 2 x 10^{-4} at 710 C) and then admitting hydrogen.
20	364	1940	80-295		Sample C	0.0958:1 ratio of hydrogen to tantalum; made from sample B by evacuating to 10^{-5} mm Hg at room temperature; heating to 700 C, and admitting additional hydrogen.

SPECIFICATION TABLE NO. 56 (continued)

Curve No.	Ref. No.	Year	Temp. Range, K	Reported Error, %	Name and Specimen Designation	Composition (weight percent), Specifications and Remarks
21	364	1940	80-295		Sample D	Prepared by pumping hydrogen from sample C at 720 C, pumping continued until pressure had fallen to 2 x 10^{-4} mm Hg at 720 C and during cooling to room temp.
22	364	1940	80-295		Sample E	0.0284:1ratio of hydrogen to tantalum; prepared by dehydrogenating sample D at 700 C.
23	364	1940	53-320		Sample E	99.85 Ta, 0.11 C, 0.037 H_2, and 0.005 Fe.
24	365	1941	1.4-4.3			>99.95 Ta, and traces of Cu, Fe, Nb and Si; sample supplied by Nederlandsche Siemens N.V.
25	366	1955	10-274			

DATA TABLE NO. 56 SPECIFIC HEAT OF TANTALUM

[Temperature, T, K; Specific Heat, C_p, Cal g^{-1} K^{-1}]

T	C_p
CURVE 1	
100	2.52 x 10^{-2}
150	2.89
200	3.10
250	3.15
300	3.30
400	3.35
500	3.42
600	3.55
800	3.67
1000	3.69
1200	3.75
1400	3.87
1600	3.98
1700	4.00
1800	4.05
1900	4.10
2000	4.15
2200	4.25
2400	4.39
2600	4.75
2800	5.10
3000	5.67
3200	6.67
CURVE 2	
100	2.74 x 10^{-2}
150	2.98
200	3.11*
250	3.19*
300	3.28*
400	3.35*
500	3.40*
600	3.45
700	3.50
800	3.55
900	3.60
1000	3.65
1100	3.70
1200	3.75*
1300	3.80
1400	3.83*
1500	3.90
CURVE 2 (cont.)	
1600	3.95 x 10^{-2}*
1700	4.00*
1800	4.05*
1900	4.11*
2000	4.17*
2100	4.24
2200	4.33
2300	4.42
2400	4.51
2500	4.60
2600	4.72*
2700	4.84
2800	5.00
2900	5.25
3000	5.58
3100	6.02
3195	6.80
CURVE 3	
1200	3.70 x 10^{-2}
1300	3.74
1400	3.77
1500	3.82
1600	3.86
1700	3.91
1800	3.97
1900	4.02
2000	4.08
2100	4.15
2200	4.22*
2300	4.29
2400	4.36*
CURVE 4	
12	4.090 x 10^{-4}
14	6.355
15	7.903
16	9.671
18	1.382 x 10^{-3}
20	1.868
22	2.448
24	3.117
CURVE 4 (cont.)	
25	3.487 x 10^{-3}
26	3.885
28	4.736
30	5.560
35	7.798
40	1.008 x 10^{-2}
45	1.231
50	1.432
55	1.623
60	1.793
65	1.945
70	2.075
75	2.195
80	2.305
85	2.403
90	2.486
95	2.555
100	2.618
105	2.675
110	2.731
115	2.778
120	2.832
125	2.861
130	2.897
135	2.930
140	2.962
145	2.986
150	3.011*
155	3.035
160	3.058
165	3.078
170	3.096
175	3.113*
180	3.129
185	3.144*
190	3.158
195	3.172*
200	3.184
205	3.196*
210	3.208
215	3.217*
220	3.227
225	3.236*
230	3.245
CURVE 4 (cont.)	
235	3.254 x 10^{-2}*
240	3.262
245	3.270*
250	3.278
255	3.284*
260	3.291*
265	3.298
270	3.305*
273.16	3.309*
275	3.311*
280	3.319
285	3.325*
290	3.332*
295	3.337
298.16	3.341*
300	3.343*
305	3.348*
310	3.353
315	3.357*
320	3.362*
325	3.367
330	3.371*
335	3.376*
340	3.381*
345	3.386
350	3.390*
355	3.395*
360	3.400*
365	3.404
370	3.408*
375	3.411*
380	3.415*
385	3.418*
390	3.421
395	3.425*
400	3.428*
405	3.431*
410	3.435*
415	3.439*
420	3.442
425	3.446*
430	3.451*
435	3.455*
440	3.459*
CURVE 4 (cont.)	
445	3.463 x 10^{-2}*
450	3.467
455	3.472*
460	3.476*
465	3.479*
470	3.484*
475	3.488*
480	3.492
485	3.496*
490	3.500*
495	3.504*
500	3.508*
505	3.512*
510	3.516
515	3.519*
520	3.523*
525	3.527*
530	3.531*
535	3.535*
540	3.539*
543.16	3.541
CURVE 5	
1200	3.20 x 10^{-2}
1300	3.51
1400	3.65
1500	3.72
1600	3.63
1700	3.94*
CURVE 6	
1273.15	3.753 x 10^{-2}*
1373.15	3.782*
1473.15	3.811*
1573.15	3.840*
1673.15	3.870*
1773.15	3.898
1873.15	3.927
1973.15	3.956
2073.15	3.985
2173.15	4.014
2273.15	4.043
CURVE 6 (cont.)	
2373.15	4.072 x 10^{-2}
2473.15	4.101
2573.15	4.130
2673.15	4.159
2773.15	4.188
2873.15	4.217*
2939.15	4.236
CURVE 7*	
1200	3.667 x 10^{-2}
1300	3.705
1400	3.743
1500	3.780
1600	3.818
1700	3.855
1800	3.893
1900	3.930
2000	3.968
2100	4.006
2200	4.043
2300	4.080
2400	4.118
2500	4.156
2600	4.193
2700	4.231
2800	4.269
2900	4.306
CURVE 8	
1088.9	4.18 x 10^{-2}
1144.4	4.02
1200.0	3.87
1255.5	3.76*
1311.1	3.79*
1366.6	3.95
1422.2	4.19
1477.8	4.41
1533.3	4.60
1588.9	4.75
1644.4	4.87
1700.0	4.93

*Not shown on plot

DATA TABLE NO. 56 (Continued)

CURVE 9

T	C_p
1088.7	3.96×10^{-2}
1144.4	4.17
1200.0	4.35
1255.5	4.49
1311.1	4.59
1366.6	4.66
1422.2	4.70
1477.8	4.72
1533.3	4.72
1588.9	4.69*
1644.4	4.63
1700.0	4.57

CURVE 10

T	C_p
1088.9	3.66×10^{-2}*
1144.4	3.94
1200.0	4.15
1255.5	4.27
1311.1	4.36
1366.6	4.40
1422.2	4.43
1477.8	4.41*
1533.3	4.36
1588.9	4.26
1644.4	4.22
1700.0	4.12

CURVE 11

T	C_p
573	3.5795×10^{-2}
623	3.6074
673	3.6352
723	3.6632
773	3.6911
823	3.7189
873	3.7468
923	3.7747
973	3.8026
1023	3.8305
1073	3.8584
1123	3.8863*
1173	3.9142*

CURVE 12*

T	C_p
273.15	3.322×10^{-2}
373.15	3.364
473.15	3.407
573.15	3.450
673.15	3.495
773.15	3.540
873.15	3.585
973.15	3.632
1073.15	3.679
1173.15	3.726
1273.15	3.774
1373.15	3.823
1473.15	3.873
1573.15	3.923
1673.15	3.974
1773.15	4.026
1873.15	4.078

CURVE 13

Series 1

T	C_p
1.255	3.1500×10^{-6}
1.380	3.1500
1.379	3.5369
1.448	3.9790
1.650	6.5765
1.875	1.0611×10^{-5}*
1.884	1.0611
2.055	1.3319
2.068	1.4811
2.209	1.7906
2.253	1.9287
2.404	2.4206
2.872	3.7580
3.200	5.0235
3.442	5.9741
3.726	7.1733
3.959	8.2344
4.109	8.9142
4.256	9.6271
4.683	5.3275
4.926	5.7641
5.168	6.1730

CURVE 13 (cont.)

Series 1 (cont.)

T	C_p
5.702	7.3502×10^{-5}
6.266	8.7705
6.732	1.0152×10^{-4}
7.365	1.2545
8.063	1.5767
8.698	1.9260
8.915	2.0719
9.365	2.3758
9.487	2.4112
9.899	2.7229
10.447	3.1236
10.490	3.1733*
11.152	3.7386
11.577	4.1299
11.700	4.2797
18.988	1.7071×10^{-3}
19.386	1.8038
20.611	2.1354
21.941	2.5670

Series 2

T	C_p
1.253	3.2053×10^{-6}
1.381	3.0948*
1.381	3.4264
1.450	3.8685
1.656	6.6317*
1.680	7.1291*
1.878	1.0887×10^{-5}*
1.887	1.0887*
2.055	1.3595*
2.068	1.5032*
2.208	1.8016*
2.252	1.9287*
2.403	2.4151*
2.870	3.8133*
3.195	5.0567*
3.437	5.9630*
3.722	7.1881*
3.957	8.1239*
4.110	8.7926*
4.259	9.5110*

CURVE 13 (cont.)

Series 2 (cont.)

T	C_p
4.445	6.4770×10^{-5}*
4.688	5.2943*
4.932	5.7530*
5.174	6.1841*
5.707	7.3833*
6.269	8.7650*
6.735	1.0163×10^{-4}*
7.366	1.2600*
11.828	4.4510
12.515	5.1601
13.041	5.7862
13.098	5.8636*
13.293	6.1344
13.864	6.9081
14.653	8.0907
15.443	9.3506
16.347	1.1020×10^{-3}
17.317	1.2971
18.353	1.5214
22.040	2.5930*
23.247	3.0799
24.323	3.4822

CURVE 14

Series 1

T	C_p
1.421	1.1274×10^{-5}
1.545	1.2269
1.714	1.3706
1.966	1.5474
2.274	1.9564
2.655	2.3266
3.126	2.8130
3.500	3.3380
3.851	3.8298
4.269	4.5814
4.685	5.1949
5.025	5.8525
5.392	6.5710

CURVE 14 (cont.)

Series 2

T	C_p
1.422	1.1219×10^{-5}*
1.551	1.1992
1.720	1.3927
1.967	1.5806*
2.274	1.9564*
2.654	2.3487*
3.122	2.8461*
3.495	3.3325*
3.848	3.7856*
4.271	4.5262*
4.690	5.1672*
5.031	5.8415*
5.397	1.0777

CURVE 15

Series 1

T	C_p
1.553	5.5817×10^{-6}
1.755	8.5660
1.960	1.2379×10^{-5}
2.185	1.7629
2.409	2.3819
2.620	2.9567
2.842	3.6475
3.067	4.4433
3.242	5.1949
3.403	7.7370
3.539	1.8547×10^{-4}
3.824	5.7475×10^{-5}
4.232	4.5096
4.519	4.9241

Series 2

T	C_p
1.558	5.4712×10^{-6}
1.761	8.7318
1.962	1.2656
2.184	1.7795*
2.408	2.3764*
2.619	2.9843*
2.840	3.6972*

* Not shown on plot

DATA TABLE NO 56 (continued)

T	C_p
CURVE 15 (cont.)	
Series 2 (cont.)	
3.064	4.4985 x 10^{-5}*
3.238	5.2225*
3.399	7.7315*
3.536	1.8447 x 10^{-4}*
3.822	5.6922 x 10^{-5}*
4.234	4.4488*
4.523	4.8909*
CURVE 16	
Series I	
1.334	5.7475 x 10^{-6}
1.500	7.9028
1.648	1.0003 x 10^{-5}
1.796	1.2435
1.954	1.5695
2.123	1.9840
2.299	2.5256
2.586	3.9348
2.707	8.4057
2.844	7.7978
3.033	2.8130
3.254	3.0948
Series II	
1.334	5.5817 x 10^{-6}
1.505	7.6818
1.654	1.0058 x 10^{-5}*
1.801	1.2711*
1.956	1.6082*
2.122	2.0116*
2.298	2.5256*
2.445	3.0230*
2.584	3.9680*
2.705	8.5107*
2.842	7.9028*
3.030	2.8517*
3.250	3.1114*
CURVE 17	
Series I	
1.583	1.3595 x 10^{-5}
1.875	1.9177
2.144	2.7301
2.285	4.0012
2.335	6.3022

T	C_p
CURVE 17 (cont.)	
Series II	
1.589	1.3374 x 10^{-5}*
1.879	1.9674*
2.143	2.7688*
2.284	4.0012*
2.334	6.2946*
CURVE 18*	
53	1.55 x 10^{-2}
55	1.63
60	1.80
65	1.96
70	2.10
75	2.23
80	2.33
85	2.43
90	2.51
95	2.59
100	2.64
105	2.69
110	2.75
115	2.80
120	2.84
125	2.88
130	2.91
135	2.95
140	2.98
145	3.01
150	3.03
155	3.05
160	3.07
165	3.08
170	3.11
175	3.12
180	3.14
185	3.15
190	3.17
195	3.18
200	3.19
205	3.21
210	3.22
215	3.23
220	3.24
225	3.25
230	3.26
235	3.27
240	3.27
245	3.28

T	C_p
CURVE 18 (cont.)*	
250	3.29 x 10^{-2}
255	3.29
260	3.30
265	3.31
270	3.32
275	3.33
280	3.33
285	3.33
290	3.34
295	3.34
CURVE 19*	
80	2.30 x 10^{-2}
85	2.40
90	2.48
95	2.56
100	2.63
105	2.70
110	2.75
115	2.81
120	2.86
125	2.91
130	2.96
135	3.01
140	3.05
145	3.09
150	3.13
155	3.17
160	3.21
165	3.26
170	3.30
175	3.34
180	3.39
185	3.44
190	3.49
195	3.55
200	3.60
205	3.67
210	3.74
215	3.82
220	3.91
225	4.02
230	4.13
235	4.27
240	4.42
245	4.58
250	4.74
255	4.84

T	C_p
CURVE 19 (cont.)*	
260	4.43 x 10^{-2}
265	4.00
270	3.58
275	3.43
280	3.41
285	3.39
290	3.38
295	3.37
CURVE 20*	
80	2.28 x 10^{-2}
85	2.39
90	2.48
95	2.55
100	2.62
105	2.69
110	2.75
115	2.80
120	2.86
125	2.91
130	2.96
135	3.01
140	3.05
145	3.09
150	3.14
155	3.18
160	3.23
165	3.27
170	3.32
175	3.36
180	3.40
185	3.45
190	3.51
195	3.57
200	3.63
205	3.70
210	3.78
215	3.86
220	3.96
225	4.07
230	4.19
235	4.32
240	4.47
245	4.63
250	4.82
255	5.03
260	5.27
265	5.42

T	C_p
CURVE 20 (cont.)*	
270	4.96 x 10^{-2}
275	4.46
280	3.96
285	3.50
290	3.47
295	3.45
CURVE 21*	
80	2.33 x 10^{-2}
85	2.42
90	2.50
95	2.58
100	2.64
105	2.69
110	2.74
115	2.79
120	2.84
125	2.87
130	2.91
135	2.95
140	2.97
145	3.00
150	3.02
155	3.04
160	3.06
165	3.08
170	3.10
175	3.12
180	3.13
185	3.15
190	3.16
195	3.17
200	3.18
205	3.19
210	3.21
215	3.22
220	3.23
225	3.24
230	3.25
235	3.26
240	3.27
245	3.27
250	3.28
255	3.29
260	3.29
265	3.30
270	3.30
275	3.32

T	C_p
CURVE 21 (cont.)*	
280	3.33 x 10^{-2}
285	3.33
290	3.34
295	3.34
CURVE 22*	
80	2.32 x 10^{-2}
85	2.42
90	2.50
95	2.58
100	2.64
105	2.70
110	2.75
115	2.81
120	2.86
125	2.91
130	2.96
135	3.01
140	3.06
145	3.10
150	3.14
155	3.18
160	3.23
165	3.27
170	3.31
175	3.35
180	3.40
185	3.44
190	3.50
195	3.55
200	3.61
205	3.68
210	3.75
215	3.82
220	3.89
225	3.72
230	3.51
235	3.33
240	3.30
245	3.30
250	3.30
255	3.31
260	3.31
265	3.32
270	3.32
275	3.32
280	3.33
285	3.33

* Not shown on plot

DATA TABLE NO. 56 (continued)

T	C_p
CURVE 22 (cont.)*	
290	3.34 x 10^{-2}
295	3.35
CURVE 23*	
53	1.50 x 10^{-2}
55	1.57
60	1.74
65	1.90
70	2.06
75	2.19
80	2.30
85	2.39
90	2.48
95	2.55
100	2.63
105	2.69
110	2.75
115	2.81
120	2.86
125	2.91
130	2.96
135	3.01
140	3.05
145	3.09
150	3.13
155	3.17
160	3.22
165	3.26
170	3.30
175	3.35
180	3.39
185	3.43
190	3.48
195	3.53
200	3.58
205	3.63
210	3.69
215	3.74
220	3.79
225	3.85
230	3.89
235	3.94
240	3.99
245	4.03
250	4.08
255	4.13
260	4.18
CURVE 23 (cont.)*	
265	4.22 x 10^{-2}
270	4.27
275	4.30
280	4.26
285	4.19
290	4.13
295	4.07
300	4.01
310	3.93
320	3.92
CURVE 24*	
Series I	
1.396	4.255 x 10^{-6}
1.432	4.753
1.467	5.140
1.502	5.305
1.621	7.019
1.671	8.787
1.712	9.727
1.717	1.000 x 10^{-5}
1.783	1.039
1.891	1.133
1.971	1.343
2.046	1.630
2.182	1.873
2.237	2.012
2.342	2.448
2.406	2.697
2.746	3.355
3.039	4.515
3.652	7.599
4.017	5.775
Series II	
3.435	5.913 x 10^{-5}
3.500	6.007
3.575	6.289
3.634	6.687
3.798	7.533
3.849	7.560
3.960	8.632
4.024	8.069
4.031	7.129
4.140	5.322
4.213	5.012
CURVE 24 (cont.)*	
4.213	5.990 x 10^{-5}
4.252	4.637
Series III	
3.247	5.051 x 10^{-5}
3.320	5.189
3.375	5.604
3.427	5.958
3.480	5.908
3.675	6.720
3.780	7.157
3.906	7.942
3.914	8.035
3.980	8.820
4.007	8.196
4.086	5.239
4.114	5.239
4.163	4.333
4.194	4.736
4.244	4.880
4.270	4.952
4.278	5.068
4.303	5.101
CURVE 25*	
10.12	2.730 x 10^{-4}
11.59	3.907
11.60	3.769
13.40	5.941
13.66	5.770
13.81	6.416
15.86	9.318
16.37	1.027 x 10^{-3}
18.21	1.364
18.61	1.460
18.84	1.517
20.92	2.016
21.27	2.145
24.67	3.266
26.63	4.019
30.47	5.687
34.12	7.278
38.40	9.295
42.54	1.119 x 10^{-2}
46.73	1.301
51.32	1.488
CURVE 25 (cont.)*	
55.04	1.623 x 10^{-2}
59.49	1.776
64.06	1.903
67.88	2.013
68.35	2.019
71.36	2.095
74.87	2.186
79.32	2.260
83.57	2.354
88.14	2.406
92.10	2.480
96.81	2.538
101.57	2.594
106.66	2.660
111.71	2.710
116.99	2.755
122.04	2.827
127.08	2.845
133.36	2.899
138.52	2.937
144.07	2.964
149.33	3.006
155.06	3.017
160.49	3.052
166.60	3.079
172.12	3.080
177.57	3.087
183.69	3.129
189.18	3.138
195.55	3.152
197.71	3.161
201.64	3.168
203.22	3.170
207.12	3.189
209.39	3.185
213.21	3.202
215.04	3.214
218.68	3.222
221.23	3.228
221.29	3.226
224.77	3.234
226.83	3.238
226.85	3.245
230.61	3.248
233.04	3.254
233.12	3.252
236.27	3.255
238.58	3.263
CURVE 25 (cont.)*	
238.69	3.266 x 10^{-2}
241.90	3.267
245.17	3.282
245.46	3.272
250.97	3.286
251.21	3.282
257.61	3.301
257.98	3.300
264.40	3.318
270.10	3.318
274.20	3.329

* Not shown on plot

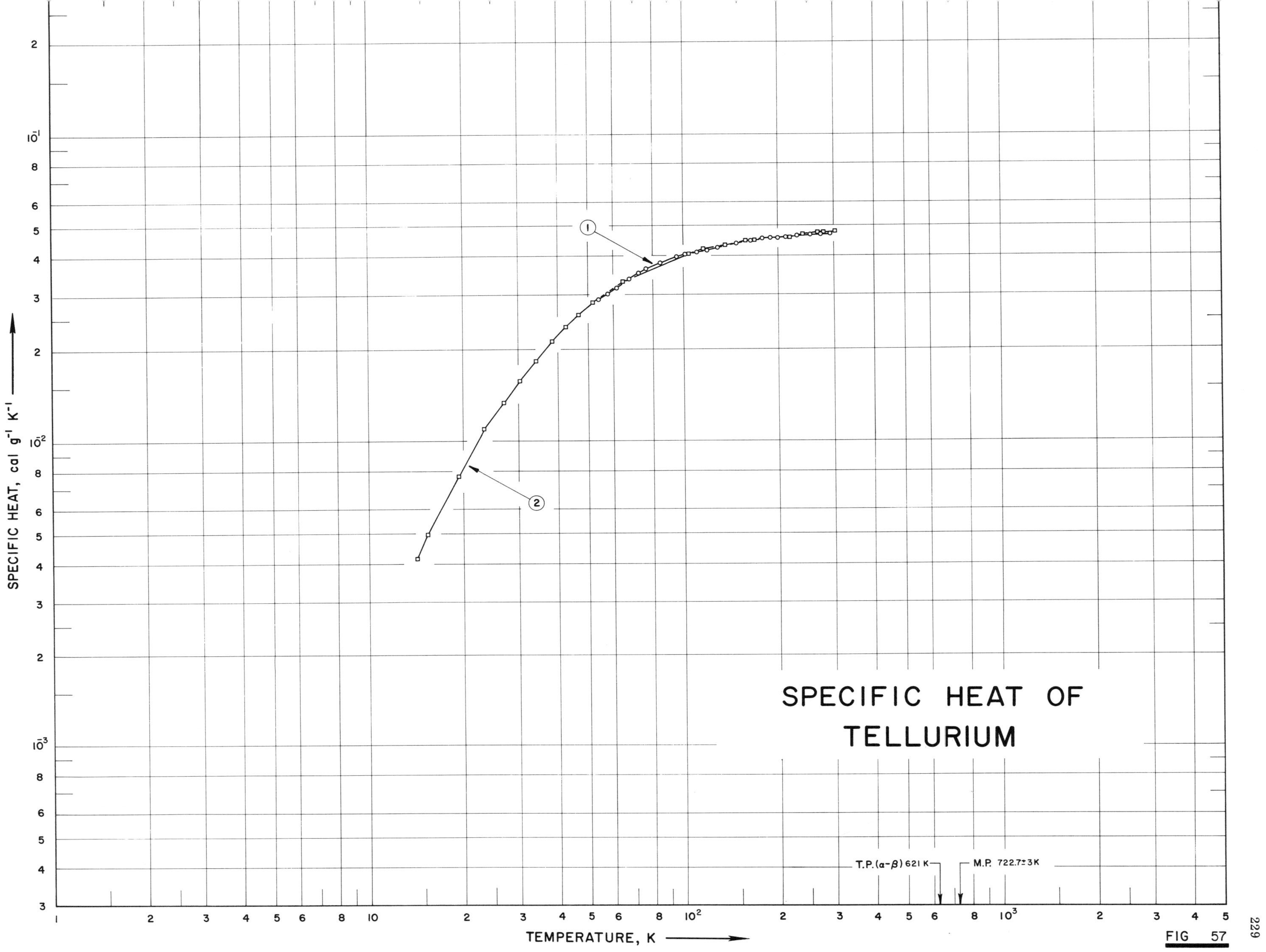
SPECIFIC HEAT OF
TELLURIUM
SPECIFIC HEAT, cal g⁻¹ K⁻¹
TEMPERATURE, K
T.P. (α−β) 621 K
M.P. 722.7±3K
1
2
FIG 57

SPECIFICATION TABLE NO. 57 SPECIFIC HEAT OF TELLURIUM

(Impurity $< 0.20\%$ each; total impurities $< 0.50\%$)

[For Data Reported in Figure and Table No. 57]

Curve No.	Ref. No.	Year	Temp. Range, K	Reported Error, %	Name and Specimen Designation	Composition (weight percent), Specifications and Remarks
1	358	1937	54-292			0.2 Se.
2	367	1939	14-301			0.2 Se (estimated).

DATA TABLE NO. 57 SPECIFIC HEAT OF TELLURIUM

[Temperature, T,K; Specific Heat, C_p, Cal $g^{-1}K^{-1}$]

T	C_p
CURVE 1	
54.0	2.922×10^{-2}
57.6	3.044
61.8	3.184
67.4	3.399
72.3	3.555
76.0	3.665
84.9	3.823
95.1	4.013
101.8	4.063
111.6	4.161
119.5	4.218
128.5	4.294
147.9	4.431
164.4	4.504
178.4	4.591
189.4	4.597
198.0	4.605
211.8	4.617
229.5	4.697
250.3	4.728
272.6	4.723
292.0	4.744
CURVE 2	
14.22	4.201×10^{-3}
15.40	5.031
19.35	7.844
23.39	1.114×10^{-2}
26.91	1.362
30.21	1.590
34.04	1.853
38.32	2.149
42.41	2.389
46.82	2.618
51.75	2.862
57.71	3.116*
64.42	3.329
104.75	4.096
117.25	4.271
127.15	4.326*
137.04	4.381
147.38	4.428*
157.85	4.514
168.48	4.538
CURVE 2 (cont.)	
178.48	4.569×10^{-2}*
188.95	4.592*
200.44	4.634*
212.10	4.655*
218.34	4.647
223.71	4.679*
228.62	4.679*
239.80	4.749
250.12	4.820*
258.28	4.781*
259.64	4.757*
267.09	4.796
278.75	4.804
289.80	4.828*
301.27	4.820

* Not shown on plot

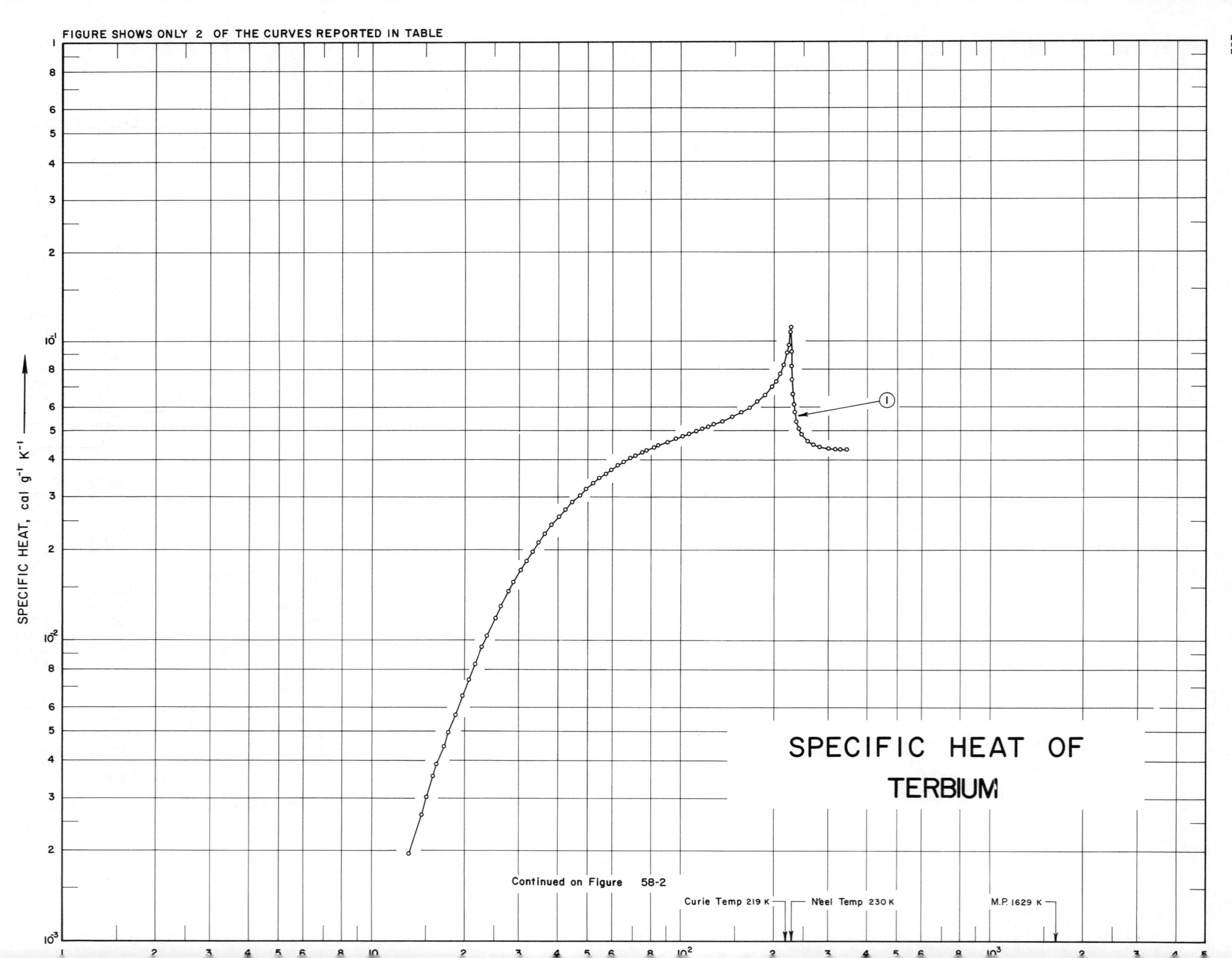

FIGURE SHOWS ONLY 2 OF THE CURVES REPORTED IN TABLE
SPECIFIC HEAT, cal g^{-1} K^{-1}
1
SPECIFIC HEAT OF TERBIUM
Continued on Figure 58-2
Curie Temp 219 K
Néel Temp 230 K
M.P. 1629 K

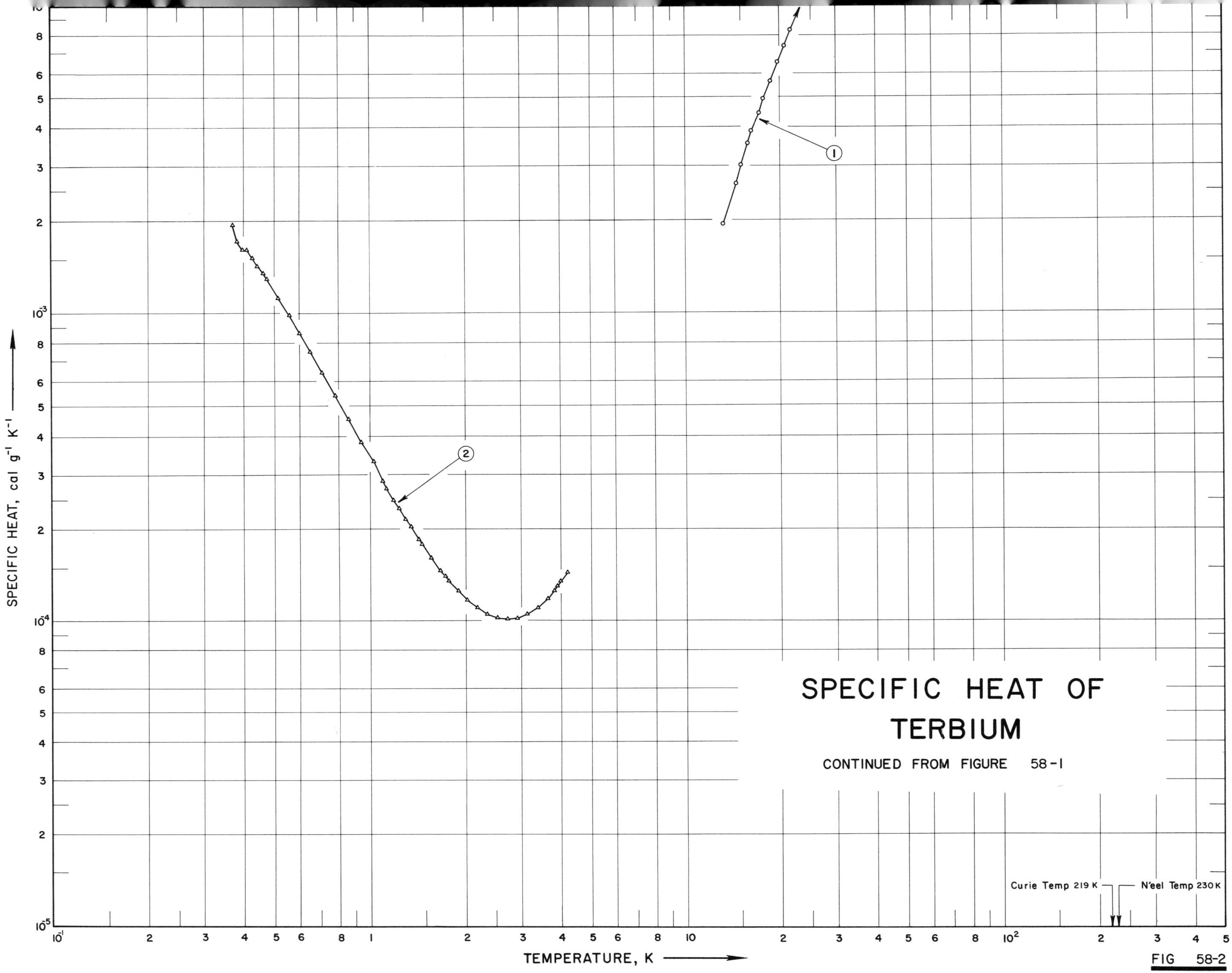

SPECIFIC HEAT OF TERBIUM
CONTINUED FROM FIGURE 58-1
Curie Temp 219 K
N'eel Temp 230 K
1
2
SPECIFIC HEAT, cal g^{-1} K^{-1}
TEMPERATURE, K
10^{-5}
2
3
4
5
6
8
10^{-4}
2
3
4
5
6
8
10^{-3}
2
3
4
5
6
8
10^{-1}
2
3
4
5
6
8
1
2
3
4
5
6
8
10
2
3
4
5
6
8
10^{2}
2
3
4
5

FIG 58-2

SPECIFICATION TABLE NO. 58 SPECIFIC HEAT OF TERBIUM

(Impurity $< 0.20\%$ each; total impurities $< 0.50\%$)

[For Data Reported in Figure and Table No. 58]

Curve No.	Ref. No.	Year	Temp. Range, K	Reported Error, %	Name and Specimen Designation	Composition (weight percent), Specifications and Remarks
1	118	1957	13-347	± 0.1		
2	89	1962	0.4-4	< 2.0		0.14 C, 0.12 O_2, 0.02 H_2, and 0.01 N_2; sample supplied by the Nuclear Corp. of America; vacuum distilled.
3	220	1964	0.05-0.9			Hexagonal closed packed.
4	301	1966	298-1900	< 2		0.06 Ca, 0.05 Si, 0.04 Fe, 0.025 Mg, 0.01 each Al, Ni, N_2, 0.004 Cu, 0.003 O_2, and 0.001 Cr; prepared by metallothermic reduction of the fluoride with calcium and purified by distillation.

DATA TABLE NO. 58 SPECIFIC HEAT OF TERBIUM

[Temperature, T, K; Specific Heat, C_p, Cal g^{-1} K^{-1}]

T	C_p
CURVE 1	
Series 1	
13.33	1.95×10^{-3}
14.57	2.636
15.88	3.549
17.26	4.449
18.76	5.644
20.72	7.393
22.85	9.457
25.21	1.182×10^{-2}
27.89	1.451
30.50	1.710
33.22	1.971
36.42	2.259
40.42	2.581
44.76	2.880
49.50	3.184
54.54	3.453
59.90	3.698
65.61	3.923
71.43	4.106
77.81	4.291
84.54	4.461
90.44	4.589
96.16	4.691
101.73	4.791
107.18	4.884
112.52	4.961
117.75	5.046
122.89	5.143
126.37	5.200*
127.62	5.218*
128.23	5.226
131.26	5.266*
136.66	5.369
141.25	5.477*
147.16	5.547
152.27	5.641*
157.30	5.739*
162.25	5.843
167.39	5.966*
172.73	6.102
177.05	6.240

T	C_p
CURVE 1 (cont.)	
Series 1 (cont.)	
183.05	6.396×10^{-2}*
188.31	6.561
193.70	6.794*
198.93	6.994
204.48	7.288
210.29	7.689
215.78	8.242
220.86	9.078
224.81	9.690
226.78	1.077×10^{-1}
227.68	1.118
228.65	9.129×10^{-2}
229.81	7.393
231.15	6.603
232.62	6.105
234.18	5.772
235.81	5.534*
237.49	5.353
239.20	5.209*
241.87	5.044
246.55	4.848
252.52	4.691*
258.65	4.591
263.39	4.533*
269.47	4.478
275.60	4.435
281.76	4.404
287.93	4.380*
294.09	4.363*
300.27	4.353
Series 2	
15.08	3.014×10^{-3}
16.37	3.889
17.87	4.946
19.77	6.525
21.68	8.312
23.76	1.036×10^{-2}
26.30	1.291
28.99	1.560

T	C_p
CURVE 1 (cont.)	
Series 2 (cont.)	
31.80	1.835×10^{-2}
34.79	2.114
38.43	2.422
42.45	2.731
47.13	3.038
52.20	3.330
57.37	3.581
62.93	3.821
68.98	4.029
75.15	4.212
82.08	4.398
87.90	4.537*
93.53	4.644*
99.02	4.742*
104.39	4.835*
109.27	4.918*
114.42	5.004*
120.20	5.097*
125.85	5.192*
130.72	5.265*
134.82	5.338*
139.53	5.416*
144.51	5.500*
149.40	5.589*
154.85	5.694*
160.21	5.802*
165.15	5.911*
169.99	6.030*
175.03	6.163*
180.25	6.312*
201.72	7.141*
207.46	7.486*
211.80	7.827*
214.02	8.038*
215.21	8.154*
216.39	8.333*
218.35	8.682*
220.48	9.063*

T	C_p
CURVE 1 (cont.)	
Series 3	
220.32	9.124×10^{-2}*
221.73	9.142*
223.28	9.333*
224.53	9.590*
225.56	9.953*
226.56	1.054×10^{-1}*
227.33	1.118*
227.68	1.124*
227.84	1.116*
228.10	1.079*
228.30	1.029*
228.45	9.821×10^{-2}*
228.60	9.488*
229.22	8.187*
230.40	7.036*
231.70	6.412*
233.34	5.901*
235.06	5.641*
236.59	5.438*
238.30	5.283*
240.19	5.141*
244.00	4.945*
249.80	4.756*
255.75	4.632*
261.80	4.550*
266.15	4.504*
272.33	4.459*
278.52	4.420*
284.74	4.391*
290.98	4.370*
297.22	4.353*
303.44	4.342*
309.67	4.336*
315.88	4.326
322.09	4.323*
328.28	4.320

T	C_p
CURVE 1 (cont.)	
Series 4*	
226.14	1.024×10^{-1}
227.12	1.102
227.74	1.129
228.05	1.089
228.55	9.482×10^{-2}
231.46	6.519
236.43	5.462
241.66	5.054
246.90	4.835
Series 5*	
210.91	7.732×10^{-2}
212.81	7.914
215.65	8.234
218.33	8.664
219.39	8.870
219.78	8.970
220.23	9.032
220.84	9.112
221.46	9.137
221.99	9.170
223.65	9.364
225.64	9.660
226.65	1.061×10^{-1}
227.21	1.111
227.43	1.128
227.60	1.133
227.77	1.131
227.93	1.110
228.53	9.583
Series 6*	
220.69	8.721×10^{-2}
220.94	8.799
221.42	8.938
221.77	9.030
222.90	9.217
224.62	9.583
225.96	1.010×10^{-1}

T	C_p
CURVE 1 (cont.)	
Series 6 (cont.)*	
226.92	1.082×10^{-1}
227.44	1.125
Series 7*	
201.91	7.140
205.71	7.359
209.07	7.588
211.42	7.786
213.56	7.985
215.65	8.212
217.53	8.517
219.33	8.883
220.50	9.083
221.07	9.121
221.64	9.142
222.20	9.161
Series 8*	
226.30	1.026×10^{-1}
226.80	1.066
Series 9*	
203.85	7.245×10^{-2}
209.68	7.634
215.19	8.163
218.61	8.730
219.59	8.938
220.26	9.059
Series 10*	
221.00	9.128×10^{-2}
221.43	9.125
221.84	9.142
222.48	9.200
223.34	9.280
225.29	9.764

*Not shown on plot

DATA TABLE NO. 58 (Continued)

T	C_p
CURVE 1 (cont.)	
Series 11*	
219.91	8.668 x 10^{-2}
220.36	8.791
220.81	8.910
221.47	9.035
222.98	9.228
225.08	9.692
Series 12	
319.40	4.331 x 10^{-2}*
326.05	4.330*
332.68	4.335*
339.27	4.331*
345.86	4.318
Series 13*	
307.81	4.331 x 10^{-2}
314.61	4.323
321.40	4.316
328.18	4.307
334.94	4.314
340.84	4.313
346.75	4.314
CURVE 2	
Series 1A	
0.4796	1.304 x 10^{-3}
0.5202	1.135
0.5625	9.907 x 10^{-4}
0.6069	8.656
0.6580	7.488
0.7173	6.399
0.7873	5.397
0.8659	4.534
0.9492	3.832
1.0425	3.231
1.1454	2.735
1.2498	2.352
1.3670	2.039
1.4755	1.793
1.5780	1.622

T	C_p
CURVE 2 (cont.)	
Series 1A (cont.)	
1.6808	1.482 x 10^{-4}
1.7927	1.364
1.9140	1.263
2.0441	1.179
2.1970	1.109
2.3723	1.056
2.5540	1.026
2.7391	1.017
2.9388	1.024
3.1645	1.053
3.4101	1.107
3.6685	1.189
3.9428	1.303
4.2322	1.451
Series 1B	
0.4681	1.360 x 10^{-3}*
0.4872	1.269*
0.5077	1.182*
0.5297	1.098*
0.5534	1.018*
0.5791	9.426 x 10^{-4}*
0.6070	8.645*
0.6373	7.922*
0.6705	7.229*
0.7069	6.569*
0.7472	5.931*
0.7918	5.335*
0.8416	4.773*
0.8974	4.245*
0.9603	3.750*
1.0318	3.292*
1.1134	2.874
1.2071	2.493
1.3152	2.156
1.4402	1.860
1.5848	1.608*
1.7511	1.403
1.9398	1.243*
2.1498	1.129*
2.3771	1.055*
2.6154	1.020*
2.8579	1.015*

T	C_p
CURVE 2 (cont.)	
Series 1B (cont.)	
3.0973	1.041 x 10^{-4}*
3.3503	1.090*
3.6120	1.163*
3.8550	1.256
4.0787	1.360
Series 2	
0.3742	1.954 x 10^{-3}
0.3873	1.732
0.4011	1.635
0.4158	1.635
0.4314	1.537
0.4480	1.442
0.4658	1.343*
0.4847	1.262*
0.5050	1.174*
CURVE 3*	
0.0459	6.11 x 10^{-3}
0.0485	6.30
0.0508	6.86
0.0549	7.07
0.0584	7.23
0.0667	8.32
0.0742	8.38
0.0877	8.42
0.0982	7.65
0.101	7.64
0.113	7.56
0.120	7.53
0.130	7.49
0.134	7.47
0.138	7.16
0.149	7.10
0.149	6.83
0.162	6.42
0.178	5.88
0.190	5.73
0.201	5.05
0.213	4.93
0.231	4.53
0.248	4.02

T	C_p
CURVE 3 (cont.)*	
0.264	3.71 x 10^{-3}
0.279	3.40
0.294	3.22
0.312	2.734
0.331	2.602
0.355	2.340
0.387	2.078
0.425	1.770
0.449	1.504
0.483	1.16
0.544	1.01
0.619	8.02 x 10^{-4}
0.750	6.15
0.894	4.47
CURVE 4*	
298.15	4.15 x 10^{-2}
300	4.15
400	4.26
500	4.39
600	4.54
700	4.71
800	4.90
900	5.11
1000	5.34
1100	5.59
1200	5.86
1300	6.15
1400	6.456
1500	6.783
1560	6.997
1560	4.172
1600	4.172
1630	4.172
1630	6.991
1700	6.991
1800	6.991
1900	6.991

*Not shown on plot

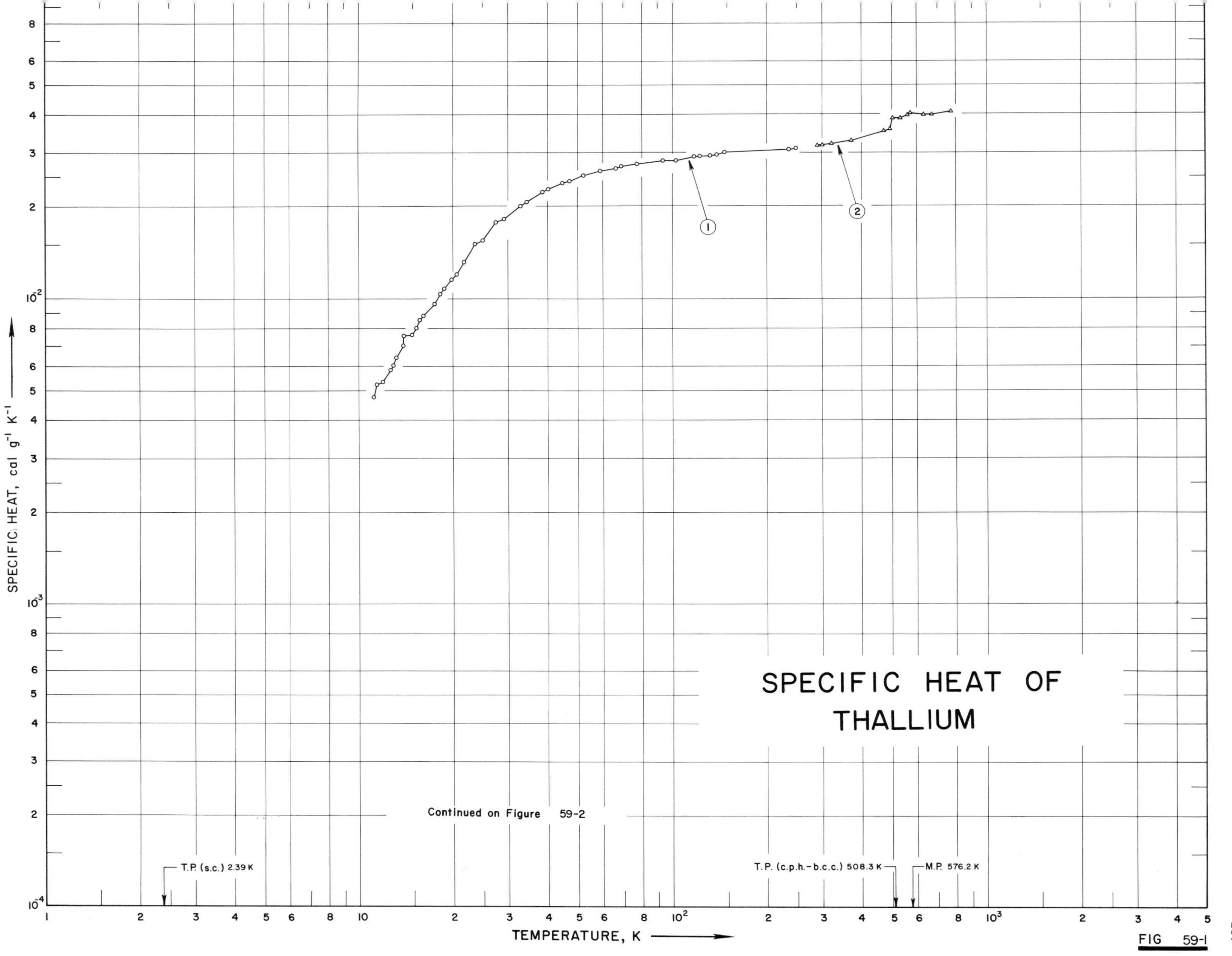
SPECIFIC HEAT OF
THALLIUM
SPECIFIC HEAT, cal g⁻¹ K⁻¹
TEMPERATURE, K
Continued on Figure 59-2
T.P. (s.c.) 2.39 K
T.P. (c.p.h.-b.c.c.) 508.3 K
M.P. 576.2 K
1
2
FIG 59-1

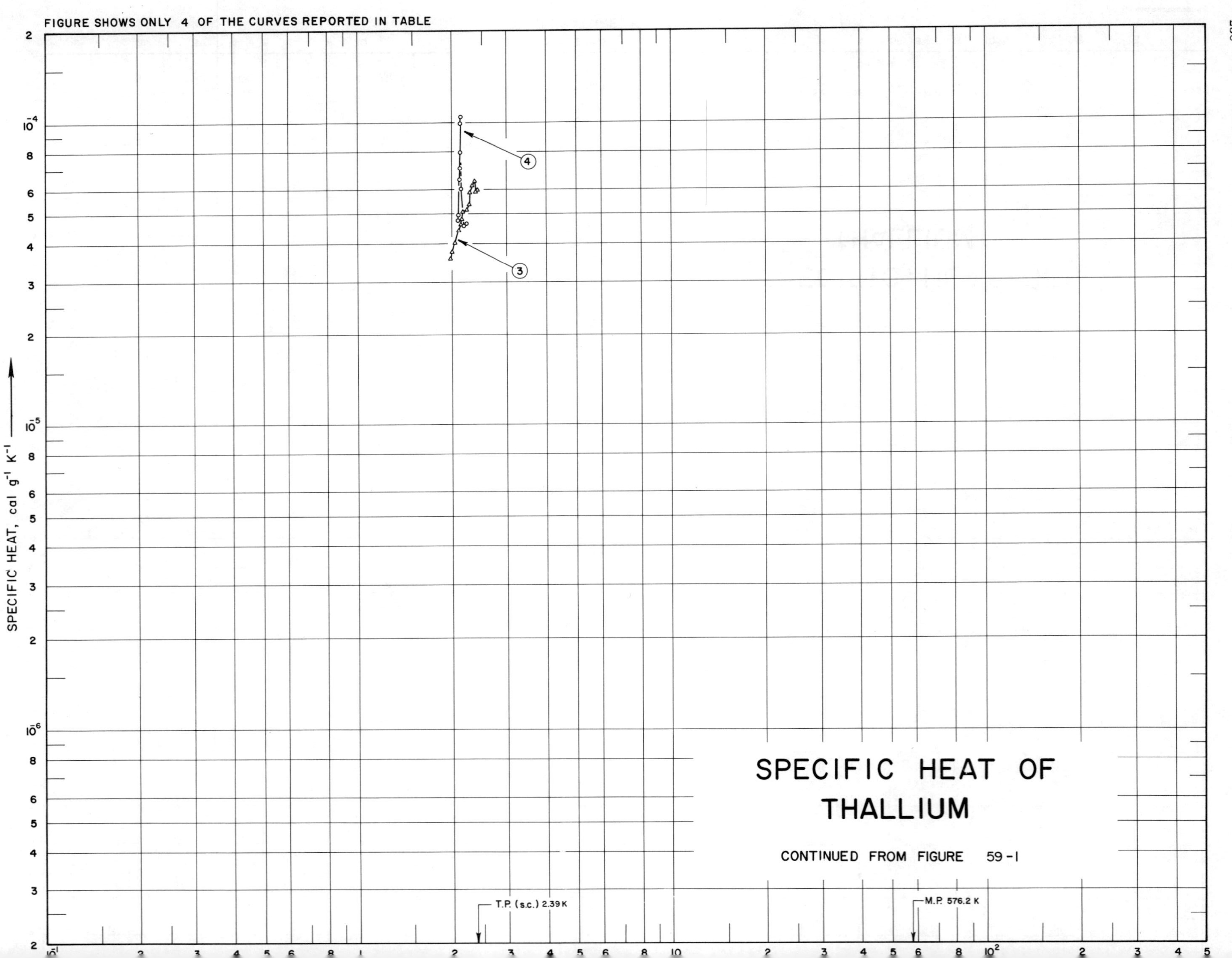

FIGURE SHOWS ONLY 4 OF THE CURVES REPORTED IN TABLE
SPECIFIC HEAT, cal g^{-1} K^{-1}
10^{-4}
10^{-5}
10^{-6}
4
3
T.P. (s.c.) 2.39 K
M.P. 576.2 K
10^{-1}
1
10
10^{2}
SPECIFIC HEAT OF THALLIUM
CONTINUED FROM FIGURE 59-1

SPECIFICATION TABLE NO. 59 SPECIFIC HEAT OF THALLIUM

(Impurity $< 0.20\%$ each; total impurities $< 0.50\%$)

[For Data Reported in Figure and Table No. 59]

Curve No.	Ref. No.	Year	Temp. Range, K	Reported Error, %	Name and Specimen Designation	Composition (weight percent), Specifications and Remarks
1	186	1930	11-249			Traces of Al and Fe.
2	187	1931	291-773	±3.0		
3	188	1934	1.9-2.4			Zero magnetic field.
4	188	1934	2.0-2.2			33.6 gauss magnetic field.
5	368	1938	15-301	≤5.0		~99.98 Tl; sample supplied by the American Smelting and Refining Co.; density = 11.878 g cm^{-3} at 20.7 C.
6	369	1934	1.3-4.1			
7	370	1957	2.4-4.1			99.999 Tl; sample supplied by the A.D. Mackay, Inc.; zero magnetic field; superconducting.
8	370	1957	1.2-4.1			Same as above; 200 oersteds; normal state.
9	370	1957	1.1-2.3			Same as above; superconducting; zero magnetic field.

DATA TABLE NO. 59 SPECIFIC HEAT OF THALLIUM

[Temperature, T, K; Specific Heat, C_p, Cal $g^{-1}K^{-1}$]

CURVE 1

T	C_p
11.18	4.776 x 10^{-3}
11.40	5.231
11.98	5.378
12.64	5.837
12.88	6.038
13.22	6.449
13.94	7.026
13.98	7.594
14.86	7.633
14.93	7.736*
15.32	8.010
15.66	8.529
16.08	8.808
17.50	9.664
17.78	9.830*
18.24	1.042 x 10^{-2}
18.36	1.030*
18.80	1.089
19.86	1.167
20.60	1.202
20.70	1.202*
21.70	1.323
23.60	1.505
25.00	1.554
27.50	1.774
29.20	1.819
33.00	2.009
34.50	2.067
38.70	2.237
40.60	2.288
44.90	2.387
47.20	2.427
52.30	2.532
54.00	2.557*
59.20	2.614
61.40	2.603*
66.10	2.668
69.00	2.706
77.50	2.756
94.00	2.820
96.20	2.801*
103.50	2.838
118.40	2.906
123.20	2.916

CURVE 1 (cont.)

T	C_p
132.80	2.941 x 10^{-2}
139.80	2.955
147.80	3.014
236.60	3.078
246.60	3.102*
249.00	3.102

CURVE 2

T	C_p
291.15	3.16 x 10^{-2}
301.15	3.17
323.15	3.21
373.15	3.30
473.15	3.53
493.15	3.58
503.15	3.90
535.15	3.90
563.15	3.99
573.15	4.06
633.15	4.00
673.15	4.00
773.15	4.11

CURVE 3

T	C_p
1.969	3.551 x 10^{-5}
1.994	3.739
2.034	3.981
2.097	4.404
2.120	4.598
2.140	4.789
2.158	5.030
2.191	5.079*
2.229	5.118
2.262	5.329
2.279	5.828
2.299	5.921*
2.315	6.146
2.332	6.239*
2.352	6.385
2.375	5.886
2.404	5.979

CURVE 4

T	C_p
2.081	4.744 x 10^{-5}
2.091	4.932
2.107	6.444
2.111	7.012
2.115	7.927
2.119	9.899
2.124	1.034 x 10^{-4}
2.130	6.004 x 10^{-5}
2.170	4.557
2.213	4.631

CURVE 5*
Series I

T	C_p
58.27	2.591 x 10^{-2}
60.77	2.625
64.13	2.638
68.05	2.700
72.01	2.733
76.25	2.754
80.60	2.766
83.85	2.786
88.38	2.816
92.81	2.834
97.05	2.858
101.85	2.866
107.10	2.883
112.57	2.894
118.65	2.905
125.24	2.927
131.77	2.925
139.04	2.969
146.73	2.957
154.41	2.976
162.04	2.980
169.85	2.996
178.25	3.001
186.34	3.015
193.64	3.006
201.99	3.007
211.03	3.004
219.11	3.031
227.22	3.031
235.43	3.023
244.28	3.035

CURVE 5 (cont.)*

T	C_p
251.97	3.053 x 10^{-2}
261.01	3.047
269.73	3.039
278.17	3.074
288.34	3.017
298.35	3.109
300.53	3.096

Series II

T	C_p
49.98	2.464 x 10^{-2}
53.13	2.536
57.10	2.567
61.19	2.601
65.07	2.666
139.14	2.933
146.45	2.958
186.23	3.020
214.21	3.024
223.42	3.024
233.72	3.035
242.04	3.064
269.33	3.055
281.50	3.093

Series III

T	C_p
14.56	7.467 x 10^{-3}
17.24	9.864
20.39	1.262 x 10^{-2}
23.99	1.509
27.85	1.742
31.87	1.949
35.99	2.092
39.33	2.231
43.82	2.314
48.04	2.408
52.81	2.495

CURVE 6*
Series I

T	C_p
1.297	1.063 x 10^{-5}
2.012	3.876

CURVE 6 (cont.)*

T	C_p
2.058	4.088 x 10^{-5}
2.073	4.344
2.140	4.702
2.169	4.918
2.188	4.837
2.204	5.241
2.270	5.612
2.291	6.053
2.328	6.268
2.353	6.297
2.898	9.757
3.145	1.199 x 10^{-4}
3.147	1.197
3.348	1.413
3.515	1.640
3.701	1.859
3.943	2.181
4.183	2.632

Series II

T	C_p
1.590	1.813 x 10^{-5}
1.957	3.553
2.098	4.367
2.190	5.059
2.208	5.402
2.230	5.260
2.247	5.661
2.272	5.749
2.288	5.784
2.304	6.092
2.320	5.911
2.335	6.219
2.350	6.503
2.368	5.926
2.384	5.764
2.401	6.121
2.420	5.896
2.439	6.204
2.456	6.493
2.479	6.513
2.500	6.845
2.519	6.973

CURVE 6 (cont.)*
Series III

T	C_p
1.310	1.038 x 10^{-5}
1.338	1.112

CURVE 7*
Series I

T	C_p
2.413	5.63 x 10^{-5}
2.443	5.92
2.481	6.02
2.575	6.80
2.659	7.58
2.706	7.73
2.766	8.42
2.814	8.81
2.881	9.20
2.949	9.93
2.977	1.01 x 10^{-4}
2.977	1.02
3.030	1.08
3.038	1.06
3.080	1.11
3.117	1.18
3.142	1.17
3.196	1.26
3.203	1.23
3.258	1.30
3.321	1.40
3.406	1.49
3.447	1.51
3.521	1.61
3.585	1.74
3.649	1.80
3.677	1.83
3.739	1.89
3.794	2.02
3.841	2.10
3.895	2.19
3.938	2.20
3.983	2.32
4.043	2.41
4.143	2.60

*Not shown on plot

DATA TABLE NO. 59 (continued)

T	C_p
CURVE 7 (cont.)*	
Series II	
2.396	5.53 x 10^{-5}
2.407	5.58
2.438	5.97
2.482	6.07
2.488	6.12
2.566	6.70
2.570	6.80
2.649	7.49
2.654	7.24
2.740	7.98
2.835	8.71
2.892	9.30
2.931	9.69
2.967	1.02 x 10^{-4}
3.012	1.03
3.054	1.09
3.101	1.13
3.188	1.25
3.302	1.37
3.425	1.51
3.568	1.69
3.655	1.79
3.824	2.05
3.933	2.28
4.054	2.50
4.186	2.79
CURVE 8*	
1.224	9.93 x 10^{-6}
1.230	1.01 x 10^{-5}
1.239	9.69 x 10^{-6}
1.247	1.03 x 10^{-5}
1.254	1.05
1.257	1.02
1.314	1.13
1.336	1.19
1.377	1.27
1.406	1.34
1.469	1.54
1.519	1.64
1.539	1.74
1.542	1.71
1.561	1.78
1.567	1.81
CURVE 8 (cont.)*	
1.568	1.80 x 10^{-5}
1.587	1.95
1.622	1.97
1.634	2.01
1.657	2.06
1.667	2.11
1.710	2.24
1.839	2.81
1.849	2.75
1.887	2.93
1.966	3.19
2.020	3.50
2.032	3.59
2.112	4.03
2.131	4.08
2.203	4.49
2.211	4.54
2.313	5.19
2.353	5.33
2.408	5.68
2.465	5.77
2.492	6.31
2.543	6.46
2.567	6.85
2.591	6.80
2.701	7.73
2.829	8.71
3.001	1.04 x 10^{-4}
3.095	1.11
3.194	1.25
3.306	1.39
3.443	1.52
3.576	1.68
3.660	1.84
3.728	1.95
3.841	2.14
3.921	2.29
4.008	2.48
4.187	2.71
CURVE 9*	
Series I	
1.160	8.71 x 10^{-6}
1.161	7.88
1.180	8.01
CURVE 9 (cont.)*	
1.194	9.00 x 10^{-6}
1.274	1.09 x 10^{-5}
1.294	1.14
1.296	1.14
1.312	1.16
1.365	1.29
1.377	1.44
1.396	1.49
1.438	1.52
1.465	1.61
1.485	1.66
1.485	1.70
1.504	1.71
1.547	1.83
1.548	1.84
1.644	2.15
1.693	2.46
1.733	2.55
1.775	2.74
1.804	2.86
1.839	2.99
1.845	2.99
1.887	3.20
1.900	3.20
1.942	3.44
1.976	3.60
1.987	3.58
2.055	3.97
2.063	3.99
2.118	4.24
2.153	4.53
2.177	4.88
2.217	5.06
2.233	5.18
2.244	5.28
2.244	5.42
2.251	5.23
2.276	5.32
2.284	5.62
2.307	5.64
2.332	5.94
Series II	
1.150	8.42 x 10^{-6}
1.150	8.71
CURVE 9 (cont.)*	
1.206	9.20 x 10^{-6}
1.230	9.93
1.281	1.14 x 10^{-5}
1.324	1.21
1.397	1.43
1.487	1.66
1.540	1.87
1.565	1.96
1.620	2.11
1.658	2.24
1.675	2.33
1.758	2.63
1.798	2.81
1.888	3.20
1.937	3.39
1.988	3.55
2.099	4.37
2.216	5.13
2.224	5.14
2.282	5.65
2.309	5.82
2.342	5.73

* Not shown on plot

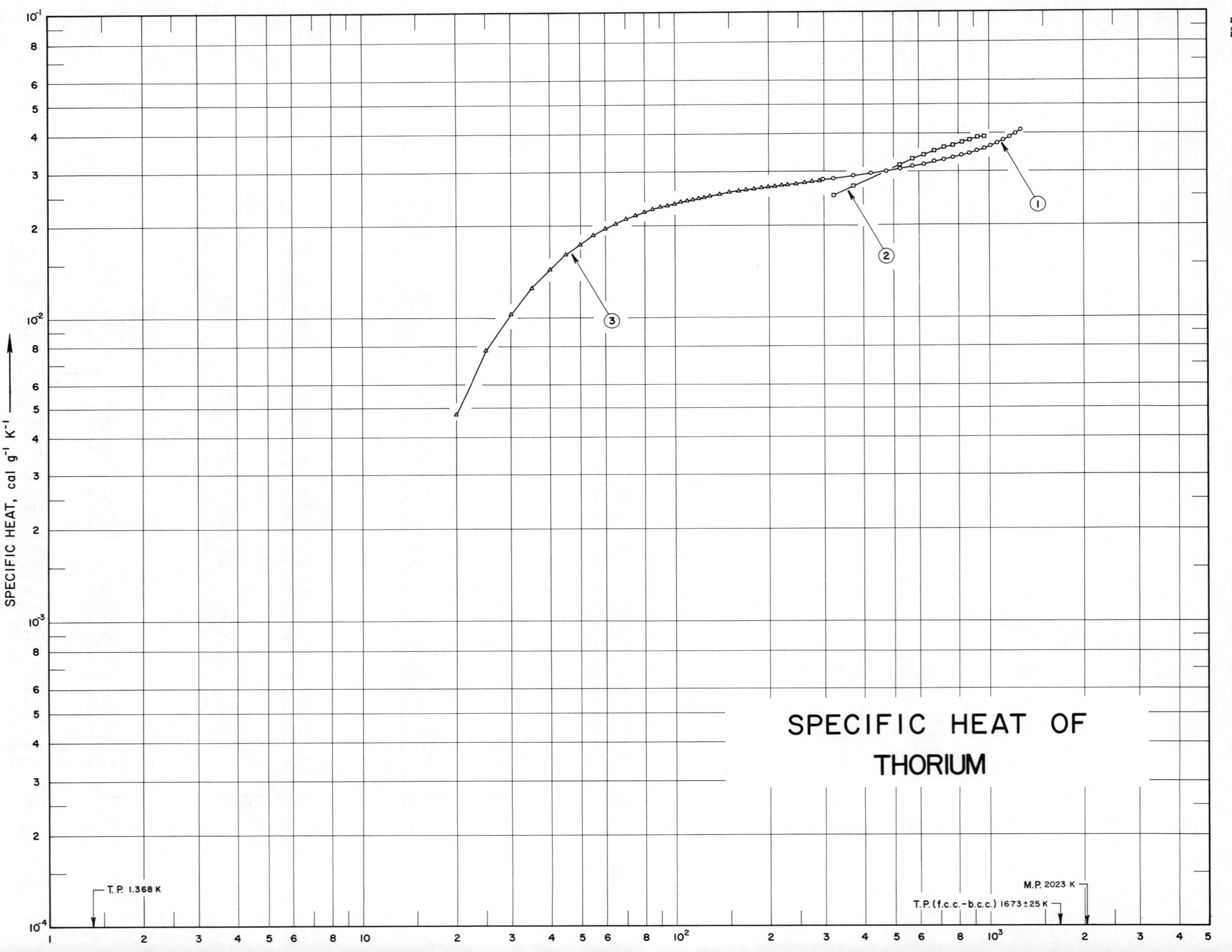

SPECIFIC HEAT OF
THORIUM
SPECIFIC HEAT, cal g⁻¹ K⁻¹
T.P. 1.368 K
M.P. 2023 K
T.P. (f.c.c.-b.c.c.) 1673 ± 25 K
1
2
3

SPECIFICATION TABLE NO. 60 SPECIFIC HEAT OF THORIUM

(Impurity < 0.20% each; total impurities < 0.50%)

[For Data Reported in Figure and Table No. 60]

Curve No.	Ref. No.	Year	Temp. Range, K	Reported Error, %	Name and Specimen Designation	Composition (weight percent), Specifications and Remarks
1	119	1960	308-1334	2.0	Crystal bar	0.02 Zr, 0.0075 C, 0.00583 O_2, < 0.005 Si, 0.003 Al, 0.002 Ca, < 0.002 each Be, Fe, Mg, Mn, and Ni, 0.00035 H_2, and 0.00033 N; annealed at 100 C for at least one hr. under 10^{-5} mm Hg pressure; cooled to room temperature at 40 C per hr.; arc melted; cleaned with hot nitric acid in sodium fluosilicate.
2	7	1959	323-973	≤ 2.0		99.81 Th.
3	120	1953	20-300			0.06 O_2, 0.04 N_2, 0.025 Si, and < 0.01 other metals.

DATA TABLE NO. 60 SPECIFIC HEAT OF THORIUM

[Temperature, T, K; Specific Heat, C_p, Cal g^{-1} K^{-1}]

T	C_p
CURVE 1	
298.15	2.83×10^{-2}
323.15	2.85
373.15	2.91
423.15	2.96
473.15	3.01
523.15	3.07
573.15	3.12
623.15	3.17
673.15	3.23
723.15	3.28
773.15	3.33
823.15	3.38
873.15	3.44
923.15	3.50
973.15	3.57
1023.15	3.64
1073.15	3.71
1123.15	3.80
1173.15	3.89
1223.15	3.99
1273.15	4.11
CURVE 2	
323.15	2.50×10^{-2}
373.15	2.68
523.15	3.15
573.15	3.30
623.15	3.40
673.15	3.50
723.15	3.60
773.15	3.65
823.15	3.74
873.15	3.80
923.15	3.88
973.15	3.90
CURVE 3	
20	4.765×10^{-3}
25	7.759
30	1.033×10^{-2}
35	1.254
40	1.445
45	1.608
50	1.744
55	1.856
60	1.951
65	2.035
70	2.101
75	2.159
80	2.211
85	2.261
90	2.301
95	2.333
100	2.362
105	2.390
110	2.416
115	2.440
120	2.463
125	2.484
130	2.505
135	2.523*
140	2.540
145	2.555*
150	2.570
155	2.583*
160	2.596
165	2.609*
170	2.621
175	2.632*
180	2.642
185	2.652*
190	2.661
195	2.670*
200	2.678
205	2.687*
210	2.695
215	2.703*
220	2.711
225	2.718*
230	2.727
CURVE 3 (cont.)	
235	2.734×10^{-2}*
240	2.741*
245	2.748
250	2.754*
255	2.760*
260	2.767
265	2.773*
270	2.780*
275	2.786
280	2.793*
285	2.799*
290	2.805
295	2.811*
298.16	2.814*
300	2.817*

* Not shown on plot

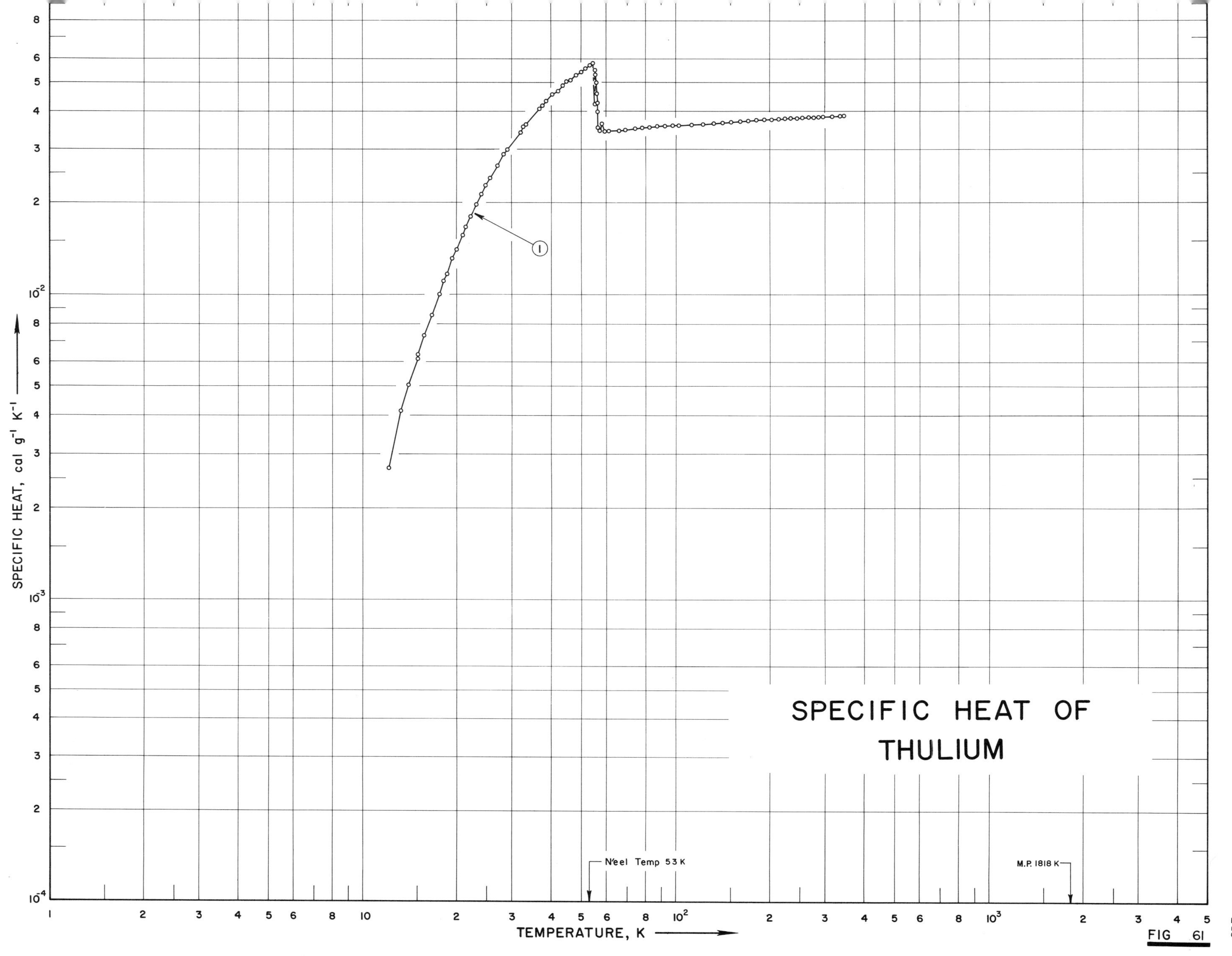
SPECIFIC HEAT OF
THULIUM
SPECIFIC HEAT, cal g⁻¹ K⁻¹
TEMPERATURE, K
Néel Temp 53 K
M.P. 1818 K
1
FIG 61

SPECIFICATION TABLE NO. 61 SPECIFIC HEAT OF THULIUM

(Impurity < 0.20% each; total impurities < 0.50%)

[For Data Reported in Figure and Table No. 61]

Curve No.	Ref. No.	Year	Temp. Range, K	Reported Error, %	Name and Specimen Designation	Composition (weight percent), Specifications and Remarks
1	126	1961	12-352	0.1		0.4 Ta, 0.02 - 0.005 Fe, <0.01 Mg, and 0.02 rare earth; cast and machined; data corrected for impurities.
2	301	1966	298-1900	<2.0		0.05 Mg, 0.02 Ca, 0.02 Cr, 0.01 Fe, and 0.0002 Si; prepared by metallothermic reduction of the fluorides with calcium and purified by distillation.

DATA TABLE NO. 61 SPECIFIC HEAT OF THULIUM

[Temperature, T, K; Specific Heat, C_p, Cal $g^{-1} K^{-1}$]

T	C_p
CURVE 1	
Series 1	
12.20	2.701×10^{-3}
13.36	4.152
15.04	6.142
16.74	8.559
18.71	1.174×10^{-2}
20.99	1.563
23.19	1.965
25.66	2.395
29.18	2.979
33.42	3.602
37.82	4.182
42.18	4.669
46.35	5.072
50.21	5.397
53.85	5.669
58.32	3.637
61.76	3.433
69.29	3.459
74.54	3.482
82.50	3.524
92.98	3.558
103.21	3.577
113.28	3.598
123.17	3.619
132.92	3.637
141.74	3.651
151.26	3.666
161.08	3.682
171.19	3.709
182.05	3.738
193.33	3.746
204.08	3.752
214.76	3.759
224.34	3.766
234.80	3.773
245.38	3.780
255.81	3.786
266.16	3.806
276.43	3.804
286.66	3.811
CURVE 1 (cont.)	
Series 1 (cont.)	
296.84	3.821×10^{-2}
306.95	3.830*
317.01	3.836
327.03	3.845*
336.99	3.852
346.89	3.860
Series 2	
14.08	5.014×10^{-3}
15.80	7.327
17.76	1.007×10^{-2}
20.04	1.410
22.23	1.802
24.93	2.277
28.49	2.883
32.94	3.542
38.69	4.301
41.74	4.639*
45.54	5.006
50.09	5.387*
53.08	5.622*
53.58	5.660*
54.08	5.696*
54.57	5.722
55.12	5.698*
56.20	4.569
56.96	3.507
57.81	3.437
58.67	3.422*
59.62	3.419
61.54	3.421*
66.60	3.447
71.65	3.468*
78.77	3.503
87.89	3.546
98.23	3.567
109.44	3.590*
120.43	3.612*
131.27	3.632*
CURVE 1 (cont.)	
Series 2 (cont.)	
139.30	3.645×10^{-2}*
147.23	3.658*
157.71	3.675*
167.47	3.700*
177.67	3.734*
188.18	3.742*
199.18	3.747*
209.97	3.754*
220.16	3.762*
230.29	3.769*
240.89	3.775*
251.43	3.793*
261.91	3.791*
272.32	3.799*
282.69	3.808*
291.21	3.813*
301.47	3.821*
311.66	3.829*
321.80	3.838*
331.87	3.846*
341.91	3.853*
351.88	3.863*
Series 3	
51.95	5.511×10^{-2}
52.79	5.59*
53.60	5.657*
54.26	5.702*
54.68	5.722*
54.97	5.728*
55.31	5.660*
55.82	5.281
56.51	3.967
Series 4	
15.09	6.345×10^{-3}
16.62	8.558*
18.21	1.110
CURVE 1 (cont.)	
Series 4 (cont.)	
19.46	1.316×10^{-2}
21.46	1.666
24.03	2.120
27.12	2.641
32.06	3.395
36.94	4.079
40.64	4.523
43.96	4.868
48.25	5.244
51.47	5.504*
52.54	5.583*
53.44	5.649*
53.81	5.678*
54.25	5.705*
54.72	5.733*
55.19	5.707*
55.50	4.211
55.73	5.450*
56.04	4.935
56.36	4.263
56.72	3.539*
Series 5*	
49.40	5.311×10^{-2}
49.88	5.351
Series 6*	
53.05	5.599×10^{-2}
53.54	5.637
54.03	5.671
Series 7*	
79.46	3.499×10^{-2}
80.43	3.505
81.97	3.521
83.52	3.529
85.04	3.534
CURVE 1 (cont.)	
Series 7 (cont.)*	
86.56	3.542×10^{-2}
88.07	3.548
89.58	3.551
91.08	3.552
92.58	3.556
Series 8*	
166.50	3.682×10^{-2}
167.91	3.688
169.99	3.698
172.06	3.704
174.12	3.711
176.18	3.717
178.23	3.722
180.28	3.732
182.32	3.735
184.36	3.738
Series 9*	
83.41	3.527×10^{-2}
86.61	3.546
89.76	3.552
93.57	3.556
97.50	3.564
101.56	3.571
Series 10*	
159.15	3.673×10^{-2}
162.03	3.680
166.09	3.689
170.12	3.704
174.12	3.722
178.10	3.729
182.07	3.735
186.02	3.740
190.11	3.742
194.31	3.744
CURVE 1 (cont.)	
Series 11*	
73.61	3.471×10^{-2}
78.49	3.497
83.01	3.525
87.66	3.544
92.38	3.553
99.49	3.567
109.23	3.588
119.42	3.610
129.63	3.629
Series 12*	
140.51	3.639×10^{-2}
146.65	3.650
152.75	3.659
157.98	3.667
163.13	3.674
168.29	3.684
173.40	3.708
178.47	3.723
185.51	3.733
188.54	3.736
196.15	3.739
206.34	3.746
216.46	3.753
226.52	3.761
Series 13*	
65.14	3.443×10^{-2}
76.92	3.488
82.99	3.519
87.81	3.538
93.69	3.551
99.59	3.559
105.24	3.571
111.02	3.586
116.68	3.607

*Not shown on plot

DATA TABLE NO. 61 (continued)

T	C_p
CURVE 1 (cont.)	
Series 14*	
164.06	3.676×10^{-2}
169.32	3.703
174.60	3.721
179.87	3.726
185.05	3.729
190.21	3.733
195.89	3.745
201.01	3.746
206.11	3.746
211.21	3.747
216.29	3.749
Series 15*	
73.40	3.461×10^{-2}
79.80	3.493
86.05	3.528
92.23	3.544
98.32	3.554
101.34	3.568
113.30	3.581
116.21	3.595
CURVE 2*	
298.15	3.59×10^{-2}
300	3.60
400	3.73
500	3.85
600	3.98
700	4.10
800	4.22
900	4.34
1000	4.45
1100	4.56
1200	4.67
1300	4.78
1400	4.88
1500	4.99
1600	5.09
1700	5.19
1800	5.29

T	C_p
CURVE 2 (cont.)*	
1818	5.30×10^{-2}
1818	5.85
1900	5.85

*Not shown on plot

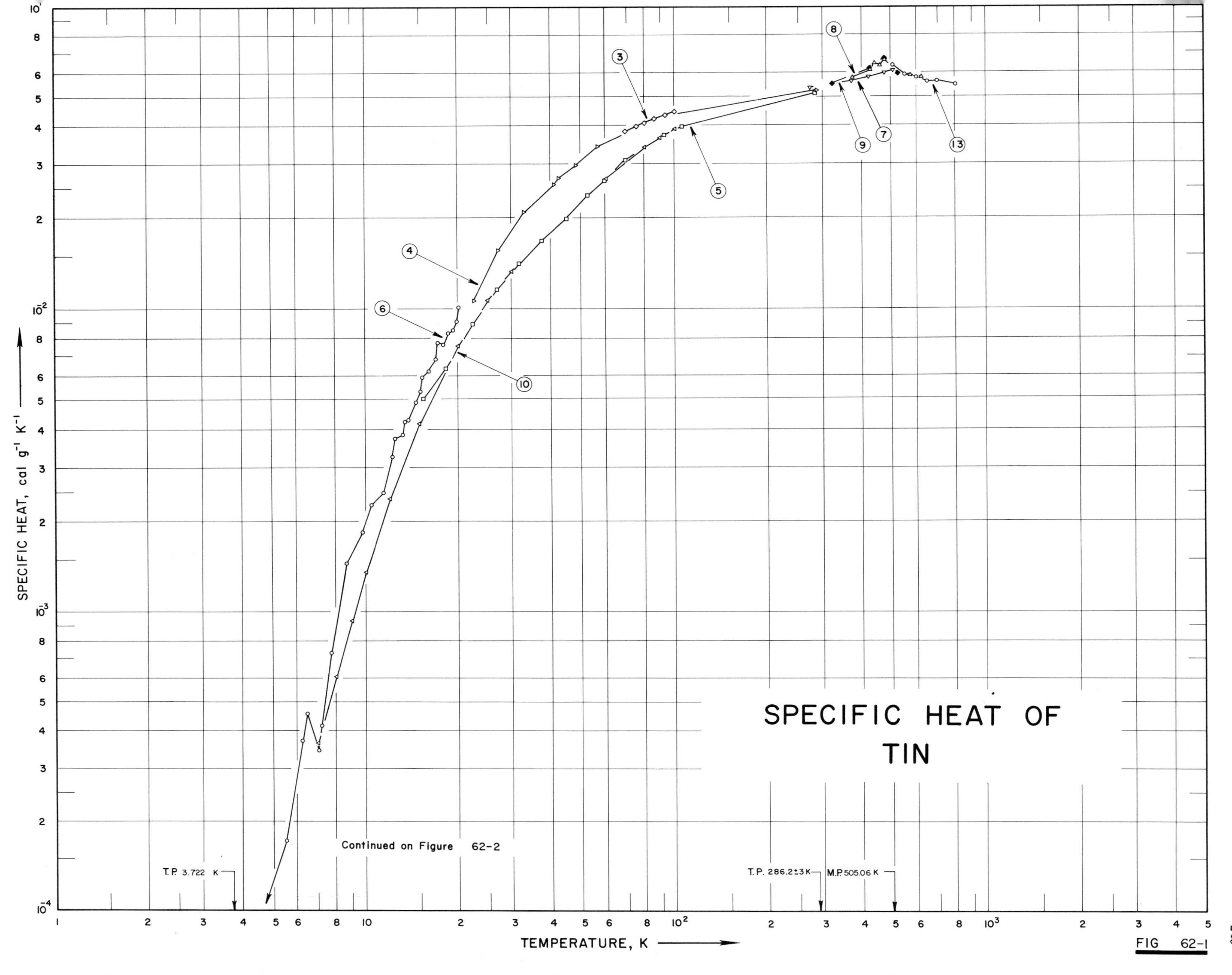
SPECIFIC HEAT OF TIN
SPECIFIC HEAT, cal g⁻¹ K⁻¹
TEMPERATURE, K
T.P. 3.722 K
T.P. 286.2±3 K
M.P. 505.06 K
Continued on Figure 62-2
3
4
5
6
7
8
9
10
13

FIG 62-1

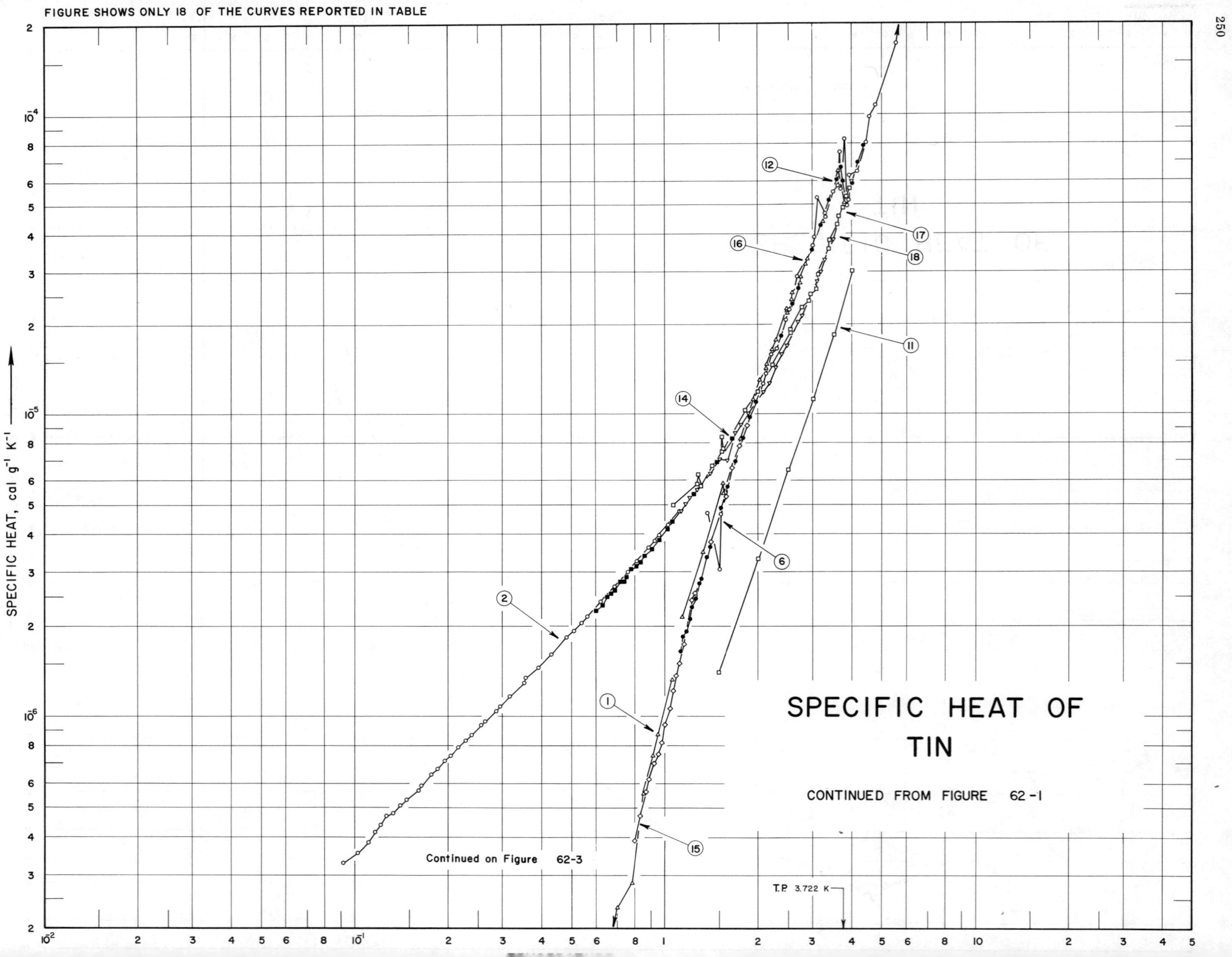
FIGURE SHOWS ONLY 18 OF THE CURVES REPORTED IN TABLE
SPECIFIC HEAT OF
TIN
CONTINUED FROM FIGURE 62-1
Continued on Figure 62-3
T.P. 3.722 K
SPECIFIC HEAT, cal g⁻¹ K⁻¹
12
17
16
18
11
14
6
2
1
15

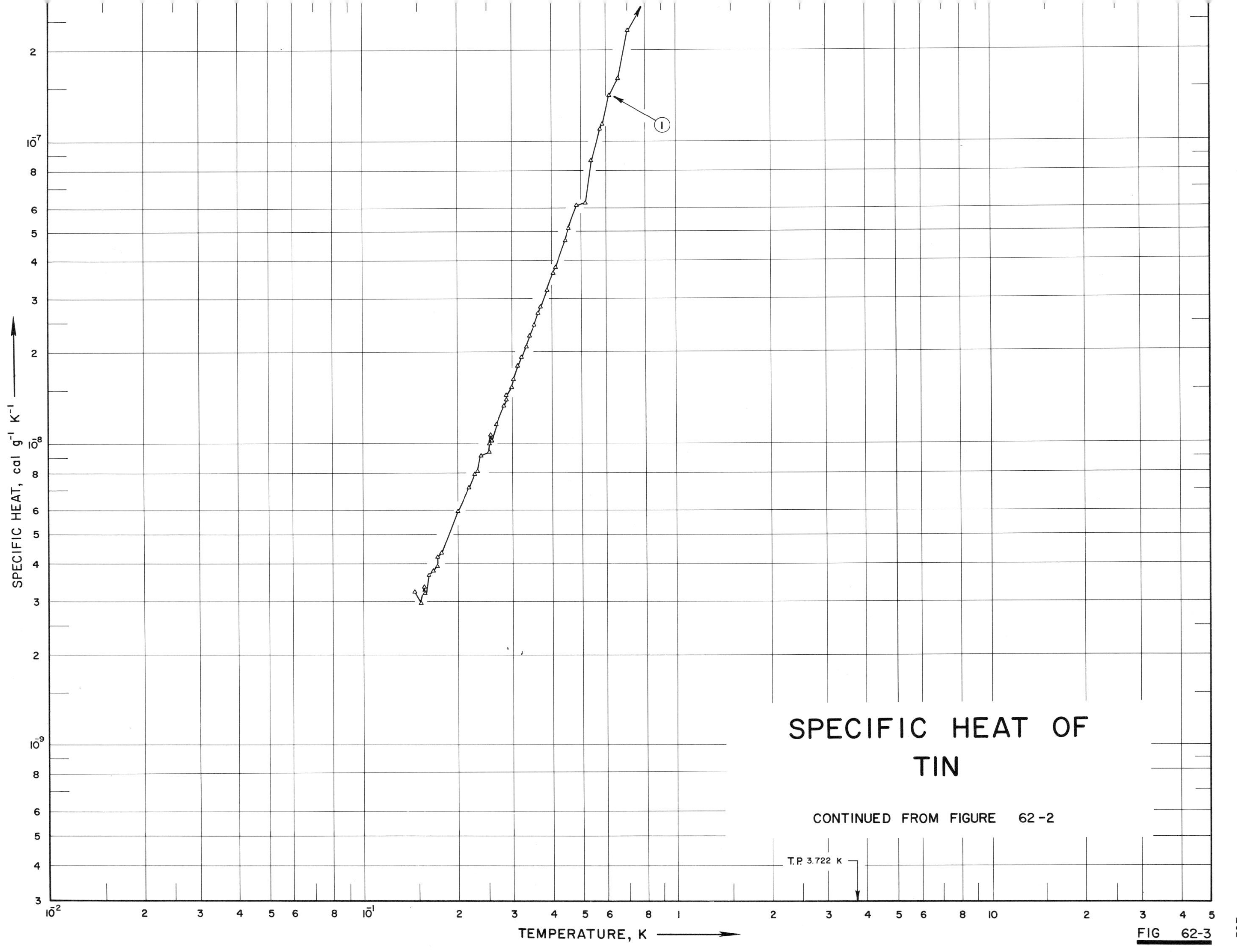
SPECIFIC HEAT OF TIN
CONTINUED FROM FIGURE 62-2
T.P. 3.722 K
1
SPECIFIC HEAT, cal g⁻¹ K⁻¹
TEMPERATURE, K
FIG 62-3

SPECIFICATION TABLE NO. 62 SPECIFIC HEAT OF TIN

(Impurity < 0.20% each; total impurities < 0.50%)

[For Data Reported in Figure and Table No. 62

Curve No.	Ref. No.	Year	Temp. Range, K	Reported Error, %	Name and Specimen Designation	Composition (weight percent), Specifications and Remarks
1	85	1965	0.1-1.0	1.0		99.9999 Sn; polycrystalline; sample supplied by Cominco Products Inc.; zero magnetic field; vacuum cast.
2	85	1965	0.1-1.1	1.0		99.9999 Sn; polycrystalline; sample supplied by Cominco Products Inc.; H = 1000 Oe, magnetic field; vacuum cast.
3	189	1923	70-101	1.0		Good quality.
4	190	1924	22-286		White tin	
5	190	1924	16-284		Grey tin	
6	191	1932	1-20			Kahlbum's purity; 0.01 Fe and trace Cu.
7	192	1932	273-505		White tin	Purest tin; sample supplied by E. Merck.
8	193	1947	323-623			Tin, heating.
9	193	1947	323-523			Tin, cooling.
10	43	1952	7-100	1.0	Grey tin	< 0.05 total impurity, coarse powder; under 10^{-5} mm Hg helium atm.
11	194	1955	1.5-4.0		Grey tin	99.8 Sn; powdered form.
12	195	1956	1-4	1.3		99.999^+ Sn; zero magnetic field; superconducting; 4.2 K the residual resistivity is 0.002 at room temperature.
13	196	1958	506-800			99.9 Sn, 0.05 Pb, 0.02 Cu, 0.01 Bi, 0.005 Sb, 0.002 Ni, 0.001 Ca, 0.001 In, 0.0005 Cd, and 0.0005 Ag; sample supplied by the American Smelting and Refining Co.; molten state.
14	197	1961	0.6-1.6	1-2		99.990± 0.002 Sn; normal state; 500 gauss magnetic field.
15	197	1961	0.8-2	1-2		99.990± 0.002 Sn; superconducting; zero magnetic field.
16	198	1938	1.1-4			99.992 Sn, 0.01 Fe, and trace Cu; superconducting; zero magnetic field.
17	198	1938	1.5-4			Same as above; normal state; 299.0 gauss magnetic field.
18	195	1956	1.1-4			99.999^+ Sn; 800 oersteds; normal state; 4.2 K the residual resistivity is 0.002 at room temperature.
19	198	1938	2.8-3.9			99.992 Sn, 0.01 Fe, and trace Cu; normal state; 138.6 gauss magnetic field.
20	371	1960	0.4-4.3		Tin I	99.999 Sn; sample supplied by the Consolidated Mining and Smelting Co. of Canada, Ltd.; self annealed at room temperature; superconducting; zero magnetic field.
21	371	1960	0.4-3.8		Tin I	Same as above; normal state; 500 oersteds.
22	371	1960	0.4-4.1		Tin II	99.999 Sn; sample supplied by the Consolidated Mining and Smelting Co. of Canada, Ltd.; annealed in air for one hr. at 200 C; superconducting; zero magnetic field.

SPECIFICATION TABLE NO. 62 (continued)

Curve No.	Ref. No.	Year	Temp. Range, K	Reported Error, %	Name and Specimen Designation	Composition (weight percent), Specifications and Remarks
23	371	1960	0.6-4.2		Tin II	Same as above; normal state; 500 oersteds.
24	1	1961	295	± 5.0		
25	268	1926	348-873			99.989 Sn, 0.008 Cu, and 0.003 Pb.

DATA TABLE NO. 62 SPECIFIC HEAT OF TIN

[Temperature, T, K; Specific Heat, C_p, Cal $g^{-1}K^{-1}$]

T	C_p
CURVE 1	
Series 1	
0.1456	3.191×10^{-9}
0.1519	2.942
0.1569	3.153
0.1667	3.766
0.1676	3.830*
0.1720	3.868
0.2305	8.051
0.2380	9.015
0.2384	8.919*
0.2523	9.321
0.2539	9.974
0.2540	1.014×10^{-8}*
0.2572	1.019
0.2575	1.017*
0.2818	1.330
0.2875	1.397
0.2884	1.449
0.2894	1.413*
0.2914	1.466*
0.2981	1.536
0.3024	1.610
0.3067	1.638*
0.3128	1.798
0.3178	1.798*
0.3195	1.851*
0.3200	1.905
0.3333	2.072
Series 2	
0.3382	2.126×10^{-8}*
0.3402	2.251
0.3537	2.452
0.3635	2.682
0.3577	2.582*
0.3672	2.737*
0.3703	2.803
0.1558	3.317×10^{-9}
0.1601	3.369*
0.1616	3.625
0.1725	4.176
CURVE 1 (cont.)	
Series 2 (cont.)	
0.1770	4.307×10^{-9}
0.2009	5.890
0.2014	5.824*
0.2174	7.076
0.2279	7.880
0.2288	8.081*
0.2333	8.315*
0.2338	8.176*
0.2340	8.415*
0.2345	8.172
0.2525	1.000×10^{-8}*
0.2564	1.067
0.2667	1.153
0.2831	1.355*
0.2840	1.362*
0.3021	1.630*
Series 3	
0.3082	1.723×10^{-8}*
0.3242	1.959*
0.3287	2.032*
0.3367	2.165*
0.3543	2.479*
0.3636	2.676*
0.3736	2.886*
0.3881	3.202
0.3900	3.262*
0.4058	3.639
0.4112	3.810
0.4419	4.686
0.4434	4.710*
0.4554	5.139
0.4837	6.106
0.5150	6.238
0.5383	8.655
0.5749	1.099×10^{-7}
0.5826	1.140
0.6170	1.410
0.6557	1.609
0.7016	2.316
0.7879	2.800
CURVE 1 (cont.)	
Series 3 (cont.)	
0.8596	5.465×10^{-7}
0.9203	7.334
0.9564	8.601
1.061	1.307×10^{-6}
CURVE 2	
Series 1	
0.09269	3.252×10^{-7}
0.09474	3.313*
0.09517	3.339*
0.10373	3.532
0.10470	3.536*
0.11136	3.800
0.11165	3.800*
0.11734	4.114
0.11868	4.199*
0.1217	4.354
0.1286	4.668
0.1315	4.770*
0.1342	4.766
0.1410	5.010
0.1453	5.145*
0.1480	5.230
0.1485	5.240*
0.1603	5.632
0.1625	5.733*
0.1652	5.840
Series 2	
0.1780	6.325×10^{-7}
0.1821	6.452*
0.1865	6.639
0.1966	7.034
0.2003	7.106*
0.2051	7.342
0.2182	7.829
0.2204	7.880*
0.2301	8.232
CURVE 2 (cont.)	
Series 2 (cont.)	
0.2402	8.590×10^{-7}
0.2438	8.685*
0.2594	9.295
0.2652	9.539
0.2888	1.037×10^{-6}
0.2925	1.049*
0.2978	1.067
0.3198	1.153
0.3592	1.339
0.3550	1.281
0.3948	1.441
Series 3	
0.3973	1.440×10^{-6}*
0.4339	1.596
0.4408	1.611*
0.4886	1.812
0.5137	1.901
0.5264	1.960*
0.5423	2.020
0.5681	2.128
0.6260	2.376
0.6374	2.424*
0.6923	2.656
0.6932	2.668*
0.7637	2.966
0.8199	3.216
0.8393	3.309*
0.8941	3.572
0.9341	3.750
0.9637	3.915
1.032	4.243
1.037	4.249*
1.126	4.708
CURVE 3	
69.63	3.850×10^{-2}
70.00	3.867*
72.39	3.909*
CURVE 3 (cont.)	
75.11	3.993×10^{-2}
77.71	4.069*
80.00	4.103
80.34	4.111*
84.00	4.195*
86.36	4.221
88.90	4.271*
90.00	4.305*
91.26	4.339*
93.56	4.356
96.21	4.431*
98.59	4.440*
100.00	4.482
101.00	4.499*
CURVE 4	
22.4	1.07×10^{-2}
26.9	1.58
32.7	2.09
40.9	2.56
42.5	2.70
48.0	2.97
56.7	3.42
92.5	4.39*
101.1	4.55*
286.3	5.28
CURVE 5	
15.5	5.90×10^{-3}
18.2	6.36
22.35	8.90
26.8	1.16×10^{-2}
31.5	1.41
37.3	1.69
44.9	1.99
52.3	2.36
59.3	2.65
69.5	3.09
92.5	3.72
106.2	3.99
283.7	5.16
CURVE 6	
Series 1	
12.520	3.749×10^{-3}
13.264	3.852
13.818	4.300
14.564	4.968
15.154	5.362
16.020	6.246
16.950	6.862
17.924	7.613
18.628	8.301
19.304	8.517
19.740	9.110
20.170	1.015×10^{-2}
Series 2	
3.652	7.456×10^{-5}
4.553	9.815
Series 3	
3.084	5.257×10^{-5}
3.910	6.226
4.755	1.077×10^{-4}
5.520	1.719
6.232	3.696
6.458	4.554
7.002	3.441
7.206	4.140
7.726	7.291
8.676	1.451×10^{-3}
9.742	1.843
10.492	2.255
11.480	2.497
12.250	3.252
13.460	4.252
15.300	5.971
17.174	7.754
Series 4	
2.200	1.580×10^{-5}
2.292	1.641

*Not shown on plot

DATA TABLE NO. 62 (continued)

T	C_p
CURVE 6 (cont.)	
Series 4 (cont.)	
2.453	2.055 x 10^{-5}
2.650	2.861
2.992	3.634
3.295	4.521
3.598	5.762
3.895	5.151
4.175	6.453
Series 5	
1.373	4.617 x 10^{-6}
1.511	4.608
1.510	3.033
2.110	1.358 x 10^{-5}
2.064	1.263
2.537	2.227
3.073	3.889
3.279	4.675
3.486	5.485
3.687	5.510
3.853	4.933
3.786	8.257
4.113	6.454*
4.420	8.072
CURVE 7	
273	5.3929 x 10^{-2}
293	5.4393*
323	5.5156
373	5.6609
423	5.8287
463	5.9792*
473	6.0191
505	6.1527
CURVE 8	
323.15	5.50 x 10^{-2}*
373.15	5.80
423.15	6.16
439.15	6.50
458.15	6.40

T	C_p
CURVE 8 (cont.)	
473.15	6.65 x 10^{-2}
523.15	6.00*
573.15	5.92
623.15	5.84
CURVE 9	
323.15	5.52 x 10^{-2}
373.15	5.80*
423.15	6.20
439.15	6.50*
458.15	6.42*
473.15	6.72
523.15	6.00
CURVE 10	
7	3.622 x 10^{-4}
8	6.066
9	9.351
10	1.356 x 10^{-3}
12	2.367
15	4.195
20	7.523
25	1.070 x 10^{-2}
30	1.348
60	2.696
70	3.126*
80	3.404
90	3.665
100	3.917
CURVE 11	
1.5	1.389 x 10^{-6}
2.0	3.281
2.5	6.462
3.0	1.117 x 10^{-5}
3.5	1.824
4.0	2.999
CURVE 12	
Series 1	
1.122	1.632 x 10^{-6}
1.148	1.820

T	C_p
CURVE 12 (cont.)	
1.177	1.892 x 10^{-6}
1.208	2.081
1.229	2.275
1.263	2.442
1.289	2.726
1.312	2.816
1.367	3.301
1.400	3.583
1.517	4.831
1.597	5.677
1.685	6.844
1.782	8.213
1.876	9.662
1.962	1.097 x 10^{-5}
1.985	1.115*
2.072	1.280*
2.270	1.620*
2.371	1.820
2.470	2.063*
2.572	2.323
2.690	2.615
2.763	2.856*
2.877	3.180*
2.980	3.523
3.082	3.905*
3.176	4.227
3.268	4.630*
3.369	5.113
3.462	5.536*
3.568	6.039
3.679	6.623
3.699	6.663*
3.711	6.663*
3.723	5.958*
3.734	4.892*
3.748	4.751*
3.760	4.791*
3.773	4.871*
3.799	4.992*
4.003	5.858*
4.112	6.462*
4.191	6.904*
4.341	7.871
Series 3	
1.122	1.633 x 10^{-6}
1.148	1.820
1.177	1.893
1.208	2.122

T	C_p
CURVE 13	
506	6.4 x 10^{-2}
510	6.06*
550	5.96
600	5.86
650	5.66
700	5.68
800	5.50
CURVE 14	
0.604	2.223 x 10^{-6}
0.632	2.300
0.655	2.457
0.673	2.537
0.692	2.590
0.710	2.688*
0.727	2.769
0.742	2.775
0.757	2.88
0.769	2.92*
0.783	3.06
0.798	3.06*
0.817	3.10
0.840	3.22
0.865	3.36
0.881	3.38*
0.912	3.54
0.930	3.58*
0.966	3.79
0.989	3.91*
1.021	4.13
1.057	4.33
1.079	4.45*
1.118	4.63*
1.183	5.01*
1.245	5.32
1.470	6.83
1.653	8.19
CURVE 15	
0.800	3.85 x 10^{-7}
0.832	4.63
0.875	5.58
0.893	6.12

T	C_p
CURVE 15 (cont.)	
0.928	6.91 x 10^{-7}
0.958	7.41
0.980	8.05
1.003	9.28
1.025	1.03 x 10^{-6}*
1.048	1.05
1.070	1.20
1.093	1.35
1.114	1.48
1.156	1.72
1.178	1.94*
1.210	2.13*
1.236	2.408
1.255	2.535
1.308	2.86*
1.405	3.71
1.570	5.24
1.640	6.54
1.741	7.75
1.845	9.04
1.937	1.08 x 10^{-5}*
2.057	1.25*
CURVE 16	
1.143	2.106 x 10^{-6}
1.145	2.022*
1.335	3.454
1.541	5.813
1.544	5.392
1.753	8.172
1.920	1.112 x 10^{-5}
1.927	1.095*
2.015	1.289
2.027	1.272*
2.111	1.407
2.125	1.466
2.273	1.761
2.456	2.249
2.474	2.165
2.484	2.148*
2.550	2.392
2.562	2.527
2.717	2.721
2.730	2.856

T	C_p
CURVE 16 (cont.)	
2.837	3.151 x 10^{-5}
2.881	3.277
3.215	4.364
3.270	4.507*
3.600	6.436
3.633	6.605*
3.775	5.021
3.794	5.063*
CURVE 17	
Series 1	
1.537	7.389 x 10^{-6}
1.548	7.549
1.068	4.912
1.067	4.920*
1.273	5.788
1.280	6.201
1.305	5.687
1.307	5.729*
1.425	6.656
1.430	6.555*
1.524	8.291
1.528	7.692*
1.810	1.018 x 10^{-5}
1.811	1.000*
1.985	1.288*
1.998	1.179
2.077	1.281*
2.094	1.296*
2.230	1.446
2.238	1.446*
Series 2	
2.521	1.904 x 10^{-5}
2.525	1.862
2.760	2.266
2.780	2.266*
2.794	2.241*
3.076	2.595
3.108	2.907
3.363	3.547
3.383	3.783
3.595	4.280
3.635	4.533
3.768	5.106*

*Not shown on plot

DATA TABLE NO. 62 (continued)

T	C_p
CURVE 17 (cont.)	
Series 3	
2.781	2.233 x 10^{-5}*
2.904	2.376
2.951	2.494
3.382	3.657*
3.402	3.631*
3.628	4.398*
3.691	4.600*
3.747	4.870
3.784	5.080*
3.848	5.291
3.940	5.611
Series 4	
1	1.47 x 10^{-6}
1.5	5.11
2.0	1.23 x 10^{-5}
2.5	2.22
3.0	3.59
3.5	5.73
3.7	7.18
Series 5	
1	3.84 x 10^{-6}
1.5	7.14
2.0	1.18 x 10^{-5}
2.5	1.80
CURVE 18	
Series 1	
1.130	4.692 x 10^{-6}
1.164	4.954
1.200	5.195
1.235	5.336*
1.270	5.518
1.306	5.739*
1.404	6.242
1.498	6.988
1.476	6.887*
1.590	6.927

T	C_p
CURVE 18 (cont.)	
Series 1 (cont.)	
1.686	8.518 x 10^{-6}
1.763	9.062
1.857	9.948
1.965	1.083 x 10^{-5}*
2.073	1.180
1.980	1.081*
2.069	1.172*
2.166	1.265
2.266	1.434
2.373	1.567
2.476	1.683
2.570	1.845*
2.681	2.016
2.767	2.122
2.886	2.350*
2.985	2.539*
3.082	2.759
3.184	2.972
3.277	3.242
3.388	3.564*
3.473	3.806
3.584	4.209*
3.689	4.571*
3.787	5.034*
3.887	5.518*
3.985	5.961
Series 2	
1.130	4.692 x 10^{-6}
1.164	4.954
1.200	5.215
1.235	5.356
1.270	5.538
1.306	5.759
1.403	6.263
1.495	7.028
1.474	6.927
1.587	6.967
1.682	8.578
1.758	9.162
1.851	1.005 x 10^{-5}
1.958	1.099

T	C_p
CURVE 18 (cont.)	
Series 2 (cont.)	
1.973	1.097 x 10^{-5}
2.065	1.200
2.061	1.192
2.156	1.291
2.259	1.389
2.369	1.528
2.474	1.657
2.569	1.820
2.680	2.002
2.766	2.116
2.885	2.354
2.983	2.555
3.080	2.783
3.180	3.004
3.272	3.282
3.382	3.605
3.466	3.826
3.577	4.229
3.682	4.551
3.781	4.934
3.880	5.397
3.982	5.840
CURVE 19*	
2.781	2.233 x 10^{-5}
2.904	2.376
2.951	2.494
3.382	3.657
3.402	3.631
3.628	4.398
3.691	4.600
3.747	4.870
3.784	5.080
3.848	5.291
3.940	5.611
CURVE 20*	
0.4374	3.884 x 10^{-8}
0.4449	4.12
0.5021	6.299
0.5084	6.552

T	C_p
CURVE 20 (cont.)*	
0.5727	1.024 x 10^{-7}
0.6274	1.397
0.6678	1.888
0.6874	1.993
0.7117	2.353
0.7407	2.798
0.7712	3.339
0.8047	4.002
0.8326	4.711
0.8804	5.975
0.9217	7.276
0.9630	8.843
1.0490	1.257 x 10^{-6}
1.1307	1.696
1.2069	2.187
1.2985	2.887
1.4172	3.944
1.5744	5.659
1.7272	7.617
1.8961	1.019 x 10^{-5}
2.028	1.250
2.138	1.464
2.404	2.004
2.655	2.673
2.837	3.228
3.007	3.833
3.153	4.381
3.280	4.881
3.393	5.336
3.495	5.813
3.589	6.367
3.729	4.806
3.839	5.311
4.154	6.801
4.389	7.945
CURVE 21*	
0.4026	1.484 x 10^{-6}
0.4237	1.604
0.4374	1.612
0.4847	1.894
0.6042	2.275
0.7062	2.588
0.7319	2.830

T	C_p
CURVE 21 (cont.)*	
0.7575	2.883 x 10^{-6}
0.7975	3.132
0.8303	3.327
0.9067	3.670
0.9778	4.036
1.0329	4.243
1.1502	4.887
1.2391	5.464
1.3461	6.072
1.5163	7.175
1.819	9.707
2.161	1.314 x 10^{-5}
2.377	1.575
2.555	1.832
2.903	2.426
3.170	3.007
3.387	3.765
3.574	4.131
3.874	5.219
CURVE 22*	
0.4060	3.156 x 10^{-8}
0.4061	3.271
0.4233	3.793
0.4798	5.652
0.5350	7.951
0.6112	1.342 x 10^{-7}
0.7021	2.265
0.7393	2.678
0.9180	7.561
0.9923	1.023 x 10^{-6}
1.0404	1.234
1.1078	1.566
1.2193	2.320
1.3571	3.389
1.4954	4.719
1.6616	6.725
1.9302	1.061 x 10^{-5}
2.221	1.628
2.511	2.291
2.744	2.934
2.938	3.544
3.101	4.128
3.244	4.703

T	C_p
CURVE 22 (cont.)*	
3.365	5.238 x 10^{-5}
3.477	5.776
3.579	6.256
3.644	6.611
3.668	6.699
3.678	6.829
3.695	6.860
3.706	5.291
3.719	4.958
3.722	4.797
3.737	5.034
3.749	5.005
3.765	4.989
3.857	5.413
3.986	6.068
4.148	6.816
CURVE 23*	
0.6182	2.365 x 10^{-6}
0.7340	2.912
0.8374	3.389
0.9474	3.930
1.0129	4.243
1.0189	4.329
1.1748	5.110
1.3582	6.208
1.5166	7.268
1.6965	8.643
1.9983	1.155 x 10^{-5}
2.596	2.316
3.383	3.631
3.665	4.583
3.878	5.373
4.069	6.432
4.240	7.138
CURVE 24*	
295	5.7 x 10^{-2}

T	C_p
CURVE 25*	
348	6.21 x 10^{-2}
373	6.41
423	6.81
448	7.03
498	7.40
523	5.20
623	5.20
723	5.20
823	5.20
873	5.20

*Not shown on plot

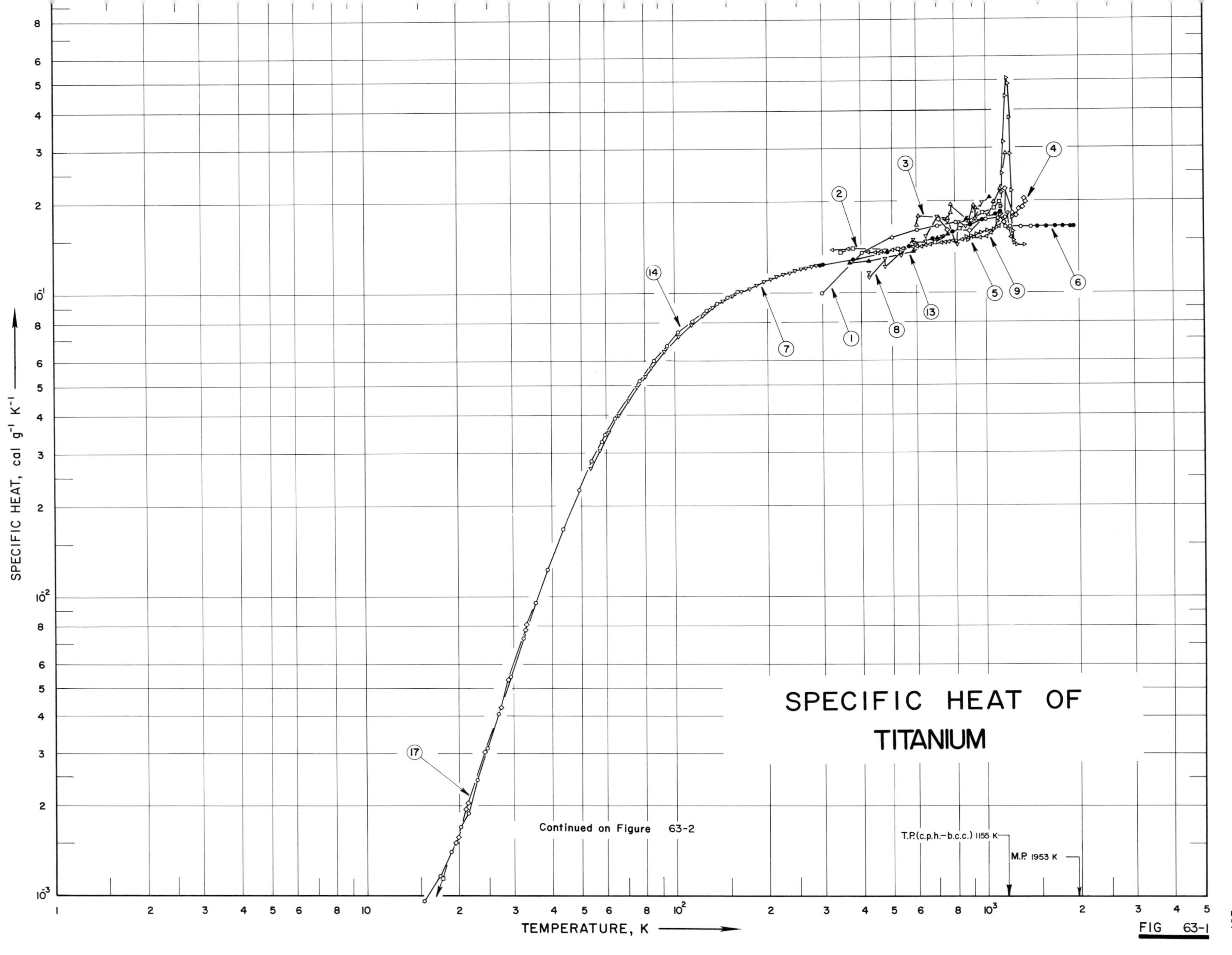
SPECIFIC HEAT OF
TITANIUM
SPECIFIC HEAT, cal g⁻¹ K⁻¹
TEMPERATURE, K
Continued on Figure 63-2
T.P.(c.p.h.–b.c.c.) 1155 K
M.P. 1953 K
FIG 63-1

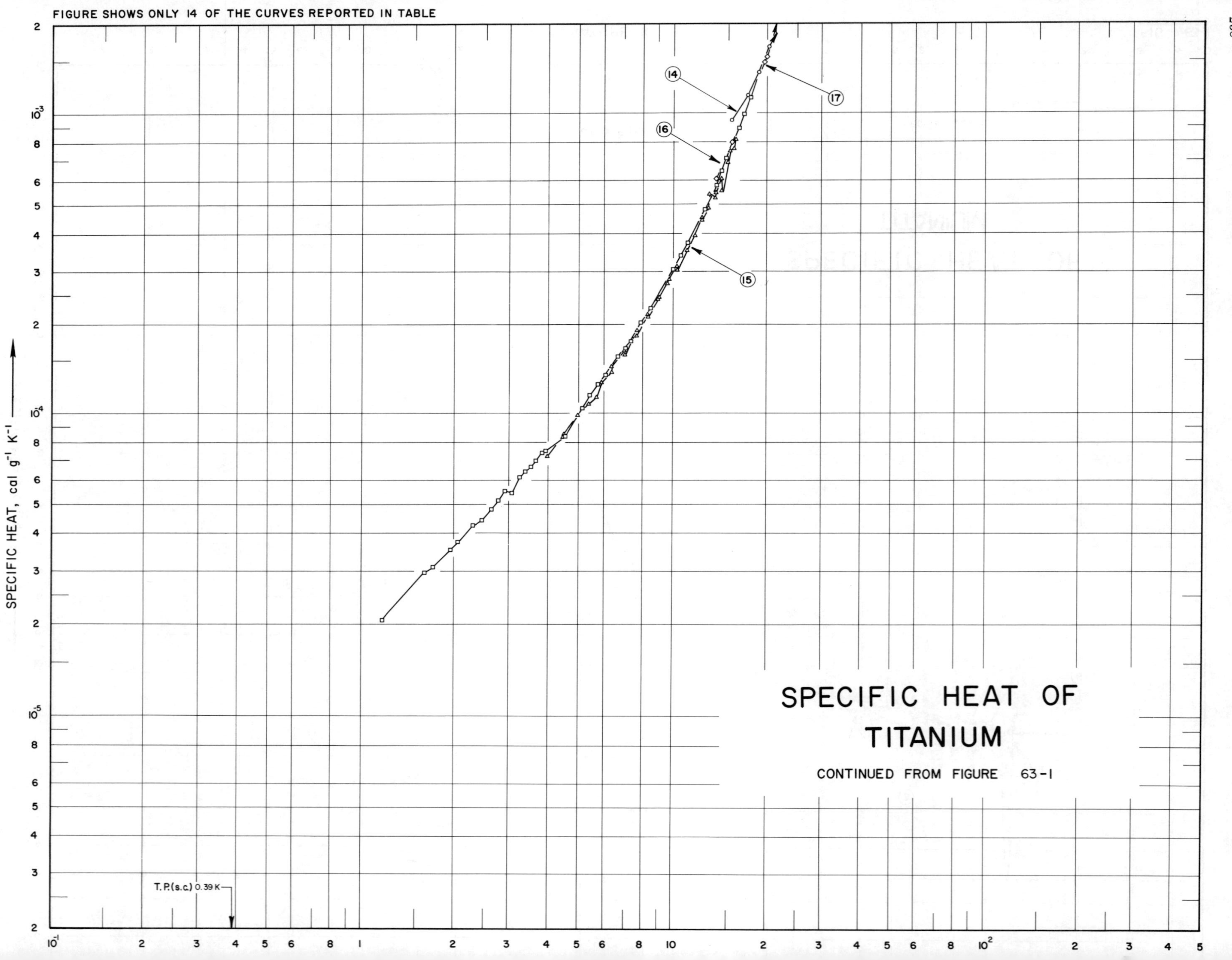
FIGURE SHOWS ONLY 14 OF THE CURVES REPORTED IN TABLE
SPECIFIC HEAT, cal g⁻¹ K⁻¹
SPECIFIC HEAT OF
TITANIUM
CONTINUED FROM FIGURE 63-1
T.P.(s.c.) 0.39 K
14
15
16
17

SPECIFICATION TABLE NO. 63 SPECIFIC HEAT OF TITANIUM

(Impurity < 0.20% each; total impurities < 0.50%)

[For Data Reported in Figure and Table No. 63]

Curve No.	Ref. No.	Year	Temp. Range, K	Reported Error, %	Name and Specimen Designation	Composition (weight percent), Specifications and Remarks
1	127	1959	298-1400	± 1.6		
2	160	1965	343-1103	± 5.0		0.0055 O_2, 0.002 other metals, 0.0015 C, 0.001 N_2, and 0.0008 H_2.
3	129	1963	599-1066	≤ 6.0	Filament No. 2	≤ 0.2 C, few tenths percent maximum O_2, N_2, and Fe; sample supplied by the Driver-Harris Co.; sealed under vacuum.
4	129	1963	758-1150	≤ 6.0	Filament No. 3	Same as above.
5	130	1958	868-1348	± 5.0		Commercial grade; sealed under vacuum.
6	81	1961	294-1923			99.705 Ti, 0.08 Fe, 0.07 Si, 0.05 C, 0.03 N_2, 0.02 O_2, 0.005 H_2, and 0.04 other impurities.
7	132	1944	54-295			98.75 Ti, 0.5 Si, 0.27 Fe, and 0.15 V; data corrected for impurities.
8	133	1956	422-978	± 18		99.9 Ti.
9	134	1957	323-1233		Iodide titanium	0.032 C, 0.03 Fe, 0.011 O_2, 0.0067 H_2, 0.001 Cu, and 0.00077 N_2; under 0.01 μ Hg vacuum.
10	134	1957	433-1223		Iodide titanium	Same as above.
11	134	1957	363-1113		Iodide titanium	Same as above.
12	134	1957	333-1033		Iodide titanium	Same as above.
13	135	1956	311-1033		Ti 75 A	99.75 Ti, 0.131 O_2, 0.07 Fe, 0.06 C, 0.048 N_2, and 0.0068 H_2.
14	136	1953	15-306			0.0082 Mn, 0.007 Si, 0.0066 Al, 0.02 total of Cu, Pb, N_2, and Te; prepared by iodide process; annealed under high vacuum at 800 C.
15	128	1956	6-16			≥ 99.95 Ti; annealed.
16	308	1957	1-21			~99.9 Ti; hexagonal close-packed structure.
17	200	1958	14-272			
18	199	1936	473-1090			Purest ductile titanium; prepared by the Van Arkel iodide process.
19	372	1956	4-15	< 2.0		99.95 - 99.99 Ti, sample supplied by the Foote Mineral Co.
20	340	1952	15-1900	0.2		99.96 Ti, 0.0082 Mn, 0.007 Si, 0.0066 Al, Cu, Pb, N_2, and Te; sample supplied by the New Jersey Zinc Co.
21	203	1956	5-18	< 1.0		99.95 - 99.99 Ti, impurities total < .005 of Al, Ca, C, Cr, Cu, Fe, Mg, Mn, Ni, N_2, Si, and Sn; prepared by admitting hydrogen to tube of titanium; annealed at temperatures < 500 C with final anneal at 180 C; cooled slowly to room temperature.
22	340	1952	298-1900	0.2		99.96 Ti, 0.0082 Mn, 0.007 Si, 0.0066 Al, Cu, Pb, N_2 and Te; sample supplied by the New Jersey Zinc Co.

DATA TABLE NO. 63 SPECIFIC HEAT OF TITANIUM

[Temperature, T, K; Specific Heat, C_p, Cal g^{-1} K^{-1}]

T	C_p
CURVE 1	
(α) 298.15	9.985 x 10^{-2}
300	1.009 x 10^{-1}*
400	1.363
500	1.529
600	1.620
700	1.677
800	1.714
900	1.740
1000	1.761
1100	1.776
1155	1.784
(β) 1155	1.663
1200	1.663
1300	1.663
1400	1.663
CURVE 2	
343.15	1.37 x 10^{-1}
376.15	1.41
527.15	1.40
688.15	1.47
822.15	1.64
977.15	1.84
1061.15	1.91
1103.15	2.00
CURVE 3	
599	1.68 x 10^{-1}
608	1.80
609	1.82*
756	1.77
771	1.85
773	1.97
884	1.73
911	1.96
925	1.91
1022	1.82
1066	2.00
1115	2.23
1154	2.92
CURVE 4	
708	1.76 x 10^{-1}
758	1.62
890	1.62
1023	1.82*
1060	1.94*
1150	1.72
1152	2.23
1223	1.85
1254	1.80
1277	1.90
1323	1.94
1338	2.07
1345	2.02
CURVE 5	
868.15	1.54 x 10^{-1}
878.15	1.49
888.15	1.52
898.15	1.55
908.15	1.56*
918.15	1.53
928.15	1.54*
938.15	1.57
948.15	1.56*
958.15	1.57*
968.15	1.59
978.15	1.58*
988.15	1.58*
998.15	1.60*
1008.15	1.62
1018.15	1.62*
1028.15	1.64*
1038.15	1 62*
1048.15	1.62*
1058.15	1.63
1068.15	1.66*
1078.15	1.67*
1088.15	1.70*
1098.15	1.74
1108.15	1.82
1118.15	1.94
CURVE 5 (cont.)	
1128.15	2.17 x 10^{-1}
1138.15	2.50
1148.15	3.19
1158.15	4.50
1168.15	5.18
1178.15	4.96
1188.15	3.82
1198.15	2.89
1208.15	2.19
1218.15	1.78
1228.15	1.61
1238.15	1.50
1248.15	1.47
1258.15	1.45*
1268.15	1.45
1278.15	1.46*
1288.15	1.44*
1298.15	1.44*
1308.15	1.44*
1318.15	1.44*
1328.15	1.46*
1338.15	1.44*
1348.15	1.45
CURVE 6	
(α) 294.09	1.246 x 10^{-1}
301.02	1.249
375.45	1.294
482.15	1.369
569.15	1.434
676.15	1.515
784.15	1.599
891.15	1.682
978.15	1.750
1075.15	1.827
1083.15	1.833*
1093.15	1.840*
1113.15	1.856*
1123.15	1.864
1153.15	1.672*
(β) 1173.15	1.672*
CURVE 6 (cont.)	
(β) 1273.15	1.672 x 10^{-1}*
1373.15	1.672*
1473.15	1.672
1573.15	1.672
1673.15	1.672
1773.15	1.672
1873.15	1.672
1923.15	1.672
CURVE 7	
53.5	2.668 x 10^{-2}
57.1	3.056
61.4	3.505
65.9	3.998
70.5	4.472
75.1	4.921
80.1	5.390
83.8	5.714
92.7	6.474
102.9	7.223
112.7	7.860
123.2	8.478
133.3	8.990
143.4	9.428
154.0	9.827
163.9	1.018 x 10^{-1}
174.8	1.047
184.2	1.073
194.4	1.098
204.7	1.118
214.6	1.137
224.9	1.156
235.0	1.171
244.9	1.186
255.5	1.201
265.7	1.217
276.1	1.232
285.8	1.242
295.1	1.249*
CURVE 8	
Series 1	
422.0	1.17 x 10^{-1}
477.5	1.30
533.1	1.37
588.7	1.51
644.2	1.54
699.8	1.78
755.3	1.613*
810.9	1.43
866.4	1.66
922.0	1.76
977.5	2.00
Series 2	
422.0	1.132 x 10^{-1}
477.5	1.221
533.1	1.341
588.7	1.469
644.2	1.468
699.8	1.672*
755.3	1.740
810.9	1.725
866.4	1.778*
922.0	1.892
977.5	1.838*
CURVE 9	
323.15	1.397 x 10^{-1}
343.15	1.387*
363.15	1.400
383.15	1.404*
403.15	1.359
423.15	1.357
443.15	1.363
463.15	1.369
483.15	1.379*
503.15	1.388
523.15	1.394*
543.15	1.404
CURVE 9 (cont.)	
563.15	1.411 x 10^{-1}*
583.15	1.415
603.15	1.424
623.15	1.431
643.15	1.440
663.15	1.447
683.15	1.451*
703.15	1.456*
723.15	1.463
743.15	1.469
763.15	1.471
783.15	1.483
803.15	1.488
823.15	1.490*
843.15	1.497
863.15	1.502*
883.15	1.507*
903.15	1.513*
923.15	1.514*
943.15	1.521*
963.15	1.526
983.15	1.536*
1003.15	1.545
1023.15	1.558*
1043.15	1.588
1063.15	1.623*
1083.15	1.652*
1103.15	1.665
1173.15	1.841
1193.15	1.636
1213.15	1.527
1233.15	1.502*
CURVE 10*	
433.15	1.370 x 10^{-1}
473.15	1.371
513.15	1.394
553.15	1.413
593.15	1.428
633.15	1.444
673.15	1.454

*Not shown on plot

DATA TABLE NO. 63 (continued)

T	C_p
CURVE 10 (cont.)*	
713.15	1.465×10^{-1}
753.15	1.479
793.15	1.494
833.15	1.508
873.15	1.527
913.15	1.532
953.15	1.539
993.15	1.552
1023.15	1.563
1043.15	1.588
1063.15	1.610
1083.15	1.632
1103.15	1.641
1103.15	1.644
1138.15	1.656
1148.15	1.830
1183.15	1.709
1203.15	1.608
1223.15	1.538
CURVE 11*	
363.15	1.336×10^{-1}
403.15	1.355
443.15	1.368
483.15	1.380
523.15	1.397
563.15	1.416
593.15	1.421
603.15	1.434
633.15	1.436
643.15	1.447
673.15	1.448
683.15	1.459
713.15	1.459
723.15	1.473
753.15	1.471
763.15	1.485
793.15	1.480
803.15	1.500
833.15	1.491
843.15	1.516
873.15	1.511
883.15	1.532
913.15	1.521
923.15	1.547
953.15	1.533
963.15	1.562

T	C_p
CURVE 11 (cont.)	
993.15	1.549×10^{-1}
1003.15	1.570
1033.15	1.573
1043.15	1.601
1073.15	1.613
1113.15	1.640
CURVE 12*	
333.15	1.312×10^{-1}
353.15	1.321
373.15	1.328
393.15	1.336
413.15	1.351
433.15	1.357
453.15	1.363
475.15	1.370
493.15	1.380
513.15	1.386
533.15	1.402
553.15	1.407
573.15	1.415
593.15	1.427
613.15	1.437
633.15	1.443
653.15	1.452
673.15	1.457
693.15	1.465
713.15	1.469
733.15	1.472
753.15	1.480
773.15	1.486
793.15	1.495
813.15	1.501
833.15	1.510
853.15	1.517
873.15	1.521
893.15	1.522
913.15	1.525
933.15	1.521
953.15	1.556
973.15	1.561
993.15	1.557
1013.15	1.557
1033.15	1.583

T	C_p
CURVE 13	
310.93	1.250×10^{-1}*
366.48	1.258
422.04	1.275*
477.59	1.302*
533.15	1.339
588.71	1.385*
644.26	1.441
699.82	1.506
755.37	1.581*
810.93	1.665*
866.48	1.759
922.04	1.863*
977.59	1.976*
1033.15	2.099
CURVE 14	
15.44	9.603×10^{-4}
17.36	1.169×10^{-3}
18.75	1.399
20.04	1.691
21.31	1.879
22.87	2.443
24.60	3.111
26.71	4.008
29.32	5.428
32.23	7.286
35.26	9.520
38.67	1.232×10^{-2}
43.54	1.687
49.04	2.265
53.89	2.818
58.00	3.282
59.33	3.451
63.95	3.910
70.27	4.537*
77.00	5.196
85.62	6.013
94.76	6.789
104.49	7.480
114.76	8.115
127.07	8.800
137.65	9.267
148.70	9.716
160.37	1.014×10^{-1}
172.74	1.048*
185.70	1.077*
198.46	1.108*

T	C_p
CURVE 14 (cont.)	
212.40	1.133×10^{-1}*
215.29	1.141*
224.52	1.156*
234.03	1.170*
248.05	1.186*
259.30	1.204*
271.73	1.224*
283.32	1.234*
293.57	1.242*
299.58	1.244*
305.51	1.254*
CURVE 15	
3.946	7.235×10^{-5}
4.450	8.532
4.917	9.880
5.381	1.093×10^{-4}
5.686	1.148
5.867	1.277
6.316	1.372
6.348	1.447
6.984	1.636
6.986	1.577
7.618	1.931
7.640	1.846
8.253	2.176
8.288	2.126
8.898	2.445
8.953	2.485
9.559	2.764
9.621	2.849
10.242	3.158
10.319	3.054
11.042	3.568
11.691	3.982*
11.692	4.077
12.336	4.591
12.344	4.481
12.978	4.925
12.998	5.497
13.619	5.328
13.648	5.535
14.233	6.187
14.287	5.618
14.899	6.986
15.129	7.634
15.555	7.734
15.757	8.332

T	C_p
CURVE 16	
1.17	2.06×10^{-5}
1.60	2.96
1.71	3.09
1.94	3.53
2.05	3.74
2.28	4.26
2.45	4.43
2.62	4.80
2.76	5.14
2.89	5.51
3.05	5.44
3.22	6.14
3.36	6.43
3.50	6.68
3.64	6.99
3.80	7.47
3.91	7.53
4.50	8.43
5.11	1.05×10^{-4}
5.40	1.16
5.74	1.26
6.05	1.35
6.66	1.56
7.00	1.66
7.30	1.76*
7.58	1.93
7.85	2.04*
8.15	2.17
8.46	2.28*
8.74	2.42
9.03	2.55*
9.35	2.69*
9.64	2.88
9.93	3.07*
10.22	3.13
10.52	3.42*
10.80	3.55
11.15	3.78*
11.52	3.99
11.90	4.32
12.28	4.53
12.65	4.89*
13.18	5.39
13.74	5.89
14.25	6.53
14.85	7.20*
15.52	8.02
16.18	9.04*
16.85	1.01×10^{-3}*
17.58	1.15*

T	C_p
CURVE 16 (cont.)	
18.29	1.29×10^{-3}*
19.00	1.44*
19.66	1.62*
20.32	1.75*
20.95	1.97*
CURVE 17	
13.72	5.97×10^{-4}*
13.76	5.93*
13.77	6.10
14.19	6.58*
15.49	8.18*
16.18	9.12*
16.57	9.67*
17.39	1.10×10^{-3}*
18.03	1.21*
18.92	1.39
19.46	1.51
19.65	1.56*
19.72	1.57
20.93	1.95
21.37	2.04*
21.57	2.102*
21.62	2.127
24.07	3.035
27.13	4.230
28.66	5.319
32.54	7.752
32.94	8.079
38.06	1.219×10^{-2}*
43.02	1.679*
48.95	2.278*
54.77	2.900*
61.16	3.587*
61.21	3.591*
66.03	4.102*
66.77	4.175*
71.34	4.641*
77.47	5.236*
82.72	5.714*
87.78	6.123*
92.71	6.549*
98.92	7.042*
104.06	7.403*
109.13	7.727*
114.09	8.073*
120.79	8.440*

* Not shown on plot

DATA TABLE NO. 63 (continued)

T	C_p
CURVE 17 (cont.)	
126.71	8.787 x 10^{-2}*
132.62	9.071*
138.42	9.359*
146.59	9.691*
152.79	9.908*
158.90	1.007 x 10^{-1}*
164.88	1.029*
173.50	1.053*
179.98	1.068*
186.20	1.084*
192.29	1.097*
198.80	1.111*
201.18	1.116*
203.83	1.122*
207.55	1.133*
210.19	1.136*
210.80	1.137*
213.57	1.144*
216.26	1.145*
218.04	1.151*
224.32	1.162*
229.97	1.170*
230.24	1.171*
236.37	1.180*
238.52	1.183*
242.38	1.189*
244.79	1.192*
248.07	1.196*
250.62	1.201*
259.03	1.211*
259.32	1.210*
265.44	1.219*
265.80	1.220*
271.76	1.227*
271.92	1.225*
CURVE 18*	
473	1.353 x 10^{-1}
573	1.440
673	1.474
773	1.492
873	1.529
973	1.622
1073	1.807
1090	1.851

T	C_p
CURVE 19*	
4.0	7.38 x 10^{-5}
4.5	8.58
5.0	9.88
5.5	1.12 x 10^{-4}
6.0	1.26
6.5	1.42
7.0	1.59
7.5	1.77
8.0	1.97
8.5	2.19
9.0	2.41
9.5	2.65
10.0	2.92
10.5	3.20
11.0	3.51
11.5	3.84
12.0	4.18
12.5	4.56
13.0	4.95
13.5	5.39
14.0	5.84
14.5	6.39
15.0	6.94
CURVE 20*	
15.439	9.541 x 10^{-4}
17.362	1.167 x 10^{-3}
18.754	1.399
20.045	1.695
21.311	1.875
22.872	2.438
24.60	3.106
26.708	4.006
29.323	5.428
32.232	7.292
35.264	9.514
38.666	1.231 x 10^{-2}
43.540	1.687
49.044	2.265
53.890	2.818
57.999	3.282
63.946	3.911
70.273	4.537
76.999	5.196
85.618	6.012

T	C_p
CURVE 20 (cont.)*	
94.758	6.789 x 10^{-2}
104.492	7.480
114.763	8.115
127.074	8.800
137.652	9.267
148.701	9.716
160.373	1.014 x 10^{-1}
172.744	1.048
185.705	1.077
198.459	1.108
212.404	1.133
215.286	1.155
224.524	1.156
234.026	1.170
248.048	1.186
259.296	1.204
271.726	1.224
283.322	1.234
293.571	1.242
299.576	1.244
305.510	1.254
CURVE 21*	
Series I	
4.837	7.135 x 10^{-4}
5.406	8.383
5.931	9.430
6.457	1.068 x 10^{-3}
6.993	1.232
7.537	1.402
8.097	1.612
8.665	1.816
9.229	2.016
9.787	2.255
10.349	2.520
10.914	2.789
11.484	3.114
12.049	3.443
12.614	3.802
13.180	4.236
13.749	4.630
14.322	5.104
14.893	5.698
15.457	6.162
16.021	6.841

T	C_p
CURVE 21 (cont.)*	
16.590	7.504 x 10^{-3}
17.157	8.228
17.722	8.971
Series II	
3.895	4.905 x 10^{-3}
4.427	6.437 x 10^{-4}
4.990	7.485
5.459	8.433
5.959	9.530
6.513	1.103 x 10^{-3}
16.672	7.744
17.236	8.567
17.802	9.186
CURVE 22*	
298.16	1.24 x 10^{-1}
300	1.24
400	1.30
500	1.36
600	1.42
700	1.48
800	1.54
900	1.60
1000	1.66
1050	1.69
1100	1.72
1154	1.75
1154	1.49
1200	1.49
1300	1.49
1400	1.49
1500	1.59
1600	1.68
1700	1.81
1800	1.98
1900	2.19

*Not shown on plot

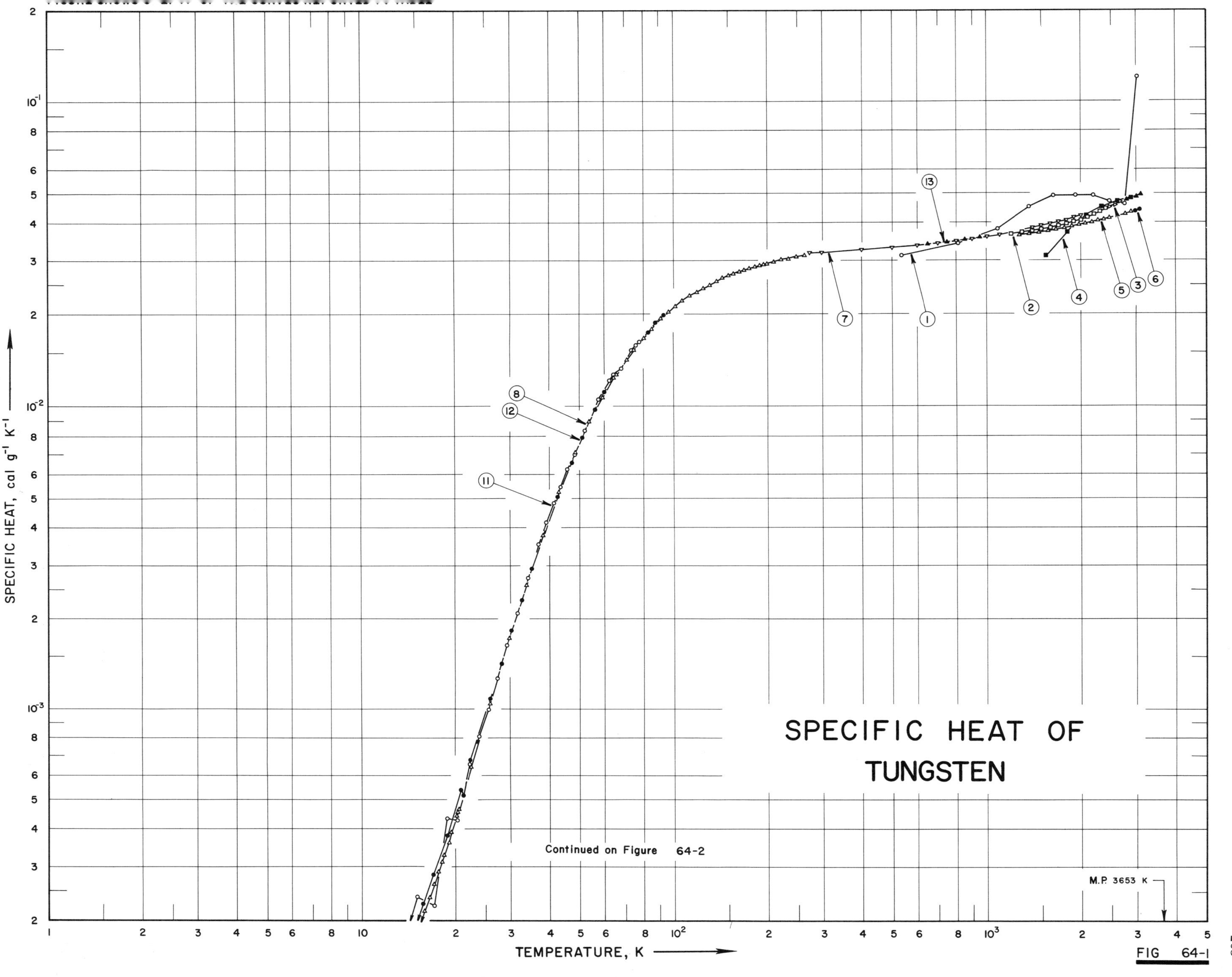
SPECIFIC HEAT OF
TUNGSTEN
SPECIFIC HEAT, cal g⁻¹ K⁻¹
TEMPERATURE, K
Continued on Figure 64-2
M.P. 3653 K

FIG 64-1

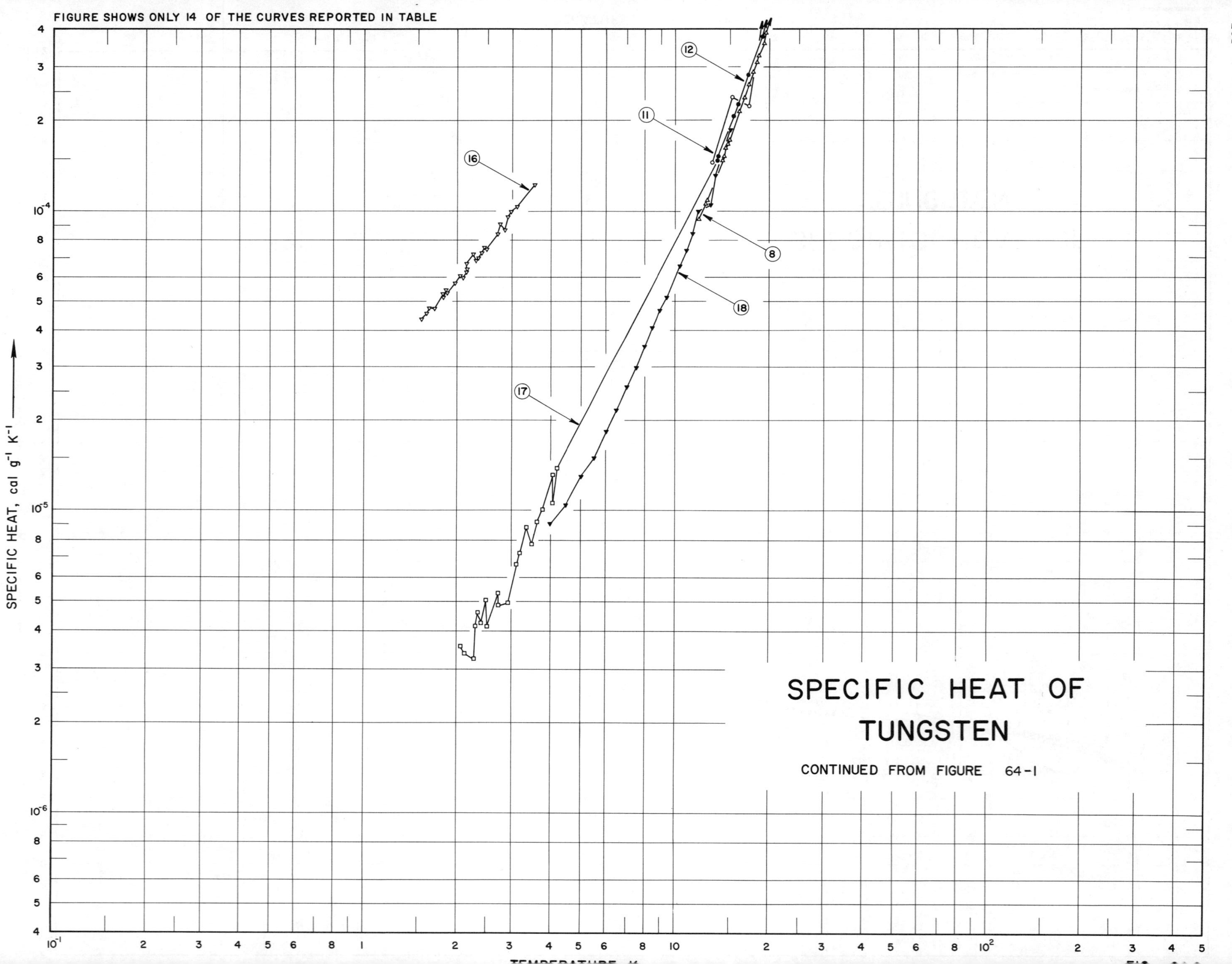
FIGURE SHOWS ONLY 14 OF THE CURVES REPORTED IN TABLE
SPECIFIC HEAT OF
TUNGSTEN
CONTINUED FROM FIGURE 64-1
SPECIFIC HEAT, cal g⁻¹ K⁻¹
12
11
16
8
18
17
4
3
2
10⁻⁴
8
6
5
10⁻⁵
10⁻⁶
10⁻¹
1
10
10²

SPECIFICATION TABLE NO. 64 SPECIFIC HEAT OF TUNGSTEN

(Impurity $<0.20\%$ each; total impurities $<0.50\%$)

[For Data Reported in Figure and Table No. 64]

Curve No.	Ref. No.	Year	Temp. Range, K	Reported Error, %	Name and Specimen Designation	Composition (weight percent), Specifications and Remarks
1	12	1962	533-3033	≤ 5.0		99.9914 W, 0.003 Fe, 0.0026 Si, 0.002 O_2, 0.001 Cu, H_2, Ni, N_2, and P; sample supplied by Union Carbide Muetals Co.
2	64	1963	1200-2400	≤ 2.0	Wire sample	99.90 W, residuals WC; out gassed and sealed in $< 1 \times 10^{-6}$ mm Hg glass envelpe.
3	137	1962	2273-2673	≤ 1.2		99.95 W; powder metallurgy; sealed under 10^{-2} - 10^{-3} mm Hg at 1.05 atm.
4	67	1962	1550-2880	$\pm$ 10.2		
5	116	1961	1273-2893	1.0		99.8 W; degassed at 2150 C for 2 hrs; sealed under vacuum.
6	138	1962	2673-3093	≤ 0.5		99.95 W, 0.05 impurities; polished surface.
7	139	1962	273-2600	≤ 1.2		Same as above.
8	69	1959	12-273			99.9917 W, 0.002-0.006 Fe, 0.002 Si, and 0.0001-0.0003 Cu, Mg, Mo, and Ni; sample rods prepared by powder metallurgy.
9	71	1960	1089-1700			99.9 W, and <0.02 R_2O_3.
10	140	1962	1500-2200	<4.0		Sealed under argon atmosphere.
11	90	1958	13-78	0.5	W-1	99.99 W, and traces of Ag, Cu, Fe, Mn, Ni, and Si.
12	90	1958	13-93	0.5	W-2	99.985 W, 0.01 Na_2O, and 0.005-0.008 Ni.
13	80	1963	600-3100	<1.2		≤ 0.05 impurities; prepared by powdered metallurgy.
14	201	1929	373-1173			
15	164	1932	273-1873			
16	202	1950	1.5-2.9	± 0.5		99.9 W.
17	175	1953	2-20			99.9 W; sample supplied by Fansteel Metallurgical Corp; under helium atmosphere.
18	203	1957	4-15	2.0		99.9 W, and ≤ 0.02 H_2O_3.

DATA TABLE NO. 64 SPECIFIC HEAT OF TUNGSTEN

[Temperature, T, K; Specific Heat, C_p, Cal g^{-1} K^{-1}]

T	C_p
CURVE 1	
533.1	3.1 x 10^{-2}
810.9	3.4
1088.7	3.8
1366.5	4.5
1644.3	4.9
1922.0	4.9
2199.8	4.9
2477.6	4.7
2755.4	4.6
3033.1	1.2 x 10^{-1}
CURVE 2	
1200	3.65 x 10^{-2}
1300	3.70
1400	3.74
1500	3.79
1600	3.83
1700	3.88
1800	3.94
1900	4.00
2000	4.06
2100	4.14
2200	4.24
2300	4.33
2400	4.44
CURVE 3	
2273	4.367 x 10^{-2}*
2373	4.432*
2473	4.498
2573	4.563
2673	4.680
CURVE 4	
1550	3.1 x 10^{-2}
1810	3.7
2080	4.2
2340	4.5
2610	4.7
2880	4.8
CURVE 5	
1273	3.606 x 10^{-2}
1373	3.650
1473	3.693
1573	3.737
1673	3.780
1773	3.824
1873	3.868
1973	3.911
2073	3.955
2173	3.998
2273	4.042
2373	4.086
2473	4.129
2573	4.173*
2673	4.216*
2773	4.260
2873	4.304*
2893	4.332
CURVE 6	
2673	4.216 x 10^{-2}*
2723	4.238*
2773	4.260*
2823	4.282*
2873	4.304*
2923	4.325*
2973	4.347
3023	4.369*
3073	4.391*
3093	4.400
CURVE 7	
273.15	3.171 x 10^{-2}
298.15	3.184
300	3.185*
400	3.241
500	3.297
600	3.353
700	3.410
800	3.467
900	3.525
1000	3.583
CURVE 7 (cont.)	
1100	3.642 x 10^{-2}
1200	3.701*
1300	3.761*
1400	3.821
1500	3.882
1600	3.943
1700	4.005
1800	4.067
1900	4.130
2000	4.193
2100	4.257*
2200	4.321*
2300	4.385*
2400	4.450*
2500	4.516*
2600	4.582*
CURVE 8	
11.81	9.410 x 10^{-5}
11.87	9.355*
12.46	1.066 x 10^{-4}
12.57	1.093
14.07	1.479
14.24	1.572
14.46	1.626
14.65	1.681
14.82	1.729
15.92	2.148
16.51	2.371
17.05	2.611
17.62	2.877
18.03	3.100
18.38	3.269
19.06	3.595
19.42	3.878
20.23	4.411
20.44	4.520
20.50	4.629
22.44	6.402
25.85	1.045 x 10^{-3}
29.74	1.707
33.74	2.568
38.22	3.761
CURVE 8 (cont.)	
43.00	5.218 x 10^{-3}
48.50	7.066
53.94	8.903
59.40	1.069 x 10^{-2}
59.80	1.084*
64.47	1.240
65.57	1.265
70.83	1.420
74.51	1.522
79.97	1.664
85.29	1.782
91.31	1.921
96.69	2.030
101.99	2.114
107.25	2.205
113.74	2.293
119.40	2.363
125.05	2.435
130.67	2.486
137.97	2.562
144.02	2.628
149.89	2.671
155.81	2.702
163.16	2.751
169.36	2.790
175.38	2.818
182.89	2.852
189.29	2.882
195.41	2.900
198.67	2.911
201.36	2.932*
202.10	2.920*
204.38	2.941*
207.70	2.949*
209.94	2.964
213.17	2.974*
215.33	2.975*
219.42	2.988*
222.53	3.006
225.11	3.014*
228.57	3.022*
230.61	3.024*
234.27	3.037
237.34	3.046*
CURVE 8 (cont.)	
239.82	3.055 x 10^{-2}*
243.10	3.056*
248.71	3.081
250.22	3.075*
254.97	3.095*
255.98	3.093*
260.78	3.102*
262.77	3.101
268.65	3.112*
273.84	3.111*
CURVE 9*	
1088.9	3.571 x 10^{-2}
1144.4	3.594
1200.0	3.617
1255.5	3.640
1311.1	3.662
1366.6	3.686
1422.2	3.709
1477.8	3.731
1533.3	3.754
1588.9	3.778
1644.4	3.800
1700.0	3.823
CURVE 10*	
1500	3.726 x 10^{-2}
1600	3.797
1700	3.867
1800	3.938
1900	4.009
2000	4.079
2100	4.150
2200	4.221
CURVE 11	
13.07	1.452 x 10^{-4}
15.16	2.382
17.03	2.225
18.80	4.313
20.33	4.259
CURVE 11 (cont.)	
22.23	6.527 x 10^{-4}
23.88	8.104
25.52	9.953
27.33	1.278 x 10^{-3}
29.35	1.621
31.60	2.089
34.01	2.714
36.91	3.524
39.05	4.144
41.43	4.803
43.74	5.461
45.90	6.222
48.02	6.935
52.09	8.322
57.66	1.055 x 10^{-2}
62.22	1.216
64.11	1.261
68.18	1.339
73.21	1.526
75.26	1.586
77.54	1.617
CURVE 12	
Series 1	
13.54	1.474 x 10^{-4}
15.73	2.257
19.22	3.894*
20.83	5.395
22.37	6.744
24.03	8.267*
25.97	1.082 x 10^{-3}
28.13	1.420
30.42	1.838
32.79	2.295
35.08	2.910
42.61	5.042
47.16	6.548
51.09	7.908
56.10	9.779
60.14	1.121 x 10^{-2}
65.07	1.260*
72.95	1.491*

*Not shown on plot

DATA TABLE NO. 64 (Continued)

CURVE 12 (cont.)

T	C_p
78.07	1.634 x 10^{-2}*
83.08	1.740
88.18	1.879
93.22	1.978
Series 2	
13.65	1.534 x 10^{-4}
15.35	2.061
17.00	2.801
18.90	3.796
21.24	5.167

CURVE 13

T	C_p
600	3.353 x 10^{-2}*
650	3.381
700	3.410*
750	3.438
800	3.467*
850	3.496
900	3.525*
950	3.554
1000	3.583*
1100	3.642*
1200	3.701*
1300	3.761*
1400	3.821*
1500	3.882*
1600	3.943*
1700	4.005*
1800	4.067*
1900	4.130*
2000	4.193*
2100	4.257*
2200	4.321*
2300	4.396*
2400	4.450*
2500	4.516*
2600	4.582*
2700	4.649*
2800	4.716
2900	4.783*
3000	4.851
3100	4.920

CURVE 14*

T	C_p
373	3.252 x 10^{-2}
423	3.274
473	3.297
523	3.320
573	3.342
623	3.365
673	3.388
723	3.410
773	3.433
823	3.456
873	3.478
923	3.501
973	3.524
1023	3.546
1073	3.569
1123	3.592
1173	3.614

CURVE 15*

T	C_p
273.15	3.199 x 10^{-2}
373.15	3.247
473.15	3.295
573.15	3.343
673.15	3.391
773.15	3.438
873.15	3.486
973.15	3.533
1073.15	3.579
1173.15	3.626
1273.15	3.672
1373.15	3.718
1473.15	3.764
1573.15	3.809
1673.15	3.855
1773.15	3.900
1873.15	3.935

CURVE 16

T	C_p
Series 1	
2.063	5.999 x 10^{-5}
2.140	6.391
2.229	7.180
2.273	6.826
2.320	6.968
2.377	7.245
2.430	7.588
2.675	8.376
2.738	9.056
2.895	9.546
2.955	9.954
Series 2	
1.686	4.73
1.793	5.30
1.849	5.33
1.953	5.744
2.027	6.076
2.128	6.282
2.135	6.641
Series 3	
1.533	4.35
1.593	4.54
1.614	4.75
1.655	4.77*
1.633	4.82*
1.823	5.35*
2.088	6.282*
2.362	7.125*
2.464	7.451
2.806	8.642
3.093	1.047 x 10^{-4}
3.505	1.229
1.792	5.16 x 10^{-5}
1.848	5.33*
1.974	5.771*
1.825	5.444

CURVE 17

T	C_p
Series 1*	
15.19	1.90 x 10^{-4}
15.27	1.75
15.37	1.77
15.47	1.81
15.54	2.19
15.64	2.25
15.73	1.89
15.84	2.15
15.94	2.09
16.04	2.17
16.14	2.31
16.24	2.30
16.30	2.27
16.40	2.18
16.53	1.88
16.68	2.03
16.83	2.53
17.50	2.82
17.67	2.69
17.83	2.34
18.02	2.47
18.26	3.22
18.57	3.29
18.86	3.40
19.17	3.46
19.51	4.37
19.84	4.71
20.13	4.42
Series 2	
2.06	3.57 x 10^{-6}
2.13	3.38
2.29	4.15
2.34	4.61
2.39	4.25
2.48	5.07
3.10	6.64
3.19	7.29
3.26	7.29*
3.34	8.81
3.47	7.78
3.61	9.25

CURVE 17 (cont.)

T	C_p
Series 2 (cont.)	
3.75	1.01 x 10^{-5}
4.02	1.32
4.18	1.40
Series 3	
2.27	3.26 x 10^{-6}
2.50	4.12
2.72	5.35
2.73	4.87
2.92	4.97
4.04	1.07 x 10^{-5}

CURVE 18

T	C_p
Series 1	
4.431	1.040 x 10^{-5}
4.980	1.300
5.484	1.495
5.980	1.833
6.473	2.185
6.958	2.574
7.433	2.997
7.906	3.523
8.383	4.080
8.857	4.641
9.334	5.148
10.316	6.578
10.814	7.371
11.304	8.398
11.828	9.412
12.359	1.049 x 10^{-4}*
12.939	1.050
13.448	1.318
13.949	1.452*
14.448	1.632*
14.926	1.863
Series 2	
3.978	9.10 x 10^{-6}
4.546	1.040 x 10^{-5}*

CURVE 18 (cont.)

T	C_p
Series 2 (cont.)	
5.187	1.326 x 10^{-5}*
5.675	1.599*
6.154	1.924*
6.628	2.275*
7.101	2.652*
7.574	3.120*
8.045	3.614*
8.519	4.212*
8.988	4.758*
9.451	5.330*
9.921	6.110*
10.435	6.773*
10.939	7.670*
11.432	8.580*
11.922	9.698*
12.413	1.062 x 10^{-4}*
12.909	1.180*
13.404	1.301*
13.901	1.431*
14.394	1.581*
14.889	1.895*

* Not shown on plot

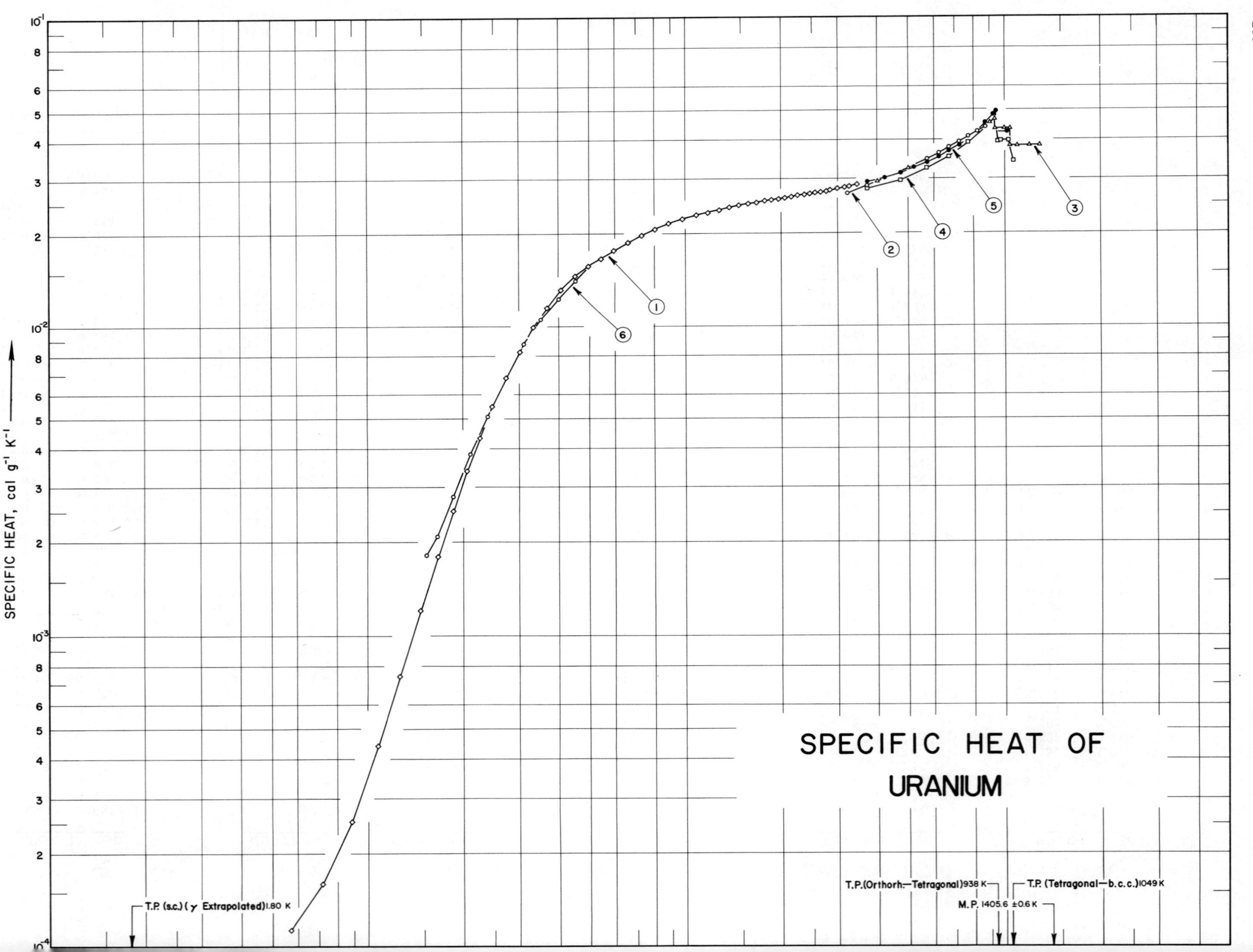

SPECIFIC HEAT OF
URANIUM
SPECIFIC HEAT, cal g⁻¹ K⁻¹
10⁻¹
8
6
5
4
3
2
10⁻²
8
6
5
4
3
2
10⁻³
8
6
5
4
3
2
10⁻⁴
T.P. (s.c.) (γ Extrapolated) 1.80 K
T.P. (Orthorh.—Tetragonal) 938 K
T.P. (Tetragonal—b.c.c.) 1049 K
M.P. 1405.6 ±0.6 K
1
2
3
4
5
6

SPECIFICATION TABLE NO. 65 SPECIFIC HEAT OF URANIUM

(Impurity < 0.20% each; total impurities < 0.50%)

[For Data Reported in Figure and Table No. 65]

Curve No.	Ref. No.	Year	Temp. Range, K	Reported Error, %	Name and Specimen Designation	Composition (weight percent), Specifications and Remarks
1	141	1960	6-348	≤ 5.0	α - uranium	~99.99 U, 0.0018 C, 0.0015 O_2, 0.0012 Si, 0.0005 Al, 0.0005 N_2, 0.0002 Cr, 0.0002 Fe, 0.0001 Cu, and 0.0001 Mg, annealed for 0.5 hr. at 600 - 650 C in a vacuo and cooled slowly to room temperature; sealed under helium atm.
2	7	1959	323-873	≤ 2.0		99.72 U.
3	142	1947	298-1300			99.71 U; sealed under helium in silica glass bulb.
4	143	1956	373-1073			0.1 C, 0.046 Si, 0.017 N_2, 0.01 Ni, 0.0095 Fe, 0.0035 Cr, 0.0007 Mn, 0.0005 Cu, and 0.0002 Co.
5	144	1947	273-1173			99.96 U, 0.015 C, 0.003 N_2, 0.002 O_2, and 0.0005 H_2; capsulated in Nichrome V.
6	145	1942	15-298		U_{238}	99.71 U; cased and cleaned.

DATA TABLE NO. 65 SPECIFIC HEAT OF URANIUM

[Temperature, T, K; Specific Heat, C_p, Cal g^{-1} K^{-1}]

T	C_p
CURVE 1	
Series 1	
300.104	2.779 x 10^{-2}
307.465	2.793*
317.478	2.815*
327.514	2.834
337.489	2.853*
347.549	2.868
Series 2	
5.703	1.134 x 10^{-4}
7.204	1.596
8.906	2.521
10.784	4.411
12.644	7.394
14.648	1.206 x 10^{-3}
16.653	1.790
18.645	2.508
20.662	3.357
22.687	4.323
24.927	5.478
27.490	6.823
30.297	8.285
33.432	9.915
36.911	1.160 x 10^{-2}
40.832	1.325
45.148	1.465
49.741	1.573
Series 3	
49.520	1.566 x 10^{-2}*
54.470	1.665
59.986	1.768
66.134	1.875
73.049	1.971
80.640	2.065
88.873	2.150
98.141	2.219
108.041	2.280
118.131	2.336
128.225	2.384
138.123	2.425
147.980	2.460
157.765	2.489
CURVE 1 (cont.)	
Series 3 (cont.)	
167.755	2.518 x 10^{-2}
177.804	2.544
186.311	2.563
196.331	2.586
206.324	2.606
216.322	2.626
226.307	2.651
236.222	2.664
246.154	2.684
256.120	2.701
266.096	2.717
278.483	2.743
285.104	2.754
294.998	2.777*
304.950	2.789*
314.904	2.806*
CURVE 2	
323.15	2.68 x 10^{-2}
373.15	2.84
573.15	3.45
623.15	3.62
673.15	3.78
723.15	3.94
773.15	4.10
823.15	4.25
873.15	4.40
CURVE 3	
(α) 298	2.758 x 10^{-2}*
300	2.760*
400	2.951
500	3.230
600	3.543*
700	3.873*
800	4.212*
900	4.555
935	4.676
(β) 935	4.360
950	4.360*
1000	4.360
1045	4.360
CURVE 3 (cont.)	
(γ) 1045	3.822 x 10^{-2}
1100	3.822
1200	3.822
1300	3.822
CURVE 4	
373.15	2.78 x 10^{-2}
473.15	2.96
573.15	3.24
673.15	3.53
773.15	3.92
873.15	4.37*
933.15	4.66*
953.15	3.94
973.15	3.96
1043.15	3.97
1063.15	3.40*
1073.15	3.40
CURVE 5	
(α) 273.15	2.75 x 10^{-2}*
323.15	2.83*
373.15	2.919
423.15	3.022
473.15	3.135
523.15	3.257
573.15	3.388
623.15	3.529
673.15	3.681
723.15	3.846*
773.15	4.031*
823.15	4.253*
873.15	4.521
923.15	4.818*
941.15	4.93
(β) 941.15	4.262
973.15	4.262
1023.15	4.262
1047.15	4.262*
(γ) 1047.15	3.843*
1073.15	3.843*
1123.15	3.843*
1173.15	3.843*
CURVE 6	
15.38	1.815 x 10^{-3}
16.61	2.083
18.55	2.793
21.01	3.839
23.91	5.087
27.55	6.885*
31.19	8.737
35.32	1.063 x 10^{-2}
40.10	1.241
45.25	1.415
50.49	1.567*
55.53	1.685*
60.63	1.777*
65.78	1.868*
71.57	1.956*
75.99	2.019*
81.52	2.081*
87.05	2.130*
92.96	2.182*
98.93	2.237*
104.81	2.283*
111.00	2.324*
117.40	2.348*
123.48	2.369*
128.23	2.393*
134.58	2.417*
140.90	2.446*
147.06	2.467*
153.35	2.488*
159.76	2.505*
166.65	2.524*
173.61	2.543*
180.04	2.558*
186.37	2.575*
192.57	2.585*
198.47	2.606*
205.12	2.618*
211.07	2.635*
217.76	2.644*
224.89	2.661*
231.88	2.666*
238.73	2.680*
245.59	2.695*
252.89	2.706*
259.36	2.720*
266.32	2.724*
CURVE 6 (cont.)	
273.02	2.740 x 10^{-2}*
280.16	2.749*
287.50	2.750*
294.88	2.754*
297.71	2.761*

*Not shown on plot

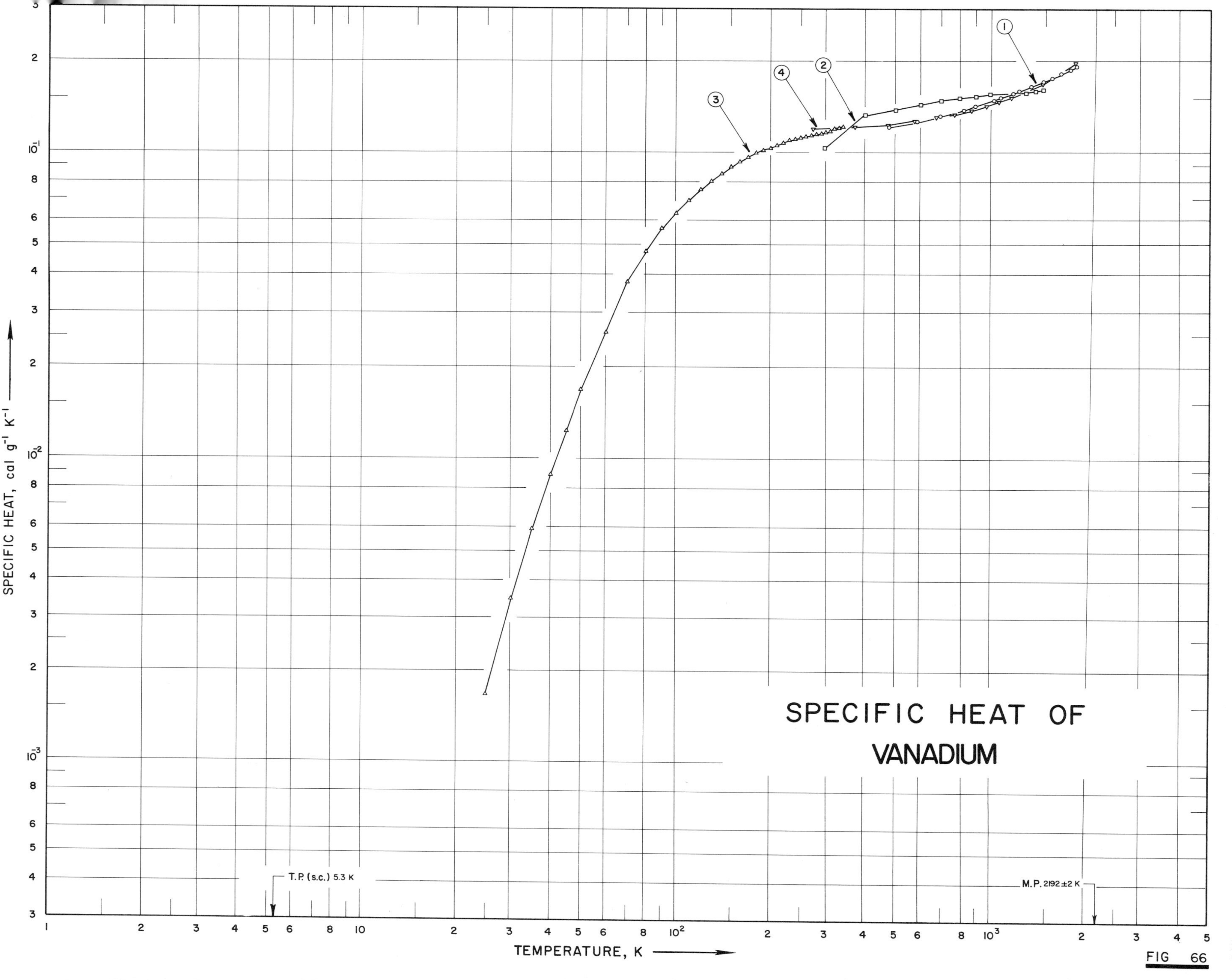

SPECIFIC HEAT OF
VANADIUM
SPECIFIC HEAT, cal g⁻¹ K⁻¹
TEMPERATURE, K
T.P. (s.c.) 5.3 K
M.P. 2192±2 K
1
2
3
4

FIG 66

SPECIFICATION TABLE NO. 66 SPECIFIC HEAT OF VANADIUM

(Impurity < 0.20% each; total impurities < 0.50%)

[For Data Reported in Figure and Table No. 66]

Curve No.	Ref. No.	Year	Temp. Range, K	Reported Error, %	Name and Specimen Designation	Composition (weight percent), Specifications and Remarks
1	146	1961	479-1894	3.0		99.74 V, 0.073 O_2, 0.048 Fe, 0.043 N, and 0.042 C; hot rolled; annealed; sealed under helium atm.; density = 378 lb ft^{-3}.
2	147	1962	298-1485	1.7		99.8 V, 0.1 C, 0.07 O_2, and 0.03 N_2.
3	148	1961	25-340			99.8 V, 0.05 Fe, 0.01 Hf, 0.01 Nb, 0.001 each Co, Cr, Mg, Ni, and Si, < 0.001 Mn, and 0.0001 Cu; carbothermic vanadium powder; annealed in vacuum for few hrs. at 800 K.
4	70	1934	273-1873	≤ 0.2		Purest vanadium; sample supplied by the Vanadium Corp. of America.
5	373	1936	54-297			≥ 99.5 V; sample supplied by the Vanadium Corp. of America; pellets of -8 to + 35 mesh; density = 6.009 g cm^{-3} at 22.5 C.
6	374	1954	1.2-5			Annealed in vacuo (pressure <3 x 10^{-6} mm Hg) for 3 hrs. at 850 C and then cooled slowly at about 50 C per hr.

DATA TABLE NO. 66 SPECIFIC HEAT OF VANADIUM

[Temperature, T, K; Specific Heat, C_p, Cal g^{-1} K^{-1}]

T	C_p
CURVE 1	
479.26	1.213 x 10^{-1}
586.48	1.267
695.93	1.322
823.15	1.386
896.48	1.423
1034.82	1.493
1084.82	1.518
1183.15	1.568
1245.37	1.599
1355.37	1.654
1484.26	1.719
1582.04	1.769
1694.82	1.825
1700.93	1.829*
1802.04	1.880
1894.26	1.926
CURVE 2	
298	1.043 x 10^{-1}
300	1.050*
400	1.328
500	1.387
600	1.448
700	1.488
700	1.488
800	1.516
900	1.537
1000	1.553
1200	1.579*
1300	1.590
1400	1.599
1485	1.607
CURVE 3	
25	1.67 x 10^{-3}
30	3.47
35	5.89
40	8.87
45	1.227 x 10^{-2}
50	1.672
60	2.591
70	3.779
80	4.780
90	5.63
CURVE 3 (cont.)	
100	6.34 x 10^{-2}
110	6.99
120	7.56
130	8.07
140	8.54
150	8.97
160	9.34
170	9.68
180	9.95
190	1.02 x 10^{-1}
200	1.04
210	1.06
220	1.08
230	1.10
240	1.11
250	1.12
260	1.13
270	1.14
280	1.15
290	1.16
300	1.174
310	1.18
320	1.20
330	1.21
340	1.22
298.16	1.172*
CURVE 4	
273.15	1.198 x 10^{-1}
373.15	1.218
473.15	1.243
573.15	1.271
673.15	1.303
873.15	1.379
973.15	1.423
1073.15	1.470
1173.15	1.521
1273.15	1.575*
1373.15	1.633*
1473.15	1.694
1573.15	1.759*
1673.15	1.826*
1773.15	1.897*
1873.15	1.971
CURVE 5*	
54.2	2.320 x 10^{-2}
56.0	2.499
59.7	2.877
69.7	3.872
74.3	4.281
80.2	4.791
91.5	5.635
101.6	6.457
110.2	6.995
120.3	7.558
135.7	8.337
151.0	9.021
166.0	8.797
185.2	1.009 x 10^{-1}
206.7	1.052
225.2	1.083
254.8	1.112
272.1	1.133
296.5	1.159
CURVE 6* Series I	
1.199	3.618 x 10^{-6}
1.255	5.142
1.311	6.892
1.391	9.299
1.583	1.799 x 10^{-5}
1.783	3.044
1.981	4.621
2.179	6.493
2.377	8.816
2.573	1.112 x 10^{-4}
2.769	1.364
2.986	1.670
3.189	1.891
3.378	2.290
3.402	2.336
3.601	2.726
3.780	3.115
3.995	3.509
4.193	3.904
4.402	4.227
4.575	4.584
4.784	5.086
4.968	5.541
5.135	2.491
CURVE 6 (cont.)* Series II	
1.214	4.035 x 10^{-6}
1.272	5.466
1.326	7.146
1.517	1.469 x 10^{-5}
1.718	2.635
1.887	4.005
2.066	5.250
2.286	7.812
2.482	9.604
2.678	1.239 x 10^{-4}
2.891	1.530
3.096	1.834
3.301	2.182
3.483	2.491
3.608	2.740
3.904	3.345
4.110	3.767
4.305	4.054
4.508	4.410
4.685	4.814
4.815	5.152
4.919	5.414
4.949	5.466
4.980	5.536
5.013	5.325
5.048	4.340
5.088	2.637
5.132	2.477
5.178	2.501
5.226	2.529
5.343	2.604
5.447	2.674

* Not shown on plot

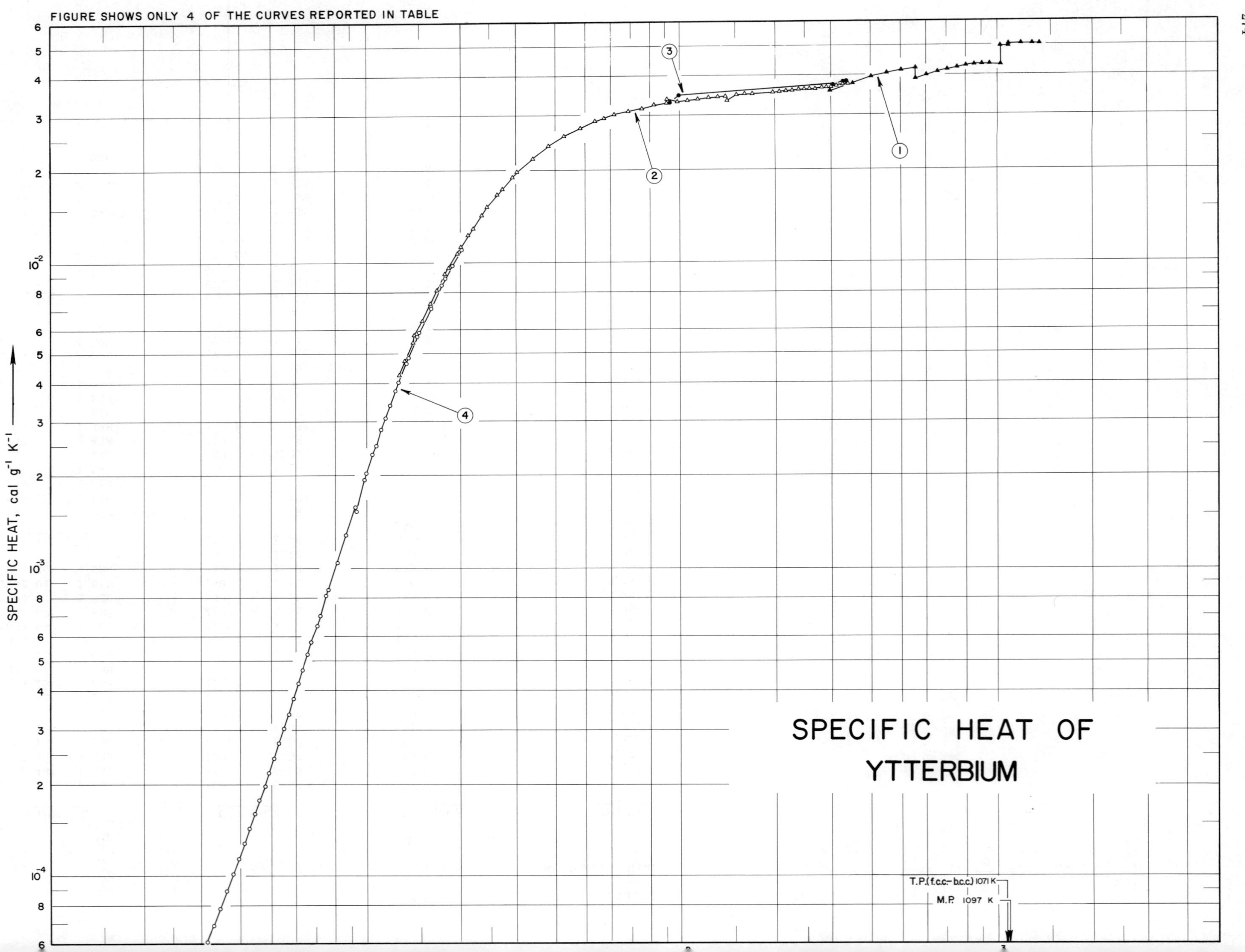
FIGURE SHOWS ONLY 4 OF THE CURVES REPORTED IN TABLE
SPECIFIC HEAT, cal g^{-1} K^{-1}
SPECIFIC HEAT OF
YTTERBIUM
T.P.(f.c.c.-b.c.c.) 1071 K
M.P. 1097 K
1
2
3
4
6
5
4
3
2
10^{-2}
8
6
5
4
3
2
10^{-3}
8
6
5
4
3
2
10^{-4}
8
6

SPECIFICATION TABLE NO. 67 SPECIFIC HEAT OF YTTERBIUM

(Impurity < 0.20% each; total impurities < 0.50%)

[For Data Reported in Figure and Table No. 67]

Curve No.	Ref. No.	Year	Temp. Range, K	Reported Error, %	Name and Specimen Designation	Composition (weight percent), Specifications and Remarks
1	36	1961	298-1373			> 99.92 Yb, < 0.06 Ca, < 0.05 Ta, 0.0225 O_2, < 0.02 Mg, 0.0117 C, < 0.01 each Fe, Si, and Y, 0.0035 H_2, < 0.002 Er, < 0.001 each Cr, Lu, Se, Tm, and 0.0001 N_2; cast into 1/2 inch dia. rods from which 3/4 inch long samples were prepared; sealed under helium atm. in tantalum crucible.
2	149	1964	13-341	< 3.0	Sample 1	99.9432 Yb, 0.026 O_2, 0.01 Y, 0.0083 H_2, 0.005 Er, 0.003 Lu, 0.002 Tm, 0.0015 F, and 0.001 Sc.
3	149	1964	14-335	< 3.0	Sample 2	99.877 Yb, 0.05 Ta, 0.032 O_2, 0.02 Er, 0.01 H_2, 0.01 Y, and 0.001 Sc.
4	205	1966	3-25	0.5		0.12 O_2, 0.06 Ca, 0.046 H_2, and 0.007 N_2.
5	375	1963	0.4-4	≤ 1.5		Same as above; specimen prepared by Research Chemicals.

DATA TABLE NO. 67 SPECIFIC HEAT OF YTTERBIUM

[Temperature, T, K; Specific Heat, C_p, Cal g^{-1} K^{-1}]

T	C_p
CURVE 1	
298.15	3.56 x 10^{-2}
300	3.57*
350	3.77
400	3.95
450	4.08
500	4.18
550	4.24*
553.15	4.24
553.15	3.91
600	4.02
650	4.11
700	4.19
750	4.25
800	4.31
850	4.34
900	4.36
950	4.36
1000	4.35*
1033.15	4.33
1033.15	4.99
1050	4.99*
1097.15	4.99
1097.15	5.08
1100	5.08*
1150	5.08*
1200	5.08
1250	5.08*
1300	5.08
1350	5.08*
1373.15	5.08
CURVE 2	
Series 1	
57.23	2.919 x 10^{-2}
61.80	3.006
68.30	3.080
75.50	3.153
82.77	3.221
90.27	3.272
97.96	3.307
105.99	3.341
114.42	3.373
123.14	3.403
CURVE 2 (cont.)	
Series 1 (cont.)	
131.35	3.430 x 10^{-2}
139.47	3.452
Series 2	
87.08	3.254 x 10^{-2}*
94.18	3.289*
101.26	3.320*
109.00	3.360*
117.92	3.387*
127.33	3.417*
Series 3	
280.19	3.663 x 10^{-2}
290.41	3.664
300.38	3.673
312.56	3.713
326.63	3.750
340.55	3.754
Series 4	
57.42	2.917 x 10^{-2}*
58.86	2.946*
60.52	2.964*
62.43	3.010*
64.50	3.041*
66.59	3.043*
68.71	3.086*
70.86	3.108*
73.05	3.130*
75.27	3.151*
77.44	3.173*
79.66	3.195*
81.93	3.218*
84.15	3.236*
86.33	3.254*
88.48	3.267*
88.57	3.268*
90.66	3.367
92.84	3.209*
95.19	3.293*
CURVE 2 (cont.)	
Series 4 (cont.)	
97.68	3.305 x 10^{-2}*
100.42	3.322*
103.36	3.356*
106.76	3.351*
111.90	3.366*
118.80	3.390*
126.22	3.413*
133.53	3.434*
140.72	3.347
Series 5	
134.15	3.434 x 10^{-2}*
142.34	3.456*
151.44	3.474
160.34	3.488
169.10	3.503
197.36	3.540
207.40	3.558
217.47	3.570
227.74	3.583
237.58	3.606
247.48	3.612
257.43	3.617
267.30	3.634
277.08	3.652*
286.80	3.669*
Series 6	
275.78	3.648 x 10^{-2}*
285.47	3.660*
295.08	3.682*
307.47	3.706*
322.57	3.739*
Series 7	
13.33	4.71 x 10^{-3}
15.14	6.40
16.85	8.11
18.42	9.57
20.08	1.128 x 10^{-2}
CURVE 2 (cont.)	
Series 7 (cont.)	
22.01	1.298 x 10^{-2}
24.39	1.517
27.26	1.736
30.47	1.961
34.12	2.174
38.36	2.384
42.98	2.559
48.06	2.714
53.97	2.852
60.48	2.974*
67.16	3.069*
Series 8	
126.13	3.413 x 10^{-2}*
133.52	3.432*
140.77	3.452*
149.21	3.467*
158.82	3.482*
168.25	3.500*
184.00	3.529*
192.65	3.536*
201.36	3.546*
210.39	3.452*
221.53	3.575*
232.49	3.588*
242.36	3.602*
252.31	3.617*
262.32	3.627*
272.25	3.645*
282.11	3.653*
292.07	3.674*
318.09	3.740*
333.29	3.778*
Series 9	
219.12	3.515 x 10^{-2}*
228.76	3.579*
238.30	3.593*
CURVE 2 (cont.)	
Series 10	
12.76	4.25 x 10^{-3}
14.28	5.77
16.06	7.33
17.86	9.13
19.56	1.075 x 10^{-2}
21.24	1.233
23.43	1.432
26.31	1.669
29.40	1.892
Series 11	
28.00	1.789 x 10^{-2}*
31.99	2.058*
35.49	2.253*
38.94	2.414*
42.75	2.548*
47.01	2.684*
51.57	2.802*
56.50	2.901*
61.80	2.997*
67.56	3.072*
74.12	3.140*
80.93	3.209*
86.97	3.254*
93.09	3.290*
CURVE 3	
Series 1	
266.96	3.566 x 10^{-2}*
273.35	3.648*
283.23	3.678*
294.02	3.657*
306.54	3.722*
319.91	3.757*
Series 2	
58.72	2.946 x 10^{-2}*
63.14	3.021*
68.91	3.088*
CURVE 3 (cont.)	
Series 2 (cont.)	
74.65	3.146 x 10^{-2}*
86.54	3.253*
92.75	3.285*
99.11	3.459
Series 3	
203.61	3.557 x 10^{-2}
212.65	3.572*
222.42	3.594*
232.08	3.605*
241.63	3.616*
251.10	3.640*
261.22	3.655*
272.07	3.669*
282.84	3.688*
293.52	3.707*
304.13	3.743
315.38	3.761*
327.29	3.827
Series 4	
14.13	5.40 x 10^{-3}
15.24	6.51 *
17.03	8.30 *
19.08	1.026 x 10^{-2}
21.37	1.242*
23.96	1.481*
26.58	1.691*
29.32	1.885*
32.07	2.061*
35.03	2.225*
38.90	2.413*
43.37	2.577*
Series 5	
298.62	3.727 x 10^{-2}*
305.57	3.750*
314.76	3.777*
325.06	3.818*
335.34	3.849

* Not shown on plot

T	C_p
CURVE 4	
Series 1	
3.099	6.082 x 10^{-5}
3.404	7.800
3.752	1.016 x 10^{-4}
4.082	1.286
4.401	1.597
4.726	1.966
5.067	2.421
5.436	3.003
5.842	3.750
6.262	4.662
6.679	5.724
7.104	6.973
7.562	8.493
8.060	1.041 x 10^{-3}
8.603	1.281
9.211	1.582
9.844	1.935
10.489	2.341
11.174	2.814
11.900	3.369
12.700	4.011
13.631	4.826
14.760	5.854
16.060	7.077
17.453	8.419
18.84	9.789
20.168	1.106 x 10^{-2}
21.522	1.233*
22.971	1.371*
24.529	1.519*
Series 2	
3.248	6.883 x 10^{-5}
3.577	8.913
3.914	1.144 x 10^{-4}
4.240	1.435
4.560	1.771
4.890	2.177
5.249	2.696
5.642	3.358
6.059	4.202
6.492	5.224
6.945	6.479
7.445	8.101
CURVE 4 (cont.)	
Series 2 (cont.)	
8.016	1.025 x 10^{-3}*
8.647	1.299*
9.303	1.630
9.986	2.027
10.720	2.494
11.510	3.063
12.401	3.771
13.403	4.615
14.573	5.676
16.127	7.137*
17.947	8.899
19.766	1.067 x 10^{-2}*
21.340	1.217*
22.606	1.335
23.871	1.456*
CURVE 5*	
Series I	
0.4275	1.959 x 10^{-6}
0.4428	2.004
0.4651	2.086
0.4959	2.214
0.5353	2.395
0.5823	2.634
0.6350	2.932
0.6915	3.282
0.7494	3.666
0.8090	4.097
0.8727	4.573
0.9403	5.126
1.0109	5.760
1.0848	6.443
1.1615	7.232
1.2422	8.117
1.3283	9.156
1.4200	1.031 x 10^{-5}
1.5201	1.180
1.6343	1.363
1.7626	1.600
1.9097	1.896
2.0682	2.266
2.2351	2.710
2.4120	3.240
2.6012	3.867
CURVE 5 (cont.)*	
Series 1 (cont.)	
2.7992	4.700 x 10^{-5}
3.0023	5.647
3.2097	6.719
3.4197	7.908
3.6341	9.336
3.8576	1.083 x 10^{-4}
Series II	
0.3697	1.802 x 10^{-6}
0.3864	1.852
0.4051	1.890
0.4268	1.950
0.4514	2.039
0.4781	2.146
0.5058	2.262
0.5360	2.412
0.5688	2.569
0.6043	2.779
0.6422	2.999
0.6811	3.215
0.7217	3.492
0.7652	3.779
0.8123	4.112
0.8645	4.503
0.9226	4.982
0.9850	5.511
1.0518	6.120
1.1238	6.825
1.2044	7.684
1.2924	8.696
1.3909	9.928
1.4957	1.143 x 10^{-5}
1.6309	1.357
1.7562	1.573
1.9045	1.883
2.0815	2.296
2.2782	2.831
2.4883	3.499
2.7076	4.312
2.9215	5.251
3.1315	6.300
3.3404	7.453
3.5487	8.736
3.7636	1.019 x 10^{-4}
3.9959	1.192

*Not shown on plot

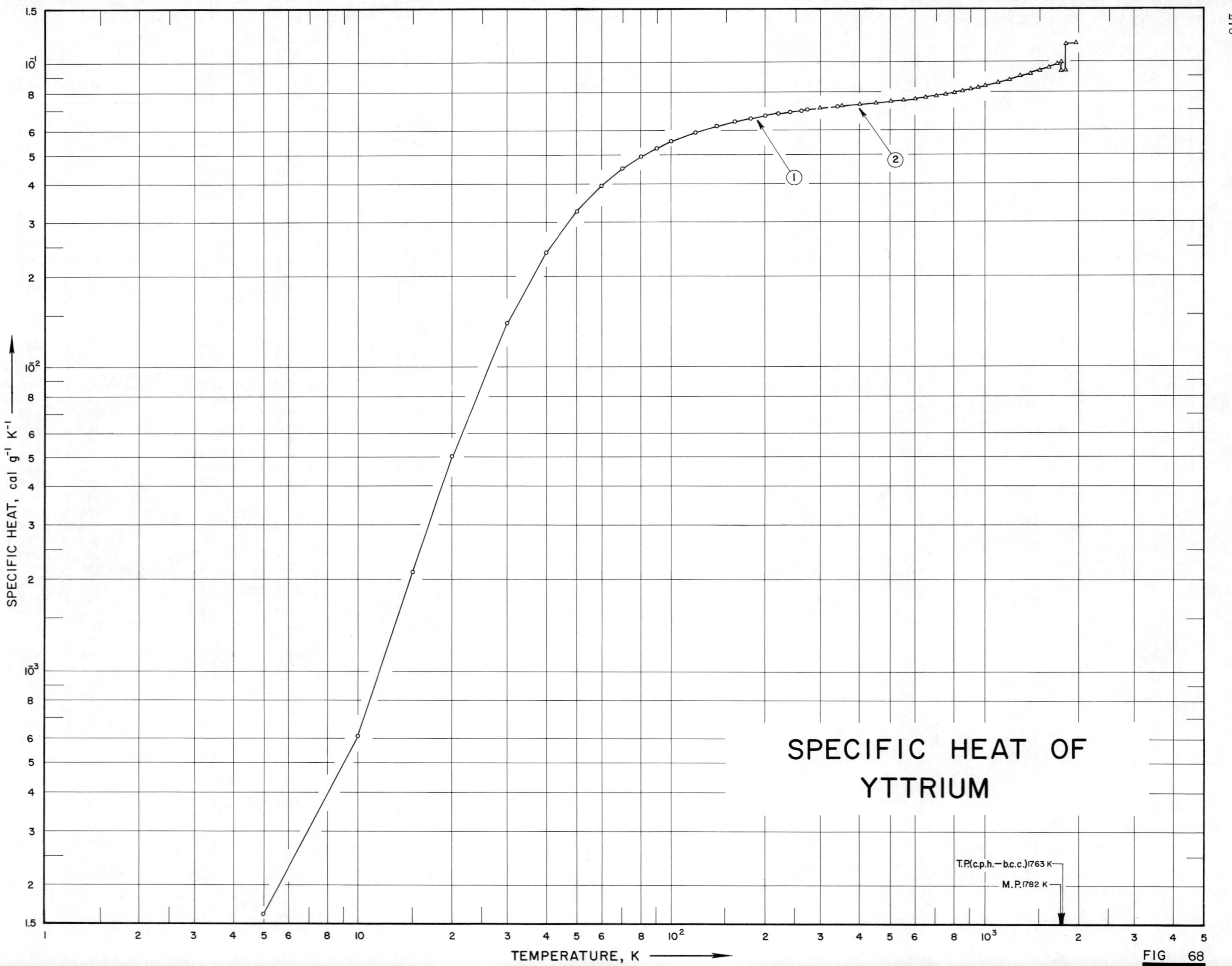

SPECIFIC HEAT OF YTTRIUM
T.P.(c.p.h.—b.c.c.) 1763 K
M.P. 1782 K
SPECIFIC HEAT, cal g⁻¹ K⁻¹
TEMPERATURE, K
1
2
FIG 68

SPECIFICATION TABLE NO. 68 SPECIFIC HEAT OF YTTRIUM

(Impurity < 0.20% each; total impurities < 0.50%)

[For Data Reported in Figure and Table No. 68]

Curve No.	Ref. No.	Year	Temp. Range, K	Reported Error, %	Name and Specimen Designation	Composition (weight percent), Specifications and Remarks
1	158	1960	5-340	0.3		0.5 total of Ca, Cr, Dy, Gd, Mg, 0.44 Ta, 0.025 N_2 and 0.015 C; after heat capacity measurements, chemical analysis showed 0.97 YOF, and 0.44 Ta; data corrected for impurities.
2	36	1961	298-1950			98.76 Y; < 1.0 Ta, < 0.05 each Ca, Er, Ho, Yb, < 0.01 each Fe, Gd, Si, < 0.005 each Dy, Mg, 0.025 O_2, 0.007 N_2, 0.0077 C, and 0.1 F; obtained as crystals by distillation; pressed into 1/2 inch dia. rod; sealed under reduced pressure of helium in two concentric tantalum crucibles.

DATA TABLE NO. 68 SPECIFIC HEAT OF YTTRIUM

[Temperature, T, K; Specific Heat, C_p, Cal $g^{-1}K^{-1}$]

T	C_p
CURVE 1	
5	1.609×10^{-4}
10	6.167
15	2.118×10^{-3}
20	5.067
30	1.418×10^{-2}
40	2.400
50	3.268
60	3.968
70	4.515
80	4.941
90	5.284
100	5.553
120	5.936
140	6.212
160	6.429
180	6.604
200	6.730
220	6.826
240	6.912
260	6.984
273.15	7.035
280	7.054*
298.15	7.110*
300	7.116*
320	7.169*
340	7.218
CURVE 2	
298.15	7.141×10^{-2}
300	7.141*
350	7.231
400	7.321
450	7.411
500	7.501
550	7.591
600	7.681
650	7.771
700	7.872
750	7.962
800	8.041
850	8.153
900	8.255
950	8.356
CURVE 2 (cont.)	
1000	8.457×10^{-2}
1050	8.547*
1100	8.648
1150	8.749*
1200	8.862
1250	8.963*
1300	9.064
1350	9.166*
1400	9.278
1450	9.402*
1500	9.492
1550	9.604*
1600	9.705
1650	9.818*
1700	9.930
1750	1.004×10^{-1}*
1758.15	1.007
1758.15	9.413×10^{-2}
1800	9.413*
1803.15	9.413
1803.15	1.158×10^{-1}
1850	1.158*
1900	1.158*
1950	1.158

*Not shown on plot

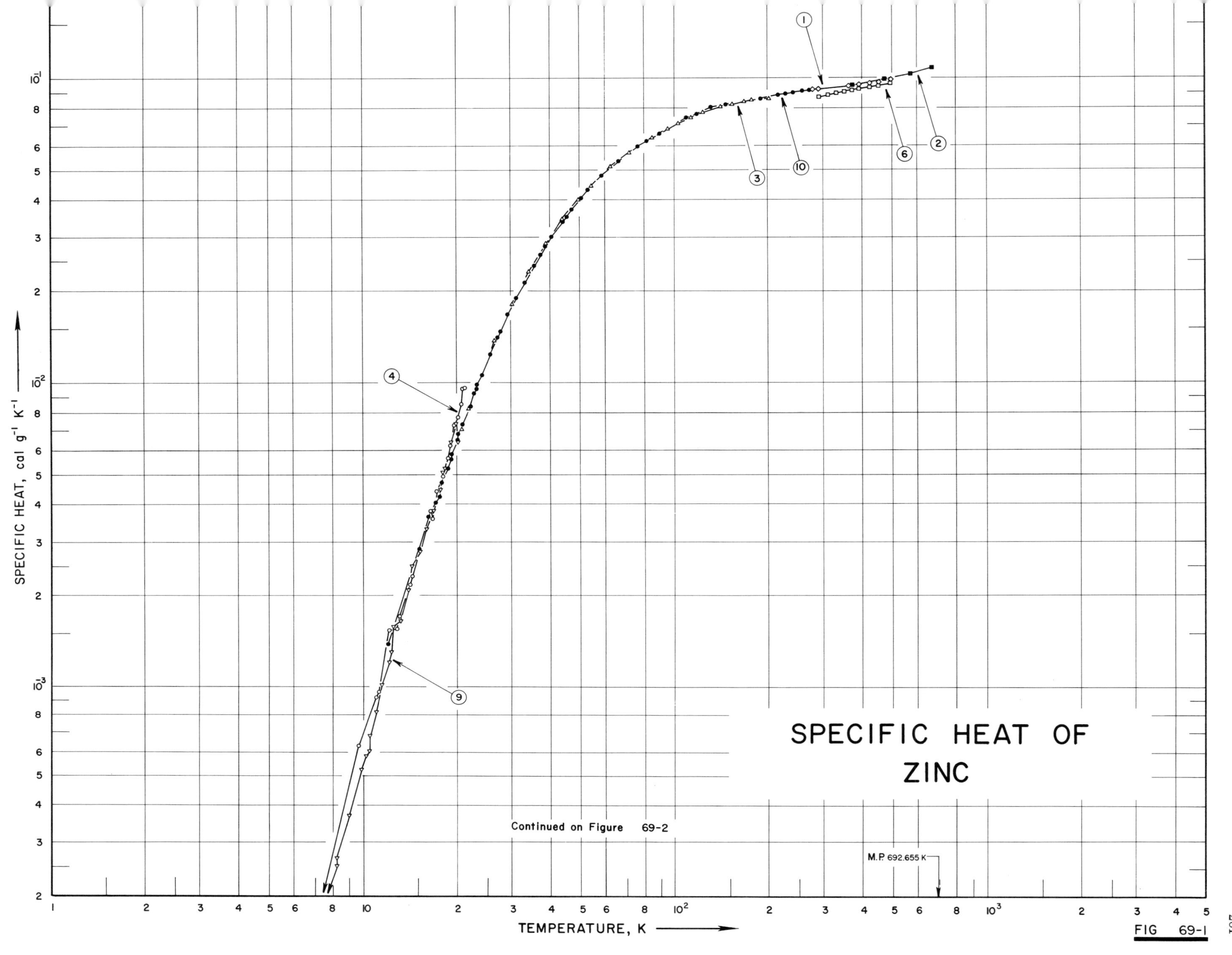
SPECIFIC HEAT OF ZINC
Continued on Figure 69-2
M.P. 692.655 K
TEMPERATURE, K
SPECIFIC HEAT, cal g⁻¹ K⁻¹
FIG 69-1

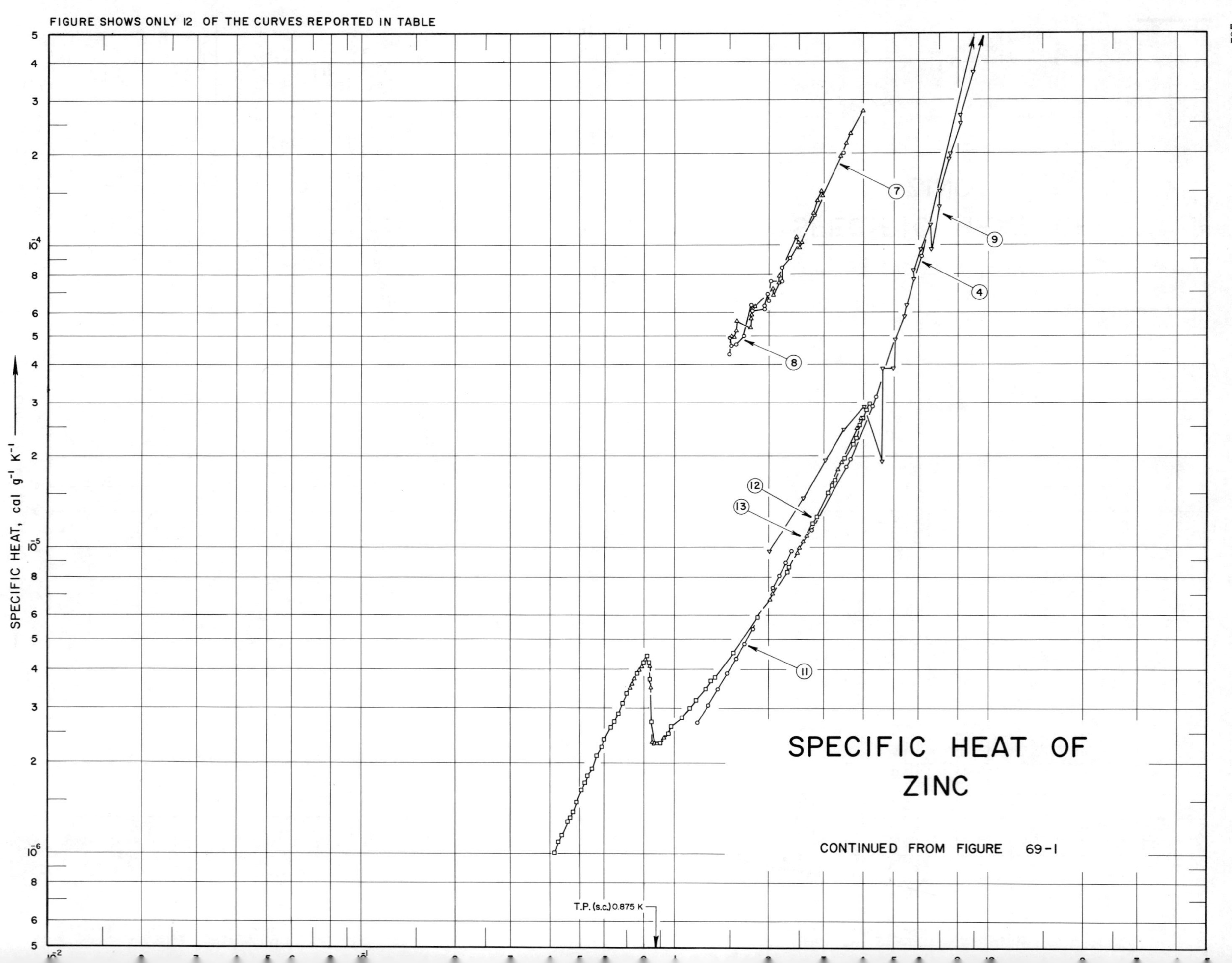
FIGURE SHOWS ONLY 12 OF THE CURVES REPORTED IN TABLE
SPECIFIC HEAT OF ZINC
CONTINUED FROM FIGURE 69-1
SPECIFIC HEAT, cal g⁻¹ K⁻¹
T.P. (s.c.) 0.875 K
4
7
8
9
11
12
13

SPECIFICATION TABLE NO. 69 SPECIFIC HEAT OF ZINC

(Impurity < 0.20% each; total impurities < 0.50%)

[For Data Reported in Figure and Table No. 69]

Curve No.	Ref. No.	Year	Temp. Range, K	Reported Error, %	Name and Specimen Designation	Composition (weight percent), Specifications and Remarks
1	206	1924	278-498			
2	179	1924	373-673			Pure commercial product; negligible amounts of Cd, Fe, and Pb.
3	207	1928	21-202			Kahlbaum's purity.
4	191	1932	1-21			Kahlbaum's purity.
5	208	1935	373-673			Pure Zn; fused in hydrogen atmosphere.
6	182	1936	193-393			99.98Zn, 0.009 Pb, 0.004 Cd, 0.0014 Fe, and 0.001 Cu.
7	202	1950	1.5-4	± 0.5		99.9 Zn, and 0.1 Pb; powder specimen.
8	202	1950	1.5-3.4	± 0.5		Same as above; fused Zn.
9	208	1955	2-20			
10	210	1959	12-273	0.2-0.7		99.995 Zn.
11	211	1959	1.2-2.4	<3.0		99.95 Zn.
12	216	1958	0.4-4.2			99.999 Zn; sample supplied by the American Smelting and Refining Co.; superconducting; zero magnetic field; sample maintained at 340 C for 40 hrs. and then returned to room temperature in 60 hrs.
13	216	1958	0.7-3.9			99.999 Zn; sample supplied by the American Smelting and Refining Co.; normal state; sample maintained at 340 C for 40 hrs. and then returned to room temperature in 60 hrs.
14	282	1924	10-373			
15	268	1926	348-1073			99.948 Zn, 0.03 Fe, and 0.022 Si.
16	376	1956	12-320			0.0003 Fe, 0.0002 Pb, 0.00005 each Cd, Cu, Sn, and 0.000006 As; melted at 650 - 700 C for 1 hr.; cooled, flushed with helium and outgassed again at 700 C for 5 min. outgassed again under vacuum for 15 min. at 300 - 380 C; under helium melted again and then permitted to solidify; annealed under helium atm. for 24 hrs. at 65 C and then for 9 days at room temperature.
17	1	1961	295	± 5.0		
18	377	1962	1.8-4.1			Large single crystals.

DATA TABLE NO. 69 SPECIFIC HEAT OF ZINC

[Temperature, T, K; Specific Heat, C_p, Cal $g^{-1}K^{-1}$]

CURVE 1

T	C_p
278	9.21×10^{-2}
291	9.27
363	9.42
393	9.53
423	9.63
438	9.69*
453	9.74
473	9.83*
498	9.92

CURVE 2

T	C_p
373	9.498×10^{-2}
473	9.972
573	1.0431×10^{-1}
673	1.0890

CURVE 3

T	C_p
20.74	7.09×10^{-3}
21.90	8.21
26.5	1.38×10^{-2}
30.3	1.806
34.3	2.319
38.9	2.863
43.9	3.454
49.2	4.006
54.2	4.458
62.8	5.188
71.8	5.736
85.5	6.447
95.6	6.841
104.1	7.146
114.7	7.482
124.7	7.782
140.9	8.109
154.8	8.290
169.2	8.440
178.6	8.502
201.9	8.706

CURVE 4

T	C_p
1.317	3.763×10^{-6}
1.358	3.931
1.399	3.747
2.149	7.235
2.159	6.975
2.771	1.155×10^{-5}
3.595	1.852
3.698	1.964
3.728	2.005*
4.313	2.921
4.419	3.152
6.200	9.246
6.260	9.474*
9.658	6.305×10^{-4}
11.002	9.202
11.270	9.604
12.122	1.542×10^{-3}
12.928	1.552
13.044	1.702
14.180	2.163
14.400	2.319
16.500	3.805
16.742	3.579
17.328	4.425
18.028	4.943
18.836	5.694
19.086	6.239
19.546	7.236
19.992	7.365*
20.160	7.750
20.762	8.530
20.916	9.529
21.374	9.645

CURVE 5*

T	C_p
373	9.443×10^{-2}
473	9.825
573	1.045×10^{-1}
673	1.133

CURVE 6

T	C_p
193	8.712×10^{-2}
203	8.793*
213	8.865
223	8.932*
233	8.992
243	9.047*
253	9.097
263	9.147*
273	9.195
283	9.238*
293	9.278
303	9.319*
313	9.360*
323	9.398
333	9.434*
343	9.472*
353	9.508
363	9.546*
373	9.582*
383	9.618*
393	9.654

CURVE 7

T	C_p
1.544	5.02×10^{-5}
1.567	4.97
1.596	5.25
1.599	5.66
1.619	5.63*
1.592	5.42*
1.766	5.37
1.772	5.77
1.780	5.92
1.820	6.30
2.097	7.21
2.098	6.87
2.174	7.57
2.178	7.95
2.479	1.07×10^{-4}
2.509	1.03

CURVE 7 (cont.)

T	C_p
2.529	9.87×10^{-5}
2.566	1.03×10^{-4}
2.799	1.294
2.881	1.412
2.967	1.524
2.992	1.461
3.422	1.958
3.483	2.003*
3.570	2.180
3.667	2.325
4.012	2.723

CURVE 8

T	C_p
1.515	4.33×10^{-5}
1.520	4.97
1.528	5.06*
1.692	5.02
1.777	6.38
1.793	6.09
1.585	5.31
1.543	4.64
1.590	4.70
1.909	6.18
1.910	6.36
1.901	6.23*
1.921	6.36*
2.005	6.95
2.008	6.55
2.052	7.63
2.221	8.44
2.230	7.66
2.363	9.10
2.831	1.29×10^{-4}*
2.823	1.27
3.034	1.48*
3.022	1.46*
3.498	2.027*
3.491	2.004

CURVE 9

Series 1

T	C_p
4.63	1.94×10^{-5}
5.02	3.89
5.47	5.83
5.85	7.777
6.17	9.71
6.62	9.71
7.07	1.35×10^{-4}
7.59	1.94
8.23	2.52
9.00	3.69
9.84	5.24
10.2	5.83
10.5	6.80
11.0	8.16
11.5	1.01×10^{-3}
12.1	1.20
12.4	1.30
13.2	1.65
12.5	1.59
14.4	2.50
14.0	2.08
15.2	2.78
16.0	3.30
16.8	3.79
17.7	4.45
18.1	5.03
18.4	5.22
19.2	6.35
20.1	6.37
19.8	7.38

Series 2

T	C_p
2.04	9.75×10^{-6}
2.61	1.46×10^{-5}
3.06	1.95
3.50	2.44
4.08	2.93

CURVE 9 (cont.)

Series 2 (cont.)

T	C_p
4.65	3.90×10^{-5}
5.10	4.88
5.54	6.34
5.86	8.29
6.31	9.26*
6.62	1.17×10^{-4}
7.07	1.51
7.64	2.00
8.22	2.68
8.92	3.75*
9.81	5.31*
10.5	6.05
10.9	8.00*
11.5	1.01×10^{-3}*

CURVE 10

T	C_p
12.05	1.381×10^{-3}
15.09	2.856
15.18	2.877*
16.34	3.644
17.06	4.065
17.61	4.248
17.90	4.716
18.17	4.898*
18.67	5.272
19.15	5.654
19.23	5.844
20.22	6.503
20.24	6.873
20.47	6.872*
20.97	7.337
21.17	7.464*
21.92	8.224*
21.99	8.366*
22.21	8.437
22.88	9.298
23.10	9.559

*Not shown on plot

DATA TABLE NO. 69 (continued)

T	C_p
CURVE 10 (cont.)	
23.32	9.882×10^{-3}
24.16	1.068×10^{-2}
24.23	1.082*
24.26	1.075*
25.68	1.247
27.09	1.412
27.75	1.485
29.34	1.684
31.14	1.901
33.18	2.137*
33.37	2.168*
35.68	2.435
37.24	2.631
38.58	2.792
40.67	3.020
44.01	3.396
45.33	3.544
47.00	3.720*
49.83	4.016*
50.47	4.054
53.14	4.317*
55.71	4.557*
56.03	4.598*
56.58	4.644*
58.84	4.825*
58.84	4.848*
61.05	4.984*
61.94	5.054*
61.97	5.075*
64.89	5.275*
65.91	5.348*
66.75	5.394
68.00	5.478*
70.10	5.622*
71.27	5.688*
72.88	5.758*
74.23	5.862*
74.34	5.750*
74.42	5.877*
74.84	5.874*
75.29	5.954*
75.77	5.917×10^{-2}*
76.24	6.013*
76.67	6.041*
77.12	6.059*
77.91	6.062*
78.46	6.081*
78.77	6.093*
79.90	6.165*
80.47	6.195*
81.08	6.223*
81.85	6.252*
82.31	6.311*
83.00	6.310*
83.51	6.255*
84.13	6.333*
84.88	6.394*
85.54	6.446*
85.82	6.443*
86.20	6.454*
86.86	6.515*
87.51	6.515*
88.16	6.581*
88.81	6.610*
90.18	6.656*
91.95	6.729*
94.97	6.856*
95.00	6.876*
98.97	7.015*
100.10	7.084*
102.81	7.178*
105.02	7.257*
106.55	7.308*
109.91	7.438*
110.22	7.413*
113.56	7.543*
114.91	7.577*
117.02	7.624*
119.71	7.708
120.63	7.743*
124.31	7.831*
124.50	7.809*
128.50	7.901×10^{-2}*
129.16	7.919*
132.63	8.043*
133.57	8.020*
137.19	8.100*
137.93	8.092*
142.12	8.192*
142.19	8.144*
146.62	8.239
147.10	8.262*
151.32	8.311*
152.06	8.328*
155.60	8.385*
157.12	8.389*
160.18	8.423*
161.95	8.484*
166.74	8.548*
169.64	8.533*
173.66	8.571*
174.06	8.583*
177.05	8.578
180.47	8.651
181.52	8.632*
185.82	8.693*
187.20	8.678*
192.16	8.700*
194.12	8.707*
196.61	8.750*
197.29	8.762*
198.47	8.744*
200.54	8.765*
201.04	8.776*
204.88	8.784*
205.76	8.781*
208.11	8.863*
210.08	8.850*
213.48	8.863*
216.83	8.902
220.05	8.927*
223.36	8.924*
226.46	8.966*
229.75	8.958×10^{-2}
232.09	8.981*
236.25	8.980*
239.31	9.041*
242.34	9.026
248.73	9.059*
251.63	9.100*
254.58	9.143*
258.27	9.131
262.28	9.185*
266.16	9.152*
269.77	9.159*
272.95	9.183
CURVE 11	
1.2	2.699×10^{-6}
1.3	3.064
1.4	3.462
1.5	3.896
1.6	4.369
1.7	4.884
1.8	5.442*
1.9	6.046*
2.0	6.700
2.1	7.405
2.2	8.165
2.3	8.981
2.4	9.856
CURVE 12	
0.4207	1.013×10^{-6}
0.4236	1.020*
0.4253	1.042*
0.4279	1.053*
0.4365	1.104
0.4458	1.159
0.4654	1.287
0.4735	1.338
0.4816	1.389
0.4900	1.448×10^{-6}*
0.4970	1.492
0.5163	1.649
0.5278	1.729
0.5387	1.824*
0.5474	1.883*
0.5556	1.923
0.5651	2.011*
0.5762	2.124
0.5982	2.263
0.6087	2.395*
0.6191	2.464*
0.6393	2.636
0.6537	2.749
0.6672	2.914*
0.6809	2.998*
0.6938	3.141*
0.7047	3.236*
0.7168	3.386*
0.7283	3.437*
0.7389	3.587*
0.7440	3.660*
0.7579	3.737*
0.7712	3.916*
0.7826	4.051*
0.7851	4.058*
0.7948	4.080*
0.7952	4.161
0.8063	4.241*
0.8064	4.230*
0.8187	4.424*
0.8232	4.391
0.8306	4.486*
0.8396	4.464*
0.8432	4.278
0.8492	3.755
0.8591	2.738*
0.8627	2.735*
0.8735	2.336*
0.8891	2.311*
0.8892	2.303*
0.9053	2.303×10^{-6}*
0.9183	2.347*
0.9462	2.431*
0.9723	2.512
0.9915	2.643
1.0790	2.837
1.1410	3.031
1.1950	3.210
1.2880	3.488
1.3310	3.729
1.3700	3.806
1.5630	4.581*
1.5770	4.625*
1.8730	5.985
1.8970	6.062*
2.3220	8.409
2.3520	8.775
2.7940	1.203×10^{-5}
2.8410	1.243*
2.8840	1.276
3.1260	1.506
3.2140	1.601
3.2950	1.671
3.5410	1.974
3.7560	2.201
3.8470	2.300
3.9400	2.537
4.0430	2.673
4.1450	2.852
4.2390	2.994
CURVE 13	
0.7331	3.525×10^{-6}
0.7450	3.645
0.7566	3.777
0.7866	4.029
0.7981	4.153
0.8196	4.369*
0.8244	4.439*
0.8352	4.486*

*Not shown on plot

DATA TABLE NO. 69 (continued)

T	C_p
CURVE 13 (cont.)	
0.8463	4.153 x 10^{-6}
0.8521	3.536
0.8689	2.369
0.8761	2.311*
0.8946	2.336
0.9104	2.362*
0.9405	2.450
2.0540	6.877
2.0780	7.009*
2.0990	7.141
2.5060	9.799
2.5480	1.002 x 10^{-5}
2.6060	1.057
2.6780	1.104
3.2700	1.693*
3.3640	1.817
3.4520	1.916
3.8510	2.475
3.9730	2.680
CURVE 14*	
10	5.507 x 10^{-4}
20	4.650 x 10^{-3}
30	1.392 x 10^{-2}
40	2.646
60	4.834
80	6.287
120	7.756
200	8.827
273.2	9.209
373.2	9.546
CURVE 15*	
348	9.40 x 10^{-2}
423	9.78
523	1.02 x 10^{-1}
623	1.07
723	1.24
773	1.24
873	1.24
973	1.24
1073	1.24

T	C_p
CURVE 16*	
Series I	
207.25	8.804 x 10^{-2}
210.13	8.809
214.02	8.816
218.93	8.847
Series II	
216.41	8.836 x 10^{-2}
221.69	8.851
226.96	8.888
232.21	8.920
237.43	8.958
242.63	8.991
247.80	9.018
252.95	9.047
258.08	9.102
263.20	9.147
268.27	9.168
273.32	9.197
278.36	9.214
283.80	9.241
289.62	9.272
Series III	
80.81	6.280 x 10^{-2}
84.43	6.493
87.94	6.650
91.89	6.816
96.31	6.976
101.12	7.146
106.33	7.313
111.40	7.463
116.37	7.594
121.24	7.715
125.47	7.810
130.18	7.912
134.83	8.006
139.42	8.085
143.96	8.163
148.74	8.246
153.77	8.311
158.74	8.365

T	C_p
CURVE 16 (cont.)*	
Series IV	
11.59	1.130 x 10^{-3}
12.93	1.590
14.21	2.150
15.90	3.320
18.44	5.270
20.95	7.530
23.36	1.007 x 10^{-2}
25.48	1.241
28.87	1.637
33.16	2.152
37.52	2.666
41.79	3.151
45.99	3.607
50.23	4.038
54.71	4.451
59.19	4.833
Series V	
65.32	5.340 x 10^{-2}
70.13	5.661
75.65	5.996
81.33	6.320
Series VI	
159.39	8.389 x 10^{-2}
164.33	8.449
169.90	8.512
177.10	8.582
185.24	8.663
193.38	8.730
201.26	8.778
Series VII	
297.55	9.298 x 10^{-2}
304.34	9.332
311.84	9.362
319.30	9.388

T	C_p
CURVE 17*	
295.15	8.8 x 10^{-2}
CURVE 18*	
1.85	5.76 x 10^{-6}
1.92	6.04
1.98	6.33
2.04	6.64
2.10	6.99
2.16	7.19
2.24	7.74
2.32	8.27
2.40	8.66
2.46	9.073
2.52	9.425
2.58	1.268 x 10^{-5}
2.63	1.027
2.70	1.067
2.80	1.158
2.90	1.244
3.00	1.329
3.11	1.432
3.22	1.544
3.34	1.674
3.44	1.782
3.55	1.927
3.71	2.129
3.89	2.390
3.99	2.569
4.03	2.619
4.09	2.739
4.15	2.839
4.16	2.878
4.18	2.890

*Not shown on plot

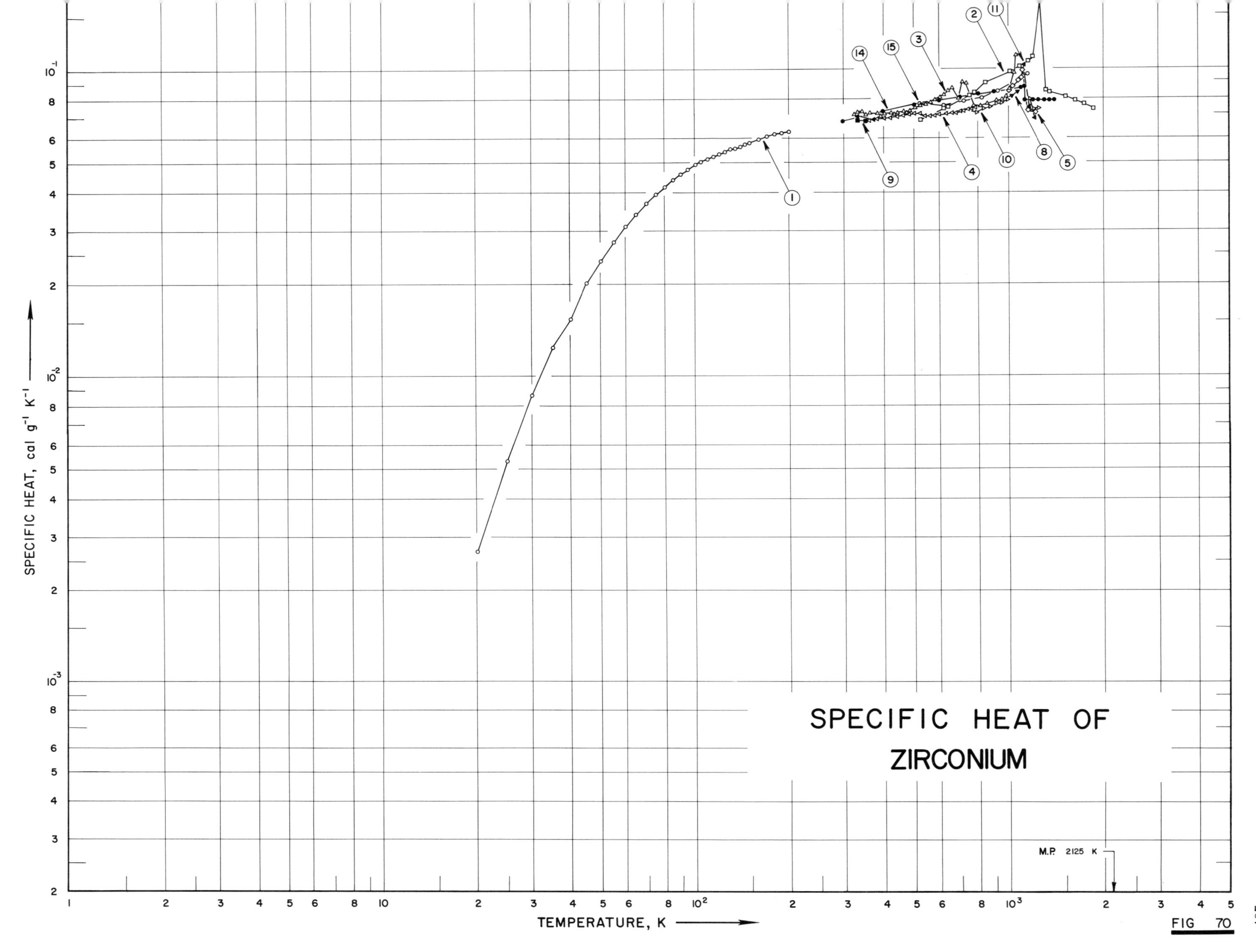
SPECIFIC HEAT OF
ZIRCONIUM
SPECIFIC HEAT, cal g^{-1} K^{-1}
TEMPERATURE, K
M.P. 2125 K
1
2
3
4
5
8
9
10
11
14
15
FIG 70

SPECIFICATION TABLE NO. 70 SPECIFIC HEAT OF ZIRCONIUM

(Impurity < 0.20% each; total impurities < 0.50%)

[For Data Reported in Figure and Table No. 70]

Curve No.	Ref. No.	Year	Temp. Range, K	Reported Error, %	Name and Specimen Designation	Composition (weight percent), Specifications and Remarks
1	128	1954	20-200	< 5.0		99.5 Zr.
2	146	1961	528-1863	3.0		99.95 Zr, 0.029 Fe, 0.017 C, 0.0045 Hf, and <0.031 all other elements; sealed under helium atmosphere; density = 405 lb ft^{-3}.
3	134	1957	323-1063		Zr - 300 ppm H alloy	99.966 Zr, and 0.03 H_2; homogenized 14 days at 1300 C; sealed under 0.01 μ Hg vacuum.
4	134	1957	363-883		Iodide zirconium	0.022 C, 0.015 O_2, 0.013 Fe, 0.0075 N_2, 0.007 Hf, 0.0004 K, 0.0035 H_2, 0.003 each Na, Ni, 0.0018 each Si, W, 0.0014 Cu, 0.0007 Cr, 0.0006 Ca, 0.0005 each Mg, Pb, 0.0004 each Al, Sn, 0.00035 Mo, 0.0002 Ti and 0.0001 Co; homogenized 14 days at 1300 C; sealed under 0.01 μ Hg vacuum.
5	134	1957	363-1223		Iodide zirconium	Same as above.
6	134	1957	543-863		Iodide zirconium	Same as above.
7	134	1957	333-743		Iodide zirconium	Same as above.
8	134	1957	353-1073		Iodide zirconium	Same as above.
9	134	1957	333-673		Iodide zirconium	Same as above.
10	134	1957	593-1253		Iodide zirconium	Same as above.
11	134	1957	333-1213		Iodide zirconium	Same as above.
12	134	1957	393-1033		Iodide zirconium	Same as above.
13	134	1957	413-973		Iodide zirconium	Same as above.
14	159	1950	298-1400			2.15 Hf; sample supplied by the Foote Mineral Co.; corrected for impurities.
15	150	1957	323-1154			99.91 Zr, 0.03 Fe, 0.02 C, and 0.0145 Hf.
16	378	1950	53-298			2.15 Hf; corrected for impurities.
17	379	1951	14-298			0.35 Fe, 0.05 Hf, 0.02 C, 0.004 N_2, and ~0.005 others; pellets; annealed under vacuum for 15 min. at 800 C; corrected for Fe impurity.
18	~~379~~ 429	1951	298-1800			Same as above.

DATA TABLE NO. 70 SPECIFIC HEAT OF ZIRCONIUM

[Temperature, T, K; Specific Heat, C_p, Cal g^{-1} K^{-1}]

T	C_p
CURVE 1	
20	2.675 x 10^{-3}
25	5.300
30	8.645
35	1.245 x 10^{-2}
40	1.646
45	2.028
50	2.393
55	2.753
60	3.093
65	3.401
70	3.694
75	3.957
80	4.199
85	4.415
90	4.605
95	4.770
100	4.930
105	5.053
110	5.161
115	5.270
120	5.362
125	5.449
130	5.522
135	5.593
140	5.671
145	5.743
150	5.814
155	5.892*
160	5.969
165	6.047*
170	6.108
175	6.165*
180	6.211
185	6.242*
190	6.263
195	6.288*
200	6.303
CURVE 2	
528.15	6.887 x 10^{-2}
625.38	7.486
643.15	7.591
749.82	8.252
780.93	8.443
CURVE 2 (cont.)	
844.27	9.079 x 10^{-2}
1009.26	9.849
1085.38	1.032 x 10^{-1}
1087.60	1.033*
1153.71	1.073
1199.27	1.102
1265.94	1.70
1288.72	1.70*
1322.05	8.589 x 10^{-2}
1352.05	8.410
1527.61	8.160
1640.94	7.923
1740.94	7.714
1862.61	7.460
CURVE 3	
Series 71	
323.15	7.175 x 10^{-2}
343.15	7.316
363.15	7.215
383.15	7.205
403.15	7.208
423.15	7.242
443.15	7.267
463.15	7.327
485.15	7.387
503.15	7.487
523.15	7.645
543.15	7.740
563.15	7.823
583.15	7.995
603.15	8.144
623.15	8.322
643.15	8.526
663.15	8.735
Series 72	
333.15	7.306 x 10^{-2}
373.15	7.201*
413.15	7.154
453.15	7.209
493.15	7.378
CURVE 3 (cont.)	
Series 72 (cont.)	
533.15	7.673 x 10^{-2}*
573.15	7.892*
613.15	8.219*
653.15	8.637*
693.15	8.077
733.15	9.077
773.15	7.661
813.15	7.638
853.15	7.740
Series 73	
353.15	7.118 x 10^{-2}
393.15	7.117
433.15	7.173
473.15	7.271
513.15	7.463*
553.15	7.749*
593.15	8.039*
633.15	8.405*
673.15	8.825*
713.15	9.178
753.15	8.355*
793.15	7.546*
833.15	7.717*
873.15	7.806*
913.15	7.917
953.15	7.982
983.15	8.185
1003.15	8.706
1023.15	9.765*
1043.15	9.795
1063.15	1.1249 x 10^{-1}
CURVE 4	
363.15	6.838 x 10^{-2}
383.15	6.903
403.15	6.926
423.15	6.993
443.15	7.041
463.15	7.079
483.15	7.160
CURVE 4 (cont.)	
503.15	7.201 x 10^{-2}
523.15	7.201
543.15	7.042
563.15	7.081
583.15	7.091
603.15	7.126
633.15	7.173
663.15	7.257
683.15	7.256
703.15	7.317
723.15	7.359
743.15	7.410
763.15	7.445
783.15	7.506
803.15	7.560*
823.15	7.590*
843.15	7.696*
863.15	7.779*
883.15	7.826*
CURVE 5	
363.15	6.849 x 10^{-2}*
403.15	7.009*
443.15	7.106*
483.15	7.232*
523.15	7.276*
563.15	7.124*
603.15	7.195*
683.15	7.237*
703.15	7.325*
723.15	7.366*
743.15	7.448*
763.15	7.512*
783.15	7.563*
803.15	7.646*
823.15	7.668*
843.15	7.699*
863.15	7.783*
883.15	7.857*
903.15	7.914*
923.15	7.991*
943.15	8.094*
973.15	8.223*
1003.15	8.543
1023.15	8.660*
CURVE 5 (cont.)	
1043.15	8.820 x 10^{-2}
1063.15	9.086*
1078.15	9.212
1088.15	9.235*
1098.15	9.296*
1108.15	9.464
1118.15	9.938
1165.15	7.361
1183.15	7.179*
1203.15	7.102*
1223.15	7.224
CURVE 6*	
543.15	7.243 x 10^{-2}
563.15	7.161
583.15	7.149
603.15	7.209
623.15	7.257
643.15	7.349
663.15	7.371
683.15	7.419
703.15	7.454
723.15	7.450
743.15	7.584
763.15	7.646
783.15	7.736
803.15	7.814
823.15	7.873
843.15	7.919
863.15	7.973
CURVE 7*	
333.15	7.012 x 10^{-2}
353.15	7.004
373.15	7.095
393.15	7.122
413.15	7.145
433.15	7.180
453.15	7.202
473.15	7.146
493.15	7.122
513.15	7.130
533.15	7.136
553.15	7.170
CURVE 7 (cont.)*	
573.15	7.178 x 10^{-2}
593.15	7.187
623.15	7.257
663.15	7.335
703.15	7.441
743.15	7.543
CURVE 8	
353.15	7.045 x 10^{-2}*
393.15	7.081*
433.15	7.152*
473.15	7.135*
513.15	7.118*
553.15	7.196*
593.15	7.232*
633.15	7.266*
673.15	7.289*
713.15	7.362*
753.15	7.448*
793.15	7.578*
833.15	7.718*
873.15	7.801*
913.15	7.937*
953.15	8.035*
993.15	8.113*
1033.15	8.201
1073.15	8.483
CURVE 9	
333.15	6.816 x 10^{-2}
353.15	6.809
373.15	6.840*
393.15	7.007*
413.15	7.100*
433.15	7.138*
453.15	7.166*
473.15	7.183*
493.15	7.233*
513.15	7.260*
533.15	7.099*
553.15	7.051*
573.15	7.092*
593.15	7.146*

*Not shown on plot

DATA TABLE NO. 70 (Continued)

T	C_p
CURVE 9 (cont.)	
613.15	7.143×10^{-2}
633.15	7.163
653.15	7.191
673.15	7.216
CURVE 10	
593.15	7.231×10^{-2}*
633.15	7.268*
673.15	7.301*
713.15	7.339*
753.15	7.341
793.15	7.263
833.15	7.477
873.15	7.600
913.15	7.726
953.15	7.814
993.15	7.970*
1033.15	8.107*
1073.15	8.493
1113.15	1.007×10^{-1}*
1173.15	8.030×10^{-2}
1193.15	7.504*
1213.15	7.418*
1233.15	7.462*
1253.15	7.497
CURVE 11	
333.15	7.008×10^{-2}
373.15	6.891
413.15	7.026*
453.15	7.127*
493.15	7.112*
533.15	7.110*
573.15	7.169*
613.15	7.213*
653.15	7.258*
693.15	7.310*
733.15	7.403*
773.15	7.506*
813.15	7.604*
853.15	7.726*
893.15	7.877*
933.15	7.996*
973.15	8.065*
1013.15	8.131*
CURVE 11 (cont.)	
1053.15	8.309×10^{-2}*
1093.15	8.472*
1123.15	1.048×10^{-1}
1173.15	7.553×10^{-2}
1193.15	6.918*
1213.15	6.901
CURVE 12*	
393.15	7.148×10^{-2}
433.15	7.175
473.15	7.175
513.15	7.109
553.15	7.178
593.15	7.248
633.15	7.265
673.15	7.299
713.15	7.370
753.15	7.406
793.15	7.484
833.15	7.573
873.15	7.700
913.15	7.821
953.15	7.911
993.15	7.980
1033.15	8.046
CURVE 13*	
413.15	7.104×10^{-2}
453.15	7.161
493.15	7.122
533.15	7.157
573.15	7.210
613.15	7.269
653.15	7.291
693.15	7.342
753.15	7.465
813.15	7.553
853.15	7.657
893.15	7.768
933.15	7.834
973.15	8.016
CURVE 14	
298	6.8×10^{-2}
300	6.8*
400	7.3
500	7.7
600	7.96
700	8.14
800	8.32
900	8.47
1000	8.62*
1100	8.76
1135	8.8
1135	7.97
1150	7.97*
1200	7.97
1250	7.97
1300	7.97
1350	7.97
1400	7.97
CURVE 15	
323.15	6.91×10^{-2}*
423.15	7.17*
523.15	7.72
623.15	7.64
723.15	7.93
823.15	8.09
923.15	8.51*
1023.15	8.60*
1098.15	9.32*
1129.15	7.90*
1154.15	9.70
CURVE 16*	
53.2	2.651×10^{-2}
56.8	2.894
60.8	3.150
65.6	3.436
70.6	3.709
75.4	3.956
79.8	4.154
84.1	4.325
94.9	4.709
104.5	5.001
115.0	5.258
124.1	5.449
CURVE 16 (cont.)*	
136.1	5.670×10^{-2}
146.2	5.802
156.0	5.930
166.1	6.040
176.0	6.146
186.2	6.218
196.1	6.289
206.3	6.360
216.4	6.425
226.2	6.473
236.4	6.521
246.0	6.557
256.7	6.624
266.4	6.668
276.4	6.717
286.6	6.741
296.8	6.762
CURVE 17*	
14.38	1.041×10^{-3}
16.51	1.644
18.86	2.280
20.93	3.376
23.06	4.440
25.46	5.722
27.98	7.093
31.07	9.198
34.07	1.151×10^{-2}
37.55	1.448
41.43	1.766
44.97	2.039
48.52	2.311
53.59	2.669
58.91	3.068
64.62	3.423
70.33	3.727
76.04	3.998
82.86	4.297
89.22	4.526
95.48	4.745
101.97	4.946
108.61	5.130
115.03	5.277
122.09	5.419
128.70	5.520
135.50	5.615
CURVE 17 (cont.)*	
142.59	5.717×10^{-2}
149.94	5.809
157.56	5.901
165.96	5.980
173.99	6.040
183.07	6.116
191.77	6.189
199.90	6.253
208.14	6.291
215.00	6.341
221.10	6.373
227.40	6.393
234.19	6.398
239.89	6.461
245.00	6.434
250.43	6.481
255.74	6.500
261.35	6.537
265.96	6.541
270.30	6.565
274.16	6.473
278.18	6.564
282.00	6.549
286.07	6.568
290.11	6.578
293.81	6.584
298.23	6.582
CURVE 18*	
298.16	6.591×10^{-2}
300	6.593
350	6.800
400	6.972
450	7.127
500	7.271
600	7.542
700	7.800
800	8.048
900	8.291
1000	8.534
1100	8.775
α 1143	8.880
β 1143	7.339
1200	7.439
1300	7.616
1400	7.791
CURVE 18 (cont.)*	
1500	7.968×10^{-2}
1600	8.144
1700	8.319
1800	8.496

*Not shown on plot

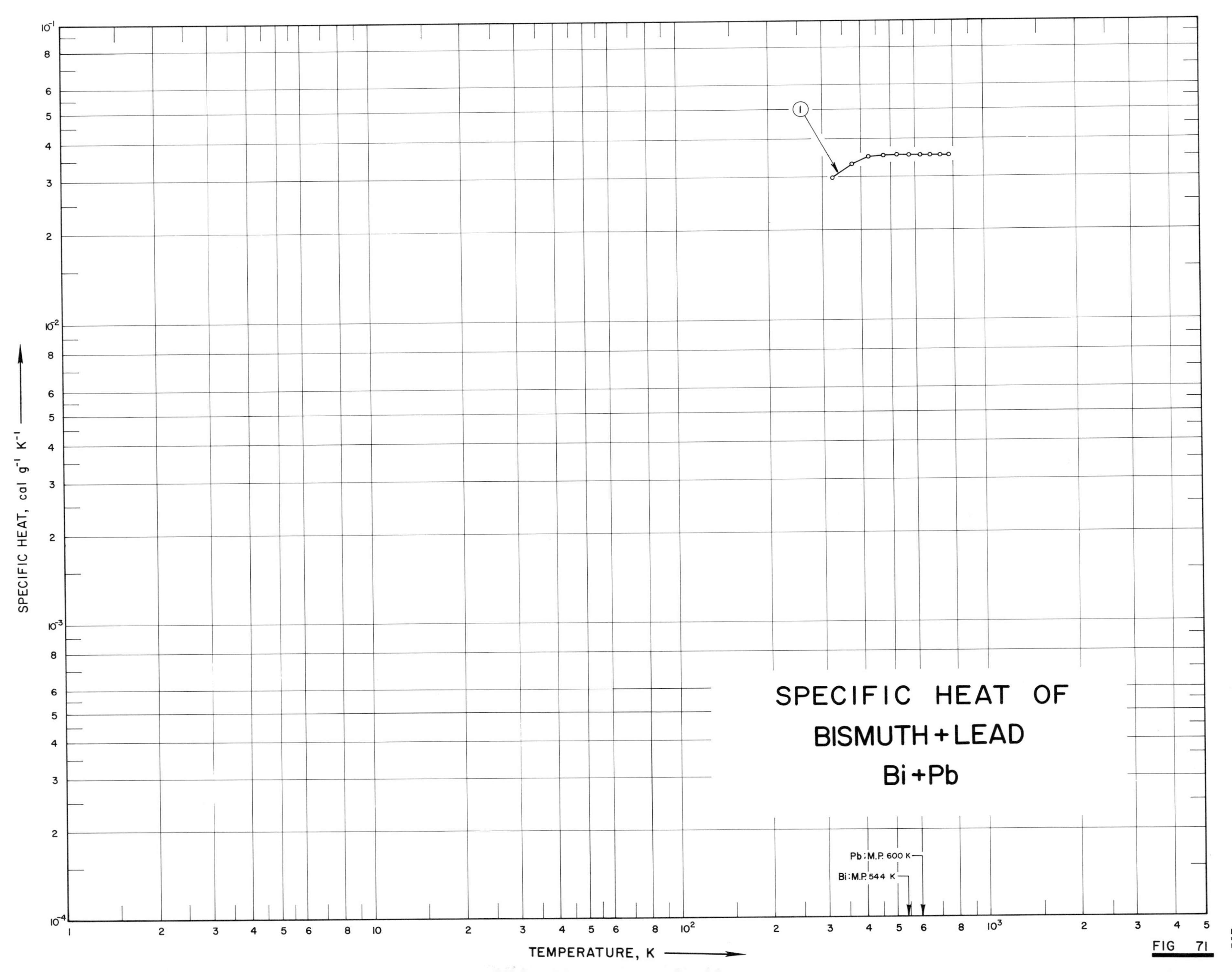
SPECIFIC HEAT OF
BISMUTH + LEAD
Bi + Pb
SPECIFIC HEAT, cal g⁻¹ K⁻¹
TEMPERATURE, K
Pb: M.P. 600 K
Bi: M.P. 544 K
1
FIG 71

SPECIFICATION TABLE NO. 71 SPECIFIC HEAT OF BISMUTH + LEAD Bi + Pb

[For Data Reported in Figure and Table No. 71]

Curve No.	Ref. No.	Year	Temp. Range, K	Reported Error, %	Name and Specimen Designation	Composition (weight percent), Specifications and Remarks
1	7	1959	323-773	1.5-2		56.5 Bi, 43.5 Pb.

DATA TABLE NO. 71 SPECIFIC HEAT OF BISMUTH + LEAD, Bi + Pb

[Temperature, T, K; Specific Heat, C_p, Cal $g^{-1}K^{-1}$]

T	C_p
CURVE 1	
323	2.98×10^{-2}
373	3.32
423	3.50
473	3.53
523	3.55
573	3.55
623	3.55
673	3.55
723	3.55
773	3.55

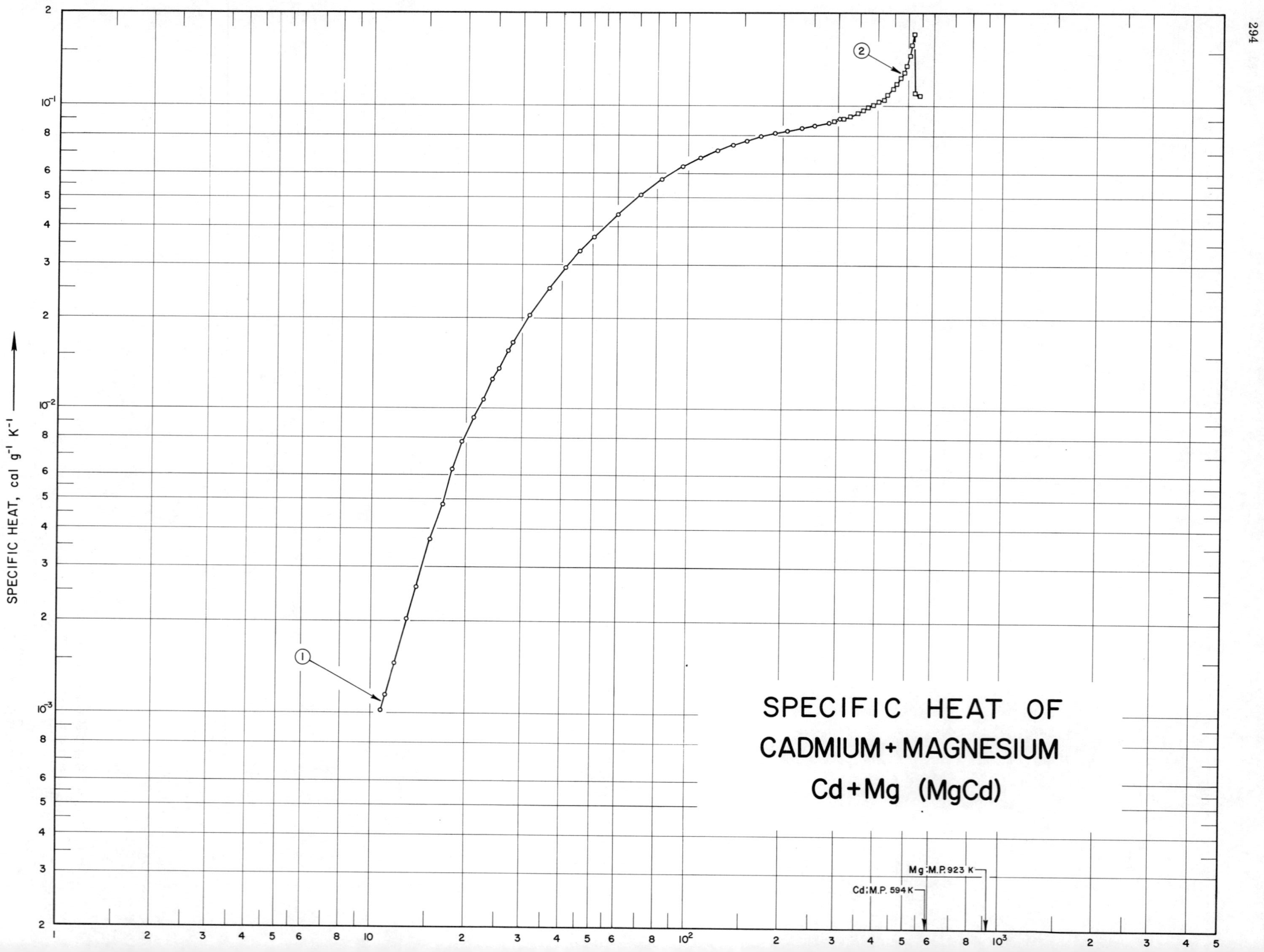

SPECIFIC HEAT OF
CADMIUM + MAGNESIUM
Cd + Mg (MgCd)
Mg: M.P. 923 K
Cd: M.P. 594 K
SPECIFIC HEAT, cal g⁻¹ K⁻¹
1
2

SPECIFICATION TABLE NO. 72 SPECIFIC HEAT OF CADMIUM + MAGNESIUM, Cd + Mg (MgCd)

[For Data Reported in Figure and Table No. 72]

Curve No.	Ref. No.	Year	Temp. Range, K	Reported Error, %	Name and Specimen Designation	Composition (weight percent), Specifications and Remarks
1	415	1952	10-304		MgCd	82.5 Cd, 17.5 Mg; sealed in Pyrex tube containing one half atm. of pure helium; held 10 days at 325 C, then held 2 days at 225 C and finally allowed to cool slowly to room temperature.
2	363	1957	293-547		MgCd	50.52 ± 0.04 at % Cd; stored in desicator for 5 yrs; before measurement series 1 MgCd sample was held at room temperature for 6 days; series 2 MgCd sample held in furnace at final temperature for 30 hrs.

DATA TABLE NO. 72 SPECIFIC HEAT OF CADMIUM + MAGNESIUM, Cd + Mg (MgCd)

[Temperature, T, K; Specific Heat, C_p, Cal $g^{-1}K^{-1}$]

T	C_p
CURVE 1	
Series 1	
10.78	1.024×10^{-3}
11.95	1.463
13.97	2.589
16.87	4.828
19.93	7.768
22.69	1.074×10^{-2}
25.29	1.361
27.92	1.650
31.67	2.047
36.37	2.519
40.93	2.939
45.48	3.340
50.08	3.719
54.83	4.086*
59.79	4.441
65.03	4.798*
70.56	5.135
76.29	5.449*
82.19	5.758
83.10	5.801*
90.30	6.119*
95.96	6.356
100.30	6.504*
104.65	6.653*
109.08	6.794
113.63	6.924*
118.29	7.047*
123.08	7.167
127.96	7.277*
132.95	7.382*
138.02	7.480
143.18	7.575*
148.44	7.666*
153.78	7.841
159.22	7.831*
164.81	7.907*
170.50	7.982
176.30	8.051*
182.19	8.116*
Series 2	
11.22	1.156×10^{-3}
CURVE 1 (cont.)	
Series 2 (cont.)	
13.08	2.033×10^{-3}
15.49	3.687
18.43	6.261
21.41	9.348
24.33	1.258×10^{-2}
27.15	1.559
31.19	1.993*
36.28	2.503*
41.12	2.962*
46.04	3.387*
50.93	3.782*
55.91	4.169*
206.15	8.339
212.07	8.386*
217.92	8.435*
223.70	8.472*
229.42	8.510
235.09	8.548*
240.72	8.587*
246.29	8.627*
251.80	8.663
262.73	8.741*
268.11	8.783*
273.45	8.829*
278.76	8.868
Series 3	
188.17	8.176×10^{-2}
194.24	8.238*
200.39	8.295*
206.59	8.345*
212.72	8.394*
218.84	8.438*
224.94	8.481*
231.03	8.522*
237.10	8.563*
243.11	8.599*
249.12	8.639*
255.09	8.681*
261.03	8.722*
266.95	8.764*
272.85	8.807*
CURVE 1 (cont.)	
Series 3 (cont.)	
278.69	8.865×10^{-2}*
284.49	8.912*
290.25	8.966*
295.97	9.015*
301.65	9.063*
Series 4	
274.62	8.832×10^{-2}*
277.79	8.858*
280.24	8.881*
283.09	8.902*
288.36	8.947*
293.60	8.993*
298.80	9.037*
303.98	9.082
CURVE 2	
Series 2	
360.14	9.65×10^{-2}
364.89	9.72*
369.62	9.78*
374.32	9.86
378.98	9.93*
383.61	9.99*
388.22	1.006×10^{-1}
392.84	1.013*
Series 3	
389.56	1.008×10^{-1}*
394.18	1.016*
398.76	1.023*
403.30	1.032
407.81	1.041*
412.33	1.049*
416.82	1.058*
421.27	1.067*
425.71	1.075*
430.15	1.086
CURVE 2 (cont.)	
Series 4	
442.28	1.112×10^{-1}*
446.58	1.124*
450.84	1.137
455.05	1.149*
459.26	1.161*
463.45	1.178
Series 5	
496.18	1.346×10^{-1}
499.96	1.371*
503.66	1.414*
507.26	1.457
Series 6	
506.90	1.443×10^{-1}*
511.35	1.508*
515.65	1.577
519.78	1.668*
528.87	1.094
534.38	1.095*
Series 8	
302.64	8.99×10^{-2}*
307.65	9.04*
312.63	9.08
317.57	9.13*
322.50	9.19*
327.41	9.24
332.31	9.29*
337.21	9.34*
342.08	9.40*
346.90	9.46
Series 9*	
350.01	9.50×10^{-2}
354.86	9.57
359.69	9.63
364.50	9.68
369.26	9.78
CURVE 2 (cont.)	
Series 9(cont.)*	
373.99	9.86×10^{-2}
378.69	9.92
383.37	9.98
388.03	1.01×10^{-1}
392.68	1.01
397.31	1.02
401.92	1.03
406.50	1.04
411.07	1.05
415.59	1.05
Series 10*	
411.36	1.046×10^{-1}
416.99	1.357
422.60	1.067
428.14	1.080
433.63	1.090
439.11	1.104
444.56	1.119
449.95	1.133
455.27	1.149
460.52	1.167
Series 11	
461.48	1.167×10^{-1}*
466.70	1.187*
471.85	1.208*
476.92	1.229
481.94	1.253*
486.89	1.279
491.77	1.308*
496.54	1.344*
501.25	1.381*
505.85	1.430*
510.34	1.480*
514.69	1.548*
518.87	1.633*
521.90	1.711
Series 13*	
529.10	1.089×10^{-1}
CURVE 2 (cont.)	
Series 13 (cont.)*	
533.60	1.083×10^{-1}
538.07	1.081
542.53	0.081
546.99	1.078
Series 14*	
298.21	8.97×10^{-2}
303.26	9.00
308.28	9.05
313.28	9.09
318.26	9.14
Series 15	
293.01	8.90×10^{-2}
298.06	8.97*
303.09	8.98*
308.12	9.04*
313.11	9.08*
318.07	9.15*

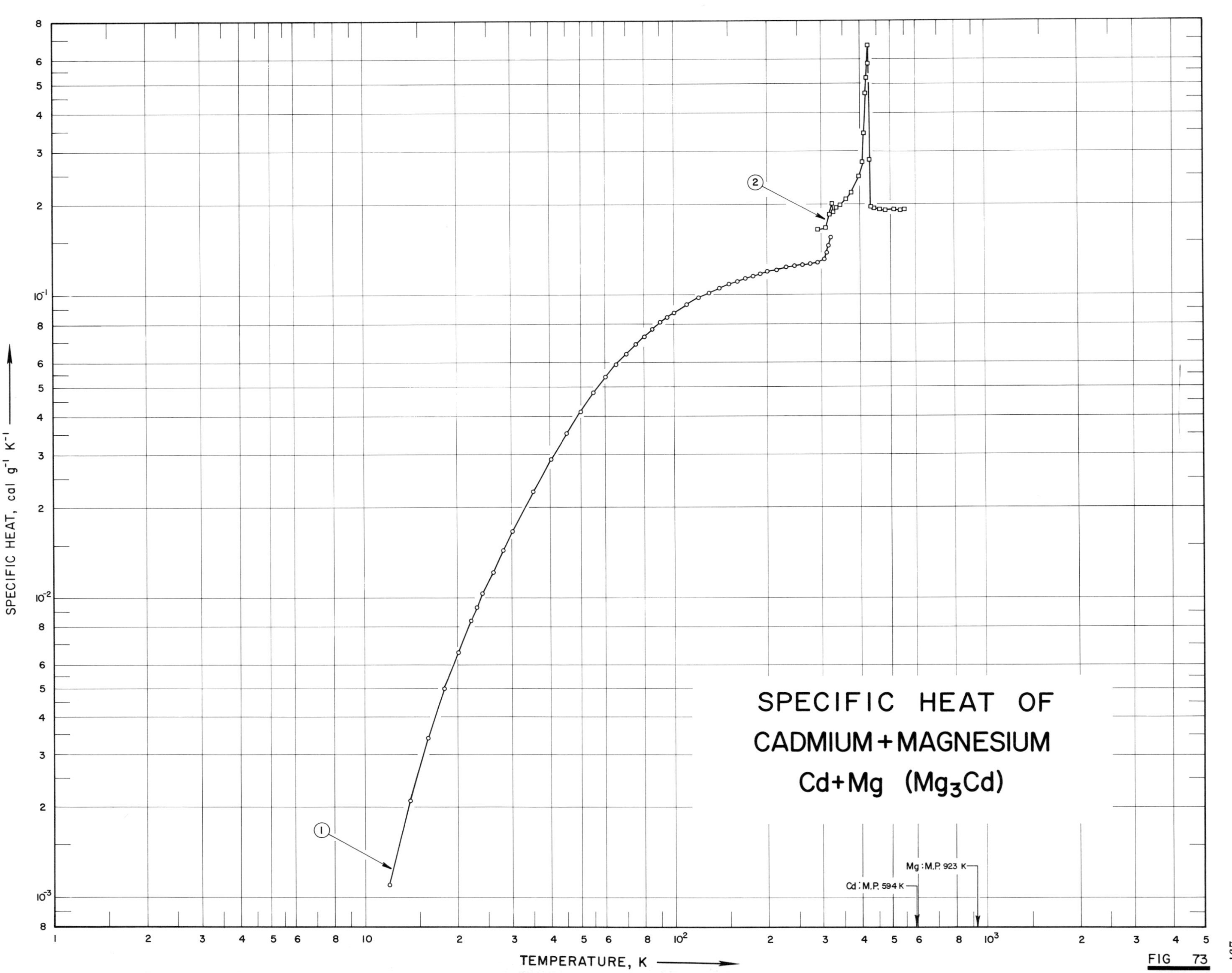

SPECIFIC HEAT OF
CADMIUM + MAGNESIUM
Cd+Mg (Mg_3Cd)
SPECIFIC HEAT, cal g^{-1} K^{-1}
TEMPERATURE, K
Cd: M.P. 594 K
Mg: M.P. 923 K
1
2

FIG 73

SPECIFICATION TABLE NO. 73 SPECIFIC HEAT OF CADMIUM + MAGNESIUM, Cd + Mg (Mg_3Cd)

[For Data Reported in Figure and Table No. 73]

Curve No.	Ref. No.	Year	Temp. Range, K	Reported Error, %	Name and Specimen Designation	Composition (weight percent), Specifications and Remarks
1	48	1954	12-320		Mg_3Cd	60.64 Cd, 39.36 Mg, impurities $<$ 0.01; annealed 47 days at 345-350 C; machined strain relieved 2 days at 350 C.
2	381	1955	291-552		Mg_3Cd	24.98 Cd.

DATA TABLE NO. 73 SPECIFIC HEAT OF CADMIUM + MAGNESIUM, Cd + Mg (Mg_3Cd)

[Temperature, T, K; Specific Heat, C_p, Cal $g^{-1}K^{-1}$]

CURVE 1

T	C_p
12	1.1×10^{-3}
14	2.1
16	3.4
18	5.0
20	6.6
22	8.4
23	9.3
24	1.03×10^{-2}
26	1.22
28	1.44
30	1.66
35	2.26
40	2.89
45	3.52
50	4.15
55	4.78
60	5.38
65	5.90
70	6.39
75	6.86
80	7.30
85	7.72
90	8.10
95	8.43
100	8.72
105	9.01*
110	9.29
115	9.55*
120	9.76
125	9.97*
130	1.018×10^{-1}
135	1.038*
140	1.055
145	1.071*
150	1.086
155	1.099*
160	1.112
165	1.124*
170	1.136
175	1.146*
180	1.156
185	1.166*
190	1.174
195	1.183*
200	1.191

CURVE 1 (cont.)

T	C_p
205	1.198×10^{-1}*
210	1.206*
215	1.213
220	1.219*
225	1.226*
230	1.231
235	1.237*
240	1.241*
245	1.247
250	1.252
255	1.257*
260	1.262
265	1.266*
270	1.270*
275	1.273
280	1.275*
285	1.277*
290	1.278
295	1.279*
298.16	1.281*
300	1.282*
305	1.322
310	1.383
315	1.466
320	1.566

CURVE 2

Series 1

T	C_p
291.57	1.641×10^{-1}
297.20	1.646*
302.81	1.652*
308.34	1.669
344.35	1.987
349.30	2.016*
354.19	2.046*
359.04	2.076

Series 2

T	C_p
358.36	2.073×10^{-1}*
363.16	2.105*
367.90	2.147*
373.15	2.189

CURVE 2 (cont.)

Series 2 (cont.)

T	C_p
378.89	2.247×10^{-1}*
384.51	2.316*
390.03	2.389*
395.42	2.480
400.66	2.601*
405.69	2.751
410.24	3.444
426.35	2.801
431.71	1.962
437.94	1.962*
444.24	1.943

Series 3

T	C_p
334.90	1.942×10^{-1}
339.87	1.966*
344.76	1.995*
415.44	4.661
418.57	5.236
421.41	5.841
423.48	6.704
442.81	1.946*
447.86	1.936*
452.95	1.924*
460.88	1.923
465.96	1.916*
471.04	1.912*
476.14	1.907*
481.21	1.906
486.93	1.903*
493.31	1.904*
506.15	1.901*
512.55	1.912
519.02	1.903*
525.48	1.908*
531.89	1.915*
538.35	1.906
544.79	1.919*
551.15	1.919

CURVE 2 (cont.)

Series 4

T	C_p
313.87	1.693×10^{-1}*
318.88	1.846
323.65	2.011
328.12	1.918*
332.54	1.933*
504.77	1.896*
511.20	1.914*
517.59	1.905*
523.96	1.899*
530.33	1.900*

Series 5

T	C_p
489.51	1.905×10^{-1}*
495.92	1.904*
502.30	1.907*
508.65	1.917*
514.97	1.903*
521.33	1.901*
527.69	1.904*
533.96	1.905*
540.30	1.911*

Series 6

T	C_p
325.03	1.875×10^{-1}

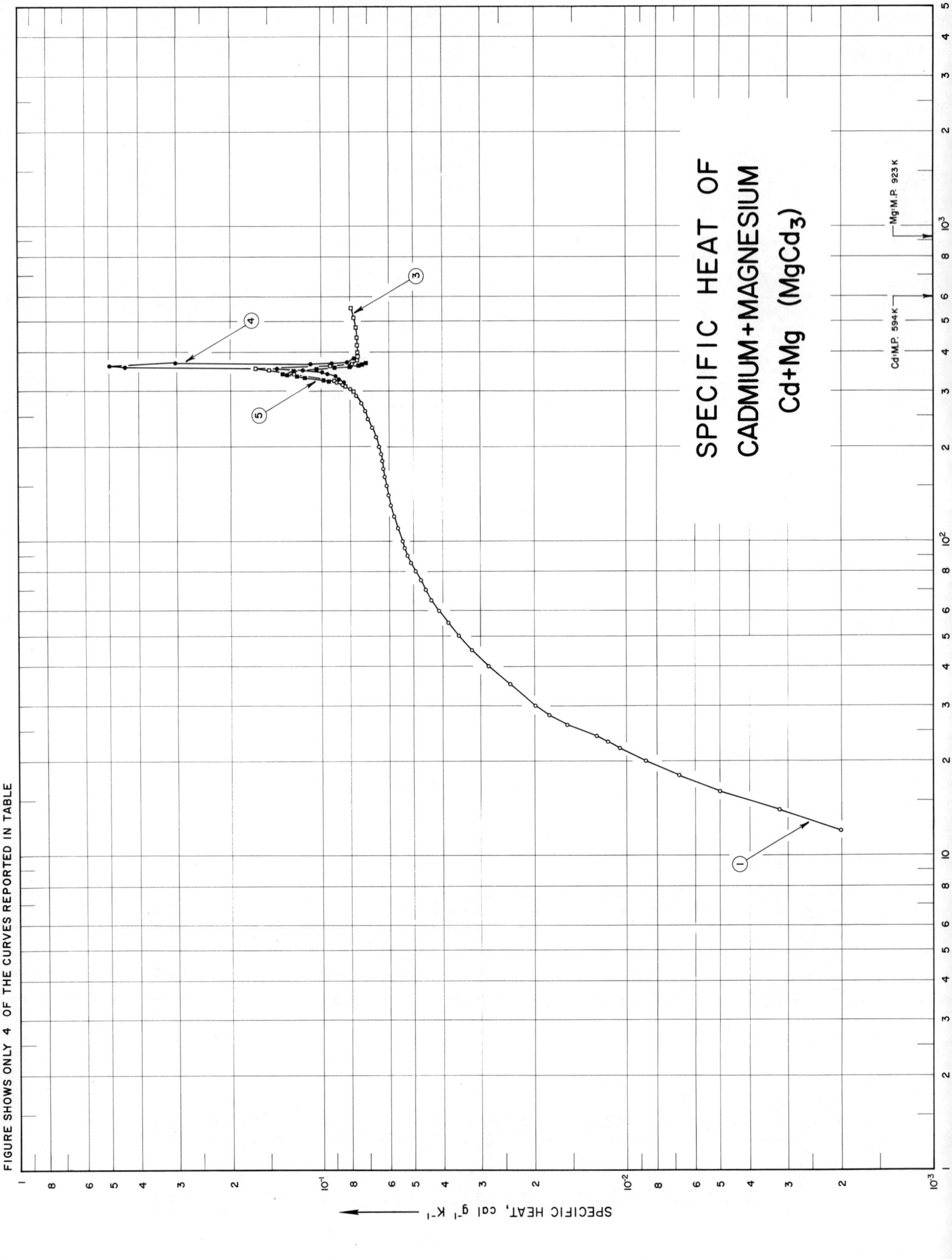

FIGURE SHOWS ONLY 4 OF THE CURVES REPORTED IN TABLE

SPECIFIC HEAT OF
CADMIUM + MAGNESIUM
Cd+Mg ($MgCd_3$)

SPECIFIC HEAT, cal g^{-1} K^{-1}

Cd:M.P. 594 K

Mg:M.P. 923 K

1

3

4

5

SPECIFICATION TABLE NO. 74 SPECIFIC HEAT OF CADMIUM + MAGNESIUM, Cd + Mg ($MgCd_3$)

[For Data Reported in Figure and Table No. 74]

Curve No.	Ref. No.	Year	Temp. Range, K	Reported Error, %	Name and Specimen Designation	Composition (weight percent), Specifications and Remarks
1	48	1954	12-320		$MgCd_3$	<0.01 impurities for each metal; magnesium prepared by National Lead Co.; cadmium prepared by Anaconda Mining Co.
2	363	1957	202-280		$MgCd_3$	74.98 ±0.04 at % Cd; stored at room temperature in desicator for 7 months under He atm.
3	381	1955	295-553		$MgCd_3$	74.98 Cd.
4	382	1958	307-384		Alloy No. 1	77.2 at % Cd.
5	382	1958	308-368		Alloy No. 1	Same as above; additional heat treatment.
6	382	1958	299-378		Alloy No. 2	75.9 at % Cd.
7	382	1958	300-370		Alloy No. 3	73.0 at % Cd.

DATA TABLE NO. 74 SPECIFIC HEAT OF CADMIUM + MAGNESIUM, Cd + Mg ($MgCd_3$)

[Temperature, T, K; Specific Heat, C_p, Cal $g^{-1}K^{-1}$]

CURVE 1

T	C_p
12	2.0 x 10^{-3}
14	3.2
16	5.0
18	6.8
20	8.7
22	1.06 x 10^{-2}
23	1.16
24	1.26
26	1.57
28	1.79
30	1.98
35	2.41
40	2.83
45	3.22
50	3.55
55	3.84
60	4.12
65	4.35
70	4.55
75	4.74
80	4.91
85	5.08
90	5.22
95	5.33
100	5.44
105	5.53*
110	5.63
115	5.71*
120	5.78
125	5.86*
130	5.92
135	5.98*
140	6.03
145	6.08*
150	6.13
155	6.18*
160	6.22
165	6.25*
170	6.29
175	6.32*
180	6.34
185	6.36*
190	6.39
195	6.43*
200	6.48 x 10^{-2}
205	6.53*
210	6.58*
215	6.64
220	6.70*
225	6.77*
230	6.83
235	6.89*
240	6.94*
245	7.00
250	7.06*
255	7.11*
260	7.17
265	7.23*
270	7.29*
275	7.36
280	7.45*
285	7.55*
290	7.66
295	7.76*
298.16	7.83*
300	7.87*
305	8.00
310	8.19*
315	8.48
320	8.86

CURVE 2*

T	C_p
202.71	6.50 x 10^{-2}
208.51	6.55
214.24	6.71
219.95	6.72
225.57	6.79
231.15	6.85
236.70	6.90
242.24	6.97
247.73	7.03
255.85	7.12
268.91	7.29
279.58	7.47
284.74	7.57
289.85	7.66
294.96	7.76
299.95	7.90 x 10^{-2}
304.88	8.02
309.80	8.20
314.67	8.44
319.43	8.81

CURVE 3

Series 1

T	C_p
295.21	7.723 x 10^{-2}*
299.72	7.830
304.16	7.988*
308.53	8.095*
312.84	8.310
317.57	8.653*
322.67	9.139*
323.99	9.029
328.81	9.814*
376.51	7.622
382.22	7.606*
387.93	7.599
393.62	7.601*
399.33	7.583
405.01	7.588*
410.73	7.589*
416.47	7.597*
422.20	7.614
427.93	7.617*
433.64	7.650*
439.34	7.671*
445.03	7.642
450.71	7.657*
456.40	7.664*
462.09	7.669*
467.77	7.680*
473.47	7.688*
479.23	7.735
484.83	7.742*
490.61	7.752*
496.35	7.754*
502.02	7.784*
507.71	7.788*
513.36	7.819
519.00	7.856 x 10^{-2}*
524.64	7.906*
530.24	7.857*
535.91	7.890*
541.55	7.895*
547.32	7.937*
552.90	7.955

Series 2

T	C_p
475.72	7.721 x 10^{-2}*
481.37	7.729*
487.02	7.733*
492.59	7.756*
498.11	7.775*

Series 3

T	C_p
341.19	1.070 x 10^{-1}*
345.23	1.372*
351.73	1.433*
355.08	1.653
357.84	1.486*
361.36	9.350 x 10^{-2}
366.61	7.922
372.23	7.662*
377.91	7.605*
383.61	7.607*
516.08	7.836*
521.66	7.790*
527.29	7.872*
532.90	7.891*

Series 4

T	C_p
347.80	1.467 x 10^{-1}*
456.31	7.668 x 10^{-2}*
462.03	7.697*
467.71	7.712*
473.40	7.705*
479.08	7.705*
484.79	7.625*

Series 5

T	C_p
341.19	1.236 x 10^{-1}
351.23	1.492

CURVE 4

T	C_p
307.2	8.1 x 10^{-2}*
310.0	8.1*
312.8	8.4*
315.7	8.4*
317.4	8.4*
320.2	8.4
322.4	8.5*
325.1	8.5*
327.3	8.7
330.2	8.7*
332.7	8.7*
335.1	8.7*
337.0	8.9
339.6	9.1*
341.4	9.5
344.4	9.5*
346.4	9.9
348.3	1.06 x 10^{-1}*
351.7	1.15
353.1	1.24*
355.9	1.40
357.8	1.63*
359.3	3.44
362.2	3.96
365.1	2.03
366.9	1.08
368.9	9.2 x 10^{-2}
372.1	8.2
374.7	8.1*
376.6	7.9*
379.0	7.9*
381.7	7.8*
384.1	7.8

CURVE 5

T	C_p
308.2	8.0 x 10^{-2}*
310.9	8.3*
312.9	8.4*
316.1	8.6*
317.8	8.9*
320.4	9.1*
322.4	9.4
325.5	9.8
327.2	1.01 x 10^{-1}*
331.5	1.13
334.9	1.20
337.4	1.25*
338.9	1.30
341.5	1.33
344.2	1.32*
346.1	1.30*
349.0	1.23
351.1	1.13*
353.4	1.03
356.1	9.0 x 10^{-2}
358.9	8.0
361.2	7.5
364.0	7.4*
365.3	7.3
368.1	7.1

CURVE 6*

T	C_p
299.7	8.3 x 10^{-2}
302.0	8.3
305.4	8.4
307.2	8.5
310.1	8.6
313.1	8.6
315.3	8.8
317.7	9.0
320.1	9.1
322.0	9.4
325.3	9.7
327.1	1.02 x 10^{-1}
330.4	1.07
332.2	1.13
334.5	1.20
336.0	1.28

DATA TABLE NO. 74 (continued)

T	C_p
CURVE 6 (cont.)*	
338.4	1.40×10^{-1}
341.7	1.51
343.3	1.70
346.3	1.68
348.2	1.53
350.7	1.35
362.8	1.21
356.1	1.08
358.1	9.8×10^{-2}
360.0	8.9
363.6	8.2
365.4	7.6
367.5	7.6
370.2	7.4
372.3	7.3
374.3	7.3
377.5	7.4
CURVE 7*	
300.4	8.5×10^{-2}
302.5	8.5
305.8	8.7
307.8	8.8
310.7	8.9
312.3	9.1
315.0	9.2
317.4	9.4
320.4	9.5
322.2	9.9
324.7	1.01×10^{-1}
327.4	1.06
330.1	1.09
332.4	1.14
334.5	1.17
337.5	1.20
339.5	1.23
342.3	1.25
345.1	1.25
346.2	1.19
349.2	1.09
351.4	9.7×10^{-2}
354.0	8.7
356.7	7.6
358.9	7.3
361.0	7.2
CURVE 7 (cont.)*	
363.6	7.1×10^{-2}
365.4	7.0
367.6	7.0
369.8	7.0

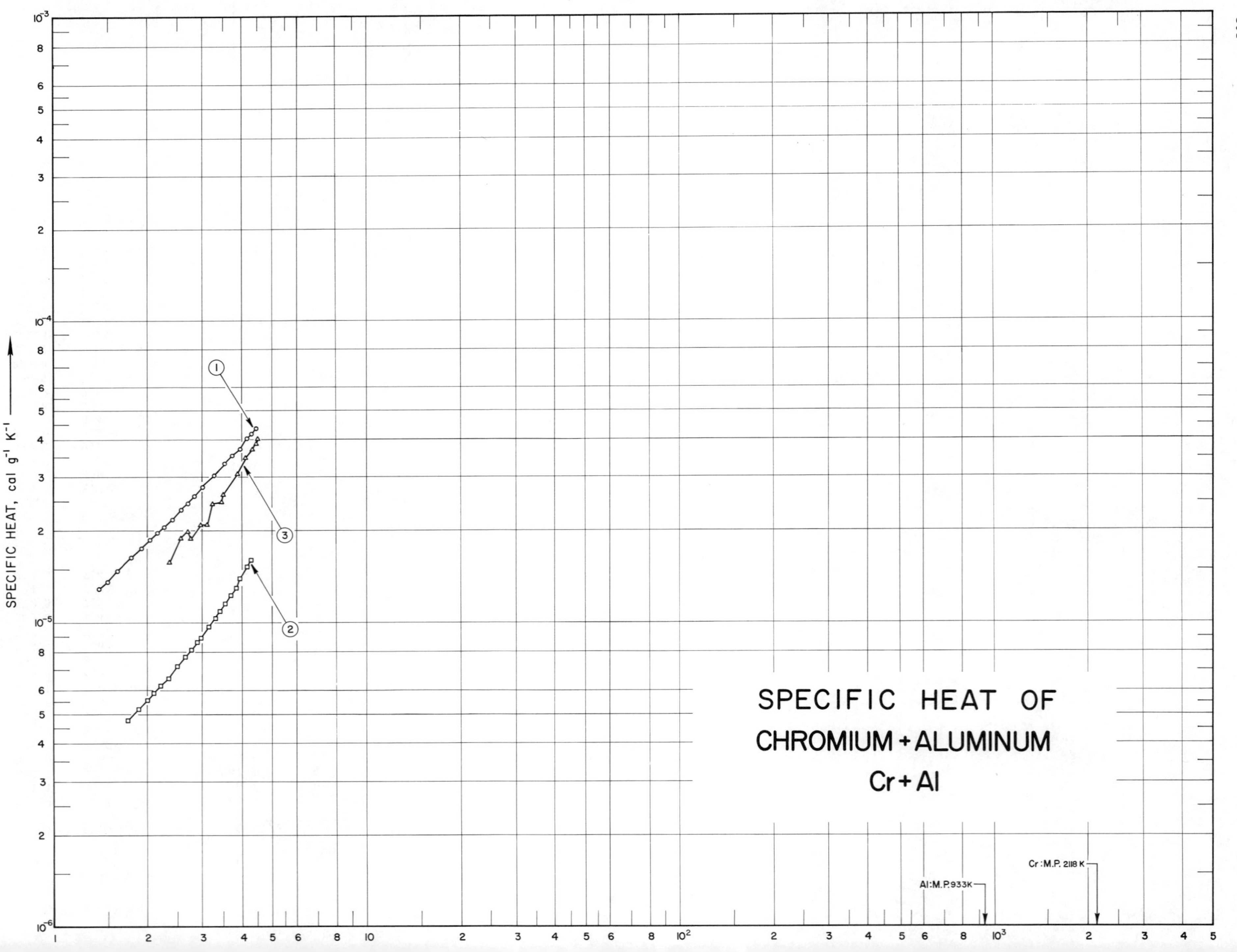
SPECIFIC HEAT OF
CHROMIUM + ALUMINUM
Cr + Al
SPECIFIC HEAT, cal g^{-1} K^{-1}
Al: M.P. 933 K
Cr: M.P. 2118 K
1
2
3

SPECIFICATION TABLE NO. 75 SPECIFIC HEAT OF CHROMIUM + ALUMINUM Cr + Al

[For Data Reported in Figure and Table No. 75]

Curve No.	Ref. No.	Year	Temp. Range, K	Reported Error, %	Name and Specimen Designation	Composition (weight percent), Specifications and Remarks
1	349	1962	1.4-4.5	≤2	Cr(90) Al(10)	91.94 Cr, 8.01 Al; annealed under vacuum at 1100 C for 72 hrs; etched with 10-20% HCl.
2	349	1962	1.7-4.3	≤2	Cr(80) Al(20)	86.15 Cr, 13.82 Al; same as above.
3	349	1962	2.3-4.6	≤2	Cr(70) Al(30)	78.36 Cr, 21.59 Al; same as above.

DATA TABLE NO. 75 SPECIFIC HEAT OF CHROMIUM + ALUMINUM, Cr + Al

[Temperature, T, K; Specific Heat, C_p, Cal $g^{-1}K^{-1}$]

T	C_p
CURVE 1	
1.401	1.30×10^{-5}
1.498	1.37
1.613	1.49
1.778	1.63
1.920	1.76
2.049	1.87
2.163	1.97
2.173	1.97*
2.268	2.06
2.412	2.187
2.437	2.228*
2.572	2.354
2.683	2.464
2.824	2.600
3.003	2.784
3.260	3.054
3.529	3.329
3.743	3.556
3.970	3.825
4.161	4.062
4.309	4.217
4.484	4.434
CURVE 2	
1.733	4.77×10^{-6}
1.885	5.21
2.001	5.53
2.098	5.88
2.207	6.20
2.211	6.14*
2.336	6.59
2.498	7.20
2.504	7.22*
2.648	7.72
2.655	7.81*
2.761	8.18
2.795	8.31*
2.886	8.66
2.963	8.94
3.141	9.72
3.288	1.04×10^{-5}
3.411	1.10
3.551	1.16
CURVE 2 (cont.)	
3.694	1.24×10^{-5}
3.841	1.31
3.953	1.41
4.057	1.45*
4.176	1.53
4.296	1.61
CURVE 3	
2.359	1.59×10^{-5}
2.557	1.89
2.691	1.99
2.751	1.88
2.950	2.11
3.095	2.11
3.220	2.466
3.449	2.487
3.485	2.635
3.875	3.080
4.117	3.482
4.344	3.742
4.459	3.918
4.567	4.051

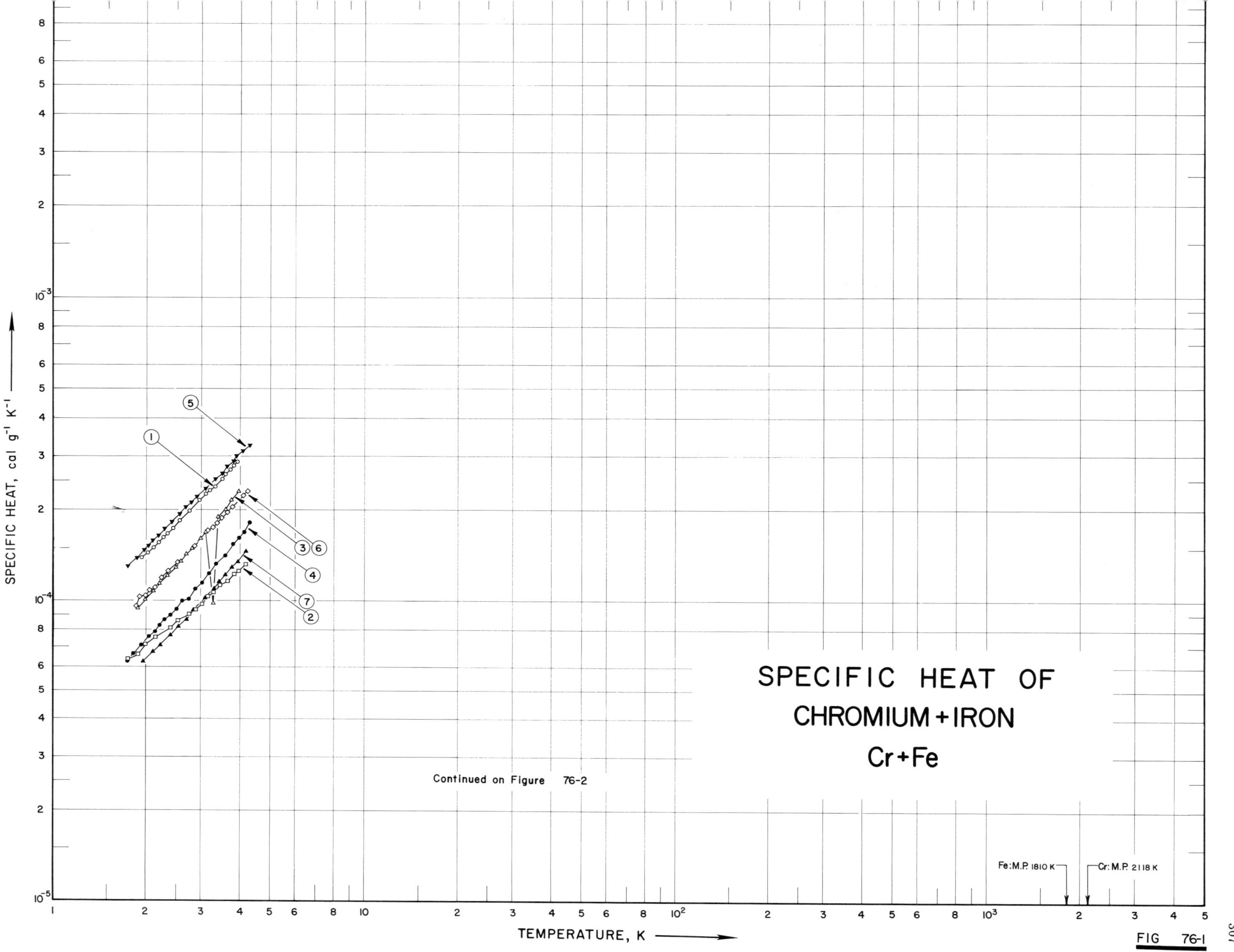

Continued on Figure 76-2

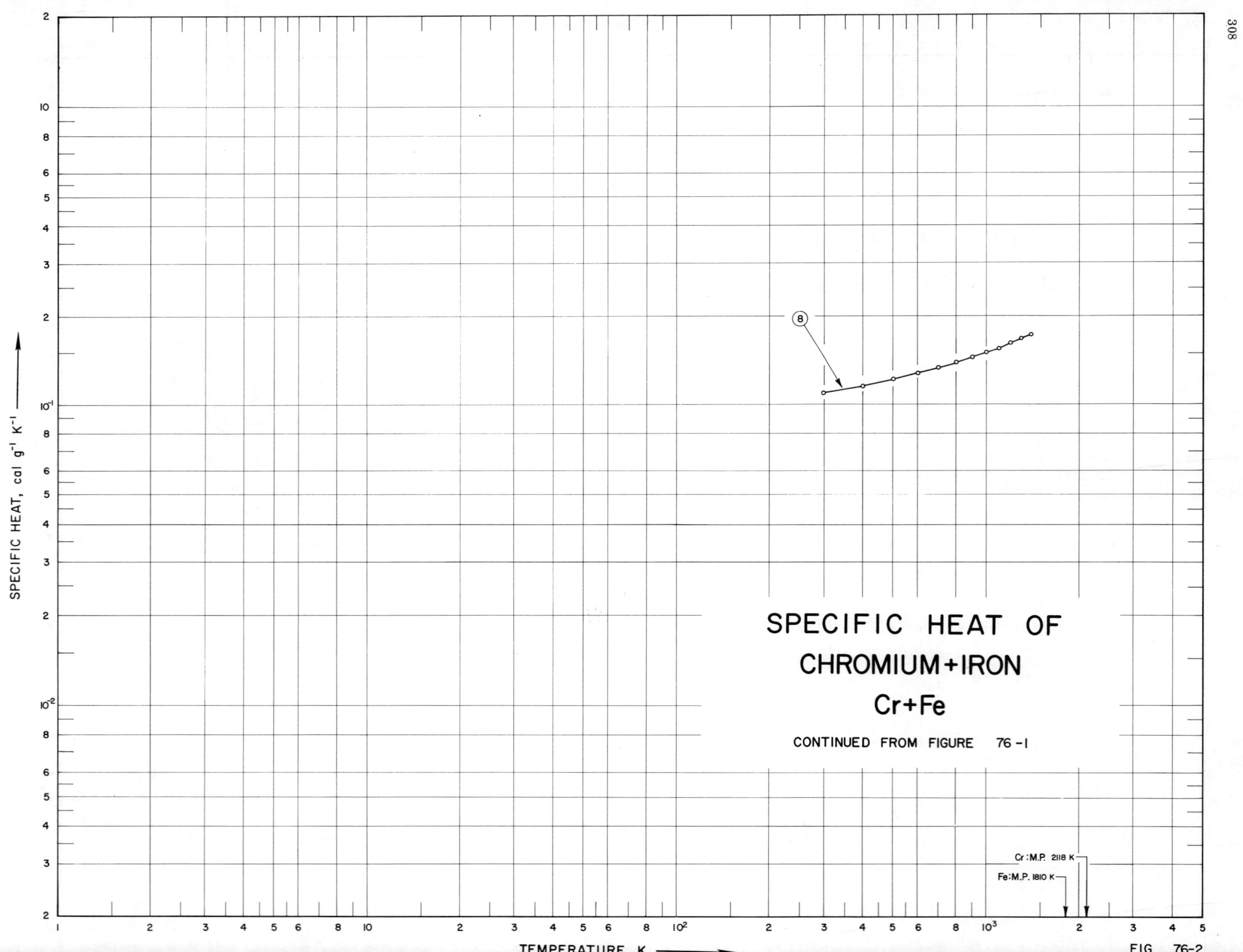
SPECIFIC HEAT OF
CHROMIUM+IRON
Cr+Fe
CONTINUED FROM FIGURE 76-1
8
Cr:M.P. 2118 K
Fe:M.P. 1810 K
SPECIFIC HEAT, cal g^{-1} K^{-1}
TEMPERATURE, K
FIG 76-2

SPECIFICATION TABLE NO. 76 SPECIFIC HEAT OF CHROMIUM + IRON Cr + Fe

[For Data Reported in Figure and Table No. 76]

Curve No.	Ref. No.	Year	Temp. Range, K	Reported Error, %	Name and Specimen Designation	Composition (weight percent), Specifications and Remarks
1	320	1959	1.9-3.9			80 at % Cr, 20 at % Fe; induction melted from electrolytic Cr and Fe flakes; alloy kept at molten state 3 min for homogenization; annealed 3 days at 1170 C under $92He-8H_2$ gas mixture.
2	320	1959	1.7-4.2			95 at % Cr, 5 at % Fe; same as above.
3	320	1959	1.8-4.0			90 at % Cr, 10 at % Fe; same as above.
4	320	1959	1.7-4.3			85 at % Cr, 15 at % Fe; same as above.
5	320	1959	1.7-4.3			82 at % Cr, 18 at % Fe; same as above.
6	320	1959	1.8-4.2			72 at % Cr, 28 at % Fe; same as above.
7	320	1959	1.9-4.2			63 at % Cr, ~~27~~ 37 at % Fe; same as above.
8	222	1959	298-1400	±0.5	$Cr_{0.784}Fe_{0.216}$ Sample No. 80 Cr	77.2 ~~Cu~~ Cr, 22.8 Fe; homogenized for 4 days at 1350 C under helium atmosphere; air cooled to room temperature.

DATA TABLE NO. 76 SPECIFIC HEAT OF CHROMIUM + IRON, Cr + Fe

[Temperature, T, K; Specific Heat, C_p Cal $g^{-1}K^{-1}$]

T	C_p
CURVE 1	
1.946	1.401 x 10^{-4}
2.023	1.452
2.117	1.520
2.199	1.576
2.267	1.631
2.345	1.681
2.438	1.755
2.565	1.853
2.755	1.985
2.968	2.160
3.113	2.268
3.205	2.325
3.310	2.400
3.378	2.463*
3.501	2.536
3.599	2.624
3.659	2.679*
3.730	2.723
3.823	2.813
3.929	2.885
CURVE 2	
1.753	6.392 x 10^{-5}
1.791	6.506*
1.877	6.616
1.920	6.761*
1.981	7.104
2.060	7.280*
2.145	7.518
2.256	7.592*
2.399	8.125
2.524	8.608
2.633	8.699*
2.753	9.017
2.883	9.372
3.021	9.785
3.172	1.042 x 10^{-4}
3.298	1.075
3.460	1.127
3.654	1.174
3.834	1.230
3.968	1.267
4.068	1.299*
4.188	1.331
CURVE 3	
1.875	9.480 x 10^{-5}
1.926	9.870*
1.986	1.017 x 10^{-4}
2.057	1.052*
2.116	1.089
2.170	1.125*
2.217	1.151
2.267	1.162*
2.335	1.221
2.403	1.242*
2.480	1.308
2.575	1.371
2.691	1.446
2.818	1.504
2.979	1.626
3.100	1.703
3.183	1.721*
3.289	9.885 x 10^{-5}
3.419	1.918 x 10^{-4}
3.580	2.014
3.761	2.158
3.964	2.313
CURVE 4	
1.744	6.347 x 10^{-5}
1.827	6.663
1.876	6.842*
1.931	7.109
1.993	7.315*
2.042	7.585
2.131	7.907
2.206	8.329
2.293	8.673
2.392	8.982
2.502	9.431
2.619	1.003 x 10^{-4}
2.744	1.019
2.874	1.109
3.026	1.167
3.191	1.249
3.367	1.336
3.580	1.426
3.817	1.552
3.985	1.630
CURVE 4 (cont.)	
4.124	1.704 x 10^{-4}
4.292	1.830
CURVE 5	
1.746	1.309 x 10^{-4}
1.813	1.362*
1.869	1.395
1.913	1.431*
1.965	1.477
2.024	1.527
2.098	1.592
2.187	1.654
2.288	1.741
2.413	1.830
2.560	1.944
2.676	2.049
2.782	2.110
2.904	2.210
3.103	2.354
3.339	2.527
3.505	2.646
3.635	2.778
3.809	2.897
3.897	2.931*
3.987	3.008
4.078	3.115
4.182	3.158*
4.320	3.255
CURVE 6	
1.865	9.571 x 10^{-5}
1.917	1.034 x 10^{-4}
1.980	1.048
2.051	1.091
2.133	1.134
2.237	1.210
2.361	1.272
2.531	1.351
2.708	1.454*
2.853	1.534
2.971	1.598*
3.153	1.724
3.288	1.762
CURVE 6 (cont.)	
3.378	1.824
3.498	1.890
3.629	1.976
3.714	2.005*
3.786	2.054
4.105	2.236
4.219	2.312
CURVE 7	
1.957	6.262 x 10^{-5}
2.016	6.433*
2.104	6.759
2.222	7.128
2.386	7.696
2.528	8.223
2.601	8.452*
2.681	8.740
2.725	8.918*
2.781	9.102*
2.847	9.378
2.919	9.586*
2.997	9.867*
3.093	1.027 x 10^{-4}
3.202	1.063*
3.313	1.114
3.426	1.171
3.515	1.209*
3.589	1.228
3.670	1.260*
3.761	1.309
3.846	1.326*
3.934	1.361
4.055	1.419*
4.187	1.479
CURVE 8	
298.15	1.11 x 10^{-1}
400	1.17
500	1.23
600	1.28
700	1.34
800	1.40
900	1.46
CURVE 8 (cont.)	
1000	1.51 x 10^{-1}
1100	1.57
1200	1.63
1300	1.69
1400	1.75

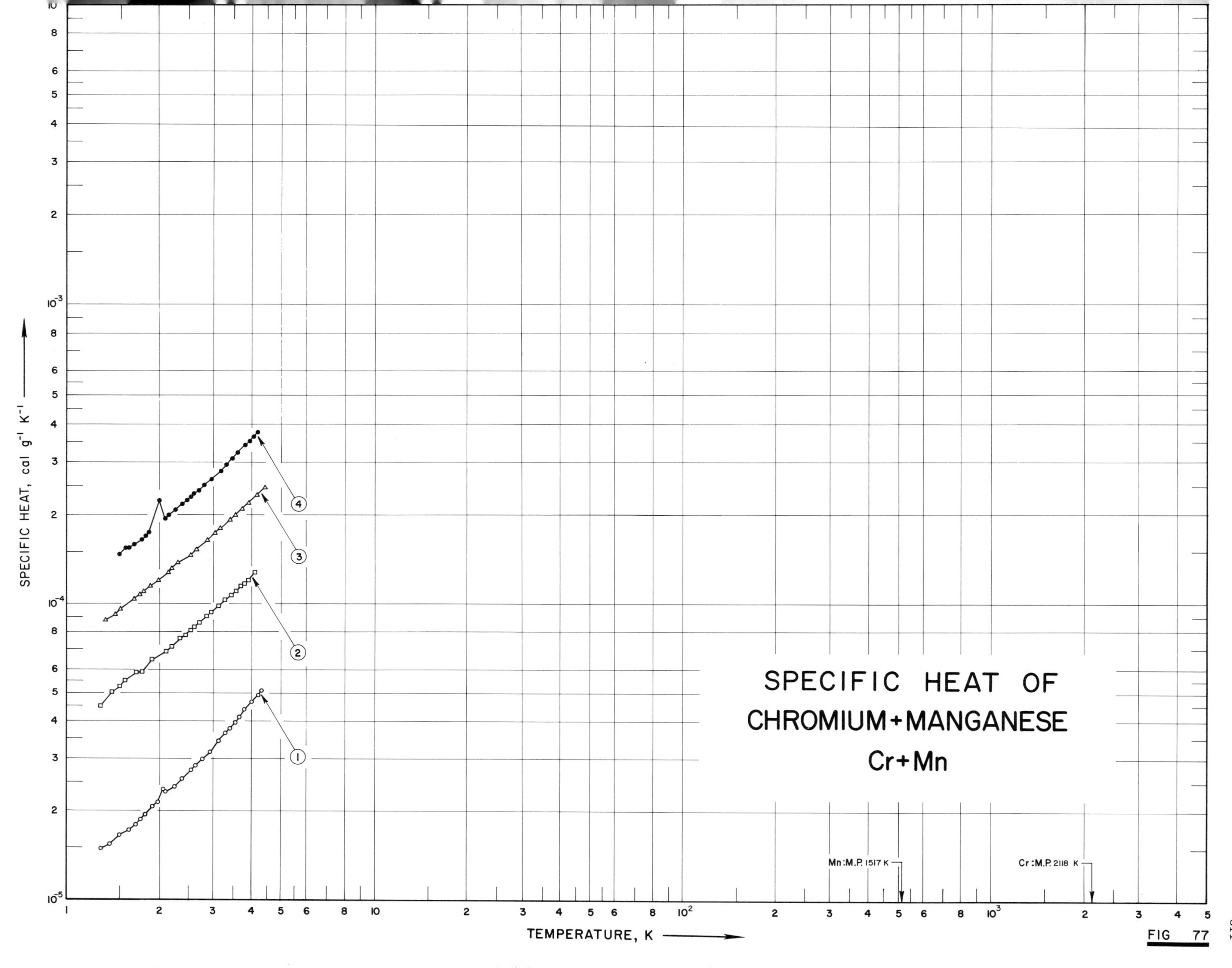
SPECIFIC HEAT OF
CHROMIUM+MANGANESE
Cr+Mn
SPECIFIC HEAT, cal g⁻¹ K⁻¹
TEMPERATURE, K
Mn: M.P. 1517 K
Cr: M.P. 2118 K
1
2
3
4
FIG 77

SPECIFICATION TABLE NO. 77 SPECIFIC HEAT OF CHROMIUM + MANGANESE Cr + Mn

[For Data Reported in Figure and Table No. 77]

Curve No.	Ref. No.	Year	Temp. Range, K	Reported Error, %	Name and Specimen Designation	Composition (weight percent), Specifications and Remarks
1	297	1959	1.3-4.3		$Cr_{0.9}Mn_{0.1}$	88.97 Cr, 10.3 Mn; induction melted.
2	297	1959	1.3-4.1		$Cr_{0.8}Mn_{0.2}$	79.6 Cr, 20.4 Mn; same as above.
3	297	1959	1.3-4.4		$Cr_{0.7}Mn_{0.3}$	68 Cr, 32 Mn; same as above.
4	297	1959	1.5-4.2		$Cr_{0.6}Mn_{0.4}$	59.6 Cr, 40.4 Mn; same as above.

DATA TABLE NO. 77 SPECIFIC HEAT OF CHROMIUM + MANGANESE, Cr + Mn

[Temperature, T, K; Specific Heat, C_p, Cal $g^{-1}K^{-1}$]

T	C_p
CURVE 1	
1.313	1.490×10^{-5}
1.389	1.543
1.499	1.657
1.529	1.662*
1.622	1.722
1.680	1.796
1.752	1.873
1.809	1.943
1.918	2.069
1.987	2.132
2.073	2.371
2.149	2.300
2.255	2.413
2.385	2.562
2.552	2.746
2.628	2.849
2.765	2.989
2.949	3.147
3.130	3.449
3.294	3.662
3.407	3.780
3.537	3.968
3.656	4.147
3.743	4.247*
3.807	4.379
3.848	4.428*
4.003	4.652
4.114	4.782*
4.218	4.935
4.320	5.113
CURVE 2	
1.296	4.497×10^{-5}
1.419	5.044
1.483	5.252
1.558	5.539
1.561	5.480*
1.683	5.862
1.752	5.877
1.914	6.489
1.969	6.482*
2.112	6.892
2.203	7.145
CURVE 2 (cont.)	
2.344	7.594×10^{-5}
2.423	7.786
2.523	8.069
2.608	8.328
2.716	8.606
2.854	9.056
2.957	9.355
3.120	9.849
3.212	1.009×10^{-4}*
3.274	1.028
3.428	1.075
3.599	1.110
3.689	1.151
3.771	1.173
3.890	1.214
4.091	1.276
CURVE 3	
1.332	8.800×10^{-5}
1.435	9.233
1.493	9.626
1.666	1.047×10^{-4}
1.724	1.076
1.775	1.101
1.872	1.152
1.907	1.165*
1.995	1.203
2.070	1.239
2.145	1.287
2.203	1.325
2.254	1.354*
2.318	1.383
2.384	1.418*
2.543	1.473
2.630	1.531
2.744	1.590*
2.875	1.654
3.046	1.745
3.164	1.811
3.242	1.841*
3.403	1.927
3.548	2.001
3.722	2.093*
CURVE 3 (cont.)	
3.721	2.110×10^{-4}
3.908	2.193
4.158	2.327
4.278	2.398*
4.407	2.481
CURVE 4	
1.484	1.476×10^{-4}
1.562	1.554
1.600	1.558
1.665	1.599
1.765	1.664
1.808	1.700
1.862	1.757
2.013	2.230
2.086	1.958
2.156	2.005
2.203	2.025*
2.264	2.087
2.326	2.136*
2.370	2.178
2.396	2.206*
2.460	2.242
2.535	2.317
2.582	2.364
2.637	2.401*
2.691	2.430
2.750	2.475*
2.812	2.537
2.882	2.589*
2.965	2.646
3.066	2.740*
3.178	2.821
3.305	2.966
3.453	3.092
3.613	3.240
3.738	3.379*
3.828	3.422
3.933	3.539
4.056	3.663
4.188	3.782

*Not shown on plot

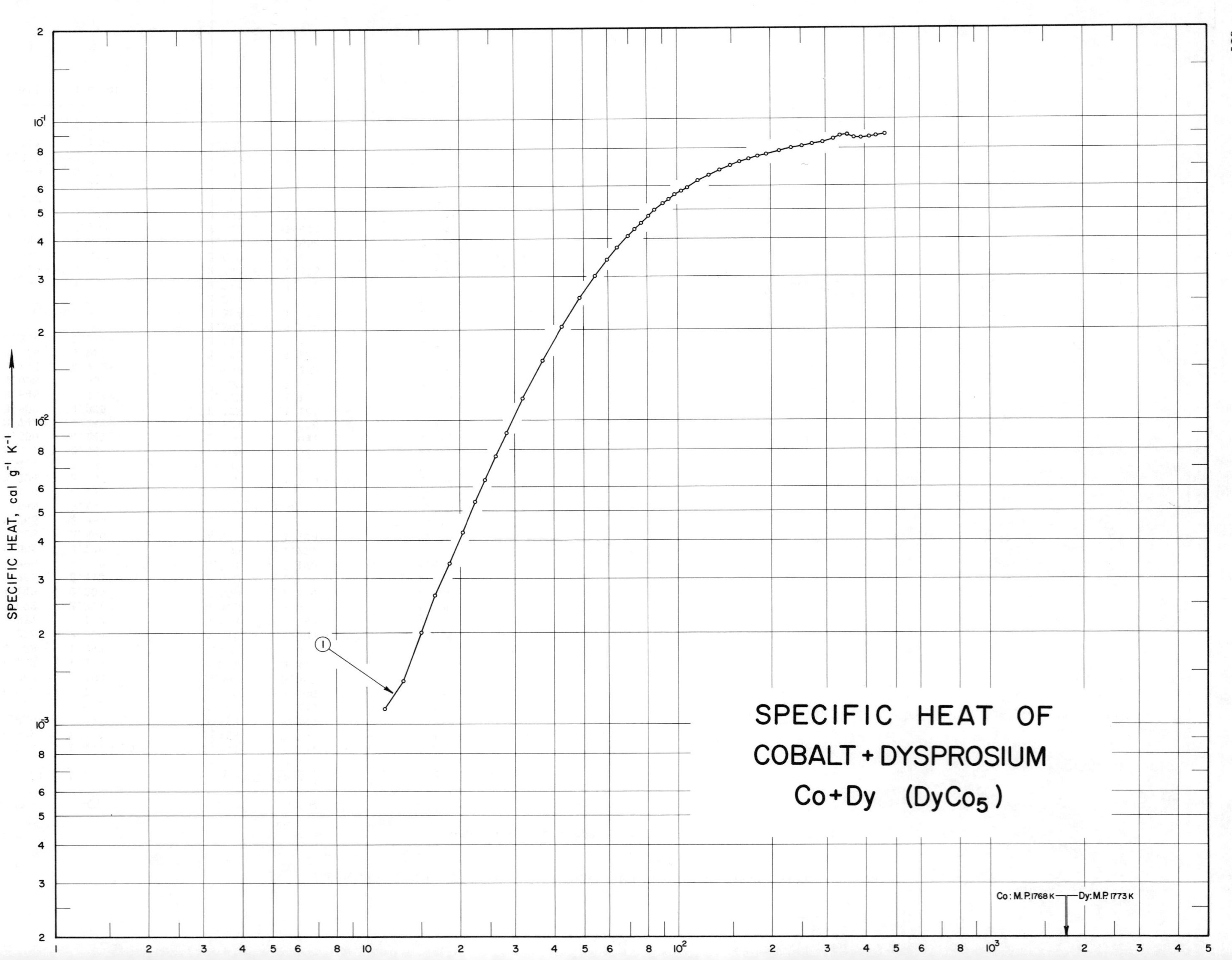
SPECIFIC HEAT OF
COBALT + DYSPROSIUM
Co+Dy (DyCo5)
SPECIFIC HEAT, cal g-1 K-1
Co : M.P. 1768 K
Dy: M.P. 1773 K
1

SPECIFICATION TABLE NO. 78 SPECIFIC HEAT OF COBALT + DYSPROSIUM, Co + Dy ($DyCo_5$)

[For Data Reported in Figure and Table No. 78]

Curve No.	Ref. No.	Year	Temp. Range, K	Reported Error, %	Name and Specimen Designation	Composition (weight percent), Specifications and Remarks
1	383	1961	11-466	≤2.0	$DyCo_5$	Almost pure $DyCo_5$; prepared by levitation melting to fuse together stoichiometric quantities of component metals; Dy sample contained 0.2 Ta, 0.1 Ca; Co sample contained spectroscopic traces of metallic impurities.

DATA TABLE NO. 78 SPECIFIC HEAT OF COBALT + DYSPROSIUM, Co + Dy ($DyCo_5$)

[Temperature, T, K; Specific Heat, C_p, Cal $g^{-1}K^{-1}$]

T	C_p
CURVE 1	
Series 1	
69.65	4.086×10^{-2}
73.03	4.300
76.74	4.541
80.77	4.764
84.70	4.987
86.53	5.074*
90.41	5.258
94.44	5.424
98.70	5.601
103.13	5.772
107.71	5.936
112.56	6.098*
117.46	6.258
122.35	6.398*
127.32	6.529
132.37	6.660*
137.52	6.776
142.77	6.890*
148.12	7.001
153.59	7.104*
159.16	7.200
164.86	7.294*
170.57	7.371
176.31	7.445*
182.15	7.537
188.08	7.587
194.01	7.682
199.89	7.736
205.72	7.802
Series 2	
11.49	1.126×10^{-3}
13.22	1.382
15.00	2.012
16.67	2.650
18.52	3.357
20.40	4.251
22.34	5.366
24.13	6.333
26.06	7.608
28.29	9.099
CURVE 1 (cont.)	
Series 2 (cont.)	
31.87	1.180×10^{-2}
36.94	1.579
42.67	2.049
48.66	2.541
54.46	3.001
59.67	3.404
64.28	3.748
68.96	4.035*
Series 3	
181.94	7.517×10^{-2}*
188.15	7.601*
194.31	7.666
200.47	7.741*
206.68	7.815*
212.98	7.861
219.42	7.924*
225.90	7.979*
232.50	8.040
239.07	8.086*
245.62	8.115*
252.12	8.154
258.59	8.209*
265.12	8.237*
271.74	8.274
278.32	8.309*
Series 4	
281.51	8.318×10^{-2}*
288.08	8.355*
294.63	8.406
301.16	8.467*
Series 5	
290.58	8.355×10^{-2}*
307.57	8.519*
312.81	8.591*
318.01	8.637
323.20	8.701*
CURVE 1 (cont.)	
Series 5 (cont.)	
328.37	8.749×10^{-2}*
333.53	8.804
Series 6	
333.81	8.801×10^{-2}*
339.92	8.836*
346.01	8.876*
352.09	8.882
358.17	8.828*
364.26	8.782*
370.36	8.685
382.69	8.659*
388.82	8.677*
Series 7	
389.04	8.655×10^{-2}
395.17	8.688*
401.29	8.699*
407.41	8.703*
413.52	8.734
425.70	8.773*
Series 8	
435.29	8.830×10^{-2}
441.36	8.825*
453.46	8.867*
459.51	8.904*
465.56	8.904

* Not shown on plot

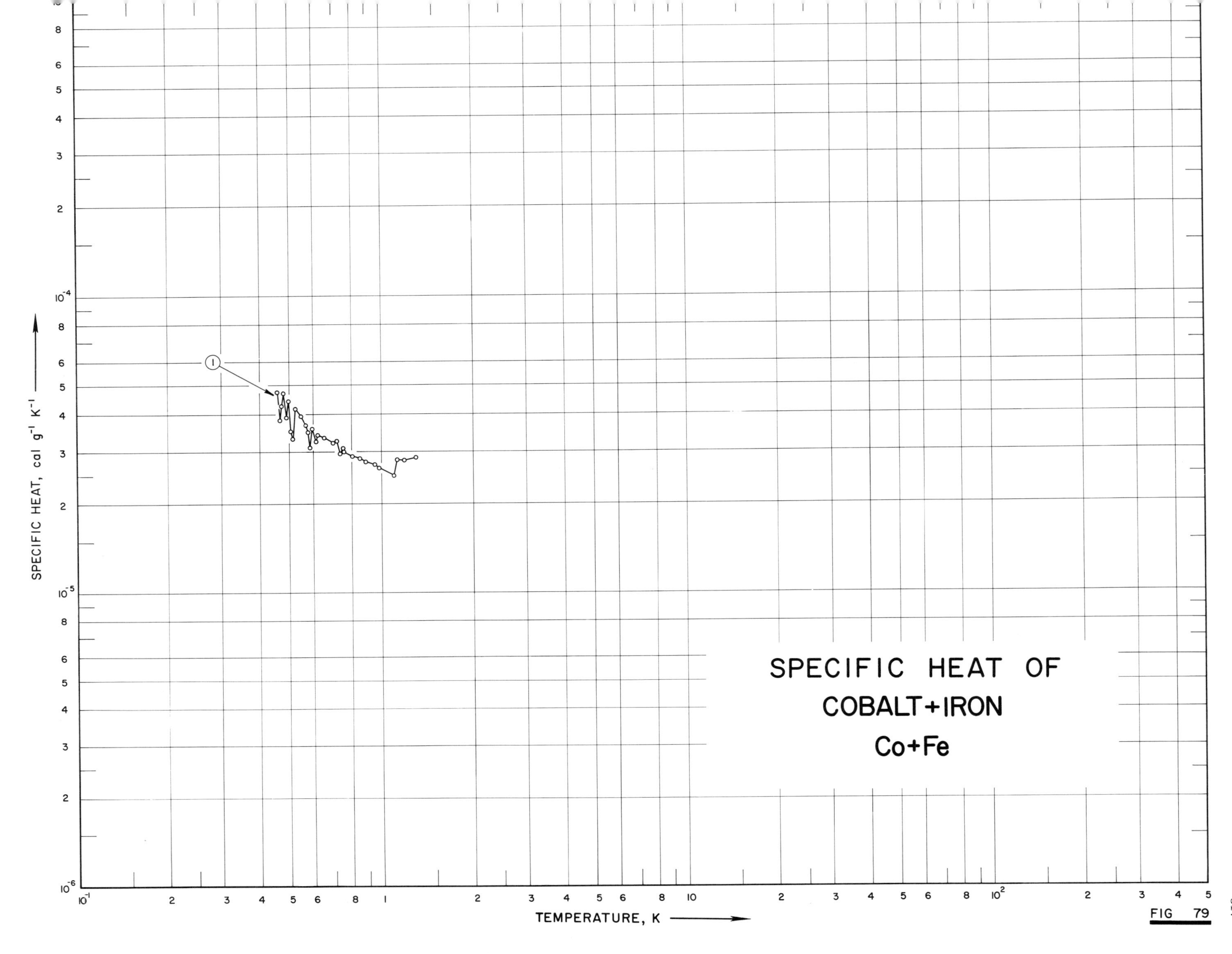
SPECIFIC HEAT OF
COBALT+IRON
Co+Fe
SPECIFIC HEAT, cal g⁻¹ K⁻¹
TEMPERATURE, K
1
FIG 79

SPECIFICATION TABLE NO. 79 SPECIFIC HEAT OF COBALT + IRON Co + Fe

[For Data Reported in Figure and Table No. 79]

Curve No.	Ref. No.	Year	Temp. Range, K	Reported Error, %	Name and Specimen Designation	Composition (weight percent), Specifications and Remarks
1	223	1959	0.4-1.4	5	Face centered cubic	93.5 Co, 6.5 Fe (99.9 Co and 99.99 Fe); prepared by electrical induction heating of constituent.

DATA TABLE NO. 79 SPECIFIC HEAT OF COBALT + IRON, Co + Fe

[Temperature, T, K; Specific Heat, C_p, Cal $g^{-1}K^{-1}$]

T	C_p
CURVE 1	
0.463	4.764 x 10^{-5}
0.470	3.827
0.478	4.283
0.484	4.735
0.494	3.937
0.500	4.434
0.505	4.418*
0.511	3.514
0.515	3.396*
0.517	3.318
0.529	4.194
0.554	3.945
0.571	3.673
0.583	3.485
0.590	3.115
0.599	3.571
0.619	3.245
0.625	3.424
0.643	3.371*
0.657	3.343
0.701	3.213
0.721	3.253
0.742	2.960
0.757	3.082
0.759	2.997
0.813	2.899
0.859	2.854
0.896	2.773
0.890	2.785*
0.962	2.712
0.965	2.687*
0.967	2.659*
0.990	2.659
1.119	2.504
1.141	2.822
1.203	2.809
1.311	2.871

*Not shown on plot

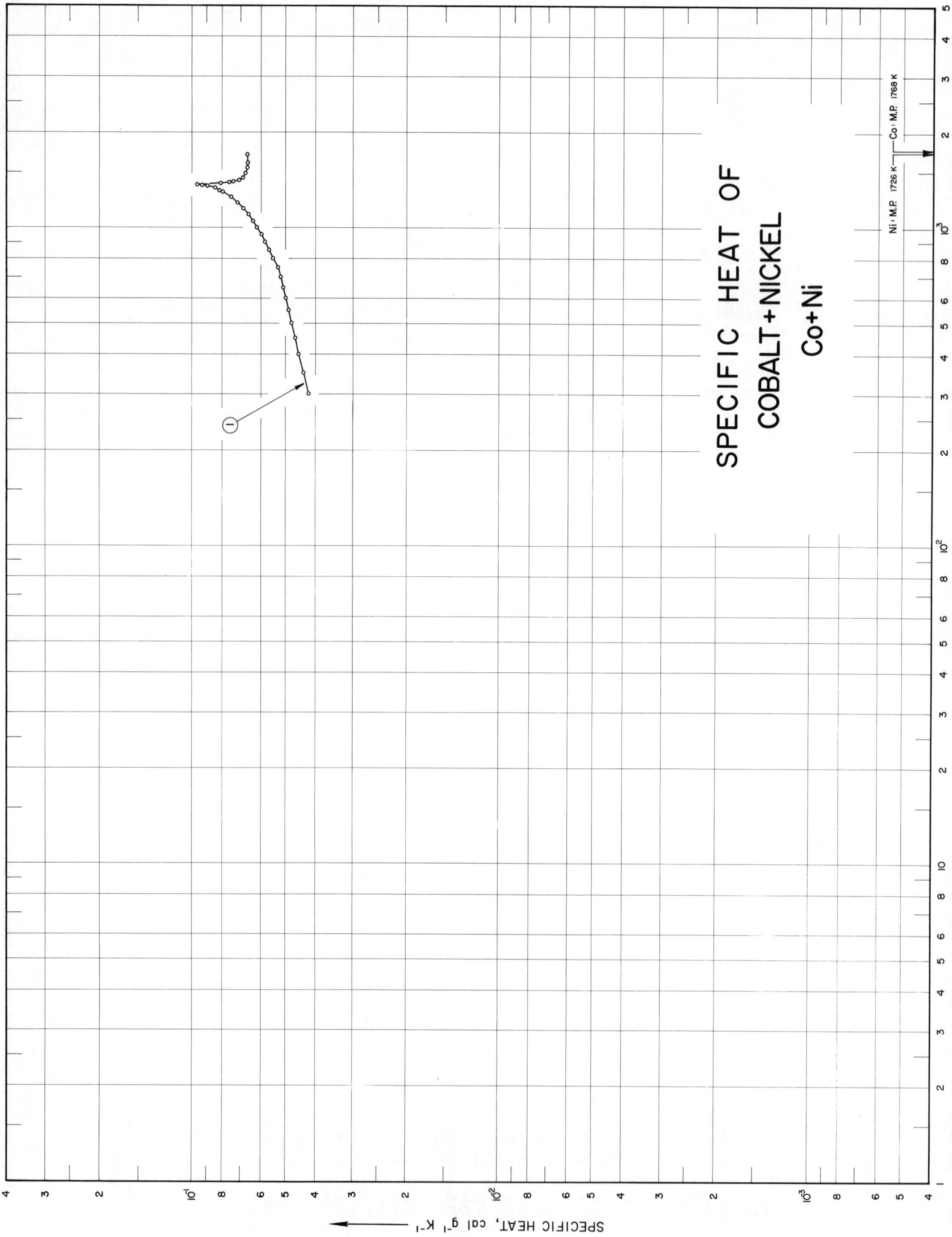

SPECIFIC HEAT OF COBALT + NICKEL Co+Ni

SPECIFIC HEAT, cal g^{-1} K^{-1}

Ni : M.P. 1726 K

Co : M.P. 1768 K

SPECIFICATION TABLE NO. 80 SPECIFIC HEAT OF COBALT + NICKEL Co + Ni

[For Data Reported in Figure and Table No. 80]

Curve No.	Ref. No.	Year	Temp. Range, K	Reported Error, %	Name and Specimen Designation	Composition (weight percent), Specifications and Remarks
1	224	1964	300-1700	2.0		99.5 Co, 0.36 Ni, 0.07 Si, 0.025 C, 0.01 Mn, 0.01 P, 0.01 S, 0.01 V, <0.01 Cr, 0.004 O_2.

DATA TABLE NO. 80 SPECIFIC HEAT OF COBALT + NICKEL, Co + Ni

[Temperature, T, K; Specific Heat, C_p, Cal $g^{-1}K^{-1}$]

T	C_p
CURVE 1	
300	4.23×10^{-1}
350	4.39
400	4.55
450	4.68
500	4.80
550	4.90
600	5.02
650	5.11
700	5.21
750	5.33
800	5.50
850	5.68
900	5.85
950	6.02
1000	6.21
1050	6.41
1100	6.63
1150	6.89
1200	7.18
1250	7.52
1300	7.98
1320	8.20
1340	8.48
1360	8.96
1370	9.33
1377	9.67
1380	8.14
1390	7.64
1400	7.40
1420	7.11
1440	6.94
1460	6.86*
1480	6.79
1500	6.74*
1550	6.67
1600	6.65
1700	6.70

*Not shown on plot

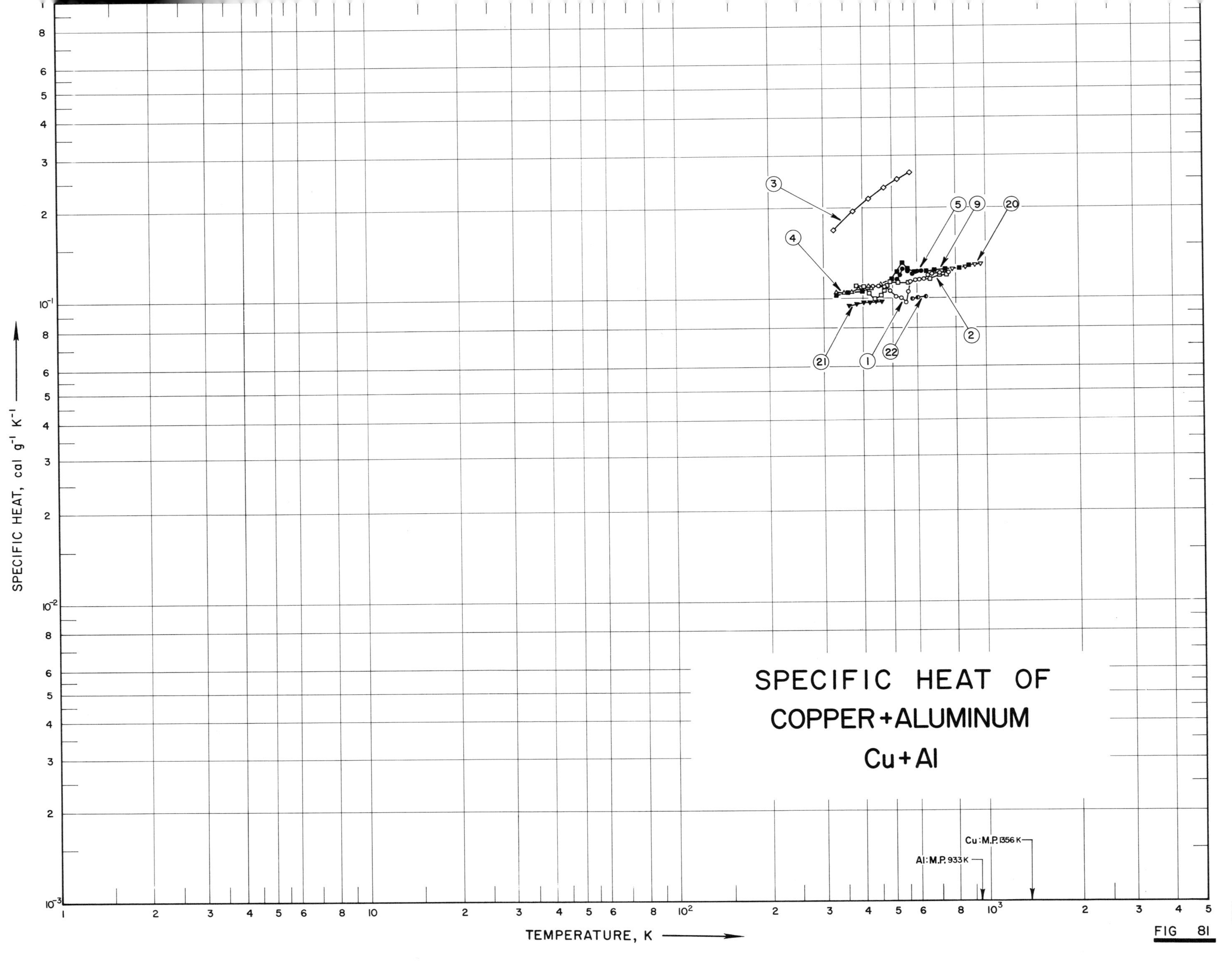
SPECIFIC HEAT OF
COPPER + ALUMINUM
Cu + Al
Cu : M.P. 1356 K
Al : M.P. 933 K
TEMPERATURE, K
SPECIFIC HEAT, cal g⁻¹ K⁻¹
FIG 81

SPECIFICATION TABLE NO. 81 SPECIFIC HEAT OF COPPER + ALUMINUM Cu + Al

[For Data Reported in Figure and Table No. 81]

Curve No.	Ref. No.	Year	Temp. Range, K	Reported Error, %	Name and Specimen Designation	Composition (weight percent), Specifications and Remarks
1	225	1959	433-757	0.8	Cu + 15.9 at.% Al alloy	92.6 Cu, 7.4 Al; quenched from 500 C.
2	225	1959	383-748	0.8	Cu + 15.9 at.% Al alloy	92.6 Cu, 7.4 Al; quenched from 900 C.
3	102	1962	323-573	3-5	$CuAl_2$	54.09 Cu, 45.91 Al; melted; kept in kiln for 1 hr; cast in steel form; preheated to 200 C.
4	384	1962	328-478	±0.5	16.8 atm % Aluminum-copper alloy	92.04 Cu, 7.91 Al; prepared from: 99.999^{+} Cu, <0.0002 As, <0.0002 Te, <0.0001 Sb, <0.0001 Pb, <0.0001 Ni, <0.0001 Se, <0.0001 Sn, <0.0001 S, <0.00007 Fe, <0.00005 Cr, <0.00003 Ag, <0.00001 Bi, <0.00001 Si; and 99.99^{+} Al; molten for about 20 min; homogenized in hydrogen 5 days at 850 C.
5	384	1962	388-618	±0.5	Same as above	Same as above.
6	384	1962	398-723	±0.5	Same as above	Same as above.
7	384	1962	378-858	±0.5	Same as above	Same as above.
8	384	1962	358-468	±0.5	Same as above	Same as above.
9	384	1962	328-883	±0.5	Same as above	Same as above.
10	384	1962	358-598	±0.5	Same as above	Same as above.
11	384	1962	328-598	±0.5	Same as above	Same as above.
12	384	1962	318-588	±0.5	Same as above	Same as above.
13	384	1962	328-588	±0.5	Same as above	Same as above.
14	384	1962	338-608	±0.5	Same as above	Same as above.
15	384	1962	348-608	±0.5	Same as above	Same as above.
16	384	1962	358-608	±0.5	Same as above	Same as above.
17	384	1962	318-638	±0.5	Same as above	Same as above.
18	384	1962	338-608	±0.5	Same as above	Same as above.
19	384	1962	883-913	±0.5	Same as above	Same as above.
20	384	1962	763-963	±0.5	Same as above	Same as above.
21	384	1962	363-463	±0.5	Same as above	Same as above.
22	384	1962	578-643	±0.5	Same as above	Same as above.

DATA TABLE NO. 81 SPECIFIC HEAT OF COPPER + ALUMINUM Cu + Al

[Temperature, T, K; Specific Heat, C_p, Cal g^{-1} K^{-1}]

CURVE 1

T	C_p
433.15	1.090×10^{-1}
453.15	1.094
473.15	1.085
479.15	1.074*
493.15	1.060
503.15	1.040*
513.15	1.020
523.15	1.010*
533.15	1.000
543.15	9.850×10^{-2}*
553.15	9.700
563.15	1.055×10^{-1}
573.15	1.130
593.15	1.150
613.15	1.155
633.15	1.170
653.15	1.195
673.15	1.200
713.15	1.210
757.15	1.220

CURVE 2

T	C_p
383.15	1.100×10^{-1}
403.15	1.091
423.15	1.041
433.15	1.005*
443.15	9.950×10^{-2}
453.15	1.006×10^{-1}*
463.15	1.026
473.15	1.070
483.15	1.109
493.15	1.130
503.15	1.130*
523.15	1.129
563.15	1.134
583.15	1.136*
623.15	1.155*
663.15	1.170
708.15	1.190
748.15	1.200

CURVE 3

T	C_p
323.15	1.691×10^{-1}
373.15	1.960×10^{-1}
423.15	2.167
473.15	2.341
523.15	2.495
573.15	2.633

CURVE 4

T	C_p
328.2	1.053×10^{-1}
338.2	1.033*
348.2	1.046
358.2	1.056*
368.2	1.057
378.2	1.078*
388.2	1.084
398.2	1.089*
408.2	1.094*
418.2	1.101
428.2	1.078
438.2	1.103*
448.2	1.104*
458.2	1.113
468.2	1.111*
478.2	1.112*

CURVE 5

T	C_p
388.2	1.087×10^{-1}*
398.2	1.091*
408.2	1.092*
418.2	1.092*
428.2	1.098*
438.2	1.101*
448.2	1.106*
458.2	1.106*
468.2	1.101*
478.2	1.111*
488.2	1.112*
498.2	1.118*
508.2	1.129*
518.2	1.162
528.2	1.190
538.2	1.252
548.2	1.241*
558.2	1.225
568.2	1.218×10^{-1}*
578.2	1.218
588.2	1.215*
603.2	1.223
618.2	1.223

CURVE 6*

T	C_p
398.2	1.086×10^{-1}
408.2	1.094
418.2	1.077
428.2	1.095
438.2	1.097
448.2	1.101
458.2	1.102
468.2	1.105
478.2	1.106
488.2	1.110
498.2	1.113
508.2	1.119
518.2	1.140
528.2	1.169
538.2	1.223
548.2	1.252
558.2	1.237
568.2	1.216
578.2	1.215
588.2	1.213
598.2	1.217
608.2	1.216
623.2	1.219
643.2	1.226
663.2	1.227
683.2	1.230
703.2	1.232
723.2	1.239

CURVE 7*

T	C_p
378.2	1.081×10^{-1}
388.2	1.083
398.2	1.081
408.2	1.087
418.2	1.090
428.2	1.088
438.2	1.093×10^{-1}
448.2	1.098
458.2	1.096
468.2	1.099
478 2	1.103
488.2	1.106
498.2	1.108
508.2	1.112
518.2	1.127
528.2	1.139
538.2	1.184
548.2	1.237
568.2	1.251
578.2	1.222
588.2	1.215
598.2	1.214
608.2	1.213
623.2	1.218
643.2	1.222
663.2	1.222
683.2	1.228
858.2	1.274

CURVE 8*

T	C_p
358.2	1.062×10^{-1}
368.2	1.058
378.2	1.069
388.2	1.079
398.2	1.073
408.2	1.084
418.2	1.082
428.2	1.088
438.2	1.093
448.2	1.092
458.2	1.096
468.2	1.101

CURVE 9

T	C_p
328.2	1.029×10^{-1}
338.2	1.032*
348.2	1.039*
358.2	1.050
368.2	1.049*
378.2	1.063×10^{-1}*
388.2	1.074*
398.2	1.071
408.2	1.085*
418.2	1.078*
428.2	1.083*
438.2	1.098*
448.2	1.105*
458.2	1.113*
468.2	1.123*
478.2	1.135*
488.2	1.147*
498.2	1.164
508.2	1.186*
518.2	1.225
528.2	1.260*
538.2	1.313
548.2	1.314*
558.2	1.255
568.2	1.218*
578.2	1.216*
588.2	1.220
598.2	1.220*
608.2	1.223*
623.2	1.225*
643.2	1.229
663.2	1.230*
683.2	1.235
703.2	1.239*
723.2	1.239*
743.2	1.250
763.2	1.250*
783.2	1.253*
803.2	1.257*
823.2	1.266
843.2	1.268*
863.2	1.260*
883.2	1.287

CURVE 10*

T	C_p
358.2	1.058×10^{-1}
368.2	1.055
398.2	1.083
408.2	1.098
418.2	1.087×10^{-1}
428.2	1.087
438.2	1.103
448.2	1.098
458.2	1.103
468.2	1.106
478.2	1.109
488.2	1.106
498.2	1.110
508.2	1.116
518.2	1.144
528.2	1.179
538.2	1.222
548.2	1.234
558.2	1.225
568.2	1.216
578.2	1.212
588.2	9.771×10^{-2}
598.2	1.217×10^{-1}

CURVE 11*

T	C_p
328.2	1.048×10^{-1}
338.2	1.047
353.2	1.054
368.2	1.059
378.2	1.073
388.2	1.083
398.2	1.085
408.2	1.089
418.2	1.089
428.2	1.093
438.2	1.102
448.2	1.103
458.2	1.106
468.2	1.108
478.2	1.112
488.2	1.114
498.2	1.121
508.2	1.134
518.2	1.170
528.2	1.212
538.2	1.260
548.2	1.271

* Not shown on plot

DATA TABLE NO. 81 (continued)

T	C_p
CURVE 11 (cont.)*	
558.2	1.233 x 10^{-1}
568.2	1.222
578.2	1.222
588.2	1.220
598.2	1.229
CURVE 12*	
318.2	1.060 x 10^{-1}
328.2	1.042
338.2	1.051
348.2	1.046
358.2	1.055
368.2	1.052
378.2	1.067
388.2	1.075
398.2	1.078
408.2	1.083
418.2	1.082
428.2	1.085
438.2	1.096
448.2	1.092
458.2	1.100
468.2	1.099
478.2	1.106
488.2	1.103
498.2	1.105
508.2	1.107
518.2	1.129
528.2	1.153
538.2	1.201
548.2	1.226
558.2	1.224
568.2	1.217
578.2	1.216
588.2	1.219
CURVE 13*	
328.2	1.043 x 10^{-1}
338.2	1.041
348.2	1.048
358.2	1.054
368.2	1.047
378.2	1.049
388.2	1.054
398.2	1.050
408.2	1.051
CURVE 13 (cont.)*	
418.2	1.044 x 10^{-1}
428.2	1.054
438.2	1.061
448.2	1.065
458.2	1.072
468.2	1.065
478.2	1.072
488.2	1.078
498.2	1.084
508.2	1.092
518.2	1.122
528.2	1.148
538.2	1.193
548.2	1.213
558.2	1.221
568.2	1.217
578.2	1.217
588.2	1.217
CURVE 14*	
338.2	1.044 x 10^{-1}
348.2	1.049
358.2	1.055
368.2	1.049
378.2	1.057
388.2	1.058
398.2	1.055
408.2	1.055
418.2	1.048
428.2	1.054
438.2	1.063
448.2	1.065
458.2	1.061
468.2	1.066
483.2	1.076
498.2	1.087
508.2	1.100
518.2	1.125
528.2	1.154
538.2	1.197
548.2	1.218
558.2	1.223
568.2	1.219
578.2	1.220
588.2	1.218
598.2	1.219
608.2	1.223
CURVE 15*	
348.2	1.025 x 10^{-1}
358.2	1.022
368.2	1.013
378.2	1.025
388.2	1.039
398.2	1.041
408.2	1.051
418.2	1.056
428.2	1.063
438.2	1.060
448.2	1.055
458.2	1.060
468.2	1.081
478.2	1.107
488.2	1.122
498.2	1.135
508.2	1.140
518.2	1.156
528.2	1.178
538.2	1.212
548.2	1.215
558.2	1.220
568.2	1.217
578.2	1.218
588.2	1.217
598.2	1.217
608.2	1.222
CURVE 16*	
358.2	1.027 x 10^{-1}
368.2	1.017
378.2	1.022
388.2	1.043
398.2	1.044
408.2	1.055
418.2	1.055
428.2	1.068
438.2	1.060
448.2	1.058
458.2	1.062
468.2	1.077
478.2	1.103
488.2	1.119
498.2	1.132
508.2	1.141
518.2	1.155
528.2	1.180
CURVE 16 (cont.)*	
538.2	1.206 x 10^{-1}
548.2	1.223
558.2	1.217
568.2	1.218
578.2	1.219
588.2	1.241
598.2	1.221
608.2	1.222
CURVE 17*	
318.2	1.051 x 10^{-1}
328.2	1.041
338.2	1.026
348.2	1.027
358.2	1.025
378.2	1.027
388.2	1.039
398.2	1.042
408.2	1.047
418.2	1.063
428.2	1.057
438.2	1.063
448.2	1.088
458.2	1.116
468.2	1.134
478.2	1.148
488.2	1.161
498.2	1.171
508.2	1.179
518.2	1.180
528.2	1.205
538.2	1.217
548.2	1.219
558.2	1.217
568.2	1.220
578.2	1.223
588.2	1.223
598.2	1.225
608.2	1.221
618.2	1.224
628.2	1.224
638.2	1.233
CURVE 18*	
338.2	1.040 x 10^{-1}
348.2	1.026
CURVE 18 (cont.)*	
358.2	1.028 x 10^{-1}
368.2	1.021
378.2	1.028
388.2	1.041
398.2	1.041
408.2	1.048
418.2	1.049
428.2	1.060
428.2	1.060
438.2	1.068
448.2	1.085
458.2	1.107
468.2	1.133
478.2	1.146
488.2	1.156
598.2	1.168
508.2	1.174
518.2	1.182
528.2	1.189
538.2	1.202
548.2	1.217
558.2	1.216
568.2	1.216
578.2	1.215
588.2	1.221
598.2	1.219
608.2	1.219
CURVE 19*	
883.2	1.258 x 10^{-1}
903.2	1.261
923.2	1.271
943.2	1.271
963.2	1.276
CURVE 20	
763.2	1.264 x 10^{-1}*
783.2	1.254
803.2	1.258*
823.2	1.264*
843.2	1.268*
863.2	1.274
883.2	1.273*
903.2	1.278*
923.2	1.293
943.2	1.291*
963.2	1.298
CURVE 21	
363.2	9.460 x 10^{-2}
383.2	9.613
403.2	9.711
423.2	9.713
443.2	9.792
463.2	9.835
CURVE 22	
578.2	1.008 x 10^{-1}
603.2	1.013
623.2	1.017*
643.2	1.021

*Not shown on plot

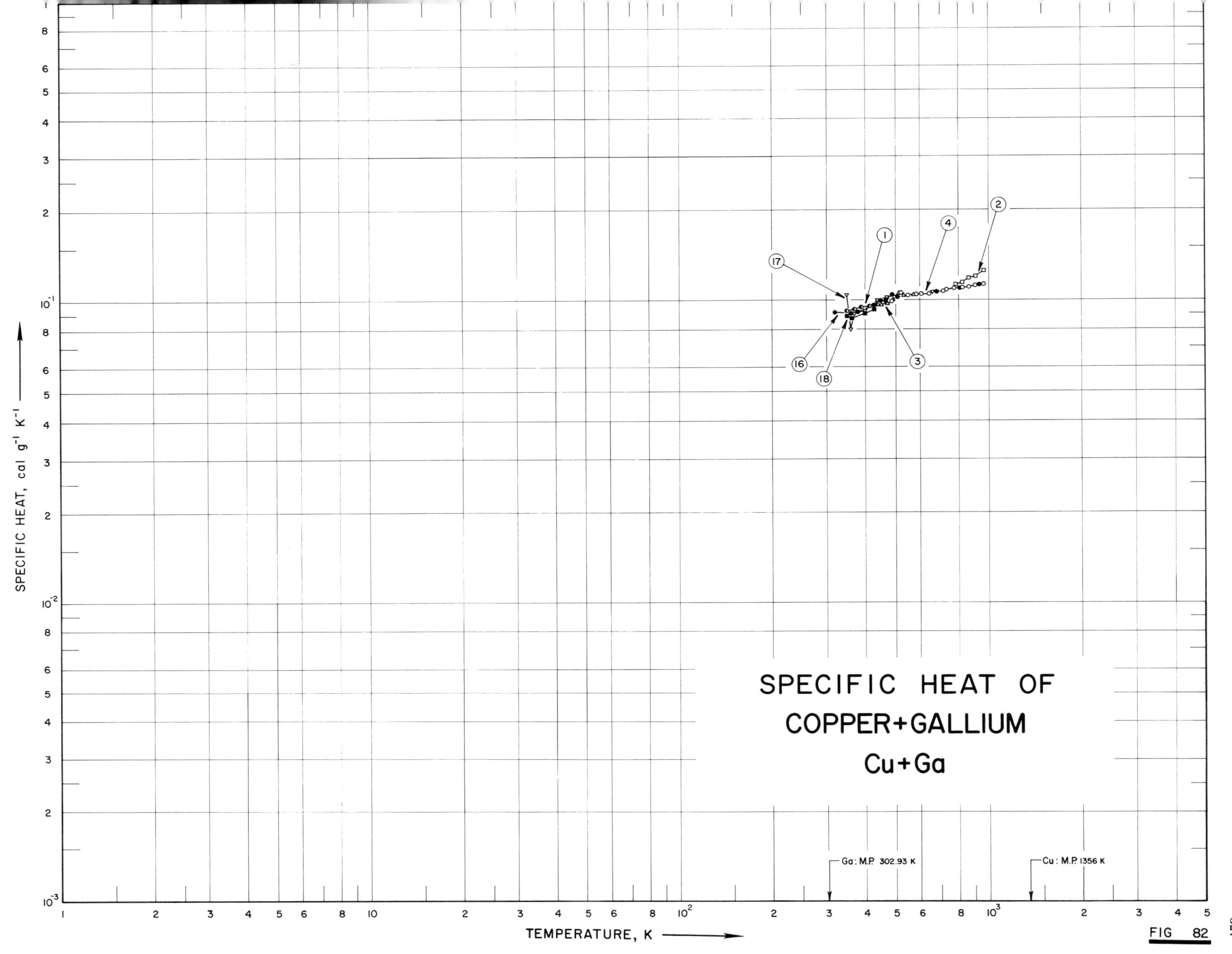
SPECIFIC HEAT OF
COPPER+GALLIUM
Cu+Ga
Ga: M.P. 302.93 K
Cu: M.P. 1356 K
SPECIFIC HEAT, cal g⁻¹ K⁻¹
TEMPERATURE, K
1
2
3
4
16
17
18

FIG 82

SPECIFICATION TABLE NO. 82 SPECIFIC HEAT OF COPPER + GALLIUM Cu + Ga

[For Data Reported in Figure and Table No. 82]

Curve No.	Ref. No.	Year	Temp. Range, K	Reported Error, %	Name and Specimen Designation	Composition (weight percent), Specifications and Remarks
1	384	1962	348-438	±0.5	17.2 atm % Gallium-copper alloy	94.18 Cu, 15.80 Ga; prepared from 99.99 Ga, 99.999^+ Cu.
2	384	1962	783-963	±0.5	Same as above	Same as above.
3	384	1962	373-573	±0.5	Same as above	Same as above.
4	384	1962	448-963	±0.5	Same as above	Same as above.
5	384	1962	403-583	±0.5	Same as above	Same as above.
6	384	1962	338-568	±0.5	Same as above	Same as above.
7	384	1962	348-583	±0.5	Same as above	Same as above.
8	384	1962	638-883	±0.5	Same as above	Same as above.
9	384	1962	338-578	±0.5	Same as above	Same as above.
10	384	1962	328-598	±0.5	Same as above	Same as above.
11	384	1962	328-568	±0.5	Same as above	Same as above.
12	384	1962	378-568	±0.5	Same as above	Same as above.
13	384	1962	338-578	±0.5	Same as above	Same as above.
14	384	1962	338-623	±0.5	Same as above	Same as above.
15	384	1962	328-578	±0.5	Same as above	Same as above.
16	384	1962	328-953	±0.5	Same as above	Same as above.
17	384	1962	348-663	±0.5	Same as above	Same as above.
18	384	1962	348-593	±0.5	Same as above	Same as above.

DATA TABLE NO. 82 SPECIFIC HEAT OF COPPER + GALLIUM Cu + Ga

[Temperature, T, K; Specific Heat, C_p, Cal g^{-1} K^{-1}]

T	C_p
CURVE 1	
348.2	9.254×10^{-2}
358.2	9.376*
368.2	9.362
378.2	9.469*
388.2	9.536
413.2	9.572
428.2	9.632*
438.2	9.711
CURVE 2	
783.2	1.134×10^{-1}
803.2	1.142*
823.2	1.155
843.2	1.168*
863.2	1.181
883.2	1.193*
903.2	1.210
923.2	1.227*
943.2	1.243*
963.2	1.261
CURVE 3	
373.2	9.398×10^{-2}*
393.2	9.572*
413.2	9.536*
433.2	9.637*
453.2	9.730
473.2	9.816
493.2	1.014×10^{-1}
513.2	1.066
533.2	1.048
553.2	1.049*
573.2	1.049
CURVE 4	
448.2	9.759×10^{-2}*
458.2	9.749*
468.2	9.787*
478.2	9.823*
488.2	9.957
498.2	1.020×10^{-1}*
508.2	1.048*
518.2	1.063*
CURVE 4 (cont.)	
528.2	1.053×10^{-1}*
538.2	1.049*
548.2	1.047
558.2	1.050*
568.2	1.052*
583.2	1.056
603.2	1.062
623.2	1.067*
643.2	1.065
663.2	1.068*
713.2	1.078
733.2	1.081
753.2	1.086*
773.2	1.091
798.2	1.096*
823.2	1.106
843.2	1.106*
863.2	1.114
883.2	1.121*
903.2	1.125
923.2	1.134*
943.2	1.132*
963.2	1.132
CURVE 5*	
403.2	9.534×10^{-2}
423.2	9.577
443.2	9.641
463.2	9.735
483.2	9.814
503.2	1.022×10^{-1}
523.2	1.067
543.2	1.049
563.2	1.045
583.2	1.047
CURVE 6*	
338.2	9.221×10^{-2}
348.2	9.235
358.2	9.400
368.2	9.381
378.2	9.467
393.2	9.541
408.2	9.663
CURVE 6 (cont.)*	
418.2	9.582×10^{-2}
428.2	9.670
438.2	9.704
448.2	9.751
458.2	9.802
468.2	9.871
478.2	9.871
488.2	1.034×10^{-1}
498.2	1.067
508.2	1.061
518.2	1.053
528.2	1.043
538.2	1.052
548.2	1.045
558.2	1.051
568.2	1.054
CURVE 7*	
348.2	9.257×10^{-2}
363.2	9.352
383.2	9.505
403.2	9.584
423.2	9.613
443.2	9.716
463.2	9.847
483.2	1.006×10^{-1}
503.2	1.049
523.2	1.068
543.2	1.048
563.2	1.049
583.2	1.054
CURVE 8*	
638.2	1.060×10^{-1}
663.2	1.063
683.2	1.068
703.2	1.073
723.2	1.076
743.2	1.082
763.2	1.087
783.2	1.088
803.2	1.097
823.2	1.099
843.2	1.104
CURVE 8 (cont.)*	
863.2	1.109×10^{-1}
883.2	1.111
CURVE 9*	
338.2	9.254×10^{-2}
348.2	9.281
358.2	9.362
368.2	9.367
378.2	9.431
388.2	9.560
398.2	9.565
408.2	9.615
418.2	9.601
428.2	9.661
438.2	9.682
448.2	9.699
458.2	1.068×10^{-1}
468.2	9.768×10^{-2}
478.2	9.823
488.2	9.983
498.2	1.027×10^{-1}
508.2	1.050
518.2	1.054
528.2	1.038
538.2	1.043
548.2	1.042
558.2	1.043
568.2	1.041
578.2	1.045
CURVE 10*	
328.2	9.321×10^{-2}
338.2	9.324*
348.2	9.371*
358.2	9.436
368.2	9.412*
378.2	9.505
388.2	9.591*
398.2	9.615
408.2	9.630
418.2	9.620
428.2	9.677
438.2	9.680
448.2	9.792
CURVE 10 (cont.)*	
458.2	9.759×10^{-2}
468.2	9.802
483.2	9.895
498.2	1.017×10^{-1}
508.2	1.042
518.2	1.065
528.2	1.055
538.2	1.049
548.2	1.046
558.2	1.049
573.2	1.053
588.2	1.057
598.2	1.059
CURVE 11*	
328.2	9.324×10^{-2}
338.2	9.359
348.2	9.398
358.2	9.469
373.2	9.491
388.2	9.653
398.2	9.610
408.2	9.637
418.2	9.620
428.2	9.677
438.2	9.718
448.2	9.732
458.2	9.768
468.2	9.771
478.2	9.826
488.2	9.945
498.2	1.010×10^{-1}
508.2	1.043
518.2	1.062
528.2	1.048
538.2	1.044
548.2	1.051
558.2	1.049
568.2	1.050
CURVE 12*	
378.2	9.610×10^{-2}
388.2	9.565
398.2	9.496
CURVE 12 (cont.)*	
408.2	9.453×10^{-2}
418.2	9.393
428.2	9.381
438.2	9.467
448.2	9.472
458.2	9.457
468.2	9.630
478.2	9.661
488.2	9.766
498.2	9.935
508.2	1.021×10^{-1}
518.2	1.045
528.2	1.048
538.2	1.047
548.2	1.045
558.2	1.048
568.2	1.051
CURVE 13*	
338.2	9.374×10^{-2}
348.2	9.331
358.2	9.352
368.2	9.324
378.2	9.321
388.2	9.319
398.2	9.247
408.2	9.338
418.2	9.290
428.2	9.386
438.2	9.448
448.2	9.489
458.2	9.551
468.2	9.658
478.2	9.517
488.2	9.924
498.2	1.013×10^{-1}
508.2	1.025
518.2	1.051
528.2	1.037
538.2	1.049
548.2	1.048
573.2	1.050
578.2	1.053

*Not shown on plot

DATA TABLE NO. 82 (continued)

T	C_p
CURVE 14*	
338.2	9.128 x 10^{-2}
348.2	9.168
358.2	9.226
368.2	9.092
378.2	9.159
388.2	9.259
398.2	9.302
408.2	9.429
418.2	9.410
428.2	9.563
438.2	9.644
448.2	9.842
458.2	9.988
468.2	1.007 x 10^{-1}
478.2	1.003
488.2	1.018
498.2	1.035
508.2	1.041
518.2	1.046
528.2	1.034
538.2	1.043
548.2	1.043
558.2	1.044
568.2	1.047
583.2	1.052
603.2	1.056
623.2	1.059
CURVE 15*	
328.2	9.238 x 10^{-2}
338.2	9.211
348.2	9.113
358.2	9.183
368.2	9.118
378.2	9.197
388.2	9.312
398.2	9.204
408.2	9.567
418.2	9.491
428.2	9.706
438.2	9.861
448.2	1.006 x 10^{-1}
458.2	1.011
468.2	1.014
478.2	1.022
488.2	1.026
CURVE 15 (cont.)*	
498.2	1.027 x 10^{-1}
508.2	1.032
518.2	1.040
528.2	1.040
538.2	1.045
548.2	1.046
558.2	1.049
568.2	1.050
578.2	1.053
CURVE 16	
328.2	9.288 x 10^{-2}
338.2	9.226*
348.2	9.159*
358.2	9.104
368.2	9.163*
378.2	9.207
388.2	9.209*
398.2	9.319*
408.2	9.536*
418.2	9.543*
428.2	9.716
438.2	9.921*
448.2	1.009 x 10^{-1}
458.2	1.017*
468.2	9.981 x 10^{-2}
478.2	1.008 x 10^{-1}*
488.2	1.055
498.2	1.034*
508.2	1.037
523.2	1.045*
543.2	1.047*
563.2	1.054*
583.2	1.058*
603.2	1.062*
623.2	1.066*
643.2	1.069*
663.2	1.071*
683.2	1.076
703.2	1.080*
723.2	1.084*
743.2	1.090*
763.2	1.092*
783.2	1.097*
803.2	1.101
823.2	1.108*
CURVE 16 (cont.)	
843.2	1.111 x 10^{-1}*
868.2	1.105*
893.2	1.125*
913.2	1.130*
933.2	1.138
953.2	1.141*
CURVE 17	
348.2	1.056 x 10^{-1}
358.2	8.054 x 10^{-2}
368.2	9.185
378.2	9.204*
388.2	9.309*
398.2	9.290
408.2	9.510*
418.2	9.587*
428.2	9.816*
438.2	1.003 x 10^{-1}
453.2	1.015*
468.2	1.021
478.2	1.024*
488.2	1.024*
498.2	1.038*
508.2	1.033*
518.2	1.047*
528.2	1.041*
538.2	1.040*
553.2	1.055*
568.2	1.057*
583.2	1.059*
603.2	1.062*
623.2	1.068*
643.2	1.069*
663.2	1.070
CURVE 18	
348.2	8.915 x 10^{-2}
363.2	8.822
398.2	9.159
428.2	9.419
438.2	9.484*
448.2	9.861*
458.2	1.009 x 10^{-1}*
468.2	1.004*
488.2	1.028*
CURVE 18 (cont.)	
508.2	1.030 x 10^{-1}*
518.2	1.049*
528.2	1.046*
538.2	1.045*
553.2	1.046*
573.2	1.050*
593.2	1.056*

*Not shown on plot

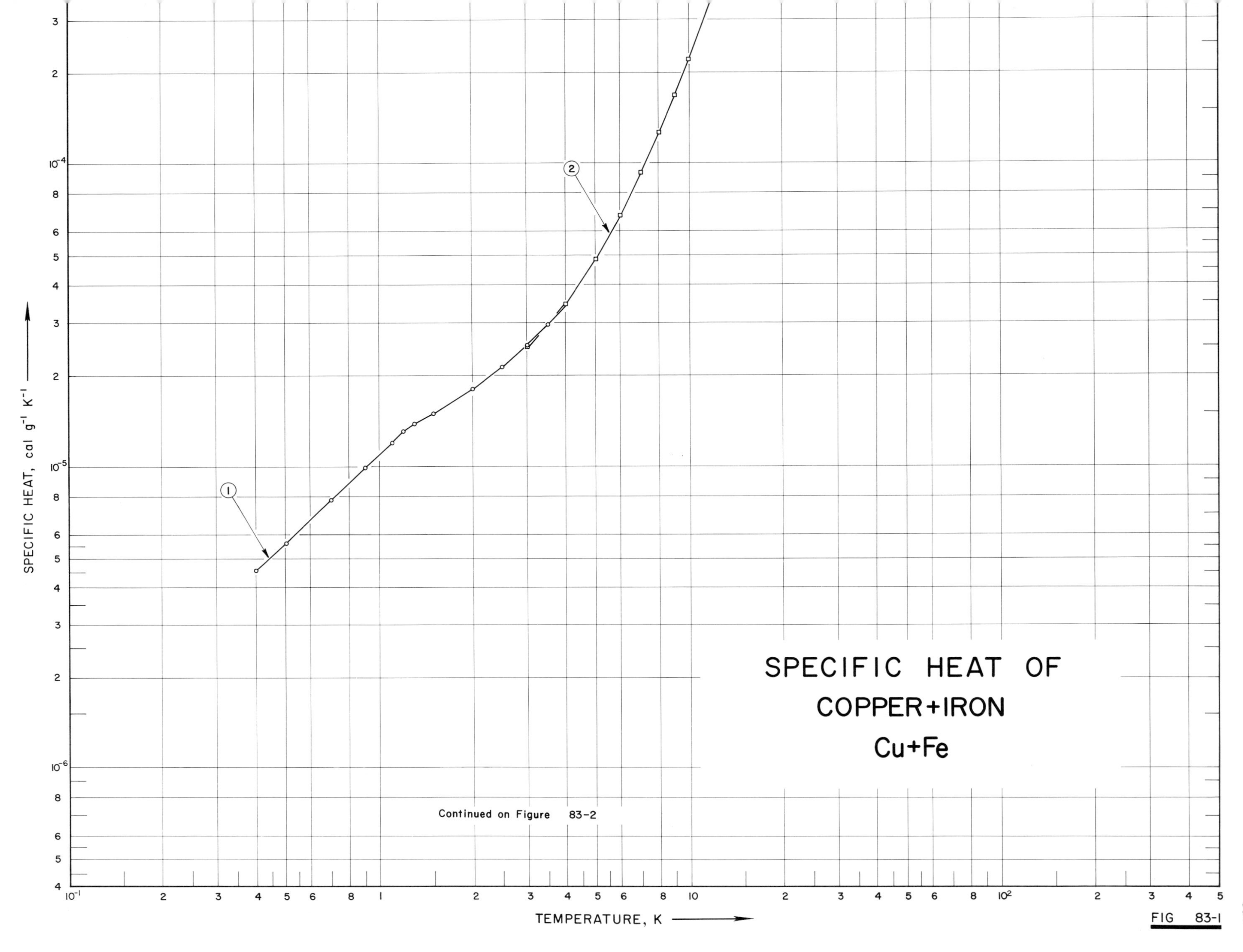
SPECIFIC HEAT OF
COPPER+IRON
Cu+Fe
SPECIFIC HEAT, cal g⁻¹ K⁻¹
TEMPERATURE, K
1
2
Continued on Figure 83-2

FIG 83-1

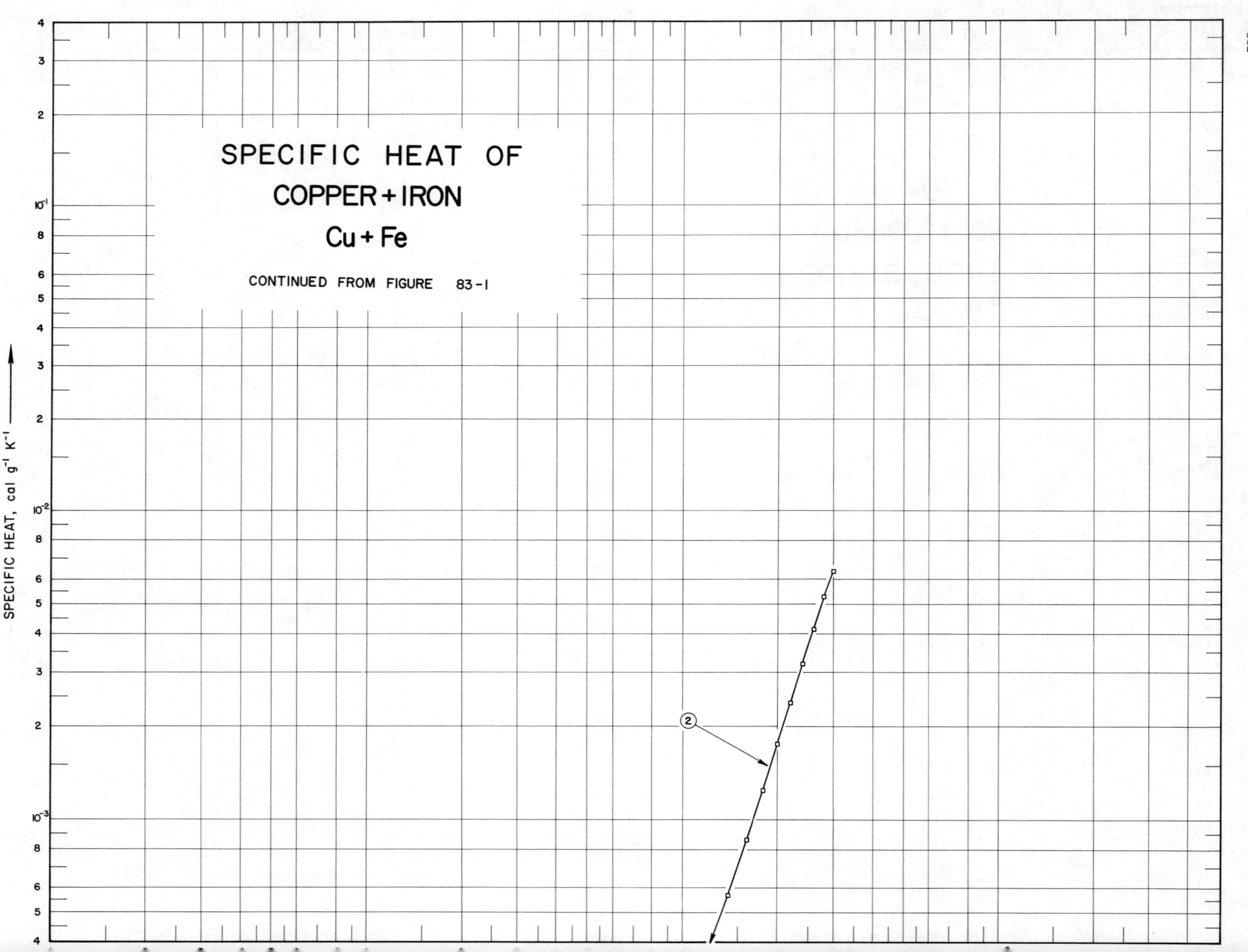
SPECIFIC HEAT OF
COPPER + IRON
Cu + Fe
CONTINUED FROM FIGURE 83-1
2
SPECIFIC HEAT, cal g⁻¹ K⁻¹

SPECIFICATION TABLE NO. 83 SPECIFIC HEAT OF COPPER + IRON Cu + Fe

[For Data Reported in Figure and Table No. 83]

Curve No.	Ref. No.	Year	Temp. Range, K	Reported Error, %	Name and Specimen Designation	Composition (weight percent), Specifications and Remarks
1	16	1961	0.4-4.0	1	0.2% Fe Dilute copper alloy	99.799 Cu, 0.2 Fe, <0.0001 Se, <0.0001 S; melted at 1300 C; annealed for 72 hrs at 870 C; cooled rapidly to room temperature.
2	16	1961	3-30	1	0.2% Fe Dilute copper alloy	Same as above.

DATA TABLE NO. 83 SPECIFIC HEAT OF COPPER + IRON Cu + Fe

[Temperature, T, K; Specific Heat, C_p, Cal g^{-1} K^{-1}]

T	C_p
CURVE 1	
0.4	4.56×10^{-6}
0.5	5.67
0.7	7.82
0.9	9.96
1.1	1.20×10^{-5}
1.3	1.39
1.5	1.50
1.2	1.32
1.5	1.52*
2.0	1.800
2.5	2.137
3.0	2.523
3.5	2.957
4.0	3.437
CURVE 2	
3.0	2.510×10^{-5}
4.0	3.451
5.0	4.860
6.0	6.752
7.0	9.367
8.0	1.271×10^{-4}
9.0	1.682
10.0	2.219
12.0	3.653
14.0	5.702
16.0	8.628
18.0	1.252×10^{-3}
20.0	1.760
22.0	2.395
24.0	3.203
26.0	4.166
28.0	5.323
30.0	6.446

* Not shown on plot

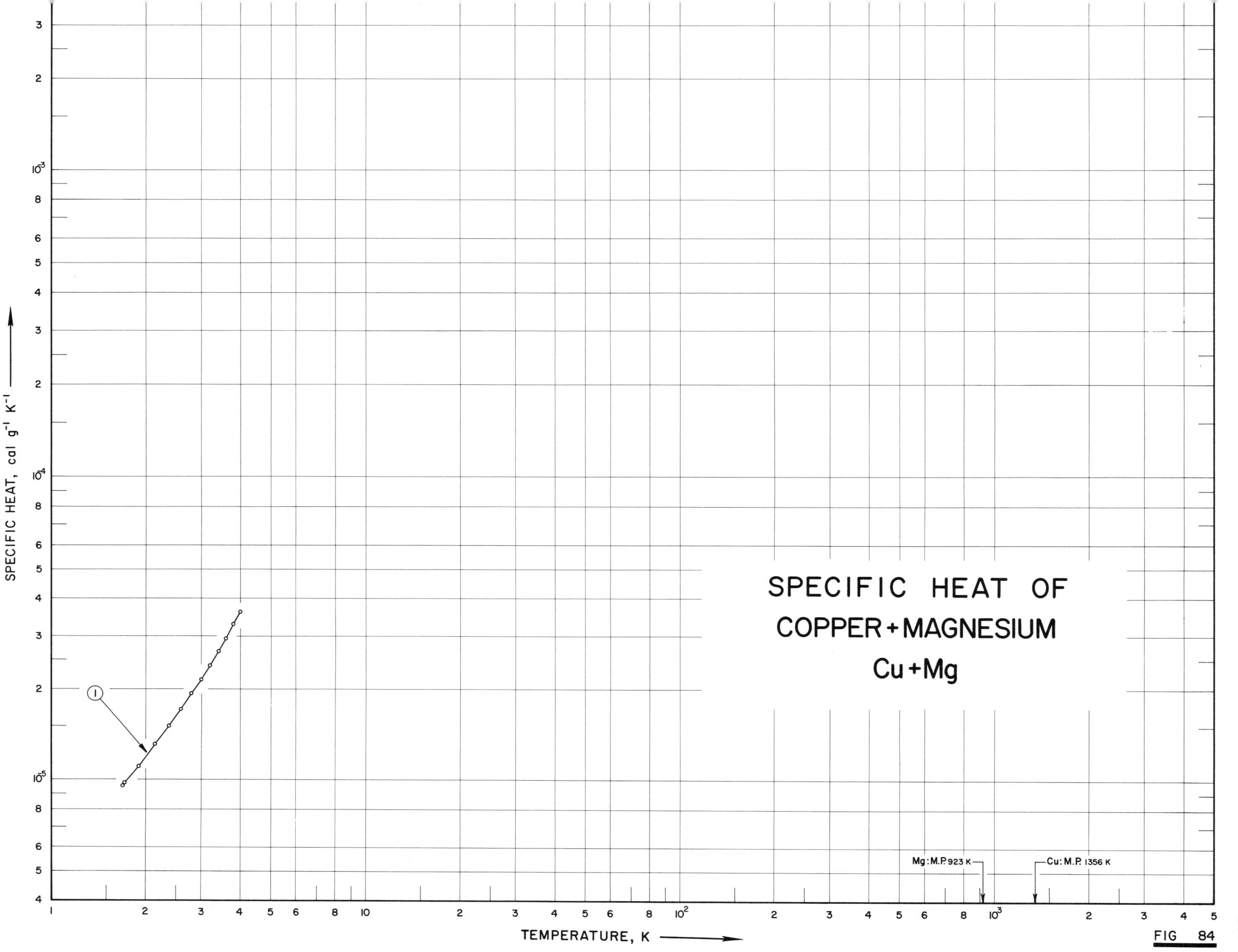
SPECIFIC HEAT OF
COPPER + MAGNESIUM
Cu + Mg
Mg: M.P. 923 K
Cu: M.P. 1356 K
TEMPERATURE, K
SPECIFIC HEAT, cal g⁻¹ K⁻¹
1

FIG 84

SPECIFICATION TABLE NO. 84 SPECIFIC HEAT OF COPPER + MAGNESIUM Cu + Mg

[For Data Reported in Figure and Table No. 84]

Curve No.	Ref. No.	Year	Temp. Range, K	Reported Error, %	Name and Specimen Designation	Composition (weight percent), Specifications and Remarks
1	385	1966	1.7-4.0		$MgCu_{2-x}$	Prepared by melting together 99.99 Cu and resublimed grade 99.98 Mg under an atmosphere of argon; after casting specimen was sealed under pure helium and held 17-24 hrs at 200-200 C below the melting temperature.

DATA TABLE NO. 84 SPECIFIC HEAT OF COPPER + MAGNESIUM Cu + Mg

[Temperature, T, K; Specific Heat, C_p, Cal g^{-1} K^{-1}]

T	C_p
CURVE 1	
1.697	9.55×10^{-6}
1.718	9.82
1.909	1.11×10^{-5}
2.148	1.31
2.373	1.51
2.583	1.72
2.799	1.94
3.010	2.15
3.207	2.39
3.407	2.66
3.606	2.95
3.811	3.30
4.008	3.62

*Not shown on plot

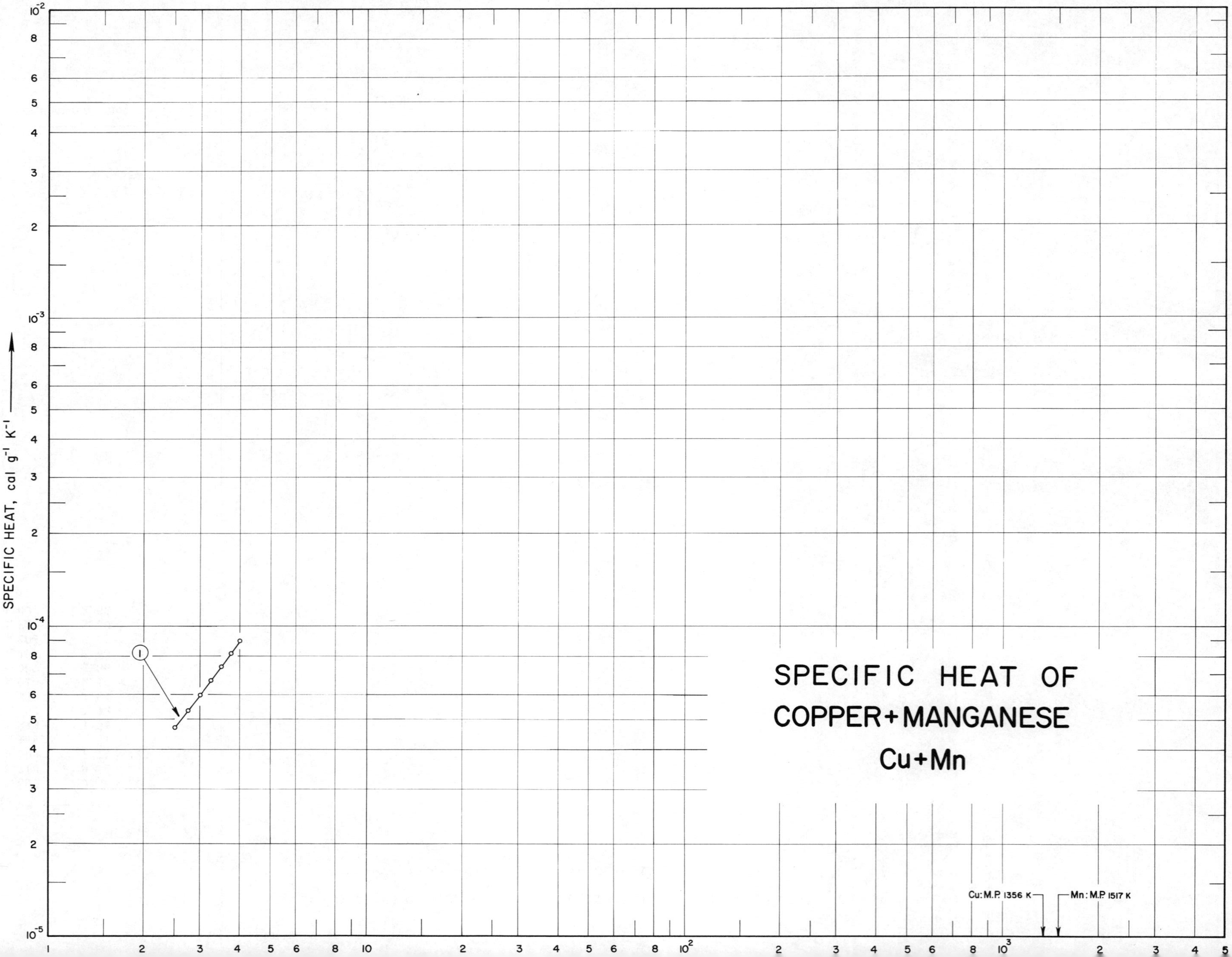

SPECIFIC HEAT OF
COPPER+MANGANESE
Cu+Mn

SPECIFIC HEAT, cal g^{-1} K^{-1}

Cu: M.P. 1356 K

Mn: M.P. 1517 K

1

SPECIFICATION TABLE NO. 85 SPECIFIC HEAT OF COPPER + MANGANESE Cu + Mn

[For Data Reported in Figure and Table No. 85]

Curve No.	Ref. No.	Year	Temp. Range, K	Reported Error, %	Name and Specimen Designation	Composition (weight percent), Specifications and Remarks
1	386	1963	2.5-4.0	1	Manganin	87 Cu, 13 Mn; from Driver-Harris Co.

DATA TABLE NO. 85 SPECIFIC HEAT OF COPPER + MANGANESE Cu + Mn

[Temperature, T, K; Specific Heat, C_p, Cal g^{-1} K^{-1}]

T	C_p
CURVE 1	
2.50	4.71×10^{-5}
2.75	5.35
3.00	6.00
3.25	6.69
3.50	7.41
3.75	8.17
4.00	8.99

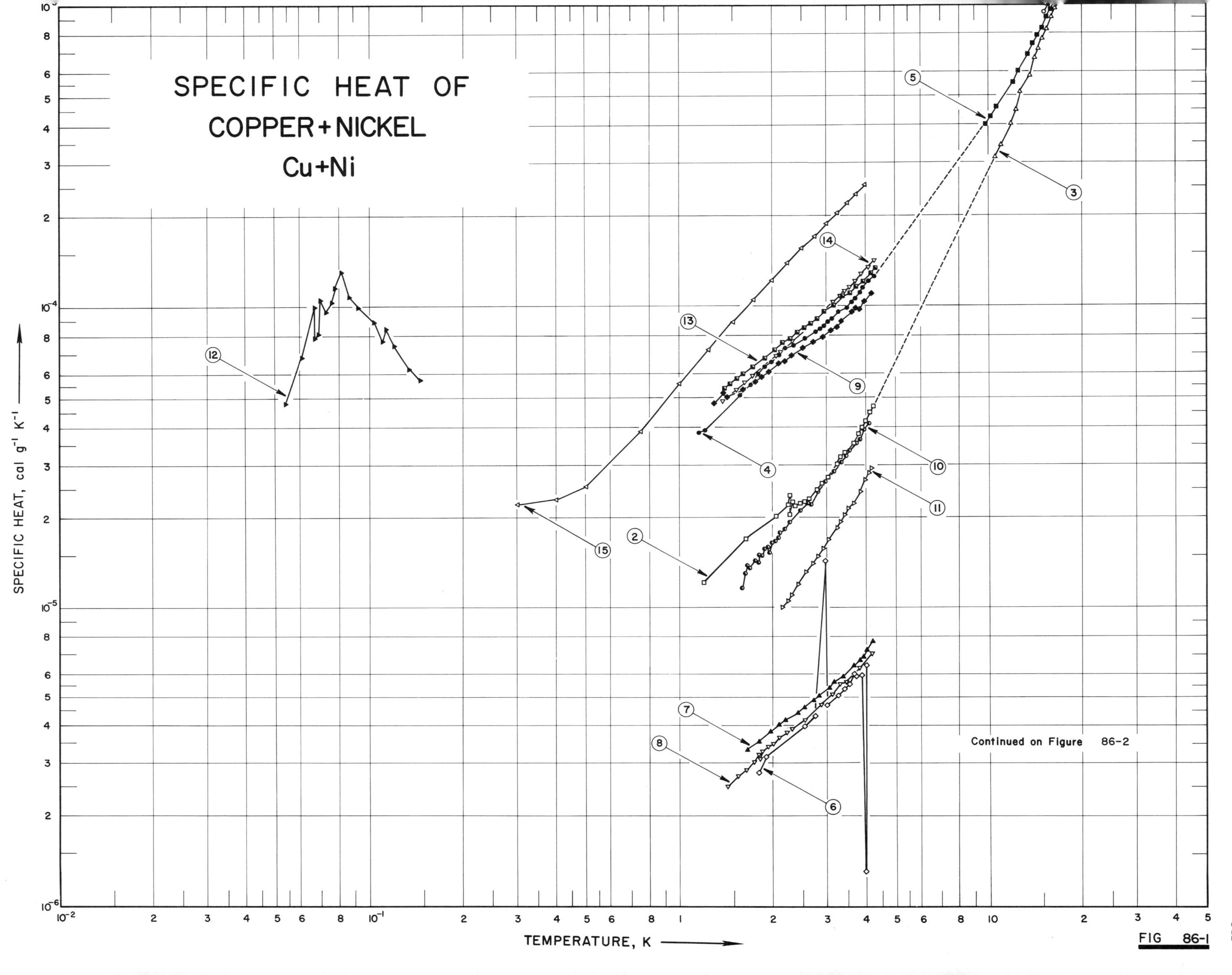
SPECIFIC HEAT OF
COPPER + NICKEL
Cu+Ni
SPECIFIC HEAT, cal g⁻¹ K⁻¹
TEMPERATURE, K
Continued on Figure 86-2
2
3
4
5
6
7
8
9
10
11
12
13
14
15

FIG 86-1

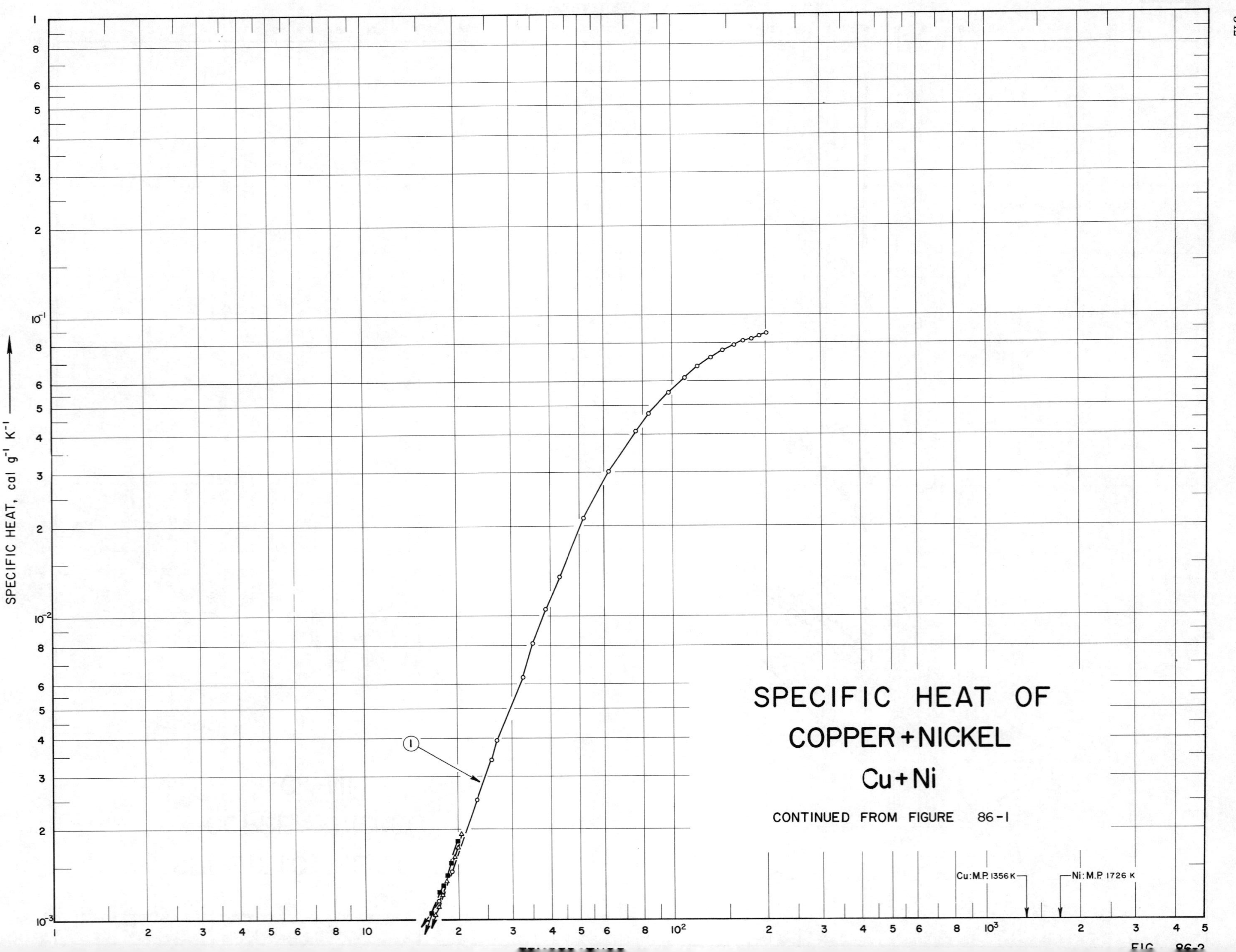

SPECIFIC HEAT OF
COPPER + NICKEL
Cu + Ni
CONTINUED FROM FIGURE 86-1
Cu: M.P. 1356 K
Ni: M.P. 1726 K
SPECIFIC HEAT, cal g⁻¹ K⁻¹

SPECIFICATION TABLE NO. 86 SPECIFIC HEAT OF COPPER + NICKEL Cu + Ni

[For Data Reported in Figure and Table No. 86]

Curve No.	Ref. No.	Year	Temp. Range, K	Reported Error, %	Name and Specimen Designation	Composition (weight percent), Specifications and Remarks
1	55	1930	15-201	1.5	Constantan	60 Cu, 40 Ni.
2	387	1940	1.2-4.2		$Cu_{80}Ni_{20}$	79.73 Cu, 20.27 Ni; prepared from 99.99 Cu and 99.98 Ni; small amounts (0.01-0.04% Al) added to alloy, melts as deoxidizers; melted; held at 200 C below melting temperature for 1 to 2 hrs and slowly cooled.
3	387	1940	10-20		$Cu_{80}Ni_{20}$	Same as above.
4	387	1940	1.1-4.3		Constantan	$Cu_{60}Ni_{40}$; 59.84 Cu and 99.98 Ni; small amounts Al added as deoxidizer; melted; held at 200 C below melting temperature for 1 to 2 hrs and slowly cooled.
5	387	1940	9-20		Constantan	Same as above.
6	388	1956	1.8-4.0	7		65 Cu, 35 Ni; specimen supplied by Westinghouse Corp; annealed and cold worked after annealing; capsule contained specimen.
7	388	1956	1.6-4.2	7		Same as above; annealed; solid sample.
8	388	1956	1.4-4.2	7		65 Cu, 35 Ni; specimen supplied by Ford Motor Co.; annealed; solid sample.
9	388	1956	1.3-4.2			60 Cu, 40 Ni; specimen supplied by Westinghouse Corp; annealed in argon atmosphere for 24 hrs at temperature 30 C below solidus curve.
10	388	1956	1.6-4.1			75 Cu, 25 Ni; same as above.
11	388	1956	2.1-4.2			90 Cu, 10 Ni; same as above.
12	389	1966	0.05-0.15			54.48 Cu, 2.67 Mn, 0.17 Fe, 0.01 C, <0.02 Zn, <0.01 Si, <0.005 Pb, remainder Ni; specimen from Driver-Harris Co.
13	349	1962	1.4-4.3	≤2	Ni(45) Cu(55)	56.92 Cu, 43.01 Ni; annealed under vacuum at 1100 C for 72 hrs; etched with 30 ml HNO_3 and 20 ml CH_3COOH.
14	349	1962	1.4-4.3	≤2	Ni(48) Cu(52)	54.09 Cu, 45.80 Ni; same as above.
15	386	1963	0.3-4.0	1	Constantan	57 Cu, 43 Ni.

DATA TABLE NO. 86 SPECIFIC HEAT OF COPPER + NICKEL Cu + Ni

[Temperature, T, K; Specific Heat, C_p, Cal $g^{-1} K^{-1}$]

T	C_p
CURVE 1	
15.20	9.395×10^{-4}
17.40	1.155×10^{-3}
19.20	1.447
23.10	2.507
25.85	3.405
26.95	3.930
32.70	6.329
35.15	8.169
38.80	1.061×10^{-2}
43.04	1.355
51.50	2.123
62.25	3.012
76.00	4.099
83.65	4.677
97.10	5.502
109.30	6.154
121.31	6.725
133.15	7.190
145.80	7.594
158.15	7.900
169.57	8.155
180.28	8.339
190.70	8.521
201.97	8.672
CURVE 2	
1.200	1.219×10^{-5}
1.638	1.706
2.05	2.021
2.25	2.202
2.27	2.373
2.27	2.048
2.29	2.366*
2.32	2.256
2.35	2.272*
2.37	2.184
2.46	2.226
2.53	2.258
2.62	2.307
2.69	2.397*
2.78	2.480
2.88	2.598
2.95	2.693*
3.02	2.728
3.09	2.762*
CURVE 2 (cont.)	
3.23	3.014×10^{-5}
3.32	3.179
3.37	3.154*
3.43	3.296
3.51	3.328*
3.67	3.536
3.76	3.776*
3.79	3.888
3.82	3.872*
3.88	3.984
3.91	4.016*
3.97	4.160*
3.99	4.208
4.06	4.352*
4.08	4.384*
4.13	4.496
4.18	4.640*
4.23	4.704
CURVE 3	
10.57	3.125×10^{-4}
10.98	3.424
11.36	3.712*
11.78	4.032
12.24	4.496
12.71	5.152
13.69	5.824
14.10	6.640
14.53	7.136
14.98	7.696
15.42	8.256
15.90	9.040
16.37	9.712
16.81	1.045×10^{-3}
17.31	1.104
17.82	1.219
18.28	1.341
18.70	1.451*
19.10	1.488*
19.58	1.621
20.07	1.741
20.66	1.930
CURVE 4	
Series 1	
1.170	3.84×10^{-5}
1.221	3.90
1.575	5.12
1.704	5.53
1.809	5.98
1.897	6.34
1.991	6.55
2.10	6.89
2.20	7.27
2.27	7.43*
2.34	7.45
2.43	7.64*
2.55	7.80
2.66	7.97*
2.76	8.24
3.26	9.51
Series 2	
2.07	6.73×10^{-6}*
2.17	6.99*
2.26	7.35*
2.33	7.46*
2.42	7.58*
2.50	7.64*
2.60	7.82*
2.69	8.05*
2.77	8.23*
2.85	8.42
2.94	8.58
3.02	8.86
3.10	9.06
3.25	9.33*
3.36	9.63*
3.49	9.80
3.60	1.02×10^{-4}
3.72	1.06
3.83	1.11
3.92	1.15
4.01	1.18*
4.10	1.22
4.19	1.24*
4.28	1.26
CURVE 5	
9.75	4.02×10^{-4}
10.19	4.26
10.66	4.59
12.07	5.53
12.51	6.00
13.46	6.81
13.98	7.40
14.43	7.87
14.84	8.34
15.37	9.04
15.91	9.64
16.47	1.06×10^{-3}
17.01	1.15*
17.55	1.23
18.07	1.30
18.61	1.40
19.10	1.54
19.68	1.65*
20.37	1.83
CURVE 6	
1.806	2.796×10^{-6}
1.844	2.893*
1.906	3.159
2.539	3.996
2.580	4.027*
2.738	4.326
2.823	4.404*
2.950	1.430
2.994	4.683
3.254	5.073
3.405	5.359
3.487	5.492*
3.531	5.550
3.534	5.610*
3.631	5.852*
3.664	6.023
3.722	5.910
3.885	5.976
3.967	6.305*
3.996	1.318
4.009	6.470
CURVE 7	
1.654	3.326×10^{-6}
1.808	3.556
1.972	3.828
2.080	4.015
2.197	4.184
2.297	4.296*
2.413	4.448
2.521	4.619
2.709	4.870
2.816	5.097
3.047	5.414
3.146	5.677
3.371	5.907
3.659	6.418
3.822	6.706
3.914	6.899
4.017	7.237
4.087	7.305*
4.171	7.702
CURVE 8	
1.427	2.521×10^{-6}
1.545	2.704
1.585	2.793*
1.630	2.843
1.708	2.979*
1.732	3.028
1.806	3.216
1.815	3.114
1.843	3.253*
1.857	3.288
1.900	3.366*
1.922	3.371*
1.937	3.405
1.977	3.433*
1.983	3.381*
2.014	3.497
2.034	3.560*
2.096	3.651
2.226	3.787
2.319	3.900
2.541	4.197
2.627	4.320*
2.745	4.519*
2.813	4.592*
CURVE 8 (cont.)	
2.868	4.713×10^{-6}
2.933	4.819*
2.982	4.865*
3.068	5.021*
3.108	5.147
3.139	5.047*
3.255	5.369*
3.285	5.562
3.428	5.549*
3.454	5.651
3.613	5.923*
3.818	6.313
4.021	6.727*
4.176	7.043
CURVE 9	
1.297	4.800×10^{-5}
1.325	4.885*
1.374	4.927*
1.383	5.192
1.427	5.002
1.469	5.083*
1.617	5.329
1.766	5.660
1.853	5.877
1.946	6.080
1.968	6.263*
2.097	6.473
2.192	6.649
2.311	6.871
2.352	6.946*
2.515	7.289
2.540	7.315*
2.718	7.632
2.735	7.687*
2.918	7.935
3.078	8.343
3.227	8.531
3.339	8.913
3.398	9.095*
3.540	9.399*
3.600	9.536
3.726	9.804
3.839	9.749
3.959	1.033×10^{-4}

*Not shown on plot

DATA TABLE NO. 86 (continued)

T	C_p
CURVE 9 (cont.)	
4.063	1.077×10^{-4}*
4.164	1.088*
4.170	1.090
CURVE 10	
Series 1	
1.581	1.172×10^{-5}
1.625	1.302
1.656	1.384
1.686	1.364
1.753	1.439
1.758	1.463*
1.819	1.501
1.819	1.491*
1.879	1.575
1.895	1.577
1.982	1.659
2.115	1.780
2.178	1.829
2.213	1.881*
2.261	1.926
2.343	2.011*
2.453	2.114
2.592	2.236
2.798	2.449
2.967	2.641
3.153	2.851
3.321	3.061
3.535	3.342
3.706	3.532*
3.835	3.635
3.953	3.877*
4.091	4.105
Series 2	
1.806	1.421×10^{-5}
1.848	1.500*
1.856	1.506
1.883	1.525*
1.893	1.527*
1.923	1.602
1.955	1.539
1.986	1.621*
2.023	1.672
CURVE 10 (cont.)	
Series 2 (cont.)	
2.054	1.698×10^{-5}*
2.055	1.686*
2.084	1.723
2.089	1.701*
2.125	1.682*
2.213	1.773*
2.322	1.911*
2.661	2.213
2.850	2.422*
3.025	2.686*
3.067	2.660*
3.216	2.901*
3.400	3.119*
3.469	3.219
3.580	3.355*
3.740	3.553
3.947	3.943
4.018	4.033*
4.070	4.007*
4.095	4.205*
CURVE 11	
2.141	1.009×10^{-5}
2.223	1.064
2.289	1.119
2.343	1.164*
2.418	1.208
2.493	1.277*
2.564	1.321
2.632	1.349*
2.697	1.415
2.797	1.490
2.916	1.589
3.024	1.695
3.125	1.777*
3.221	1.852
3.311	1.933
3.400	2.047
3.511	2.146
3.663	2.321
3.828	2.442
3.970	2.672
4.090	2.821
4.168	2.922
CURVE 12	
0.054	4.78×10^{-5}
0.061	6.84
0.067	9.97
0.067	7.89
0.069	8.08
0.070	1.06×10^{-4}
0.073	9.58×10^{-5}
0.076	1.03×10^{-4}
0.078	1.16
0.082	1.30
0.087	1.07
0.093	9.92×10^{-5}
0.104	8.89
0.112	7.70
0.113	8.41
0.122	7.41
0.135	6.21
0.147	5.74
CURVE 13	
1.406	5.392×10^{-5}
1.464	5.554
1.537	5.786
1.615	5.993
1.732	6.333
1.893	6.766
2.044	7.188
2.165	7.572
2.278	7.813
2.404	8.186
2.540	8.484
2.663	8.754
2.788	9.064
2.943	9.593
3.152	1.008×10^{-4}
3.377	1.076
3.567	1.108
3.733	1.164
3.923	1.210
4.072	1.254*
4.182	1.285
4.304	1.330
CURVE 14	
1.384	4.89×10^{-5}
CURVE 14 (cont.)	
1.444	5.067×10^{-5}*
1.528	5.341
1.620	5.604
1.726	5.907
1.813	6.134*
1.884	6.375*
1.981	6.648*
2.062	6.889
2.125	7.090
2.207	7.358*
2.319	7.697
2.454	8.131*
2.571	8.511*
2.678	8.868*
2.790	9.132*
2.943	9.625*
3.157	1.032×10^{-4}
3.312	1.078
3.423	1.119
3.555	1.161
3.716	1.214
3.885	1.275
3.998	1.309*
4.087	1.345
4.191	1.385*
4.295	1.419
CURVE 15	
0.30	2.22×10^{-5}
0.40	2.29
0.50	2.56
0.75	3.90
1.00	5.57
1.25	7.24
1.50	8.89
1.75	1.06×10^{-4}
2.00	1.22
2.25	1.39
2.50	1.55
2.75	1.71
3.00	1.88
3.25	2.04
3.50	2.21
3.75	2.37
4.00	2.53

* Not shown on plot

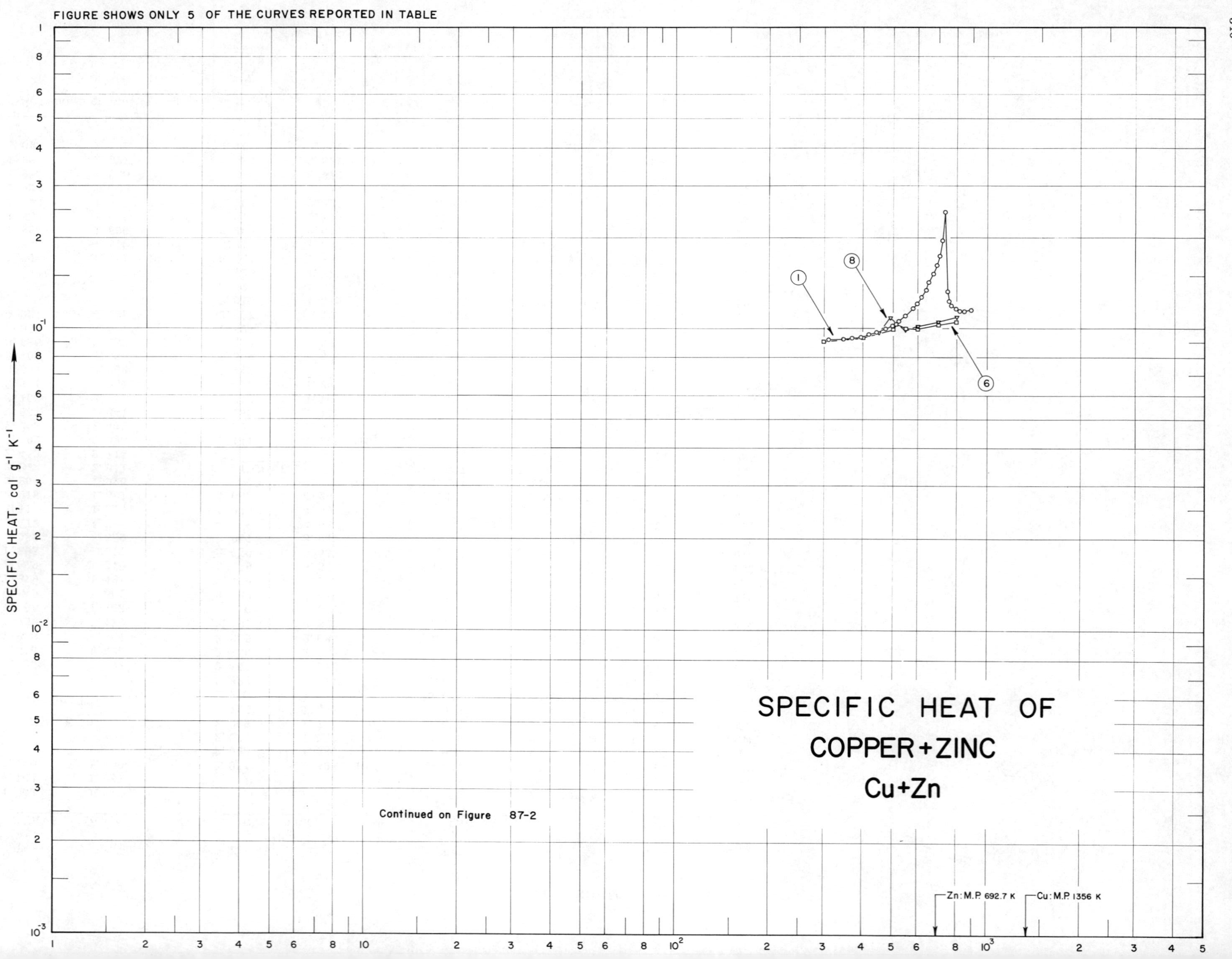
FIGURE SHOWS ONLY 5 OF THE CURVES REPORTED IN TABLE
SPECIFIC HEAT, cal g⁻¹ K⁻¹
SPECIFIC HEAT OF
COPPER + ZINC
Cu+Zn
Continued on Figure 87-2
Zn: M.P. 692.7 K
Cu: M.P. 1356 K
1
6
8

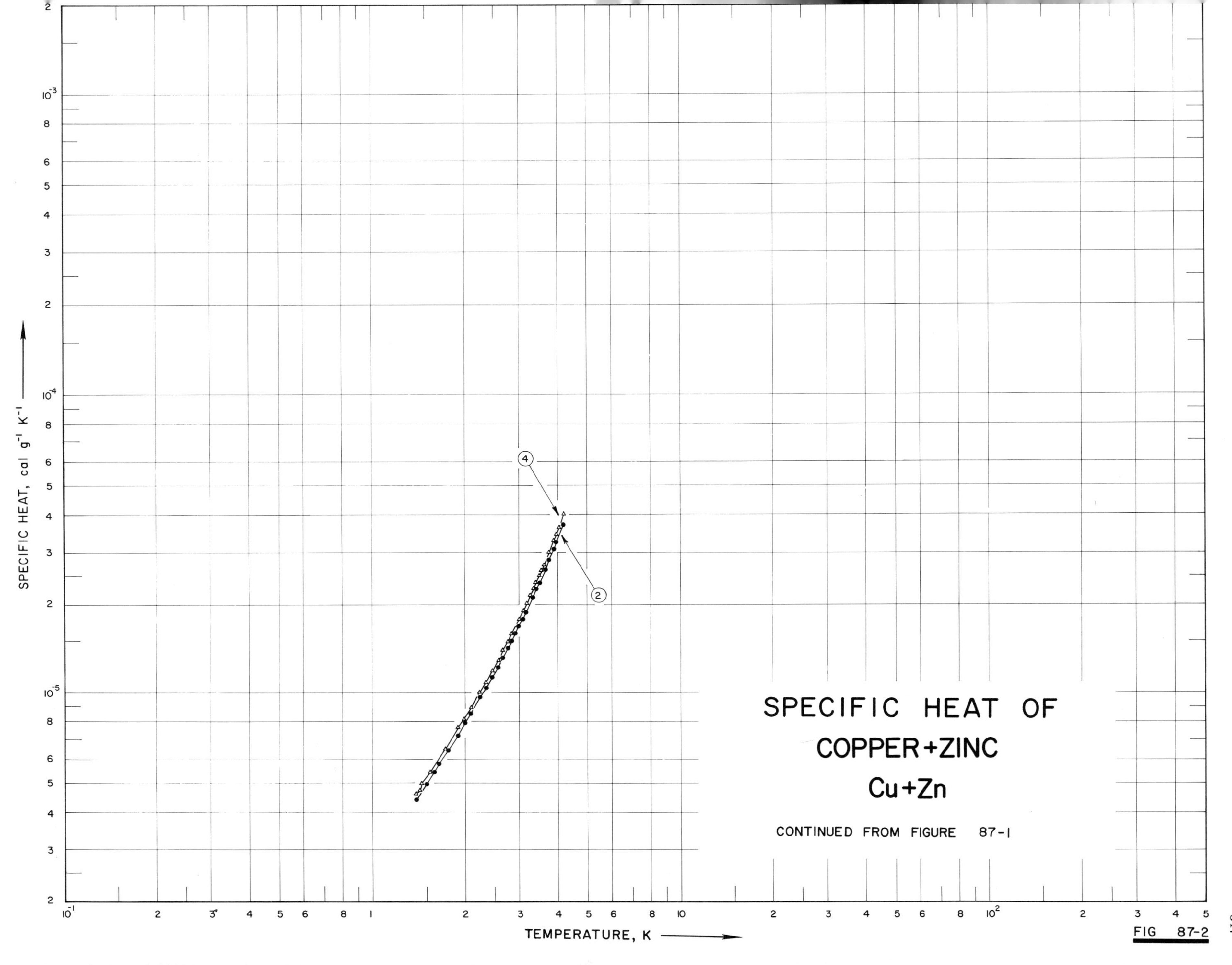

SPECIFIC HEAT OF
COPPER+ZINC
Cu+Zn
CONTINUED FROM FIGURE 87-1
SPECIFIC HEAT, cal g⁻¹ K⁻¹
TEMPERATURE, K
4
2

FIG 87-2

SPECIFICATION TABLE NO. 87 SPECIFIC HEAT OF COPPER + ZINC Cu + Zn

[For Data Reported in Figure and Table No. 87]

Curve No.	Ref. No.	Year	Temp. Range, K	Reported Error, %	Name and Specimen Designation	Composition (weight percent), Specifications and Remarks
1	52	1941	310-892		Brass	51. 8 Cu, 48. 17 An, 0. 03 Pb, and traces of Fe; β-phase; measured in argon atmosphere at reduced pressure.
2	390	1962	1. 4-4. 2			51. 45 Cu, 48. 65 Zn; prepared from 99. 999 Cu and 99. 999 Zn by induction melting under argon atmosphere; annealed and quenched to insure presence of β-phase.
3	390	1962	1. 4-4. 2			53. 09 Cu, 46. 92 Zn; same as above.
4	390	1962	1. 4-4. 2			54. 51 Cu, 45. 49 Zn; same as above.
5	390	1962	1. 4-4. 1			56. 76 Cu, 43. 24 Zn; same as above; annealed for 20 min at 830 C; quenched twice from 810 C.
6	391	1967	298-800		Alpha Brass Alloy No. 1	79. 75 Cu, 20. 22 Zn, 0. 015 Fe, 0. 01 Ni, 0. 003 Pb.
7	391	1967	298-800		Alpha Brass Alloy No. 2	70. 42 Cu, 29. 445 Zn, 0. 05 Pb, 0. 045 Sn, 0. 02 Fe, 0. 02 Ni.
8	391	1967	298-800		Alpha Brass Alloy No. 3	65. 18 Cu, 34. 815 Zn, 0. 002 Fe, 0. 002 Ni, 0. 001 Pb.
9	418	1959	328-590	<±4. 0		64. 80 Cu and 35. 20 Zn; annealed and homogenized for several days; cooled from 700 C; measured under H_2 atmosphere.
10	418	1959	331-595	<±4. 0		64. 80 Cu and 35. 20 Zn; annealed and homogenized for several days; heated for 200 hrs below 200 C; measured under H_2 atmosphere.
11	418	1959	332-641	<±4. 0		69. 89 Cu and 30. 11 Zn; annealed and homogenized for several days; cooled from 700 C; measured under H_2 atmosphere.
12	418	1959	325-629	<±4. 0		69. 89 Cu and 30. 11 Zn; annealed and homogenized for several days; heated for 200 hrs below 200 C; measured under H_2 atmosphere.
13	418	1959	332-632	<±4. 0		69. 89 Cu and 30. 11 Zn; annealed and homogenized for several days; cooled from 700 C; measured under H_2 atmosphere.
14	418	1959	326-686	<±4. 0		69. 89 Cu and 30. 11 Zn; annealed and homogenized for several days; heated for 300 hrs below 230 C; measured under H_2 atmosphere.
15	418	1959	367-669	<±4. 0		75. 52 Cu and 24. 48 Zn; annealed and homogenized for several days; cooled from 700 C; measured under H_2 atmosphere.
16	418	1959	323-662	<±4. 0		75. 52 Cu and 24. 48 Zn; annealed and homogenized for several days; heated for 200 hrs below 200 C; measured under H_2 atmosphere.
17	418	1959	325-661	<±4. 0		80. 29 Cu and 19. 71 Zn; annealed and homogenized for several days; cooled from 700 C; measured under H_2 atmosphere.
18	418	1959	330-673	<±4. 0		80. 29 Cu and 19. 71 Zn; annealed and homogenized for several days; heated for 200 hrs below 200 C; measured under H_2 atmosphere.

SPECIFICATION TABLE NO. 87 (continued)

Curve No.	Ref. No.	Year	Temp. Range, K	Reported Error, %	Name and Specimen Designation	Composition (weight percent), Specifications and Remarks
19	418	1959	380-684	<±4.0		80.29 Cu and 19.71 Zn; annealed and homogenized for several days; heated for 300 hrs below 200 C; measured under H_2 atmosphere.
20	418	1959	327-626	<±4.0		89.72 Cu and 10.28 Zn; annealed and homogenized for several days; cooled from 700 C, measured under H_2 atmosphere.
21	418	1959	337-598	<±4.0		89.72 Cu and 10.28 Zn; annealed and homogenized for several days; heated for 200 hrs below 200 C; measured under H_2 atmosphere.
22	417	1961	16-301	±0.5	Alloy No. 1	90.05 Cu, 9.93 Zn, 0.005 Fe, 0.003 Pb, and 0.001 Bi; vacuum annealed for 24 hrs at 500 C and furnace cooled.
23	417	1961	17-298	±0.5	Alloy No. 2	79.75 Cu, 20.22 Zn, 0.015 Fe, 0.01 Ni and 0.003 Pb; same as above.
24	417	1961	16-303	±0.5	Alloy No. 4	65.18 Cu, 34.81 Zn, 0.002 Fe, 0.002 Ni, and 0.001 Pb; same as above.

DATA TABLE NO. 87 SPECIFIC HEAT OF COPPER + ZINC Cu + Zn

[Temperature, T, K; Specific Heat, C_p, Cal g^{-1} K^{-1}]

CURVE 1

T	C_p
310.4	9.20 x 10^{-2}
346.1	9.26
369.3	9.34
394.9	9.44
419.8	9.60
444.1	9.76
473.5	1.006 x 10^{-1}
497.8	1.028
523.6	1.070
547.2	1.111
580.0	1.176
597.1	1.221
617.8	1.278
640.7	1.351
652.3	1.441
678.6	1.528
694.3	1.632
709.5	1.762
724.7	1.984
736.0	2.457
749.3	1.344
757.6	1.248
774.4	1.200
796.3	1.177
818.7	1.159
848.4	1.150
873.3	1.154*
892.1	1.160

CURVE 2

T	C_p
1.380	4.466 x 10^{-6}
1.505	5.028
1.594	5.505
1.657	5.881
1.776	6.525
1.900	7.292
2.011	8.082
2.099	8.622
2.258	9.765
2.359	1.065 x 10^{-5}
2.466	1.153
2.568	1.244
2.667	1.331
2.754	1.428
2.838	1.523

CURVE 2 (cont.)

T	C_p
2.922	1.615 x 10^{-5}
2.982	1.692*
2.998	1.704
3.083	1.808
3.160	1.902
3.239	2.032*
3.325	2.140
3.400	2.284
3.401	2.266*
3.429	2.263*
3.512	2.394
3.578	2.518*
3.650	2.667
3.716	2.801*
3.764	2.870
3.764	2.882*
3.827	2.966*
3.892	3.128
3.965	3.283
3.995	3.342*
4.027	3.369*
4.091	3.553*
4.180	3.762

CURVE 3*

T	C_p
1.443	4.881 x 10^{-6}
1.533	5.308
1.663	6.025
1.755	6.533
1.911	7.513
1.991	8.059
2.026	8.274
2.115	8.913
2.238	9.837
2.345	1.072 x 10^{-5}
2.460	1.181
2.581	1.294
2.658	1.372
2.761	1.474
2.840	1.564
2.912	1.657
2.990	1.745
3.078	1.860
3.158	1.969
3.245	2.087

CURVE 3 (cont.)*

T	C_p
3.301	2.153 x 10^{-5}
3.370	2.268
3.433	2.359
3.611	2.673
3.685	2.825
3.686	2.997
3.882	3.191
3.955	3.382
4.035	3.544
4.126	3.738

CURVE 4

T	C_p
1.394	4.690 x 10^{-6}
1.425	4.839
1.455	5.062
1.482	5.155*
1.554	5.507
1.737	6.633
1.917	7.780
1.925	7.710*
1.987	8.241*
1.995	8.256
2.108	9.039
2.243	1.023 x 10^{-5}
2.347	1.106
2.470	1.211
2.573	1.315
2.667	1.420
2.756	1.514
2.832	1.608
3.005	1.810
3.093	1.939
3.178	2.053
3.262	2.187
3.338	2.299
3.399	2.401
3.476	2.534
3.536	2.642
3.602	2.765
3.675	2.863*
3.743	3.034
3.891	3.351
3.966	3.520
4.039	3.676
4.192	4.093

CURVE 5*

T	C_p
1.441	5.098 x 10^{-6}
1.466	5.232
1.490	5.369
1.596	5.923
1.694	6.529
1.906	7.896
2.056	9.026
2.120	9.561
2.120	9.494
2.242	1.049 x 10^{-5}
2.345	1.149
2.410	1.210
2.455	1.263
2.567	1.375
2.660	1.474
2.751	1.583
2.834	2.048
2.916	1.800
3.002	1.905
3.097	2.042
3.175	2.153
3.237	2.266
3.324	2.422
3.385	2.526
3.456	2.610
3.519	2.735
3.587	2.882
3.655	3.032
3.656	3.032
3.734	3.175
3.814	3.400
3.888	3.511
3.961	3.746
4.039	3.943
4.116	4.106

CURVE 6

T	C_p
298.15	9.08 x 10^{-2}
400	9.41
450	9.72*
500	1.01 x 10^{-1}
513	1.05
550	1.01
600	1.00
700	1.04
800	1.06

CURVE 7*

T	C_p
298.15	9.05 x 10^{-2}
400	9.43
450	9.65
500	1.11 x 10^{-1}
550	1.00
600	1.01
700	1.05
800	1.06

CURVE 8

T	C_p
298.15	9.05 x 10^{-2}*
400	9.41*
450	9.76
483	1.09 x 10^{-1}
500	1.05*
550	1.00
600	1.02
700	1.06
800	1.09

CURVE 9*

T	C_p
328.05	9.9 x 10^{-2}
340.75	9.9
350.75	9.8
357.15	9.7
363.05	9.7
373.45	9.6
377.95	9.5
382.55	9.4
389.85	9.2
396.85	9.1
405.05	9.2
410.55	9.5
415.75	9.8
417.35	9.9
424.65	1.02 x 10^{-1}
428.35	1.05
434.35	1.06
437.95	1.06
442.85	1.06
450.05	1.06
460.15	1.06
466.75	1.06
473.65	1.06
482.55	1.06

CURVE 9 (cont.)*

T	C_p
490.65	1.07 x 10^{-1}
501.15	1.07
510.45	1.07
521.85	1.07
536.45	1.07
546.45	1.07
556.65	1.07
571.35	1.07
590.15	1.07

CURVE 10*

T	C_p
331.25	9.9 x 10^{-2}
339.05	9.9
347.45	1.00 x 10^{-1}
356.45	1.00
362.45	1.00
373.25	1.00
382.45	1.00
391.05	1.00
401.45	1.00
410.85	1.00
422.25	1.01
432.55	1.01
446.65	1.04
455.55	1.06
464.35	1.09
470.75	1.14
480.25	1.16
485.75	1.15
493.65	1.13
500.95	1.09
507.05	1.08
516.85	1.07
531.65	1.07
543.45	1.07
557.25	1.07
571.35	1.07
584.05	1.07
594.55	1.07

CURVE 11*

T	C_p
332.15	9.9 x 10^{-2}
342.55	9.8
356.15	9.8
367.75	9.8
380.55	9.6

*Not shown on plot

DATA TABLE NO. 87 (continued)

T	C_p
CURVE 11 (cont.)*	
392.65	9.2×10^{-2}
400.25	9.0
408.35	8.9
416.05	9.1
421.75	9.4
423.75	9.8
428.65	1.03×10^{-1}
436.15	1.05
443.35	1.08
450.35	1.09
457.05	1.08
465.85	1.08
473.55	1.07
481.95	1.08
491.15	1.07
498.25	1.07
506.35	1.08
513.35	1.08
522.35	1.08
529.35	1.08
537.95	1.08
547.75	1.08
557.45	1.08
566.55	1.08
572.85	1.08
582.25	1.08
595.75	1.08
608.45	1.08
617.35	1.08
629.65	1.07
640.65	1.08
CURVE 12*	
325.45	9.9×10^{-2}
331.95	9.9
342.35	1.0×10^{-1}
346.75	9.9×10^{-2}
357.35	9.9
366.25	1.0×10^{-1}
373.65	1.0
384.25	1.0
393.85	1.0
404.95	1.0
413.15	1.0
425.05	1.0
435.15	9.9×10^{-2}
CURVE 12 (cont.)*	
447.45	1.0×10^{-1}
458.05	1.01
466.85	1.02
474.85	1.05
483.75	1.08
491.75	1.13
499.85	1.18
507.25	1.19
516.35	1.13
522.65	1.09
541.05	1.07
548.65	1.07
556.95	1.06
566.45	1.07
575.25	1.07
583.45	1.07
592.65	1.07
602.35	1.07
611.75	1.06
619.45	1.07
629.45	1.08
CURVE 13*	
332.05	1.00×10^{-1}
349.95	1.00
362.25	1.00
381.15	1.00
390.55	1.00
400.15	9.9×10^{-2}
410.05	9.8
422.05	9.6
430.25	9.5
443.55	9.5
450.55	9.5
456.95	9.8
463.15	1.00×10^{-1}
473.75	1.03
479.65	1.03
487.35	1.04
495.65	1.04
502.35	1.04
519.15	1.04
536.25	1.05
542.75	1.05
560.15	1.05
573.35	1.05
585.15	1.05
CURVE 13 (cont.)*	
599.25	1.05×10^{-1}
609.95	1.05
623.05	1.05
631.75	1.06
CURVE 14*	
326.35	9.9×10^{-2}
339.05	9.9
346.55	9.9
356.85	9.9
371.95	1.00×10^{-1}
393.05	1.00
400.05	1.00
409.55	1.00
422.35	1.00
431.75	1.00
442.55	1.00
453.25	1.00
472.95	1.00
490.75	1.03
496.35	1.05
504.55	1.08
514.15	1.10
518.45	1.10
529.25	1.09
537.75	1.07
554.25	1.06
560.45	1.06
570.25	1.06
577.65	1.06
587.85	1.06
619.25	1.06
633.75	1.06
645.55	1.06
658.75	1.07
673.45	1.07
685.55	1.07
CURVE 15*	
366.75	9.8×10^{-2}
374.55	9.8
386.45	9.7
396.35	9.6
405.25	9.5
413.95	9.3
425.25	9.4
CURVE 15 (cont.)*	
434.85	9.6×10^{-2}
440.95	1.00×10^{-1}
443.55	1.02
449.65	1.05
455.35	1.06
462.75	1.06
468.75	1.06
475.85	1.06
483.05	1.06
489.35	1.06
498.75	1.06
504.85	1.06
512.55	1.07
520.35	1.07
527.45	1.07
535.55	1.07
541.95	1.07
581.45	1.07
598.05	1.07
612.85	1.07
630.25	1.08
651.55	1.08
668.75	1.09
CURVE 16*	
323.15	9.8×10^{-2}
330.25	9.8
346.05	9.8
357.95	9.9
366.55	9.9
375.05	9.9
387.25	9.9
401.75	9.9
407.85	9.9
419.35	9.9
432.55	9.9
444.15	9.9
457.05	9.9
468.75	9.9
473.95	9.9
478.65	1.00×10^{-1}
481.75	1.01
488.65	1.03
489.65	1.05
498.45	1.08
505.45	1.11
CURVE 16 (cont.)*	
514.95	1.14×10^{-1}
522.05	1.13
532.85	1.09
541.75	1.06
549.95	1.05
559.05	1.05
568.65	1.06
578.55	1.06
588.95	1.06
599.05	1.06
607.45	1.06
614.55	1.06
623.55	1.06
643.85	1.07
661.95	1.07
CURVE 17*	
324.55	9.8×10^{-2}
334.35	9.8
346.55	9.8
357.85	9.8
385.15	9.8
391.75	9.8
398.15	9.7
407.35	9.6
411.35	9.6
420.15	9.5
426.35	9.4
433.95	9.3
440.45	9.3
447.55	9.4
453.65	9.7
460.35	9.9
471.15	1.01×10^{-1}
476.15	1.02
485.35	1.02
492.75	1.02
501.35	1.02
507.85	1.02
516.25	1.03
522.95	1.03
532.35	1.04
541.35	1.04
548.35	1.04
558.75	1.04
571.35	1.04
CURVE 17 (cont.)*	
581.45	1.04×10^{-1}
597.15	1.04
612.05	1.04
628.05	1.04
637.55	1.05
652.95	1.05
661.45	1.06
CURVE 18*	
330.15	9.8×10^{-2}
343.85	9.8
353.15	9.8
361.35	9.8
370.25	9.8
379.55	9.9
388.45	9.9
396.65	9.9
406.95	9.8
419.45	9.9
428.15	9.9
439.55	9.9
450.35	9.8
460.45	9.9
471.55	9.9
482.75	9.9
487.15	1.01×10^{-1}
494.25	1.03
500.95	1.06
510.25	1.08
516.55	1.08
524.45	1.07
534.55	1.05
541.75	1.04
550.35	1.04
558.65	1.04
566.95	1.04
574.85	1.04
585.05	1.03
594.05	1.04
600.75	1.04
609.25	1.04
618.55	1.04
631.35	1.04
641.95	1.05
655.95	1.05
673.05	1.05

*Not shown on plot

DATA TABLE NO. 87 (continued)

T	C_p
CURVE 19*	
380.25	9.9 x 10^{-2}
389.85	9.9
398.25	9.9
407.75	9.9
418.05	9.9
429.65	9.9
441.55	9.9
451.65	9.9
462.05	9.9
472.35	9.9
482.85	1.00 x 10^{-1}
487.45	1.01
495.25	1.04
503.45	1.07
510.85	1.09
518.65	1.09
524.95	1.07
534.85	1.06
542.45	1.05
552.55	1.05
559.65	1.05
567.45	1.04
576.25	1.04
588.05	1.04
596.25	1.04
603.45	1.04
611.25	1.04
620.25	1.05
633.55	1.05
646.05	1.05
659.15	1.06
673.15	1.06
684.45	1.06
CURVE 20*	
326.65	9.8 x 10^{-2}
342.35	9.8
355.65	9.9
363.55	9.9
368.45	9.8
382.05	9.8
394.65	9.8
408.55	9.8
431.15	9.9
442.25	1.00 x 10^{-1}
CURVE 20 (cont.)*	
452.95	1.00 x 10^{-1}
458.35	1.01
466.05	1.01
475.35	1.01
482.75	1.01
489.95	1.01
497.05	1.02
508.05	1.02
516.85	1.02
523.85	1.03
529.75	1.03
538.35	1.04
548.05	1.04
556.95	1.04
565.05	1.04
572.75	1.04
588.55	1.04
602.65	1.04
609.65	1.04
613.75	1.04
625.75	1.05
CURVE 21*	
336.85	9.8 x 10^{-2}
355.65	9.8
375.05	9.9
385.15	9.9
396.35	9.9
406.95	9.9
418.55	9.9
429.45	9.9
439.55	9.9
453.25	9.9
464.05	9.9
476.35	9.9
487.75	9.9
496.15	9.9
503.25	1.00 x 10^{-1}
513.55	1.00
518.85	1.01
528.05	1.02
537.35	1.02
545.95	1.03
553.95	1.03
562.55	1.02
CURVE 21 (cont.)*	
569.75	1.02 x 10^{-1}
577.65	1.02
589.45	1.02
598.35	1.02
CURVE 22*	
Series 1	
15.62	9.6 x 10^{-4}
17.02	1.1 x 10^{-3}
18.3	1.5
19.40	1.76
20.53	2.13
21.90	2.65
23.43	3.34
23.49	3.34
25.46	4.25
27.55	5.52
29.82	7.06
32.51	9.01
39.60	1.50 x 10^{-2}
50.24	2.51
59.84	3.403
71.71	4.391
79.85	4.977
89.56	5.614
99.86	6.179
109.94	6.647
119.72	6.970
128.90	7.254
138.99	7.532
149.59	7.750
159.88	7.934
169.81	8.108
179.74	8.263
190.22	8.375
200.31	8.477
228.33	8.753
240.33	8.836
250.59	8.899
260.22	8.977
270.08	8.983
279.78	9.007
301.19	9.103
CURVE 22 (cont.)*	
Series 2	
16.92	7.8 x 10^{-2}
17.33	8.7
18.50	1.05 x 10^{-1}
20.26	1.45
21.96	1.91
24.62	2.78
30.59	5.34
33.31	6.77
37.38	9.03
41.27	1.138 x 10^{0}
44.99	1.369
48.62	1.585
54.98	1.980
64.57	2.568
78.34	3.220
89.99	3.673
101.11	4.019
109.13	4.244
119.50	4.512
129.60	4.679
140.06	4.877
150.95	5.000
160.44	5.129
169.29	5.210
190.71	5.383
199.99	5.462
220.22	5.559
233.33	5.635
243.67	5.665
256.00	5.718
266.64	5.738
279.86	5.773
298.17	5.795
CURVE 23*	
16.92	1.2 x 10^{-3}
17.33	1.4
18.50	1.64
20.26	2.27
21.96	2.99
24.62	4.35
30.59	8.36 x 10^{-2}
33.31	1.06
37.38	1.41
CURVE 23 (cont.)*	
41.27	1.781 x 10^{-2}
44.99	2.142
48.62	2.480
54.98	3.098
64.57	4.019
78.34	5.039
89.99	5.748
101.11	6.289
109.13	6.641
119.50	7.061
129.60	7.322
140.06	7.632
150.95	7.824
160.44	8.026
169.29	8.153
190.71	8.424
199.99	8.547
220.22	8.699
233.33	8.818
243.67	8.865
256.00	8.948
266.64	8.979
279.86	9.034
298.17	9.069
CURVE 24*	
15.94	1.0 x 10^{-3}
18.25	1.87
21.05	2.98
23.26	4.19
26.56	6.36
29.54	8.56
32.65	1.13 x 10^{-2}
35.82	1.42
39.15	1.739
43.01	2.124
47.47	2.576
55.88	3.394
61.99	3.962
70.57	4.679
76.80	5.146
86.73	5.765
94.83	6.192
108.02	6.767
121.17	7.214
148.94	7.881
CURVE 24 (cont.)*	
161.91	8.123 x 10^{-2}
174.47	8.293
188.87	8.447
203.85	8.592
216.12	8.698
229.86	8.785
232.71	7.541
243.30	8.858
257.41	8.910
271.75	8.944
303.33	9.067

*Not shown on plot

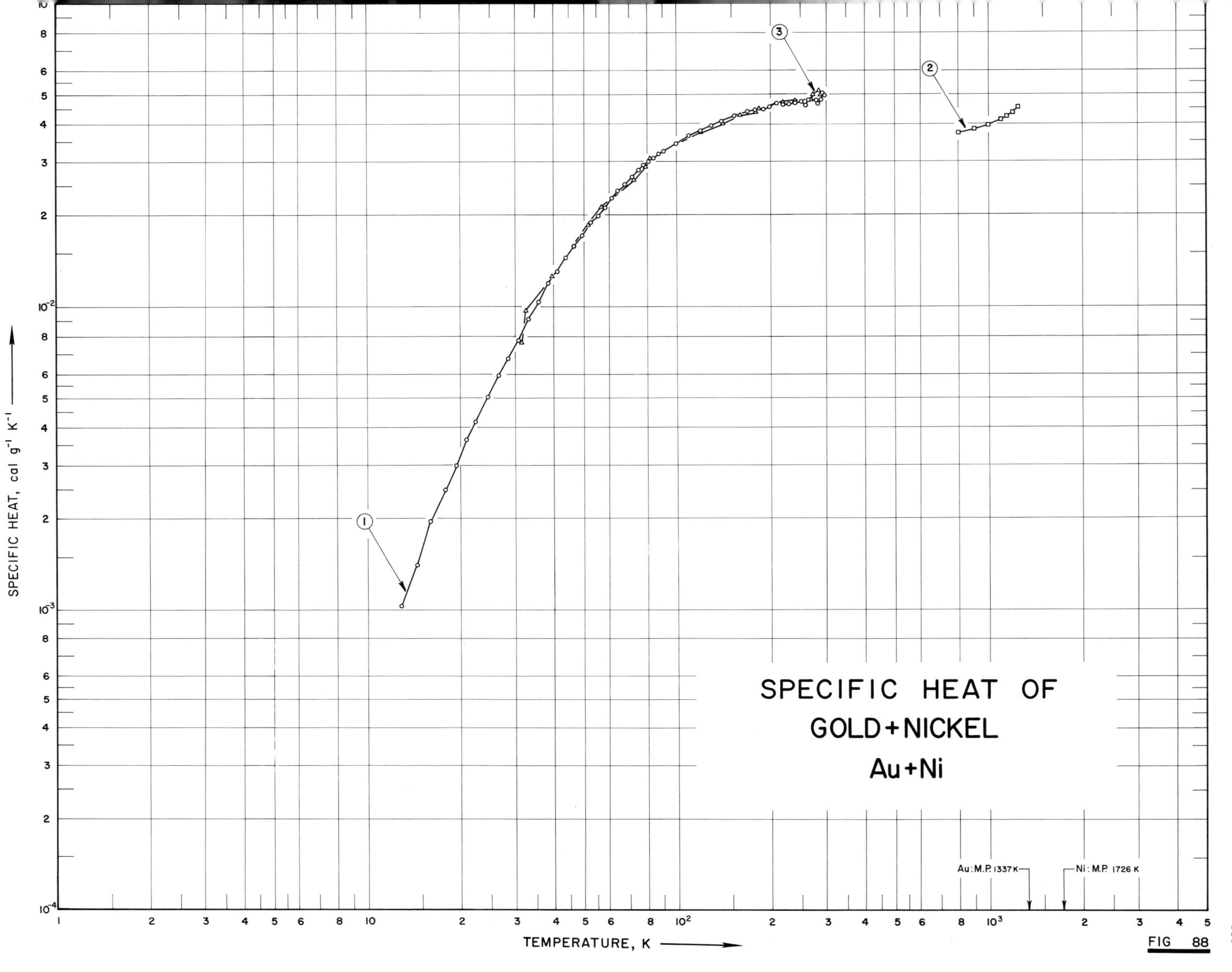
SPECIFIC HEAT OF
GOLD+NICKEL
Au+Ni
1
2
3
Au: M.P. 1337 K
Ni: M.P. 1726 K
SPECIFIC HEAT, cal g⁻¹ K⁻¹
TEMPERATURE, K
FIG 88

SPECIFICATION TABLE NO. 88 SPECIFIC HEAT OF GOLD + NICKEL Au + Ni

[For Data Reported in Figure and Table No. 88]

Curve No.	Ref. No.	Year	Temp. Range, K	Reported Error, %	Name and Specimen Designation	Composition (weight percent), Specifications and Remarks
1	226	1955	12-299			78.24 Au, 21.76 Ni; machined filings homogenized by heating several hrs at approx 900 C and water quenched.
2	227	1962	800-1250			99.95 $Au_{0.9}Ni_{0.10}$; homogenized for more than one wk at temperatures above 50 C.
3	226	1955	21-299			78.24 Au, 21.76 Ni; machined filings homogenized by heating several hrs at approx 900 C and water quenched.

DATA TABLE NO. 88 SPECIFIC HEAT OF GOLD + NICKEL Au – Ni

[Temperature, T, K; Specific Heat, C_p, Cal g^{-1} K^{-1}]

T	C_p
CURVE 1	
12.89	1.028 x 10^{-3}
14.53	1.412
16.12	1.972
17.83	2.502
19.45	3.008
20.88	3.661
22.47	4.190
24.56	5.050
26.65	5.963
28.61	6.823
30.85	7.820
33.25	9.163
35.78	1.044 x 10^{-2}
38.56	1.200
41.05	1.315
43.89	1.459
46.63	1.577
49.65	1.715
52.71	1.886
55.73	1.979
55.86	1.974*
58.68	2.122
58.73	2.127*
61.67	2.278
61.73	2.298*
64.91	2.411
68.33	2.526
71.81	2.677
75.11	2.821
78.28	2.936
80.13	2.962*
80.90	3.007
84.29	3.098
86.20	3.151*
87.57	3.182
90.79	3.267
99.90	3.457
109.70	3.672
119.85	3.813
129.83	3.962
139.77	4.111
153.65	4.271
159.65	4.331*
169.73	4.416
179.91	4.487
190.73	4.497
CURVE 1 (cont.)	
199.13	4.573 x 10^{-2}
203.06	4.547*
205.67	4.559*
210.70	4.690
210.77	4.645*
220.86	4.665*
230.57	4.673
240.49	4.709
251.22	4.759
260.15	4.621
265.11	4.820
270.99	4.799*
281.24	4.817
284.56	4.684
292.51	4.839
294.96	5.108
297.14	4.998
298.46	5.112*
CURVE 2	
800	3.734 x 10^{-2}
900	3.852
1000	3.975
1100	4.155
1150	4.255
1200	4.390
1250	4.580
CURVE 3	
20.98	3.638 x 10^{-3}*
31.58	7.659
32.57	9.724
39.55	1.269 x 10^{-2}
51.89	1.855
57.20	2.127
72.69	2.619
78.91	2.912*
79.13	2.898
81.89	3.072
99.14	3.432*
119.41	3.768
141.84	4.041
159.58	4.298
179.55	4.391
CURVE 3 (cont.)	
184.21	4.503 x 10^{-2}
199.73	4.566*
219.42	4.777
240.61	4.784
258.71	4.734*
267.37	4.799*
272.89	4.837
273.19	5.048
277.61	5.055*
279.00	5.040*
286.48	5.180
289.47	5.000
294.13	5.089*
298.80	4.981*

*Not shown on plot

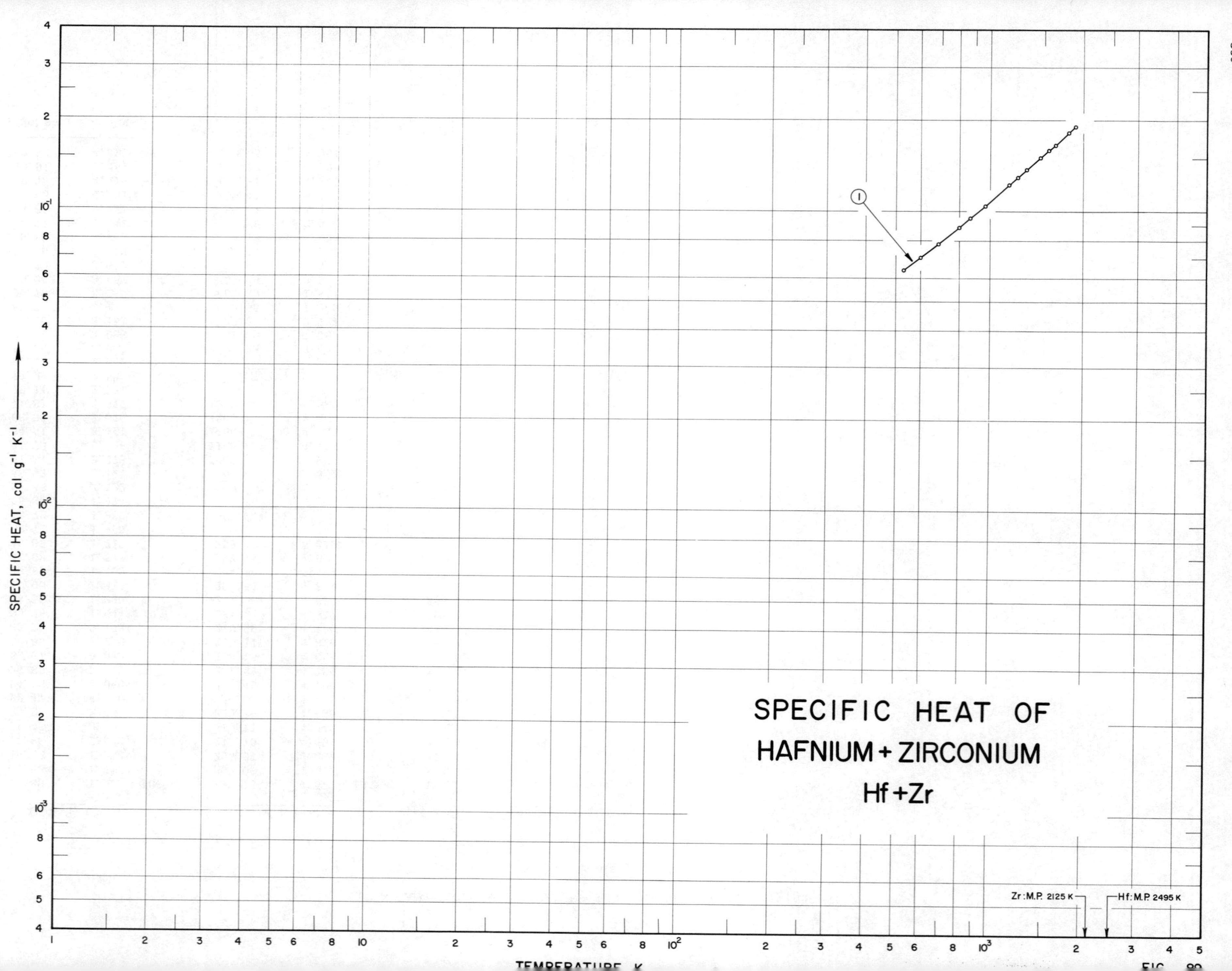
SPECIFIC HEAT OF
HAFNIUM + ZIRCONIUM
Hf + Zr
SPECIFIC HEAT, cal g⁻¹ K⁻¹
Zr: M.P. 2125 K
Hf: M.P. 2495 K
1

SPECIFICATION TABLE NO. 89 SPECIFIC HEAT OF HAFNIUM + ZIRCONIUM Hf + Zr

[For Data Reported in Figure and Table No. 89]

Curve No.	Ref. No.	Year	Temp. Range, K	Reported Error, %	Name and Specimen Designation	Composition (weight percent), Specifications and Remarks
1	146	1961	534-1884	3.0		99.0 Hf; 1 max Zr, 0.1 max Ti + Si, 0.01 max Fe + V + Cu, and 0.0001 max Mg; density = 815 lb ft^{-3}.

DATA TABLE NO. 89 SPECIFIC HEAT OF HAFNIUM + ZIRCONIUM Hf + Zr

[Temperature, T, K; Specific Heat, C_p, Cal $g^{-1} K^{-1}$]

T	C_p
CURVE 1	
533	6.385×10^{-2}
600	7.014
688	7.836
798	8.866
869	9.526
969	1.047×10^{-1}
1163	1.228
1237	1.297
1320	1.375
1472	1.517
1552	1.592
1628	1.663
1799	1.823
1883	1.902

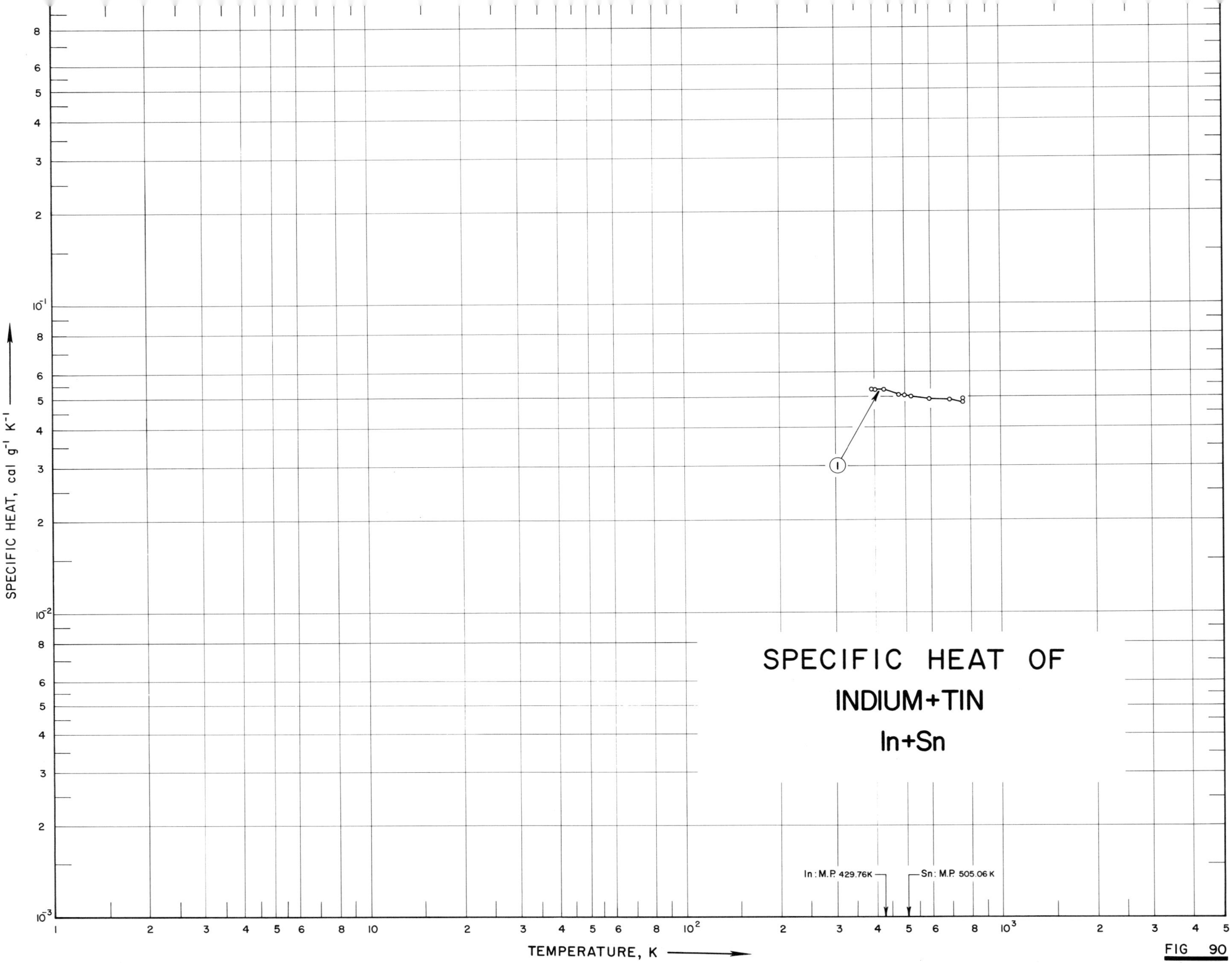

SPECIFIC HEAT OF INDIUM+TIN

In+Sn

SPECIFIC HEAT, cal g^{-1} K^{-1}

TEMPERATURE, K

In : M.P. 429.76 K

Sn : M.P. 505.06 K

FIG 90

SPECIFICATION TABLE NO. 90 SPECIFIC HEAT OF INDIUM + TIN In + Sn

[For Data Reported in Figure and Table No. 90]

Curve No.	Ref. No.	Year	Temp. Range, K	Reported Error, %	Name and Specimen Designation	Composition (weight percent), Specifications and Remarks
1	392	1962	393-758			51.95 In, 48.05 Sn; prepared by melting 99.99 In and 99.998 Sn under reducting atmosphere of N_2 plus H_2.

DATA TABLE NO. 90 SPECIFIC HEAT OF INDIUM + TIN In + Sn

[Temperature, T, K; Specific Heat, C_p, Cal g^{-1} K^{-1}]

T	C_p
CURVE 1	
393.8	6.32 x 10^{-2}
394.4	6.37*
400.7	6.34*
400.8	6.28
428.2	6.28
428.3	6.24*
475.7	6.11
497.3	6.07
524.1	6.02
524.1	6.03*
527.0	6.00*
527.0	6.02*
596.2	5.96
597.5	5.98*
693.3	5.94
693.4	5.95*
757.9	5.84
757.9	5.97

*Not shown on plot

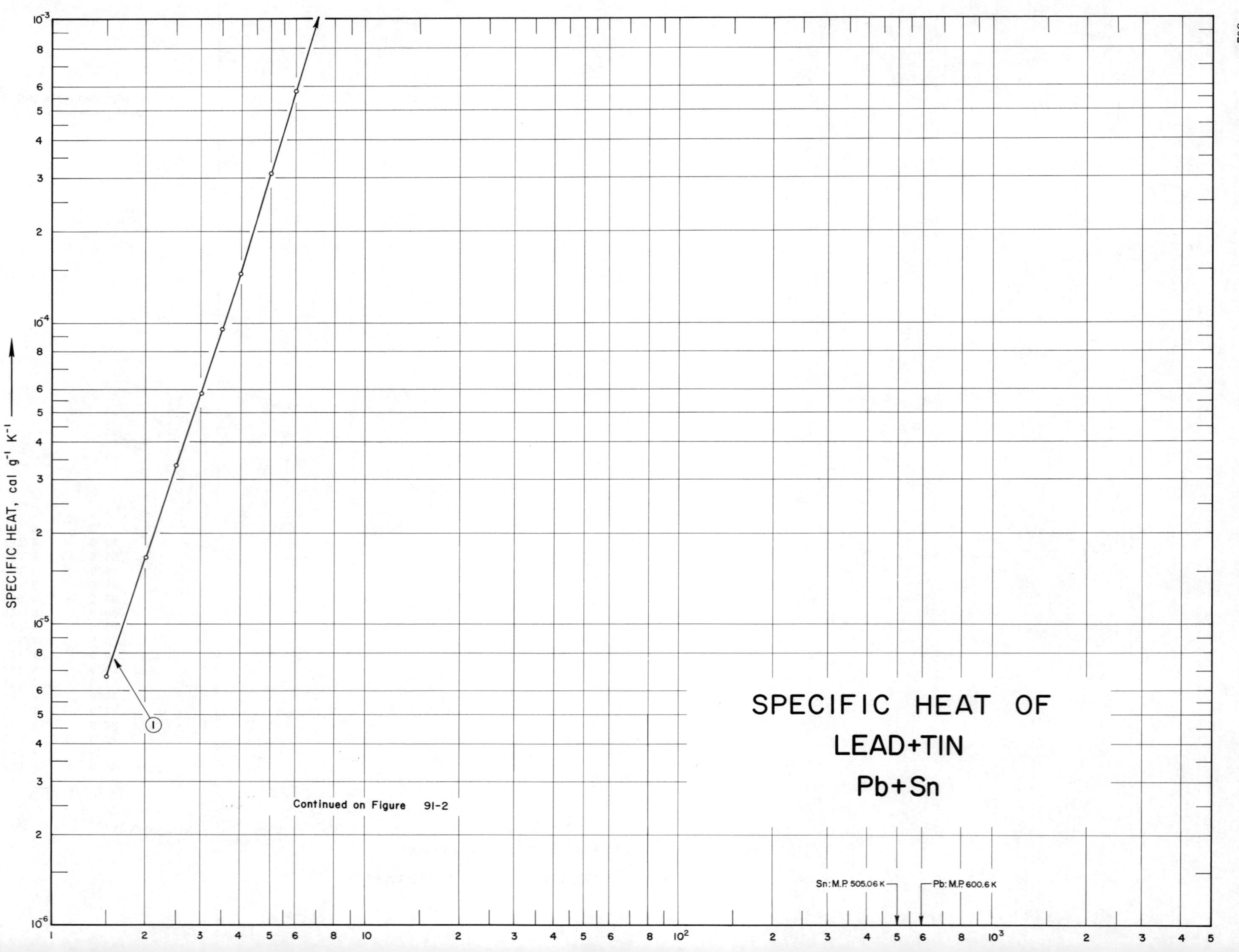
SPECIFIC HEAT OF
LEAD+TIN
Pb+Sn
SPECIFIC HEAT, cal g⁻¹ K⁻¹
Sn: M.P. 505.06 K
Pb: M.P. 600.6 K
Continued on Figure 91-2
1

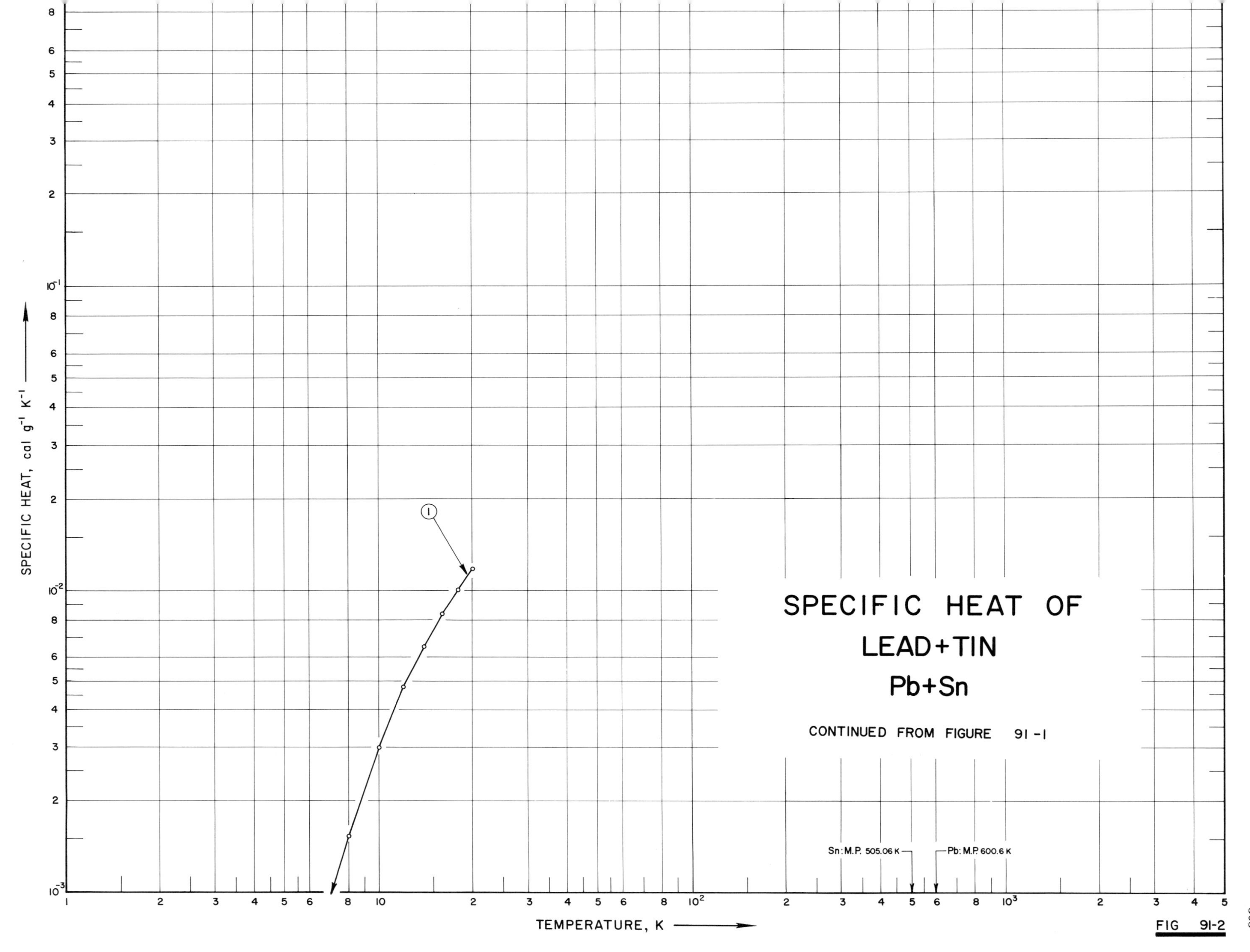

SPECIFIC HEAT OF
LEAD+TIN
Pb+Sn
CONTINUED FROM FIGURE 91-1
Sn: M.P. 505.06 K
Pb: M.P. 600.6 K
SPECIFIC HEAT, cal g⁻¹ K⁻¹
TEMPERATURE, K
1

FIG 91-2

SPECIFICATION TABLE NO. 91 SPECIFIC HEAT OF LEAD + TIN Pb + Sn

[For Data Reported in Figure and Table No. 91]

Curve No.	Ref. No.	Year	Temp. Range, K	Reported Error, %	Name and Specimen Designation	Composition (weight percent), Specifications and Remarks
1	393	1963	1.5-20			60 Sn, 40 Pb.

DATA TABLE NO. 91 SPECIFIC HEAT OF LEAD + TIN Pb + Sn

[Temperature, T, K; Specific Heat, C_p, Cal g^{-1} K^{-1}]

T	C_p
CURVE 1	
1.5	6.698×10^{-6}
2.0	1.663×10^{-5}
2.5	3.357
3.0	5.844
3.5	9.545
4.0	1.454×10^{-4}
5.0	3.102
6.0	5.814
8.0	1.534×10^{-3}
10.0	2.997
12.0	4.801
14.0	6.545
16.0	8.439
18.0	1.017×10^{-2}
20.0	1.190

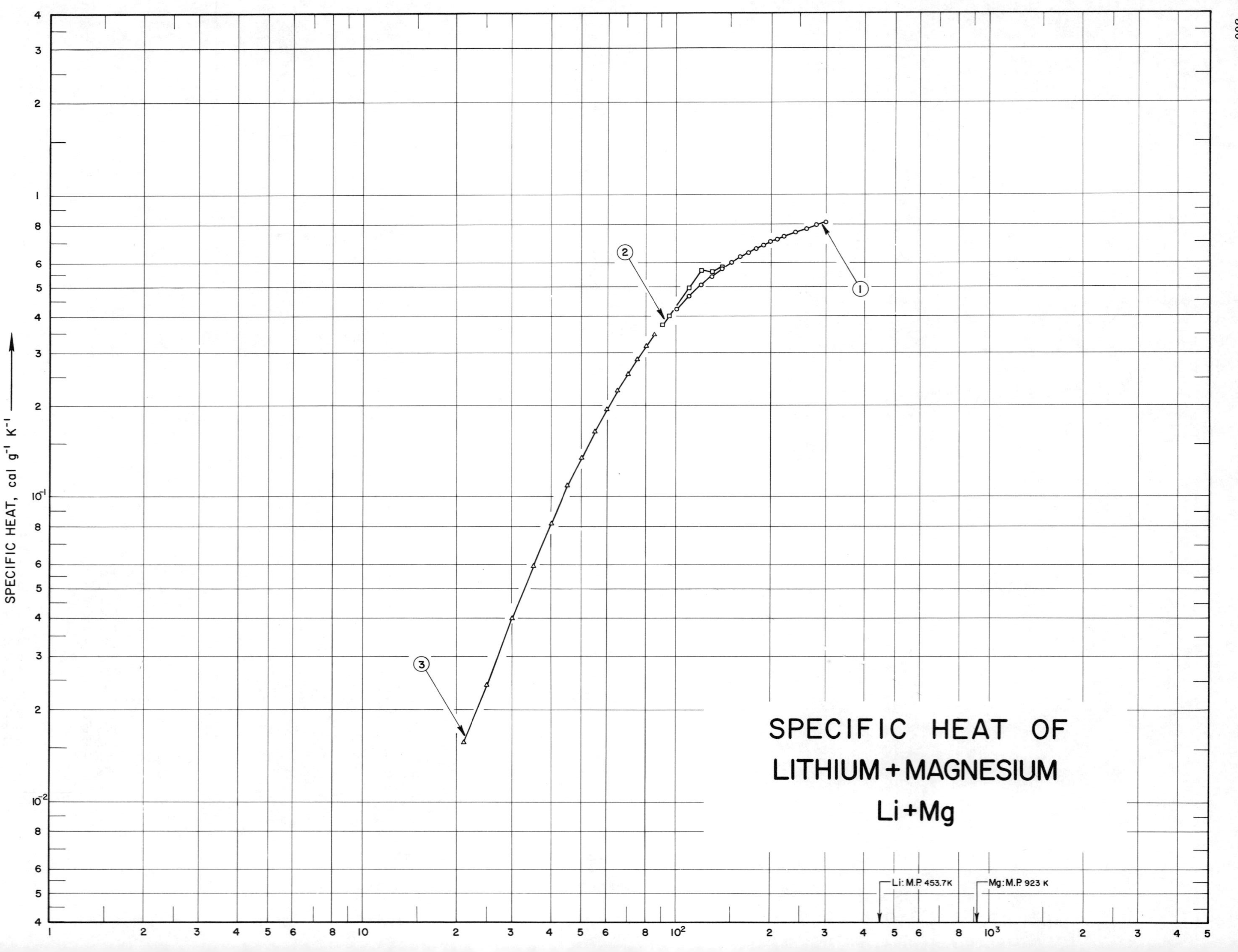
SPECIFIC HEAT OF
LITHIUM + MAGNESIUM
Li+Mg
Li: M.P. 453.7 K
Mg: M.P. 923 K
SPECIFIC HEAT, cal g⁻¹ K⁻¹
1
2
3

SPECIFICATION TABLE NO. 92 SPECIFIC HEAT OF LITHIUM + MAGNESIUM Li + Mg

[For Data Reported in Figure and Table No. 92]

Curve No.	Ref. No.	Year	Temp. Range, K	Reported Error, %	Name and Specimen Designation	Composition (weight percent), Specifications and Remarks
1	228	1960	100-300	0.3-2.0	0.95 at. % magnesium alloy	96.74 Li, 3.26 Mg; body centered cubic phase; prepared by heating lithium (Li sample impurities: 0.058 Ca, 0.056 K, 0.040 N, 0.017 Na, 0.008 Fe), and "spectroscopically pure" magnesium to 800 C in low carbon steel crucible.
2	228	1960	90-160	0.3-2.0	Same as above	Same as above; cooled to 20 K; annealed.
3	228	1960	21-85	0.3-2.0	Same as above	Same as above; not annealed.

DATA TABLE NO. 92 SPECIFIC HEAT OF LITHIUM + MAGNESIUM Li + Mg

[Temperature, T, K; Specific Heat, C_p, Cal $g^{-1} K^{-1}$]

T	C_p
CURVE 1	
100	4.256×10^{-1}
110	4.682
120	5.079
130	5.443
140	5.764
150	6.045
160	6.300
170	6.536
180	6.734
190	6.917
200	7.089
210	7.242
220	7.374
230	7.513*
240	7.634
250	7.734*
260	7.847
270	7.952*
273.15	7.982*
280	8.048
290	8.136*
298.15	8.204*
300	8.218
CURVE 2	
90	3.751×10^{-1}
95	4.021
100	4.283*
110	4.996
120	5.692
130	5.633
140	5.824
150	6.070*
160	6.304*
CURVE 3	
21	1.56×10^{-2}
25	2.43
30	4.01
35	5.96
40	8.26
45	1.08×10^{-1}
50	1.35
55	1.646
CURVE 3 (cont.)	
60	1.952×10^{-1}
65	2.262
70	2.564
75	2.869
80	3.170
85	3.467

*Not shown on plot

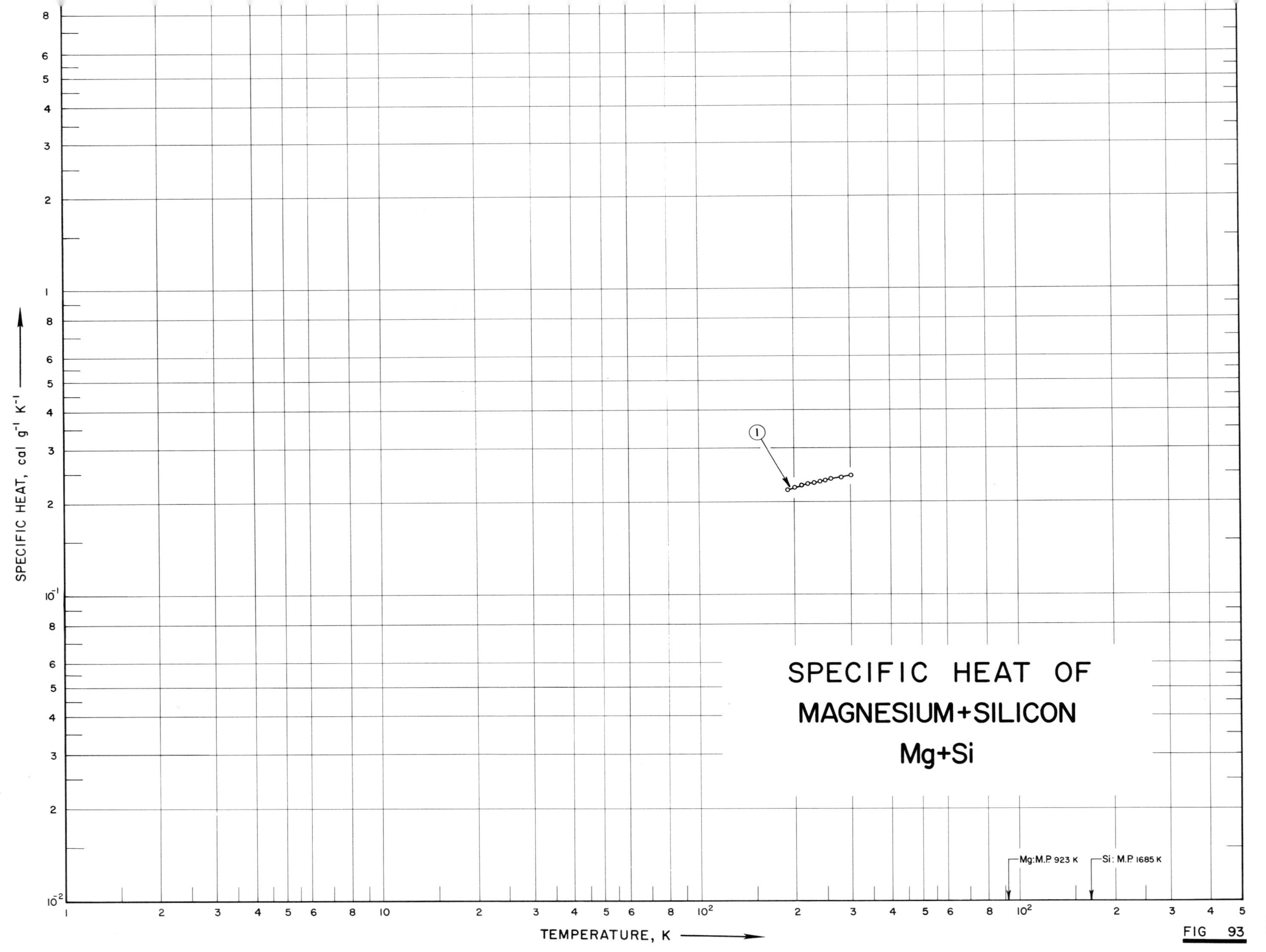

SPECIFIC HEAT OF
MAGNESIUM+SILICON
Mg+Si
Mg: M.P. 923 K
Si: M.P. 1685 K
1
SPECIFIC HEAT, cal g⁻¹ K⁻¹
TEMPERATURE, K

FIG 93

SPECIFICATION TABLE NO. 93 SPECIFIC HEAT OF MAGNESIUM + SILICON Mg + Si

[For Data Reported in Figure and Table No. 93]

Curve No.	Ref. No.	Year	Temp. Range, K	Reported Error, %	Name and Specimen Designation	Composition (weight percent), Specifications and Remarks
1	49	1960	190-300			99.80 Mg, 0.20 Si.

DATA TABLE NO. 93 SPECIFIC HEAT OF MAGNESIUM + SILICON Mg + Si

[Temperature, T, K; Specific Heat, C_p, Cal $g^{-1} K^{-1}$]

T	C_p
CURVE 1	
190	2.188 x 10^{-1}
200	2.227
210	2.260
220	2.290
230	2.316
240	2.338
250	2.359
260	2.379
280	2.415
300	2.443

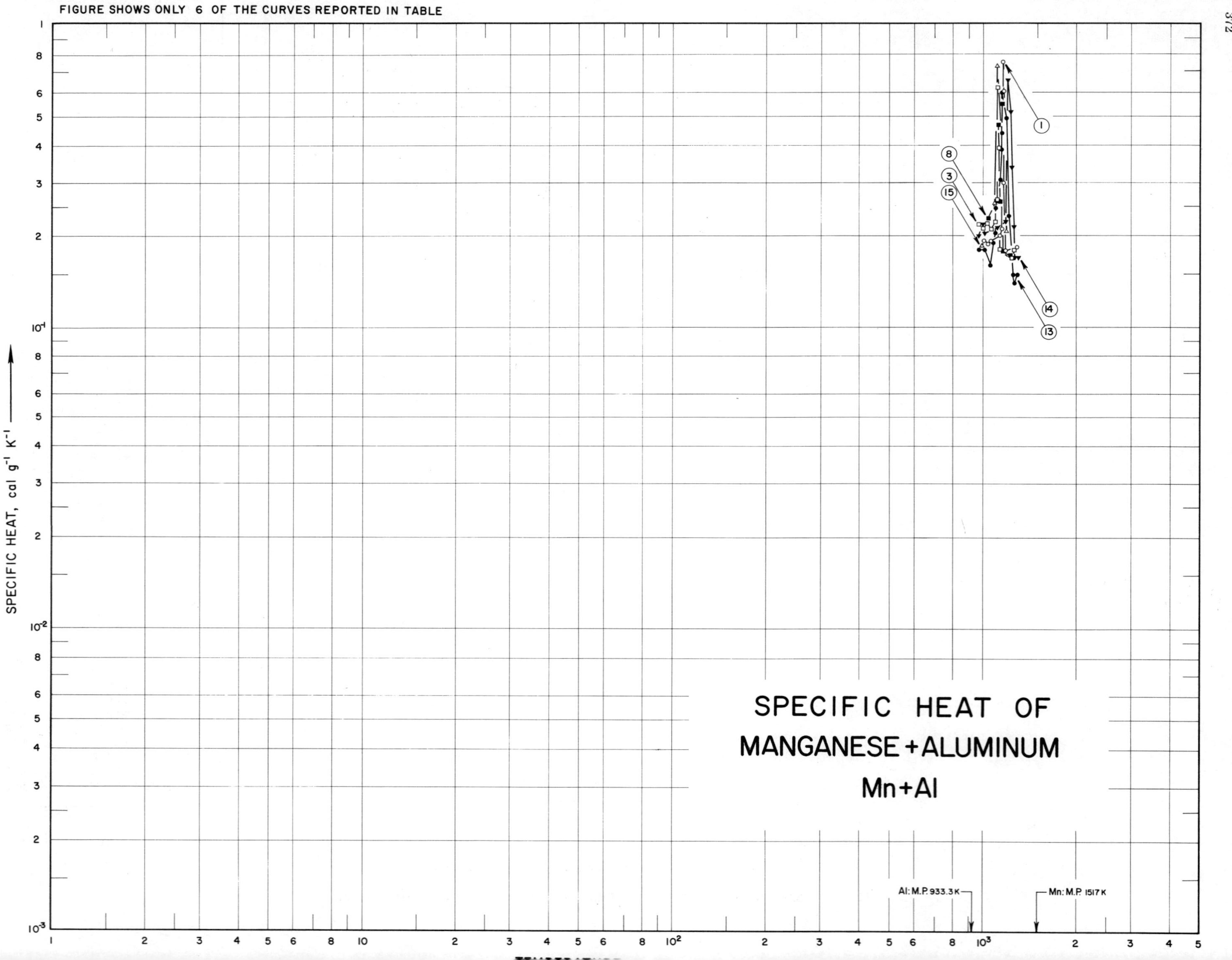
FIGURE SHOWS ONLY 6 OF THE CURVES REPORTED IN TABLE
SPECIFIC HEAT OF
MANGANESE + ALUMINUM
Mn+Al
SPECIFIC HEAT, cal g⁻¹ K⁻¹
Al: M.P. 933.3 K
Mn: M.P. 1517 K
1
8
3
15
14
13

SPECIFICATION TABLE NO. 94 SPECIFIC HEAT OF MANGANESE + ALUMINUM Mn + Al

[For Data Reported in Figure and Table No. 94]

Curve No.	Ref. No.	Year	Temp. Range, K	Reported Error, %	Name and Specimen Designation	Composition (weight percent), Specifications and Remarks
1	229	1958	1010-1283		A-47	64. 4 Mn, 35. 6 Al; prepared from desired proportions of 99. 9 Mn and 99. 99 Al; melted in an induction furnace; annealed for 1 hr at 950 C; slowly cooled to 700 C; annealed again at 950 C for 5 hrs then slowly cooled to room temperature in a vacuum.
2	229	1958	970-1208		A-48	65. 3 Mn, 34. 7 Al; same as above.
3	229	1958	972-1272		A-49	66. 30 Mn, 33. 70 Al; same as above.
4	229	1958	970-1233		A-50	69. 1 Mn, 32. 90 Al; same as above.
5	229	1958	1045-1258		A-51	69. 0 Mn, 31. 0 Al; same as above.
6	229	1958	970-1270		A-52	68. 8 Mn, 31. 2 Al; same as above.
7	229	1958	970-1258		A-53	70. 4 Mn, 29. 6 Al; same as above.
8	229	1958	1033-1233		A-54	71. 4 Mn, 28. 6 Al; same as above.
9	229	1958	970-1258		A-55	71. 8 Mn, 28. 2 Al; same as above.
10	229	1958	1020-1260		A-56	73. 2 Mn, 26. 8 Al; same as above.
11	229	1958	970-1283		A-57	73. 4 Mn, 26. 6 Al; same as above.
12	229	1958	970-1283		A-58	73. 7 Mn, 26. 3 Al; same as above.
13	229	1958	970-1295		A-59	77. 0 Mn, 23. 0 Al; same as above.
14	229	1958	970-1308		A-60	77. 3 Mn, 22. 7 Al; same as above.
15	229	1958	993-1189		A-49. 5	66. 30 Mn, 33. 70 Al; same as above.
16	229	1958	1040-1189		A-50. 5	67. 1 Mn, 32. 90 Al; same as above.

DATA TABLE NO. 94 SPECIFIC HEAT OF MANGANESE + ALUMINUM Mn + Al

[Temperature, T, K; Specific Heat, C_p, Cal g^{-1} K^{-1}]

T	C_p
CURVE 1	
1010	1.94 x 10^{-1}
1020.6	1.90*
1033	1.87*
1045.6	1.90
1053	1.91*
1070.6	1.94
1083	1.95*
1095.6	1.95*
1108	2.00*
1120.6	2.00*
1133	2.02
1145.6	2.07*
1153	2.05*
1155.6	1.95*
1158	2.12
1160.6	2.07*
1163	2.27*
1165.6	7.591
1168	6.067
1172	3.021
1174	1.90*
1176	1.80
1183	1.80*
1208	1.81*
1222	1.82*
1233	1.82*
1247	1.85*
1258	1.87*
1272	1.85*
1283	1.8
CURVE 2*	
970.65	2.060 x 10^{-1}
983.15	2.060
995.65	2.060
1008.15	2.035
1020.65	1.886
1033.15	1.960
1045.65	1.911
1058.15	1.911
1070.65	1.960
1083.15	1.935
1095.65	1.886
1108.15	1.935
1120.65	1.935
CURVE 2 (cont.)*	
1128.15	1.960 x 10^{-1}
1135.65	1.985
1138.15	2.035
1140.65	2.829
1143.15	7.940
1145.65	3.598
1148.15	3.127
1150.65	2.705
1158.15	1.774
1170.65	1.787
1183.15	1.787
1195.65	1.787
1208.15	1.762
CURVE 3	
972.15	2.187 x 10^{-1}
983.15	2.199*
997.15	2.138
1008.15	2.187*
1022.15	2.187*
1033.15	2.211
1047.15	2.211*
1058.15	2.162*
1071.15	2.113
1113.15	2.211*
1096.15	2.236
1108.15	2.408*
1115.15	2.654
1120.15	6.265
1122.15	6.167*
1124.15	3.956
1126.15	2.801*
1128.15	1.843*
1136.15	1.818
1147.15	1.867*
1158.15	1.843*
1172.15	1.843*
1183.15	1.843*
1197.15	1.769*
1208.15	1.843*
1222.15	1.769*
1233.15	1.769*
1247.15	1.695
1258.15	1.818*
1272.15	1.818
CURVE 4*	
970.65	1.832 x 10^{-1}
983.15	1.783
995.65	1.905
1008.15	1.905
1020.65	2.051
1033.15	2.027
1045.65	2.125
1070.65	2.369
1083.15	2.540
1095.65	2.686
1103.15	2.857
1105.65	2.759
1108.15	3.101
1115.65	3.565
1118.15	2.857
1120.65	2.100
1123.15	1.954
1125.65	1.905
1133.15	1.758
1145.65	1.783
1158.15	1.783
1170.65	1.880
1183.15	1.880
1195.65	1.807
1208.15	1.807
1220.65	1.856
1233.15	1.880
CURVE 5*	
1045.65	2.410 x 10^{-1}
1058.15	2.507
1070.65	2.651
1083.15	2.796
1095.65	2.989
1103.15	2.916
1105.65	2.965
1108.15	3.230
1113.15	4.604
1115.65	3.567
1118.15	2.940
1120.65	2.507
1123.15	2.410
1128.15	2.603
1130.65	2.434
1133.15	2.507
CURVE 5 (cont.)*	
1138.15	2.386 x 10^{-1}
1140.65	2.483
1143.15	2.651
1145.65	2.868
1148.15	3.374
1150.65	3.133
1153.15	2.989
1155.65	2.820
1158.15	2.724
1160.65	2.410
1163.15	2.169
1167.15	2.169
1168.15	1.904
1173.15	1.952
1175.65	1.904
1183.15	1.856
1195.65	1.880
1208.15	1.832
1220.65	1.759
1233.15	1.928
1245.65	1.928
1258.15	1.808
CURVE 6*	
970.65	2.217 x 10^{-1}
983.15	2.241
995.65	2.217
1008.15	2.241
1020.65	2.265
1033.15	2.386
1045.65	2.434
1058.15	2.506
1070.65	2.578
1083.15	2.699
1095.65	2.819
1103.15	2.771
1105.65	2.771
1108.15	2.940
1110.65	3.157
1113.15	4.289
1118.15	3.277
1120.65	3.157
1123.15	2.530
1125.65	2.530
1128.15	2.675
CURVE 6 (cont.)*	
1130.65	2.723 x 10^{-1}
1133.15	2.795
1135.65	2.723
1138.15	2.675
1140.65	2.843
1143.15	2.843
1145.65	2.988
1153.15	5.012
1155.65	4.386
1158.15	3.831
1160.65	3.301
1163.15	2.313
1165.65	2.145
1168.15	2.000
1170.65	1.880
1175.65	1.880
1183.15	1.831
1195.65	1.783
1208.15	1.711
1220.65	1.735
1233.15	1.687
1245.65	1.687
1258.15	1.663
1270.65	1.639
CURVE 7*	
970.65	2.142 x 10^{-1}
983.15	2.142
995.65	2.237
1008.15	2.261
1022.15	2.356
1033.15	2.451
1045.65	2.427
1058.15	2.451
1070.65	2.570
1083.15	2.618
1095.65	2.689
1103.15	2.713
1105.65	2.618
1108.15	2.665
1110.65	2.903
1115.65	3.427
1118.15	2.856
1120.65	2.713
1123.15	2.618
CURVE 7 (cont.)*	
1128.15	2.642 x 10^{-1}
1130.65	2.713
1133.15	2.808
1135.65	2.856
1138.15	3.022
1141.15	3.808
1145.65	3.832
1148.15	3.094
1150.65	2.570
1153.15	2.142
1155.56	1.975
1158.15	2.023
1163.15	2.023
1170.65	1.951
1183.15	1.880
1195.65	1.787
1208.15	1.809
1220.65	1.761
1223.15	1.761
1245.65	1.737
1258.15	1.737
CURVE 8	
1033.15	2.246 x 10^{-1}*
1045.65	2.340
1058.15	2.317*
1070.65	2.364*
1083.15	2.482*
1095.65	2.530*
1103.15	2.553*
1105.65	2.624*
1108.15	2.530*
1110.65	2.600
1113.15	4.728
1118.15	3.664*
1120.65	3.546*
1123.15	2.908*
1125.65	2.671*
1128.15	2.600
1130.65	2.931*
1133.15	3.286*
1135.65	3.168*
1150.65	5.532
1153.15	4.208*
1155.65	3.593*
1158.15	3.144*

* Not shown on plot

DATA TABLE NO. 94 (continued)

T	C_p
CURVE 8 (cont.)	
1160.65	2.364 x 10^{-1}*
1163.15	2.033*
1170.65	1.797
1183.15	1.820*
1195.65	1.773*
1208.15	1.797*
1220.65	1.773*
1233.15	1.726
CURVE 9*	
970.65	2.471 x 10^{-1}
983.15	2.471
995.65	2.494
1008.15	2.541
1020.65	2.471
1033.15	2.541
1045.65	2.587
1058.15	2.634
1070.65	2.611
1083.15	2.681
1095.65	2.681
1103.15	2.657
1105.65	2.681
1108.15	2.821
1110.65	2.681
1113.15	2.797
1118.15	7.459
1120.15	4.359
1121.15	3.380
1123.15	2.774
1125.65	2.727
1128.15	2.774
1130.65	2.867
1133.15	3.054
1135.65	3.263
1145.15	6.154
1145.65	4.429
1148.15	3.193
1150.65	3.124
1153.15	2.890
1155.65	2.681
1158.15	2.611
1160.65	2.471
1163.15	2.098
1170.65	1.958
1183.15	1.772
CURVE 9 (cont.)*	
1195.65	1.655 x 10^{-1}
1208.15	1.678
1220.65	1.702
1233.15	1.632
1245.65	1.632
1258.65	1.632
CURVE 10*	
1020.65	2.100 x 10^{-1}
1033.15	2.100
1045.65	2.217
1058.15	2.193
1070.35	2.263
1083.15	2.333
1095.65	2.357
1103.15	2.427
1105.35	2.380
1108.15	2.473
1110.65	2.520
1113.15	5.320
1115.65	4.433
1118.15	3.196
1120.65	2.636
1123.15	2.497
1125.65	2.753
1128.15	2.823
1130.65	3.126
1133.15	3.686
1138.15	6.836
1148.15	6.393
1150.65	5.600
1153.15	4.060
1155.65	3.290
1158.15	3.150
1160.65	2.916
1163.15	2.753
1165.65	2.706
1173.15	2.310
1185.65	1.937
1198.15	1.843
1210.65	1.727
1223.15	1.633
1235.65	1.727
1248.15	1.703
1260.65	1.680
CURVE 11*	
970.65	2.253 x 10^{-1}
983.15	2.229
995.65	2.253
1008.15	2.299
1020.65	2.299
1033.15	2.346
1045.65	2.392
1058.15	2.415
1070.65	2.462
1083.15	2.485
1095.65	2.601
1103.15	2.671
1105.65	2.624
1108.15	2.624
1110.65	2.694
1113.15	2.694
1115.65	4.412
1118.15	3.902
1120.65	2.926
1123.15	2.671
1125.65	2.810
1128.15	2.717
1133.15	2.880
1135.65	3.205
1138.15	3.321
1140.65	4.714
1148.15	7.408
1150.65	6.874
1153.15	6.456
1155.65	6.386
1158.15	6.270
1160.65	6.038
1163.15	5.783
1170.15	5.365
1183.15	3.762
1195.65	2.671
1208.15	1.765
1220.65	1.556
1233.15	1.579
1245.65	1.626
1258.15	1.510
1270.65	1.510
1283.15	1.556
CURVE 12*	
830.65	4.219 x 10^{-1}
CURVE 12 (cont.)*	
928.15	4.149 x 10^{-1}
938.15	5.911
970.65	2.017
983.15	1.970
995.65	2.063
1008.15	2.086
1020.65	2.017
1038.15	2.086
1045.65	2.063
1053.15	2.179
1055.65	2.202
1058.15	2.133
1060.65	2.225
1063.15	2.318
1065.65	2.689
1068.15	2.967
1070.65	3.639
1073.15	3.083
1075.65	2.967
1083.15	2.898
1095.65	2.944
1103.15	2.990
1105.65	3.083
1108.15	3.199
1110.65	3.245
1113.15	3.570
1115.65	3.964
1118.15	3.639
1120.65	3.523
1123.15	3.686
1125.65	3.941
1133.15	4.682
1135.65	5.795
1140.65	5.957
1143.15	5.331
1145.65	5.007
1148.15	4.937
1150.65	5.192
1153.15	5.123
1155.65	4.358
1158.15	4.358
1163.15	4.312
1170.65	4.080
1183.15	3.199
1195.65	2.411
1208.15	1.947
1220.65	1.762
CURVE 12 (cont.)*	
1233.15	1.692 x 10^{-1}
1245.65	1.576
1258.15	1.692
1270.65	1.692
1283.15	1.715
CURVE 13	
970.65	1.806 x 10^{-1}
983.15	1.806*
995.65	1.806*
1008.15	1.806*
1020.65	1.806
1033.15	1.828*
1045.65	1.851*
1058.15	1.828*
1070.65	1.603
1083.15	1.919
1090.65	2.009*
1093.15	2.054
1095.65	2.144*
1098.15	2.257*
1100.65	2.483
1103.15	2.709*
1105.65	2.664*
1108.15	2.596*
1110.65	2.438*
1113.15	2.506*
1120.65	2.438*
1133.15	2.709*
1140.65	3.093
1143.15	3.093*
1145.65	3.454*
1148.15	3.318*
1150.65	3.905
1153.15	4.402
1155.65	6.005
1158.15	6.072*
1160.65	5.621*
1163.15	5.282*
1170.65	5.169*
1183.15	4.944
1195.65	4.582*
1208.15	3.567*
1220.65	2.348
1233.15	1.874*
1245.65	1.603*
CURVE 13 (cont.)	
1258.15	1.490 x 10^{-1}
1270.65	1.400
1283.15	1.512*
1295.65	1.490
CURVE 14	
970.65	2.005 x 10^{-1}
983.15	2.117*
995.65	2.207
1008.15	2.095*
1020.65	2.050
1033.15	2.095*
1045.65	2.072*
1058.15	2.095*
1070.65	2.072*
1083.15	2.072*
1095.65	2.095*
1103.15	2.072*
1120.65	2.410
1133.15	2.162*
1145.65	2.207*
1158.15	2.185*
1170.65	2.207*
1183.15	2.252
1195.65	2.230*
1203.15	2.387*
1208.15	6.644
1210.65	5.878*
1218.15	5.473*
1230.65	5.180
1238.15	4.414*
1240.65	4.324*
1243.15	3.739*
1245.65	3.716*
1248.15	3.401
1250.35	3.153*
1253.15	2.658*
1255.65	2.613*
1258.15	2.162*
1260.65	2.162
1270.65	1.712
1283.15	1.712*
1295.65	1.712*
1308.15	1.712

*Not shown on plot

DATA TABLE NO. 94 (continued)

T	C_p
CURVE 15	
993.15	1.867×10^{-1}
1040.15	2.211*
1089.15	2.580
1108.15	7.371
1138.15	1.990*
1188.15	2.088
CURVE 16	
1040.15	2.442×10^{-1}
1089.15	2.906
1127.15	2.784
1139.15	2.027
1189.15	2.051

* Not shown on plot

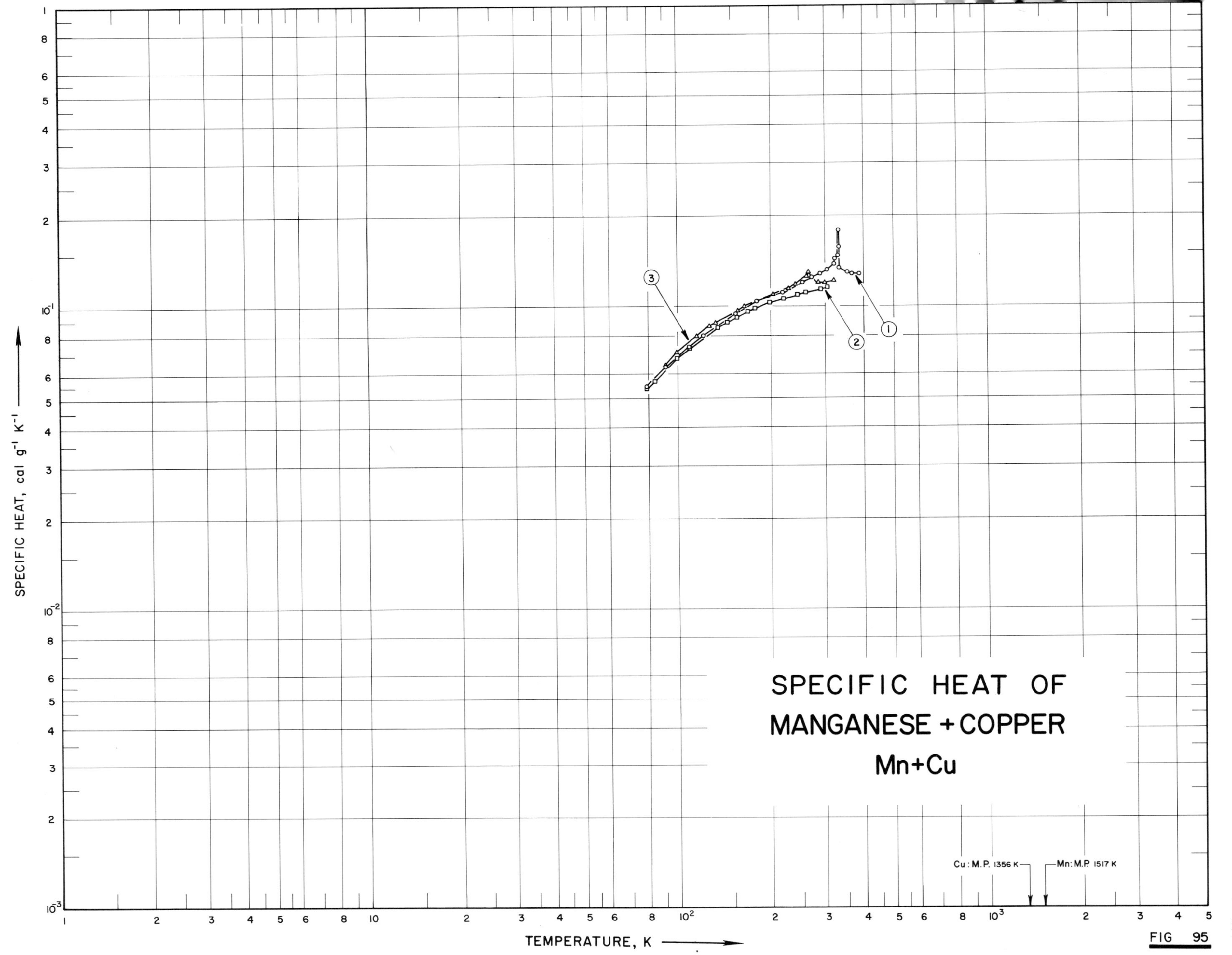
SPECIFIC HEAT OF
MANGANESE + COPPER
Mn+Cu
SPECIFIC HEAT, cal g⁻¹ K⁻¹
TEMPERATURE, K
Cu: M.P. 1356 K
Mn: M.P. 1517 K
1
2
3

FIG 95

SPECIFICATION TABLE NO. 95 SPECIFIC HEAT OF MANGANESE + COPPER Mn + Cu

[For Data Reported in Figure and Table No. 95]

Curve No.	Ref. No.	Year	Temp. Range, K	Reported Error, %	Name and Specimen Designation	Composition (weight percent), Specifications and Remarks
1	230	1961	79-386		Cu-Mn alloy 85 at. % Mn	83 Mn, 17 Cu; prepared by melting Analar grade manganese and copper together in argon arc furnace.
2	230	1961	79-307		Cu-Mn alloy 65 at. % Mn	61. 6 Mn, 38. 4 Cu; same as above.
3	230	1961	79-320		Cu-Mn alloy 80 at. % Mn	77. 5 Mn, 22. 5 Cu; same as above.

DATA TABLE NO. 95 SPECIFIC HEAT OF MANGANESE + COPPER Mn + Cu

[Temperature, T, K; Specific Heat, C_p, Cal g^{-1} K^{-1}]

T	C_p
CURVE 1	
79	5.450×10^{-2}
91	6.322
108	7.373
121	8.014
154	9.457
181	1.044×10^{-1}
218	1.118
253	1.202
272	1.250
287	1.288
292	1.286*
305	1.325
310	1.336*
321	1.384
326	1.446
331	1.475
334	1.524*
335	1.587*
337	1.647*
339	1.799
341	1.583
342	1.489*
342	1.423*
346	1.343
354	1.305*
362	1.298*
366	1.284
374	1.297*
386	1.284
CURVE 2	
79	5.389×10^{-2}
84	5.699
92	6.287*
99	6.788
109	7.340
121	7.962*
135	8.515
145	8.877
156	9.240
169	9.637
178	9.862
197	1.028×10^{-1}
221	1.060
244	1.093
260	1.117
CURVE 2 (cont.)	
291	1.145×10^{-1}
307	1.162
CURVE 3	
Series 1	
79	5.467×10^{-2}*
91	6.411×10^{-1}
99	7.070
115	7.979
127	8.602
132	8.833
157	9.671
164	9.991
181	1.047*
204	1.099
218	1.134*
228	1.150
240	1.181
250	1.215*
262	1.264
281	1.243*
261	1.289*
264	1.297*
265	1.305
268	1.305*
273	1.273*
275	1.254*
285	1.218
287	1.218*
295	1.222*
299	1.215*
300	1.207
306	1.215*
304	1.213*
310	1.209*
320	1.225
314	1.220*
294	1.231*
286	1.234*
Series 2*	
256	1.259×10^{-1}
266	1.318
272	1.279
CURVE 3 (cont.)	
270	1.288×10^{-1}
267	1.318

*Not shown on plot

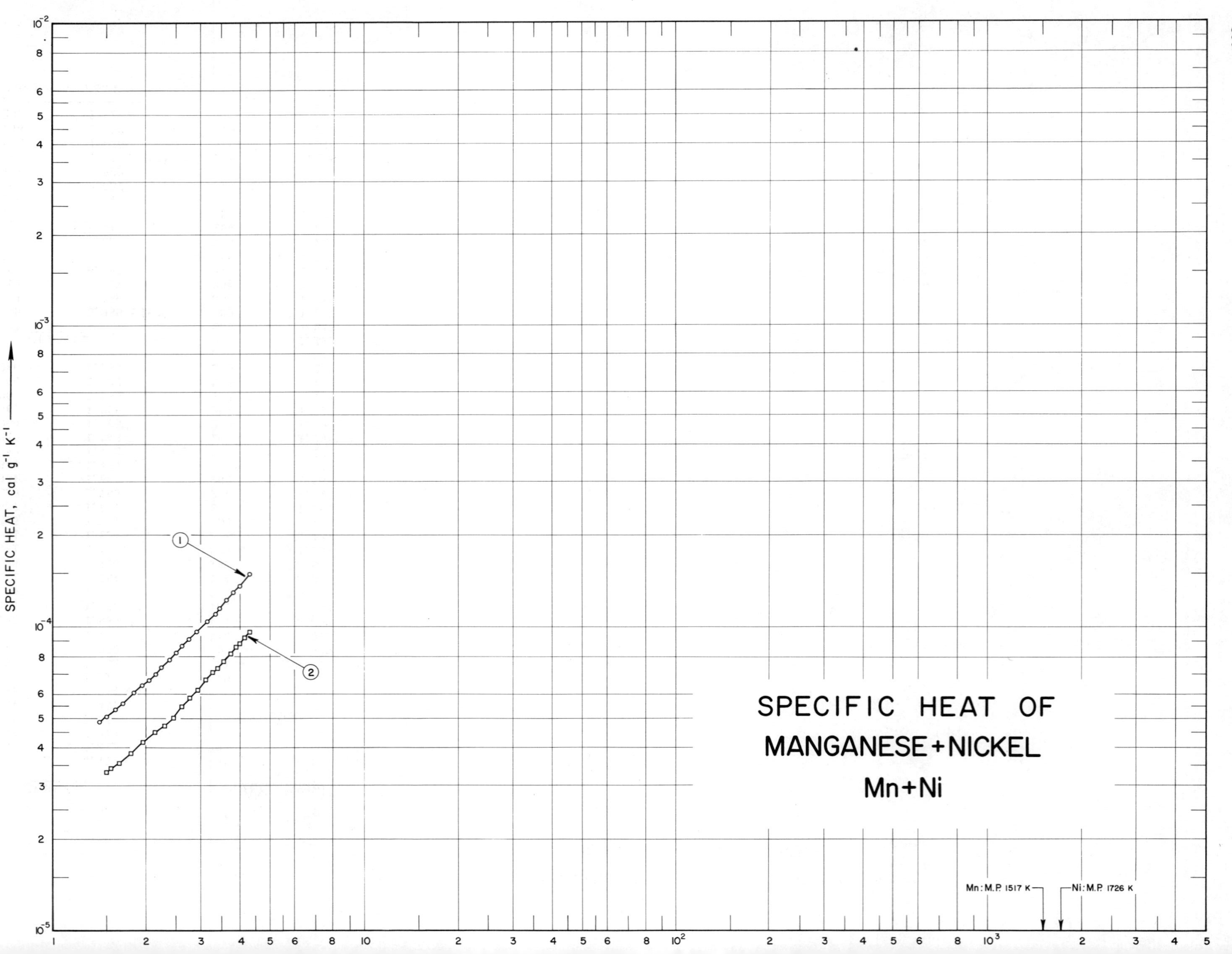

SPECIFIC HEAT OF
MANGANESE + NICKEL
Mn+Ni
SPECIFIC HEAT, cal g⁻¹ K⁻¹
Mn: M.P. 1517 K
Ni: M.P. 1726 K
1
2

SPECIFICATION TABLE NO. 96 SPECIFIC HEAT OF MANGANESE + NICKEL Mn + Ni

[For Data Reported in Figure and Table No. 96]

Curve No.	Ref. No.	Year	Temp. Range, K	Reported Error, %	Name and Specimen Designation	Composition (weight percent), Specifications and Remarks
1	349	1962	1.4-4.3	≤2	Ni(25) Mn(75)	75 Mn, 25 Ni; annealed under He + 8% H_2 gas atmosphere at 1000 C for 72 hrs; etched with 30-50% HNO_3.
2	349	1962	1.5-4.3	≤2	Ni(40) Mn(60)	60 Mn, 40 Ni; annealed under He + 8% H_2 gas atmosphere at 980 C for 72 hrs; etched with 30-50% HNO_3.

DATA TABLE NO. 96 SPECIFIC HEAT OF MANGANESE + NICKEL Mn + Ni

[Temperature, T, K; Specific Heat, C_p, Cal g^{-1} K^{-1}]

T	C_p
CURVE 1	
1.422	4.862 x 10^{-5}
1.497	5.070
1.510	5.091*
1.587	5.351
1.676	5.623
1.706	5.703*
1.829	6.075
1.946	6.430
1.947	6.410*
2.047	6.680
2.131	6.993
2.239	7.353
2.375	7.842
2.497	8.262
2.606	8.656
2.742	9.118
2.911	9.667
3.132	1.043 x 10^{-4}
3.312	1.108
3.313	1.106*
3.447	1.161
3.617	1.224
3.814	1.307
3.980	1.371
4.115	1.417*
4.270	1.499
CURVE 2	
1.481	3.306 x 10^{-5}
1.542	3.401
1.644	3.564
1.785	3.837
1.797	3.860*
1.950	4.184
2.128	4.486
2.289	4.746
2.432	5.044
2.607	5.478
2.767	5.850
2.925	6.216
3.113	6.718
3.270	7.100
3.397	7.336
3.548	7.726
3.734	8.174
CURVE 2 (cont.)	
3.883	8.599 x 10^{-5}
3.995	8.840
4.136	9.223
4.293	9.625

*Not shown on plot

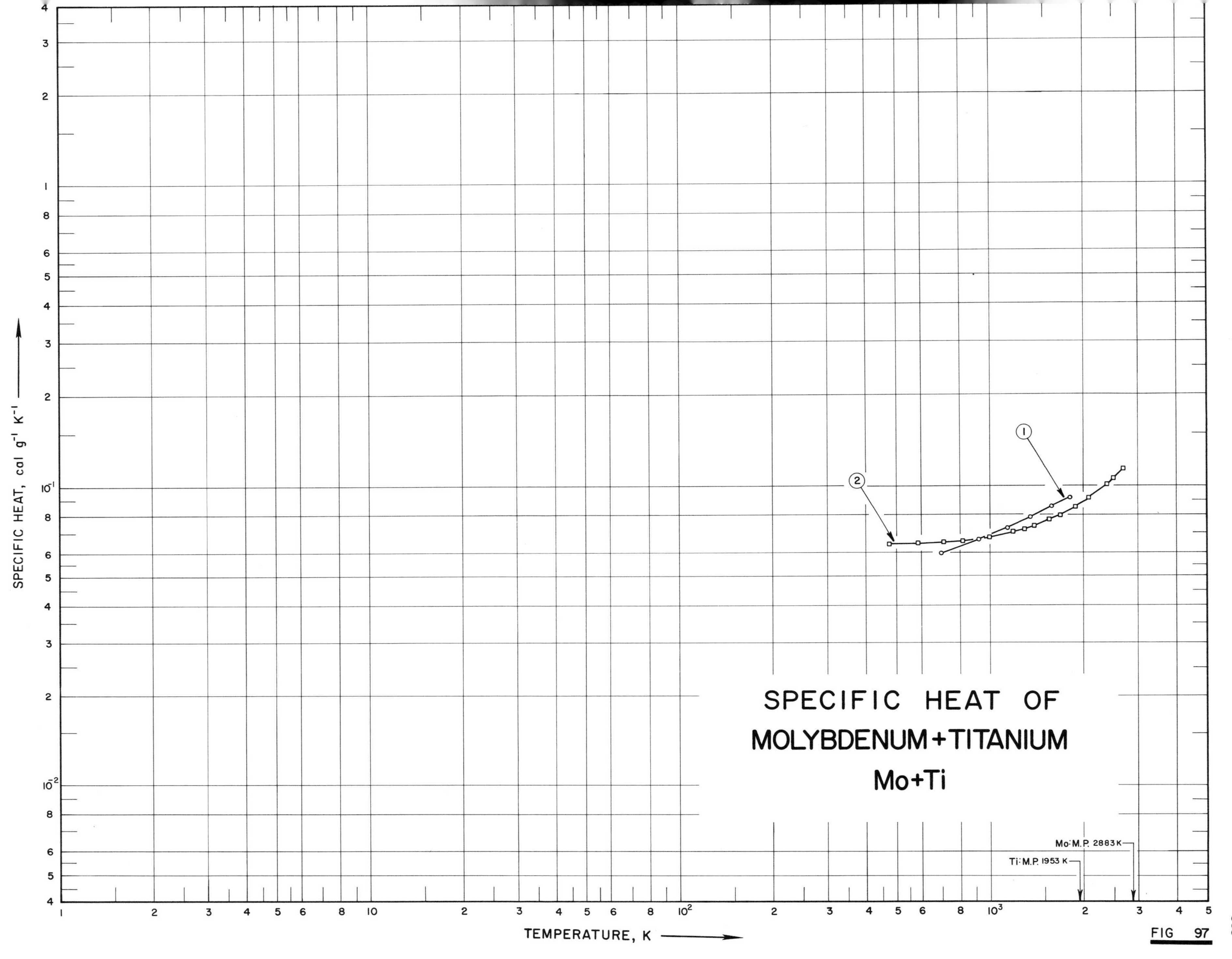
SPECIFIC HEAT OF
MOLYBDENUM+TITANIUM
Mo+Ti
SPECIFIC HEAT, cal g⁻¹ K⁻¹
TEMPERATURE, K
Mo: M.P. 2883 K
Ti: M.P. 1953 K
1
2
FIG 97

SPECIFICATION TABLE NO. 97 SPECIFIC HEAT OF MOLYBDENUM + TITANIUM Mo + Ti

[For Data Reported in Figure and Table No. 97]

Curve No.	Ref. No.	Year	Temp. Range, K	Reported Error, %	Name and Specimen Designation	Composition (weight percent), Specifications and Remarks
1	231	1960	700-1810	0.7-2.9	0.5% Ti alloy of Mo	Helium atmosphere.
2	232	1963	475-2697	±5.0		Mo - 0.5 Ti - 0.08 Zr Alloy Climax molybdenum, bal. Mo, 0.5 Ti, 0.07 Zr, 0.0290 C, <0.005 Si, <0.002 Fe, <0.001 Ni, 0.0005 O_2, 0.0003 N_2, 0.0001 H_2; density = 622 lb ft^{-3}.

DATA TABLE NO. 97 SPECIFIC HEAT OF MOLYBDENUM + TITANIUM Mo + Ti

[Temperature, T, K; Specific Heat, C_p, Cal $g^{-1}K^{-1}$]

T	C_p
CURVE 1	
700	6.00×10^{-2}
922	6.70
1144	7.30
1366	7.90
1589	8.60
1811	9.20
CURVE 2	
475	6.484×10^{-2}
481	6.485*
537	6.496*
589	6.512
650	6.537*
711	6.569
740	6.587*
818	6.642
854	6.672*
929	6.741*
945	6.758
997	6.814
1134	6.987*
1155	7.016*
1194	7.072
1271	7.194*
1285	7.218
1354	7.337*
1365	7.358*
1394	7.412
1406	7.434*
1490	7.601*
1569	7.770
1614	7.870*
1689	8.048
1694	8.061*
1791	8.307*
1880	8.550
2091	9.185
2391	1.023×10^{-1}
2522	1.074
2697	1.147

* Not shown on plot

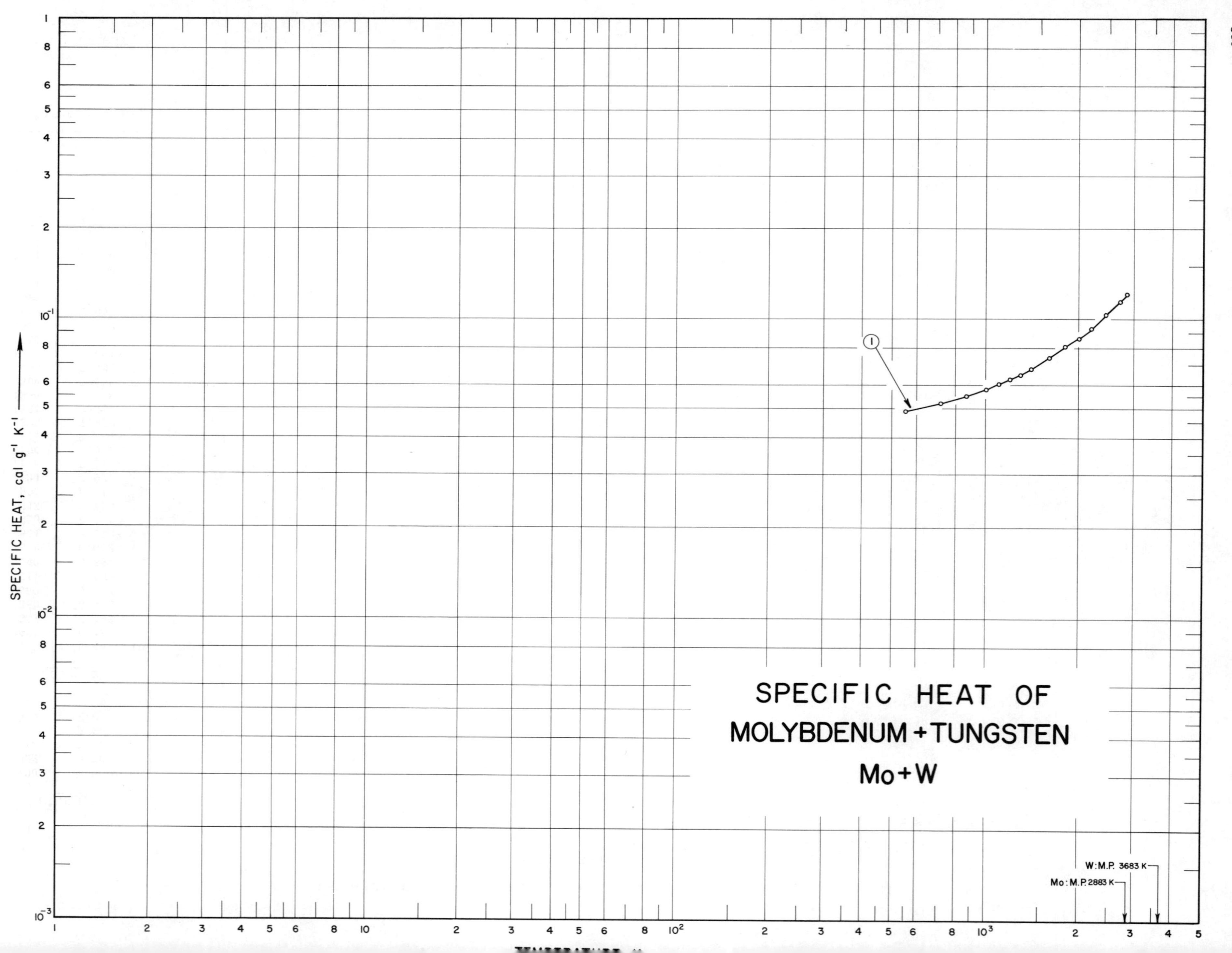
SPECIFIC HEAT OF
MOLYBDENUM + TUNGSTEN
Mo+W
W: M.P. 3683 K
Mo: M.P. 2883 K
1
SPECIFIC HEAT, cal g⁻¹ K⁻¹

SPECIFICATION TABLE NO. 98 SPECIFIC HEAT OF MOLYBDENUM + TUNGSTEN Mo + W

[For Data Reported in Figure and Table No. 98]

Curve No.	Ref. No.	Year	Temp. Range, K	Reported Error, %	Name and Specimen Designation	Composition (weight percent), Specifications and Remarks
1	232	1963	520-2855	±5		Bal. Mo; 29.83 W, 0.07 Zr, 0.012 C; density = 620 lb ft^{-3}.

DATA TABLE NO. 98 SPECIFIC HEAT OF MOLYBDENUM + TUNGSTEN Mo + W

[Temperature, T, K; Specific Heat, C_p, Cal g^{-1} K^{-1}]

T	C_p
CURVE 1	
553	4.924 x 10^{-2}
719	5.225
866	5.520
1001	5.814
1110	6.067
1198	6.282
1294	6.530
1350	6.679*
1410	6.842
1494	7.078*
1607	7.409
1703	7.701*
1819	8.073
1925	8.424
1988	8.640
2183	9.341
2305	9.804*
2436	1.032 x 10^{-1}
2591	1.096*
2716	1.149
2855	1.211

*Not shown on plot

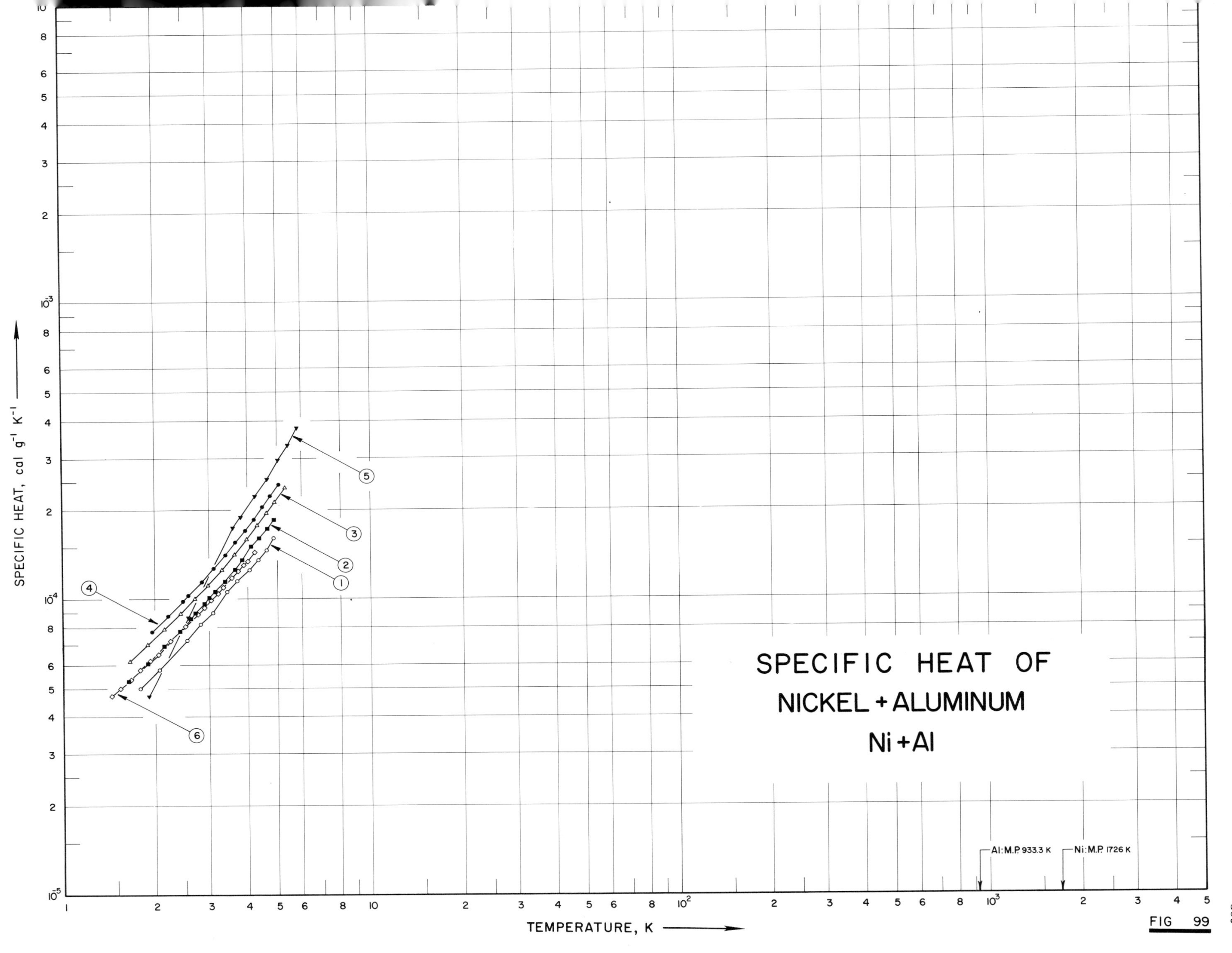
SPECIFIC HEAT OF
NICKEL + ALUMINUM
Ni + Al
Al: M.P. 933.3 K
Ni: M.P. 1726 K
TEMPERATURE, K
SPECIFIC HEAT, cal g⁻¹ K⁻¹
1
2
3
4
5
6
FIG 99

SPECIFICATION TABLE NO. 99 SPECIFIC HEAT OF NICKEL + ALUMINUM, Ni + Al (NiAl)

[For Data Reported in Figure and Table No. 99]

Curve No.	Ref. No.	Year	Temp. Range, K	Reported Error, %	Name and Specimen Designation	Composition (weight percent), Specifications and Remarks
1	394	1959	1.8-4.9	0.5		64.96 Ni, 35.04 Al; prepared from Aluminum Co. of America sample with 99.99 Al, 0.002 Cu, 0.002 Fe, 0.001 Si; and Vacuum Metals Corp sample with 99.90 Ni, 0.12 Co, 0.005 C, 0.001 N_2, 0.001 S, 0.001 P, 0.001 Mg, 0.005 Si, 0.005 Cu, 0.010 Fe, 0.001 Mn, 0.010 Ca, 0.010 Pb, 0.005 Al; alloys formed quenched to room temperature; annealed 3 days at 850-900 C under helium; then brought to room temperature from 6-8 hrs.
2	394	1959	1.6-4.9	0.5		67.64 Ni, 32.36 Al; same as above.
3	394	1959	1.6-5.3	0.5		69.37 Ni, 30.63 Al; same as above.
4	394	1959	1.9-5.0	0.5		71.05 Ni, 28.95 Al; same as above.
5	394	1959	1.9-5.8	0.5		73.47 Ni, 26.53 Al; same as above.
6	349	1962	1.4-4.2	≤2	Ni(92) Al(8)	95.20 Ni, 4.72 Al; annealed under He + 8% H_2 gas atmosphere at 1200 C for 72 hrs; etched with 30 ml HNO_3 and 20 ml CH_3COOH.

DATA TABLE NO. 99 SPECIFIC HEAT OF NICKEL + ALUMINUM Ni + Al

[Temperature, T, K; Specific Heat, C_p, Cal $g^{-1} K^{-1}$]

CURVE 1

T	C_p
1.791	5.000×10^{-5}
2.070	5.809
2.540	7.259
2.812	8.204
3.084	8.991
3.439	1.071×10^{-4}
3.704	1.152
4.053	1.255
4.358	1.361
4.633	1.470
4.896	1.618

CURVE 2

T	C_p
1.637	5.312×10^{-5}
1.881	6.060
2.138	6.920
2.401	7.757
2.618	8.552
2.692	8.954
2.882	9.615
3.006	1.013×10^{-4}
3.130	1.068
3.388	1.148
3.658	1.262
3.849	1.352
4.114	1.502
4.378	1.607
4.637	1.726
4.886	1.861

CURVE 3

T	C_p
1.664	6.209×10^{-5}
1.883	7.015
2.144	7.933
2.423	8.961
2.698	1.006×10^{-4}
2.978	1.122
3.300	1.261
3.631	1.420
3.979	1.595
4.313	1.776
4.629	1.952
4.930	2.137
5.304	2.391

CURVE 4

T	C_p
1.955	7.722×10^{-5}
2.202	8.788
2.466	9.834
2.567	1.027×10^{-4}
2.832	1.148
3.106	1.273
3.384	1.417
3.658	1.567
3.928	1.714
4.201	1.869
4.492	2.049
4.757	2.233
5.030	2.436

CURVE 5

T	C_p
1.898	4.711×10^{-5}
2.562	8.673
3.573	1.741×10^{-4}
3.808	1.897
4.249	2.225
4.657	2.559
5.048	2.950
5.428	3.323
5.824	3.786

CURVE 6

T	C_p
1.443	4.728×10^{-5}
1.486	4.847*
1.541	5.035
1.616	5.262*
1.668	5.384
1.795	5.790
1.925	6.205
2.037	6.523
2.136	6.891*
2.239	7.218
2.382	7.696*
2.511	8.091
2.632	8.549*
2.751	8.912
2.884	9.392
3.059	9.969
3.209	1.053×10^{-4}
3.314	1.094
3.432	1.131×10^{-4}*
3.567	1.179
3.723	1.247
3.882	1.302
4.009	1.349
4.119	1.397*
4.242	1.446

*Not shown on plot

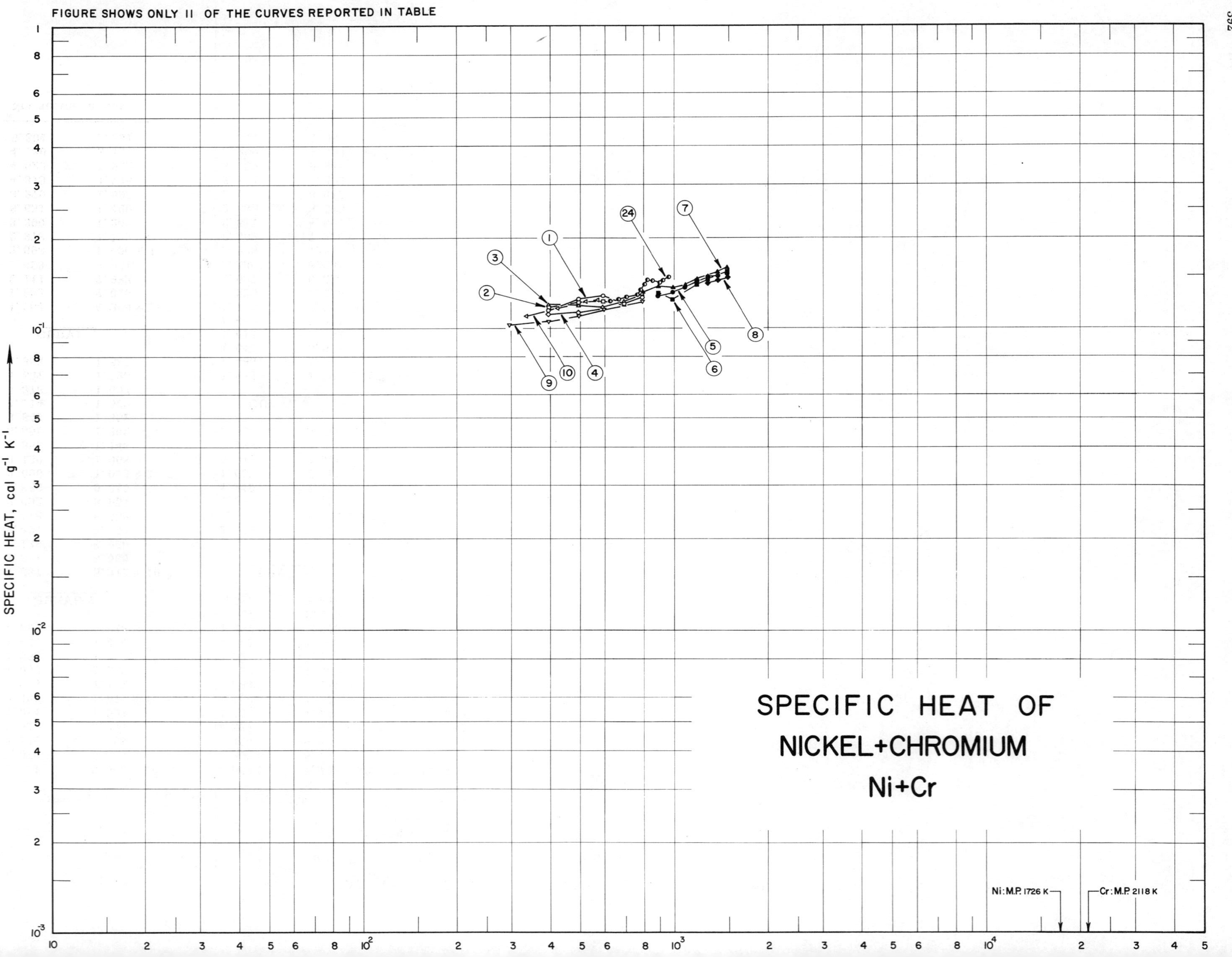
FIGURE SHOWS ONLY 11 OF THE CURVES REPORTED IN TABLE
SPECIFIC HEAT, cal g^{-1} K^{-1}
SPECIFIC HEAT OF
NICKEL+CHROMIUM
Ni+Cr
Ni: M.P. 1726 K
Cr: M.P. 2118 K
1
2
3
4
5
6
7
8
9
10
24

SPECIFICATION TABLE NO. 100 SPECIFIC HEAT OF NICKEL + CHROMIUM Ni + Cr

[For Data Reported in Figure and Table No. 100]

Curve No.	Ref. No.	Year	Temp. Range, K	Reported Error, %	Name and Specimen Designation	Composition (weight percent), Specifications and Remarks
1	233	1957	400-800		1.8 at. %	98.4 Ni, 1.60 Cr; prepared from 99.96 Ni, 0.002 S, 0.001 Cu, 0.003 Si, 0.03 Fe; and 99.94 Cr, 0.01 Fe, 0.001 Cu, 0.005 Mo, 0.04 Si, 0.009 Sn; alloys homogenized 2 wks at 1200 C.
2	233	1957	400-800		2.5 at. %	97.78 Ni, 2.22 Cr; same as above.
3	233	1957	400-800		4.4 at. %	96.10 Ni, 3.90 Cr; same as above.
4	233	1957	400-800		11 at. %	90.13 Ni, 9.87 Cr; same as above.
5	20	1957	800-1500	±0.3	$Ni_{0.9753}Cr_{0.0247}$	97.81 Ni, 2.19 Cr; argon atmosphere.
6	20	1959	800-1500	±0.3	$Ni_{0.9562}Cr_{0.0438}$	96.09 Ni, 3.91 Cr; argon atmosphere.
7	20	1959	800-1500	±0.3	$Ni_{0.8897}Cr_{0.1103}$	90.08 Ni, 9.92 Cr; argon atmosphere.
8	20	1959	800-1500	±0.3	$Ni_{0.9818}Cr_{0.0182}$	98.39 Ni, 1.61 Cr; argon atmosphere.
9	234	1963	298-1600		Chromel P	90.0 Ni, 10 Cr; sample supplied by Hoskins Mfg. Co.
10	395	1963	338-598			90.00 Ni, 10.00 Cr; specimen slowly cooled in calorimeter from about 400 C.
11	395	1963	368-618			Same as above; specimen slowly cooled from above 200 C in calorimeter.
12	395	1963	383-843			Same as above; specimen slowly cooled in calorimeter from above 350 C.
13	395	1963	428-898			Same as above; specimen slowly cooled in calorimeter from above 600 C; held at 120 C for approx 1 hr prior to data.
14	395	1963	528-968			Same as above; specimen slowly cooled in calorimeter from above 600 C.
15	395	1963	608-938			Same as above; specimen held at 300 C for 7.5 hrs prior to data.
16	395	1963	658-928			Same as above; specimen slowly cooled in calorimeter from above 600 C.
17	395	1963	598-968			Same as above; specimen slowly cooled in calorimeter from above 600 C.
18	395	1963	338-608			80.0 Ni, 20.0 Cr; specimen slowly cooled in calorimeter from about 200 C.
19	395	1963	398-658			Same as above; specimen slowly cooled in calorimeter from above 400 C.
20	395	1963	528-978			Same as above; specimen slowly cooled in calorimeter from above 640 C.
21	395	1963	578-978			Same as above; specimen slowly cooled in calorimeter from 700 C.
22	395	1963	628-968			Same as above; specimen slowly cooled in calorimeter from 620 C.
23	395	1963	473-953			Same as above; specimen slowly cooled in calorimeter from above 600 C.
24	395	1963	513-973			Same as above; specimen slowly cooled in calorimeter from above 600 C.
25	395	1963	358-668			95 Ni, 5 Cr; initial run of alloy homogenized at 1200 F for 1 wk prior to machining into specimen.
26	395	1963	338-618			Same as above; specimen slowly cooled in calorimeter from above 600 C.

SPECIFICATION TABLE NO. 100 (continued)

Curve No.	Ref. No.	Year	Temp. Range, K	Reported Error, %	Name and Specimen Designation	Composition (weight percent), Specifications and Remarks
27	395	1963	608-968			Same as above; specimen slowly cooled in calorimeter from above 600 C.
28	395	1963	578-848			Same as above; specimen slowly cooled in calorimeter from above 250 C.
29	395	1963	608-968			Same as above; specimen slowly cooled in calorimeter from above 575 C.

DATA TABLE NO. 100 SPECIFIC HEAT OF NICKEL + CHROMIUM Ni + Cr

[Temperature, T, K; Specific Heat, C_p, Cal g^{-1} K^{-1}]

T	C_p
CURVE 1	
400	1.16 x 10^{-1}
500	1.27
600	1.29
700	1.22
800	1.28
CURVE 2	
400	1.18 x 10^{-1}
500	1.23
600	1.23
700	1.21
800	1.25
CURVE 3	
400	1.21
500	1.20
600	1.18
700	1.23
800	1.32
CURVE 4	
400	1.13 x 10^{-1}
500	1.15
600	1.18
700	1.23
800	1.29
CURVE 5	
800	1.252 x 10^{-1}
900	1.281
1000	1.325
1100	1.378
1200	1.431
1300	1.476
1400	1.503
1500	1.522
CURVE 6	
800	1.283 x 10^{-1}*
900	1.312
1000	1.264
1100	1.384*
1200	1.413
1300	1.454
1400	1.506*
1500	1.559
CURVE 7	
800	1.338 x 10^{-1}
900	1.382
1000	1.377
1100	1.415
1200	1.460
1300	1.503
1400	1.555
1500	1.617
CURVE 8	
800	1.279 x 10^{-1}*
900	1.310*
1000	1.344*
1100	1.368*
1200	1.396*
1300	1.430
1400	1.457
1500	1.486
CURVE 9	
298.15	1.023 x 10^{-1}
300	1.024*
400	1.066
500	1.109
600	1.152
700	1.195*
800	1.237
900	1.280*
1000	1.323*
1100	1.366*
1200	1.408*
1300	1.451*
CURVE 9 (cont.)	
1400	1.490 x 10^{-1}*
1600	1.580*
CURVE 10	
338.15	1.107 x 10^{-1}
348.15	1.112*
358.15	1.132*
368.15	1.128*
378.15	1.136*
388.15	1.143*
398.15	1.150*
408.15	1.166*
418.15	1.155*
428.15	1.175
438.15	1.179*
448.15	1.187*
458.15	1.190*
468.15	1.193*
478.15	1.197*
488.15	1.204*
498.15	1.209*
508.15	1.212*
518.15	1.223*
528.15	1.228
538.15	1.242*
548.15	1.247*
558.15	1.240*
568.15	1.241*
578.15	1.252
588.15	1.255*
598.15	1.255*
CURVE 11*	
368.15	1.128 x 10^{-1}
378.15	1.130
388.15	1.133
398.15	1.148
408.15	1.165
418.15	1.158
428.15	1.168
438.15	1.187
448.15	1.180
458.15	1.192
CURVE 11 (cont.)*	
468.15	1.196 x 10^{-1}
478.15	1.201
488.15	1.206
498.15	1.209
508.15	1.216
518.15	1.224
528.15	1.218
538.15	1.244
548.15	1.235
558.15	1.236
568.15	1.249
578.15	1.255
588.15	1.252
598.15	1.261
608.15	1.257
618.15	1.264
CURVE 12*	
383.15	1.126 x 10^{-1}
403.15	1.138
423.15	1.139
443.15	1.157
463.15	1.166
483.15	1.179
503.15	1.163
523.15	1.200
543.15	1.210
563.15	1.222
583.15	1.230
603.15	1.241
623.15	1.248
643.15	1.260
663.15	1.261
683.15	1.275
703.15	1.287
723.15	1.296
743.15	1.308
763.15	1.324
783.15	1.360
803.15	1.406
823.15	1.391
843.15	1.395
CURVE 13*	
428.15	1.150 x 10^{-1}
438.15	1.155
448.15	1.168
458.15	1.167
468.15	1.173
478.15	1.176
488.15	1.185
498.15	1.188
508.15	1.192
518.15	1.205
528.15	1.201
538.15	1.211
548.15	1.216
558.15	1.225
568.15	1.228
578.15	1.228
588.15	1.240
598.15	1.241
608.15	1.250
618.15	1.249
628.15	1.256
638.15	1.262
648.15	1.268
658.15	1.270
668.15	1.273
678.15	1.275
688.15	1.287
698.51	1.286
708.15	1.295
718.15	1.300
728.15	1.297
538.15	1.308
748.15	1.320
758.15	1.326
768.15	1.342
778.15	1.354
788.15	1.380
798.15	1.397
808.15	1.401
818.15	1.403
828.15	1.395
838.15	1.406
848.15	1.403
858.15	1.411
868.15	1.403
CURVE 13 (cont.)*	
878.15	1.411 x 10^{-1}
888.15	1.421
898.15	1.412
CURVE 14*	
528.15	1.208 x 10^{-1}
538.15	1.213
548.15	1.221
558.15	1.231
568.15	1.229
578.15	1.235
588.15	1.239
598.15	1.247
608.15	1.251
618.15	1.253
628.15	1.257
638.15	1.264
648.15	1.270
658.15	1.272
668.15	1.277
678.15	1.280
688.15	1.286
698.15	1.293
708.15	1.293
718.15	1.297
728.15	1.295
738.15	1.309
748.15	1.313
758.15	1.319
768.15	1.328
778.15	1.353
788.15	1.362
798.15	1.384
808.15	1.382
818.15	1.380
828.15	1.374
838.15	1.392
848.15	1.394
858.15	1.393
868.15	1.398
878.15	1.398
888.15	1.397
898.15	1.393
908.15	1.405

*Not shown on plot

DATA TABLE NO. 100 (continued)

T	C_p
CURVE 14 (cont.)*	
918.15	1.407×10^{-1}
928.15	1.410
938.15	1.414
948.15	1.409
958.15	1.419
968.15	1.429
CURVE 15*	
608.15	1.237×10^{-1}
618.15	1.235
628.15	1.246
638.15	1.247
648.15	1.241
658.15	1.252
668.15	1.260
678.15	1.262
688.15	1.265
698.15	1.270
708.15	1.273
718.15	1.274
728.15	1.262
738.15	1.282
748.15	1.287
758.15	1.294
768.15	1.304
778.15	1.303
788.15	1.341
798.15	1.359
808.15	1.362
818.15	1.349
828.15	1.363
838.15	1.368
848.15	1.366
858.15	1.371
868.15	1.381
878.15	1.359
888.15	1.377
898.15	1.372
908.15	1.372
918.15	1.386
928.15	1.392
938.15	1.390
CURVE 16*	
658.15	1.247×10^{-1}
668.15	1.255
678.15	1.251
698.15	1.262
708.15	1.266
728.15	1.258
738.15	1.270
748.15	1.280
758.15	1.284
768.15	1.297
778.15	1.319
788.15	1.332
798.15	1.352
808.15	1.346
818.15	1.339
828.15	1.353
838.15	1.356
848.15	1.351
858.15	1.359
868.15	1.357
878.15	1.357
888.15	1.359
898.15	1.363
908.15	1.352
918.15	1.365
928.15	1.370
CURVE 17*	
598.15	1.237×10^{-1}
608.15	1.240
618.15	1.241
628.15	1.244
638.15	1.248
648.15	1.239
658.15	1.252
668.15	1.261
678.15	1.260
688.15	1.268
698.15	1.270
708.15	1.273
718.15	1.274
728.15	1.263
738.15	1.281
748.15	1.290
758.15	1.293
768.15	1.304
CURVE 17 (cont.)*	
778.15	1.326×10^{-1}
788.15	1.339
198.15	1.362
808.15	1.361
818.15	1.353
828.15	1.366
838.15	1.370
848.15	1.371
858.15	1.373
868.15	1.372
878.15	1.376
888.15	1.375
898.15	1.383
908.15	1.374
918.15	1.389
928.15	1.396
938.15	1.400
948.15	1.391
958.15	1.399
968.15	1.405
CURVE 18*	
338.15	1.126×10^{-1}
348.15	1.126
358.15	1.115
368.15	1.117
378.15	1.138
388.15	1.143
398.15	1.150
408.15	1.153
418.15	1.157
428.15	1.163
438.15	1.172
448.15	1.177
458.15	1.179
468.15	1.187
478.15	1.195
488.15	1.188
498.15	1.206
508.15	1.203
518.15	1.207
528.15	1.220
538.15	1.231
548.15	1.235
558.15	1.237
568.15	1.239
CURVE 18 (cont.)*	
578.15	1.244×10^{-1}
588.15	1.254
598.15	1.260
608.15	1.259
CURVE 19*	
398.15	1.147×10^{-1}
408.15	1.155
418.15	1.114
428.15	1.157
438.15	1.164
448.15	1.178
458.15	1.178
468.15	1.179
478.15	1.187
488.15	1.190
498.15	1.198
508.15	1.202
518.15	1.213
528.15	1.213
538.15	1.228
548.15	1.230
558.15	1.242
568.15	1.236
578.15	1.248
588.15	1.248
598.15	1.269
608.15	1.254
618.15	1.264
628.15	1.270
638.15	1.271
648.15	1.273
658.15	1.283
CURVE 20*	
528.15	1.205×10^{-1}
538.15	1.215
548.15	1.223
558.15	1.230
568.15	1.234
578.15	1.242
588.15	1.241
598.15	1.247
608.15	1.251
618.15	1.253
CURVE 20 (cont.)*	
628.15	1.261×10^{-1}
638.15	1.267
648.15	1.272
658.15	1.276
668.15	1.281
678.15	1.283
688.15	1.292
698.15	1.291
708.15	1.301
718.15	1.300
728.15	1.303
738.15	1.304
748.15	1.303
758.15	1.313
768.15	1.315
778.15	1.333
788.15	1.354
798.15	1.401
808.15	1.440
818.15	1.470
828.15	1.474
838.15	1.476
848.15	1.468
858.15	1.436
868.15	1.456
878.15	1.462
888.15	1.461
898.15	1.456
908.15	1.463
918.15	1.462
928.15	1.460
938.15	1.477
948.15	1.475
958.15	1.472
968.15	1.462
978.15	1.481
CURVE 21*	
578.15	1.241×10^{-1}
588.15	1.234
598.15	1.250
608.15	1.245
618.15	1.256
628.15	1.255
638.15	1.271
648.15	1.263
CURVE 21 (cont.)*	
658.15	1.267×10^{-1}
668.15	1.265
678.15	1.286
688.15	1.288
698.15	1.290
708.15	1.299
718.15	1.290
728.15	1.311
738.15	1.300
748.15	1.307
758.15	1.313
768.15	1.320
778.15	1.337
788.15	1.352
798.15	1.395
808.15	1.435
818.15	1.472
828.15	1.478
838.15	1.486
848.15	1.467
858.15	1.462
868.15	1.455
878.15	1.457
888.15	1.453
898.15	1.442
908.15	1.467
918.15	1.450
928.15	1.477
938.15	1.462
948.15	1.471
958.15	1.468
968.15	1.477
978.15	1.478
CURVE 22*	
628.15	1.254×10^{-1}
638.15	1.254
648.15	1.251
658.15	1.262
668.15	1.270
678.15	1.273
688.15	1.280
698.15	1.284
708.15	1.290
718.15	1.291
728.15	1.277

*Not shown on plot

DATA TABLE NO. 100 (continued)

T	C_p
CURVE 22 (cont.)*	
738.15	1.299 x 10^{-1}
748.15	1.297
758.15	1.305
768.15	1.317
778.15	1.340
788.15	1.369
798.15	1.418
808.15	1.452
818.15	1.469
828.15	1.492
838.15	1.490
848.15	1.479
858.15	1.477
868.15	1.478
878.15	1.479
888.15	1.481
898.15	1.483
908.15	1.469
918.15	1.477
928.15	1.488
938.15	1.485
948.15	1.480
958.15	1.482
968.15	1.485
CURVE 23*	
473.15	1.175 x 10^{-1}
493.15	1.183
513.15	1.194
533.15	1.198
553.15	1.216
573.15	1.229
593.15	1.237
613.15	1.245
633.15	1.252
653.15	1.251
673.15	1.272
713.15	1.288
733.15	1.288
753.15	1.304
773.15	1.315
793.15	1.358
813.15	1.434
833.15	1.483
853.15	1.472
873.15	1.470

T	C_p
CURVE 23 (cont.)*	
893.15	1.467 x 10^{-1}
913.15	1.469
933.15	1.496
953.15	1.510
CURVE 24	
513.15	1.192 x 10^{-1}*
533.15	1.190*
553.15	1.213*
573.15	1.223*
593.15	1.230*
613.15	1.239*
633.15	1.243
653.15	1.247*
673.15	1.262
693.15	1.264*
713.15	1.277
733.15	1.274*
753.15	1.280*
773.15	1.304
793.15	1.350*
813.15	1.424
833.15	1.470
853.15	1.462*
873.15	1.455
893.15	1.446*
913.15	1.434
933.15	1.468
953.15	1.477*
973.15	1.484
CURVE 25*	
358.15	1.129 x 10^{-1}
368.15	1.136
378.15	1.134
388.15	1.147
398.15	1.158
408.15	1.165
418.15	1.170
428.15	1.174
438.15	1.184
448.15	1.193
458.15	1.194
468.15	1.204
478.15	1.203

T	C_p
CURVE 25 (cont.)*	
488.15	1.206 x 10^{-1}
498.15	1.210
508.15	1.212
518.15	1.221
528.15	1.206
538.15	1.228
548.15	1.232
558.15	1.241
578.15	1.249
588.15	1.251
598.15	1.255
608.15	1.258
618.15	1.259
628.15	1.266
638.15	1.262
648.15	1.259
658.15	1.275
668.15	1.286
CURVE 26*	
338.15	1.138 x 10^{-1}
348.15	1.142
358.15	1.156
368.15	1.157
378.15	1.161
388.15	1.160
398.15	1.170
408.15	1.175
418.15	1.146
428.15	1.174
438.15	1.182
448.15	1.187
458.15	1.192
468.15	1.196
478.15	1.195
488.15	1.204
498.15	1.206
508.15	1.209
518.15	1.215
528.15	1.225
538.15	1.227
548.15	1.233
558.15	1.231
568.15	1.239
578.15	1.244
588.15	1.248

T	C_p
CURVE 26 (cont.)*	
598.15	1.251 x 10^{-1}
608.15	1.255
618.15	1.260
CURVE 27*	
608.15	1.247 x 10^{-1}
618.15	1.249
628.15	1.252
638.15	1.254
648.15	1.258
658.15	1.264
668.15	1.269
678.15	1.271
688.15	1.282
698.15	1.287
708.15	1.288
718.15	1.292
728.15	1.281
738.15	1.300
748.15	1.307
758.15	1.309
768.15	1.313
778.15	1.311
788.15	1.310
798.15	1.321
808.15	1.316
818.15	1.313
828.15	1.327
838.15	1.331
848.15	1.332
858.15	1.335
868.15	1.335
878.15	1.341
888.15	1.332
898.15	1.345
908.15	1.354
918.15	1.351
928.15	1.348
938.15	1.351
948.15	1.348
958.15	1.338
968.15	1.357

T	C_p
CURVE 28*	
578.15	1.219 x 10^{-1}
588.15	1.222
598.15	1.227
608.15	1.229
618.15	1.231
628.15	1.239
638.15	1.238
648.15	1.237
658.15	1.250
668.15	1.261
678.15	1.261
688.15	1.275
698.15	1.277
708.15	1.287
718.15	1.286
728.15	1.272
738.15	1.292
748.15	1.295
758.15	1.300
768.15	1.304
778.15	1.310
788.15	1.309
798.15	1.317
808.15	1.314
818.15	1.314
828.15	1.328
838.15	1.335
848.15	1.332
CURVE 29*	
608.15	1.226 x 10^{-1}
618.15	1.229
628.15	1.233
638.15	1.231
648.15	1.230
658.15	1.244
668.15	1.255
678.15	1.256
688.15	1.271
698.15	1.273
708.15	1.281
718.15	1.282
728.15	1.269
738.15	1.284
748.15	1.292
758.15	1.297

T	C_p
CURVE 29 (cont.)*	
768.15	1.297 x 10^{-1}
778.15	1.305
788.15	1.305
798.15	1.312
808.15	1.309
818.15	1.302
828.15	1.319
838.15	1.325
848.15	1.326
858.15	1.330
868.15	1.334
878.15	1.333
888.15	1.334
898.15	1.341
908.15	1.330
918.15	1.342
928.15	1.350
938.15	1.348
948.15	1.343
958.15	1.350
968.15	1.351

*Not shown on plot

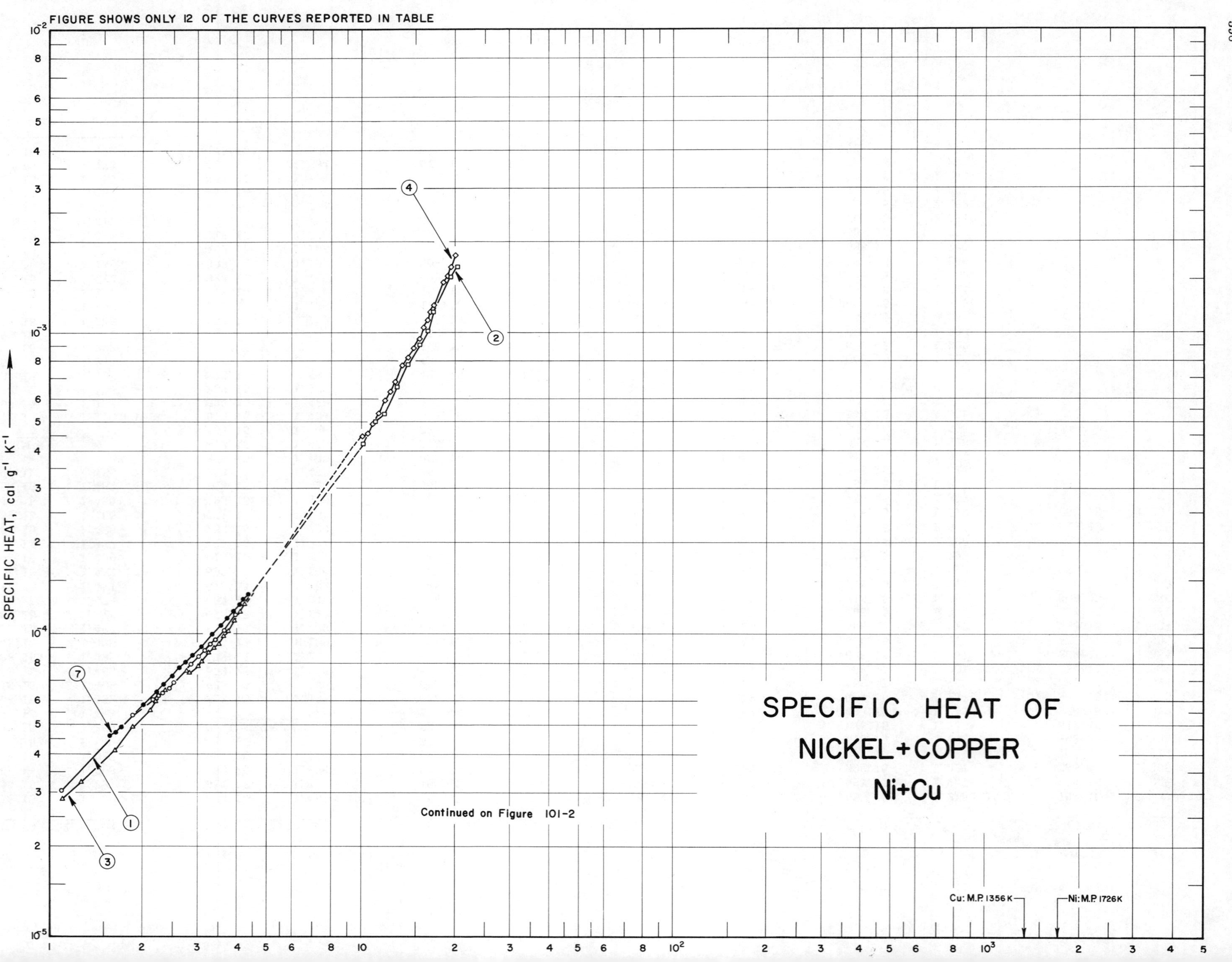
FIGURE SHOWS ONLY 12 OF THE CURVES REPORTED IN TABLE
SPECIFIC HEAT, cal g⁻¹ K⁻¹
SPECIFIC HEAT OF
NICKEL+COPPER
Ni+Cu
Continued on Figure 101-2
Cu: M.P. 1356 K
Ni: M.P. 1726 K
1
2
3
4
7

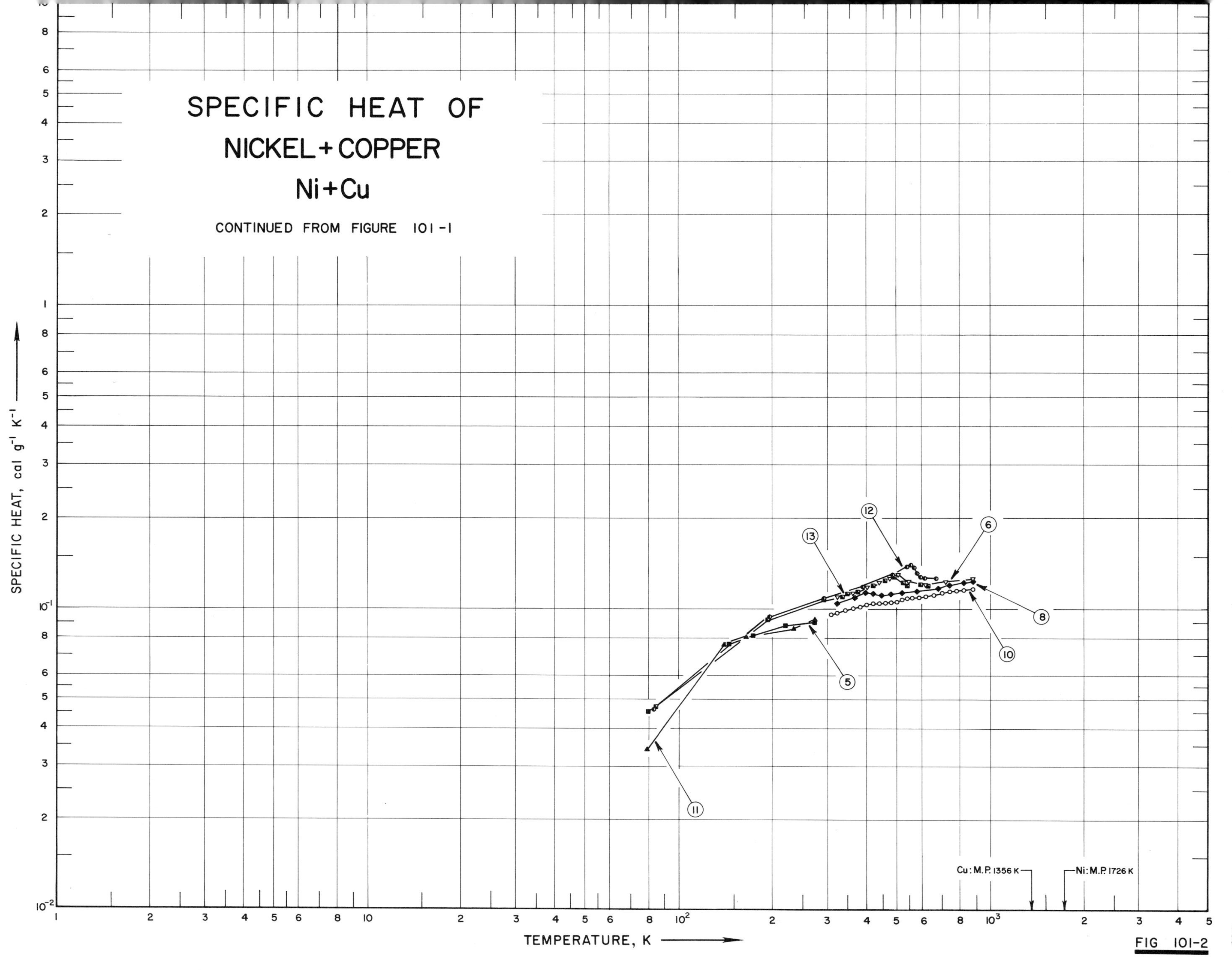
SPECIFIC HEAT OF
NICKEL + COPPER
Ni + Cu
CONTINUED FROM FIGURE 101-1
SPECIFIC HEAT, cal g⁻¹ K⁻¹
TEMPERATURE, K
Cu: M.P. 1356 K
Ni: M.P. 1726 K
FIG 101-2

SPECIFICATION TABLE NO. 101 SPECIFIC HEAT OF NICKEL + COPPER Ni + Cu

[For Data Reported in Figure and Table No. 101]

Curve No.	Ref. No.	Year	Temp. Range, K	Reported Error, %	Name and Specimen Designation	Composition (weight percent), Specifications and Remarks
1	387	1940	1. 1-4. 0		$Cu_{20}Ni_{80}$	80. 39 Ni, 19. 61 Cu; prepared from 99. 98 Ni and 99. 99 Cu; small amounts of Al added to alloy melts as deoxidizers; melted; held at 200 C below melting temperature for 1 to 2 hrs and slowly cooled.
2	387	1940	10-20		Same as above	Same as above.
3	387	1940	1. 1-4. 2		$Cu_{40}Ni_{60}$	60. 09 Ni, 39. 91 Cu; prepared from 99. 99 Cu and 99. 98 Ni; small amounts of Al added to alloy melts as deoxidizers; melted; held at 200 C below melting temperature for 1 to 2 hrs and slowly cooled.
4	387	1940	10-20		Same as above	Same as above.
5	293	1941	79-273			33. 6 Ni, 66. 4 Cu.
6	18	1956	323-883	±0. 5	90% Nickel alloy	90. 05 Ni, bal. Cu.
7	349	1962	1. 6-4. 4	≤2	Ni(90) Cu(10)	89. 25 Ni, 10. 66 Cu; annealed under vacuum at 1100 C for 72 hrs; etched in 30 ml HNO_3 and 20 ml CH_3COOH.
8	18	1956	323-883	±0. 5	75% Nickel alloy	75. 07 Ni, bal. Cu.
9	349	1962	1. 5-4. 3	≤2	Ni(55) Cu(45)	53. 01 Ni, 46. 92 Cu; annealed under vacuum at 1100 C for 72 hrs, etched with 30 ml HNO_3 and 20 ml CH_3COOH.
10	18	1956	308-863	±0. 5	50% Nickel alloy	50. 04 Ni, bal. Cu.
11	293	1941	79-273			50. 4 Ni, 49. 6 Cu.
12	343	1934	83-670	1. 5-2	Alloy 1	94. 0 Ni.
13	343	1934	84-634	1. 5-2	Alloy 2	87. 2 Ni.
14	343	1934	81-660	1. 5-2	Alloy 3	78. 8 Ni.

DATA TABLE NO. 101 SPECIFIC HEAT OF NICKEL + COPPER Ni + Cu

[Temperature, T, K; Specific Heat, C_p, Cal $g^{-1} K^{-1}$]

T	C_p
CURVE 1	
1.109	3.037 x 10^{-5}
1.109	3.007*
1.860	5.353
2.170	6.024
2.200	6.091*
2.220	6.158*
2.250	6.225
2.260	6.242*
2.280	6.275*
2.320	6.359
2.360	6.477*
2.360	6.460*
2.430	6.594
2.490	6.728*
2.520	6.896
2.580	7.047*
2.860	7.903
3.020	8.373
3.170	8.792
3.280	9.245
3.420	9.547
3.630	1.025 x 10^{-4}
3.650	1.027*
3.940	1.159*
3.970	1.164
CURVE 2	
10.12	4.25 x 10^{-4}
11.12	5.00
11.79	5.34
13.05	6.58
14.22	7.79
15.41	9.06
16.36	1.02 x 10^{-3}
17.07	1.18
19.36	1.53
19.84	1.58*
20.36	1.67
CURVE 3	
1.117	2.866 x 10^{-5}
1.275	3.245
1.635	4.11
1.864	4.92
2.12	5.60
CURVE 3 (cont.)	
2.20	5.98 x 10^{-5}
2.36	6.49
2.40	6.52*
2.44	6.57*
2.76	7.53
2.80	7.58*
2.82	7.43
3.00	7.85
3.08	8.11
3.25	8.65
3.38	8.99
3.50	9.23
3.64	9.83
3.76	1.02 x 10^{-4}
3.92	1.11
4.03	1.15*
4.12	1.18
4.24	1.26
CURVE 4	
10.15	4.48 x 10^{-4}
10.52	4.56
10.88	4.94
11.38	5.37
11.96	5.95
12.40	6.34
12.84	6.84
13.63	7.75
14.21	8.24
14.78	8.82
15.41	9.55
15.78	1.05 x 10^{-3}
16.29	1.12
16.68	1.18
17.18	1.23
18.43	1.48
18.94	1.55
19.49	1.65
20.14	1.81
CURVE 5	
79.85	4.53 x 10^{-2}
145.61	7.60
173.55	8.15
CURVE 5 (cont.)	
220.10	8.80 x 10^{-2}
273.20	9.03
CURVE 6	
323.15	1.100 x 10^{-1}
343.15	1.112*
363.15	1.128
383.15	1.154*
403.15	1.181
423.15	1.210*
443.15	1.224
463.15	1.250*
478.15	1.270
488.15	1.276*
498.15	1.298*
508.15	1.311
518.15	1.317*
528.15	1.316*
538.15	1.284*
548.15	1.245
558.15	1.229*
568.15	1.225*
578.15	1.219*
588.15	1.217*
598.15	1.213*
608.15	1.214*
623.15	1.212
643.15	1.217*
663.15	1.218*
683.15	1.219*
703.15	1.224*
723.15	1.230
743.15	1.234*
763.15	1.239*
783.15	1.242*
803.15	1.251*
823.15	1.251*
843.15	1.255*
863.15	1.258*
883.15	1.261
CURVE 7	
1.576	4.58 x 10^{-5}
1.638	4.73
CURVE 7 (cont.)	
1.711	4.94 x 10^{-5}
1.802	5.22*
1.894	5.47*
2.017	5.84
2.130	6.16*
2.221	6.40
2.332	6.80
2.479	7.26
2.623	7.70
2.747	8.05
2.889	8.48
3.080	9.02
3.347	9.84
3.572	1.07 x 10^{-4}
3.721	1.13
3.912	1.19
4.078	1.25
4.209	1.31
4.351	1.35
CURVE 8	
323.15	1.045 x 10^{-1}
338.15	1.061*
343.15	1.065*
348.15	1.075*
358.15	1.088*
363.15	1.089*
368.15	1.091
378.15	1.105*
383.15	1.113*
388.15	1.115*
398.15	1.130
403.15	1.124*
408.15	1.126*
418.15	1.129*
423.15	1.126
428.15	1.126*
438.15	1.122*
443.15	1.122*
448.15	1.119
458.15	1.122*
463.15	1.123*
468.15	1.122*
478.15	1.124*
483.15	1.124
CURVE 8 (cont.)	
503.15	1.133 x 10^{-1}*
523.15	1.139
543.15	1.141*
563.15	1.145*
583.15	1.150
603.15	1.153*
623.15	1.155*
643.15	1.166*
663.15	1.173*
683.15	1.176
703.15	1.187*
723.15	1.195*
743.15	1.203
763.15	1.211*
783.15	1.217*
803.15	1.223*
823.15	1.226
843.15	1.231*
863.15	1.237*
883.15	1.243
CURVE 9*	
1.464	4.25 x 10^{-5}
1.515	4.40
1.580	4.55
1.670	4.82
1.776	5.12
1.879	5.41
2.004	5.76
2.132	6.08
2.239	6.39
2.365	6.73
2.504	7.18
2.635	7.62
2.756	8.01
2.896	8.42
3.068	8.97
3.208	9.31
3.306	9.58
3.418	9.97
3.560	1.04 x 10^{-4}
3.736	1.10
3.911	1.16
4.042	1.21
4.158	1.25
4.295	1.31
CURVE 10	
308.15	9.590 x 10^{-2}
323.15	9.700
343.15	9.930
363.15	1.008 x 10^{-1}
383.15	1.022
403.15	1.036
423.15	1.042
443.15	1.048
463.15	1.054
483.15	1.059
503.15	1.068
523.15	1.076
543.15	1.081
563.15	1.086
583.15	1.091
630.15	1.096*
623.15	1.103
643.15	1.113*
663.15	1.122
683.15	1.128*
703.15	1.136
723.15	1.143*
743.15	1.149
763.15	1.153*
783.15	1.156
803.15	1.160*
823.15	1.163
843.15	1.167*
863.15	1.170*
883.15	1.173
CURVE 11	
79.33	3.39 x 10^{-2}
140.56	7.55
156.80	8.01
230.30	8.64
273.20	9.19
CURVE 12	
83.15	4.600 x 10^{-2}
196.55	9.400
293.55	1.080 x 10^{-1}
296.55	1.081*
325.55	1.115*

* Not shown on plot

DATA TABLE NO. 101 (continued)

T	C_p
CURVE 12 (cont.)	
391.45	1.186×10^{-1}
486.55	1.318
519.55	1.364*
541.55	1.391
556.45	1.416
561.75	1.416*
567.05	1.413*
573.35	1.388
582.15	1.322
592.75	1.293*
596.25	1.282
618.45	1.278
655.35	1.279*
670.05	1.276
CURVE 13	
84.85	4.700×10^{-2}
194.55	9.200
293.85	1.075×10^{-1}
300.45	1.075*
336.95	1.107
349.95	1.128
376.15	1.148
421.25	1.202
422.05	1.205*
460.95	1.256
478.95	1.286*
493.65	1.289
505.35	1.284*
526.25	1.228
543.85	1.205
565.05	1.208*
599.85	1.212
633.85	1.206
CURVE 14*	
81.25	4.600×10^{-2}
197.15	9.400
294.55	1.061×10^{-1}
297.45	1.056
321.05	1.088
354.55	1.130
390.65	1.167
396.85	1.172
407.65	1.171
CURVE 14 (cont.)*	
414.75	1.178×10^{-1}
422.65	1.174
444.65	1.174
470.25	1.155
518.45	1.166
587.05	1.182
628.95	1.202
659.95	1.192

*Not shown on plot

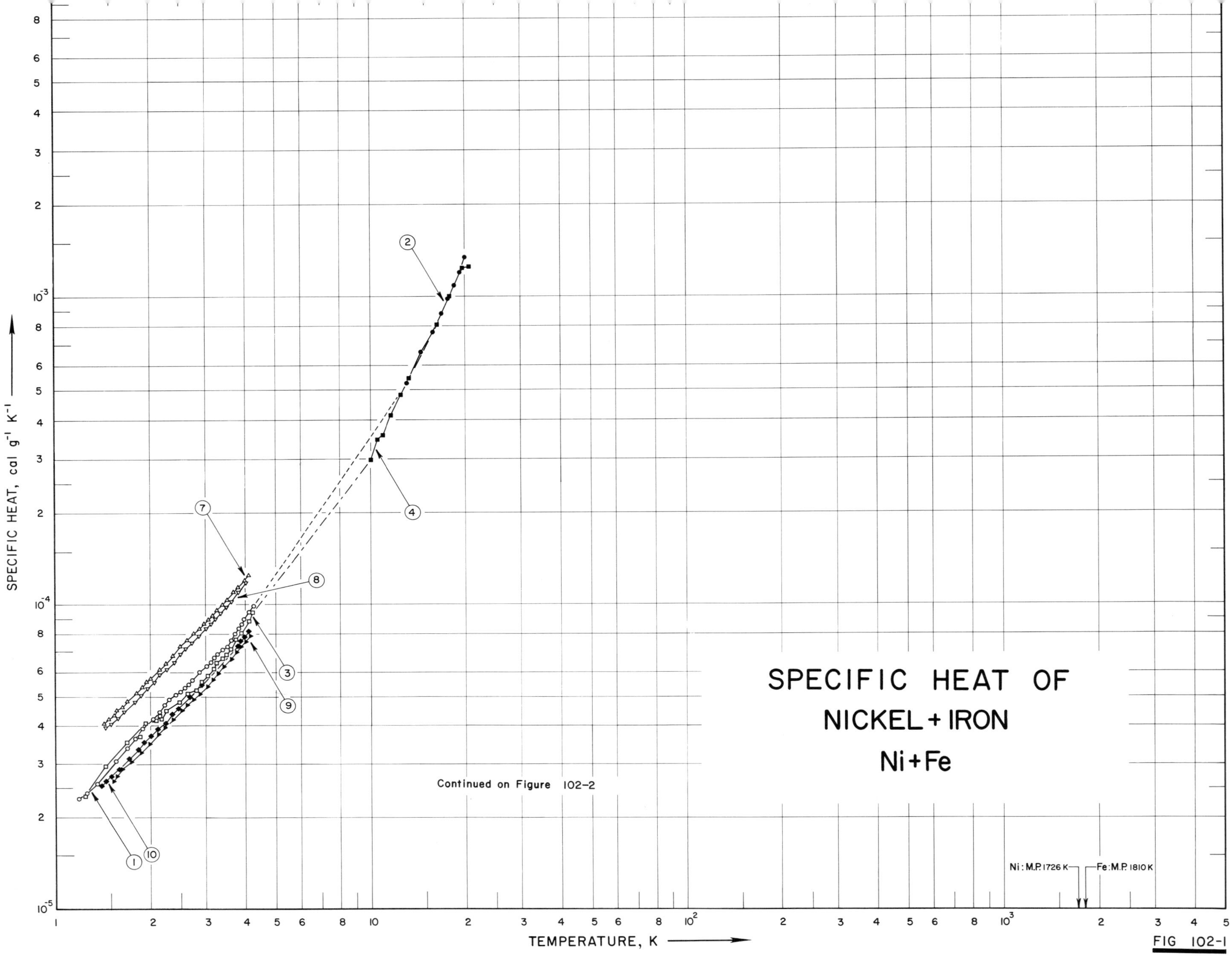

SPECIFIC HEAT OF NICKEL + IRON Ni+Fe

SPECIFIC HEAT, cal g^{-1} K^{-1}

TEMPERATURE, K

Ni: M.P. 1726 K

Fe: M.P. 1810 K

Continued on Figure 102-2

FIG 102-1

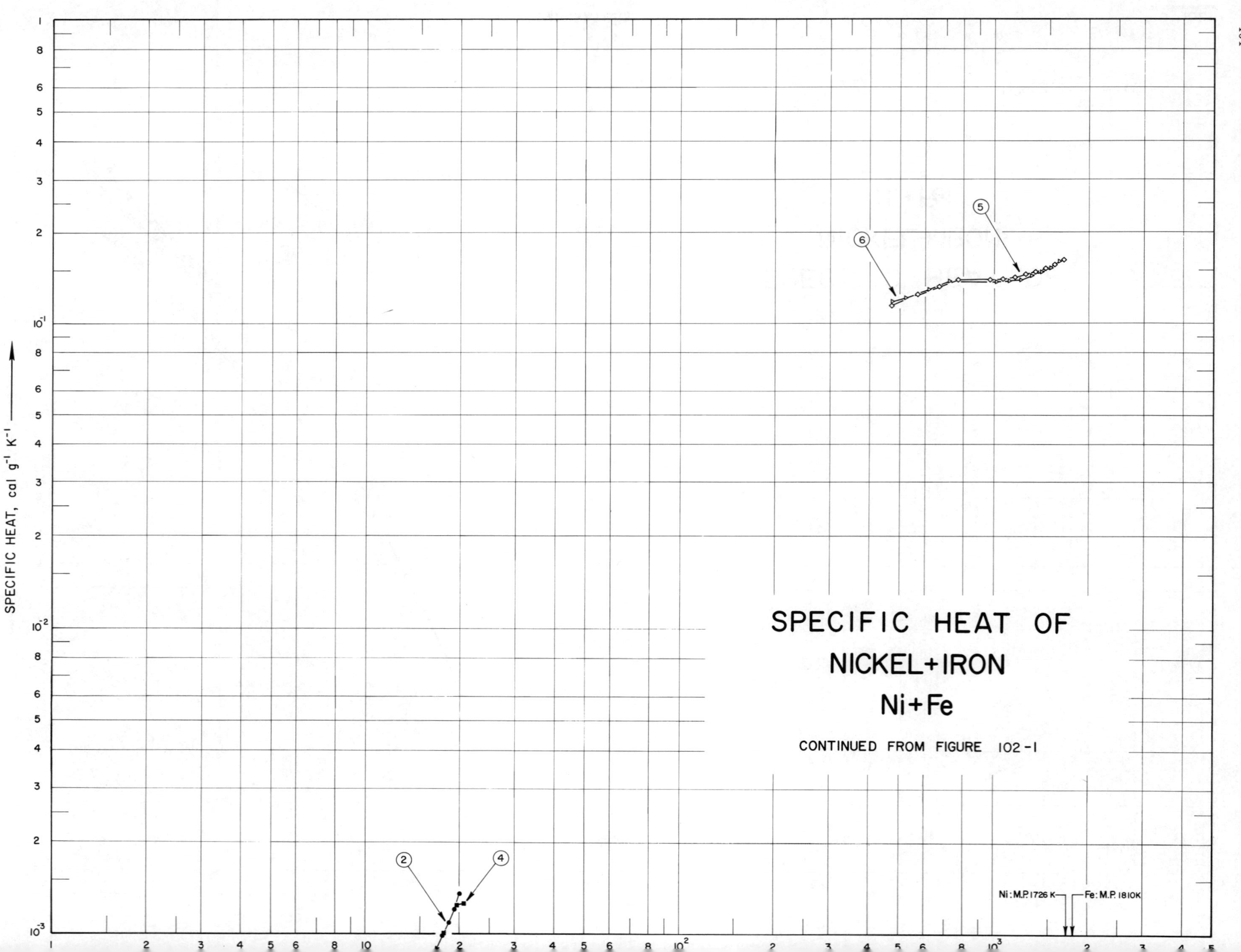
SPECIFIC HEAT OF
NICKEL+IRON
Ni+Fe
CONTINUED FROM FIGURE 102-1
SPECIFIC HEAT, cal g-1 K-1
Ni: M.P. 1726 K
Fe: M.P. 1810K
6
5
2
4

SPECIFICATION TABLE NO. 102 SPECIFIC HEAT OF NICKEL + IRON Ni + Fe

[For Data Reported in Figure and Table No. 102]

Curve No.	Ref. No.	Year	Temp. Range, K	Reported Error, %	Name and Specimen Designation	Composition (weight percent), Specifications and Remarks
1	387	1940	1.2-4.3		$Fe_{15}Ni_{85}$	85.11 Ni, 14.89 Fe; prepared from 99.98 Ni, >99.98 Fe, 0.001 C, <0.01 Si, <0.01 P, <0.01 Mn; sintered; prepared by Molybdenum Co. of Reutte, Germany.
2	387	1940	13-20		Same as above	Same as above.
3	387	1940	1.2-4.2		$Fe_{20}Ni_{80}$	80.37 Ni, 19.63 Fe; prepared from 99.98 Ni, >99.98 Fe, 0.001 C, <0.01 Si, <0.01 P, <0.01 Mn; sintered; prepared by Molybdenum Co. of Reutte, Germany.
4	387	1940	10-20		Same as above	Same as above.
5	236	1940	473-1673			79.3 Ni, 20.7 Fe.
6	236	1940	473-1673			69.76 Ni, 30.24 Fe.
7	349	1962	1.4-4.1	≤2	Ni(97.5) Fe(2.5)	97.5 Ni, 2.38 Fe; annealed under He + 8% H_2 gas atmosphere at 1100 C for 72 hrs; etched with 30 ml HNO_3 and 20 ml CH_3COOH.
8	349	1962	1.4-4.0	≤2	Ni(95) Fe(5)	95.15 Ni, 4.75 Fe; same as above.
9	349	1962	1.5-4.2	≤2	Ni(68) Fe(32)	68.59 Ni, 31.35 Fe; same as above.
10	349	1962	1.4-4.1	≤2	Ni(55) Fe(45)	55.72 Ni, 44.16 Fe; same as above.

DATA TABLE NO. 102 SPECIFIC HEAT OF NICKEL + IRON Ni + Fe

[Temperature, T, K; Specific Heat, C_p, Cal $g^{-1} K^{-1}$]

T	C_p
CURVE 1	
1.177	2.320 x 10^{-5}
1.261	2.405
1.363	2.580
1.565	3.065
1.685	3.385
1.787	3.64
1.887	3.91
2.03	4.22
2.05	4.34*
2.14	4.46
2.14	4.48*
2.22	4.72
2.25	4.77*
2.28	4.94
2.34	4.99*
2.34	4.89*
2.41	5.08
2.43	5.06*
2.49	5.20
2.54	5.29*
2.57	5.37
2.64	5.49
2.71	5.68
2.72	5.65*
2.86	6.04
2.94	6.16*
3.02	6.30
3.11	6.49
3.18	6.71
3.26	6.87
3.39	7.11
3.50	7.29
3.61	7.64
3.72	8.05
3.81	8.39
3.88	8.65
3.96	8.99
4.04	9.13*
4.12	9.49
4.19	9.65*
4.26	9.92
CURVE 2	
13.13	5.29 x 10^{-4}
14.52	6.66
15.82	7.74
16.85	8.87
17.71	9.89
18.51	1.10 x 10^{-3}
19.26	1.21
20.06	1.36
CURVE 3	
1.246	2.363 x 10^{-5}
1.446	2.961
1.691	3.54
1.858	3.70
1.928	4.09
2.03	4.18*
2.09	4.18
2.15	4.33
2.16	4.23
2.24	4.50
2.29	4.68*
2.47	4.85
2.48	4.86*
2.62	5.16
2.78	5.28
2.91	5.60
2.93	5.67*
3.05	5.88
3.18	6.19
3.25	6.48
3.30	6.45*
3.38	6.67
3.48	6.86
3.61	7.17
3.73	7.60
3.88	8.06
3.95	8.28*
4.14	8.85
4.21	9.11*
4.22	9.45
CURVE 4	
10.00	2.98 x 10^{-4}
10.64	3.49
CURVE 4 (cont.)	
10.98	3.59 x 10^{-4}
11.72	4.16
12.56	4.83
13.35	5.47
16.31	8.15
17.86	1.00 x 10^{-3}
19.74	1.24
20.71	1.26
CURVE 5	
473	1.167 x 10^{-1}
523	1.210*
573	1.252
623	1.291*
673	1.329
723	1.365*
773	1.400
973	1.402
1023	1.408*
1073	1.416
1123	1.425*
1173	1.437
1223	1.450*
1273	1.464
1323	1.480*
1373	1.498
1423	1.518*
1473	1.539
1523	1.562*
1573	1.587
1623	1.612*
1673	1.640
CURVE 6	
473	1.191 x 10^{-1}
523	1.226
573	1.263*
623	1.303
673	1.345*
723	1.390
773	1.437*
1023	1.388
1073	1.391*
1123	1.398
CURVE 6 (cont.)	
1173	1.407 x 10^{-1}*
1223	1.419
1273	1.435*
1323	1.453
1373	1.474*
1423	1.499
1473	1.526*
1523	1.556
1573	1.589*
1623	1.625
1673	1.665*
CURVE 7	
1.423	4.07 x 10^{-5}
1.475	4.22
1.530	4.35
1.568	4.51
1.595	4.57*
1.632	4.65
1.681	4.86
1.736	4.97*
1.802	5.16
1.877	5.40
1.935	5.62
1.948	5.61*
2.007	5.76
2.146	6.18
2.239	6.46
2.355	6.84
2.489	7.29
2.617	7.67
2.730	8.06
2.862	8.39
2.968	8.68
3.055	8.97
3.151	9.28
3.259	9.65
3.379	1.00 x 10^{-4}
3.514	1.05
3.668	1.10
3.784	1.14
3.870	1.17*
3.981	1.21
4.108	1.26
CURVE 8	
1.436	3.96 x 10^{-5}
1.475	3.99*
1.515	4.05
1.576	4.26
1.657	4.46
1.720	4.68*
1.781	4.84
1.862	5.07
1.967	5.36
2.060	5.60
2.142	5.93
2.255	6.15
2.388	6.52
2.494	6.89
2.588	7.18
2.702	7.49
2.851	7.90
3.010	8.39
3.122	8.69
3.221	8.97
3.352	9.42
3.491	9.83
3.623	1.03 x 10^{-4}
3.716	1.06*
3.806	1.10
3.918	1.12*
4.028	1.18
CURVE 9	
1.533	2.64 x 10^{-5}
1.565	2.75
1.625	2.89
1.741	3.06
1.868	3.27
1.998	3.50
2.114	3.76
2.222	3.94
2.371	4.20
2.525	4.51
2.617	4.71
2.741	4.95
2.876	5.14
3.035	5.45
3.170	5.76
3.281	5.97
CURVE 9 (cont.)	
3.427	6.29 x 10^{-5}
3.603	6.65
3.754	6.98
3.874	7.28
4.012	7.56
4.159	7.95
CURVE 10	
1.398	2.55 x 10^{-5}
1.449	2.64
1.512	2.74
1.594	2.90
1.711	3.13
1.831	3.36
1.915	3.55
1.926	3.54*
2.011	3.72
2.114	3.91
2.228	4.13
2.359	4.39
2.468	4.60
2.572	4.80*
2.674	5.04
2.806	5.28*
2.912	5.53
2.992	5.62*
3.096	5.91*
3.206	6.20*
3.340	6.40*
3.479	6.73*
3.653	7.12*
3.782	7.34
3.871	7.64
3.987	7.86
4.110	8.25

* Not shown on plot

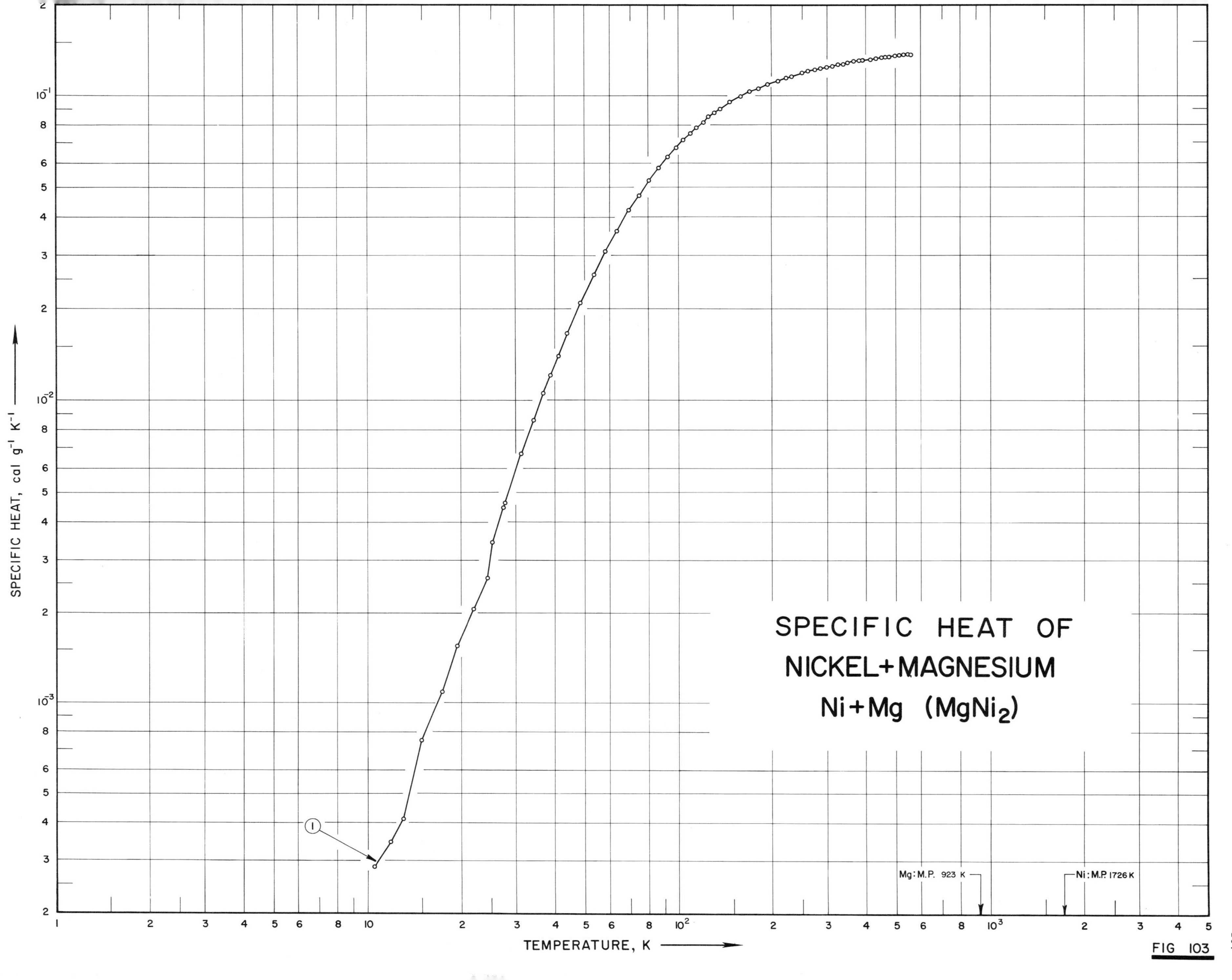

SPECIFIC HEAT OF
NICKEL+MAGNESIUM
Ni+Mg (MgNi$_2$)
SPECIFIC HEAT, cal g^{-1} K^{-1}
TEMPERATURE, K
Mg: M.P. 923 K
Ni: M.P. 1726 K
1

FIG 103

SPECIFICATION TABLE NO. 103 SPECIFIC HEAT OF NICKEL + MAGNESIUM, Ni + Mg ($MgNi_2$)

[For Data Reported in Figure and Table No. 103]

Curve No.	Ref. No.	Year	Temp. Range, K	Reported Error, %	Name and Specimen Designation	Composition (weight percent), Specifications and Remarks
1	396	1960	10-545	0.1-1.0	$MgNi_2$	~99.92 $MgNi_2$, ~0.01 Fe, ~0.01 Si, ~0.01 Ca, ~0.05 (Al + Cr + Mo); prepared from high purity elements by fusion at 1200 C (composition of Ni: 99.9 Ni, 0.065 O, 0.014 C, 0.0018 Co; composition of Mg: 99.95 Mg).

DATA TABLE NO. 103 SPECIFIC HEAT OF NICKEL + MAGNESIUM, Ni + Mg ($MgNi_2$)

[Temperature, T, K; Specific Heat, C_p, Cal g^{-1} K^{-1}]

CURVE 1

Series 1

T	C_p
10.55	2.843 x 10^{-4}
11.77	3.436
13.10	4.128
14.90	7.486
17.32	1.089 x 10^{-3}
19.46	1.542
21.87	2.059
24.24	2.613
25.11	3.427
27.29	4.462
27.63	4.647
31.19	6.689
34.21	8.594
36.68	1.050 x 10^{-2}
38.72	1.209
41.02	1.399
43.93	1.664
48.39	2.083
53.54	2.599
58.29	3.079
63.44	3.616
69.46	4.211
74.66	4.726
80.43	5.271
86.16	5.797
92.40	6.320
98.30	6.768
103.93	7.171
109.40	7.544
114.73	7.882
120.03	8.186
125.62	8.503
131.34	8.791
136.90	9.057
142.39	9.296*
147.85	9.521
153.25	9.669
158.65	9.926
164.09	1.011 x 10^{-1}*
169.64	1.029
175.31	1.045*
181.18	1.061
187.21	1.076*
193.22	1.090

CURVE 1 (cont.)

Series 1 (cont.)

T	C_p
199.29	1.103 x 10^{-1}*
209.70	1.123
216.07	1.135*
217.80	1.138*
222.33	1.146
224.02	1.149*
230.27	1.159
231.53	1.170*
242.82	1.178*
249.09	1.187
255.35	1.195*
261.61	1.201
267.90	1.209*
274.21	1.214
280.54	1.222*
286.90	1.228
293.29	1.234*
299.83	1.240
306.18	1.245*
312.50	1.249
318.81	1.255*

Series 2

T	C_p
306.92	1.250 x 10^{-1}*
308.74	1.246*
312.83	1.253*
314.67	1.254*
320.56	1.259*
326.45	1.264
332.33	1.269*
338.18	1.271
344.02	1.278*
349.84	1.281
360.54	1.287*
366.34	1.291
372.12	1.295*
374.73	1.295*
380.51	1.300
386.27	1.302*
392.02	1.307
403.51	1.315*
409.25	1.318*
414.97	1.319

CURVE 1 (cont.)

Series 2 (cont.)

T	C_p
420.68	1.320 x 10^{-1}*
426.38	1.324*
432.08	1.327
437.76	1.330*
443.46	1.332*
449.15	1.333
454.83	1.336*
456.10	1.336*
461.78	1.338
467.43	1.366*
473.07	1.343*
478.76	1.345
484.44	1.348*
490.11	1.350*
495.78	1.353
501.44	1.355*
507.12	1.358*
512.79	1.365
518.45	1.360*
524.11	1.357*
529.77	1.369
535.44	1.364*
541.11	1.365*
546.75	1.371
552.36	1.370*
558.03	1.365

Series 3*

T	C_p
310.68	1.250 x 10^{-1}
319.55	1.257
328.37	1.264
461.76	1.339
465.92	1.419
469.63	1.340
511.63	1.360
517.22	1.365
522.83	1.365
528.42	1.365
534.01	1.367
539.59	1.371
545.17	1.368

* Not shown on plot

FIGURE SHOWS ONLY 3 OF THE CURVES REPORTED IN TABLE
SPECIFIC HEAT, cal g⁻¹ K⁻¹
SPECIFIC HEAT OF
NICKEL + MANGANESE
Ni+Mn
Mn: M.P. 1517 K
Ni: M.P. 1726 K
1
2
3

SPECIFICATION TABLE NO. 104 SPECIFIC HEAT OF NICKEL + MANGANESE Ni + Mn

[For Data Reported in Figure and Table No. 104]

Curve No.	Ref. No.	Year	Temp. Range, K	Reported Error, %	Name and Specimen Designation	Composition (weight percent), Specifications and Remarks
1	349	1962	1.4-4.1	≤2	Ni(80) Mn(20)	80 Ni, 20 Mn; annealed under H + 8% H_2 gas atmosphere at 1100 C for 72 hrs; etched with 30 ml HNO_3 and 20 ml CH_3COOH.
2	349	1962	1.4-4.1	≤2	Ni(70) Mn(30)	70 Ni, 30 Mn; same as above.
3	349	1962	1.4-4.1	≤2	Ni(60) Mn(40)	60 Ni, 40 Mn; annealed under He + 8% H_2 gas atmosphere at 1000 C for 72 hrs; etched with 30-50% HNO_3.
4	420	1967	0.26-3.78		MnNi	

DATA TABLE NO. 104 SPECIFIC HEAT OF NICKEL + MANGANESE Ni + Mn

[Temperature, T, K; Specific Heat, C_p, Cal g^{-1} K^{-1}]

T	C_p
CURVE 1	
1.350	5.68 x 10^{-5}
1.382	5.77*
1.412	5.86
1.447	5.96*
1.501	6.18
1.577	6.48*
1.661	6.79
1.756	7.18
1.850	7.54
1.919	7.86
1.980	8.11
2.059	8.45
2.160	8.89
2.269	9.39
2.439	1.01 x 10^{-4}
2.651	1.10
2.819	1.18
2.970	1.25
3.121	1.32
3.273	1.37
3.374	1.43
3.574	1.53
3.743	1.61
3.887	1.68
4.013	1.75
4.134	1.80
CURVE 2	
1.363	4.54 x 10^{-5}
1.391	4.64*
1.439	4.77
1.500	4.94
1.555	5.11
1.601	5.27
1.655	5.42*
1.716	5.62
1.796	5.92
1.903	6.19
1.994	6.49
2.065	6.71
2.154	6.99
2.249	7.32
2.380	7.74
2.570	8.43
2.711	8.96
CURVE 2 (cont.)	
2.904	9.63 x 10^{-5}
3.148	1.05 x 10^{-4}
3.334	1.12
3.467	1.17
3.637	1.23
3.768	1.29
3.839	1.32*
3.941	1.36
4.073	1.41
CURVE 3	
1.353	2.87 x 10^{-5}
1.450	2.97
1.511	3.06*
1.578	3.16
1.668	3.30
1.790	3.49
1.919	3.71
2.053	3.95
2.184	4.16
2.286	4.36
2.432	4.64
2.569	4.92
2.686	5.15
2.823	5.42
2.982	5.74
3.132	6.08
3.250	6.33
3.372	6.60
3.513	6.93
3.671	7.29
3.839	7.69
3.974	8.06
4.091	8.39
CURVE 4*	
Series I	
0.700	9.325 x 10^{-6}
0.734	8.654
0.769	8.158
0.806	7.737
0.844	7.190
0.883	6.879
0.920	6.652
CURVE 4 (cont.)*	
0.957	6.372 x 10^{-6}
0.993	6.261
1.029	6.105
1.062	6.126
1.102	6.143
1.153	5.825
1.210	5.754
1.268	5.636
1.320	5.642
1.404	5.342
1.496	5.708
1.524	5.788
1.591	5.867
1.703	5.895
1.805	6.099
1.897	6.309
1.973	6.429
2.027	6.850
2.146	7.062
2.259	7.152
2.373	7.520
2.582	8.114
2.828	8.824
3.016	9.663
3.173	1.022 x 10^{-5}
3.196	1.001
3.308	1.058
3.370	1.059
Series II	
0.565	1.305 x 10^{-5}
0.594	1.231
0.622	1.135
0.653	1.032
0.688	9.405 x 10^{-6}
0.723	8.843
0.761	8.231
0.800	7.806
0.841	7.285
0.882	6.894
0.924	6.530
0.965	6.307
1.003	6.143
1.041	6.015
1.077	6.265
CURVE 4 (cont.)*	
1.119	5.859 x 10^{-6}
1.181	5.680
1.255	5.571
1.393	5.508
1.473	5.626
1.619	5.785
1.701	6.080
1.792	5.968
1.913	6.469
2.139	6.704
2.233	7.062
2.346	8.057
2.466	8.397
2.511	8.734
2.662	9.169
2.789	9.541
2.898	9.899
2.994	1.044 x 10^{-5}
3.081	1.073
3.229	1.013
3.407	1.095
3.553	1.140
3.673	1.196
3.775	1.230
Series III	
1.244	5.544 x 10^{-6}
1.342	5.544
1.437	5.567
1.617	5.773
1.703	5.865
1.785	6.038
1.886	6.265
1.962	6.391
2.079	6.721
2.187	6.879
2.342	7.304
2.519	7.863
2.680	8.151
2.826	8.662
2.958	9.127
3.122	9.651
3.312	1.035 x 10^{-5}
3.485	1.076
3.633	1.153
3.776	1.266

* Not shown on plot

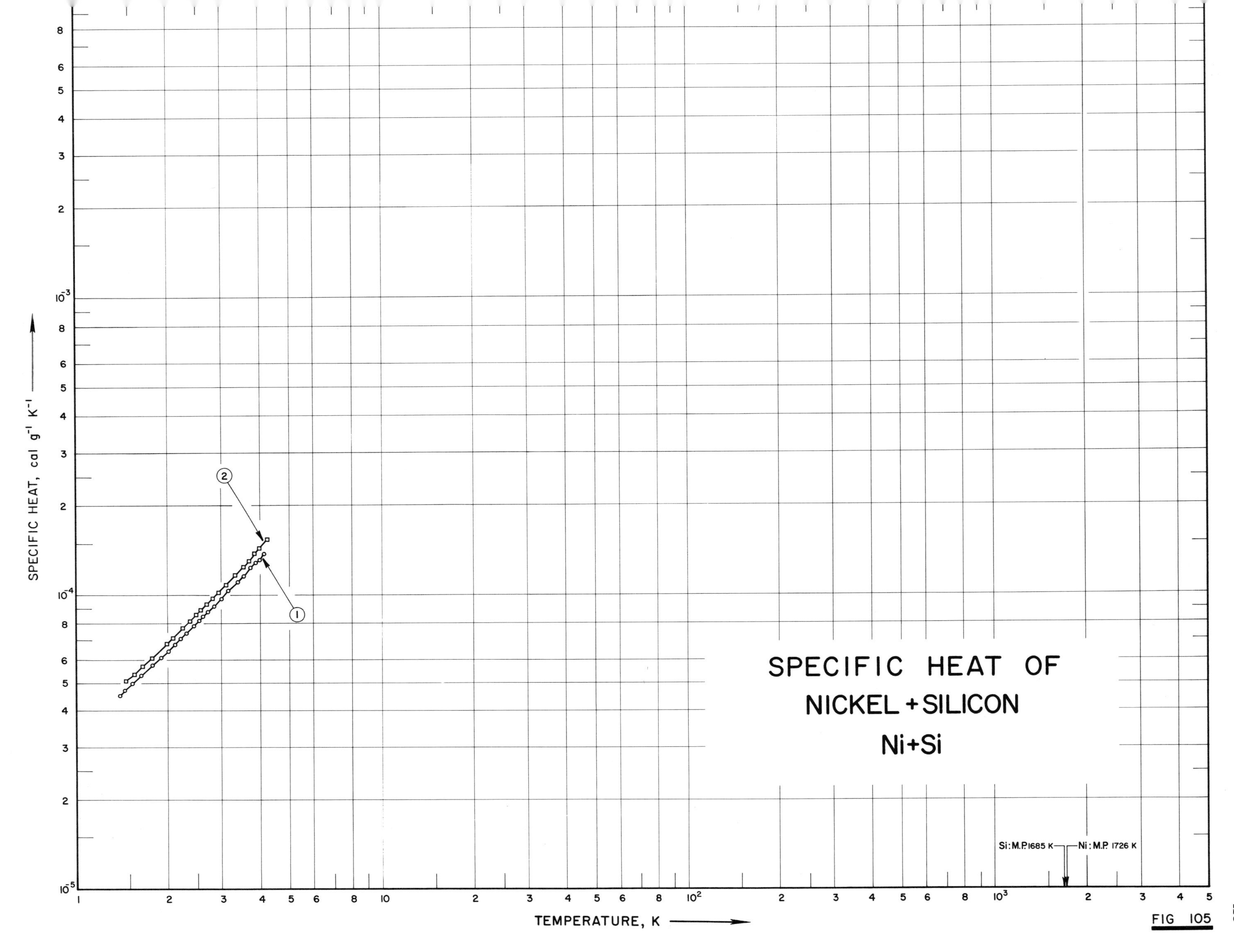
SPECIFIC HEAT OF
NICKEL + SILICON
Ni+Si
SPECIFIC HEAT, cal g⁻¹ K⁻¹
TEMPERATURE, K
Si: M.P. 1685 K
Ni: M.P. 1726 K
1
2

FIG 105

SPECIFICATION TABLE NO. 105 SPECIFIC HEAT OF NICKEL + SILICON Ni + Si

[For Data Reported in Figure and Table No. 105]

Curve No.	Ref. No.	Year	Temp. Range, K	Reported Error, %	Name and Specimen Designation	Composition (weight percent), Specifications and Remarks
1	349	1962	1. 4–4. 1	≤2	Ni(96) Si(4)	97. 95 Ni, 1. 92 Si; annealed under vacuum at 1200 C for 72 hrs; etched with 30 ml HNO_3 and 20 ml CH_3COOH.
2	349	1962	1. 5–4. 3	≤2	Ni(92) Si(8)	95. 82 Ni, 3. 97 Si; same as above.

DATA TABLE NO. 105 SPECIFIC HEAT OF NICKEL + SILICON Ni + Si

[Temperature, T, K; Specific Heat, C_p, Cal $g^{-1} K^{-1}$]

T	C_p
CURVE 1	
1.390	4.52×10^{-5}
1.451	4.72
1.535	4.97
1.648	5.32
1.791	5.75
1.911	6.12
2.022	6.42
2.119	6.76
2.207	7.07
2.316	7.39
2.437	7.85
2.537	8.17
2.616	8.45
2.711	8.78
2.836	9.17
3.010	9.73
3.171	1.03×10^{-4}
3.292	1.07*
3.415	1.11
3.558	1.16
3.740	1.23
3.898	1.29
4.013	1.33
4.146	1.39
CURVE 2	
1.472	5.098×10^{-5}
1.561	5.365
1.655	5.707
1.779	6.097
1.996	6.815
2.079	7.117
2.149	7.344*
2.246	7.702
2.368	8.151
2.478	8.581
2.571	8.994
2.682	9.333
2.806	9.757
2.938	1.024×10^{-4}
3.117	1.086
3.345	1.174
3.547	1.249
3.700	1.314
3.861	1.382
CURVE 2 (cont.)	
3.992	1.441×10^{-4}
4.112	1.489*
4.255	1.548

*Not shown on plot

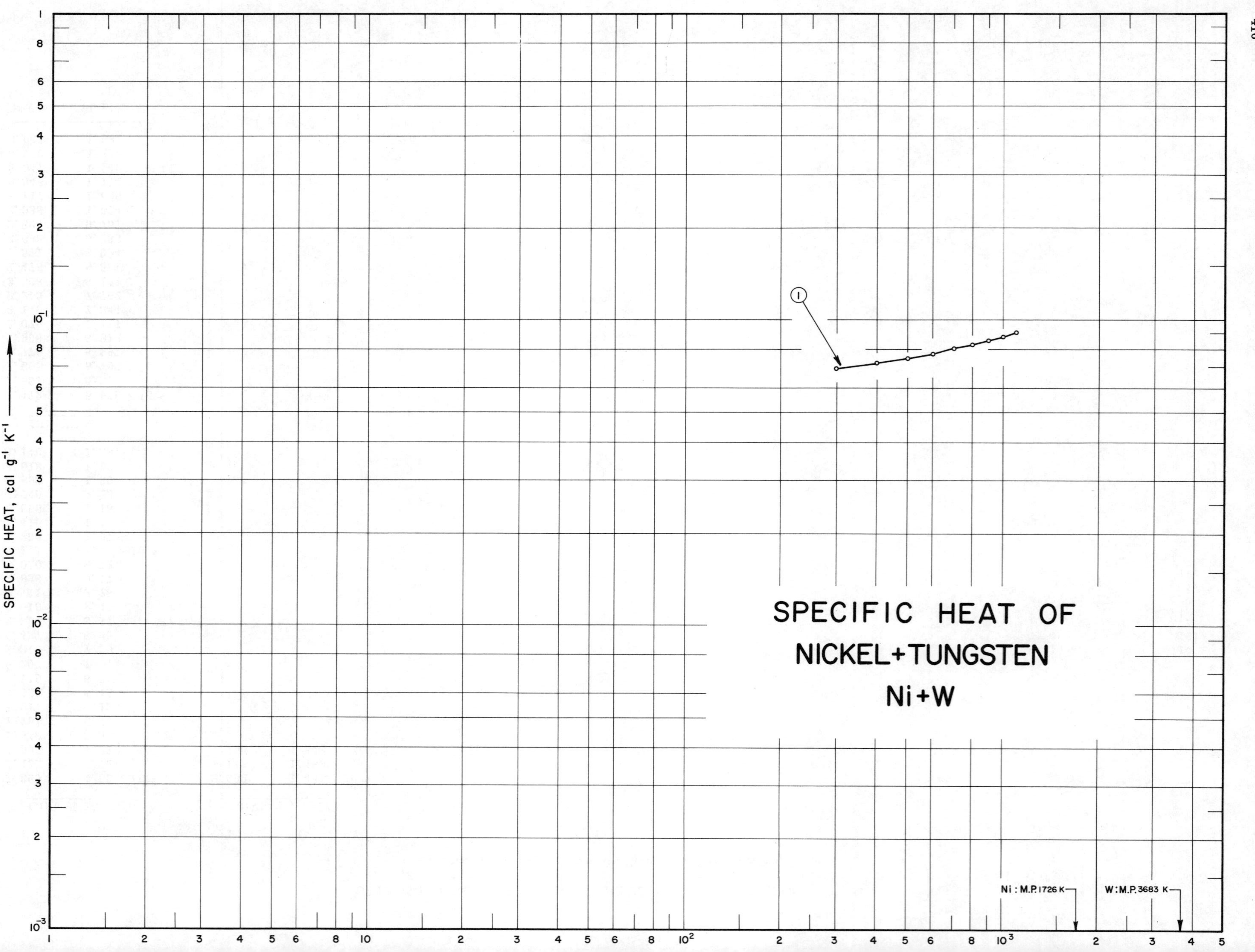
SPECIFIC HEAT OF
NICKEL+TUNGSTEN
Ni+W
SPECIFIC HEAT, cal g⁻¹ K⁻¹
Ni : M.P. 1726 K
W: M.P. 3683 K
1

SPECIFICATION TABLE NO. 106 SPECIFIC HEAT OF NICKEL + TUNGSTEN, Ni + W (Ni_4W)

[For Data Reported in Figure and Table No. 106]

Curve No.	Ref. No.	Year	Temp. Range, K	Reported Error, %	Name and Specimen Designation	Composition (weight percent), Specifications and Remarks
1	397	1962	298-1100	0.4	Ni_4W	Prepared by reduction of mixture of nickel and tungsten oxides.

DATA TABLE NO. 106 SPECIFIC HEAT OF NICKEL + TUNGSTEN, Ni + W (Ni_4W)

[Temperature, T, K; Specific Heat, C_p, Cal g^{-1} K^{-1}]

T	C_p
CURVE 1	
298.15	6.931 x 10^{-2}
300	6.936*
400	7.202
500	7.469
600	7.735
700	8.001
800	8.267
900	8.533
1000	8.799
1100	9.065

*Not shown on plot

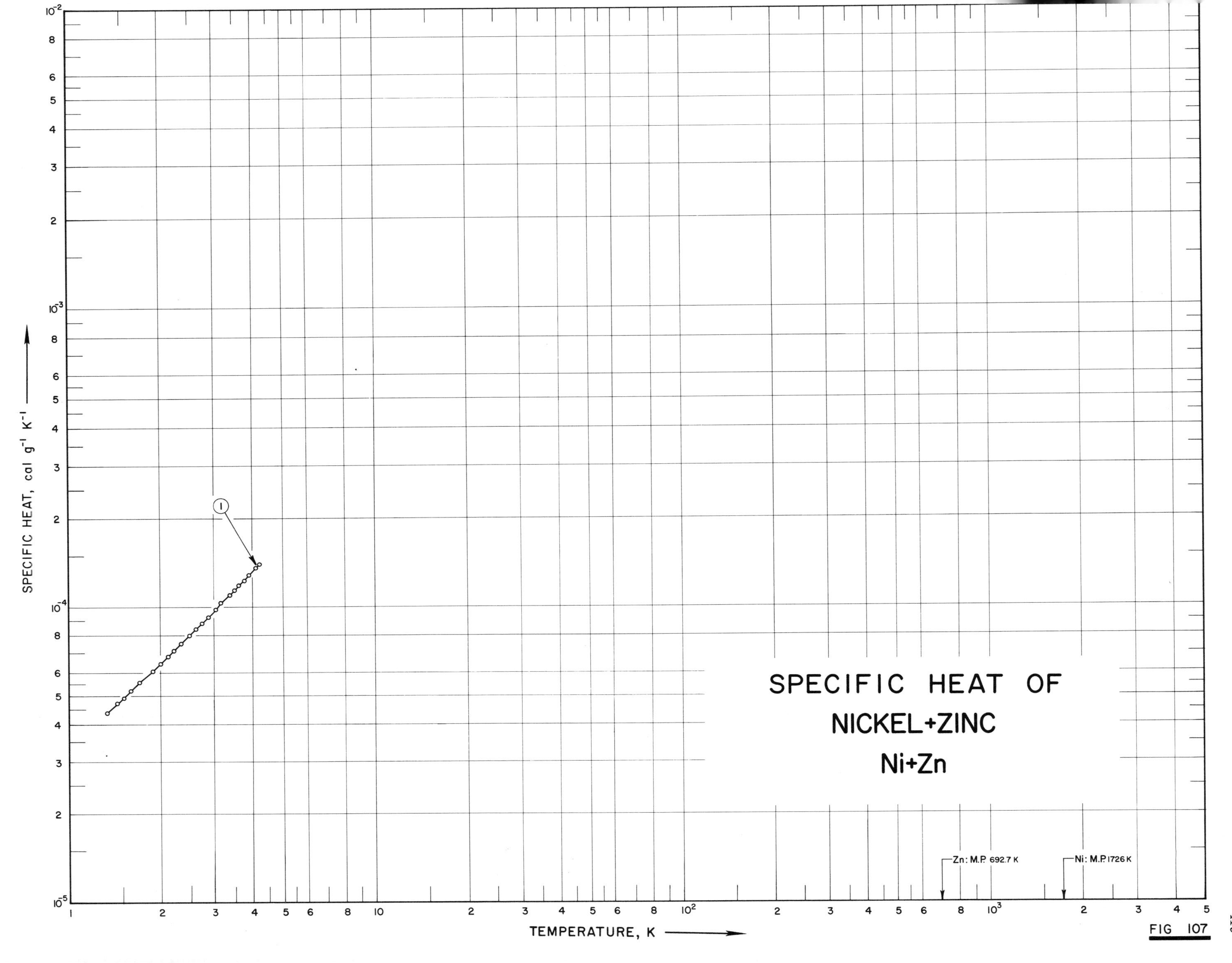
SPECIFIC HEAT OF
NICKEL+ZINC
Ni+Zn
Zn: M.P. 692.7 K
Ni: M.P. 1726 K
TEMPERATURE, K
SPECIFIC HEAT, cal g⁻¹ K⁻¹
1
FIG 107

SPECIFICATION TABLE NO. 107 SPECIFIC HEAT OF NICKEL + ZINC Ni + Zn

[For Data Reported in Figure and Table No. 107]

Curve No.	Ref. No.	Year	Temp. Range, K	Reported Error, %	Name and Specimen Designation	Composition (weight percent), Specifications and Remarks
1	349	1962	1.3-4.2	≤2	Ni(80) Zn(20)	79.80 Ni, 20.2 Zn; etched in 30 ml HNO_3 and 20 ml CH_3COOH.

DATA TABLE NO. 107 SPECIFIC HEAT OF NICKEL + ZINC Ni + Zn

[Temperature, T, K; Specific Heat, Cal g^{-1} K^{-1}]

T	C_p
CURVE 1	
1.332	4.38 x 10^{-5}
1.385	4.54*
1.447	4.71
1.522	4.95
1.607	5.21
1.716	5.55
1.889	6.09
2.013	6.46
2.119	6.82
2.212	7.12
2.334	7.52
2.485	8.00
2.616	8.44
2.728	8.83
2.867	9.26
3.032	9.82
3.166	1.03 x 10^{-4}
3.270	1.07*
3.380	1.10
3.490	1.14
3.611	1.18
3.770	1.24
3.885	1.28
3.973	1.31*
3.987	1.32*
4.101	1.36
4.139	1.37*
4.223	1.41

* Not shown on plot

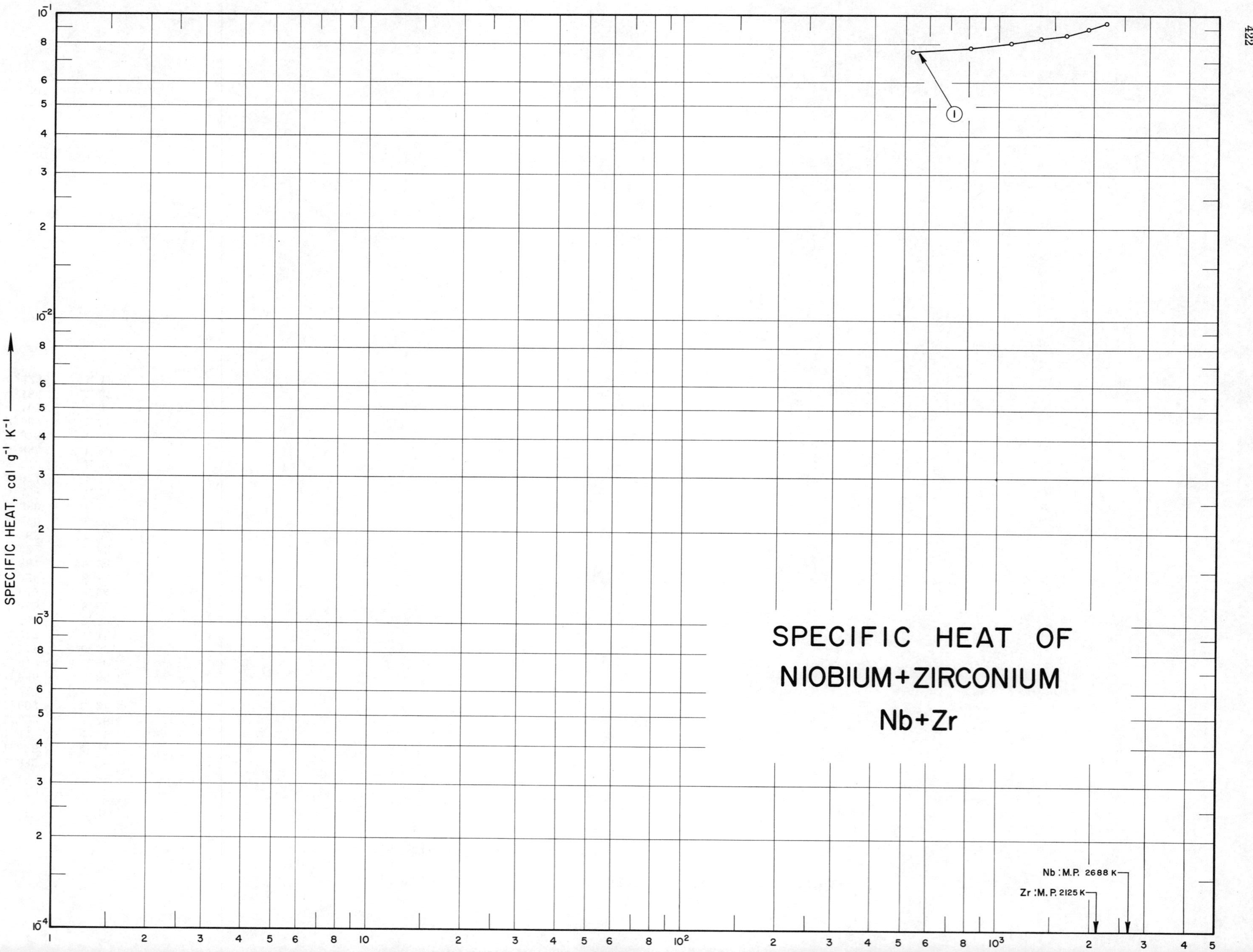
SPECIFIC HEAT OF
NIOBIUM+ZIRCONIUM
Nb+Zr
SPECIFIC HEAT, cal g⁻¹ K⁻¹
Nb : M.P. 2688 K
Zr : M.P. 2125 K
1

SPECIFICATION TABLE NO. 108 SPECIFIC HEAT OF NIOBIUM + ZIRCONIUM Nb + Zr

[For Data Reported in Figure and Table No. 108]

Curve No.	Ref. No.	Year	Temp. Range, K	Reported Error, %	Name and Specimen Designation	Composition (weight percent), Specifications and Remarks
1	237	1962	533-2200	≤5		Before exposure: 99.2 Nb, 0.5 Zr, < 0.1 total elements by semi-quantitative emission spectrography; after exposure: 99.5 Nb, 0.41 C; sample supplied by General Astrometals Corp; crushed in a hardened steel mortar to pass 100-mesh screen; hot pressed; density at 25 C, apparent density (ASTM method B311-58) 492 lb ft^{-3}; true density (by immersion in xylene) 505 lb ft^{-3}; after exposure: apparent density = 502 lb ft^{-3}, true density = 529 lb ft^{-3}.

DATA TABLE NO. 108 SPECIFIC HEAT OF NIOBIUM + ZIRCONIUM Nb + Zr

[Temperature, T, K; Specific Heat, C_p, Cal g^{-1} K^{-1}]

T	C_p
CURVE 1	
533	7.600 x 10^{-2}
811	7.800
1089	8.100
1366	8.400
1644	8.600
1922	9.000
2200	9.400

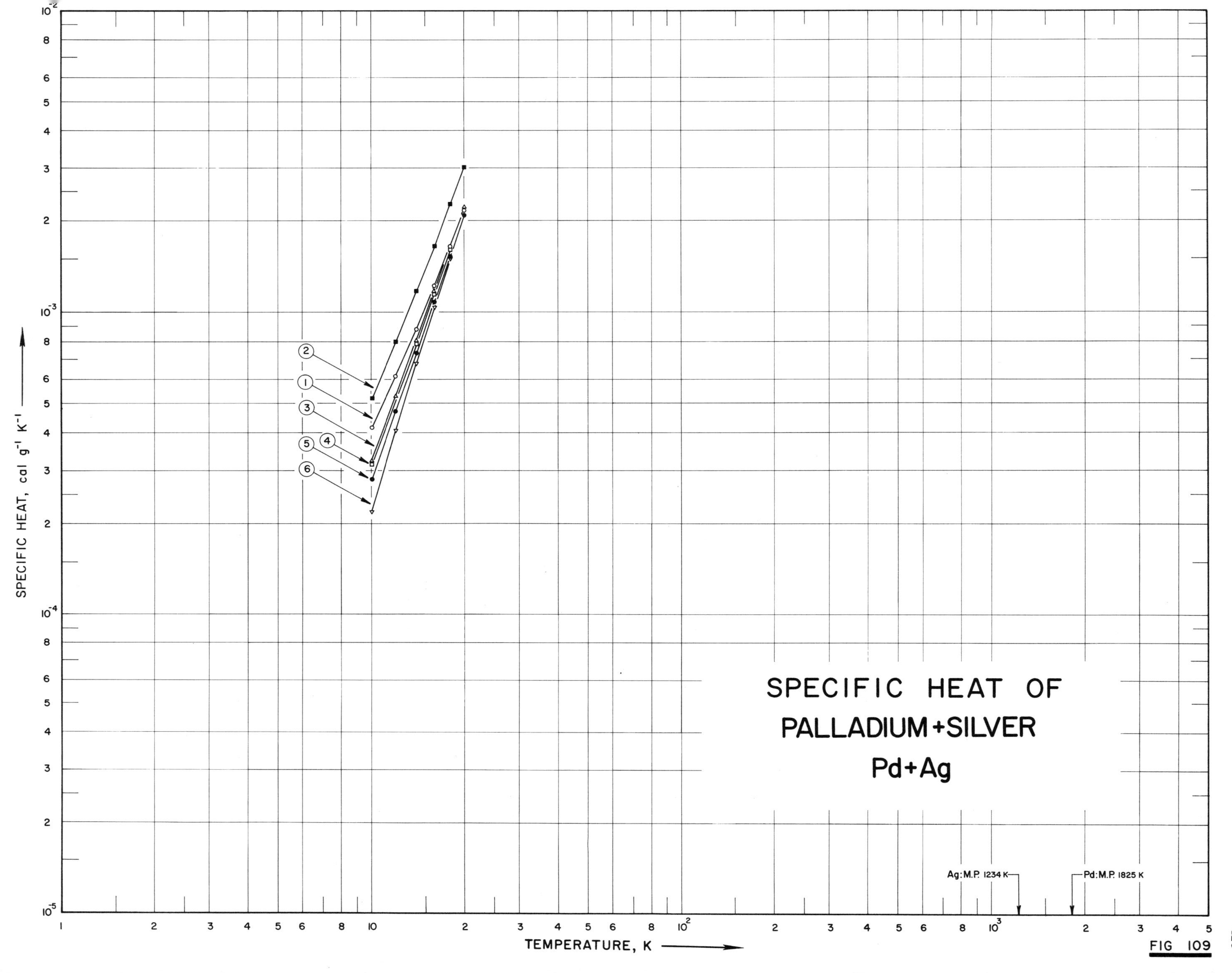
SPECIFIC HEAT OF
PALLADIUM+SILVER
Pd+Ag
SPECIFIC HEAT, cal g⁻¹ K⁻¹
TEMPERATURE, K
Ag: M.P. 1234 K
Pd: M.P. 1825 K
1
2
3
4
5
6

FIG 109

SPECIFICATION TABLE NO. 109 SPECIFIC HEAT OF PALLADIUM + SILVER Pd + Ag

[For Data Reported in Figure and Table No. 109]

Curve No.	Ref. No.	Year	Temp. Range, K	Reported Error, %	Name and Specimen Designation	Composition (weight percent), Specifications and Remarks
1	398	1953	10-20		PdAg(4.7)	Prepared by fusion using high frequency induction furnace; treated with boiling hydrochloric acid to remove surface contamination; degassed by heating to a dull red heat in vacuo for 3 hrs.
2	398	1953	10-20		PdAg(7.8)	Same as above.
3	398	1953	10-20		PdAg(14.1)	Same as above.
4	398	1953	10-20		PdAg(26.1)	Same as above.
5	398	1953	10-20		PdAg(32.1)	Same as above.
6	398	1953	10-20		PdAg(49.6)	Same as above.

DATA TABLE NO. 109 SPECIFIC HEAT OF PALLADIUM + SILVER Pd + Ag

[Temperature, T, K; Specific Heat, Cp, Cal $g^{-1}K^{-1}$]

T	Cp	T	Cp
CURVE 1		CURVE 6	
10	4.168×10^{-4}	10	2.186×10^{-4}
12	6.186	12	4.082
14	8.846	14	6.804
16	1.226×10^{-3}	16	1.044×10^{-3}
18	1.652	18	1.515
20	2.175	20	2.106*
CURVE 2			
10	5.239×10^{-3}		
12	8.013		
14	1.177×10^{-2}		
16	1.661		
18	2.274		
20	3.026		
CURVE 3			
10	3.263×10^{-4}		
12	5.328		
14	8.147		
16	1.185×10^{-3}		
18	1.656*		
20	2.244		
CURVE 4			
10	3.135×10^{-4}		
12	5.136*		
14	7.916		
16	1.156×10^{-3}		
18	1.623		
20	2.204*		
CURVE 5			
10	2.804×10^{-4}		
12	4.718		
14	7.366		
16	1.082×10^{-3}		
18	1.538		
20	2.099		

* Not shown on plot

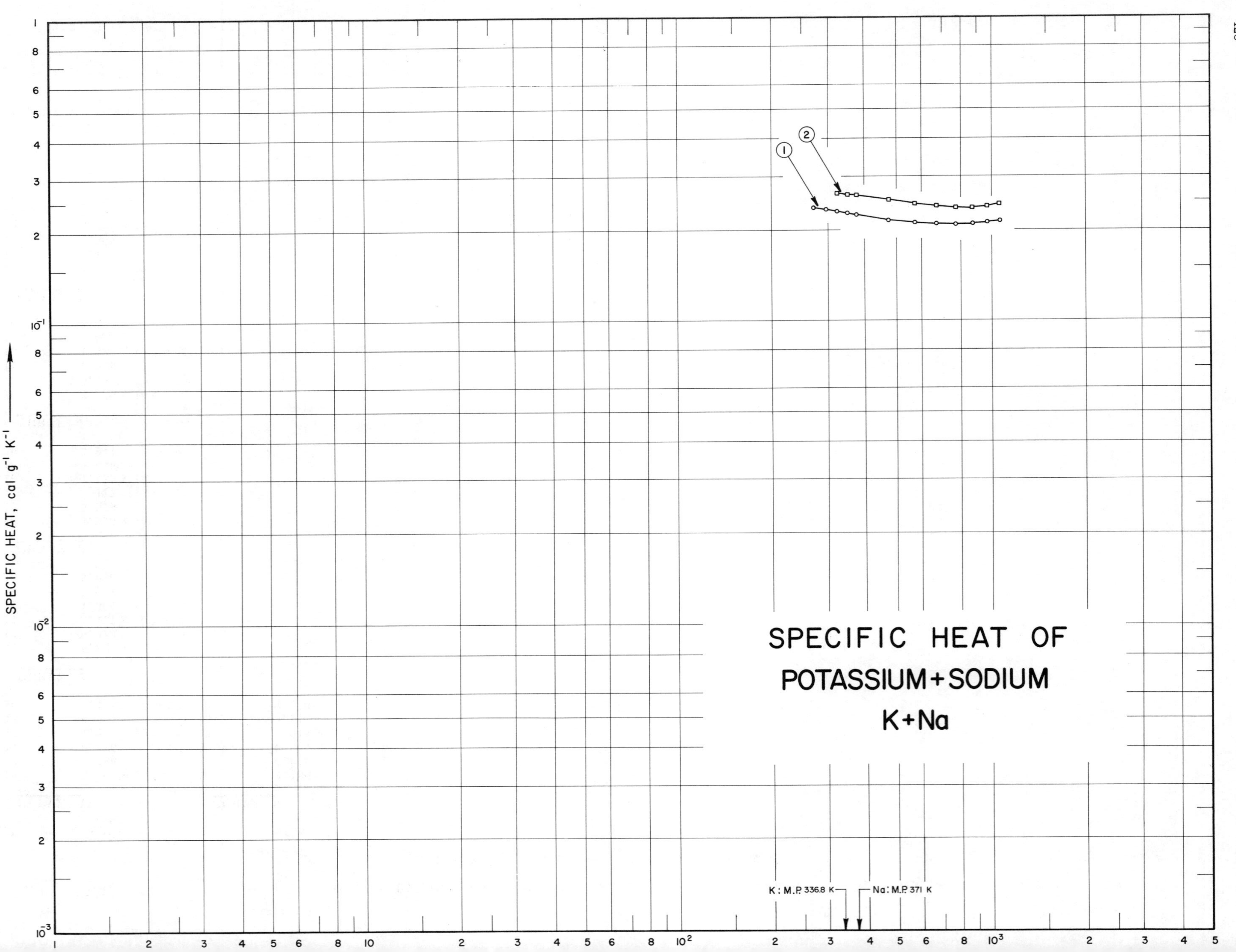
SPECIFIC HEAT OF
POTASSIUM+SODIUM
K+Na
K: M.P. 336.8 K
Na: M.P. 371 K
SPECIFIC HEAT, cal g^{-1} K^{-1}
1
2
1 2 3 4 5 6 8 10 2 3 4 5 6 8 10^2 2 3 4 5 6 8 10^3 2 3 4 5
10^{-3} 2 3 4 5 6 8 10^{-2} 2 3 4 5 6 8 10^{-1} 2 3 4 5 6 8 1

SPECIFICATION TABLE NO. 110 SPECIFIC HEAT OF POTASSIUM + SODIUM K + Na

[For Data Reported in Figure and Table No. 110]

Curve No.	Ref. No.	Year	Temp. Range, K	Reported Error, %	Name and Specimen Designation	Composition (weight percent), Specifications and Remarks
1	353	1952	273-1073	0.4	Eutectic mixture	78.0 K, 22.0 Na; 0.009 Cl_2, 0.006 S, 0.005 Ca, <0.01 alkali oxides; after measurement: 78.26 K; (5.6419 g sample).
2	353	1952	323-1073	0.4		54.0 K, 46.0 Na; after measurement: 43.64 K; (4.9182 g sample).

* Not shown on plot

DATA TABLE NO. 110 SPECIFIC HEAT OF POTASSIUM + SODIUM K + Na

[Temperature, T, K; Specific Heat, C_p, Cal g^{-1} K^{-1}]

T	C_p
CURVE 1	
273.15	2.378 x 10^{-1}
298.15	2.340
323.15	2.306
348.15	2.276
373.15	2.249
473.14	2.170
573.15	2.123
673.15	2.098
773.15	2.089
873.15	2.093
973.15	2.109
1073.15	2.135
CURVE 2	
323.15	2.657 x 10^{-1}
348.15	2.629
373.15	2.603
473.15	2.512
573.15	2.444
673.15	2.397
773.15	2.373
873.15	2.371
973.15	2.391
1073.15	2.433

* Not shown on plot

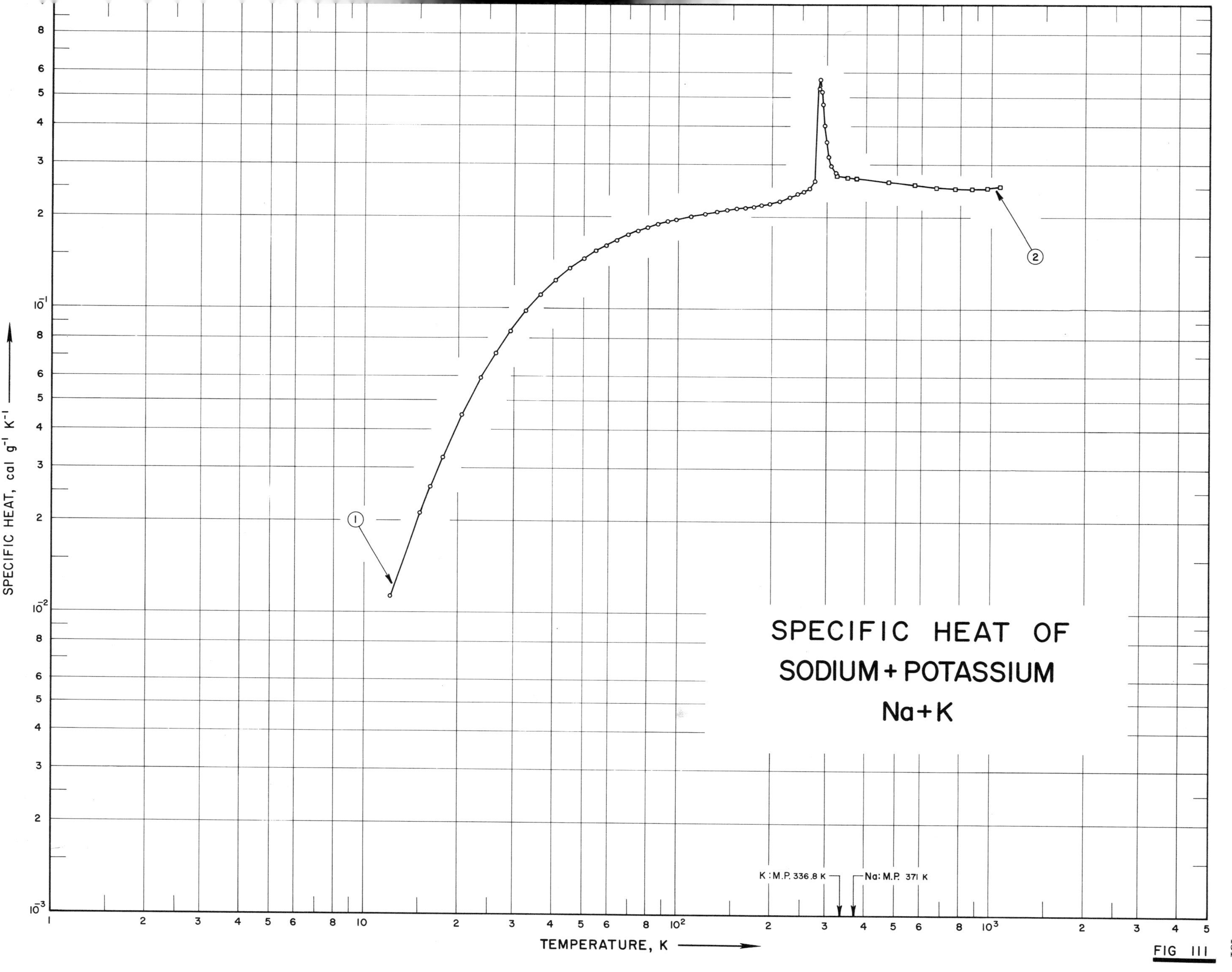
SPECIFIC HEAT OF
SODIUM + POTASSIUM
Na+K
SPECIFIC HEAT, cal g⁻¹ K⁻¹
TEMPERATURE, K
K : M.P. 336.8 K
Na: M.P. 371 K
1
2
FIG 111

SPECIFICATION TABLE NO. 111 SPECIFIC HEAT OF SODIUM + POTASSIUM Na + K

[For Data Reported in Figure and Table No. 111]

Curve No.	Ref. No.	Year	Temp. Range, K	Reported Error, %	Name and Specimen Designation	Composition (weight percent), Specifications and Remarks
1	259	1957	12-321		Na_2K	46.07 K (45.96 theo.).
2	353	1952	323-1073	0.5		55.0 Na and 45.0 K; after measurement: 44.80 K; (4.0182 g sample).

DATA TABLE NO. 111 SPECIFIC HEAT OF SODIUM + POTASSIUM Na + K

[Temperature, T, K; Specific Heat, C_p, Cal g^{-1} K^{-1}]

T	C_p
CURVE 1	
Series 1	
71.22	1.761 x 10^{-1}
80.48	1.843
86.62	1.886
93.16	1.926
99.43	1.956
105.51	1.981
111.43	2.005
117.20	2.025
122.86	2.044
128.41	2.063
133.87	2.077
139.27	2.092
144.58	2.104
149.81	2.115
154.95	2.127
160.24	2.138
165.66	2.148
171.00	2.160
176.31	2.171
181.59	2.183
186.82	2.192
192.12	2.207
197.51	2.220
202.84	2.231
Series 2	
12.19	1.128 x 10^{-2}
15.15	2.116
16.35	2.581
17.84	3.230
20.56	4.461
23.66	5.906
26.34	7.116
29.34	8.392
32.53	9.633
32.86	9.781
36.60	1.106 x 10^{-1}
40.98	1.231
45.57	1.352
50.37	1.462
54.97	1.549
59.43	1.549
CURVE 1 (cont.)	
Series 2 (cont.)	
59.43	1.618 x 10^{-1}
64.31	1.673
69.54	1.744
74.94	1.795
Series 3	
197.88	2.225 x 10^{-1}*
202.63	2.239*
207.56	2.250*
212.59	2.265
217.70	2.288*
222.90	2.307*
228.11	2.323
233.34	2.340*
238.60	2.362*
243.89	2.385
249.21	2.407*
254.49	2.427
Series 4	
240.21	2.363 x 10^{-1}*
245.52	2.386*
250.80	2.395*
256.13	2.443*
265.49	2.487
268.68	2.518*
271.89	2.552*
275.89	2.552
284.91	5.337
288.17	5.230
291.55	4.736
295.33	4.028
299.71	3.556
304.50	3.172
309.73	2.963
315.51	2.835*
321.42	2.819
CURVE 1 (cont.)	
Series 5*	
303.57	2.873 x 10^{-1}
308.01	2.861
312.12	2.850
316.32	2.836
321.06	2.823
Series 6	
282.27	5.632 x 10^{-1}*
284.12	5.697*
286.14	5.734
288.17	5.640*
290.42	5.245*
293.19	4.451*
296.65	3.670*
300.87	3.224*
CURVE 2	
323	2.736 x 10^{-1}
348	2.713
373	2.691
473	2.613
573	2.554
673	2.513
773	2.490
873	2.485
973	2.498
1073	2.530

* Not shown on plot

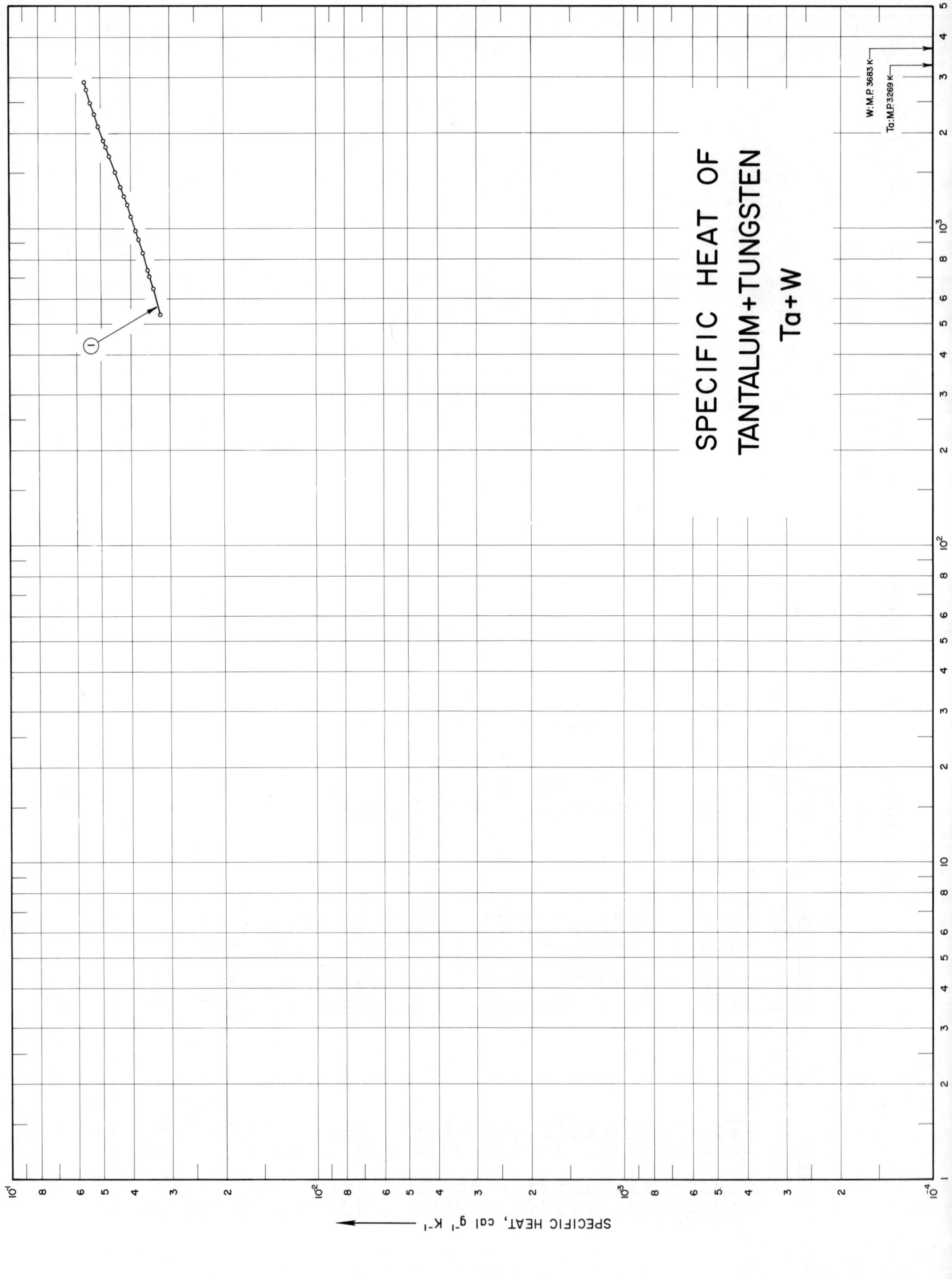

SPECIFIC HEAT OF
TANTALUM + TUNGSTEN
Ta + W

SPECIFIC HEAT, cal g^{-1} K^{-1}

W: M.P. 3683 K
Ta: M.P. 3269 K

1

SPECIFICATION TABLE NO. 112 SPECIFIC HEAT OF TANTALUM + TUNGSTEN Ta + W

[For Data Reported in Figure and Table No. 112]

Curve No.	Ref. No.	Year	Temp. Range, K	Reported Error, %	Name and Specimen Designation	Composition (weight percent), Specifications and Remarks
1	232	1963	537-2890	±5.0	Ta-10W alloy	Bal. Ta, 9.50 W, 0.087 Nb, 0.02 Si, 0.02 Ti, 0.015 Mo, 0.005 Fe, 0.001 C, 0.005 O_2, 0.003 N_2; sample supplied by the Fansteel Metallurgical Corp; density = 1035 lb ft^{-3}.

DATA TABLE NO. 112 SPECIFIC HEAT OF TANTALIUM + TUNGSTEN Ta + W

[Temperature, T, K; Specific Heat, Cp, Cal $g^{-1}K^{-1}$]

T	Cp
CURVE 1	
537	3.233×10^{-2}
648	3.399
709	3.487
740	3.533
836	3.668
851	3.689*
923	3.787
941	3.812*
983	3.868
988	3.876*
1089	4.009
1140	4.073*
1184	4.129
1267	4.232
1357	4.340
1399	4.390*
1420	4.415*
1515	4.524
1565	4.579*
1597	4.613*
1695	4.719
1803	4.830
1885	4.912
2094	5.106
2283	5.268
2480	5.421
2737	5.597
2888	5.689

* Not shown on plot

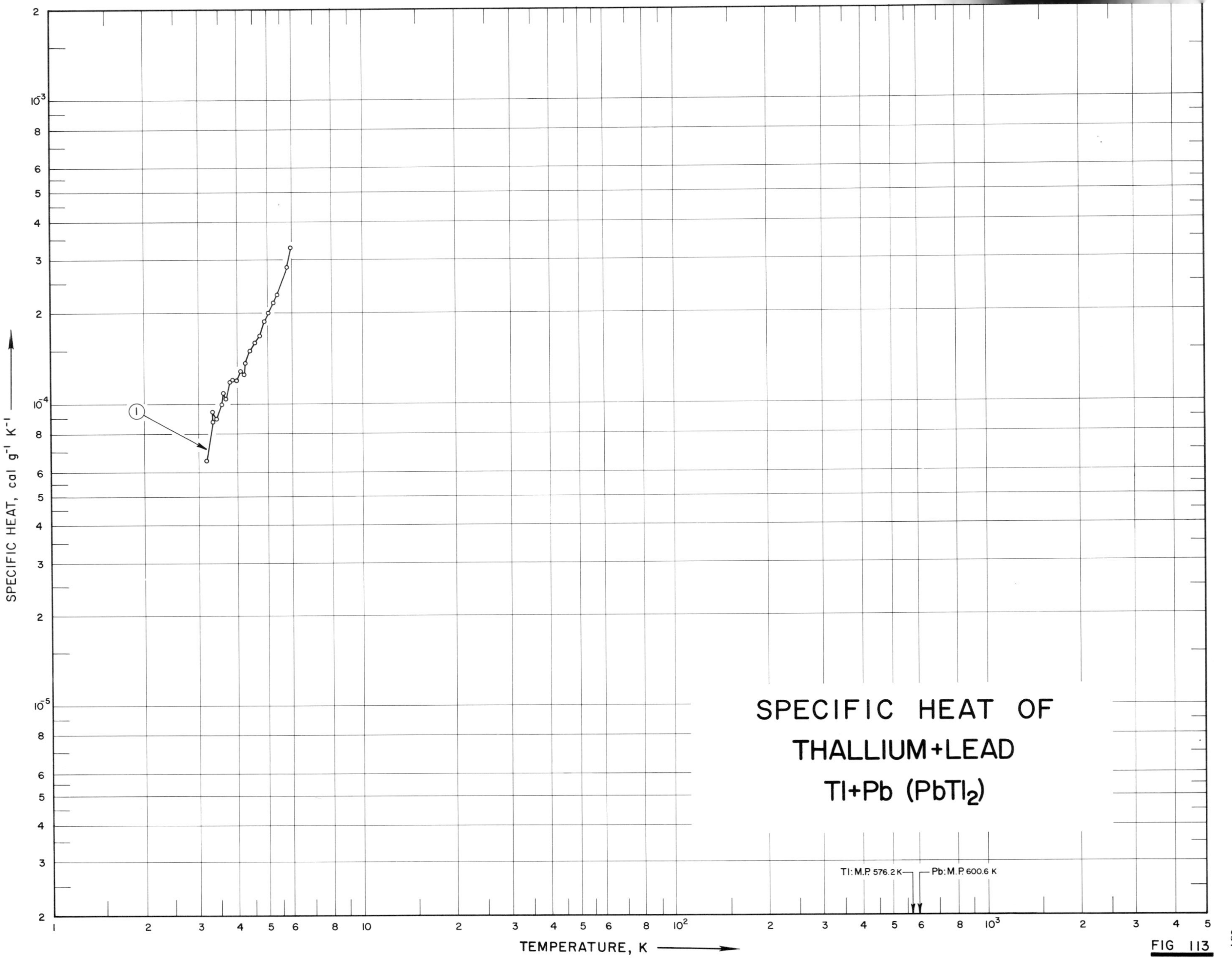
SPECIFIC HEAT OF
THALLIUM+LEAD
Tl+Pb ($PbTl_2$)
SPECIFIC HEAT, cal g⁻¹ K⁻¹
TEMPERATURE, K
Tl: M.P. 576.2 K
Pb: M.P. 600.6 K
1

FIG 113

SPECIFICATION TABLE NO. 113 SPECIFIC HEAT OF THALLIUM + LEAD Tl + Pb ($PbTl_2$)

[For Data Reported in Figure and Table No. 113]

Curve No.	Ref. No.	Year	Temp. Range, K	Reported Error, %	Name and Specimen Designation	Composition (weight percent), Specifications and Remarks
1	421	1935	3.2-5.9		$PbTl_2$	55.2 Tl and 44.87 Pb; cast in pyrex tube, annealed for 3 days in CO_2 atmosphere at a temperature slightly below melting point.

DATA TABLE NO. 113 SPECIFIC HEAT OF THALLIUM + LEAD $Tl + Pb$ ($PbTl_2$)

[Temperature, T, K; Specific Heat, Cp, Cal $g^{-1}K^{-1}$]

T	Cp
CURVE 1	
3.170	6.575×10^{-5}
3.320	8.759
3.320	9.451
3.410	8.997
3.540	9.948
3.580	1.032×10^{-4}*
3.590	1.090
3.670	1.045
3.740	1.071*
3.770	1.181
3.830	1.213
3.900	1.189*
3.960	1.202
4.030	1.274*
4.090	1.287
4.200	1.269
4.220	1.375
4.390	1.514
4.430	1.512*
4.540	1.609
4.700	1.685
4.720	1.700*
4.870	1.879
5.020	2.016
5.040	1.996*
5.210	2.173
5.360	2.312
5.660	2.855
5.940	3.320

* Not shown on plot

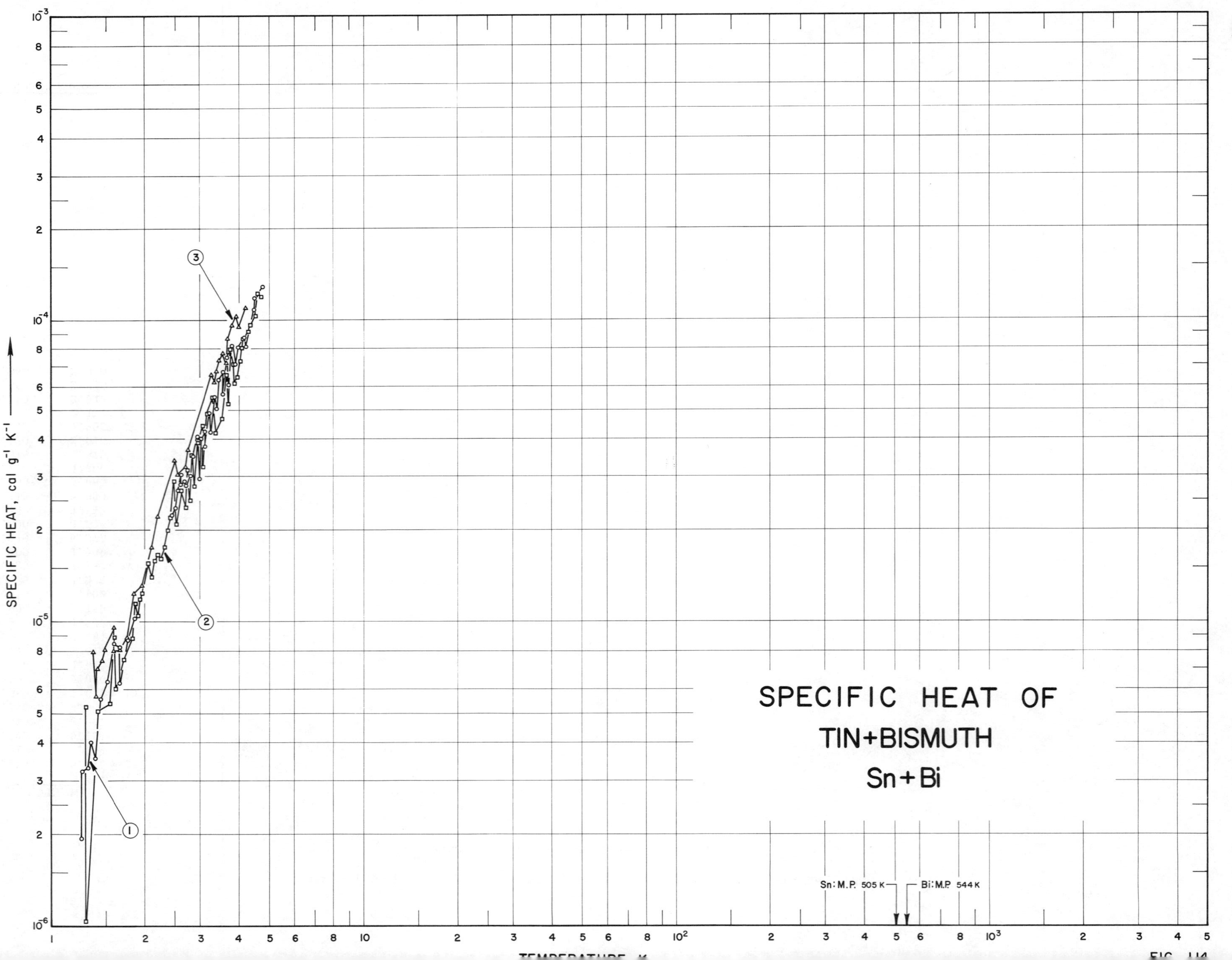
SPECIFIC HEAT OF
TIN+BISMUTH
Sn+Bi
Sn: M.P. 505 K
Bi: M.P. 544 K
SPECIFIC HEAT, cal g⁻¹ K⁻¹
1
2
3

SPECIFICATION TABLE NO. 114 SPECIFIC HEAT OF TIN + BISMUTH Sn + Bi

[For Data Reported in Figure and Table No. 114]

Curve No.	Ref. No.	Year	Temp. Range, K	Reported Error, %	Name and Specimen Designation	Composition (weight percent), Specifications and Remarks
1	239	1955	1.3-4.8	<8	6 at.% Bi	99.999 Sn, 99.95 Bi; sample supplied by the McKay Co.; annealed 2 wks at 135 C.
2	239	1955	1.3-4.6	<8	3 at.% Bi	Same as above.
3	239	1955	1.4-4.2	<8	9 at.% Bi	Same as above.

DATA TABLE NO. 114 SPECIFIC HEAT OF TIN + BISMUTH Sn + Bi

[Temperature, T, K; Specific Heat, Cp, Cal $g^{-1}K^{-1}$]

T	Cp
CURVE 1	
Series I	
2.427	2.23 x 10^{-5}
2.481	2.40*
2.549	2.70
2.605	3.04
2.667	2.89
2.760	3.00*
2.848	3.52
2.938	4.07
3.014	4.02
3.103	4.27
3.226	4.93
3.335	5.59
3.435	6.35
3.544	6.71
3.668	7.49
3.793	8.18
3.963	8.10
4.219	9.15
4.468	1.19 x 10^{-4}
4.750	1.29
Series II	
1.260	3.24 x 10^{-6}
1.260	1.94
1.320	3.32
1.380	3.56
1.310	3.40*
1.340	4.05
1.430	5.67*
1.670	6.31
1.760	8.74
1.850	1.03 x 10^{-5}
2.526	2.36
2.614	2.82
2.704	2.79
2.786	3.02
2.979	2.95
3.104	3.78
3.245	4.24
3.395	5.07
3.546	5.70
3.717	6.12
3.887	7.15
4.150	8.75
4.453	1.08 x 10^{-4}
CURVE 1 (cont.)	
Series III	
1.430	5.59 x 10^{-6}
1.510	6.40
1.580	8.50
1.670	8.26
CURVE 2	
Series I	
3.528	6.55 x 10^{-5}*
3.642	7.34*
3.730	7.97
3.838	6.93*
3.947	6.47
4.021	7.31
4.093	7.65*
4.145	8.67
4.283	9.23
4.514	1.03 x 10^{-4}
4.721	1.20
Series II	
1.881	1.05 x 10^{-5}
1.919	1.19
1.959	1.17*
2.041	1.55
2.095	1.41
2.138	1.58
2.186	1.66
2.243	1.61
2.297	1.76
2.357	2.00
2.414	2.19
2.472	2.90
2.534	2.55*
2.606	2.69
Series III	
1.280	5.28 x 10^{-6}
1.290	1.04 x 10^{-5}
1.420	5.11 x 10^{-6}
1.540	5.44
1.620	6.02
1.710	7.50
1.820	8.82
CURVE 2 (cont.)	
2.516	2.09 x 10^{-5}
2.596	2.14*
2.684	2.37
2.780	2.50
2.869	2.79
2.968	3.19*
3.064	3.24
3.360	4.21
3.518	4.70
3.675	5.26
3.864	6.18
Series IV	
2.558	2.74 x 10^{-5}*
2.638	2.89*
2.731	3.17
2.834	3.54
2.935	3.92
3.050	4.45
3.159	4.86
3.286	5.53
3.440	6.27*
3.644	6.57
3.839	7.14
4.065	8.09
4.329	9.73
4.592	1.23 x 10^{-4}
Series V	
1.590	8.91 x 10^{-6}
1.690	8.82*
1.870	1.15 x 10^{-5}
1.950	1.25
1.990	1.25*
3.518	4.70*
3.675	5.26*
3.864	6.18*
Series VI	
2.558	2.74 x 10^{-5}*
2.638	2.89*
2.731	3.17*
2.834	3.54*
2.935	3.92*
CURVE 2 (cont.)	
3.050	4.45 x 10^{-5}*
3.159	4.86*
3.286	5.53*
3.440	6.27*
3.644	6.57*
3.839	7.14*
4.065	8.09*
4.329	9.73*
4.592	1.23 x 10^{-4}*
Series VII	
1.590	8.91 x 10^{-6}*
1.690	8.82*
1.870	1.15 x 10^{-5}
1.950	1.25
1.990	1.25
CURVE 3	
Series I	
3.265	6.59 x 10^{-5}
3.374	6.75
3.449	7.33
3.548	7.75
3.668	8.68
3.792	9.62
3.905	1.03 x 10^{-4}
Series II	
2.479	3.41 x 10^{-5}
2.532	3.04
2.612	3.10*
2.686	3.23
2.761	3.48*
2.843	3.68
3.344	6.21
3.493	6.67*
3.635	7.18
3.987	9.50
4.197	1.10 x 10^{-4}
Series III	
1.399	5.68 x 10^{-6}
1.476	8.12
CURVE 3 (cont.)	
1.586	8.04 x 10^{-6}
1.673	8.12
1.757	8.91
1.836	1.24 x 10^{-5}
1.952	1.32
2.096	1.76
2.198	2.22
2.031	2.26*
Series IV	
1.362	7.97 x 10^{-6}
1.405	7.02
1.452	7.49
1.503	8.04*
1.577	9.62

* Not shown on plot

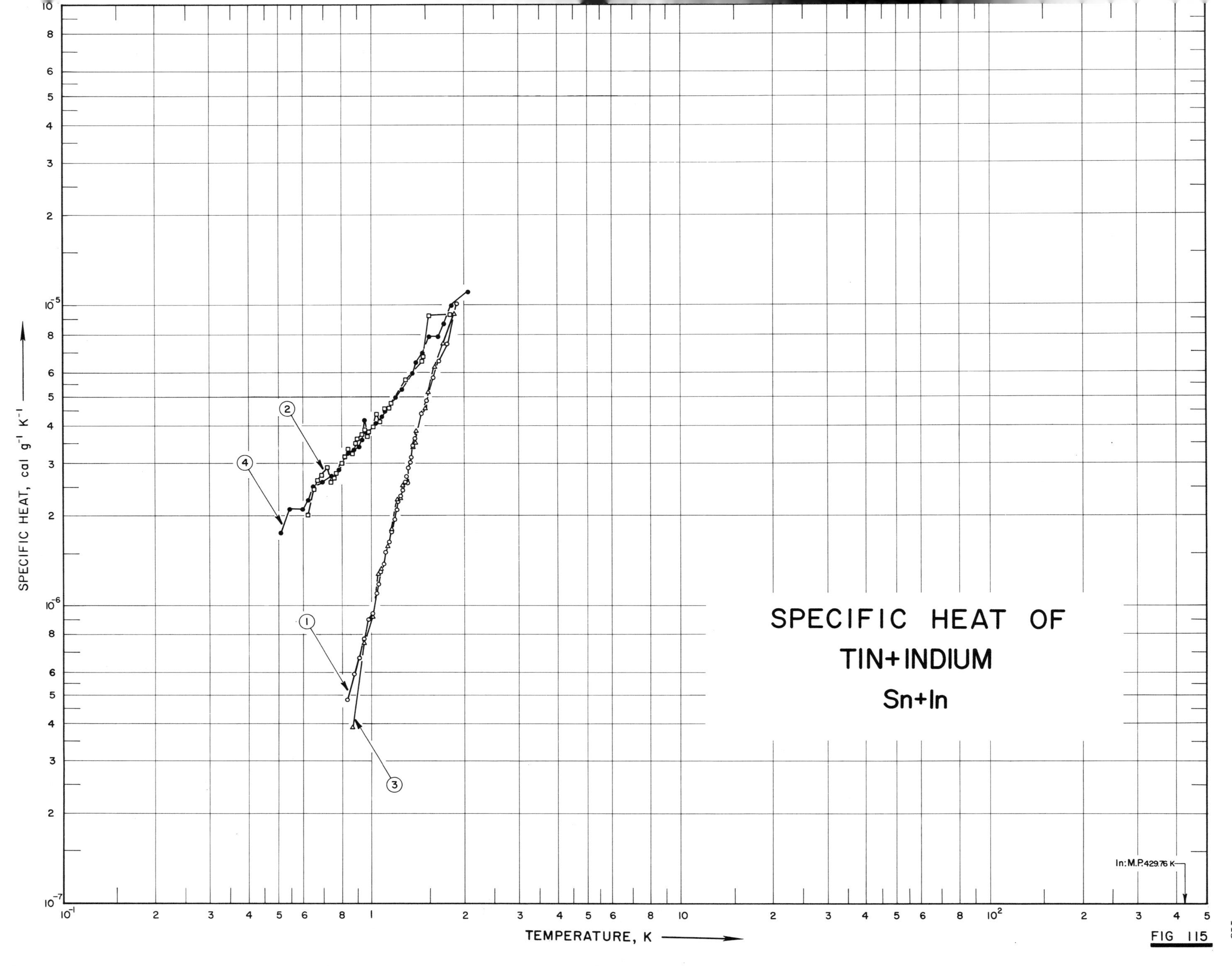
SPECIFIC HEAT OF
TIN+INDIUM
Sn+In
In: M.P. 429.76 K
SPECIFIC HEAT, cal g^{-1} K^{-1}
TEMPERATURE, K
1
2
3
4
FIG 115

SPECIFICATION TABLE NO. 115 SPECIFIC HEAT OF TIN + INDIUM Sn + In

[For Data Reported in Figure and Table No. 115]

Curve No.	Ref. No.	Year	Temp. Range, K	Reported Error, %	Name and Specimen Designation	Composition (weight percent), Specifications and Remarks
1	197	1961	0.8-1.9		Sn 2% In	2.0 In; superconducting.
2	197	1961	0.6-1.8		Sn 2% In	2.0 In; normal state, 500 gauss magnetic field.
3	197	1961	0.8-1.8		Sn 1% In	1.0 In; superconducting.
4	197	1961	0.5-2.0		Sn 1% In	1.0 In; normal state, 500 gauss magnetic field.

DATA TABLE NO. 115 SPECIFIC HEAT OF TIN + INDIUM Sn + In

[Temperature, T, K; Specific Heat, C_p, Cal $g^{-1}K^{-1}$]

T	C_p
CURVE 1	
0.837	4.82 x 10^{-7}
0.875	5.93
0.909	6.72
0.940	7.76
0.974	9.01
1.000	9.40
1.029	1.12 x 10^{-6}
1.052	1.18
1.072	1.30
1.091	1.38
1.110	1.52
1.138	1.64
1.147	1.63*
1.162	1.77
1.178	1.95
1.193	2.01*
1.207	2.097
1.220	2.227
1.233	2.297*
1.245	2.331
1.258	2.440
1.275	2.608
1.291	2.727
1.304	2.585
1.316	2.92
1.329	3.03
1.342	3.15
1.354	3.16*
1.366	3.45
1.377	3.63
1.388	3.73*
1.455	4.42
1.510	4.86
1.585	5.83
1.655	6.57
1.764	7.46
1.896	1.02 x 10^{-5}
CURVE 2	
0.622	2.01 x 10^{-6}
0.652	2.460
0.669	2.642
0.688	2.735
0.704	2.76*
0.719	2.92 x 10^{-6}
0.737	2.60
0.751	2.68
0.765	2.78
0.778	2.82
0.800	3.01
0.817	3.17
0.827	3.19*
0.838	3.35
0.848	3.39*
0.858	3.35*
0.868	3.23
0.879	3.31*
0.889	3.51
0.899	3.63
0.909	3.67*
0.930	3.75
0.946	3.89
0.957	3.83*
0.966	3.69
0.977	3.85
0.988	3.85
1.011	3.99
1.024	4.09*
1.038	4.42
1.054	4.34*
1.070	4.15
1.085	4.40*
1.100	4.58
1.116	4.63*
1.132	4.63*
1.147	4.60
1.162	4.78
1.296	5.73
1.468	6.57
1.478	6.80
1.738	9.26
1.800	9.34
CURVE 3	
0.864	3.89 x 10^{-7}
0.943	7.53
1.005	9.24
1.030	1.14 x 10^{-6}*
1.052	1.28 x 10^{-6}
1.073	1.33
1.098	1.41*
1.125	1.59
1.161	1.81
1.183	1.88*
1.200	2.099*
1.217	2.276
1.244	2.299
1.266	2.538
1.279	2.567*
1.292	2.72*
1.312	2.88*
1.328	3.00*
1.348	3.08*
1.369	3.40
1.382	3.52
1.394	3.87
1.480	4.61
1.524	5.22
1.603	6.34
1.713	7.53
1.855	9.42
CURVE 4	
0.505	1.77 x 10^{-6}
0.542	2.119
0.597	2.115
0.623	2.262
0.647	2.518
0.693	2.613
0.741	2.724
0.780	2.86
0.806	3.10*
0.828	3.24*
0.835	3.24
0.875	3.34
0.909	3.42
0.928	3.60
0.945	4.19
0.963	3.81
1.006	3.97*
1.035	4.11
1.078	4.31
1.110	4.49 x 10^{-6}
1.148	4.73*
1.200	4.99
1.260	5.32
1.368	6.00
1.396	6.50
1.471	6.99
1.530	7.91
1.549	8.07*
1.641	7.93
1.722	8.72
1.872	9.99
1.882	1.02 x 10^{-5}*
2.055	1.16

* Not shown on plot

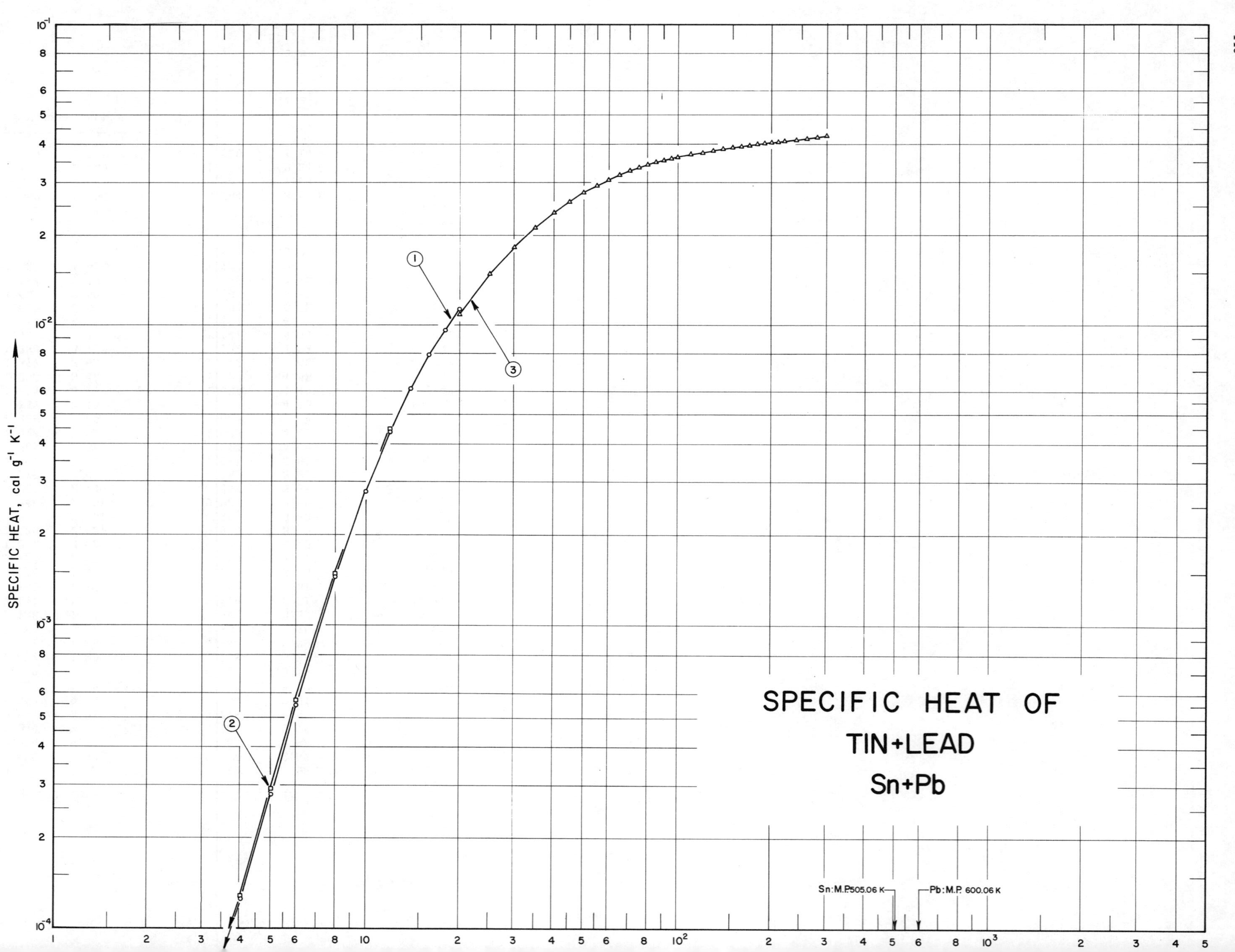
SPECIFIC HEAT OF
TIN+LEAD
Sn+Pb
Sn: M.P. 505.06 K
Pb: M.P. 600.06 K
SPECIFIC HEAT, cal g⁻¹ K⁻¹
1
2
3

SPECIFICATION TABLE NO. 116 SPECIFIC HEAT OF TIN + LEAD Sn + Pb

[For Data Reported in Figure and Table No. 116]

Curve No.	Ref. No.	Year	Temp. Range, K	Reported Error, %	Name and Specimen Designation	Composition (weight percent), Specifications and Remarks
1	393	1963	1.5-20		Pb(28) Sn(72)	72 at.% Sn, 28 at.% Pb.
2	393	1963	1.5-20		Pb(36) Sn(64)	64 at.% Sn, 36 at.% Pb.
3	399	1964	20-300		50-50 Lead-tin solder	49.9 Sn, 48.8 Pb, <0.25 Sb, <0.15 Bi, <0.05 As, <0.001 Al, <0.001 Cd, <0.001 Fe, <0.001 Ni, <0.001 Zn; sample supplied by National Lead Co.

DATA TABLE NO. 116 SPECIFIC HEAT OF TIN + LEAD $Sn + Pb$

[Temperature, T, K; Specific Heat, Cp, Cal $g^{-1}K^{-1}$]

T	Cp
CURVE 1	
1.5	6.00×10^{-6}
2	1.53×10^{-5}
2.5	3.00
3	5.15
3.5	8.23
4	1.25×10^{-4}
5	2.80
6	5.48
8	1.45×10^{-3}
10	2.80
12	4.44
14	6.18
16	7.97
18	9.68
20	1.14×10^{-2}
CURVE 2	
1.5	6.17×10^{-6}
2	1.52×10^{-5}
2.5	3.00*
3	5.24
3.5	8.44
4	1.28×10^{-4}
5	2.92
6	5.68
8	1.49×10^{-3}
10	2.82*
12	4.51
14	6.22*
16	8.02*
18	9.74*
20	1.15×10^{-2}*
CURVE 3	
20	1.097×10^{-2}
25	1.493
30	1.834
35	2.128
40	2.379
45	2.593
50	2.776
55	2.932
60	3.064
65	3.178
CURVE 3 (cont.)	
70	3.275
75	3.359
80	3.431
85	3.494
90	3.549
95	3.597
100	3.637
110	3.708
120	3.769
130	3.822
140	3.869
150	3.910
160	3.946
170	3.978
180	4.007
190	4.034
200	4.059
210	4.082
220	4.105
230	4.127*
240	4.148
250	4.170*
260	4.191
270	4.212*
280	4.233
290	4.253*
300	4.272

* Not shown on plot

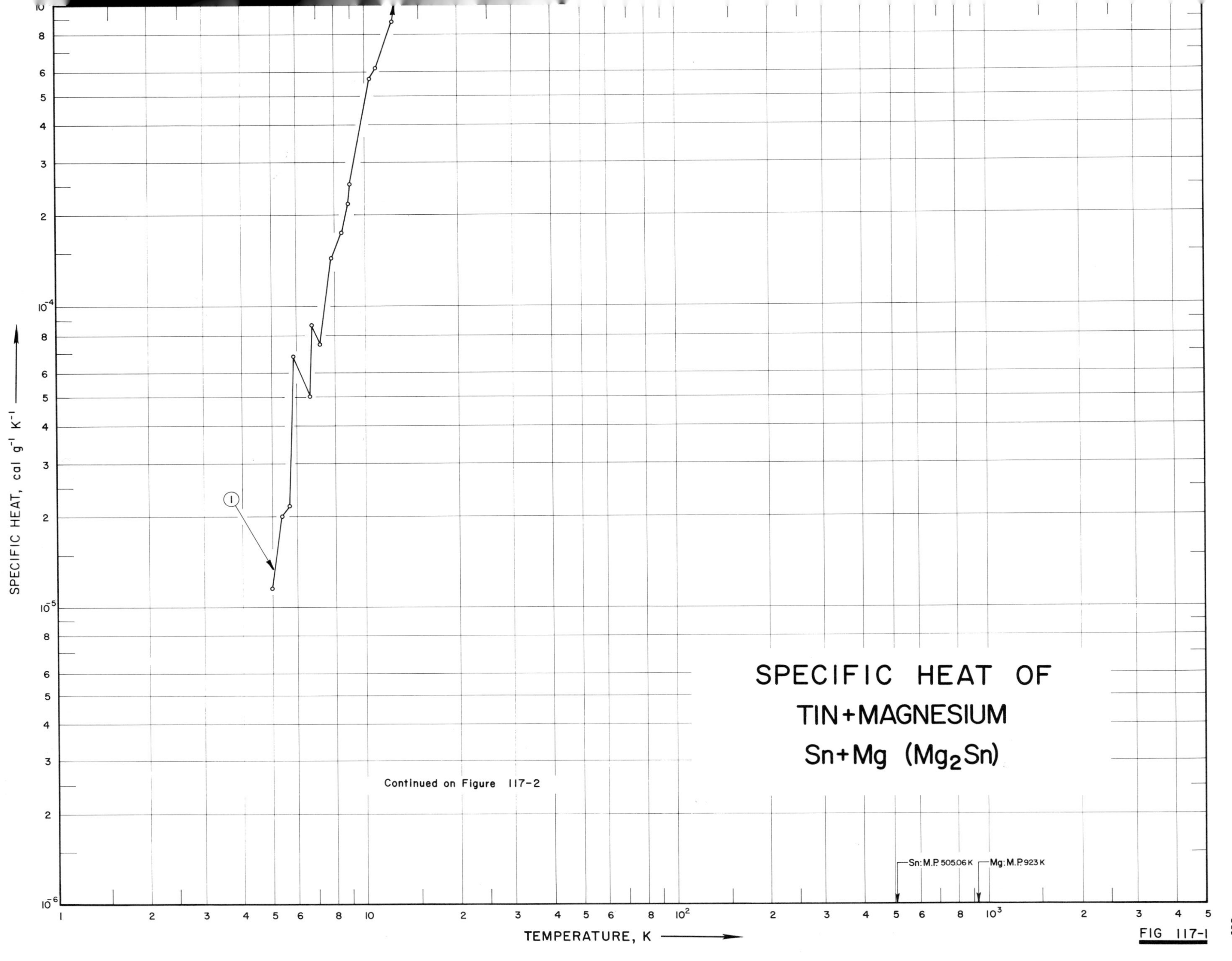
SPECIFIC HEAT OF
TIN+MAGNESIUM
Sn+Mg (Mg_2Sn)
Continued on Figure 117-2
Sn: M.P. 505.06 K
Mg: M.P. 923 K
1
TEMPERATURE, K
SPECIFIC HEAT, cal g^{-1} K^{-1}
FIG 117-1

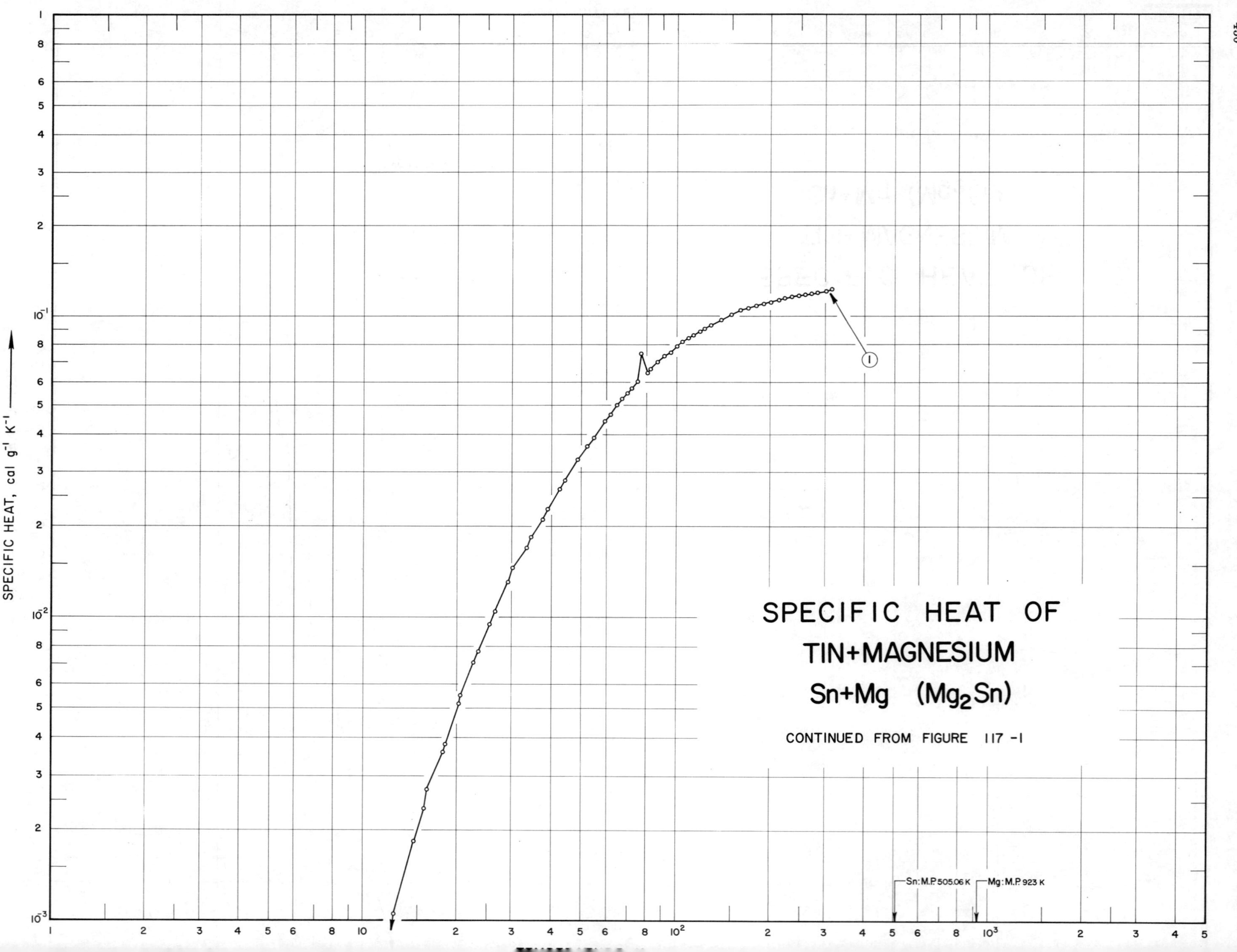

SPECIFIC HEAT OF
TIN+MAGNESIUM
Sn+Mg (Mg₂Sn)
CONTINUED FROM FIGURE 117 -1
SPECIFIC HEAT, cal g⁻¹ K⁻¹
Sn: M.P. 505.06 K
Mg: M.P. 923 K
1

SPECIFICATION TABLE NO. 117 SPECIFIC HEAT OF TIN + MAGNESIUM, Sn + Mg (Mg_2Sn)

[For Data Reported in Figure and Table No. 117]

Curve No.	Ref. No.	Year	Temp. Range, K	Reported Error, %	Name and Specimen Designation	Composition (weight percent), Specifications and Remarks
1	391	1967	5-314		Mg_2Sn	Prepared by melting stoichiometric proportions of 99. 99 Mg samples from Dow Chemical Co. and 99. 9999 Sn samples from Vulcan Materials Co. , and cooling slowly while a temperature gradient is maintained over the length of a spectroscopically pure graphite crucible.

DATA TABLE NO. 117 SPECIFIC HEAT OF TIN + MAGNESIUM Sn + Mg (Mg_2Sn)

[Temperature, T, K; Specific Heat, Cp, Cal $g^{-1}K^{-1}$]

T	Cp
CURVE 1	
Series I	
4.97	1.170×10^{-5}
5.66	2.173
7.14	7.521
8.93	2.540×10^{-4}
10.85	6.201
12.70	1.050×10^{-3}
14.58	1.832
16.43	2.721
18.36	3.822
20.57	5.514
23.31	7.708
26.46	1.048×10^{-2}
30.06	1.471
34.41	1.833
38.98	2.270
44.13	2.820
Series II	
52.14	3.66×10^{-2}
59.21	4.41
64.73	5.00
69.97	5.53
75.19	6.01
Series III	
82.90	6.65×10^{-2}
91.58	7.33
100.60	7.94
109.40	8.41
118.30	8.87
128.30	9.34
139.00	9.74
150.00	1.01×10^{-1}
160.90	1.05
Series IV	
170.0	1.06×10^{-1}
180.4	1.08
190.8	1.10
201.7	1.12
212.8	1.13
CURVE 1 (cont.)	
222.4	1.15×10^{-1}
234.0	1.16
246.1	1.17
258.2	1.18
270.2	1.19
282.2	1.20
294.3	1.22*
308.4	1.22*
Series V	
5.35	2.006×10^{-5}
6.63	5.014
8.42	1.755×10^{-4}
10.35	5.716
12.39	8.808
Series VI	
5.85	6.853×10^{-5}
6.74	8.691
7.77	1.454×10^{-4}
8.82	2.189
Series VII	
15.74	2.358×10^{-3}
18.00	3.595
20.23	5.171
22.55	7.070
25.40	9.460
29.20	1.315×10^{-2}
33.36	1.693
37.52	2.111
42.39	2.634
48.69	3.314
Series VIII	
54.64	3.92×10^{-2}
61.67	4.66
67.19	5.26
72.34	5.73
77.25	7.49
CURVE 1 (cont.)	
Series IX	
80.57	6.45×10^{-2}
87.05	7.00
96.10	7.50
104.90	8.17
113.40	8.63
122.40	9.07
132.10	9.50*
142.60	9.86*
154.10	1.02×10^{-1}*
Series X	
167.4	1.06×10^{-1}*
183.1	1.09*
186.4	1.10*
197.0	1.11*
207.3	1.13*
218.1	1.14*
229.2	1.16*
240.3	1.17*
Series XI	
253.6	1.18×10^{-1}*
265.7	1.19*
277.7	1.20*
289.5	1.21*
301.0	1.22
314.3	1.23

* Not shown on plot

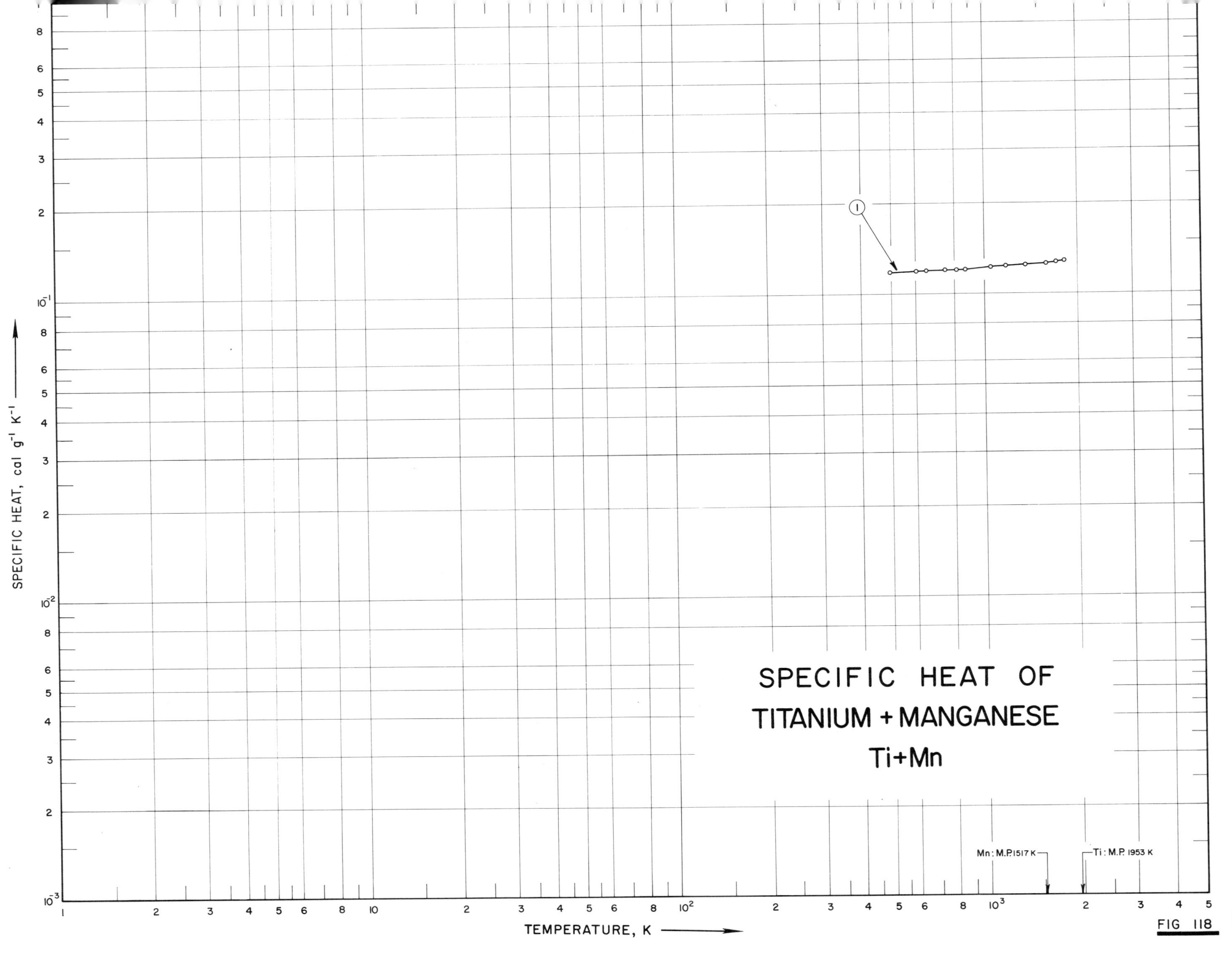

SPECIFIC HEAT OF
TITANIUM + MANGANESE
Ti+Mn
Mn: M.P. 1517 K
Ti: M.P. 1953 K
SPECIFIC HEAT, cal g⁻¹ K⁻¹
TEMPERATURE, K
1

FIG 118

SPECIFICATION TABLE NO. 118 SPECIFIC HEAT OF TITANIUM + MANGANESE Ti + Mn

[For Data Reported in Figure and Table No. 118]

Curve No.	Ref. No.	Year	Temp. Range, K	Reported Error, %	Name and Specimen Designation	Composition (weight percent), Specifications and Remarks
1	146	1961	497-1816	3.0	Ti C110M	91.81 Ti, 7.9 Mn, 0.15 O, 0.03 C, 0.01 W; measured under a helium atmosphere; density = 286 lb ft^{-3}.

DATA TABLE NO. 118 SPECIFIC HEAT OF TITANIUM + MANGANESE Ti + Mn

[Temperature, T, K; Specific Heat, Cp, Cal $g^{-1}K^{-1}$]

T	Cp
CURVE 1	
497	1.189×10^{-1}
603	1.197
650	1.201
749	1.208
812	1.213
868	1.217
1051	1.232
1176	1.241
1246	1.247
1365	1.256
1467	1.264
1588	1.273
1705	1.282
1816	1.291

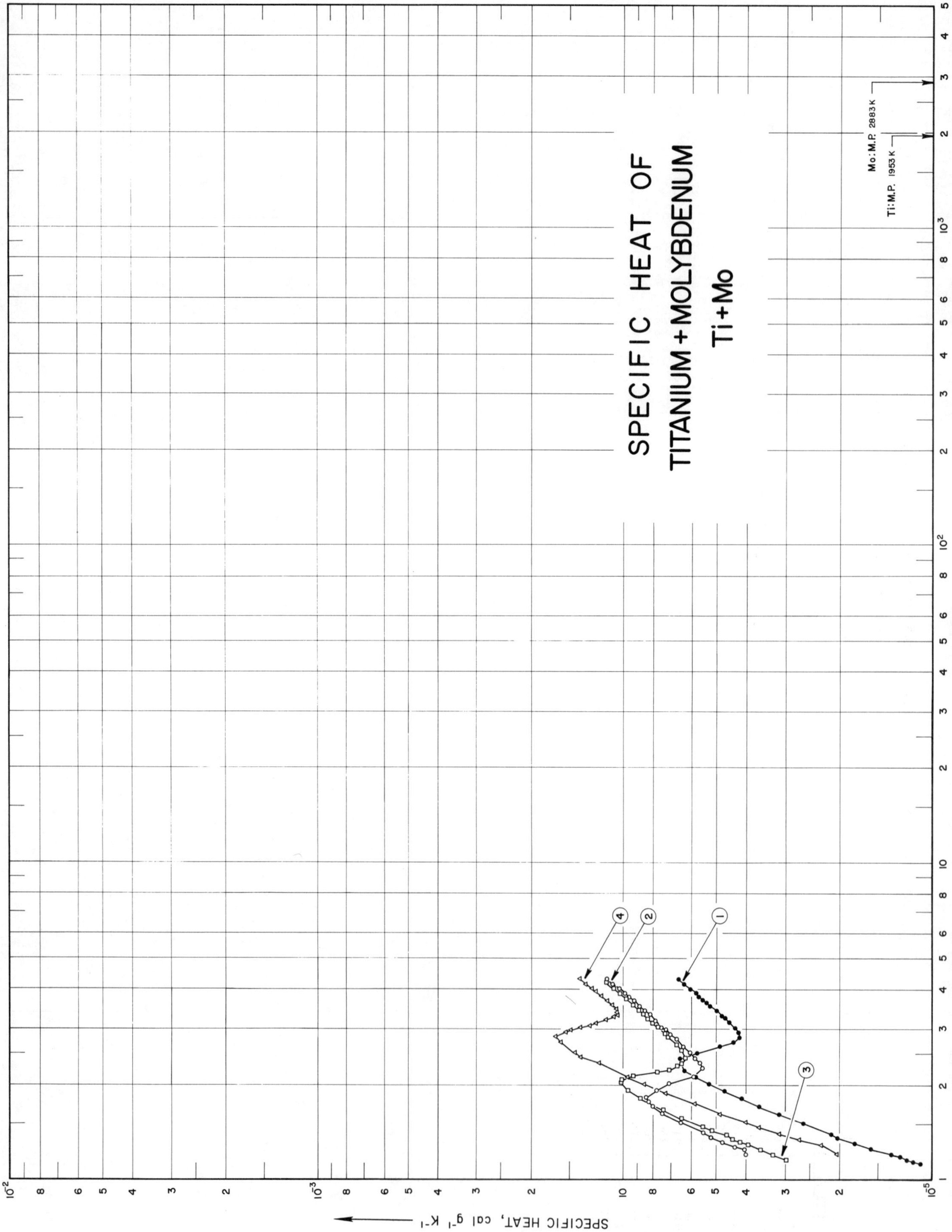

SPECIFIC HEAT OF TITANIUM + MOLYBDENUM Ti + Mo

SPECIFICATION TABLE NO. 119 SPECIFIC HEAT OF TITANIUM + MOLYBDENUM, Ti + Mo

[For Data Reported in Figure and Table No. 119]

Curve No.	Ref. No.	Year	Temp. Range, K	Reported Error, %	Name and Specimen Designation	Composition (weight percent), Specifications and Remarks
1	401	1961	1.1-4.3	± 1.0	M-6	7.54 at % Mo; prepared from 99.92 Ti and 99.9 Mo by melting together in a furnace using "gettered" argon atmosphere; remelted at least six times to promote homogeneity; quenched from region of solid solubility to room temperature.
2	401	1961	1.2-4.3	± 1.0	M-8	6.25 at % Mo. same as above.
3	401	1961	1.2-4.2	± 1.0	M-9	6.50 at % Mo. same as above.
4	401	1961	1.2-4.3	± 1.0	M-10	8.60 at % Mo; same as above.

DATA TABLE NO. 119 SPECIFIC HEAT OF TITANIUM + MOLYBDENUM, Ti + Mo

[Temperature, T, K; Specific Heat, C_p, Cal $g^{-1}K^{-1}$]

CURVE 1

T	C_p
1.127	1.084 x 10^{-5}
1.144	1.153
1.161	1.209
1.181	1.273
1.203	1.361
1.260	1.580
1.316	1.793
1.369	2.028
1.391	2.131
1.511	2.624
1.603	3.133
1.618	3.155*
1.713	3.625
1.808	4.147
1.909	4.702
2.010	5.250
2.115	5.828
2.207	6.345
2.226	6.321*
2.324	6.598*
2.408	6.542
2.502	5.755
2.623	4.863
2.716	4.403
2.820	4.212
2.921	4.242
3.017	4.332
3.126	4.522
3.175	4.595*
3.225	4.686
3.275	4.786
3.403	4.981
3.466	5.101*
3.523	5.248
3.575	5.328*
3.621	5.398
3.675	5.505
3.727	5.578*
3.777	5.679
3.828	5.767*
3.874	5.842
3.928	5.930*
3.980	6.050
4.027	6.105*
4.077	6.232 x 10^{-5}*
4.126	6.340
4.169	6.375*
4.227	6.577*
4.272	6.611

CURVE 2

T	C_p
1.209	4.005 x 10^{-5}
1.240	4.059
1.272	4.350
1.319	4.795
1.366	5.175
1.416	5.507
1.513	6.510
1.619	7.465
1.711	8.022
1.766	8.270
1.819	8.445
1.917	7.822
2.019	6.616
2.116	5.889
2.231	5.545
2.324	5.665
2.415	5.855
2.505	6.050
2.569	6.262*
2.620	6.396
2.672	6.520*
2.766	6.717
2.863	7.004
2.918	7.178*
2.960	7.286
3.011	7.507
3.061	7.697*
3.115	7.772*
3.164	7.866
3.208	8.090*
3.259	8.161*
3.306	8.241
3.400	8.533
3.501	8.888
3.606	9.133*
3.657	9.246
3.704	9.417 x 10^{-5}*
3.753	9.579
3.804	9.726*
3.859	9.904
3.898	1.008 x 10^{-4}*
3.949	1.023*
3.994	1.035
4.049	1.056*
4.090	1.077*
4.140	1.089
4.190	1.113*
4.238	1.125*
4.259	1.132*
4.285	1.135

CURVE 3

T	C_p
1.167	2.987 x 10^{-5}
1.206	3.281
1.247	3.604
1.289	3.949
1.317	4.207
1.348	4.438
1.383	4.633
1.426	5.160
1.470	5.544
1.565	6.465
1.666	7.440
1.763	8.378*
1.812	8.791
1.910	9.655
1.943	9.805*
2.017	1.018 x 10^{-4}
2.069	1.017
2.121	9.281 x 10^{-5}
2.177	7.751
2.218	7.062
2.282	6.678
2.317	6.459
2.380	6.304*
2.419	6.296
2.485	6.344*
2.561	6.474
2.663	6.710
2.717	6.864 x 10^{-5}*
2.817	7.163
2.875	7.316
2.913	7.416*
2.979	7.597*
3.023	7.722
3.120	8.037
3.218	8.321
3.324	8.642
3.408	8.917
3.450	8.980*
3.551	9.306
3.605	9.469*
3.704	9.771
3.752	9.971*
3.806	1.010 x 10^{-4}*
3.851	1.029
3.899	1.023*
3.949	1.061*
3.995	1.076
4.048	1.098*
4.095	1.119*
4.190	1.146
4.232	1.162*

CURVE 4

T	C_p
1.212	2.042 x 10^{-5}
1.281	2.307
1.342	2.720
1.396	3.136
1.460	3.648
1.520	4.012
1.620	4.891
1.737	5.898
1.874	7.292
1.996	8.546
2.107	9.733
2.323	1.214 x 10^{-4}
2.427	1.385
2.519	1.449
2.703	1.615
2.775	1.648*
2.813	1.670
2.846	1.658 x 10^{-4}*
2.910	1.549
2.935	1.510*
2.958	1.492
2.961	1.478*
3.004	1.381
3.062	1.297
3.103	1.268*
3.110	1.239
3.194	1.147
3.219	1.110*
3.261	1.077
3.283	1.090*
3.309	1.056
3.354	1.051
3.399	1.055*
3.431	1.059*
3.449	1.064
3.503	1.079*
3.515	1.087*
3.554	1.095
3.586	1.108*
3.610	1.116*
3.666	1.139
3.712	1.155*
3.716	1.159*
3.784	1.180
3.860	1.214*
3.904	1.232
3.936	1.241*
3.945	1.248*
3.994	1.272
4.045	1.292*
4.091	1.307*
4.139	1.332
4.163	1.339*
4.189	1.352*
4.202	1.360*
4.248	1.376*
4.296	1.396

* Not shown on plot

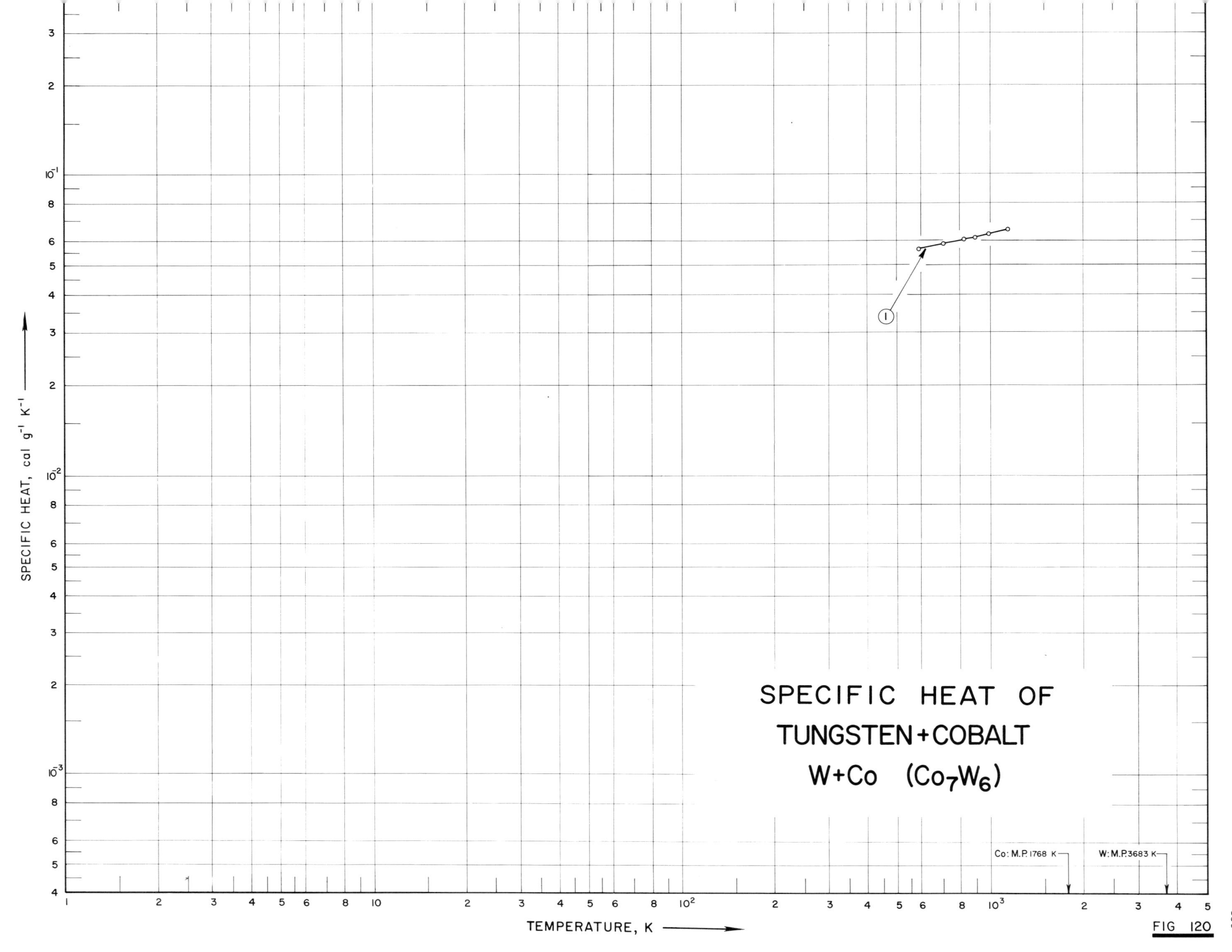
SPECIFIC HEAT OF
TUNGSTEN+COBALT
W+Co (Co7W6)
SPECIFIC HEAT, cal g-1 K-1
TEMPERATURE, K
Co: M.P. 1768 K
W: M.P. 3683 K
1

FIG 120

SPECIFICATION TABLE NO. 120 SPECIFIC HEAT OF TUNGSTEN + COBALT, W + Co (Co_7W_6)

[For Data Reported in Figure and Table No. 120]

Curve No.	Ref. No.	Year	Temp. Range, K	Reported Error, %	Name and Specimen Designation	Composition (weight percent), Specifications and Remarks
1	402	1962	589-1146	± 0.4	Co_7W_6	

DATA TABLE NO. 120 SPECIFIC HEAT OF TUNGSTEN + COBALT W + Co (Co_7W_6)

[Temperature, T, K; Specific Heat, Cp, Cal $g^{-1}K^{-1}$]

T	C_p
CURVE 1	
589.6	5.689 x 10^{-2}
705.7	5.873
704.9	5.872*
823.8	6.060
895.5	6.174
994.2	6.330
1145.4	6.570

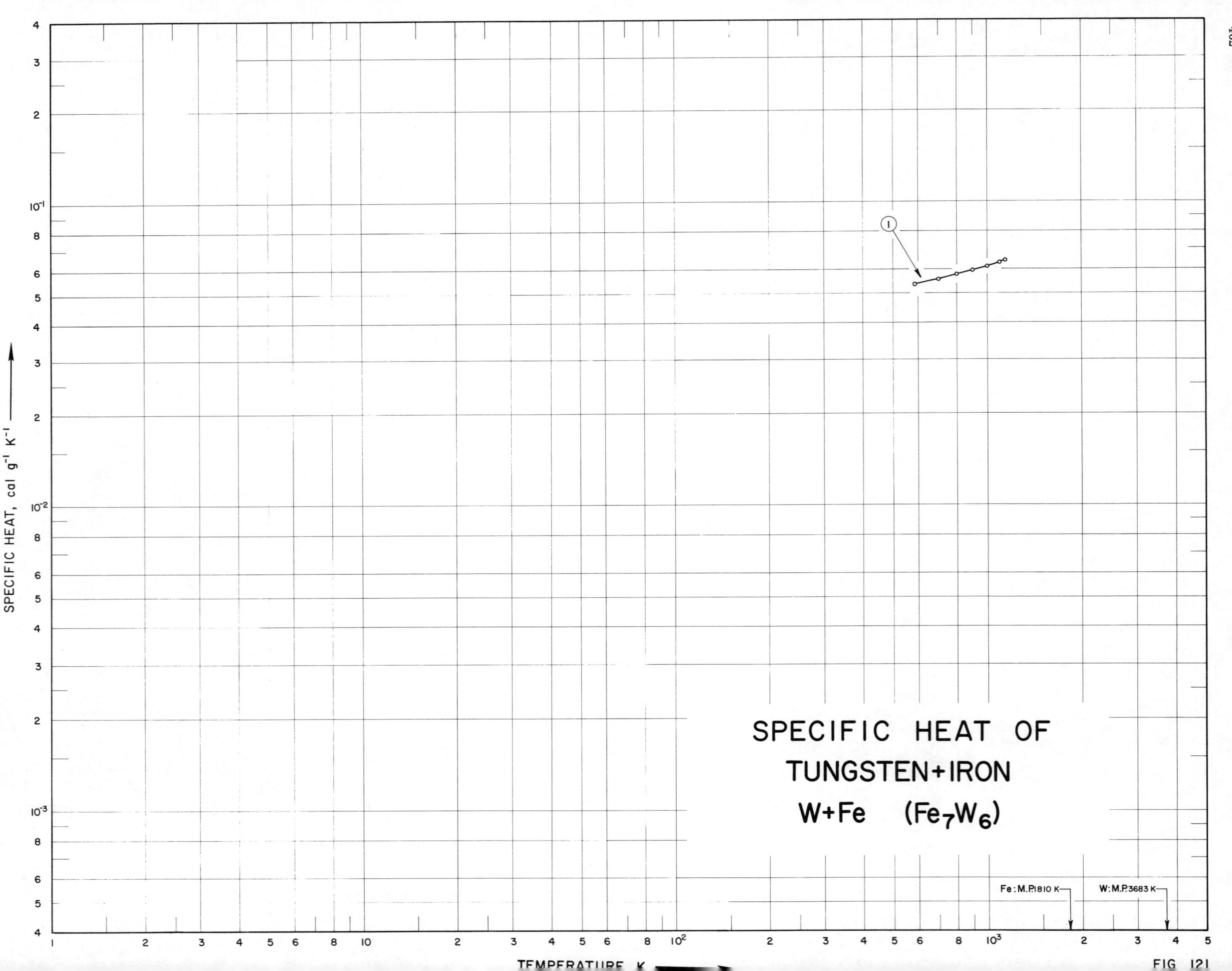
SPECIFIC HEAT OF
TUNGSTEN+IRON
W+Fe (Fe7W6)
SPECIFIC HEAT, cal g-1 K-1
TEMPERATURE, K
Fe: M.P. 1810 K
W: M.P. 3683 K
1

FIG 121

SPECIFICATION TABLE NO. 121 SPECIFIC HEAT OF TUNGSTEN + IRON, W + Fe

[For Data Reported in Figure and Table No. 121]

Curve No.	Ref. No.	Year	Temp. Range, K	Reported Error, %	Name and Specimen Designation	Composition (weight percent), Specifications and Remarks
1	402	1962	590-1145		Fe_7W_6	

DATA TABLE NO. 121 SPECIFIC HEAT OF TUNGSTEN + IRON W + Fe

[Temperature, T, K; Specific Heat, Cp, Cal $g^{-1}K^{-1}$]

T	Cp
CURVE 1	
590	5.4192×10^{-2}
600	5.4378*
700	5.6241
800	5.8104
900	5.9968
1000	6.1831
1100	6.3694
1145	6.4532

* Not shown on plot

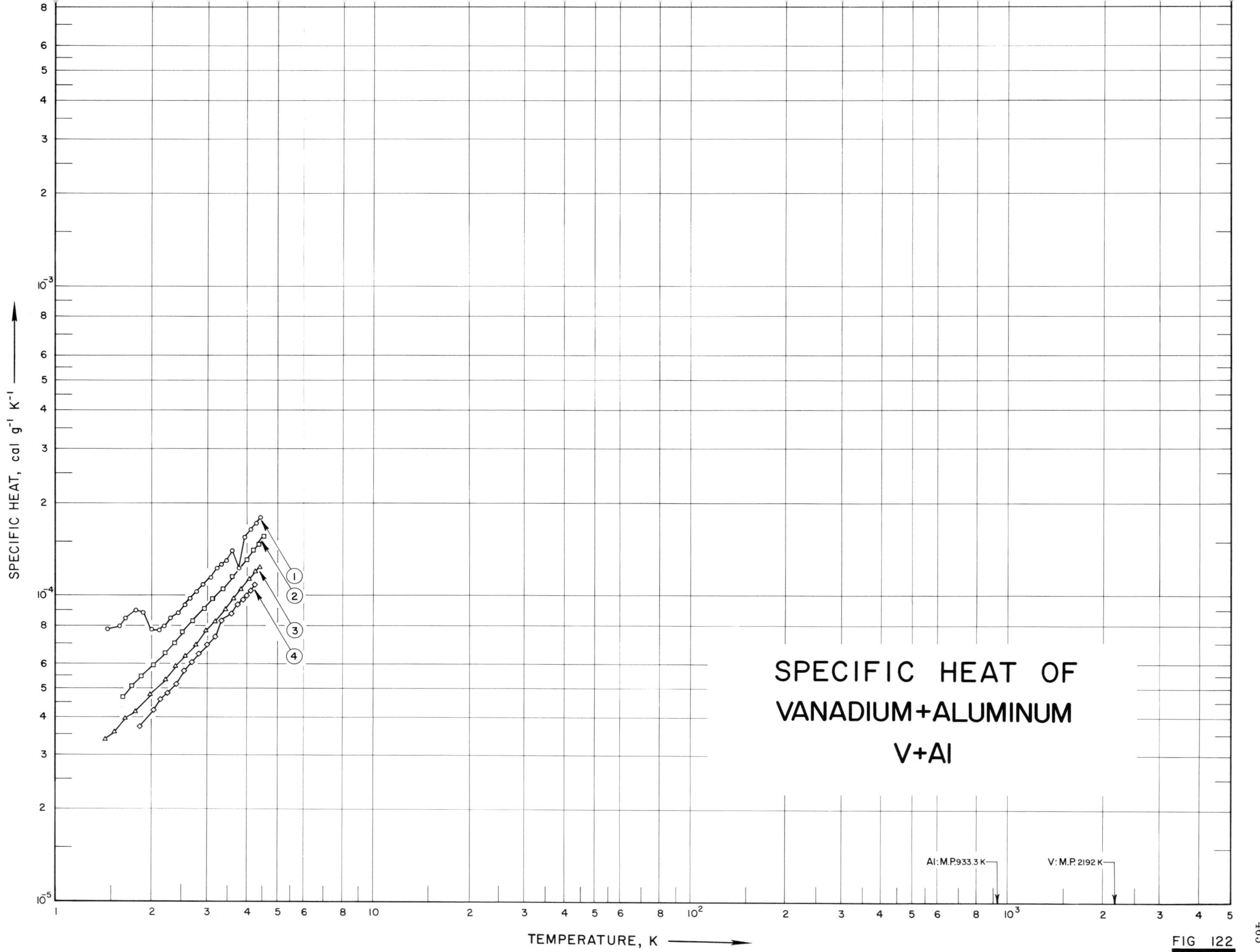
SPECIFIC HEAT OF
VANADIUM+ALUMINUM
V+Al
Al: M.P. 933.3 K
V: M.P. 2192 K
1
2
3
4
SPECIFIC HEAT, cal g⁻¹ K⁻¹
TEMPERATURE, K
FIG 122

SPECIFICATION TABLE NO. 122 SPECIFIC HEAT OF VANADIUM + ALUMINUM, V + Al

[For Data Reported in Figure and Table No. 122]

Curve No.	Ref. No.	Year	Temp. Range, K	Reported Error, %	Name and Specimen Designation	Composition (weight percent), Specifications and Remarks
1	349	1962	1.5–4.4	≤ 2	V(90)Al(10)	93.66 V, 5.98 Al; annealed under He + 8% H_2 gas atmosphere at 1100 C for 72 hrs; etched with 50% HNO_3.
2	349	1962	1.6–4.6	≤ 2	V(80)Al(20)	88.89 V, 10.90 Al; same as above.
3	349	1962	1.4–4.4	≤ 2	V(70)Al(30)	80.70 V, 18.99 Al; same as above.
4	349	1962	1.9–4.3	≤ 2	V(60)Al(40)	73.74 V, 26.2 Al; same as above.

DATA TABLE NO. 122 SPECIFIC HEAT OF VANADIUM + ALUMINUM V + Al

[Temperature, T, K; Specific Heat, Cp, Cal $g^{-1}K^{-1}$]

T	Cp
CURVE 1	
1.465	7.81×10^{-5}
1.518	7.82*
1.586	7.95
1.662	8.46
1.678	8.47*
1.785	8.98
1.897	8.87
2.020	7.72
2.125	7.72
2.213	7.97
2.318	8.41
2.439	8.82
2.556	9.32
2.666	9.78
2.788	1.03×10^{-4}
2.915	1.08
3.083	1.15
3.236	1.23
3.343	1.27
3.465	1.32
3.612	1.40
3.785	1.23
3.956	1.55
4.149	1.64
4.298	1.72
4.438	1.79
4.389	1.47×10^{-4}
4.549	1.56
CURVE 2	
1.638	4.66×10^{-5}
1.747	5.05
1.870	5.45
1.889	5.49*
2.043	5.94
2.218	6.49
2.375	7.02
2.515	7.60
2.717	8.28
2.948	9.05
2.959	9.13*
3.141	9.75
3.372	1.06×10^{-4}
3.602	1.16
4.019	1.32
4.213	1.40
CURVE 3	
1.434	3.37×10^{-5}
1.539	3.57
1.661	3.95
1.792	4.18
1.997	4.75
2.226	5.32
2.396	5.88
2.563	6.35
2.764	6.94
2.980	7.71
3.190	8.27
3.429	9.02
3.647	9.79
3.844	1.05×10^{-4}
4.072	1.13
4.262	1.20
4.418	1.25
CURVE 4	
1.850	3.80×10^{-5}
2.048	4.25
2.149	4.58
2.191	4.64*
2.260	4.81
2.400	5.15
2.554	5.70
2.697	6.05
2.847	6.45
3.017	6.91
3.192	7.35
3.434	8.29
3.579	8.76
3.755	9.39
3.904	9.74
4.017	1.00×10^{-4}
4.133	1.05
4.250	1.09

* Not shown on plot

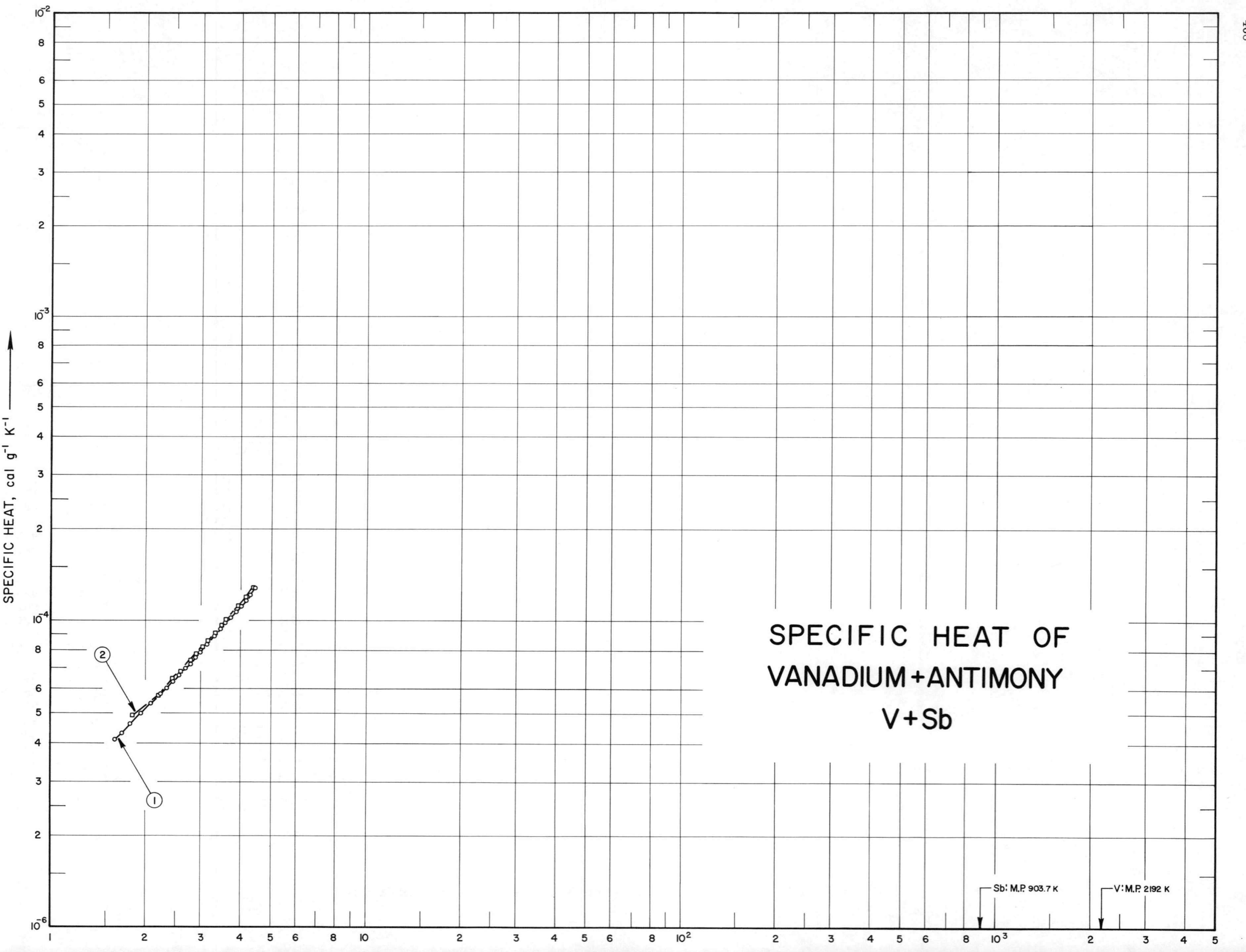
SPECIFIC HEAT OF
VANADIUM+ANTIMONY
V+Sb
SPECIFIC HEAT, cal g⁻¹ K⁻¹
Sb: M.P. 903.7 K
V: M.P. 2192 K
1
2

SPECIFICATION TABLE NO. 123 SPECIFIC HEAT OF VANADIUM + ANTIMONY, V + Sb

[For Data Reported in Figure and Table No. 123]

Curve No.	Ref. No.	Year	Temp. Range, K	Reported Error, %	Name and Specimen Designation	Composition (weight percent), Specifications and Remarks
1	349	1962	1.6-4.5	≤ 2	V(98)Sb(2)	Annealed under He + 8% H_2 gas atmosphere at 1300 C for 72 hrs; etched with 50 % HNO_3.
2	349	1962	1.8-4.4	≤ 2	V(96)Sb(4)	91.20 V, 8.77 Sb; same as above.

DATA TABLE NO. 123 SPECIFIC HEAT OF VANADIUM + ANTIMONY V + Sb

[Temperature, T, K; Specific Heat, Cp, Cal $g^{-1}K^{-1}$]

T	Cp
CURVE 1	
1.603	4.10×10^{-5}
1.685	4.32
1.785	4.61
1.929	4.98
2.076	5.39
2.108	5.46*
2.237	5.75
2.339	6.01
2.442	6.31
2.557	6.64
2.676	6.98
2.779	7.23
2.882	7.57
2.994	7.89
3.137	8.36
3.311	8.88
3.467	9.37
3.591	9.83
3.728	1.02×10^{-4}
3.876	1.07
4.028	1.12
4.183	1.17
4.320	1.22
4.450	1.28
CURVE 2	
1.823	4.907×10^{-5}
1.937	5.000*
2.080	5.395*
2.204	5.743
2.334	6.082*
2.446	6.468
2.532	6.673*
2.596	6.812
2.670	7.084*
2.733	7.235*
2.796	7.408
2.851	7.643*
2.907	7.775
2.996	7.997*
3.032	8.189
3.163	8.570
3.178	8.503*
3.341	9.053
CURVE 2 (cont.)	
3.346	9.101×10^{-5}*
3.499	9.643
3.620	1.009×10^{-4}
3.778	1.045*
3.959	1.123
4.156	1.191
4.403	1.280

* Not shown on plot

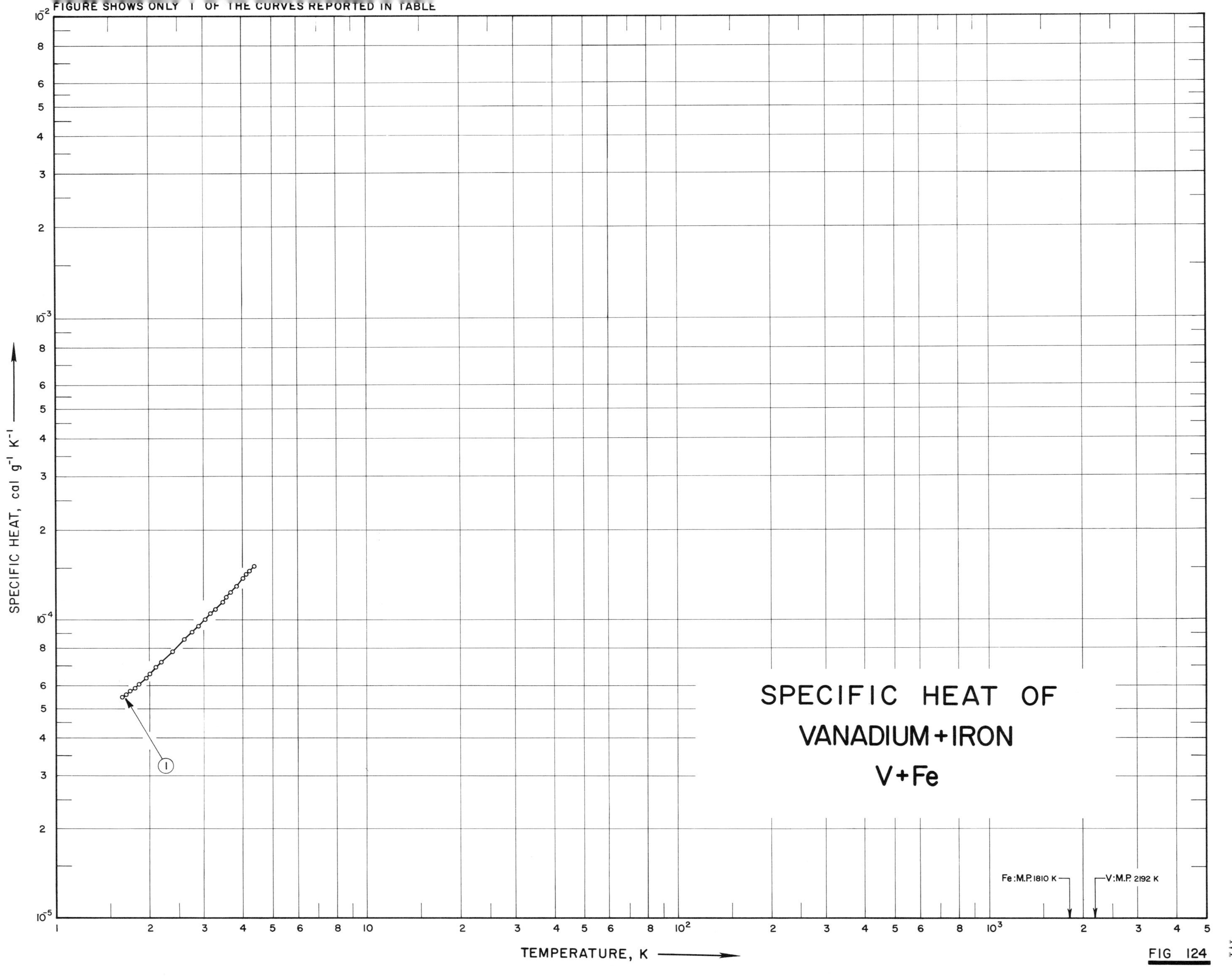
FIGURE SHOWS ONLY 1 OF THE CURVES REPORTED IN TABLE
SPECIFIC HEAT OF
VANADIUM + IRON
V + Fe
SPECIFIC HEAT, cal g⁻¹ K⁻¹
TEMPERATURE, K
Fe: M.P. 1810 K
V: M.P. 2192 K
1

FIG 124

SPECIFICATION TABLE NO. 124 SPECIFIC HEAT OF VANADIUM + IRON, V + Fe

[For Data Reported in Figure and Table No. 124]

Curve No.	Ref. No.	Year	Temp. Range, K	Reported Error, %	Name and Specimen Designation	Composition (weight percent), Specifications and Remarks
1	297	1959	1.6-4.4		$Fe_{0.33}V_{0.67}$	63.4 V, 36.6 Fe; arc melted.
2	419	1964	0.1-0.9		$Fe_{.464}V_{.536}$	

DATA TABLE NO. 124 SPECIFIC HEAT OF VANADIUM + IRON V + Fe

[Temperature, T, K; Specific Heat, Cp, Cal $g^{-1}K^{-1}$]

T	Cp
CURVE 1	
1.631	5.494 x 10^{-5}
1.680	5.611
1.734	5.765
1.800	5.877
1.858	6.097
1.907	6.099*
1.961	6.367
2.024	6.594
2.102	6.917
2.190	7.202
2.377	7.817
2.599	8.577
2.750	9.097
2.877	9.549
3.041	1.005 x 10^{-4}
3.161	1.053
3.267	1.084
3.451	1.159
3.543	1.195
3.608	1.214*
3.664	1.246
3.743	1.272*
3.825	1.303
3.877	1.328*
4.013	1.382
4.118	1.425
4.226	1.471
4.356	1.521
CURVE 2*	
0.100	3.867 x 10^{-7}
0.110	4.211
0.115	4.392
0.121	4.590
0.127	4.786
0.133	5.001
0.139	5.231
0.147	5.475
0.154	5.722
0.161	5.967
0.169	6.228
0.177	6.485
0.186	6.790
0.195	7.116
CURVE 2 (cont.)	
0.205	7.403 x 10^{-7}
0.215	7.741
0.226	8.097
0.237	8.465
0.248	8.807
0.260	9.211
0.273	9.640
0.287	1.005 x 10^{-6}
0.300	1.050
0.314	1.095
0.329	1.142
0.345	1.197
0.360	1.245
0.375	1.293
0.391	1.348
0.458	1.657
0.474	1.777
0.491	1.883
0.509	1.972
0.527	2.059
0.546	2.144
0.566	2.240
0.588	2.337
0.611	2.435
0.637	2.544
0.664	2.657
0.693	2.774
0.725	2.891
0.758	3.018
0.793	3.148
0.831	3.278
0.870	3.416
0.914	3.555
0.959	3.704

* Not shown on plot

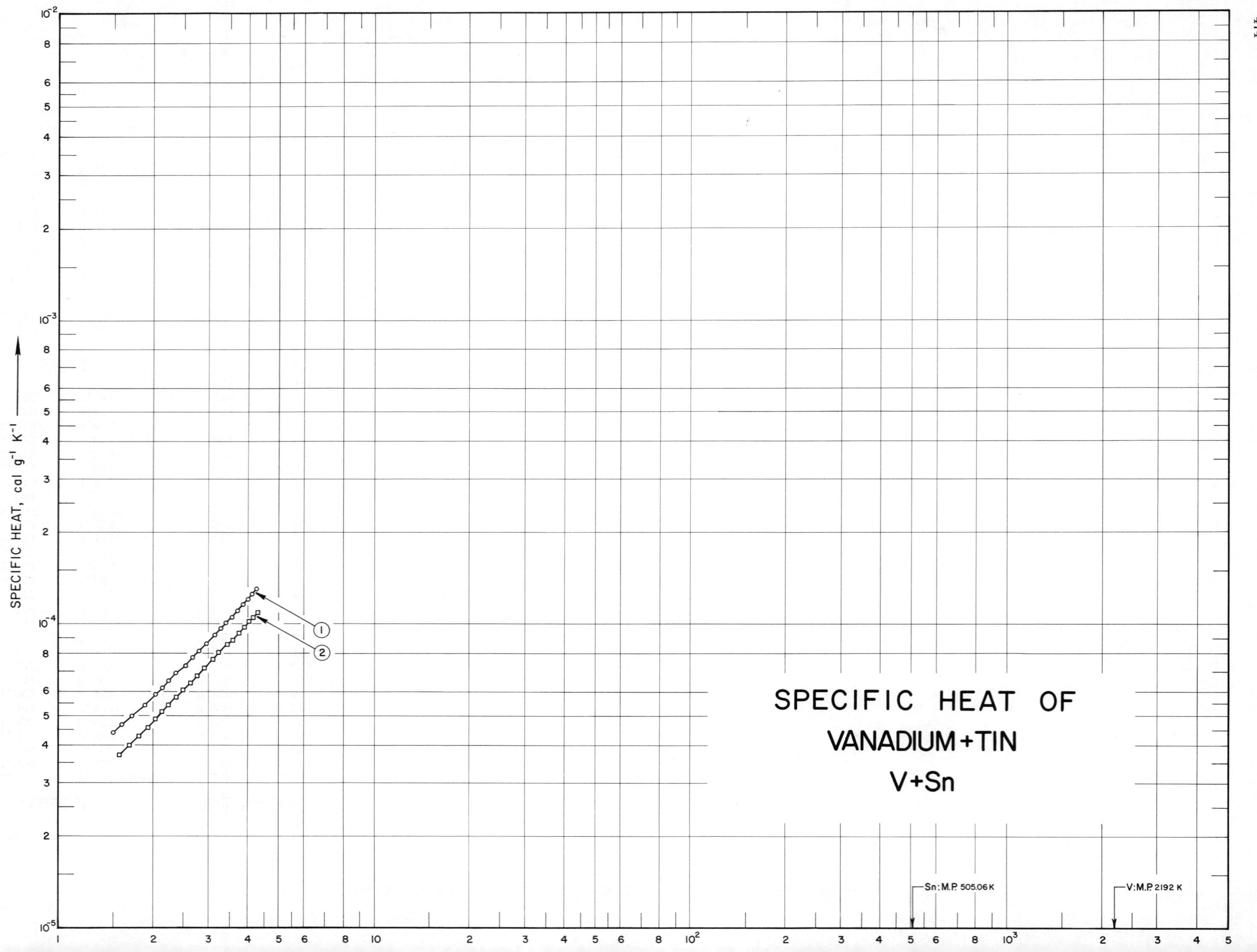

SPECIFIC HEAT OF
VANADIUM+TIN
V+Sn
SPECIFIC HEAT, cal g⁻¹ K⁻¹
Sn: M.P. 505.06 K
V: M.P. 2192 K
1
2

SPECIFICATION TABLE NO. 125 SPECIFIC HEAT OF VANADIUM + TIN, V+ Sn

[For Data Reported in Figure and Table No. 125]

Curve No.	Ref. No.	Year	Temp. Range, K	Reported Error, %	Name and Specimen Designation	Composition (weight percent), Specifications and Remarks
1	349	1962	1.5–4.3	≤ 2	V(95)Sn(5)	87.57 V, 12.33 Sn.
2	349	1962	1.6–4.3	≤ 2	V(90)Sn(10)	78.15 V, 21.77 Sn.

DATA TABLE NO. 125 SPECIFIC HEAT OF VANADIUM + TIN V + Sn

[Temperature, T,K; Specific Heat, Cp, Cal $g^{-1}K^{-1}$]

T	Cp
CURVE 1	
1.497	4.40×10^{-5}
1.597	4.66
1.598	4.63*
1.721	5.00
1.882	5.48
2.036	5.87
2.043	5.94*
2.145	6.17
2.238	6.54
2.370	6.90
2.527	7.35
2.668	7.80
2.798	8.16
2.954	8.65
3.133	9.20
3.286	9.69
3.413	1.02×10^{-4}
3.553	1.06
3.699	1.11
3.857	1.17
3.997	1.22
4.116	1.26
4.252	1.32
CURVE 2	
1.574	3.71×10^{-5}
1.682	3.99
1.813	4.30
1.927	4.56
2.035	4.87
2.147	5.16
2.245	5.43
2.370	5.76
2.499	6.10
2.625	6.44
2.754	6.76
2.914	7.19
3.088	7.65
3.226	8.07
3.325	8.34*
3.440	8.57
3.581	8.87
3.729	9.31
3.893	9.77
CURVE 2 (cont.)	
4.037	1.02×10^{-4}
4.152	1.05
4.277	1.10

* Not shown on plot

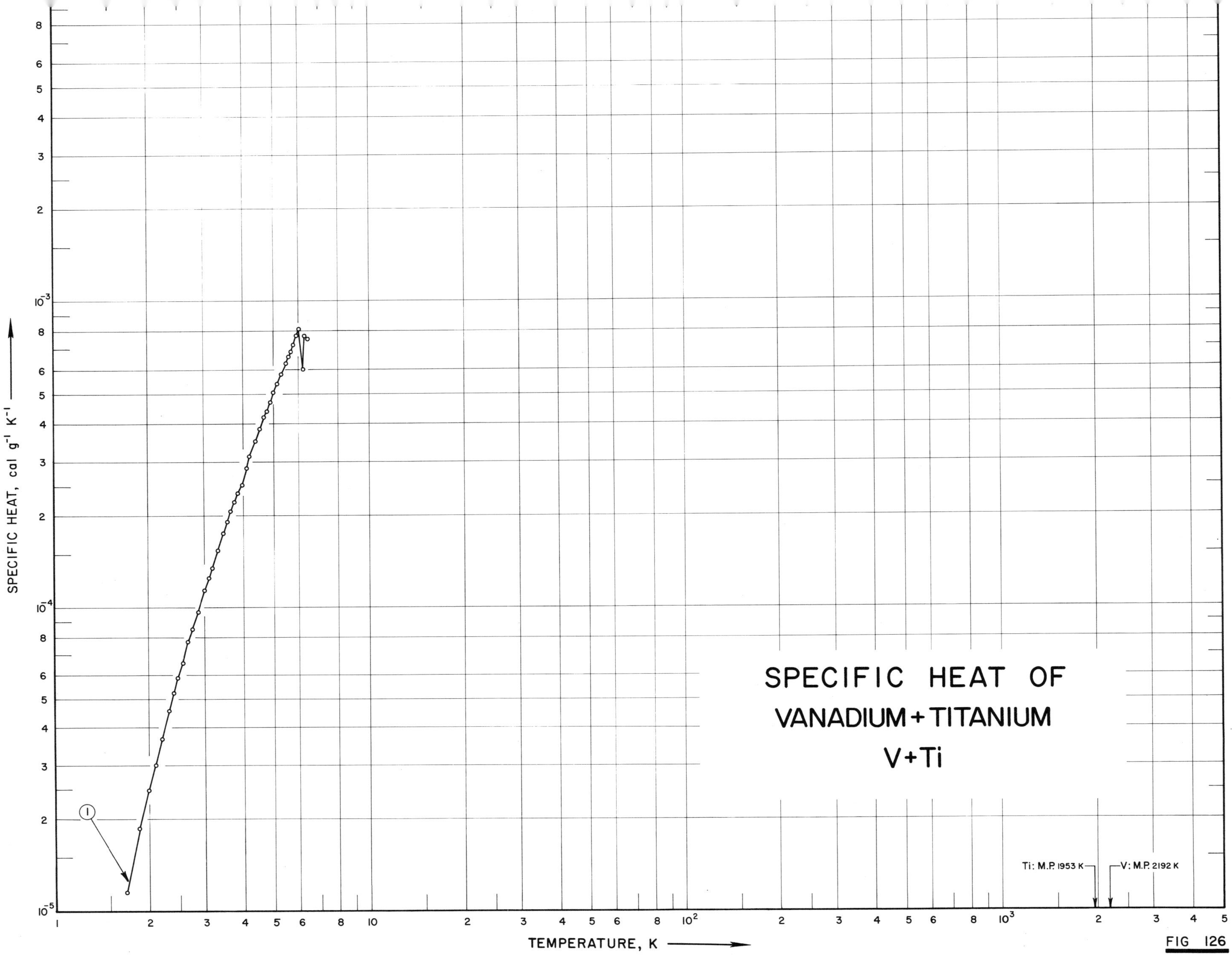

SPECIFIC HEAT OF
VANADIUM + TITANIUM
V+Ti

SPECIFIC HEAT, cal g^{-1} K^{-1}

TEMPERATURE, K

Ti: M.P. 1953 K
V: M.P. 2192 K

FIG 126

SPECIFICATION TABLE NO. 126 SPECIFIC HEAT OF VANADIUM + TITANIUM, V + Ti

[For Data Reported in Figure and Table No. 126]

Curve No.	Ref. No.	Year	Temp. Range, K	Reported Error, %	Name and Specimen Designation	Composition (weight percent), Specifications and Remarks
1	297	1959	1.7-6.4		$Ti_{0.5}V_{0.5}$	51.5 V, 48.5 Ti; arc melted.

DATA TABLE NO. 126 SPECIFIC HEAT OF VANADIUM + TITANIUM V + Ti

[Temperature, T, K; Specific Heat, Cp, Cal $g^{-1}K^{-1}$]

T	Cp
CURVE 1	
Series I	
1.690	1.155×10^{-5}
1.858	1.861
1.995	2.481
2.080	3.001
2.198	3.666
2.317	4.575
2.399	5.250
2.469	5.892
2.550	6.563
2.662	7.749
2.749	8.542
2.869	9.657
3.016	1.146×10^{-4}
3.105	1.252
3.185	1.369
3.321	1.548
3.454	1.753
3.565	1.921
3.645	2.074
3.738	2.224
3.834	2.375
3.960	2.527
4.110	2.873
4.383	3.514
Series II	
3.963	2.584×10^{-4}*
4.073	2.831*
4.199	3.146
4.347	3.456*
4.524	3.854
4.666	4.204
4.763	4.433
4.868	4.726
4.963	4.894*
4.996	5.070
5.061	5.293*
5.142	5.439
5.218	5.627*
5.297	5.815
5.463	6.282
5.564	6.622
CURVE 1 (cont.)	
5.684	6.871×10^{-4}
5.726	7.140*
5.782	7.205
5.840	7.431*
5.900	7.737
5.953	7.846*
6.008	8.115
6.076	8.257*
6.147	8.142
6.220	6.035
6.296	7.737
6.371	7.711*
6.446	7.555

* Not shown on plot

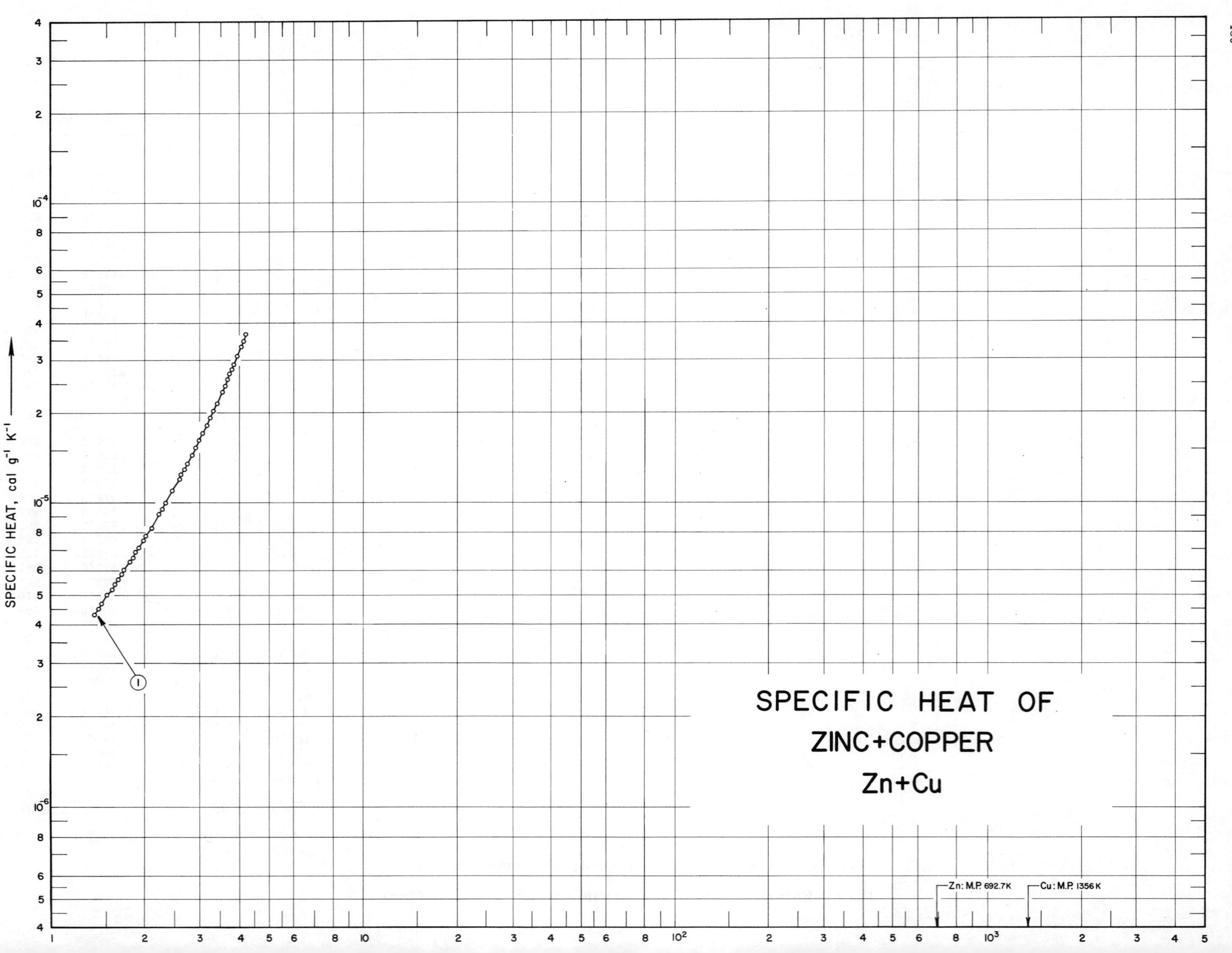
SPECIFIC HEAT OF
ZINC+COPPER
Zn+Cu
Zn: M.P. 692.7K
Cu: M.P. 1356 K
SPECIFIC HEAT, cal g⁻¹ K⁻¹
1

SPECIFICATION TABLE NO. 127 SPECIFIC HEAT OF ZINC + COPPER, Zn + Cu

[For Data Reported in Figure and Table No. 127]

Curve No.	Ref. No.	Year	Temp. Range, K	Reported Error, %	Name and Specimen Designation	Composition (weight percent), Specifications and Remarks
1	390	1962	1.4–4.2		49.92 at % Zn	50.64 Zn, 49.36 Cu; annealed 20 min at 810 C, twice quenched from 810 C.

DATA TABLE NO. 127 SPECIFIC HEAT OF ZINC + COPPER Zn + Cu

[Temperature, T, K; Specific Heat, Cp, Cal $g^{-1}K^{-1}$]

T	Cp	T	Cp
CURVE 1		CURVE 1 (cont.)	
Series I		Series II	
1.392	4.335×10^{-6}	1.400	4.424×10^{-6}*
1.424	4.535	1.424	4.542*
1.456	4.691	1.469	4.732*
1.491	4.769*	1.480	4.772*
1.524	5.021	1.531	5.039*
1.583	5.247	1.619	5.444
1.724	6.022	1.650	5.618×10^{-5}
1.877	6.908	1.706	5.848×10^{-6}
1.981	7.520	1.804	6.400
1.998	7.494*	1.838	6.604
2.014	7.691*	1.919	7.120
2.015	7.791	2.016	7.694*
2.120	8.299	2.025	7.813*
2.235	9.222	2.283	9.571
2.340	9.993	2.390	4.402*
2.463	1.104×10^{-5}	2.620	1.244×10^{-5}
2.584	1.201	2.889	1.501*
2.676	1.299	3.122	1.774*
2.748	1.367	3.356	2.083*
2.833	1.453	3.557	2.368*
2.916	1.533	3.775	2.741*
2.994	1.623	4.012	3.198*
3.071	1.720	4.202	3.585*
3.160	1.820		
3.241	1.928		
3.323	2.045		
3.422	2.169		
3.554	2.367		
3.619	2.473		
3.683	2.611		
3.748	2.705		
3.809	2.800		
3.867	2.917		
3.898	2.967*		
3.963	3.101		
4.015	3.186*		
4.080	3.343		
4.148	3.486		
4.223	3.670		

* Not shown on plot

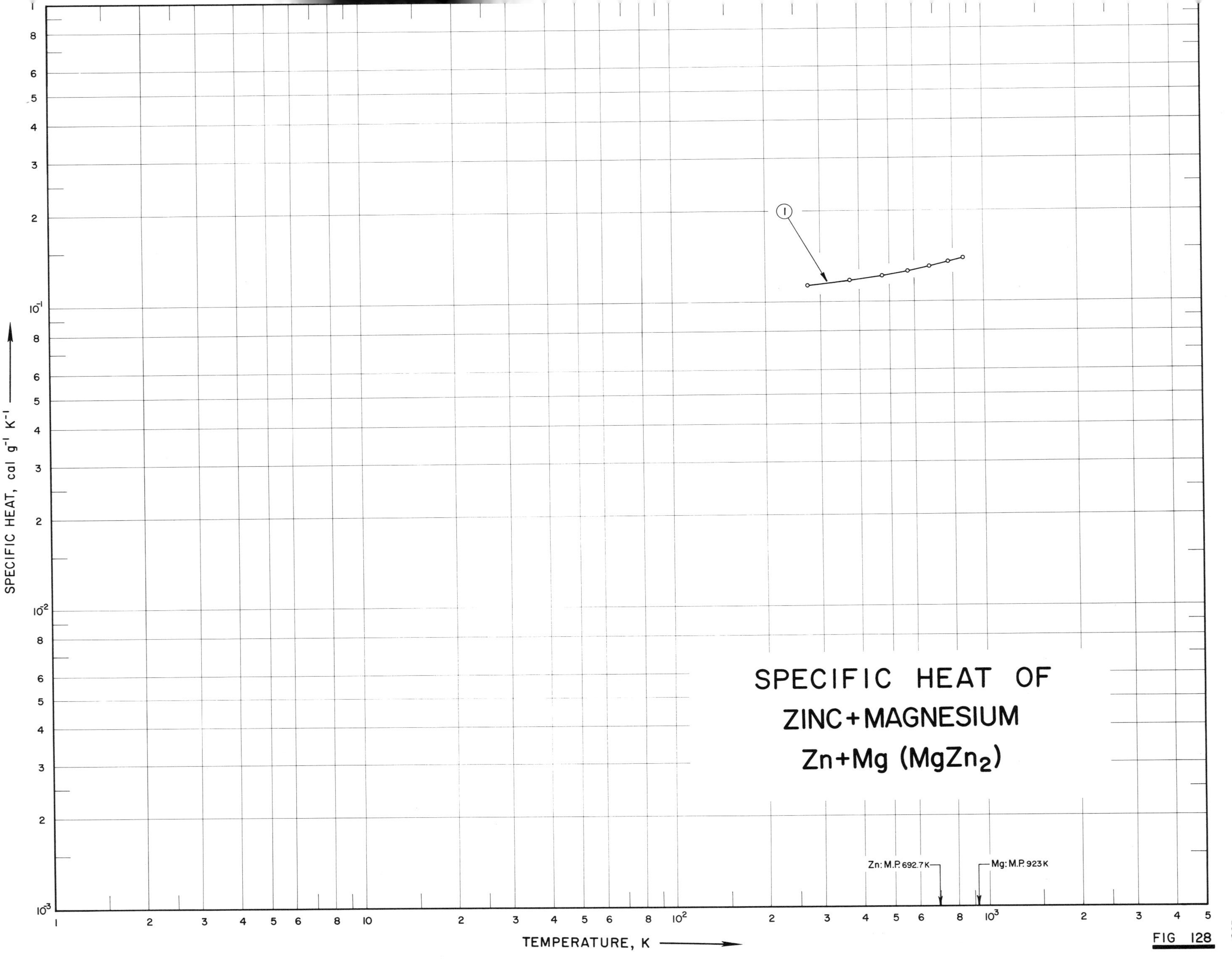
SPECIFIC HEAT OF
ZINC+MAGNESIUM
Zn+Mg ($MgZn_2$)
1
Zn: M.P. 692.7 K
Mg: M.P. 923 K
TEMPERATURE, K
SPECIFIC HEAT, cal g^{-1} K^{-1}
FIG 128

SPECIFICATION TABLE NO. 128 SPECIFIC HEAT OF ZINC + MAGNESIUM, Zn + Mg ($MgZn_2$)

[For Data Reported in Figure and Table No. 128]

Curve No.	Ref. No.	Year	Temp. Range, K	Reported Error, %	Name and Specimen Designation	Composition (weight percent), Specifications and Remarks
1	208	1935	273-590		$MgZn_2$	84.20 Zn, 15.80 Mg (84.3 Zn, 15.7 Mg theorectically); obtained by melting stoichiometric quantities in hydrogen atmosphere.

DATA TABLE NO. 128 SPECIFIC HEAT OF ZINC + MAGNESIUM Zn + Mg ($MgZn_2$)

[Temperature, T, K; Specific Heat, Cp, Cal $g^{-1}K^{-1}$]

T	Cp
CURVE 1	
273.15	1.135×10^{-1}
100	1.179
200	1.223
300	1.267
400	1.311
500	1.355
590	1.394

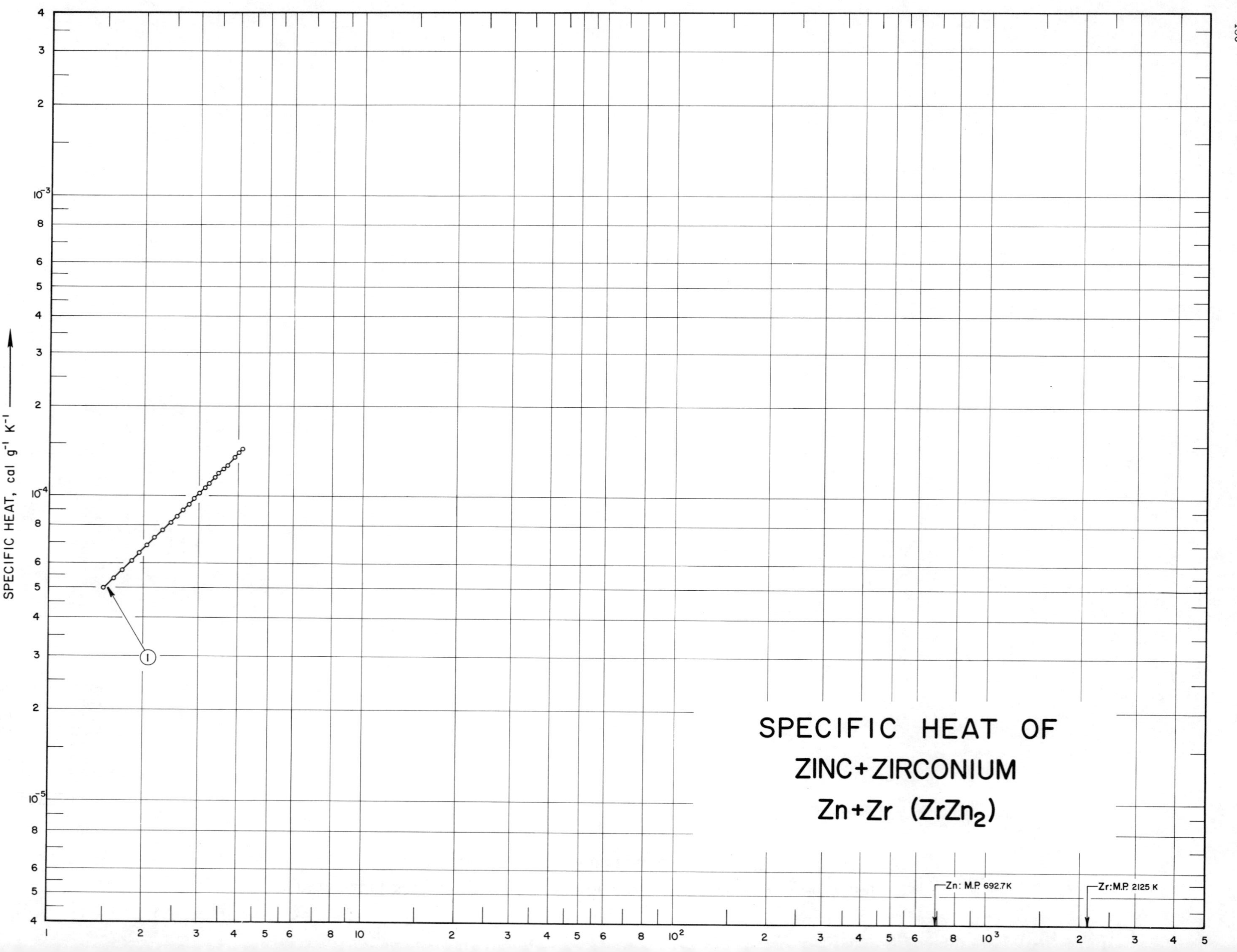

SPECIFIC HEAT OF
ZINC+ZIRCONIUM
Zn+Zr (ZrZn$_2$)
Zn: M.P. 692.7 K
Zr: M.P. 2125 K
SPECIFIC HEAT, cal g^{-1} K^{-1}

SPECIFICATION TABLE NO. 129 SPECIFIC HEAT OF ZINC + ZIRCONIUM, Zn + Zr

[For Data Reported in Figure and Table No. 129]

Curve No.	Ref. No.	Year	Temp. Range, K	Reported Error, %	Name and Specimen Designation	Composition (weight percent), Specifications and Remarks
1	403	1966	1.4-4.1		$ZrZn_2$	

DATA TABLE NO. 129 SPECIFIC HEAT OF ZINC + ZIRCONIUM Zn + Zr

[Temperature, T, K; Specific Heat, Cp, Cal $g^{-1}K^{-1}$]

T	Cp
CURVE 1	
Series I	
1.496	4.973×10^{-5}
1.607	5.364
1.639	5.456*
1.719	5.723
1.751	5.861*
1.831	6.114
1.876	6.282*
1.942	6.484
2.056	6.862
2.172	7.273
2.302	7.727
2.435	8.192
2.551	8.591
2.666	8.988
2.781	9.397
2.898	9.807
3.014	1.023×10^{-4}
3.126	1.064*
3.239	1.103
3.351	1.147*
3.469	1.191
3.581	1.233*
3.686	1.272
3.793	1.316*
3.896	1.356
4.001	1.397*
4.114	1.439*
Series II	
1.998	6.686×10^{-5}*
2.114	7.083*
2.219	7.476*
2.331	7.830*
2.449	8.232*
2.564	8.644*
2.681	9.059*
2.796	9.456*
2.911	9.877*
3.027	1.029×10^{-4}*
3.143	1.071
3.265	1.117*
3.375	1.158
3.480	1.197*
CURVE 1 (cont.)	
3.591	1.237×10^{-4}
3.698	1.280*
3.911	1.362*
4.019	1.405
4.125	1.446

* Not shown on plot

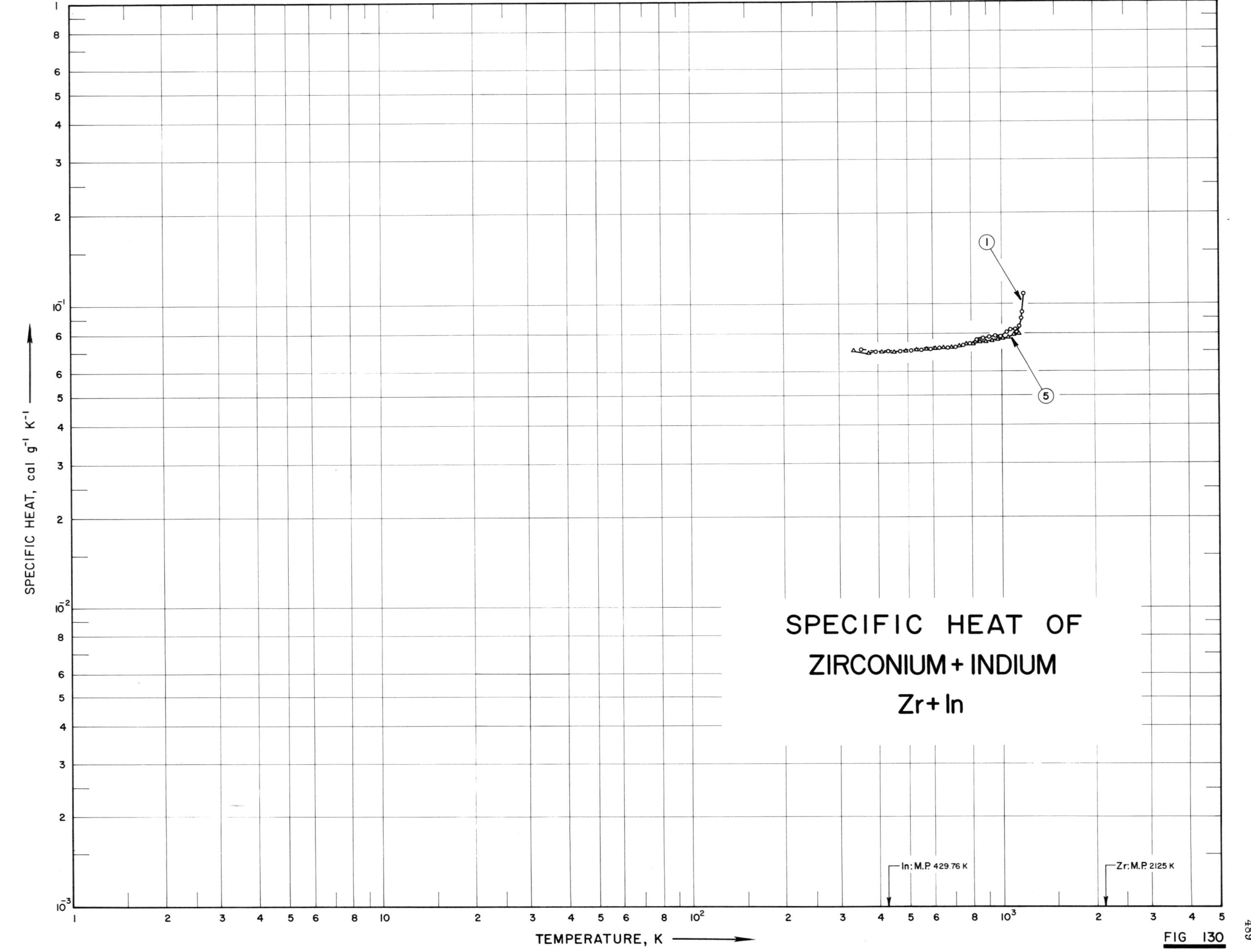
SPECIFIC HEAT OF
ZIRCONIUM + INDIUM
Zr + In
SPECIFIC HEAT, cal g⁻¹ K⁻¹
TEMPERATURE, K
In: M.P. 429.76 K
Zr: M.P. 2125 K
1
5

FIG 130

SPECIFICATION TABLE NO. 130 SPECIFIC HEAT OF ZIRCONIUM + INDIUM, Zr + In

[For Data Reported in Figure and Table No. 130]

Curve No.	Ref. No.	Year	Temp. Range, K	Reported Error, %	Name and Specimen Designation	Composition (weight percent), Specifications and Remarks
1	134	1957	343-1178		Zirconium 7.77% Indium Alloy	92.23 Zr, 7.77 In, 0.21 Fe, 0.016 O_2, 0.0067 C, 0.003 N_2, 0.00051 H_2; arc melted; homogenized 14 days at 1300 C in vacuum.
2	134	1957	353-1153		same as above	Same as above.
3	134	1957	343-1173		same as above	Same as above.
4	134	1957	333-1013		same as above	Same as above.
5	134	1957	333-1133		same as above	Same as above.

DATA TABLE NO. 130 SPECIFIC HEAT OF ZIRCONIUM + INDIUM Zr + In

[Temperature, T, K; Specific Heat, Cp, Cal $g^{-1}K^{-1}$]

CURVE 1

T	Cp
353.15	7.082 x 10^{-2}
393.15	6.970
433.15	6.980
473.15	6.991
513.15	7.032
553.15	7.052
593.15	7.109
633.15	7.175
673.15	7.184
713.15	7.215
753.15	7.310
793.15	7.438
833.15	7.626
873.15	7.705
913.15	7.772
953.15	7.863
993.15	7.845
1033.15	8.053
1073.15	8.209
1103.15	8.211
1123.15	8.071
1143.15	8.465
1158.15	8.982
1168.15	9.436
1178.15	1.0867 x 10^{-1}

CURVE 2

T	Cp
353.15	6.924 x 10^{-2}
393.15	6.960
433.15	6.969
473.15	6.991
513.15	7.065
553.15	7.153
593.15	7.185
633.15	7.227
673.15	7.247
713.15	7.296
753.15	7.397
793.15	7.529
833.15	7.582
873.15	7.642
913.15	7.653
953.15	7.668
993.15	7.880
1033.15	8.039 x 10^{-2}
1073.15	8.024
1113.15	8.113
1153.15	8.643

CURVE 3

T	Cp
343.15	7.034 x 10^{-2}
383.15	6.965
423.15	6.976
463.15	6.980
503.15	7.041
543.15	7.169
583.15	7.168
623.15	7.220
663.15	7.244
703.15	7.300
743.15	7.381
783.15	7.503
823.15	7.611
863.15	7.692
903.15	7.748
943.15	7.802
983.15	7.938
1023.15	8.024
1063.15	8.175
1103.15	8.403
1143.15	8.465
1173.15	1.1912 x 10^{-1}

CURVE 4

T	Cp
333.15	6.999 x 10^{-2}
373.15	6.903
413.15	6.964
453.15	6.986
493.15	7.044
533.15	7.103
573.15	7.142
613.15	7.197
653.15	7.261
693.15	7.303
733.15	7.409
773.15	7.540
813.15	7.669
853.15	7.758
893.15	7.837
933.15	7.886
983.15	8.011
1033.15	8.111
1073.15	8.074
1113.15	8.028

CURVE 5

T	Cp
333.15	7.017 x 10^{-2}
373.15	6.916
413.15	6.964
453.15	6.959
493.15	7.015
533.15	7.081
573.15	7.116
613.15	7.169
653.15	7.199
693.15	7.226
733.15	7.285
773.15	7.384
813.15	7.444
853.15	7.501
893.15	7.538
933.15	7.591
973.15	7.659
1013.15	7.712
1053.15	7.773
1093.15	7.922
1133.15	7.992

* Not shown on plot

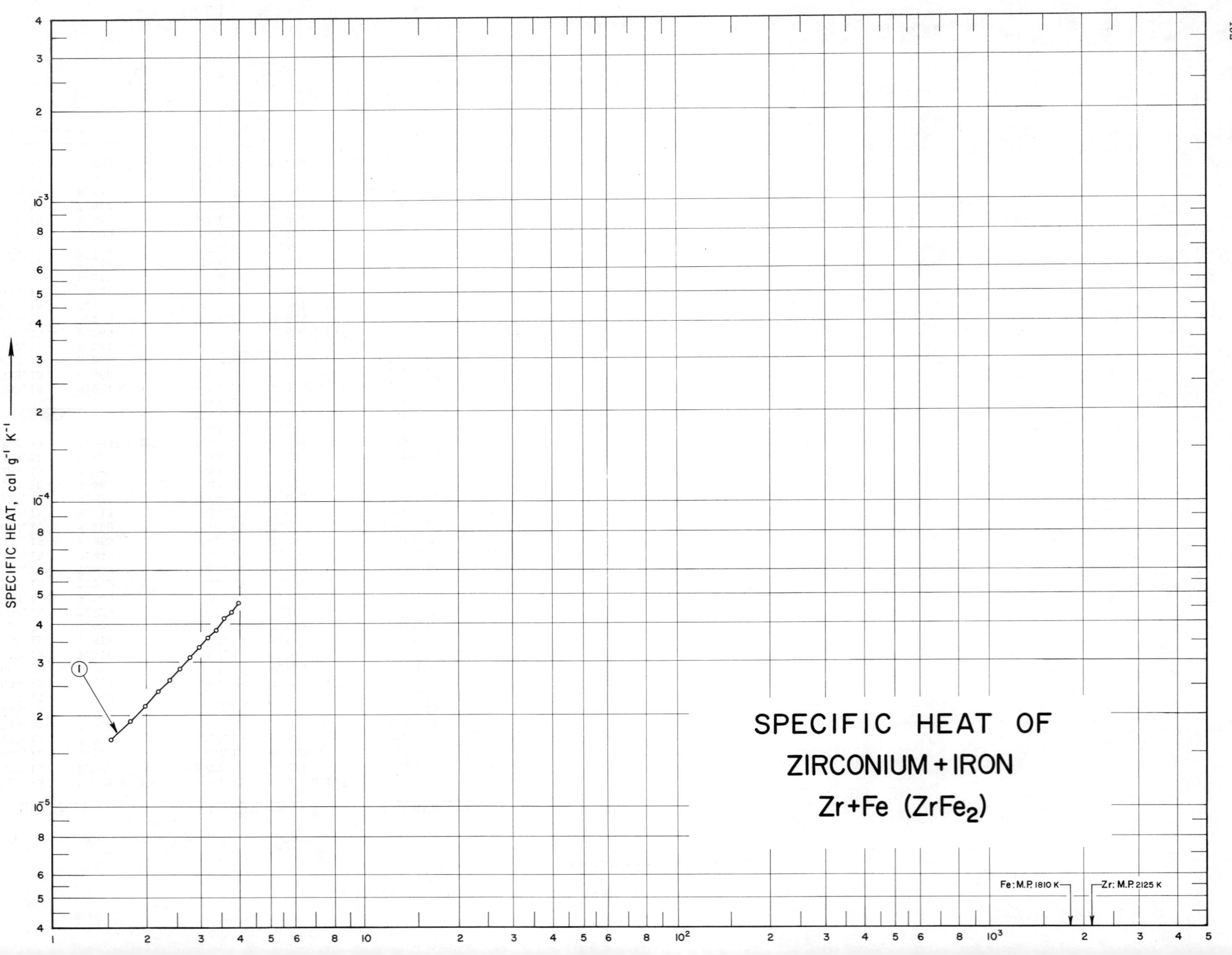
SPECIFIC HEAT OF
ZIRCONIUM + IRON
Zr+Fe (ZrFe$_2$)
Fe: M.P. 1810 K
Zr: M.P. 2125 K
SPECIFIC HEAT, cal g^{-1} K^{-1}
1

SPECIFICATION TABLE NO. 131 SPECIFIC HEAT OF ZIRCONIUM + IRON, Zr + Fe

[For Data Reported in Figure and Table No. 131]

Curve No.	Ref. No.	Year	Temp. Range, K	Reported Error, %	Name and Specimen Designation	Composition (weight percent), Specifications and Remarks
1	240	1962	1.6-4.0		$ZrFe_2$	Prepared from: Zirconium. 99.95 Zr, 0.01 Hf, 0.005 Si, 0.005 Al, 0.005 Mg, 0.005 Fe, 0.005 Ti, 0.005 Ni, 0.005 Ca, 0.0005 Cu, 0.015 n (sample supplied by Foote Mineral Co). and Ferrovac E Iron. 99.95 Fe, 0.024 C, 0.001-0.005 Mn, 0.0023 O_2, 0.0004 N_2, 0.007 Si, 0.005 Ni, 0.006 Sn, 0.001-0.004 Mo, 0.003-0.006 Co, 0.001-0.003 Cu, 0.001-0.006 Al, 0.001 Pb (sample supplied by Crucible Steel Corp).

DATA TABLE NO. 131 SPECIFIC HEAT OF ZIRCONIUM + IRON Zr + Fe

[Temperature, T, K; Specific Heat, Cp, Cal $g^{-1}K^{-1}$]

T	Cp
CURVE 1	
1.559	1.656×10^{-5}
1.787	1.918
1.992	2.135
2.188	2.392
2.381	2.605
2.572	2.845
2.770	3.094
2.969	3.340
3.170	3.607
3.372	3.822
3.571	4.161
3.773	4.351
3.972	4.671

*Not shown on plot

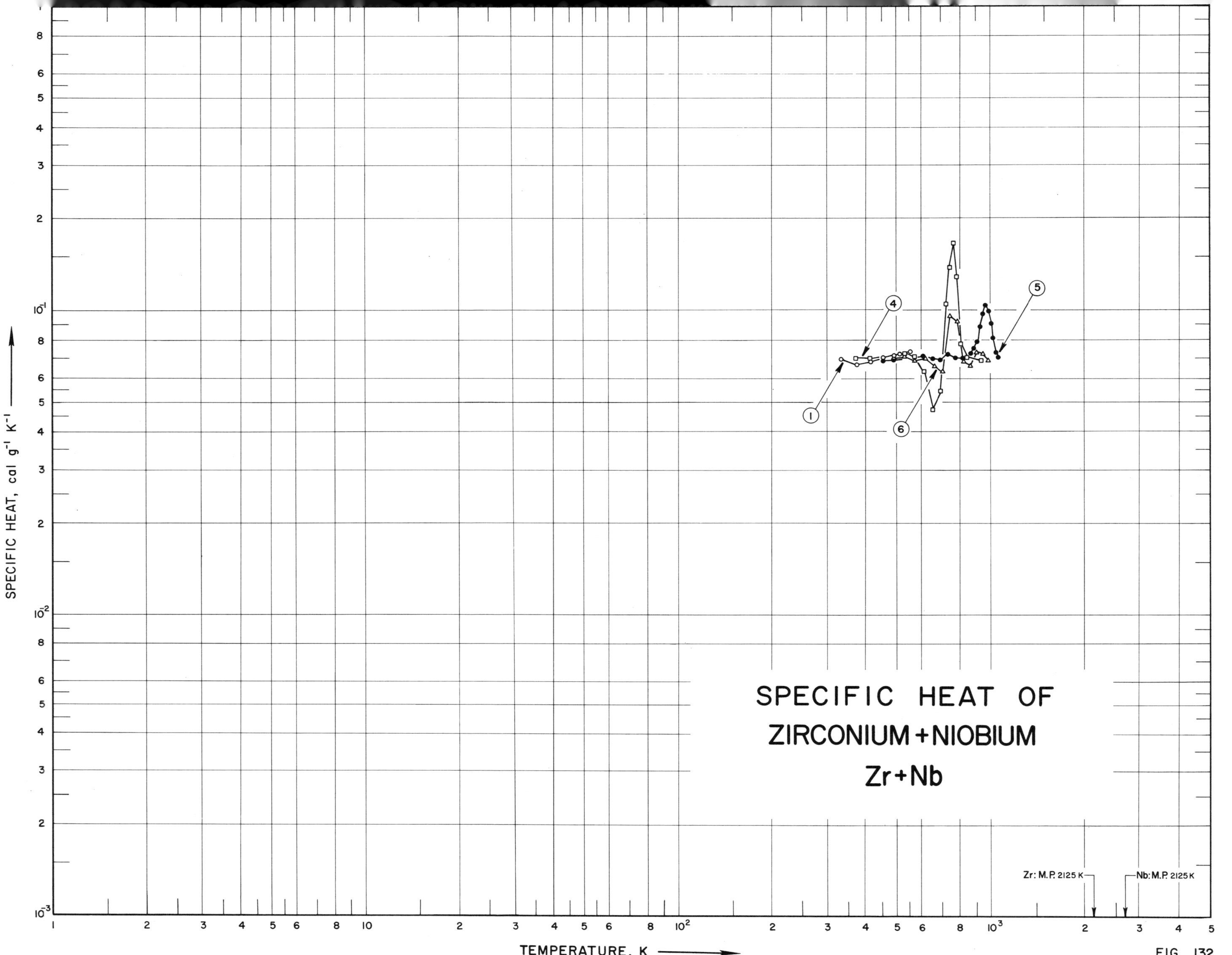

SPECIFIC HEAT OF
ZIRCONIUM + NIOBIUM
Zr + Nb
Zr: M.P. 2125 K
Nb: M.P. 2125 K
TEMPERATURE, K
SPECIFIC HEAT, cal g⁻¹ K⁻¹
1
4
5
6

FIG 132

SPECIFICATION TABLE NO. 132 SPECIFIC HEAT OF ZIRCONIUM + NIOBIUM, Zr + Nb

[For Data Reported in Figure and Table No. 132]

Curve No.	Ref. No.	Year	Temp. Range, K	Reported Error, %	Name and Specimen Designation	Composition (weight percent), Specifications and Remarks
1	134	1957	333-553		Zirconium 17.5% Niobium Alloy	82.5 Zr, 17.5 Nb; arc melted from iodide process Zr and Nb eutectoid composition homogenized 14 days at 1300 C in vacuum; tested in vacuum; 2 samples.
2	134	1957	333-553		same as above	Same as above.
3	134	1957	343-1063		same as above	Same as above.
4	134	1957	373-933		same as above	Same as above.
5	134	1957	333-1063		same as above	Same as above.
6	134	1957	373-983		same as above	Same as above.

DATA TABLE NO. 132 SPECIFIC HEAT OF ZIRCONIUM + NIOBIUM Zr + Nb

[Temperature, T, K; Specific Heat, Cp, Cal $g^{-1}K^{-1}$]

T	Cp
CURVE 1	
333.15	6.948×10^{-2}
353.15	6.713*
373.15	6.671
393.15	6.714*
413.15	6.813
433.15	6.932*
453.15	6.990
473.15	7.029*
493.15	7.136
513.15	7.228
533.15	7.308*
553.15	7.369
CURVE 2*	
333.15	6.940×10^{-2}
353.15	6.940
373.15	6.830
393.15	7.058
413.15	7.047
433.15	7.052
453.15	7.069
473.15	7.075
493.15	7.136
513.15	7.160
533.15	7.264
553.15	7.363
CURVE 3*	
343.15	6.987×10^{-2}
383.15	6.943
423.15	6.896
463.15	6.888
503.15	7.008
543.15	7.253
583.15	7.672
623.15	7.180
663.15	5.491
703.15	6.347
743.15	1.0815×10^{-1}
783.15	9.580×10^{-2}
823.15	6.921
863.15	6.854
903.15	6.829
CURVE 3* (cont.)	
943.15	6.847×10^{-2}
983.15	6.897
1023.15	6.800
1063.15	6.729
CURVE 4	
373.15	6.998×10^{-2}
413.15	6.991
453.15	6.991*
493.15	7.113*
533.15	7.269
573.15	7.073
613.15	6.304
653.15	4.733
693.15	5.463
723.15	1.0552×10^{-1}
743.15	1.3819
763.15	1.6627
783.15	1.2829
803.15	7.760×10^{-2}
823.15	7.134*
843.15	7.038
863.15	7.001*
893.15	6.949*
933.15	6.894
CURVE 5	
333.15	6.960×10^{-2}*
373.15	6.956*
413.15	6.979*
453.15	6.869
493.15	6.890
533.15	7.039*
573.15	7.156
613.15	7.110
653.15	6.968
693.15	6.938
733.15	7.213
773.15	7.041
813.15	7.036
843.15	7.070*
863.15	7.259
883.15	7.513
CURVE 5 (cont.)	
903.15	7.929×10^{-2}
923.15	8.820
943.15	9.802
963.15	1.0367×10^{-1}
983.15	9.923×10^{-2}
1003.15	9.015
1023.15	8.110
1043.15	7.255
1063.15	7.041
CURVE 6	
373.15	6.913×10^{-2}*
413.15	6.892*
453.15	6.905*
493.15	7.020
533.15	7.073
573.15	6.868
618.15	6.973
663.15	6.589
703.15	6.340
743.15	9.606
783.15	9.220
823.15	6.806
863.15	6.620
903.15	7.337
943.15	7.235
983.15	6.855

*Not shown on plot

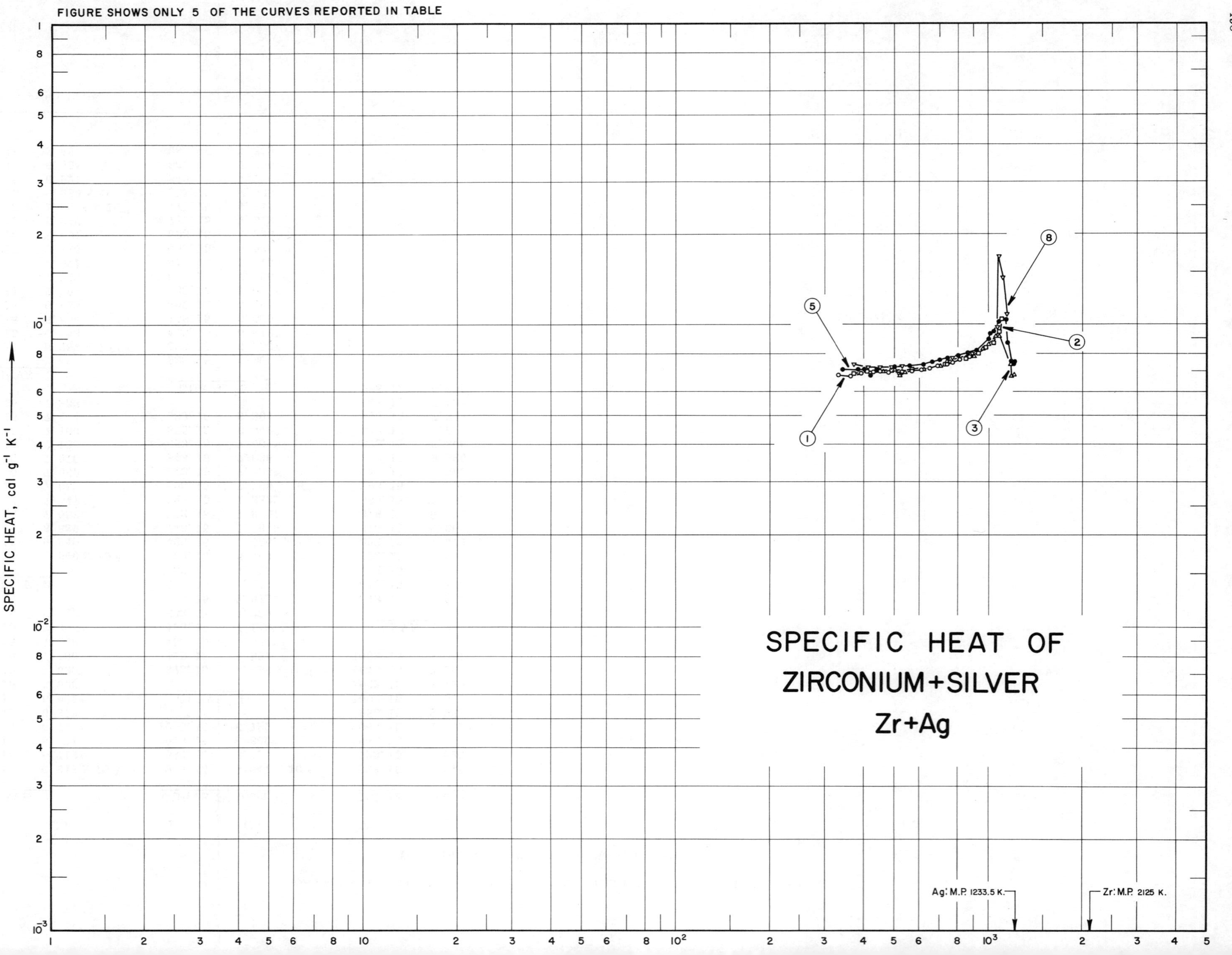
FIGURE SHOWS ONLY 5 OF THE CURVES REPORTED IN TABLE
SPECIFIC HEAT OF
ZIRCONIUM+SILVER
Zr+Ag
SPECIFIC HEAT, cal g⁻¹ K⁻¹
Ag: M.P. 1233.5 K.
Zr: M.P. 2125 K.
1
2
3
5
8

SPECIFICATION TABLE NO. 133 SPECIFIC HEAT OF ZIRCONIUM + SILVER Zr + Ag

[For Data Reported in Figure and Table No. 133]

Curve No.	Ref. No.	Year	Temp. Range, K	Reported Error, %	Name and Specimen Designation	Composition (weight percent), Specifications and Remarks
1	134	1957	333-873		Zirconium 0.881% Silver alloy	0.88Ag, 0.03 Fe, 0.015 O_2, 0.014C, 0.004 Cu, 0.0008N_2, 0.00044H_2; arc melted; homogenized 14 days at 1300 C in vacuum; measured under 0.01 μ Hg.
2	134	1957	373-1103		same as above	Same as above.
3	134	1957	343-1218		same as above	Same as above.
4	134	1957	353-1233		same as above	Same as above.
5	134	1957	343-1223		Zirconium 5.37% Silver alloy	5.37Ag, 0.028Fe, 0.022O_2, 0.013C, 0.002Cu, 0.011H_2, 0.00049N_2; arc melted; homogenized 14 days at 1300 C in vacuum; measured under 0.01 μ Hg.
6	134	1957	383-1078		same as above	Same as above.
7	134	1957	633-1033		same as above	Same as above.
8	134	1957	373-1153		same as above	Same as above.

DATA TABLE NO. 133 SPECIFIC HEAT OF ZIRCONIUM + SILVER Zr + Ag

[Temperature, T, K; Specific Heat, Cp, Cal $g^{-1}K^{-1}$]

T	Cp
CURVE 1	
333.15	6.814×10^{-2}
363.15	6.764
373.15	6.847*
393.15	6.932
413.15	6.985*
433.15	6.989
453.15	7.025
483.15	6.960
513.15	6.996*
533.15	6.994
553.15	7.031*
573.15	7.039
593.15	7.082*
613.15	7.131
633.15	7.153*
653.15	7.186
673.15	7.232*
693.15	7.302
713.15	7.351*
733.15	7.420
753.15	7.467*
773.15	7.527
793.15	7.612*
813.15	7.664
833.15	7.719*
853.15	7.766*
873.15	7.843
CURVE 2	
373.15	6.894×10^{-2}
413.15	7.000
453.15	7.038*
493.15	7.005
533.15	7.014*
573.15	7.102
613.15	7.161*
653.15	7.218*
693.15	7.295*
733.15	7.422
773.15	7.558*
813.15	7.638*
853.15	7.764
893.15	7.885*
933.15	8.044
CURVE 2 (cont.)	
963.15	8.201×10^{-2}*
983.15	8.436
1003.15	8.521*
1023.15	8.590*
1043.15	8.632
1063.15	8.782*
1083.15	9.477
1103.15	1.0544×10^{-1}
CURVE 3	
343.15	6.802×10^{-2}*
363.15	6.795*
383.15	6.918*
403.15	6.911*
423.15	7.068*
443.15	7.043*
463.15	7.028
483.15	6.972*
503.15	7.079
523.15	6.819
543.15	6.996
563.15	7.015*
583.15	7.072*
603.15	7.121*
623.15	7.146
643.15	7.197*
663.15	7.242*
683.15	7.242*
703.15	7.332
723.15	7.353*
743.15	7.554
763.15	7.488*
783.15	7.541*
803.15	7.609*
823.15	7.666*
843.15	7.710*
863.15	7.779*
883.15	7.885*
903.15	7.875
923.15	8.008*
943.15	8.196*
963.15	8.357
983.15	8.555*
1003.15	8.588
CURVE 3 (cont.)	
1023.15	8.649×10^{-2}*
1043.15	8.891*
1063.15	9.153
1083.15	9.160
1178.15	7.397
1188.15	6.833
1198.15	6.772
1208.15	6.819*
1218.15	6.857
CURVE 4*	
353.15	7.405×10^{-2}
393.15	7.237
433.15	7.212
473.15	7.165
513.15	7.192
553.15	7.254
593.15	7.299
633.15	7.349
673.15	7.390
713.15	7.427
753.15	7.488
793.15	7.555
833.15	7.677
873.15	7.828
913.15	7.966
953.15	8.093
993.15	8.238
1033.15	8.403
1073.15	8.811
1193.15	7.290
1233.15	7.030
CURVE 5	
343.15	7.111×10^{-2}
363.15	7.010*
383.15	7.088
403.15	7.135*
423.15	7.173
443.15	7.140
463.15	7.148*
483.15	7.183*
503.15	7.226
CURVE 5 (cont.)	
523.15	7.223×10^{-2}*
543.15	7.286*
563.15	7.338
583.15	7.337*
603.15	7.390*
623.15	7.434
643.15	7.461*
663.15	7.523
683.15	7.585*
703.15	7.653
723.15	7.701*
743.15	7.773
763.15	7.808*
783.15	7.888*
803.15	7.946
823.15	7.982*
843.15	8.048*
863.15	8.092
883.15	8.137*
903.15	8.174*
923.15	8.224
943.15	8.343*
963.15	8.560*
983.15	8.650*
1003.15	8.975
1023.15	9.359
1043.15	9.552
1063.15	9.802*
1078.15	1.0276×10^{-1}
1123.15	1.0586*
1143.15	1.0045
1163.15	8.728×10^{-2}
1183.15	7.588
1203.15	7.436
1223.15	7.561
CURVE 6*	
383.15	7.163×10^{-2}
413.15	7.244
453.15	7.261
493.15	7.295
533.15	7.346
573.15	7.440
613.15	7.453
CURVE 6 (cont.)	
653.15	7.526×10^{-2}
693.15	7.587
733.15	7.661
773.15	7.743
813.15	7.791
853.15	7.863
893.15	7.907
933.15	8.057
973.15	8.171
1013.15	8.554
1053.15	9.061
1078.15	9.386
CURVE 7*	
633.15	7.490×10^{-2}
673.15	7.566
713.15	7.638
753.15	7.689
793.15	7.750
833.15	7.796
873.15	7.894
913.15	7.981
953.15	8.216
993.15	8.711
1033.15	9.049
CURVE 8	
373.15	7.414×10^{-2}
413.15	7.251
453.15	7.234
493.15	7.229
533.15	7.291
573.15	7.364*
613.15	7.451*
653.15	7.527*
693.15	7.595*
733.15	7.692*
773.15	7.756
813.15	7.855*
853.15	7.910*
893.15	8.006
933.15	8.110*
973.15	8.373*
CURVE 8 (cont.)	
1013.15	8.887×10^{-2}*
1053.15	9.432*
1078.15	9.862
1088.15	1.6941×10^{-1}
1123.15	1.4361
1153.15	1.0803

* Not shown on plot

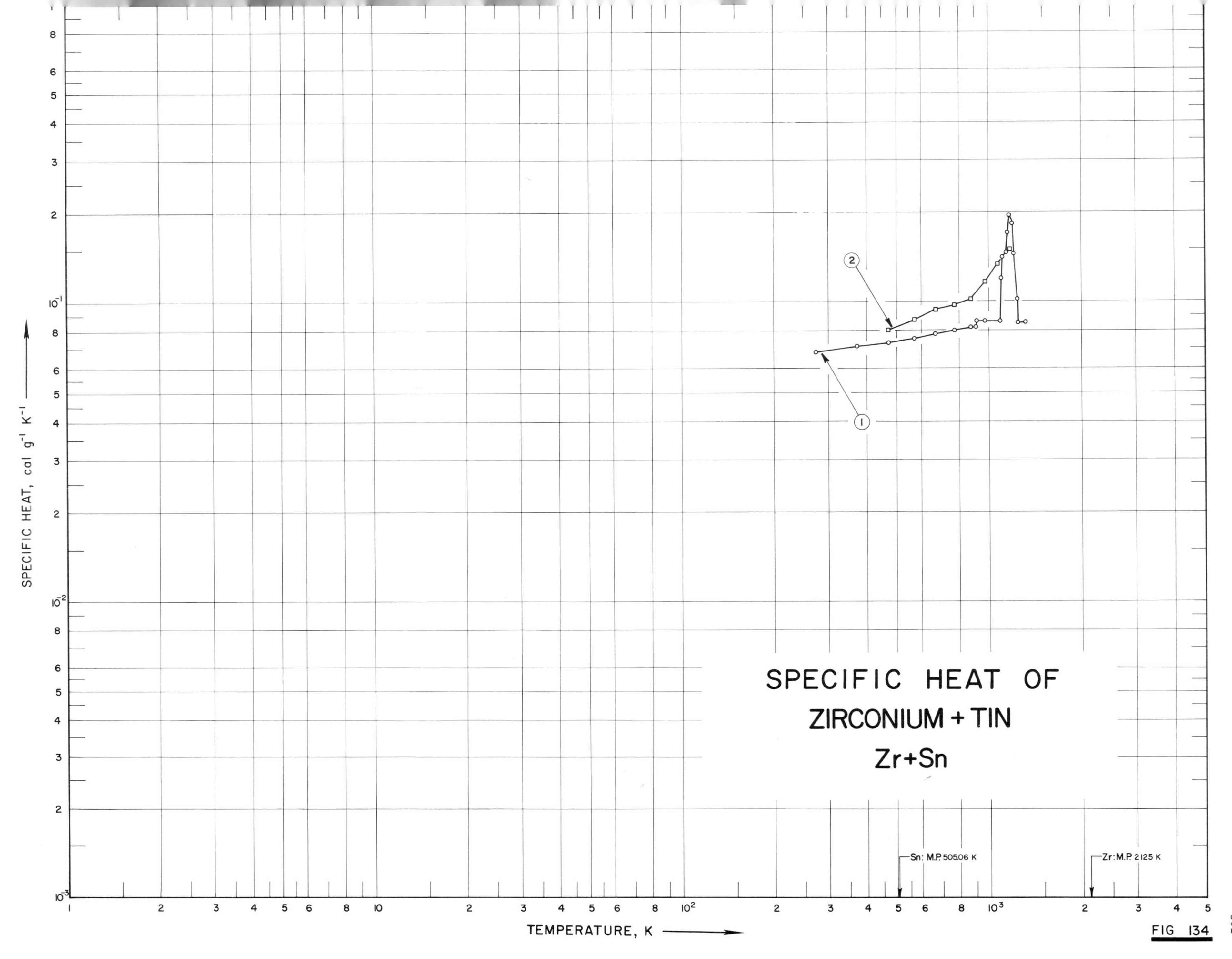

SPECIFIC HEAT OF
ZIRCONIUM + TIN
Zr+Sn
Sn: M.P. 505.06 K
Zr: M.P. 2125 K
TEMPERATURE, K
SPECIFIC HEAT, cal g⁻¹ K⁻¹
1
2

FIG 134

SPECIFICATION TABLE NO. 134 SPECIFIC HEAT OF ZIRCONIUM + TIN, Zr + Sn

[For Data Reported in Figure and Table No. 134]

Curve No.	Ref. No.	Year	Temp. Range, K	Reported Error, %	Name and Specimen Designation	Composition (weight percent), Specifications and Remarks
1	241	1963	273-1323		Zircaloy-2	
2	416	1953	473-1173		Zr(95)Sn(5)	94.7 Zr, 5.3 Sn.

DATA TABLE NO. 134 SPECIFIC HEAT OF ZIRCONIUM + TIN $Zr + Sn$

[Temperature, T, K; Specific Heat, Cp, Cal $g^{-1}K^{-1}$]

T	Cp
CURVE 1	
273.15	6.8×10^{-2}
373.15	7.1
473.15	7.3
573.15	7.5
673.15	7.8
773.15	8.0
873.15	8.2
903.15	8.2
913.15	8.6
973.15	8.6
1073.15	8.6*
1083.15	8.6
1093.15	1.20×10^{-1}
1113.15	1.41
1133.15	1.47
1153.15	1.72
1173.15	1.95
1193.15	1.84
1213.15	1.45
1233.15	1.12
1248.15	8.5×10^{-2}
1273.15	8.5*
1323.15	8.5
CURVE 2	
473.15	8.0×10^{-2}
573.15	8.7
673.15	9.4
773.15	9.7
873.15	1.02×10^{-1}
973.15	1.16
1073.15	1.34
1173.15	1.50

* Not shown on plot

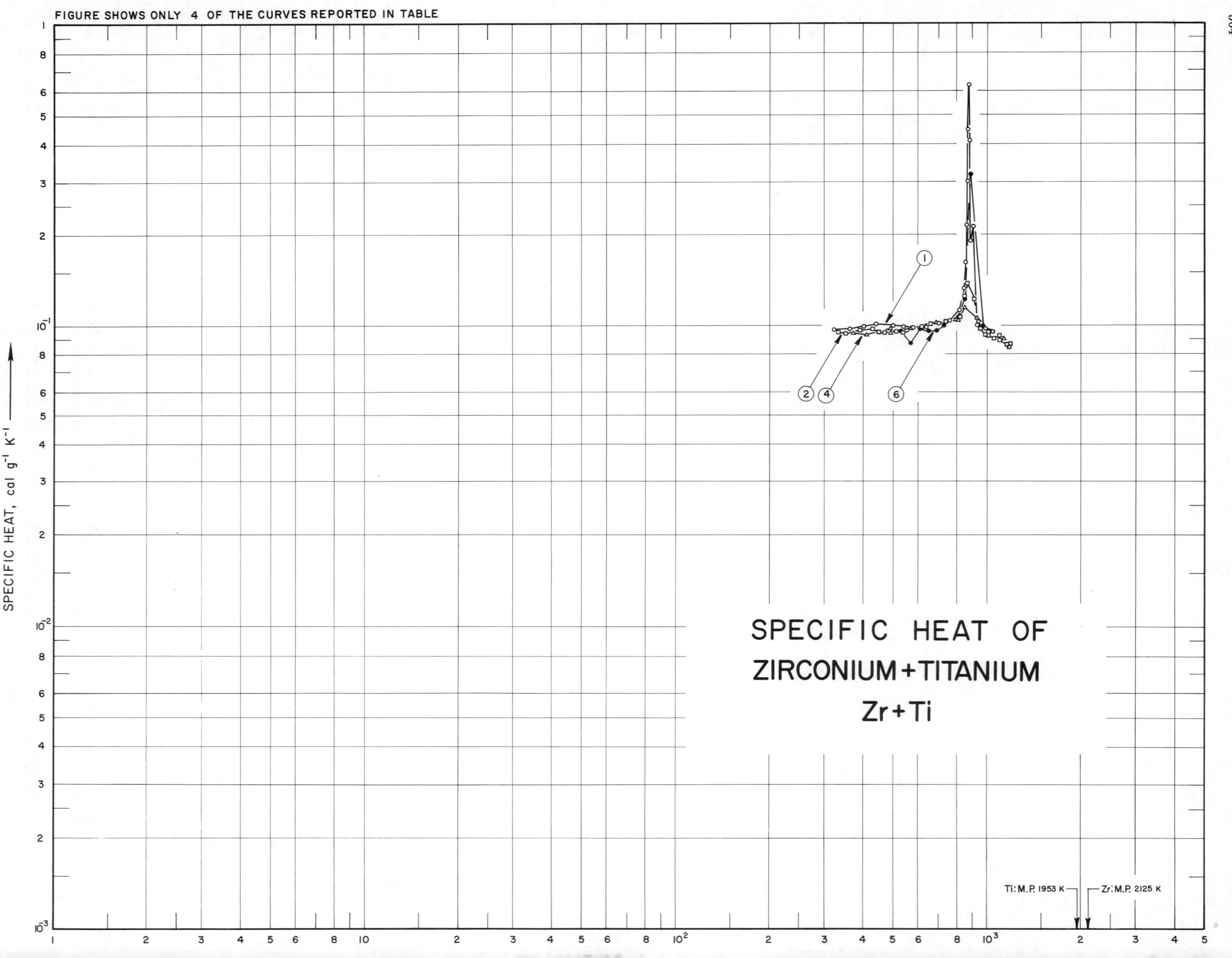
FIGURE SHOWS ONLY 4 OF THE CURVES REPORTED IN TABLE
SPECIFIC HEAT, cal g⁻¹ K⁻¹
SPECIFIC HEAT OF
ZIRCONIUM + TITANIUM
Zr + Ti
Ti: M.P. 1953 K
Zr: M.P. 2125 K
1
2
4
6

SPECIFICATION TABLE NO. 135 SPECIFIC HEAT OF ZIRCONIUM + TITANIUM, Zr + Ti

[For Data Reported in Figure and Table No. 135]

Curve No.	Ref. No.	Year	Temp. Range, K	Reported Error, %	Name and Specimen Designation	Composition (weight percent), Specifications and Remarks
1	134	1957	323-1163		Zirconium 34.4% Titanium Alloy	65.6Zr, 34.4 Ti; eutectoid composition; arc melted from iodide process Zr and Ti; homogenized 14 days at 1300 C in vacuum.
2	134	1957	333-1183		same as above	Same as above.
3	134	1957	403-1153		same as above	Same as above.
4	134	1957	373-1133		same as above	Same as above.
5	134	1957	353-1163		same as above	Same as above.
6	134	1957	333-1053		same as above	Same as above.

DATA TABLE NO. 135 SPECIFIC HEAT OF ZIRCONIUM + TITANIUM Zr + Ti

[Temperature, T, K; Specific Heat, Cp, Cal $g^{-1}K^{-1}$]

CURVE 1

T	Cp
323.15	9.685 x 10^{-2}
343.15	9.787*
363.15	9.741
383.15	9.826*
403.15	9.952
423.15	1.0099 x 10^{-1}*
443.15	1.0186
463.15	1.0243*
483.15	1.0182*
503.15	1.0084
523.15	9.966 x 10^{-2}*
543.15	9.938
563.15	9.893*
583.15	9.889
603.15	9.972*
623.15	9.877*
643.15	9.882
663.15	9.942*
683.15	1.0088 x 10^{-1}*
703.15	1.0230
723.15	1.0314*
743.15	1.0369*
763.15	1.0475
783.15	1.0620*
803.15	1.0957*
818.15	1.1257
828.15	1.1741*
838.15	1.2112*
848.15	1.3263
858.15	1.6226
865.65	2.1651
870.65	3.0203
875.65	4.4915
880.65	6.3227
885.65	4.1325
890.65	1.9066
908.15	2.1256
933.15	1.0063
953.15	9.974 x 10^{-2}*
973.15	9.885*
993.15	9.789
1013.15	9.671*
1033.15	9.554*
1053.15	9.557

CURVE 2

Series I

T	Cp
333.15	9.502 x 10^{-2}
353.15	9.434
373.15	9.532*
393.15	9.669
413.15	9.762*
433.15	9.754
453.15	9.523
473.15	9.471
493.15	9.516*
513.15	9.597
543.15	9.498
583.15	9.810*
623.15	9.929
663.15	1.0169 x 10^{-1}
703.15	1.0304*
743.15	1.0374
783.15	1.0420*
823.15	1.0737
853.15	1.2580
868.15	1.3778
913.15	1.2261
933.15	9.948 x 10^{-2}*
953.15	9.816
973.15	9.680*
993.15	9.634*
1023.15	9.541
1063.15	9.427*
1103.15	9.274
1143.15	8.932*
1173.15	8.640
1193.15	8.695

Series II

T	Cp
953.15	9.682 x 10^{-2}*
973.15	9.573*
993.15	9.340
1023.15	9.236
1063.15	9.067
1103.15	8.961
1143.15	8.605*
1183.15	8.464

CURVE 3*

Series I

T	Cp
403.15	9.399 x 10^{-2}
443.15	9.443
483.15	9.493
523.15	9.644
563.15	9.794
603.15	9.905
643.15	1.0052 x 10^{-1}
683.15	1.0313
723.15	1.0396
763.15	1.0459
803.15	1.0628
833.15	1.0728
863.15	1.2245
868.15	1.2838
913.15	1.5559
943.15	1.0334
983.15	9.983 x 10^{-2}
1023.15	9.803
1063.15	9.560
1103.15	9.011
1143.15	8.844

Series II

T	Cp
953.15	9.841 x 10^{-2}
993.15	9.501
1033.15	9.128
1073.15	8.679
1113.15	8.456
1153.15	8.338

CURVE 4

Series I

T	Cp
373.15	9.456 x 10^{-2}
413.15	9.317
453.15	9.354*
493.15	9.469
533.15	9.634*
573.15	9.820
613.15	9.941*
653.15	1.0121 x 10^{-1}*
693.15	1.0251

CURVE 4 (cont.)

T	Cp
733.15	1.0313 x 10^{-1}*
773.15	1.0351*
813.15	1.0466
853.15	1.1581

Series II

T	Cp
493.15	9.812 x 10^{-2}
533.15	9.769*
573.15	9.839*
613.15	9.940*
653.15	1.0131 x 10^{-1}*
693.15	1.0270*
733.15	1.0320*
773.15	1.0329*
813.15	1.0442*
853.15	1.1632*
933.15	1.0646
973.15	9.817 x 10^{-2}*
1013.15	9.697*
1053.15	9.548*
1093.15	9.325*
1133.15	9.069

CURVE 5*

Series I

T	Cp
353.15	9.547 x 10^{-2}
393.15	9.386
433.15	9.319
473.15	9.365
513.15	9.496
553.15	9.639
593.15	9.748
633.15	9.872
673.15	1.0099 x 10^{-1}
713.15	1.0206
753.15	1.0273
793.15	1.0314
833.15	1.0659
933.15	1.0528
973.15	9.720 x 10^{-2}
1013.15	9.610
1053.15	9.568

CURVE 5 (cont.)

T	Cp
1093.15	9.400 x 10^{-2}
1133.15	9.248

Series II

T	Cp
963.15	9.421 x 10^{-2}
1003.15	9.064
1043.15	8.922
1083.15	8.675
1123.15	8.488
1163.15	8.253

CURVE 6

T	Cp
333.15	9.515 x 10^{-2}*
373.15	9.362*
413.15	9.401*
453.15	9.416*
493.15	9.479*
533.15	9.615
573.15	8.764
613.15	9.807
653.15	9.635
693.15	9.635
733.15	1.0049 x 10^{-1}
773.15	1.0398*
813.15	1.0728
853.15	1.2285
893.15	3.1808
933.15	1.0640*
973.15	9.950 x 10^{-2}
1013.15	9.756*
1053.15	9.613*

* Not shown on plot

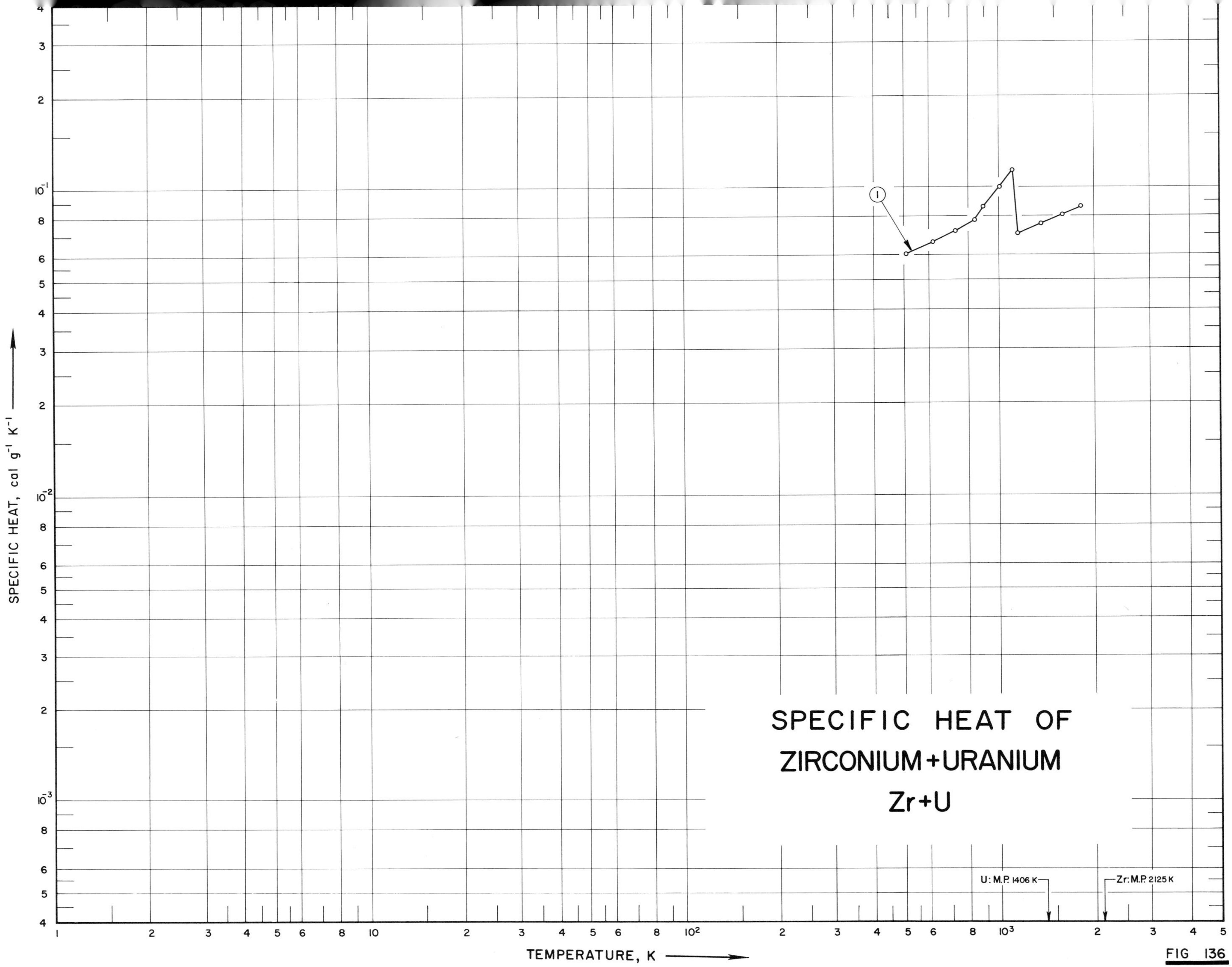

SPECIFIC HEAT OF
ZIRCONIUM+URANIUM
Zr+U
SPECIFIC HEAT, cal g^{-1} K^{-1}
TEMPERATURE, K
U: M.P. 1406 K
Zr: M.P. 2125 K
1

FIG 136

SPECIFICATION TABLE NO. 136 SPECIFIC HEAT OF ZIRCONIUM + URANIUM, Zr + U

[For Data Reported in Figure and Table No. 136]

Curve No.	Ref. No.	Year	Temp. Range, K	Reported Error, %	Name and Specimen Designation	Composition (weight percent), Specifications and Remarks
1	242	1963	505-1811	± 2.0	unhydrided zirconium-10.48% Uranium alloy	89.52 Zr, 10.48 U; measured under argon atmosphere; density = 430 lb ft^{-3}.

DATA TABLE NO. 136 SPECIFIC HEAT OF ZIRCONIUM + URANIUM Zr + U

[Temperature, T, K; Specific Heat, Cp, Cal $g^{-1}K^{-1}$]

T	Cp
CURVE 1	
506	6.050×10^{-2}
617	6.640
728	7.240
839	7.830
894	8.650
1006	1.000
1117	1.140
1144	7.060
1367	7.590
1589	8.120
1811	8.650

* Not shown on plot

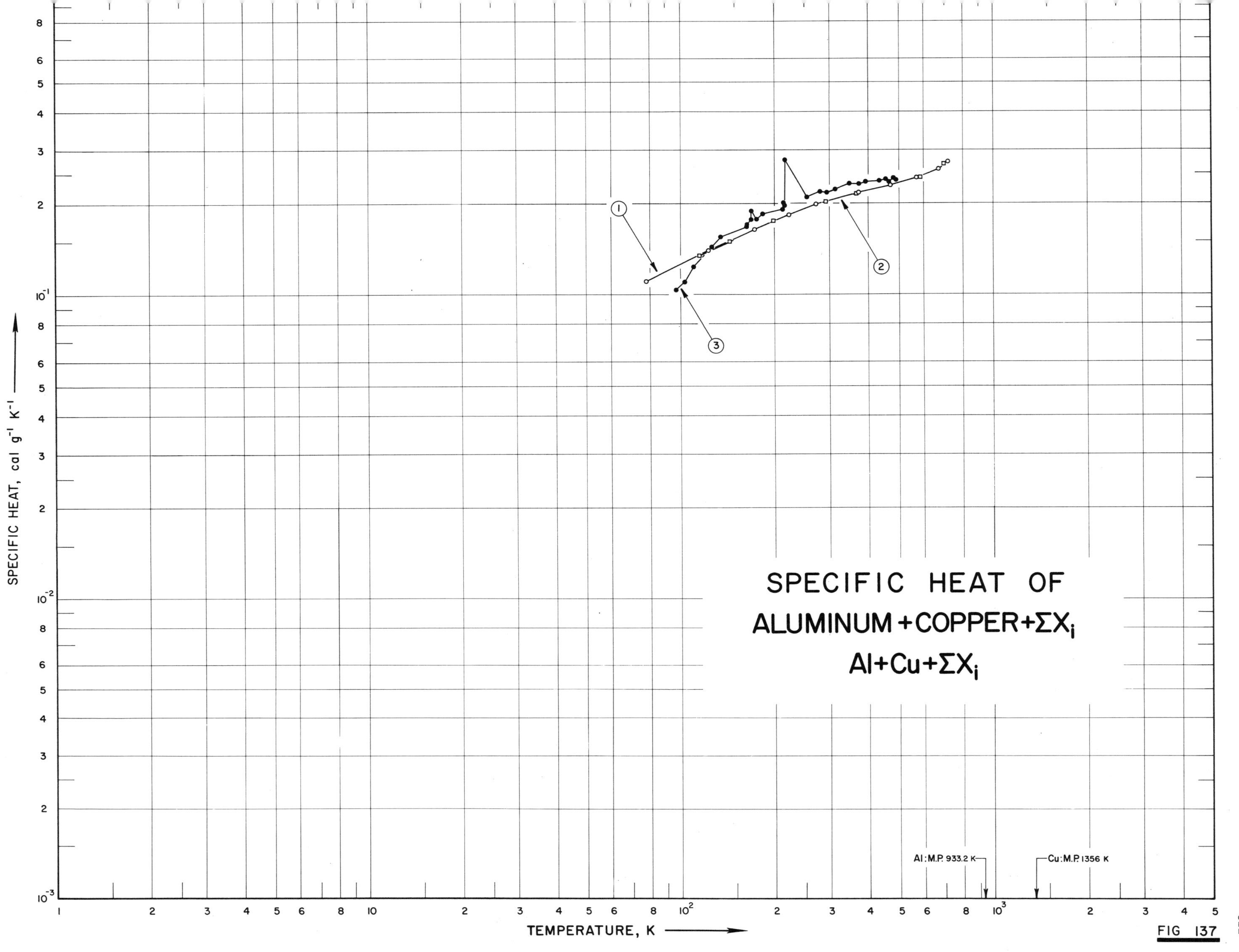
SPECIFIC HEAT OF
ALUMINUM+COPPER+ΣX_i
Al+Cu+ΣX_i
Al: M.P. 933.2 K
Cu: M.P. 1356 K
SPECIFIC HEAT, cal g^{-1} K^{-1}
TEMPERATURE, K
1
2
3
FIG 137

SPECIFICATION TABLE NO. 137 SPECIFIC HEAT OF ALUMINUM + COPPER + ΣX_i Al + Cu + ΣX_i

[For Data Reported in Figure and Table No. 137]

Curve No.	Ref. No.	Year	Temp. Range, K	Reported Error, %	Name and Specimen Designation	Composition (weight percent), Specifications and Remarks
1	243	1954	73-723		Al alloy 24S-T4	93.9 Al, 4.5 Cu, 1.5 Mg, 0.6 Mn.
2	10	1958	116-700		Al alloy 2024-T4	93.4 Al, 4.5 Cu, 1.5 Mg, 0.6 Mn; sample supplied by the Aluminum Company of America; specimen sealed in a helium capsule; density (32 F) = 174 lb ft^{-3}.
3	1	1962	97-218		Al alloy 2024	90.0 Al, 4.5 Cu, 1.5 Mg, 0.6 Mn; Hanovia liquid platinum applied on specimen's front surface for opaqueness and applied on specimen's rear surface to obtain good conductive surface; front surface painted with Parson's black for constant absorbtivity.

DATA TABLE NO. 137 SPECIFIC HEAT OF ALUMINUM + COPPER + ΣX_i Al + Cu + ΣX_i

(Temperature, T, K; Specific Heat, C_p, Cal $g^{-1} K^{-1}$)

T	C_p
CURVE 1	
73	1.12 x 10^{-1}
123	1.41
173	1.65
223	1.84
273	1.98
373	2.18
473	2.31
573	2.43
673	2.62
723	2.76
CURVE 2	
116	1.35 x 10^{-1}
144	1.50
200	1.76
293	2.03
366	2.17
478	2.31*
589	2.45
700	2.70
CURVE 3	
97	1.03 x 10^{-1}
99	1.06*
103	1.11
105	1.13*
111	1.24
115	1.28*
123	1.40*
127	1.45
133	1.48*
135	1.57
139	1.60*
163	1.69
163	1.72
165	1.75*
169	1.78
169	1.91
177	1.79
185	1.86
190	1.87
214	1.93
215	2.02
217	1.98

T	C_p
CURVE 3 (cont.)	
217	2.80 x 10^{-1}
218	2.00*
255	2.11
257	2.13*
261	2.16*
281	2.20
295	2.18
299	2.22*
315	2.24
331	2.26*
334	2.29*
343	2.29*
349	2.34
375	2.33
393	2.36
399	2.35*
411	2.38*
435	2.38
439	2.34*
440	2.37*
439	2.41*
443	2.41*
445	2.35*
455	2.42
466	2.42*
468	2.36
473	2.35*
483	2.44
487	2.38*
492	2.42

* Not shown on plot

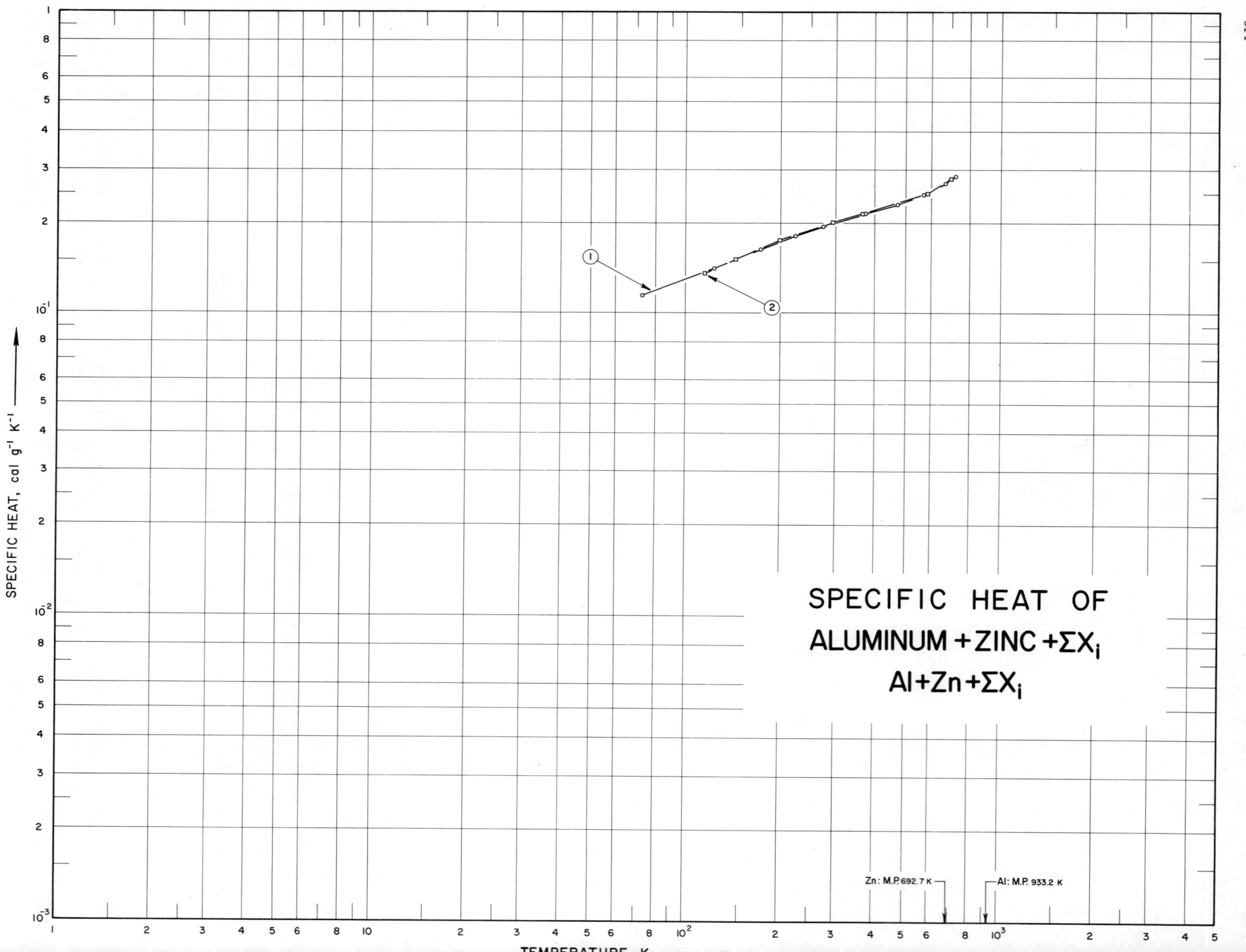
SPECIFIC HEAT OF
ALUMINUM + ZINC + ΣX_i
Al+Zn+ΣX_i
Zn: M.P. 692.7 K
Al: M.P. 933.2 K
SPECIFIC HEAT, cal g^{-1} K^{-1}
1
2

SPECIFICATION TABLE NO. 138 SPECIFIC HEAT OF ALUMINUM + ZINC + ΣX_i Al + Zn + ΣX_i

[For Data Reported in Figure and Table No. 138]

Curve No.	Ref. No.	Year	Temp. Range, K	Reported Error, %	Name and Specimen Designation	Composition (weight percent), Specifications and Remarks
1	243	1954	73-723		Al alloy 75S-T6	90 Al, 5.5 Zn, 2.5 Mg, 1.5 Cu, 0.3 Cr, 0.2 Mn.
2	10	1958	116-700		Al alloy 7075-T6	90.2 Al, 5.5 Zn, 2.5 Mg, 1.5 Cu, 0.3 Cr; sample supplied by the Aluminum Company of America; specimen sealed in a helium capsule; density (32 F) = 175 lb ft^{-3}.

DATA TABLE NO. 138 SPECIFIC HEAT OF ALUMINUM + ZINC + ΣX_i Al + Zn + ΣX_i

(Temperature, T, K; Specific Heat, C_p, Cal $g^{-1} K^{-1}$)

T	C_p
CURVE 1	
73	1.15×10^{-1}
123	1.42
173	1.64
223	1.82
273	1.96
373	2.16
473	2.31
573	2.48
673	2.70
723	2.86
CURVE 2	
116	1.37×10^{-1}
144	1.52
200	1.76
293	2.01
366	2.15
478	2.30*
589	2.52
700	2.80

* Not shown on plot

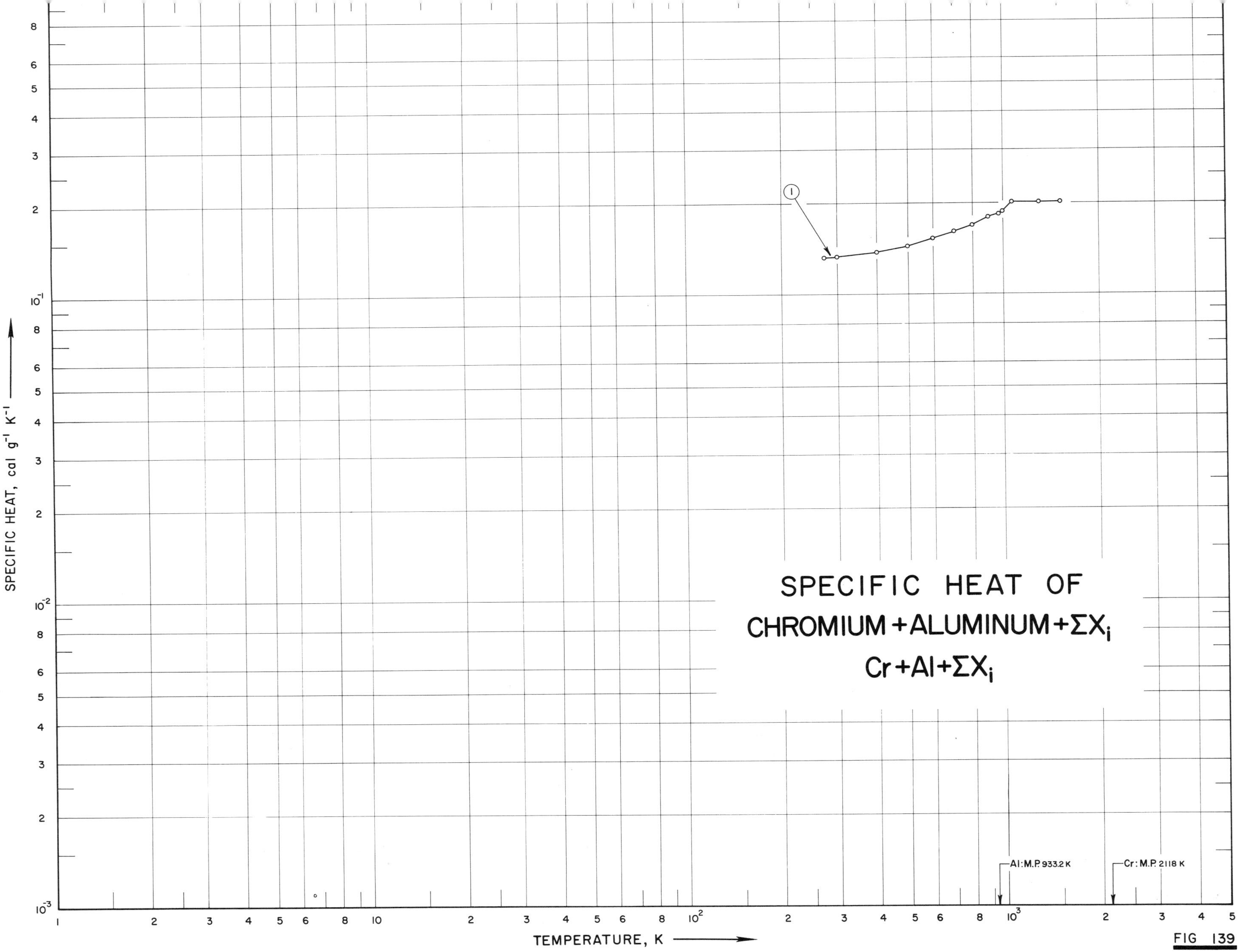

SPECIFIC HEAT OF CHROMIUM + ALUMINUM + ΣX_i

$Cr + Al + \Sigma X_i$

SPECIFIC HEAT, cal g^{-1} K^{-1}

TEMPERATURE, K

Al: M.P. 933.2 K

Cr: M.P. 2118 K

FIG 139

SPECIFICATION TABLE NO. 139 SPECIFIC HEAT OF CHROMIUM + ALUMINUM + ΣX_i Cr + Al + ΣX_i

[For Data Reported in Figure and Table No. 139]

Curve No.	Ref. No.	Year	Temp. Range, K	Reported Error, %	Name and Specimen Designation	Composition (weight percent), Specifications and Remarks
1	244	1960	273-1523	0.4		63.91 Cr, 18.11 Al, 16.55 Fe, 0.67 Si, 0.024 C, 0.006 S.

DATA TABLE NO. 139 SPECIFIC HEAT OF CHROMIUM + ALUMINUM + ΣX_i Cr + Al + ΣX_i

(Temperature, T, K; Specific Heat, C_p, Cal $g^{-1} K^{-1}$)

T	C_p
CURVE 1	
273	1.333 x 10^{-1}
300	1.340
400	1.391
500	1.463
600	1.543
700	1.627
800	1.713
900	1.800
973	1.864
1000	1.888
1073	2.03
1100	2.03*
1200	2.03*
1300	2.03
1400	2.03*
1500	2.03*
1523	2.03

* Not shown on plot

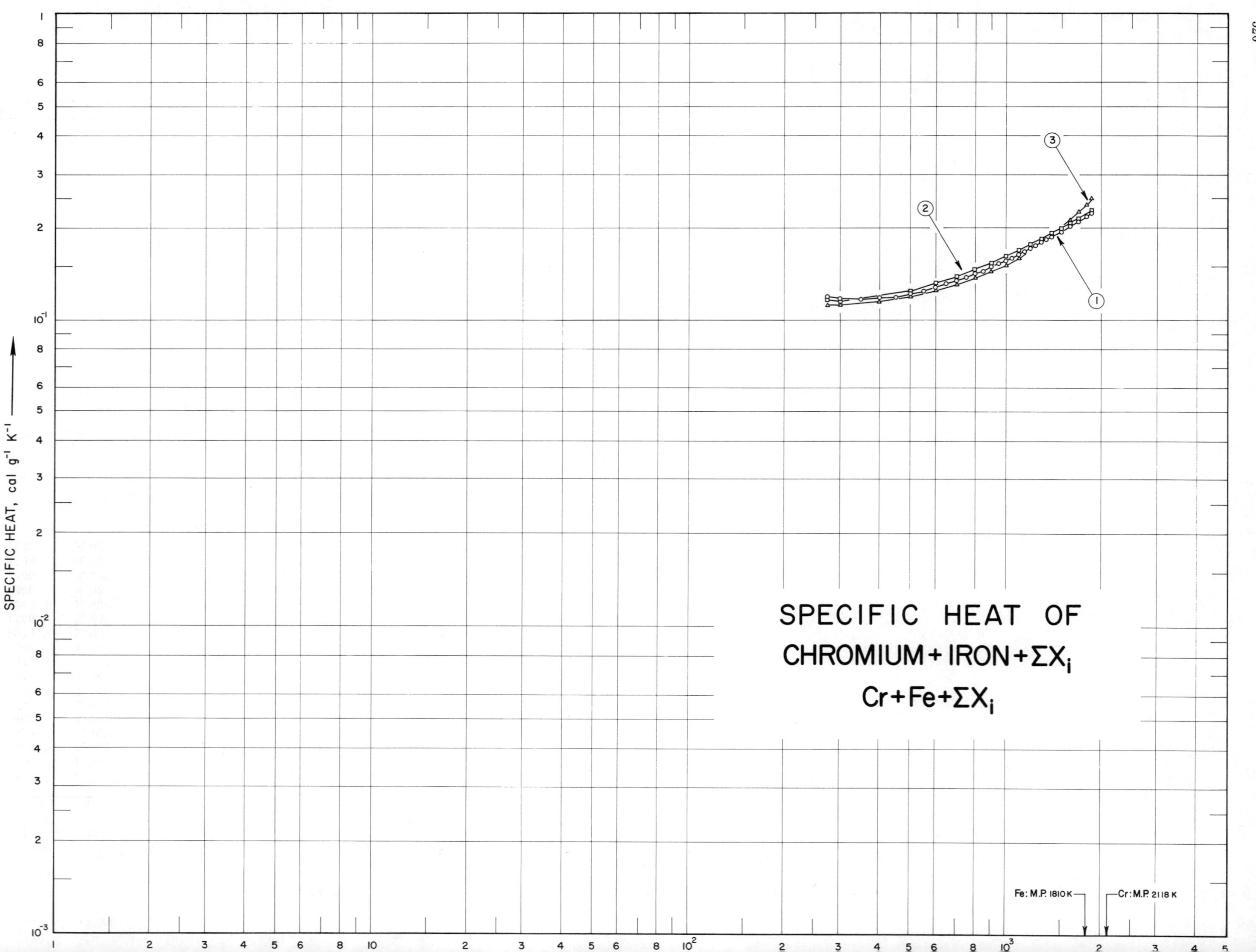
SPECIFIC HEAT OF
CHROMIUM + IRON + ΣX_i
Cr + Fe + ΣX_i
SPECIFIC HEAT, cal g^{-1} K^{-1}
Fe: M.P. 1810 K
Cr: M.P. 2118 K
1
2
3

SPECIFICATION TABLE NO. 140 SPECIFIC HEAT OF CHROMIUM + IRON + ΣX_i Cr + Fe + ΣX_i

[For Data Reported in Figure and Table No. 140]

Curve No.	Ref. No.	Year	Temp. Range, K	Reported Error, %	Name and Specimen Designation	Composition (weight percent), Specifications and Remarks
1	244	1960	273-1873	1.5	Carbonless Ferrochromium	76.45 Cr, 0.35 Si, 0.26 C, 0.14 Al, 0.008 S, bal. Fe.
2	244	1960	273-1873	1	Nitrated Ferrochromium	77.75 Cr, 1.20 N_2, 0.70 Al, 0.52 Si, 0.028 C.
3	244	1960	273-1873	0.8-1.2	Aluminothermic chromium	98.66 Cr, 0.64 Fe, 0.43 Al, 0.20 Si, 0.036 C, 0.007 P.

DATA TABLE NO. 140 SPECIFIC HEAT OF CHROMIUM + IRON + ΣX_i $Cr + Fe + \Sigma X_i$

[Temperature, T, K; Specific Heat, C_p, Cal $g^{-1} K^{-1}$]

T	C_p
CURVE 1	
273	1.214 x 10^{-1}
300	1.198
350	1.189
400	1.197
450	1.215
500	1.238
550	1.266
600	1.296
650	1.328
700	1.361
750	1.396
800	1.431
850	1.467
900	1.503
950	1.504
1000	1.577
1050	1.614
1100	1.651
1150	1.689
1200	1.727
1250	1.765
1300	1.802
1350	1.840
1400	1.879
1500	1.955
1600	2.031
1700	2.108
1800	2.185
1873	2.241
CURVE 2	
273	1.176 x 10^{-1}
300	1.175
400	1.207*
500	1.263
600	1.329
700	1.400
800	1.474
900	1.548
1000	1.624
1100	1.700
1200	1.777
1300	1.854
1400	1.931
CURVE 2 (cont.)	
1500	2.009 x 10^{-1}
1600	2.086
1700	2.163
1800	2.241*
1873	2.298
CURVE 3	
273	1.134 x 10^{-1}
300	1.142
400	1.179
500	1.221
600	1.271
700	1.327
800	1.391
900	1.461
1000	1.539
1100	1.623
1200	1.715*
1300	1.814*
1400	1.920*
1500	2.033*
1600	2.153
1700	2.280
1800	2.415
1873	2.517

* Not shown on plot

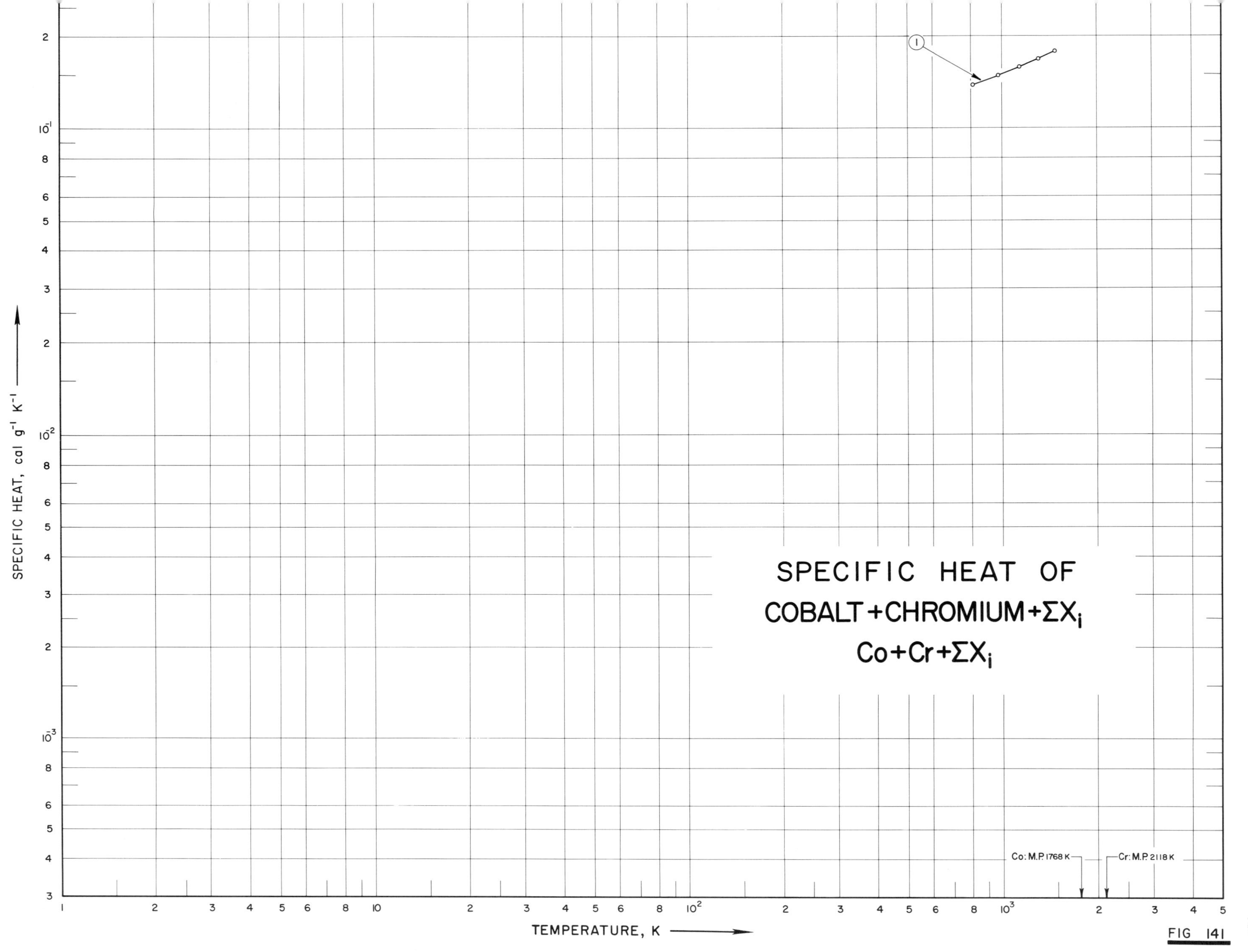
SPECIFIC HEAT OF
COBALT+CHROMIUM+ΣX_i
Co+Cr+ΣX_i
1
Co: M.P. 1768 K
Cr: M.P. 2118 K
SPECIFIC HEAT, cal g⁻¹ K⁻¹
TEMPERATURE, K

FIG 141

SPECIFICATION TABLE NO. 141 SPECIFIC HEAT OF COBALT + CHROMIUM + ΣX_i Co + Cr + ΣX_i

[For Data Reported Figure and Table No. 141]

Curve No.	Ref. No.	Year	Temp. Range, K	Reported Error, %	Name and Specimen Designation	Composition (weight percent), Specifications and Remarks
1	245	1958	810-1477		Stellite 21	Before test: 60.49 Co, 26.69 Cr, 5.42 Mo, 2.38 Ni, 1.54 Fe, 0.258 C; after test: 62.27 Co, 26.74 Cr, 5.42 Mo, 2.42 Ni, 1.23 Fe, 0.264 C, density = 511.2 lb ft^{-3}.

DATA TABLE NO. 141 SPECIFIC HEAT OF COBALT + CHROMIUM + ΣX_i — $Co + Cr + \Sigma X_i$

[Temperature, T, K; Specific Heat, C_p, Cal g^{-1} K^{-1}]

T	C_p
CURVE 1	
811	1.38×10^{-1}
978	1.48
1144	1.58
1311	1.68
1478	1.78

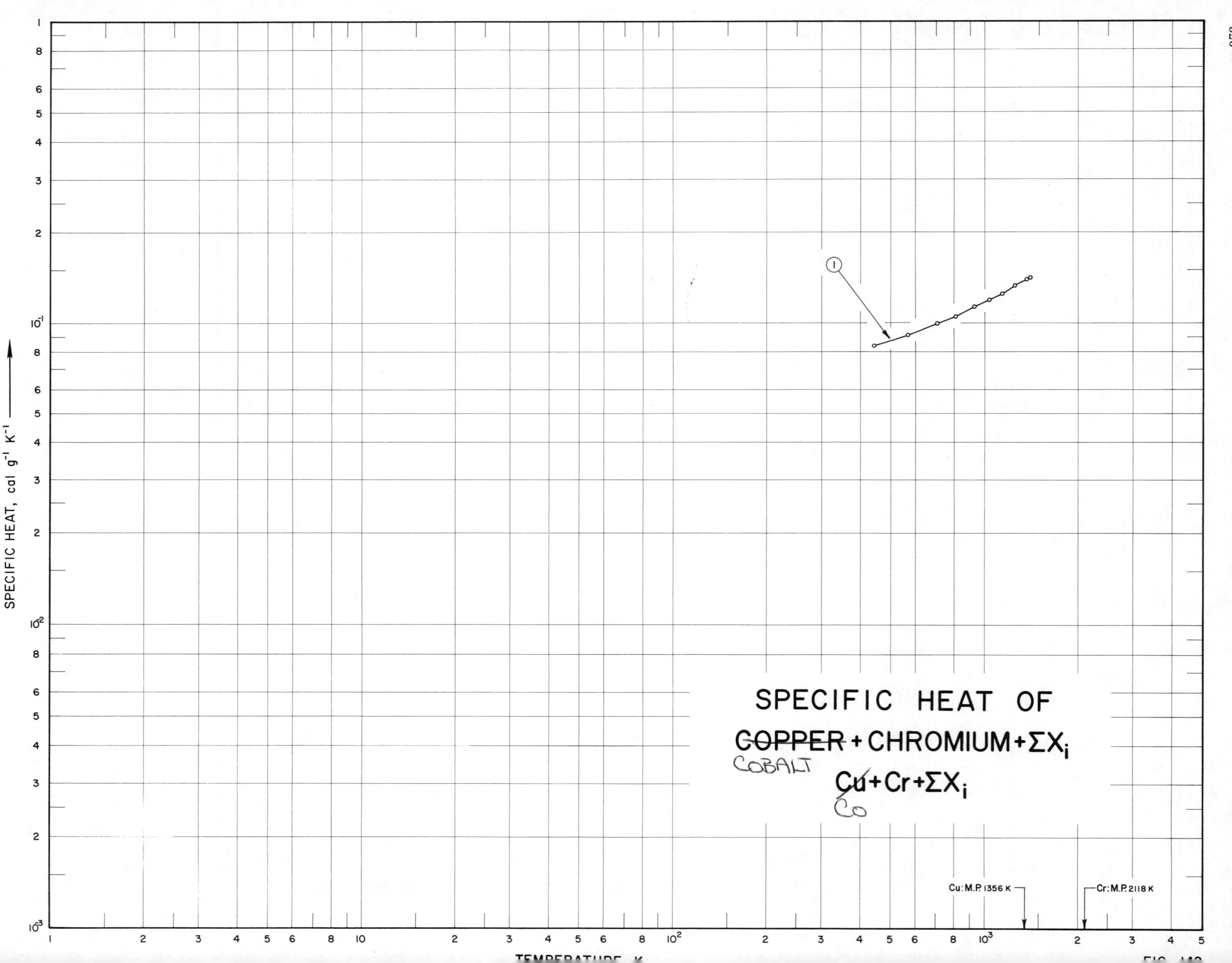
SPECIFIC HEAT OF
COPPER + CHROMIUM + ΣXi
COBALT
Cu+Cr+ΣXi
Co
Cu: M.P. 1356 K
Cr: M.P. 2118 K
SPECIFIC HEAT, cal g⁻¹ K⁻¹
1

SPECIFICATION TABLE NO. 142 SPECIFIC HEAT OF COPPER + CHROMIUM + ΣX_i Cu + Cr + ΣX_i

[For Data Reported in Figure and Table No. 142]

Curve No.	Ref. No.	Year	Temp. Range, K	Reported Error, %	Name and Specimen Designation	Composition (weight percent), Specifications and Remarks
1	146	1961	444-1412	3.0	Haynes stellite HE 1049	43.6 Cu, 26.0 Cr, 15.0 W, 10.0 Ni, 3.0 Fe, 0.8 Mn, 0.8 Si, 0.4 B; measured in helium atmosphere; density = 552 lb ft^{-3}.

DATA TABLE NO. 142 SPECIFIC HEAT OF ~~COPPER~~ COBALT + CHROMIUM + ΣX_i ~~Cu~~ Co + Cr + ΣX_i

[Temperature, T, K; Specific Heat, C_p, Cal g^{-1} K^{-1}]

T	C_p
CURVE 1	
444	8.419×10^{-2}
568	9.155
706	9.980
809	1.059×10^{-1}
928	1.129
1036	1.194
1141	1.256
1260	1.327
1375	1.395
1411	1.417

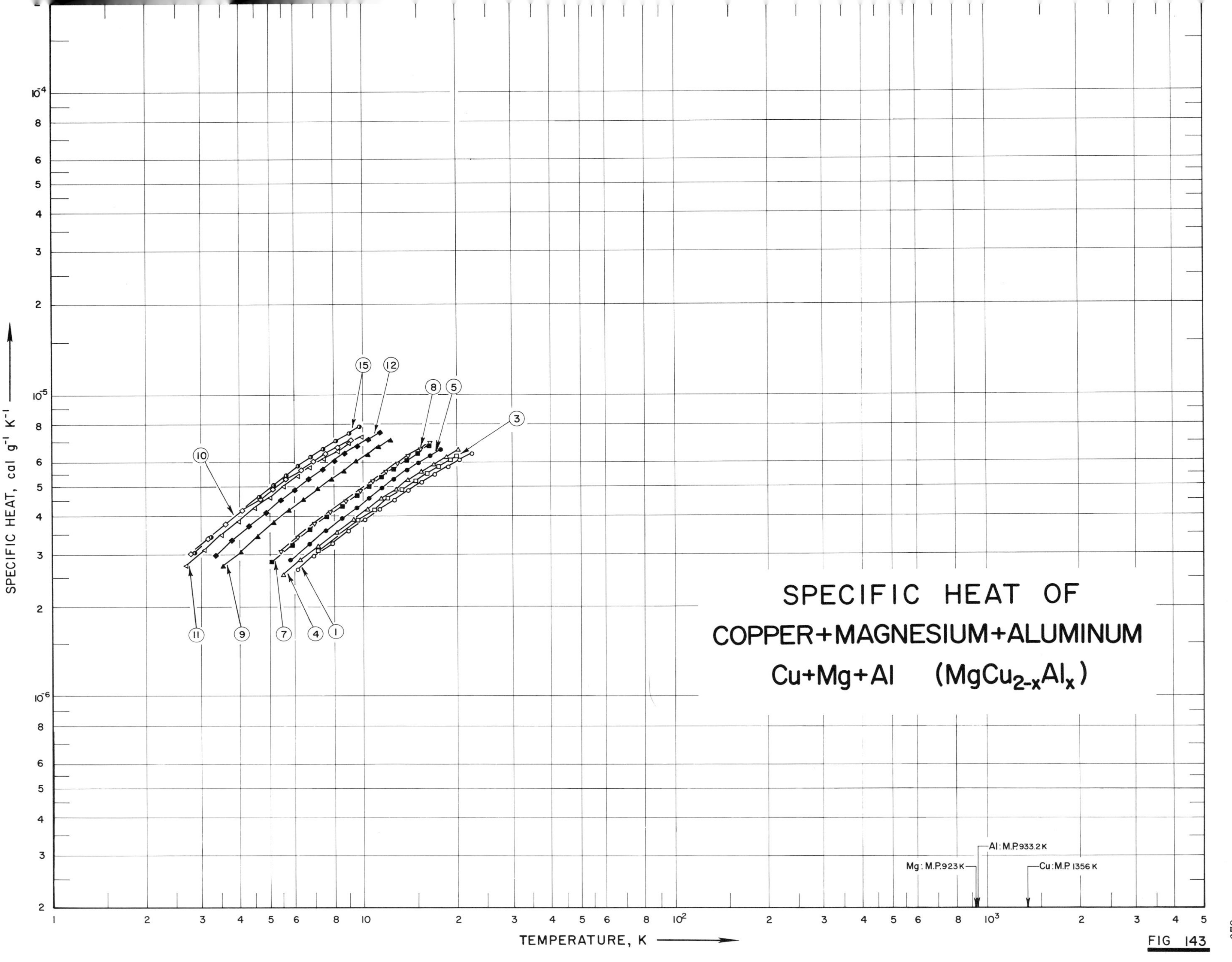
SPECIFIC HEAT OF
COPPER+MAGNESIUM+ALUMINUM
Cu+Mg+Al (MgCu$_{2-x}$Al$_x$)
Mg: M.P. 923 K
Al: M.P. 933.2 K
Cu: M.P. 1356 K
TEMPERATURE, K
SPECIFIC HEAT, cal g^{-1} K^{-1}
1
3
4
5
7
8
9
10
11
12
15

FIG 143

SPECIFICATION TABLE NO. 143 SPECIFIC HEAT OF COPPER + MAGNESIUM + ALUMINUM Cu + Mg + Al ($MgCu_{2-x}Al_x$)

[For Data Reported in Figure and Table No. 143]

Curve No.	Ref. No.	Year	Temp. Range, K	Reported Error, %	Name and Specimen Designation	Composition (weight percent), Specifications and Remarks
1	385	1965	6-23		$MgCu_{1.92}Al_{0.08}$	Prepared from: 99.99$^+$ Cu, sample supplied by American Smelting and Refining Co.; 99.99$^+$ Al, sample supplied by Aluminum Co. of America; 99.98 Mg, 0.001-0.003 Si, 0.001 Al, 0.001 Cu, 0.001 Fe, 0.001 Mn, resublimed grade sample supplied by Dow Chemical Co.; after casting each sample sealed pure helium and held 17-24 hrs at 200-250 C below melting temperature.
2	385	1965	6-23		$MgCu_{1.84}Al_{0.16}$	Same as above.
3	385	1965	7-20		$MgCu_{1.8}Al_{0.2}$	Same as above.
4	385	1965	5-20		$MgCu_{1.76}Al_{0.24}$	Same as above.
5	385	1965	5-18		$MgCu_{1.68}Al_{0.32}$	Same as above.
6	385	1965	5-18		$MgCu_{1.65}Al_{0.35}$	Same as above.
7	385	1965	5-17		$MgCu_{1.62}Al_{0.38}$	Same as above.
8	385	1965	5-17		$MgCu_{1.6}Al_{0.4}$	Same as above.
9	385	1965	3-13		$MgCu_{1.49}Al_{0.51}$	Same as above.
10	385	1965	3-9		$MgCu_{1.425}Al_{0.575}$	Same as above.
11	385	1965	2-10		$MgCu_{1.4}Al_{0.6}$	Same as above.
12	385	1965	3-12		$CuMg_{1.37}Al_{0.63}$	Same as above.
13	385	1965	3-12		$CuMg_{1.34}Al_{0.66}$	Same as above.
14	385	1965	3-12		$CuMg_{1.25}Al_{0.75}$	Same as above.
15	385	1965	3-10		$CuMg_{1.15}Al_{0.85}$	Same as above.

DATA TABLE NO. 143 SPECIFIC HEAT OF COPPER + MAGNESIUM + ALUMINUM, Cu + Mg + Al ($MgCu_{2-x}Al_x$)

[Temperature, T, K; Specific Heat, C_p, Cal g^{-1} K^{-1}]

T	C_p
CURVE 1	
6.20	2.677 x 10^{-6}
6.97	2.972
8.00	3.273
9.02	3.580
10.21	3.899
11.45	4.223
12.74	4.545
14.07	4.866
15.60	5.184
17.13	5.494
18.91	5.805
20.59	6.104
22.51	6.404
CURVE 2*	
5.99	2.560 x 10^{-6}
6.90	2.887
7.93	3.227
9.08	3.562
10.21	3.897
11.39	4.234
12.74	4.569
14.09	4.902
15.58	5.234
17.12	5.567
18.92	5.897
20.93	6.229
22.92	6.565
CURVE 3	
7.21	3.080 x 10^{-6}
9.67	3.860
10.97	4.199
12.15	4.565
13.47	4.895
14.89	5.207
16.21	5.512
17.61	5.811
19.23	6.119
20.12	6.290
CURVE 4	
5.58	2.567 x 10^{-6}
6.32	2.864
7.23	3.192
8.25	3.538
9.35	3.883
10.42	4.228
11.59	4.572
12.86	4.917
14.16	5.260
15.54	5.596
17.01	5.929
18.74	6.263
20.33	6.594
CURVE 5	
5.87	2.878 x 10^{-6}
6.79	3.251
7.62	3.593
8.56	3.933
9.57	4.272
10.53	4.611
11.60	4.955
12.73	5.300
13.99	5.665
15.22	5.992
16.53	6.312
17.88	6.633
CURVE 6*	
4.99	2.617 x 10^{-6}
5.77	2.997
6.65	3.367
7.56	3.728
8.41	4.084
9.38	4.440
10.41	4.792
11.50	5.143
12.54	5.497
13.83	5.853
15.09	6.206
16.48	6.555
18.04	6.907
CURVE 7	
5.13	2.826 x 10^{-6}
5.95	3.219
6.78	3.602
7.66	3.974
8.64	4.328
9.56	4.679
10.50	5.029
11.51	5.376
12.68	5.727
13.95	6.096
15.10	6.444
16.46	6.796
CURVE 8	
5.49	3.070 x 10^{-6}
6.20	3.425
7.00	3.780
7.86	4.131
8.84	4.489
9.83	4.860
10.75	5.236
11.91	5.608
12.92	5.957
14.11	6.302
15.32	6.646
16.57	6.988
CURVE 9	
3.57	2.744 x 10^{-6}
4.06	3.066
4.60	3.437
5.18	3.804
5.80	4.161
6.49	4.526
7.19	4.903
7.96	5.274
8.72	5.646
9.51	6.010
10.41	6.368
11.26	6.727
12.27	7.085
CURVE 10	
2.80	3.016 x 10^{-6}
3.19	3.377
3.63	3.767
4.12	4.162
4.71	4.548
5.16	4.915
5.69	5.286
6.37	5.650
6.96	6.009
7.60	6.371
8.33	6.734
9.18	7.098
CURVE 11	
2.72	2.762 x 10^{-6}
3.11	3.103
3.54	3.476
4.02	3.864
4.52	4.250
5.08	4.636
5.63	5.022
6.24	5.402
6.84	5.784
7.55	6.165
8.36	6.543
9.08	6.921
9.98	7.293
CURVE 12	
3.39	2.974 x 10^{-6}
3.81	3.340
4.33	3.713
4.90	4.106
5.48	4.500
6.07	4.897
6.74	5.290
7.45	5.672
8.12	6.050
8.76	6.422
9.62	6.787
10.49	7.148
11.38	7.517
CURVE 13*	
3.38	2.821 x 10^{-6}
3.93	3.228
4.57	3.660
5.15	4.053
5.72	4.440
6.34	4.830
6.97	5.206
7.69	5.582
8.39	5.982
9.19	6.368
10.07	6.761
10.93	7.150
11.94	7.550
CURVE 14*	
3.54	3.079 x 10^{-6}
3.63	3.133
4.17	3.543
4.76	3.960
5.38	4.357
6.05	4.746
6.67	5.142
7.29	5.510
7.92	5.888
8.71	6.287
9.49	6.692
10.36	7.098
11.35	7.501
CURVE 15	
2.89	3.051 x 10^{-6}
3.26	3.425
3.67	3.822*
4.16	4.235*
4.66	4.654
5.18	5.067
5.71	5.473
6.23	5.872
6.80	6.273
7.49	6.675
8.20	7.076
9.02	7.475
9.75	7.874

*Not shown on plot

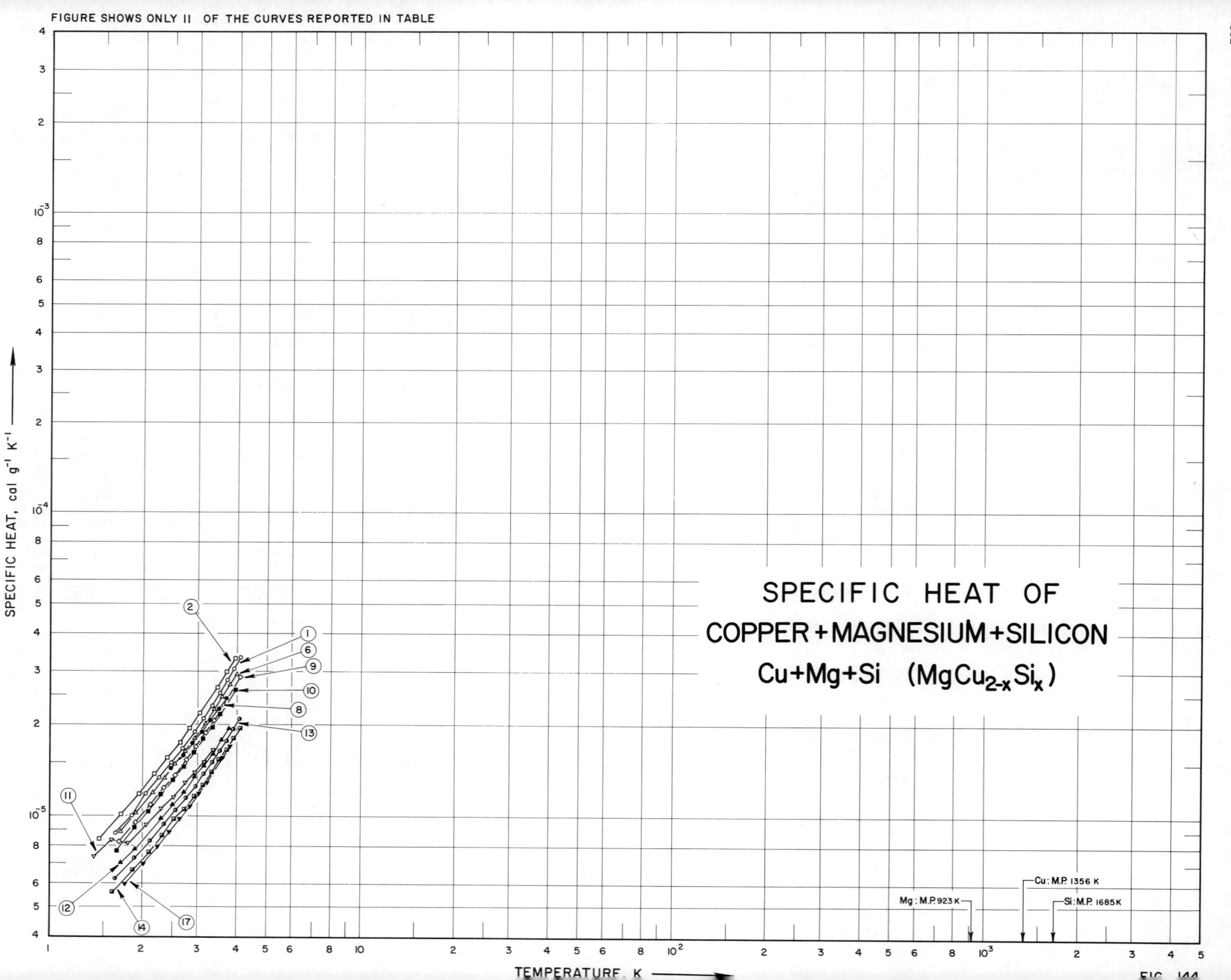
FIGURE SHOWS ONLY 11 OF THE CURVES REPORTED IN TABLE
SPECIFIC HEAT OF
COPPER + MAGNESIUM + SILICON
Cu+Mg+Si ($MgCu_{2-x}Si_x$)
SPECIFIC HEAT, cal g^{-1} K^{-1}
TEMPERATURE, K
Mg : M.P. 923 K
Cu : M.P. 1356 K
Si : M.P. 1685 K
1
2
6
9
10
8
13
11
12
14
17
4
3
2
10^{-3}
8
6
5
4
3
2
10^{-4}
8
6
5
4
3
2
10^{-5}
8
6
5
4
1
2
3
4
5
6
8
10
2
3
4
5
6
8
10^2
2
3
4
5
6
8
10^3
2
3
4
5

SPECIFICATION TABLE NO. 144 SPECIFIC HEAT OF COPPER + MAGNESIUM + SILICON, Cu + Mg + Si ($MgCu_{2-x}Si_x$)

[For Data Reported in Figure and Table No. 144]

Curve No.	Ref. No.	Year	Temp. Range, K	Reported Error, %	Name and Specimen Designation	Composition (weight percent), Specifications and Remarks
1	404	1966	1.6-4.1		$MgCu_{1.943}Si_{0.057}$	Prepared from: 99.99$^+$ Cu, sample supplied by American Smelting and Refining Co.; 99.98 Mg, 0.001-0.003 Si, 0.001 Al, 0.001 Cu, 0.001 Fe, 0.001 Mn, resublimed grade sample supplied by Dow Chemical Co.; melted; stirred for several minutes; quenched; after casting each sample held in Ar or He gas at 600-700 C for 72-98 hrs.
2	404	1966	1.5-4.0		$MgCu_{1.897}Si_{0.103}$	Same as above.
3	404	1966	1.6-4.1		$MgCu_{1.873}Si_{0.127}$	Same as above.
4	404	1966	1.5-4.0		$MgCu_{1.833}Si_{0.167}$	Same as above.
5	404	1966	1.7-3.7		$MgCu_{1.816}Si_{0.184}$	Same as above.
6	404	1966	1.7-4.0		$MgCu_{1.793}Si_{0.207}$	Same as above.
7	404	1966	1.7-4.1		$MgCu_{1.778}Si_{0.222}$	Same as above.
8	404	1966	2.5-3.7		$MgCu_{1.757}Si_{0.243}$	Same as above.
9	404	1966	1.7-4.1		$MgCu_{1.746}Si_{0.254}$	Same as above.
10	404	1966	1.7-4.0		$MgCu_{1.733}Si_{0.267}$	Same as above.
11	404	1966	1.4-3.4		$MgCu_{1.683}Si_{0.317}$	Same as above.
12	404	1966	1.7-3.8		$MgCu_{1.617}Si_{0.383}$	Same as above.
13	404	1966	1.6-4.1		$MgCu_{1.518}Si_{0.482}$	Same as above.
14	404	1966	1.6-4.1		$MgCu_{1.415}Si_{0.585}$	Same as above.
15	404	1966	1.7-4.0		$MgCu_{1.344}Si_{0.656}$	Same as above.
16	404	1966	1.5-4.0		$MgCu_{1.308}Si_{0.692}$	Same as above.
17	404	1966	1.8-4.0		$MgCu_{1.248}Si_{0.752}$	Same as above.

DATA TABLE NO. 144 SPECIFIC HEAT OF COPPER + MAGNESIUM + SILICON, Cu + Mg + Si ($MgCu_{2-x}Si_x$)

[Temperature, T, K; Specific Heat, C_p, Cal $g^{-1} K^{-1}$]

T	C_p
CURVE 1	
1.642	8.796×10^{-6}
1.852	1.004×10^{-5}
2.058	1.181
2.264	1.335
2.474	1.493
2.684	1.673
2.937	1.895
3.133	2.090
3.331	2.319
3.530	2.542
3.728	2.801
3.921	3.067
4.102	3.341
CURVE 2	
1.470	8.413×10^{-6}
1.717	1.013×10^{-5}
1.962	1.184
2.185	1.375
2.405	1.558
2.623	1.744
2.840	1.944
3.060	2.175
3.470	2.660
3.719	2.995
3.965	3.316
CURVE 3*	
1.627	9.070×10^{-6}
1.828	1.048×10^{-5}
2.048	1.223
2.241	1.375
2.433	1.525
2.632	1.694
2.838	1.887
3.040	2.104
3.252	2.332
3.465	2.593
3.669	2.843
3.869	3.125
4.071	3.416
CURVE 4*	
1.534	8.161×10^{-6}
1.695	9.139
1.881	1.035×10^{-5}
2.081	1.170
2.287	1.337
2.498	1.507
2.711	1.678
2.926	1.873
3.139	2.078
3.353	2.305
3.568	2.536
3.776	2.798
3.971	3.045
CURVE 5*	
1.679	8.828×10^{-6}
1.898	1.059×10^{-5}
2.111	1.218
2.317	1.393
2.518	1.574
2.719	1.673
3.128	2.072
3.325	2.363
3.517	2.676
3.706	2.980
CURVE 6	
1.717	8.921×10^{-6}
1.777	9.473*
1.915	1.025×10^{-5}
2.164	1.193
2.351	1.331
2.542	1.489
2.732	1.629
2.930	1.801
3.178	2.015
3.380	2.242
3.590	2.470
3.805	2.709
4.008	2.945
CURVE 7*	
1.663	8.931×10^{-6}
1.879	1.038×10^{-5}
2.077	1.190
2.277	1.341
2.480	1.491
2.687	1.662
2.893	1.846
3.099	2.056
3.304	2.281
3.508	2.502
3.713	2.750
3.914	3.023
4.097	3.335
CURVE 8	
2.461	1.438×10^{-5}
2.688	1.589
2.895	1.734
3.091	1.881
3.296	2.062
3.502	2.252
3.703	2.438
CURVE 9	
1.693	8.251×10^{-6}
1.902	9.541
2.120	1.094×10^{-5}
2.335	1.248
2.551	1.370
2.764	1.529
2.974	1.696
3.180	1.881
3.382	2.070
3.588	2.283*
3.828	2.510*
4.113	2.889
CURVE 10	
1.660	7.721×10^{-6}
1.877	9.159
CURVE 10 (cont.)	
2.083	1.039×10^{-5}
2.287	1.179
2.493	1.323
2.701	1.460
2.910	1.621
3.120	1.790
3.331	1.969
3.543	2.173
3.752	2.403*
3.956	2.607
CURVE 11	
1.393	7.351×10^{-6}
1.591	8.376
1.817	8.145
2.066	9.378
2.295	1.067×10^{-5}
2.522	1.159
2.732	1.283
2.937	1.392
3.146	1.508
3.363	1.664
CURVE 12	
1.717	7.016×10^{-6}
1.893	7.808
2.296	9.886
2.505	1.093×10^{-5}
2.715	1.205
2.929	1.363
3.147	1.472
3.363	1.611
3.571	1.781
3.772	1.933
CURVE 13	
1.632	6.230×10^{-6}
1.885	7.283
2.110	8.309
2.340	9.425
2.551	1.053×10^{-5}
CURVE 13 (cont.)	
2.755	1.154×10^{-5}
2.957	1.261
3.145	1.377
3.336	1.501
3.525	1.637
3.710	1.774
3.901	1.931
4.086	2.086
CURVE 14	
1.601	5.602×10^{-6}
1.866	6.661
2.101	7.639
2.320	8.674
2.524	9.822
2.721	1.062×10^{-5}
2.925	1.172
3.126	1.274
3.323	1.396
3.520	1.527
3.714	1.654
3.913	1.805
4.106	1.949
CURVE 15*	
1.701	5.921×10^{-6}
1.931	6.919
2.154	7.905
2.357	8.874
2.558	9.904
2.775	1.098×10^{-5}
2.997	1.218
3.206	1.349
3.394	1.465
3.570	1.581
3.761	1.723
3.970	1.889
CURVE 16*	
1.542	6.067×10^{-6}
CURVE 16 (cont.)*	
1.767	6.937×10^{-6}
1.977	7.486
2.197	7.852
2.410	9.860
2.619	1.085×10^{-5}
2.820	1.194
3.024	1.325
3.230	1.481
3.434	1.622
3.635	1.756
3.835	1.911
4.036	2.095
CURVE 17	
1.763	5.991×10^{-6}
2.019	6.970
2.240	7.965
2.444	8.888
2.638	9.815
2.837	1.077×10^{-5}
3.031	1.185
3.220	1.290
3.413	1.407*
3.607	1.553
3.801	1.692
3.993	1.811*

* Not shown on plot

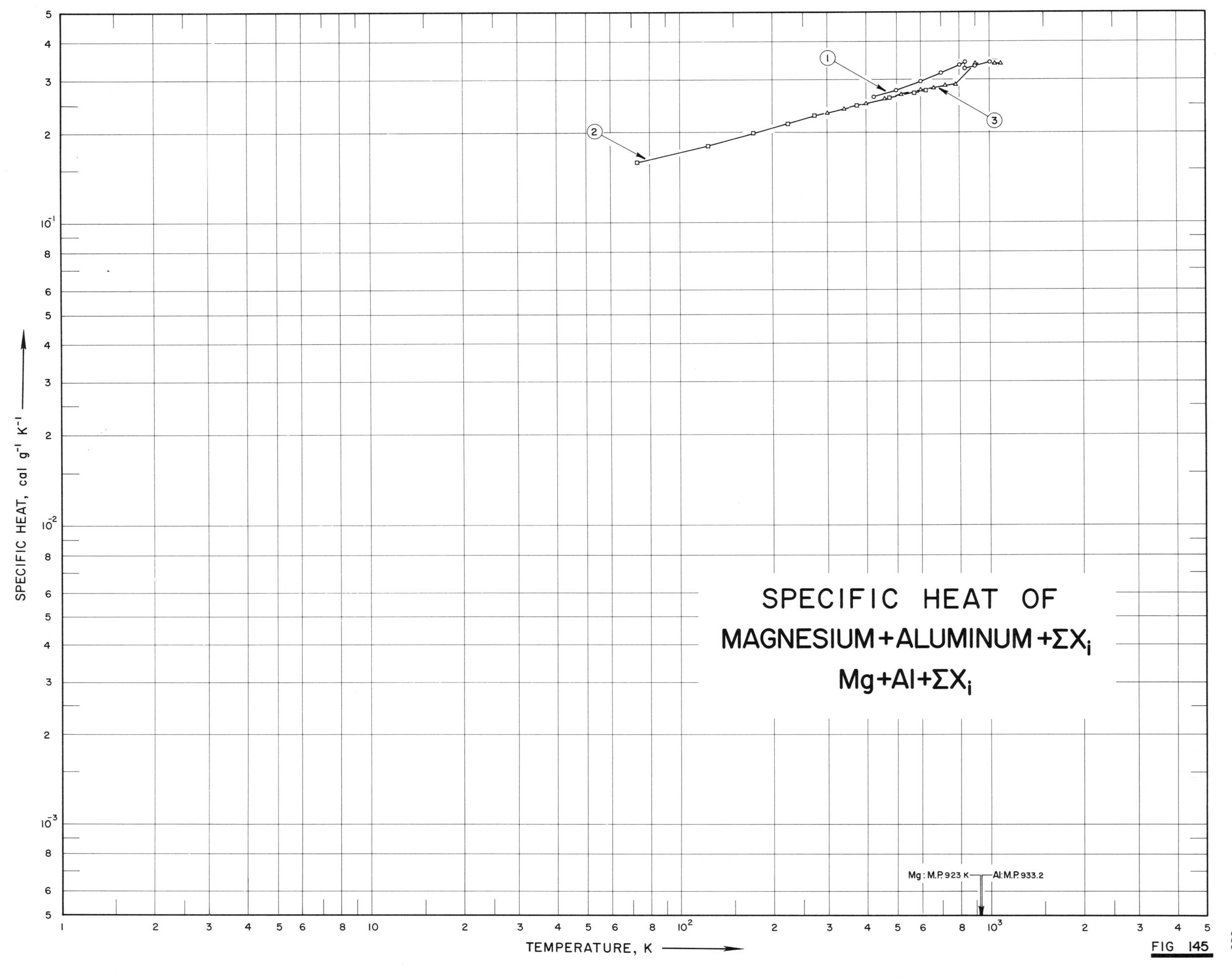
SPECIFIC HEAT OF
MAGNESIUM+ALUMINUM+ΣXi
Mg+Al+ΣXi
Mg: M.P. 923 K
Al: M.P. 933.2
TEMPERATURE, K
SPECIFIC HEAT, cal g-1 K-1
1
2
3

FIG 145

SPECIFICATION TABLE NO. 145 SPECIFIC HEAT OF MAGNESIUM + ALUMINUM + ΣX_i Mg + Al + ΣX_i

[For Data Reported in Figure and Table No. 145]

Curve No.	Ref. No.	Year	Temp. Range, K	Reported Error, %	Name and Specimen Designation	Composition (weight percent), Specifications and Remarks
1	46	1957	425-1000		Mg Alloy AZ31B	95.5 Mg, 3.0 Al, 1.0 Zn, 0.5 Mn; machined from permanent mold cast material.
2	243	1958	73-623		Mg Alloy AN-M-29	95.7 Mg, 3.0 Al, 1.0 Zn, 0.3 Mn.
3	246	1961	300-1080	0.5-3	Mg Alloy AZ-80	Bal Mg, 8.0 Al, 0.55 Zn, 0.14 Mn; measured in a helium atmosphere.

DATA TABLE NO. 145 SPECIFIC HEAT OF MAGNESIUM + ALUMINUM + ΣX_i Mg + Al + ΣX_i

[Temperature, T, K; Specific Heat, C_p, Cal $g^{-1} K^{-1}$]

	T	C_p
	CURVE 1	
	425	2.650 x 10^{-1}
	500	2.789
	600	2.982
	700	3.179
	800	3.378
(s)	838	3.454
(ℓ)	838	3.281
	900	3.352
	1000	3.463
	CURVE 2	
	73	1.60 x 10^{-1}
	123	1.82
	173	2.00
	223	2.16
	273	2.29
	373	2.48
	473	2.63
	573	2.74
	623	2.79
	CURVE 3	
	300	2.34 x 10^{-1}
	320	2.38*
	340	2.42
	360	2.45*
	380	2.49*
	400	2.52
	420	2.56*
	440	2.59*
	460	2.62
	480	2.65*
	500	2.68*
	520	2.70
	540	2.73*
	560	2.75*
	580	2.77*
	600	2.79
	620	2.81*
	640	2.83*
	660	2.84
	680	2.86*
	700	2.88*
	CURVE 3 (cont.)	
	720	2.89 x 10^{-1}
	740	2.90*
	760	2.91*
(s)	780	2.92
(ℓ)	900	3.41
	920	3.41*
	1040	3.41
	1060	3.41*
	1080	3.41

*Not shown on plot

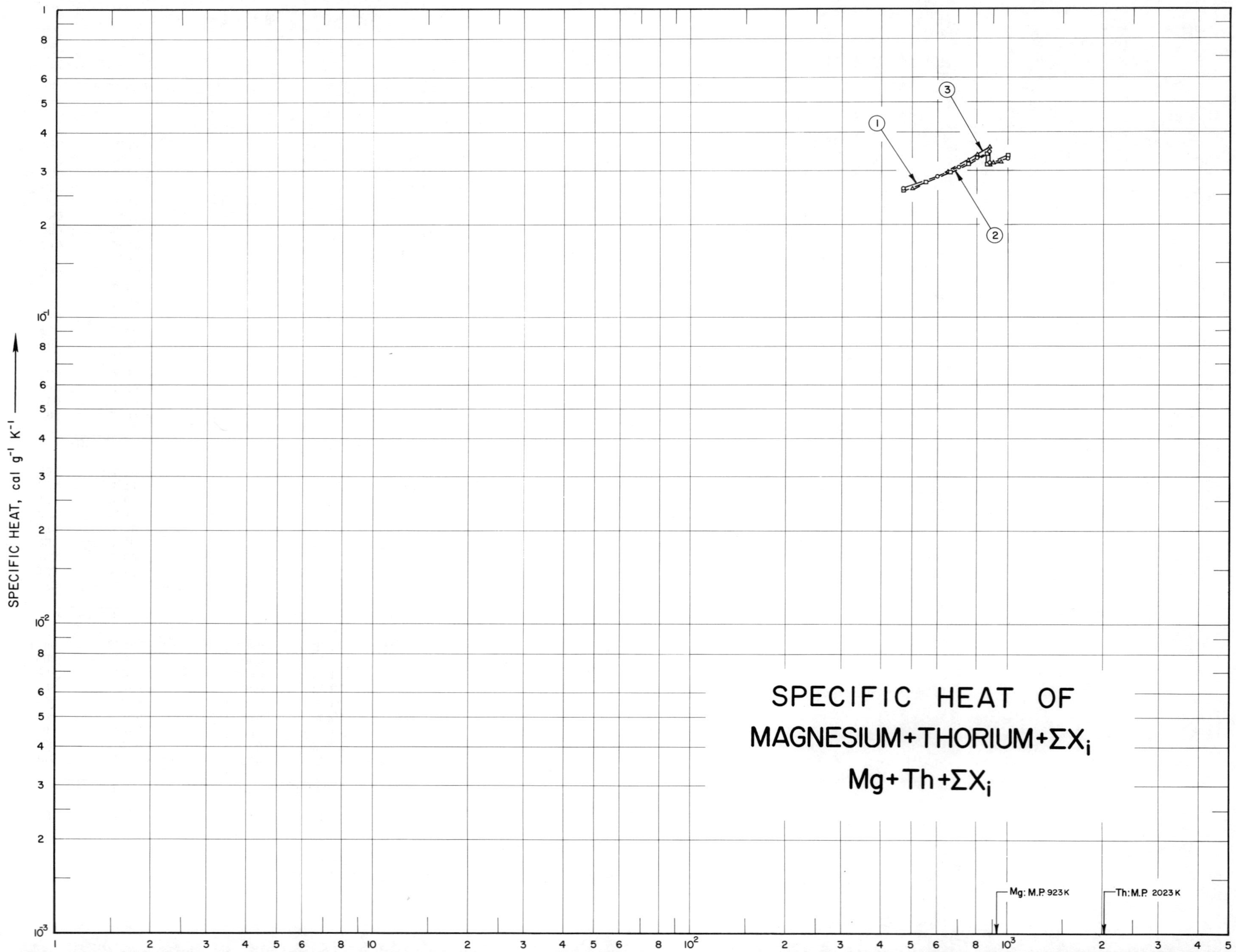

SPECIFIC HEAT OF MAGNESIUM+THORIUM+ΣX_i

Mg+Th+ΣX_i

SPECIFIC HEAT, cal g^{-1} K^{-1}

Mg: M.P. 923 K

Th: M.P. 2023 K

1

2

3

SPECIFICATION TABLE NO. 146 SPECIFIC HEAT OF MAGNESIUM + THORIUM + ΣX_i Mg + Th + ΣX_i

[For Data Reported in Figure and Table No. 146]

Curve No.	Ref. No.	Year	Temp. Range, K	Reported Error, %	Name and Specimen Designation	Composition (weight percent), Specifications and Remarks
1	46	1957	470-1000		Mg alloy HM 21XA	2.0 Th, 0.5 Mn.
2	46	1957	470-1000		Mg alloy HK 31A	96.3 Mg, 3.0 Th, 0.7 Zr.
3	46	1957	470-1000		Mg alloy HM 31Xa	2.98 total rare earth, 1.40 Mn, 0.05 Zn, 0.03 Al.

DATA TABLE NO. 146 SPECIFIC HEAT OF MAGNESIUM + THORIUM + ΣX_i Mg + Th + ΣX_i

[Temperature, T, K; Specific Heat, C_p, Cal $g^{-1} K^{-1}$]

T	C_p
CURVE 1	
470	2.629 x 10^{-1}
500	2.682*
550	2.775*
600	2.874
650	2.975*
700	3.080
750	3.187*
800	3.295
(s) 878	3.466
(ℓ) 878	3.180
900	3.200*
950	3.243*
1000	3.286
CURVE 2	
470	2.610 x 10^{-1}
500	2.661*
550	2.753
600	2.851*
660	2.973
700	3.057*
750	3.163
800	3.271*
(s) 861	3.404
(ℓ) 861	3.161
900	3.217*
950	3.288*
1000	3.359
CURVE 3	
470	2.582 x 10^{-1}*
500	2.641
550	2.748*
600	2.865*
650	2.988
700	3.115*
750	3.246
800	3.380
(s) 878	3.591
(ℓ) 878	3.164
900	3.180
950	3.218
1000	3.254*

*Not shown on plot

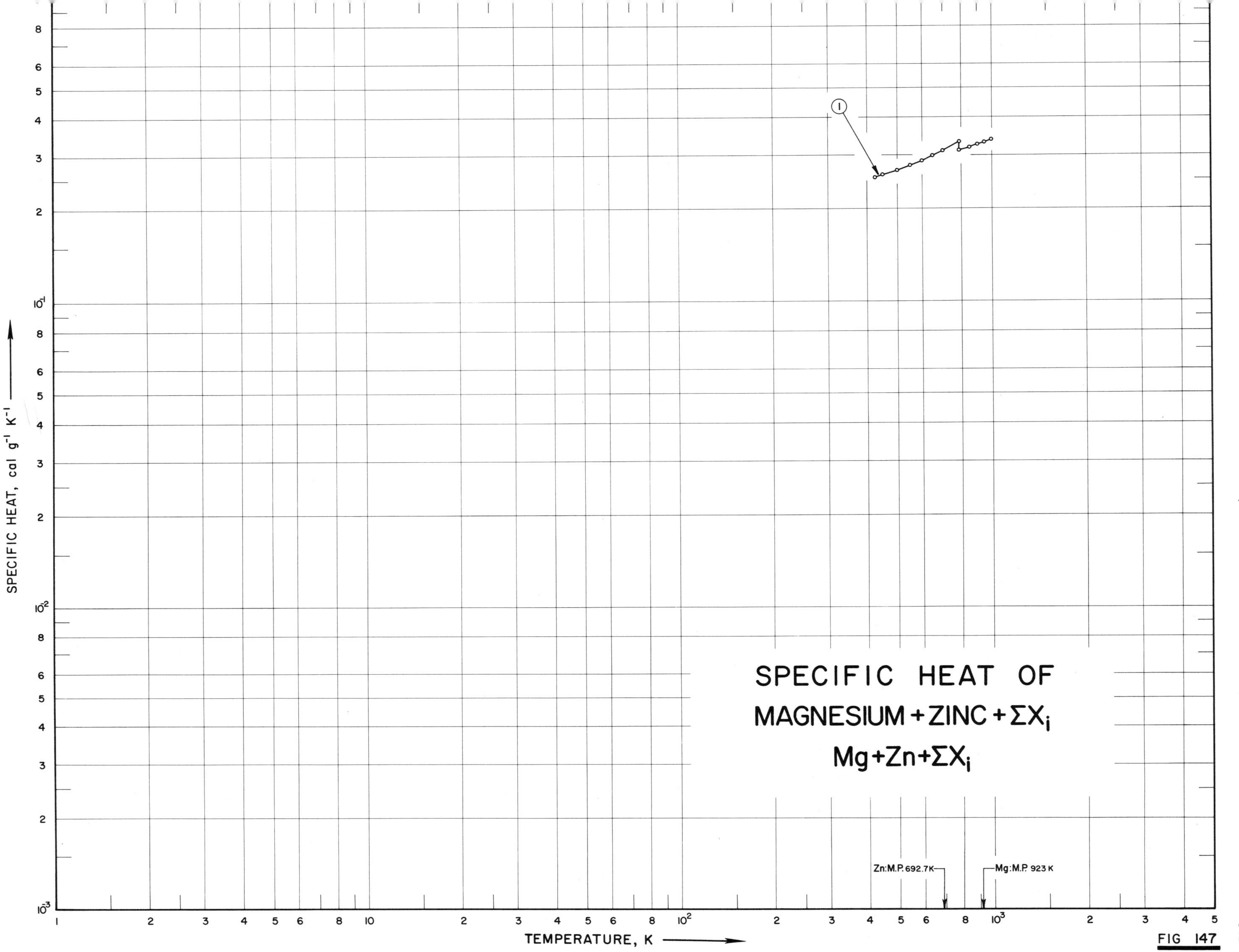
SPECIFIC HEAT OF
MAGNESIUM + ZINC + ΣX_i
Mg+Zn+ΣX_i
1
Zn: M.P. 692.7 K
Mg: M.P. 923 K
TEMPERATURE, K
SPECIFIC HEAT, cal g^{-1} K^{-1}
FIG 147

SPECIFICATION TABLE NO. 147 SPECIFIC HEAT OF MAGNESIUM + ZINC + ΣX_i Mg + Zn + ΣX_i

[For Data Reported in Figure and Table No. 147]

Curve No.	Ref. No.	Year	Temp. Range, K	Reported Error, %	Name and Specimen Designation	Composition (weight percent), Specifications and Remarks
1	46	1957	425-1000		Mg alloy ZK 60A	5.78 Zn, 0.74 Zr, 0.05 Mn, 0.03 Al.

DATA TABLE NO. 147 SPECIFIC HEAT OF MAGNESIUM + ZINC + ΣX_i $Mg + Zn + \Sigma X_i$

[Temperature, T, K; Specific Heat, C_p, Cal $g^{-1} K^{-1}$]

T	C_p
CURVE 1	
425	2.542×10^{-1}
450	2.582
500	2.674
550	2.774
600	2.882
650	2.994
700	3.110
(s) 793	3.330
(ℓ) 793	3.122
800	3.131*
850	3.193
900	3.256
950	3.318
1000	3.380

* Not shown on plot

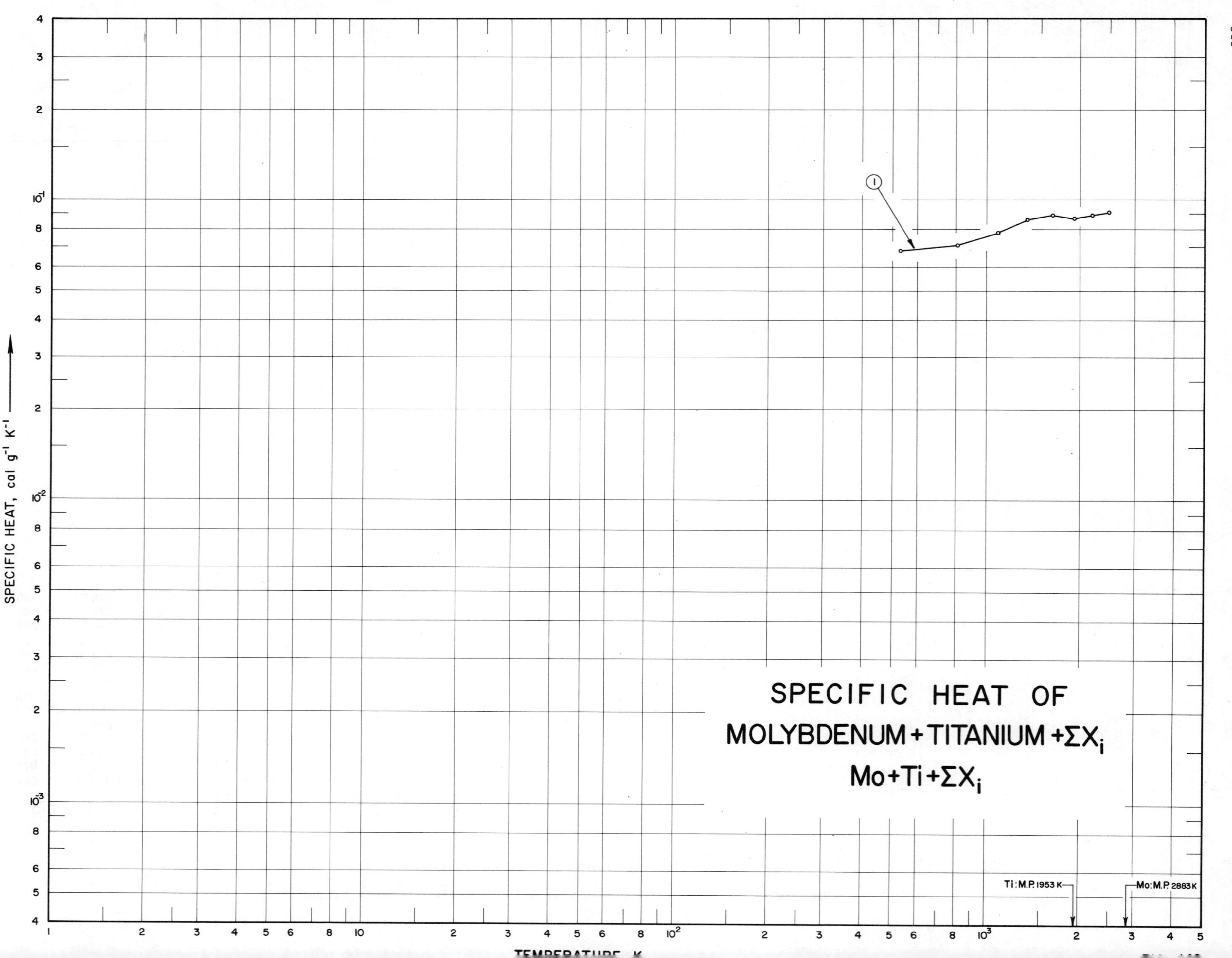

SPECIFIC HEAT OF
MOLYBDENUM + TITANIUM + ΣX_i
Mo + Ti + ΣX_i
Ti: M.P. 1953 K
Mo: M.P. 2883 K
SPECIFIC HEAT, cal g^{-1} K^{-1}
1

SPECIFICATION TABLE NO. 148 SPECIFIC HEAT OF MOLYBDENUM + TITANIUM + ΣX_i Mo + Ti + ΣX_i

[For Data Reported in Figure and Table No. 148]

Curve No.	Ref. No.	Year	Temp. Range, K	Reported Error, %	Name and Specimen Designation	Composition (weight percent), Specifications and Remarks
1	237	1962	533-2478	≤5		Before exposure: 98.6 Mo, 0.7 Ti, 0.3 Fe, 0.2 Al, 0.2 Ni, 0.1 Si; after exposure: 98.3 Mo, 0.2 C; sample supplied by General Astrometals Corporation; crushed in hardened steel mortar to pass 100-mesh screen; hot pressed; density at 25 C before exposure: apparent density (wax coated specimen) 592 lb ft^{-3}, true density (by immersion in xylene) 585 lb ft^{-3}, after exposure: apparent density (ASTM method B311-58) 539 lb ft^{-3}, true density = 565 lb ft^{-3}.

DATA TABLE NO. 148 SPECIFIC HEAT OF MOLYBDENUM + TITANIUM + ΣX_i Mo + Ti + ΣX_i

[Temperature, T, K; Specific Heat, C_p, Cal $g^{-1} K^{-1}$]

T	C_p
CURVE 1	
533	6.8×10^{-2}
811	7.1
1089	7.8
1366	8.6
1644	8.9
1922	8.7
2200	8.9
2478	9.1

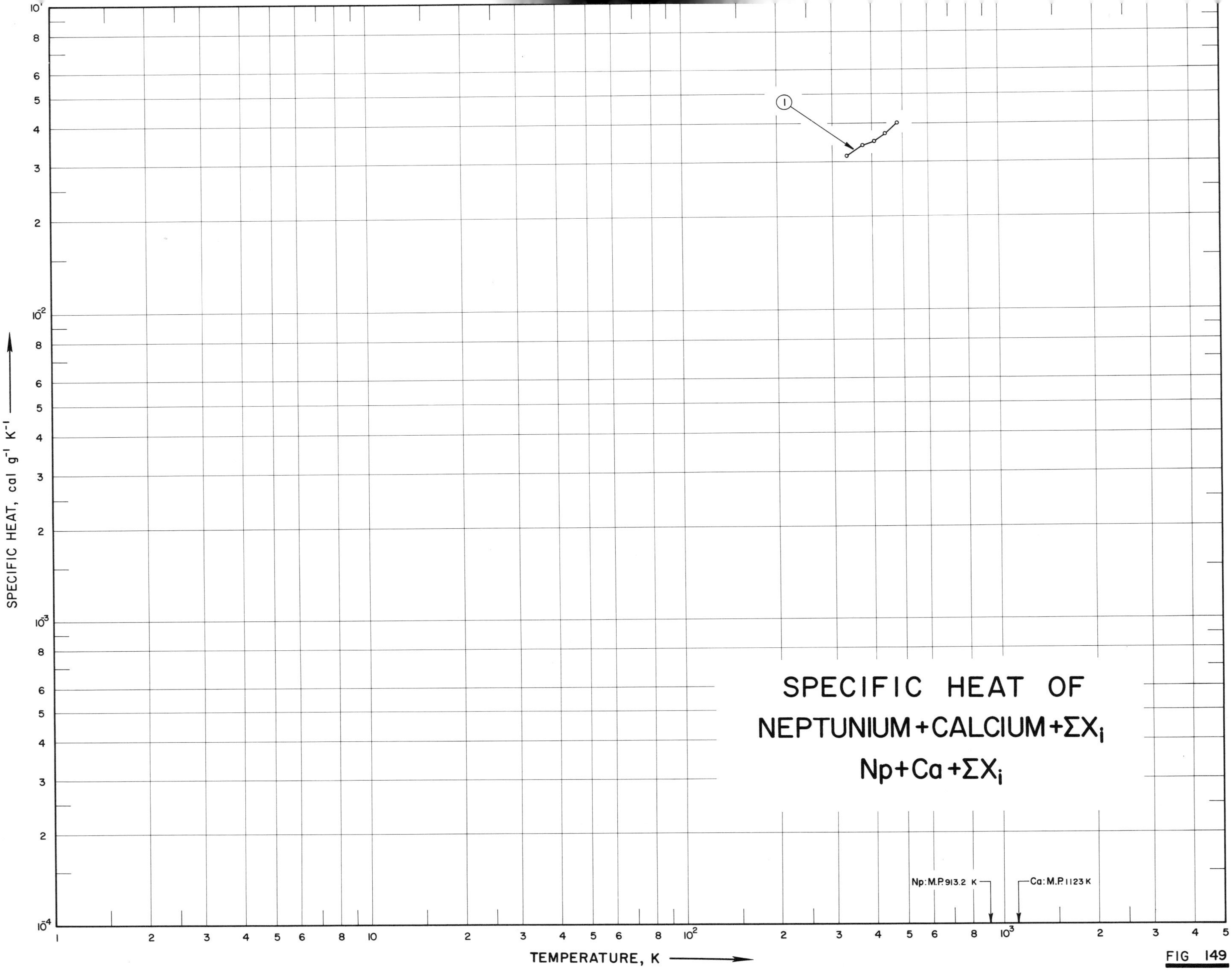

SPECIFIC HEAT OF
NEPTUNIUM + CALCIUM + ΣX_i
Np + Ca + ΣX_i

Np: M.P. 913.2 K
Ca: M.P. 1123 K

1

SPECIFIC HEAT, cal g^{-1} K^{-1}

TEMPERATURE, K

FIG 149

SPECIFICATION TABLE NO. 149 SPECIFIC HEAT OF NEPTUNIUM + CALCIUM + ΣX_i Np + Ca + ΣX_i

[For Data Reported in Figure and Table No. 149]

Curve No.	Ref. No.	Year	Temp. Range, K	Reported Error, %	Name and Specimen Designation	Composition (weight percent), Specifications and Remarks
1	247	1958	333-480	≤2.0		99.44 Np, 0.34 Ca, 0.22 U.

DATA TABLE NO. 149 SPECIFIC HEAT OF NEPTUNIUM + CALCIUM + ΣX_i $Np + Ca + \Sigma X_i$

[Temperature, T, K; Specific Heat, C_p, Cal $g^{-1} K^{-1}$]

T	C_p
CURVE 1	
333	3.14×10^{-2}
375	3.38
407	3.49
442	3.70
480	4.02

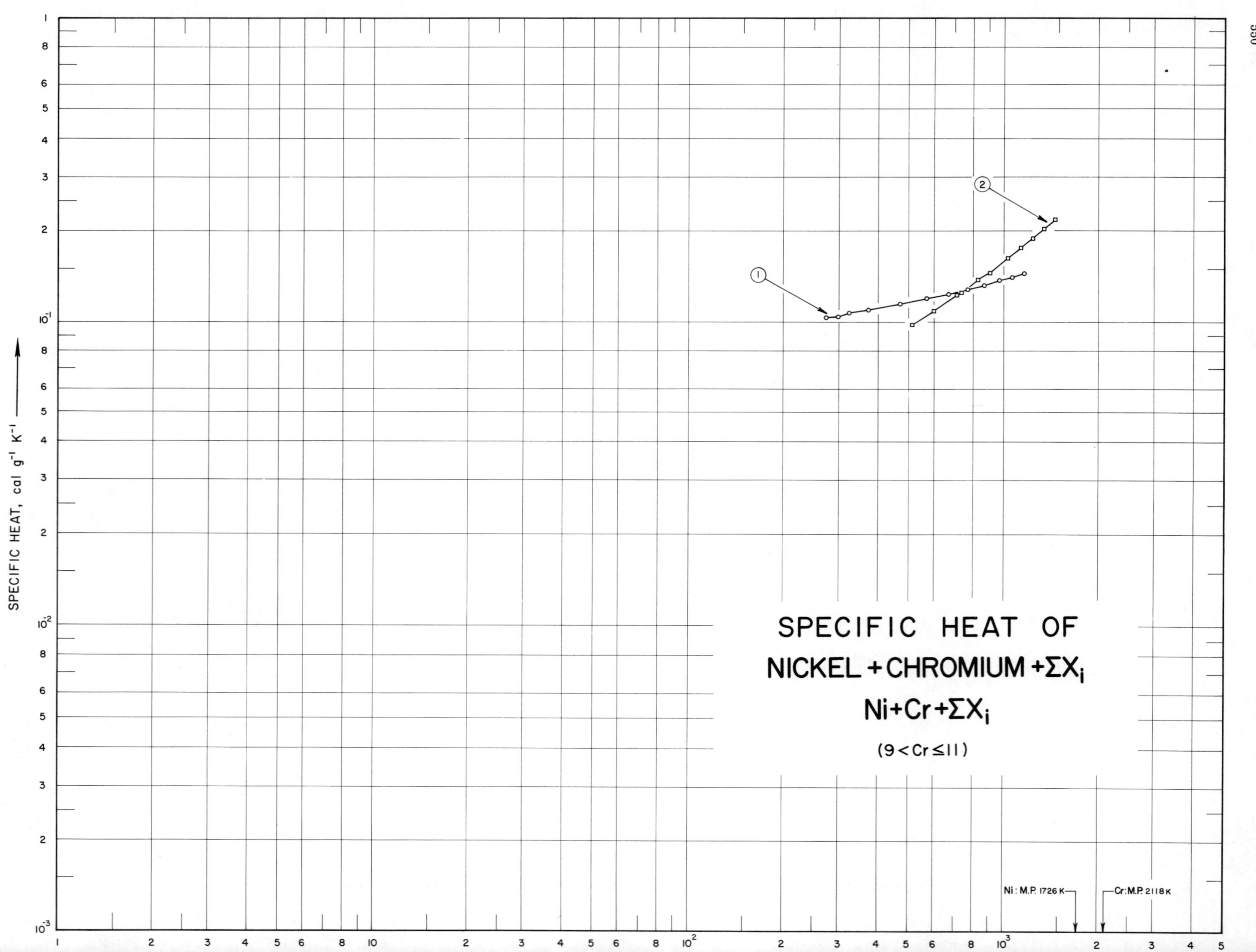
SPECIFIC HEAT OF
NICKEL + CHROMIUM + ΣX_i
Ni+Cr+ΣX_i
($9 < Cr \leq 11$)
SPECIFIC HEAT, cal g^{-1} K^{-1}
Ni: M.P. 1726 K
Cr: M.P. 2118 K
1
2

SPECIFICATION TABLE NO. 150 SPECIFIC HEAT OF NICKEL + CHROMIUM + ΣX_i, Ni + Cr + ΣX_i (9< Cr≤11)

[For Data Reported in Figure and Table No. 150]

Curve No.	Ref. No.	Year	Temp. Range, K	Reported Error, %	Name and Specimen Designation	Composition (weight percent), Specifications and Remarks
1	248	1960	273-1173	±.30		89.1 Ni, 9.6 Cr, 0.63 Fe, 0.42 Si, 0.12 Zr, 0.08 Co, 0.01 Cu, 0.01 Mn; Sample A unannealed, under helium atmosphere.
2	146	1961	513-1473	3.0	Inco 713 C	71.53 Ni, 11.0 Cr, 6.5 Al, 5.0 Fe, 3.5 Mo, 1.0 Nb + Ta, 1.0 Mn, Si, 0.25 Ti, 0.2 C; measured under helium atmosphere; density = 576 lb ft^{-3}.

DATA TABLE NO. 150 SPECIFIC HEAT OF NICKEL + CHROMIUM + ΣX_i, Ni + Cr + ΣX_i ($9 < Cr \leq 11$)

[Temperature, T, K; Specific Heat, C_p, Cal $g^{-1} K^{-1}$]

T	C_p
CURVE 1	
273	1.040×10^{-1}
298	1.058
323	1.076
373	1.107
473	1.160
573	1.206
673	1.248
773	1.288
873	1.326
973	1.380
1073	1.420
1173	1.460
CURVE 2	
513	9.821×10^{-2}
604	1.095×10^{-1}
715	1.234
738	1.262
833	1.381
901	1.465
1035	1.632
1138	1.760
1140	1.763*
1245	1.893
1254	1.905*
1353	2.028
1357	2.033*
1473	2.178

* Not shown on plot

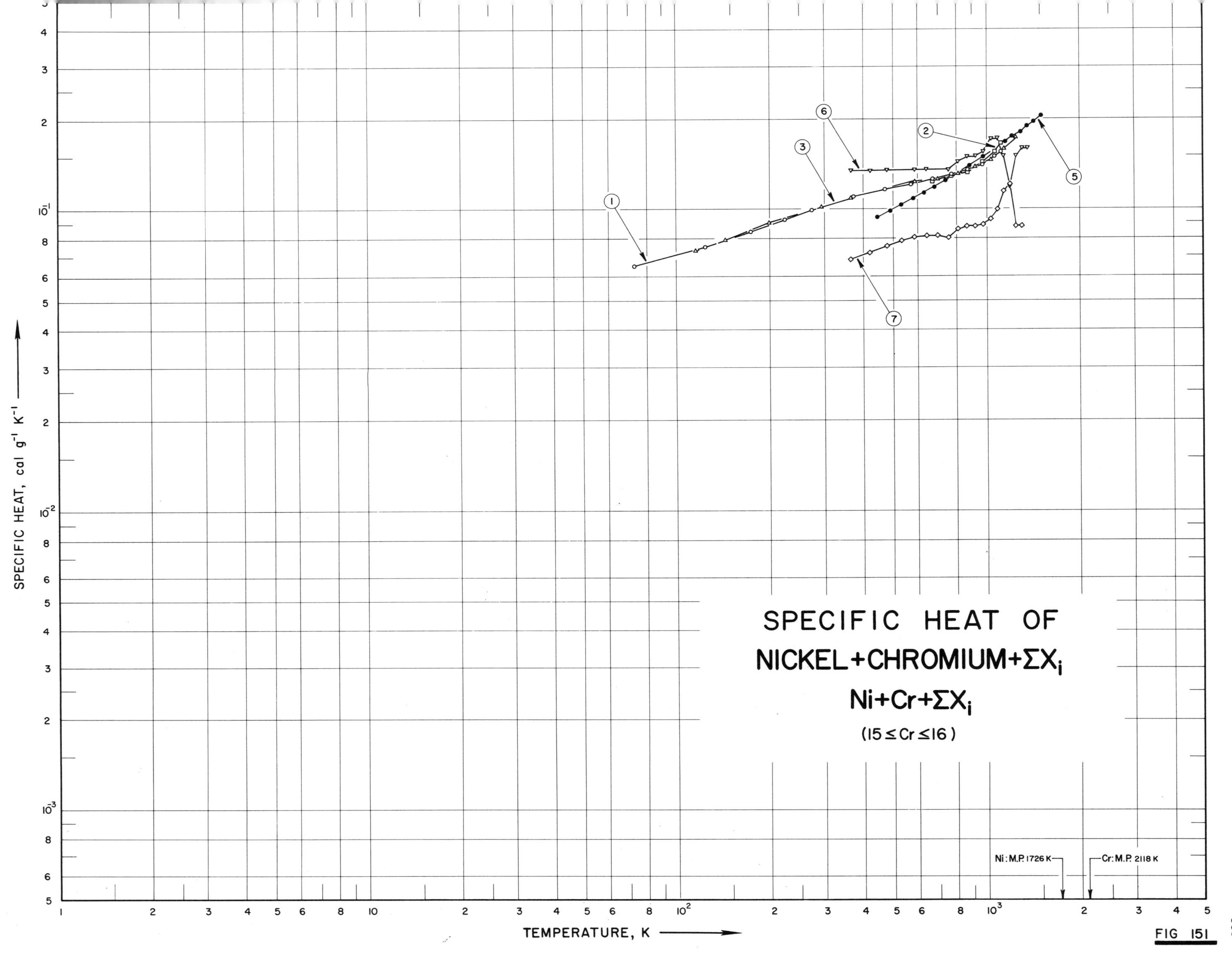
SPECIFIC HEAT OF
NICKEL+CHROMIUM+ΣX$_i$
Ni+Cr+ΣX$_i$
(15 ≤ Cr ≤ 16)
SPECIFIC HEAT, cal g^{-1} K^{-1}
TEMPERATURE, K
Ni: M.P. 1726 K
Cr: M.P. 2118 K
1
2
3
5
6
7

FIG 151

SPECIFICATION TABLE NO. 151 SPECIFIC HEAT OF NICKEL + CHROMIUM + ΣX_i, Ni + Cr + ΣX_i ($15 \leq Cr \leq 16$)

[For Data Reported in Figure and Table No. 151]

Curve No.	Ref. No.	Year	Temp. Range, K	Reported Error, %	Name and Specimen Designation	Composition (weight percent), Specifications and Remarks
1	243	1952	73-1123		Inconel	Nominal composition: 77 Ni, 15 Cr, 7 Fe.
2	243	1952	73-1123		Inconel X	Nominal composition: 73 Ni, 15 Cr, 7 Fe, 2.5 Ti.
3	10	1958	116-1255		Annealed Inconel	Nominal composition: 78 Ni, 15 Cr, 7 Fe, 0.35 Mn, 0.2 Si, 0.04 C; sample supplied by the International Nickel Co., Inc.; specimen sealed in helium capsule; annealed for 3 hours at 1600 F, held for 15 min at 1800 F and air cooled; density (32 F) = 390 lb ft^{-3}.
4	10	1958	116-1255		Inconel X	Nominal composition: 73 Ni, 15 Cr, 7 Fe, 2.5 Ti, 1 Nb, 0.9 Al, 0.7 Mn, 0.4 Si, 0.04 C; sample supplied by the International Nickel Co., Inc.; specimen sealed in helium in capsule; solution treated by heating for 3 hrs at 2100 F and air cooled; double aged for 24 hrs at 1550 F and air cooled; then held for 20 hrs at 1300 F and air cooled; density (32 F) = 380 lb ft^{-3}.
5	75	1958	445-1517	0.66-2.9	Hastelloy R-235	Nominal composition: 66.85 Ni (bal.), 15.5 Cr, 10 Fe, 5 Mo, 2.5 Ti, 0.15 C; measured in a helium atmosphere.
6	249	1959	366-1311	5-10	Inconel 702	Nominal: 80.0 Ni, 15.0 Cr, 3.0 Al, 0.5 Ti, 0.35 Fe, 0.05 C; heated to 1975 F for 0.5 hr; air cooled; heated to 1400 F for 5 hrs; air cooled.
7	249	1959	366-1311	5-10	Inconel X	Nominal: $\geq$70 Ni, 15.0 Cr, 7.0 Fe, 2.5 Ti, 0.95 Nb, 0.70 Al, $\leq$0.2 Cu, $\leq$0.08 C; heated at 2100 F for 2 hrs; air cooled; heated to 1550 F for 24 hrs; air cooled; heated to 1300 F for 20 hrs; air cooled.
8	248	1960	273-1173	±0.3	Sample 1	76 Ni, 15 Cr, 9 Fe.
9	248	1960	273-1173	±0.3	Sample 2	Same as above.
10	248	1960	273-1173	±0.3	Sample 3	Same as above.

DATA TABLE NO. 151 SPECIFIC HEAT OF NICKEL + CHROMIUM + ΣX_i, Ni + Cr + ΣX_i (15 ≤ Cr ≤ 16)

[Temperature, T, K; Specific Heat, C_p, Cal g^{-1} K^{-1}]

T	C_p
CURVE 1	
73	6.5 x 10^{-2}
123	7.5
173	8.4
223	9.2
273	9.9
373	1.10 x 10^{-1}
473	1.17
573	1.22
673	1.26
773	1.30
873	1.35
973	1.41
1073	1.50
1123	1.56
CURVE 2	
73	6.4 x 10^{-2}*
123	7.5*
173	8.5*
223	9.3*
273	1.00 x 10^{-1}*
373	1.10*
473	1.16*
573	1.20*
673	1.24
773	1.28
873	1.33
973	1.43
1073	1.56
1123	1.67
CURVE 3	
116	7.3 x 10^{-2}
144	7.9
200	9.0
293	1.02 x 10^{-1}
366	1.09
478	1.17*
589	1.23
700	1.27
811	1.32
922	1.38
1033	1.47
1144	1.59

T	C_p
CURVE 4*	
116	7.3 x 10^{-2}
144	8.0
200	9.0
293	1.03 x 10^{-1}
366	1.09
478	1.16
589	1.20
700	1.25
811	1.30
922	1.37
1033	1.51
1144	1.71
1255	1.97
CURVE 5	
446	9.41 x 10^{-2}
493	9.90
533	1.03 x 10^{-1}
567	1.07*
581	1.08
634	1.14
685	1.19
739	1.25
811	1.32*
818	1.33*
884	1.40
911	1.42*
977	1.49
1016	1.53*
1078	1.60*
1089	1.61*
1101	1.62*
1132	1.65*
1155	1.68
1174	1.70*
1221	1.75
1264	1.79*
1291	1.82
1307	1.83*
1339	1.87*
1366	1.90
1378	1.91*
1389	1.92*
1426	1.96
1434	1.97*

T	C_p
CURVE 5 (cont.)	
1474	2.01 x 10^{-1}*
1494	2.03*
1517	2.05
CURVE 6	
366	1.34 x 10^{-1}
422	1.34
478	1.35
533	1.35*
589	1.35
644	1.36
700	1.36*
755	1.36
811	1.43
866	1.48
922	1.48
978	1.55
1033	1.70
1089	1.72
1144	1.50
1200	1.18
1255	1.50
1311	1.59
CURVE 7	
366	6.8 x 10^{-2}
422	7.2
478	7.6
533	7.9
589	8.1
644	8.2
700	8.2
755	8.05
811	8.6
866	8.8
922	8.8
978	8.9
1033	9.3
1089	1.00 x 10^{-1}
1144	1.16
1200	1.215
1255	8.8 x 10^{-2}
1311	8.8

T	C_p
CURVE 8*	
273	1.05 x 10^{-1}
298	1.068
323	1.084
373	1.112
473	1.160
573	1.202
673	1.240
773	1.276
873	1.379
973	1.421
1073	1.462
1173	1.500
CURVE 9*	
273	1.04 x 10^{-1}
298	1.061
323	1.077
373	1.108
473	1.160
573	1.207
673	1.249
773	1.290
873	1.377
973	1.422
1073	1.467
1173	1.510
CURVE 10	
273	1.05 x 10^{-1}
298	1.066
323	1.082
373	1.111
473	1.162
573	1.208
673	1.251
773	1.292
873	1.396
973	1.430
1073	1.465
1173	1.500

*Not shown on plot

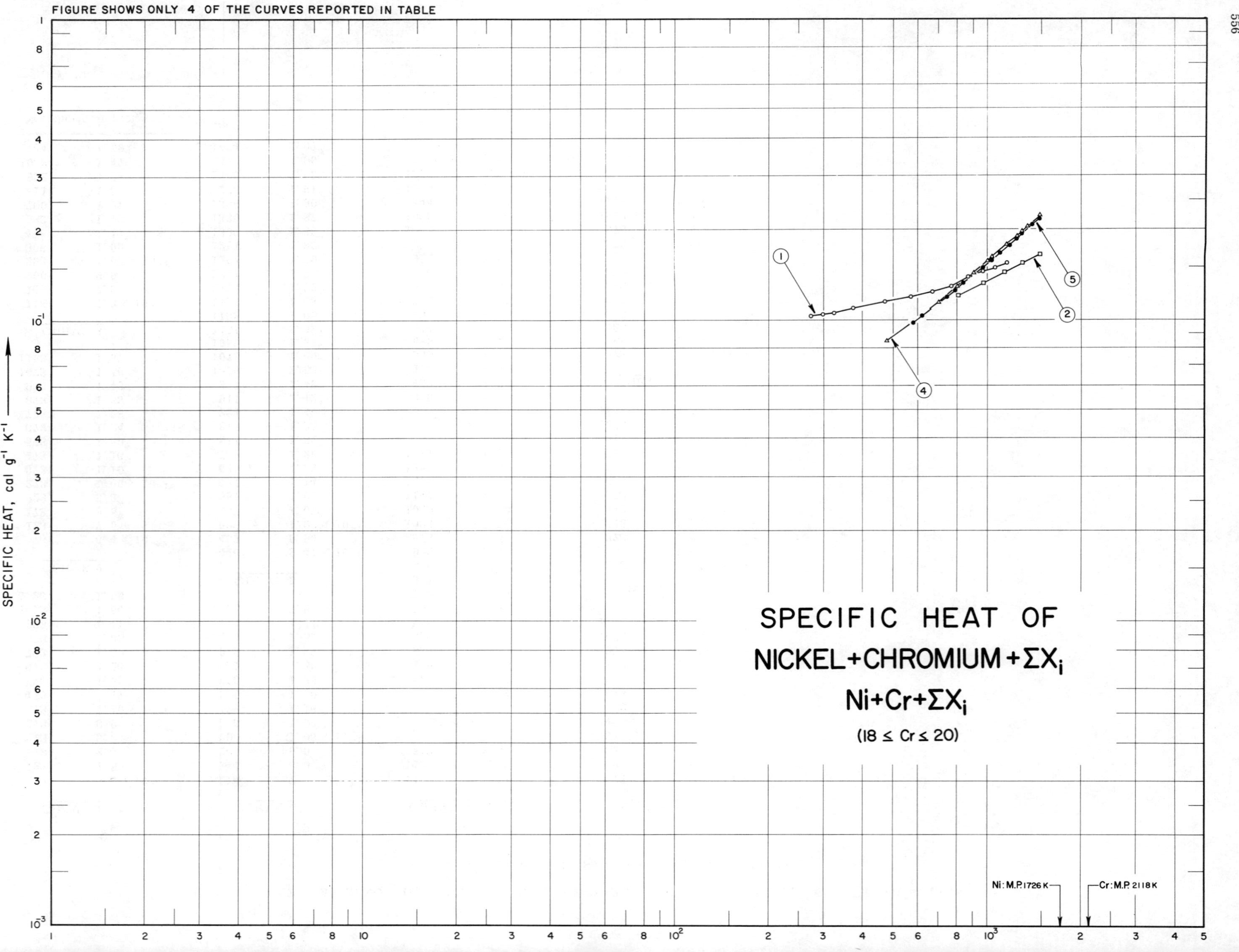
FIGURE SHOWS ONLY 4 OF THE CURVES REPORTED IN TABLE
SPECIFIC HEAT, cal g^-1 K^-1
SPECIFIC HEAT OF
NICKEL+CHROMIUM+ΣX_i
Ni+Cr+ΣX_i
(18 ≤ Cr ≤ 20)
Ni: M.P. 1726 K
Cr: M.P. 2118 K
1
2
4
5

SPECIFICATION TABLE NO. 152 SPECIFIC HEAT OF NICKEL + CHROMIUM + ΣX_i, Ni + Cr + ΣX_i $(18 \leq Cr \leq 20)$

[For Data Reported in Figure and Table No. 152]

Curve No.	Ref. No.	Year	Temp. Range, K	Reported Error, %	Name and Specimen Designation	Composition (weight percent), Specifications and Remarks
1	250, 251	1955	273-1173	±2.0	Nichrome V	77.4 Ni, 19.5 Cr, 1.4 Si, 0.59 Mn, 0.45 Fe, 0.04 C.
2	245	1958	811-1478	3	Hastelloy C	As received: 56.07 Ni, 18.83 Cr, 14.57 Mo, 4.94 Fe, 4.41 W, 0.07 C; after test: 56.00 Ni, 15.82 Cr, 14.53 Mo, 5.04 Fe, 4.49 W, 0.068 C; density = 556.9 lb ft^{-3}.
3	248	1960	273-1173	±0.3	80 Ni-20 Cr	7.74 Ni, 19.5 Cr, 1.4 Si, 0.59 Mn, 0.45 Fe, 0.04 C.
4	146	1961	479-1486	3.0	M252; Ge-J1500	57.15 Ni, 18.65 Cr, 9.98 Mo, 9.75 Cu, 2.74 Ti, 1.17 Al, <0.2 Fe, 0.12 C, 0.07 Mn, 0.06 Si; solutioned 1950 F; air cooled; measured in a helium atmosphere.
5	146	1961	479-1483	3.0	Rene 41; Ge-J1610	54.60 Ni, 18.6 Cr, 10.73 Cu, 9.63 Mo, 3.14 Ti, 1.54 Fe, 1.49 Al, 0.11 C, 0.08 Mn, 0.07 Si; solutioned 1975 F; water quenched; measured in a helium atmosphere.

DATA TABLE NO. 152 SPECIFIC HEAT OF NICKEL + CHROMIUM + ΣX_i, Ni + Cr + ΣX_i ($18 \leq Cr \leq 20$)

[Temperature, T, K; Specific Heat, C_p, Cal $g^{-1} K^{-1}$]

T	C_p
CURVE 1	
273	1.033×10^{-1}
298	1.052
323	1.071
373	1.109
473	1.171
573	1.217
673	1.264
773	1.310
873	1.400
973	1.470
1073	1.515
1173	1.563
CURVE 2	
811	1.22×10^{-1}
978	1.33
1144	1.45
1311	1.56
1478	1.67
CURVE 3*	
273	1.03×10^{-1}
298	1.05
323	1.07
373	1.11
473	1.17
573	1.22
673	1.26
773	1.31
873	1.40
973	1.47
1073	1.52
1173	1.56
CURVE 4	
479	8.521×10^{-2}
703	1.159×10^{-1}
806	1.300
915	1.449
1010	1.581
1045	1.628
1064	1.654*
CURVE 4 (cont.)	
1168	1.797×10^{-1}
1253	1.914
1301	1.980
1360	2.060
1373	2.077*
1486	2.233
CURVE 5	
479	8.463×10^{-2}*
582	9.843
624	1.041×10^{-1}
712	1.158*
749	1.207
795	1.268
842	1.332
912	1.425*
975	1.510
1031	1.585
1105	1.683
1179	1.781
1252	1.879
1305	1.950
1376	2.045*
1411	2.092
1483	2.188

*Not shown on plot

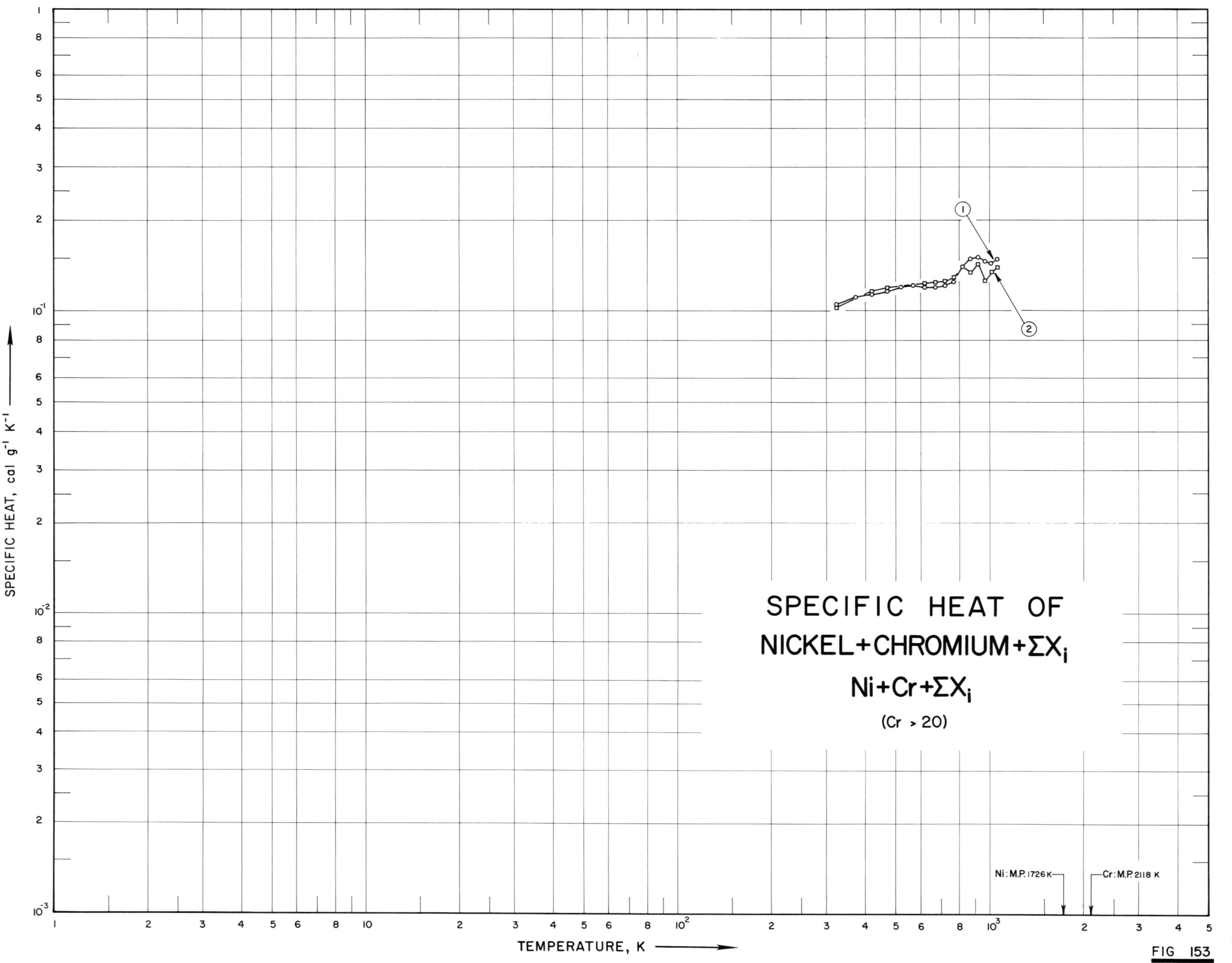
SPECIFIC HEAT OF
NICKEL+CHROMIUM+ΣXi
Ni+Cr+ΣXi
(Cr > 20)
SPECIFIC HEAT, cal g-1 K-1
TEMPERATURE, K
Ni: M.P. 1726 K
Cr: M.P. 2118 K
1
2

FIG 153

SPECIFICATION TABLE NO. 153 SPECIFIC HEAT OF NICKEL + CHROMIUM + ΣX_i, Ni + Cr + ΣX_i (Cr >20)

[For Data Reported in Figure and Table No. 153]

Curve No.	Ref. No.	Year	Temp. Range, K	Reported Error, %	Name and Specimen Designation	Composition (weight percent), Specifications and Remarks
1	252	1964	323-1073	±1	OKh21N78T [EI -435]	77.229 Ni, 21.1 Cr, 0.56 Fe, 0.49 Mn, 0.32 Si, 0.23 Ti, 0.06 C, 0.006 S, 0.005 P, trace of Cu; quenched in water from 1100 C.
2	252	1964	323-1073	±1	OKh20N60B	59.64 Ni, 20.4 Cr, 17.7 Fe, 1.59 Mn, 0.58 Nb, 0.25 Si, 0.06 C, 0.004 S; quenched in water from 1050 C; tempered 1 hr in air at 720 C.

DATA TABLE NO. 153 SPECIFIC HEAT OF NICKEL + CHROMIUM + ΣX_i, Ni + Cr + ΣX_i (Cr >20)

[Temperature, T, K; Specific Heat, C_p, Cal g^{-1} K^{-1}]

T	C_p
CURVE 1	
323	1.06 x 10^{-1}
373	1.12
423	1.14
473	1.17
523	1.20
573	1.22
623	1.20
673	1.20
723	1.22
773	1.25
823	1.40
873	1.49
923	1.51
973	1.47
1023	1.45
1073	1.48
CURVE 2	
323	1.03 x 10^{-1}
373	1.12*
423	1.17
473	1.19
523	1.21*
573	1.23*
623	1.24
673	1.25
723	1.26
773	1.29
823	1.40
873	1.34
923	1.43
973	1.26
1023	1.35
1073	1.39

*Not shown on plot

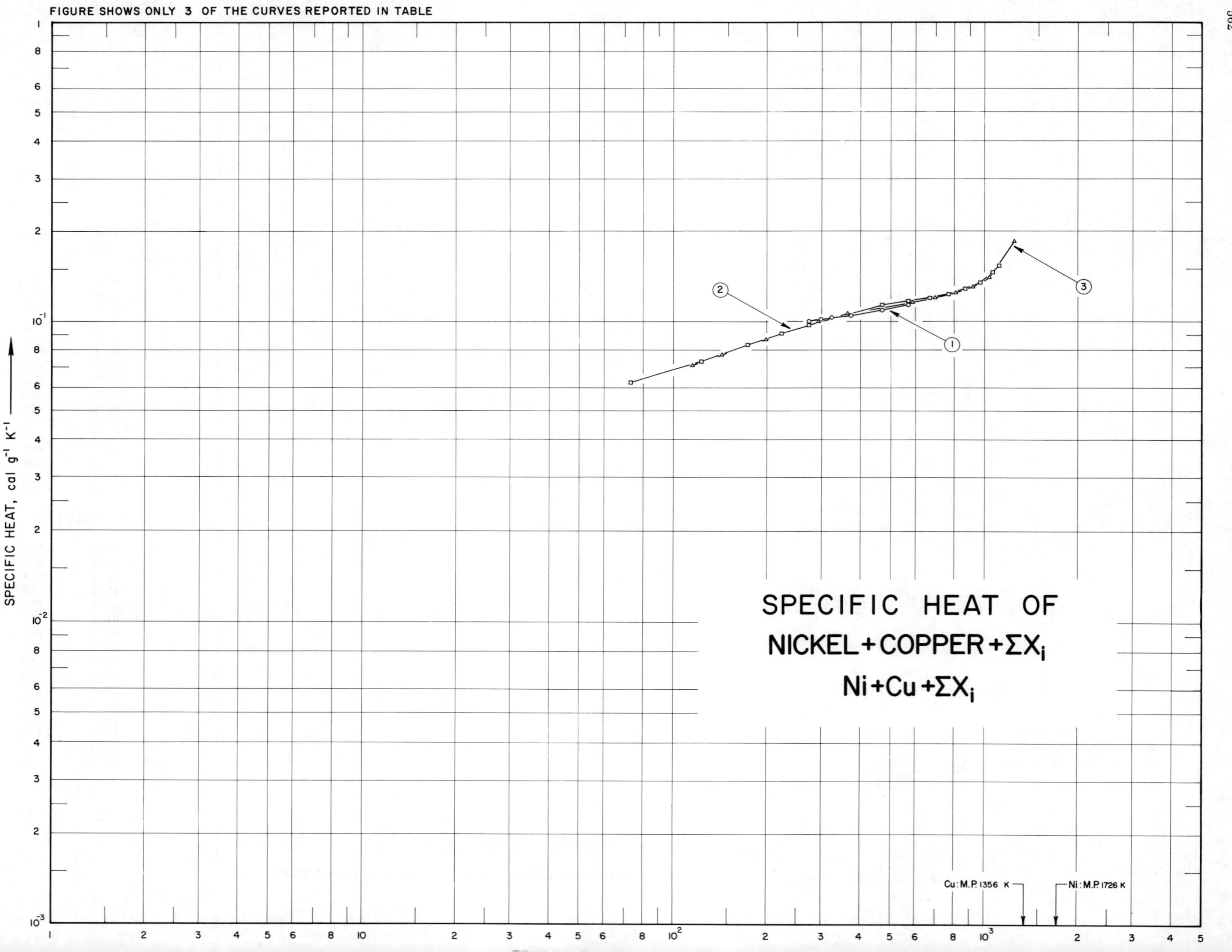

FIGURE SHOWS ONLY 3 OF THE CURVES REPORTED IN TABLE
SPECIFIC HEAT OF
NICKEL+COPPER+ΣXi
Ni+Cu+ΣXi
SPECIFIC HEAT, cal g⁻¹ K⁻¹
Cu: M.P. 1356 K
Ni: M.P. 1726 K
1
2
3

SPECIFICATION TABLE NO. 154 SPECIFIC HEAT OF NICKEL + COPPER + ΣX_i Ni + Cu + ΣX_i

[For Data Reported in Figure and Table No. 154]

Curve No.	Ref. No.	Year	Temp. Range, K	Reported Error, %	Name and Specimen Designation	Composition (weight percent), Specifications and Remarks
1	250, 251	1953	273-573	±2	Monel	67. 1 Ni, 29. 3 Cu, 1. 8 Fe, 1. 0 Mn, 0. 18 C, 0. 07 Si.
2	10	1954	73-1123		K Monel	Nominal composition: 66 Ni, 29 Cu, 2. 75 Al, 0. 9 Fe; hot rolled; annealed 1 hr at 1650 F; water quenched.
3	10	1958	116-1144		K Monel	Nominal composition: 66 Ni, 29 Cu, 3 Al; sample supplied by the International Nickel Co.; sealed in helium in capsule; annealed 1 hr at 1650 F and water quenched; density (32 F) = 527 lb ft^{-3}.
4	248	1960	273-573	±0. 3	Monel	66. 9 Ni, 29. 8 Cu, 1. 6 Fe, 1. 0 Mn, 0. 15 C, 0. 07 Si.

DATA TABLE NO. 154 SPECIFIC HEAT OF NICKEL + COPPER + ΣX_i Ni + Cu + ΣX_i

[Temperature, T, K; Specific Heat, C_p, Cal $g^{-1} K^{-1}$]

T	C_p
CURVE 1	
273	1.009×10^{-1}
298	1.021
323	1.033
373	1.054
473	1.097
573	1.142
CURVE 2	
73	6.2×10^{-2}
123	7.3
173	8.3
223	9.1
273	9.7
373	1.07×10^{-1}*
473	1.14
573	1.17
673	1.20
773	1.23
873	1.28
973	1.35
1073	1.46
1123	1.53
CURVE 3	
116	7.1×10^{-2}
144	7.7
200	8.7
293	1.00×10^{-1}
366	1.07
478	1.14*
589	1.17
700	1.20
811	1.25
922	1.32
1033	1.41
1144	1.57
CURVE 4*	
273	1.01×10^{-1}
298	1.02
323	1.03
CURVE 4 (cont.)*	
373	1.05×10^{-1}
473	1.10
573	1.14

* Not shown on plot

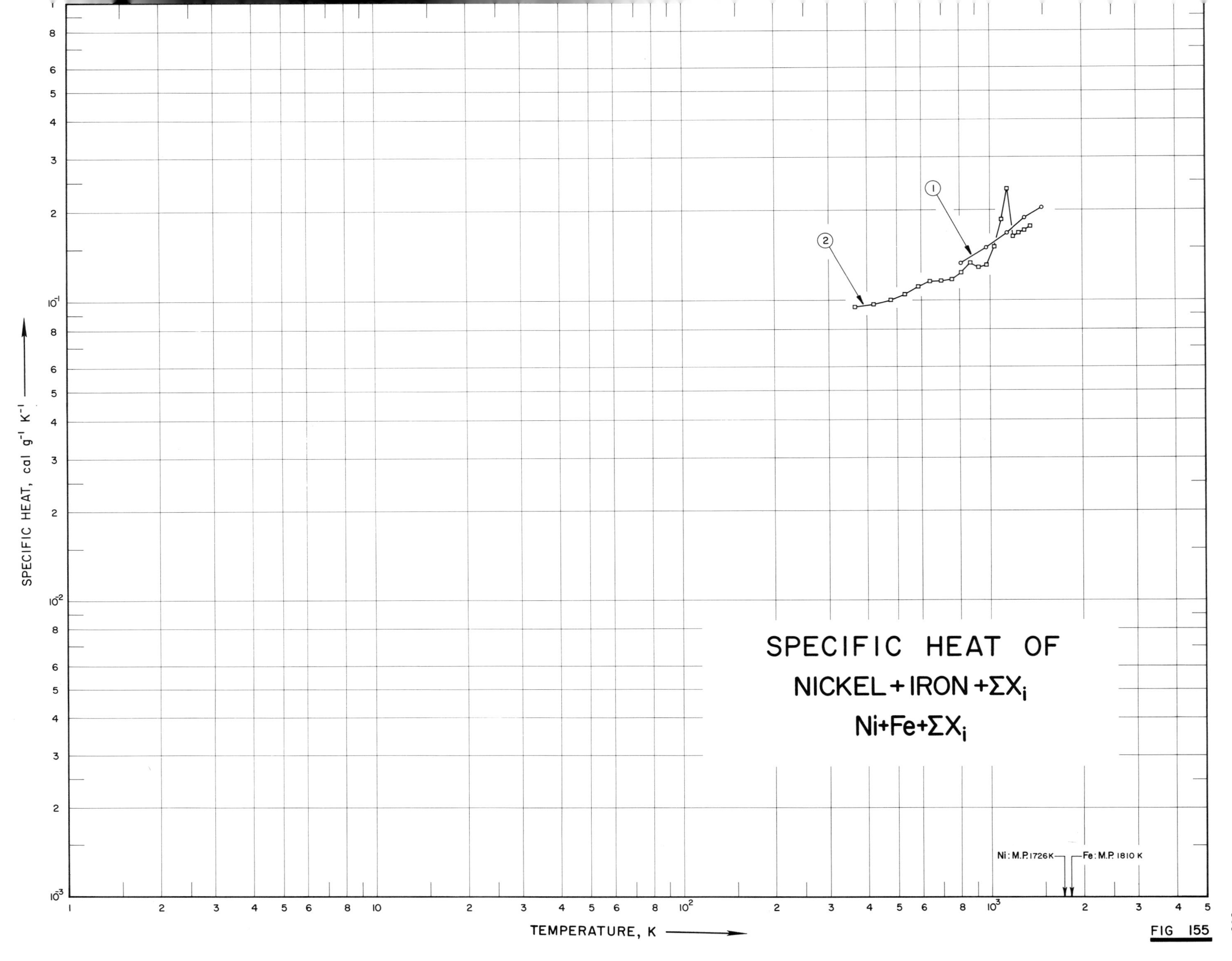
SPECIFIC HEAT OF
NICKEL + IRON + ΣXi
Ni+Fe+ΣXi
Ni: M.P. 1726 K
Fe: M.P. 1810 K
1
2
TEMPERATURE, K
SPECIFIC HEAT, cal g⁻¹ K⁻¹
FIG 155

SPECIFICATION TABLE NO. 155 SPECIFIC HEAT OF NICKEL + IRON + ΣX_i Ni + Fe + ΣX_i

[For Data Reported in Figure and Table No. 155]

Curve No.	Ref. No.	Year	Temp. Range, K	Reported Error, %	Name and Specimen Designation	Composition (weight percent), Specifications and Remarks
1	245	1958	805-1477	3	60-15 Cr (ASTM B83-46)	As received: 57. 70 Ni, 23. 92 Fe, 15. 73 Cr, 1. 14 Si, 0. 052 C, 0. 03 Mo; after test: 57. 76 Ni, 23. 91 Fe, 15. 80 Cr, 1. 33 Si, 0. 050 C, 0. 03 Mo; density = 508. 9 lb ft^{-3}.
2	249	1959	366-1255	5-10	Incoloy 901	40. 0 Ni, 35. 0 Fe, 13. 0 Cr, 6. 0 Mo, 2. 4 Ti, 0. 05 C; heated to 2050 F for 2 hrs; oil quenched; heated to 1375 F for 24 hrs; air cooled.

DATA TABLE NO. 155 SPECIFIC HEAT OF NICKEL + IRON + ΣX_i Ni + Fe + ΣX_i

[Temperature, T, K; Specific Heat, C_p, Cal g^{-1} K^{-1}]

T	C_p
CURVE 1	
811	1.33×10^{-1}
978	1.50
1144	1.68
1311	1.89
1478	2.04
CURVE 2	
366	9.5×10^{-2}
422	9.7
478	1.00×10^{-1}
533	1.05
589	1.12
644	1.16
700	1.16
755	1.18
811	1.24
866	1.33
922	1.29
978	1.32
1033	1.52
1089	1.87
1144	2.35
1200	1.64
1255	1.68

*Not shown on plot

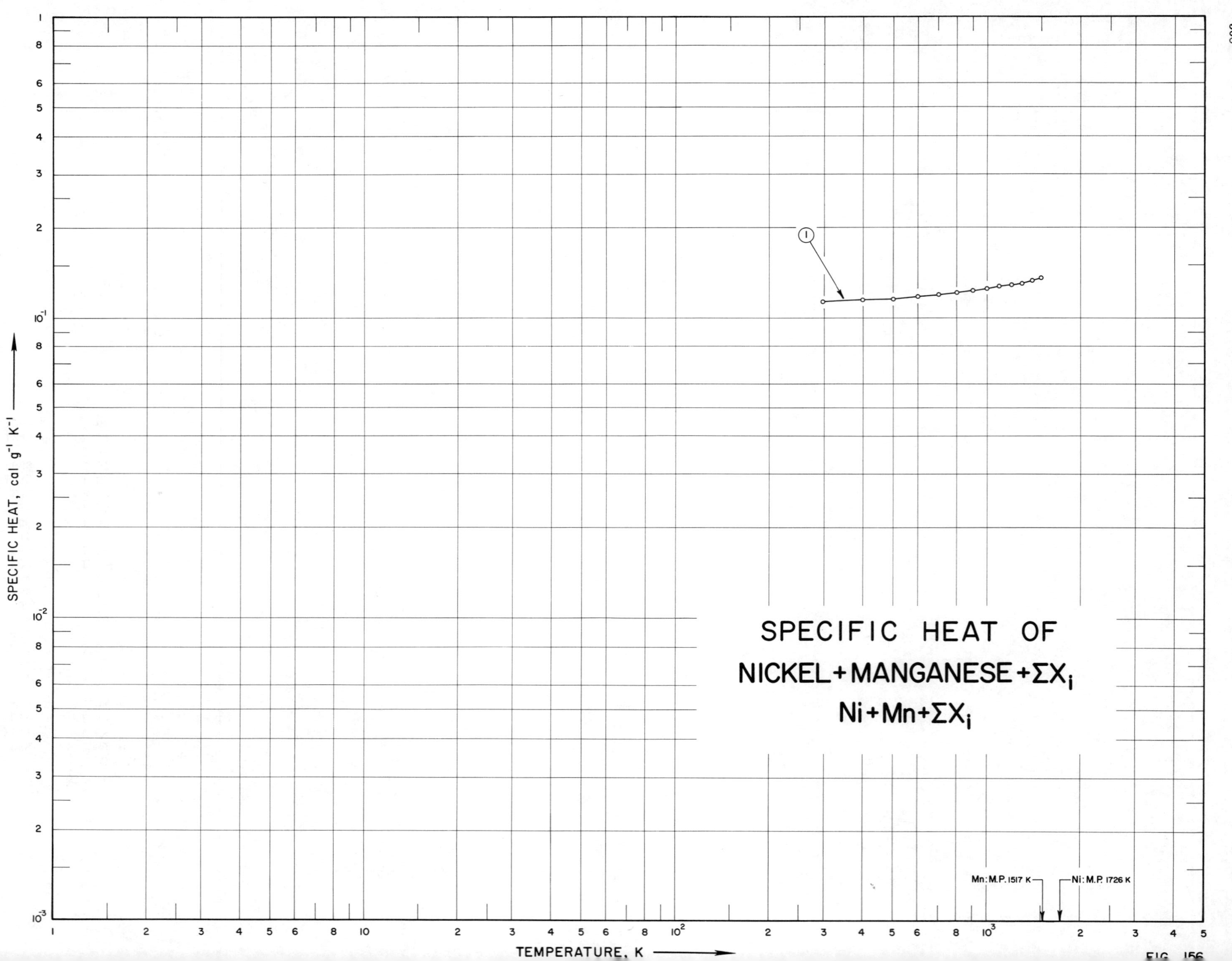

SPECIFIC HEAT OF
NICKEL+MANGANESE+ΣX_i
Ni+Mn+ΣX_i
Mn: M.P. 1517 K
Ni: M.P. 1726 K
1
TEMPERATURE, K
SPECIFIC HEAT, cal g^{-1} K^{-1}

FIG. 156

SPECIFICATION TABLE NO. 156 SPECIFIC HEAT OF NICKEL + MANGANESE + ΣX_i Ni + Mn + ΣX_i

[For Data Reported in Figure and Table No. 156]

Curve No.	Ref. No.	Year	Temp. Range, K	Reported Error, %	Name and Specimen Designation	Composition (weight percent), Specifications and Remarks
1	234	1963	298-1600		Alumel	72 Ni, 25 Mn, 2 Al, 1 Si; sample supplied by the Haskins Mfg. Co.

DATA TABLE NO. 156 SPECIFIC HEAT OF NICKEL + MANGANESE + ΣX_i Ni + Mn + ΣX_i

[Temperature, T, K; Specific Heat, C_p, Cal $g^{-1} K^{-1}$]

T	C_p
CURVE 1	
298	1.134 x 10^{-1}
300	1.135*
400	1.153
500	1.171
600	1.189
700	1.207
800	1.225
900	1.243
1000	1.261
1100	1.279
1200	1.296
1300	1.314
1400	1.33
1600	1.37

*Not shown on plot

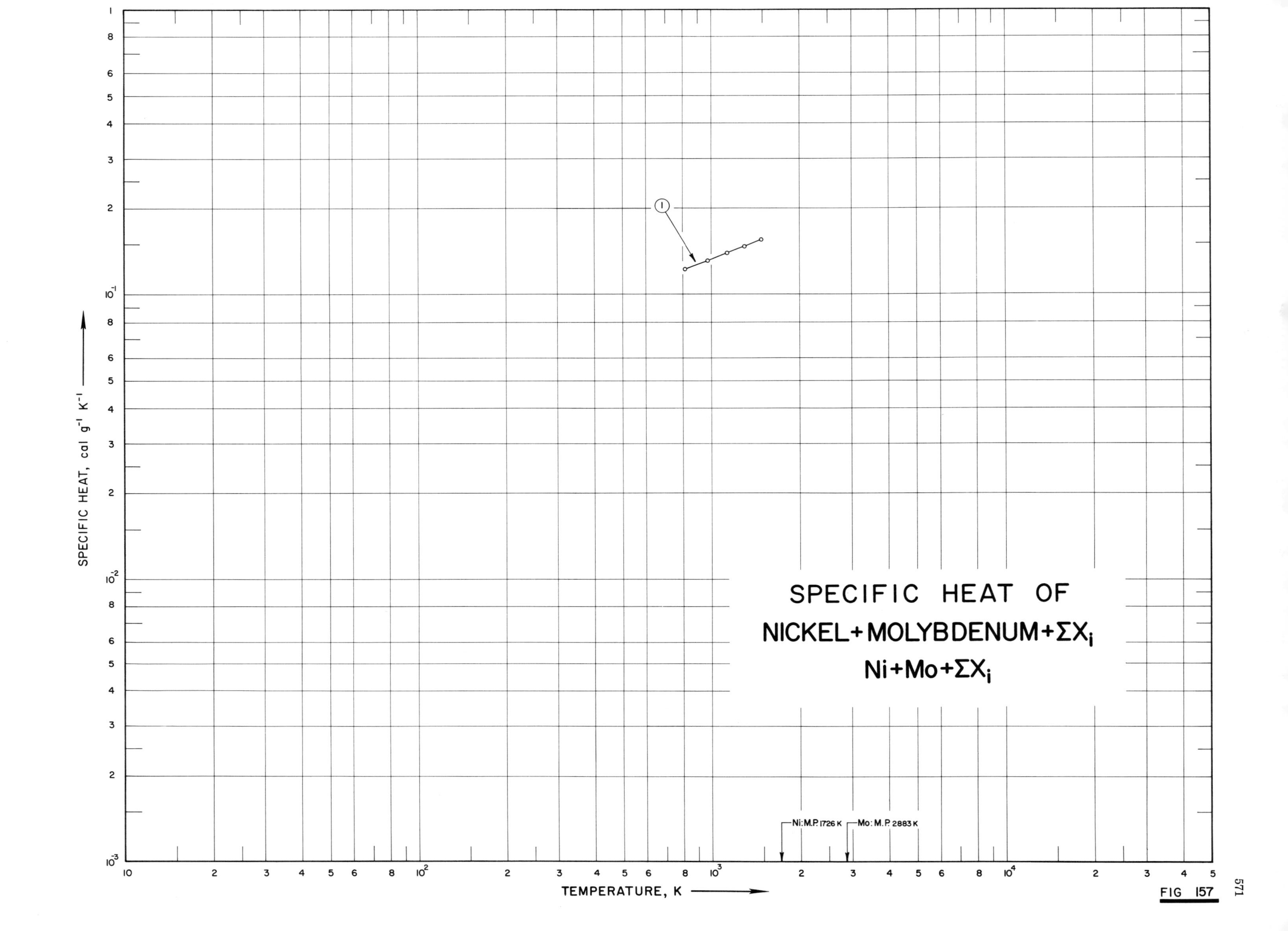
SPECIFIC HEAT OF
NICKEL+MOLYBDENUM+ΣXi
Ni+Mo+ΣXi
SPECIFIC HEAT, cal g⁻¹ K⁻¹
TEMPERATURE, K
Ni: M.P. 1726 K
Mo: M.P. 2883 K
1

FIG 157

SPECIFICATION TABLE NO. 157 SPECIFIC HEAT OF NICKEL + MOLYBDENUM + ΣX_i Ni + Mo + ΣX_i

[For Data Reported in Figure and Table No. 157]

Curve No.	Ref. No.	Year	Temp. Range, K	Reported Error, %	Name and Specimen Designation	Composition (weight percent), Specifications and Remarks
1	245	1958	784-1375	3	Hastelloy B	As received: 65. 57 Ni, 23. 78 Mo, 5. 05 Fe, 0. 020 C, after test: 65. 55 Ni, 24. 00 Mo, 4. 96 Fe, 0. 023 C; density = 585. 5 lb ft^{-3}.

DATA TABLE NO. 157 SPECIFIC HEAT OF NICKEL + MOLYBDENUM + ΣX_i Ni + Mo + ΣX_i

[Temperature, T, K; Specific Heat, C_p, Cal $g^{-1} K^{-1}$]

T	C_p
CURVE 1	
811	1.22×10^{-1}
978	1.30
1144	1.38
1311	1.47
1478	1.55

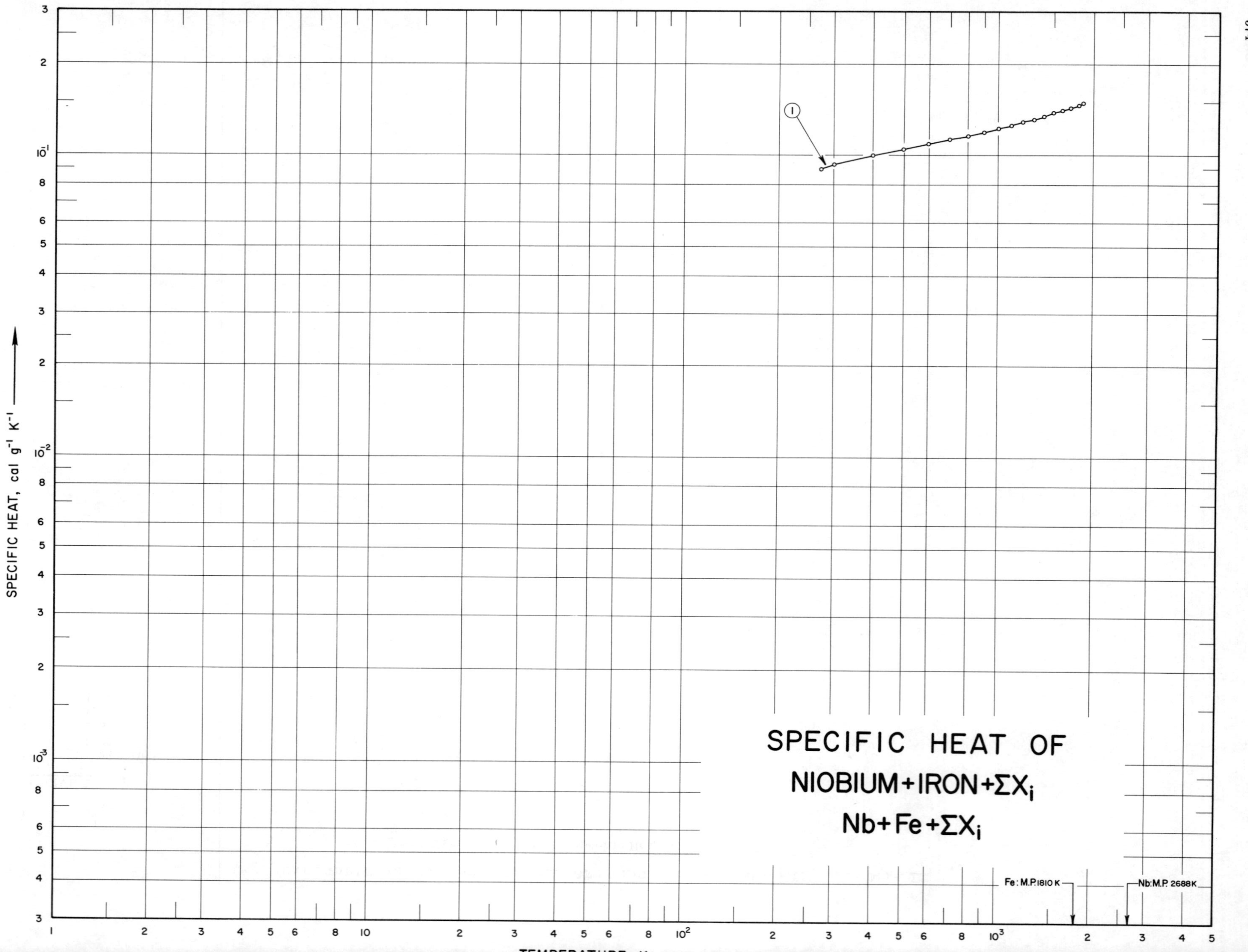

SPECIFIC HEAT OF
NIOBIUM+IRON+ΣX$_i$
Nb+Fe+ΣX$_i$
Fe: M.P. 1810 K
Nb: M.P. 2688 K
1
SPECIFIC HEAT, cal g^{-1} K^{-1}

SPECIFICATION TABLE NO. 158 SPECIFIC HEAT OF NIOBIUM + IRON + ΣX_i $Nb + Fe + \Sigma X_i$

[For Data Reported in Figure and Table No. 158]

Curve No.	Ref. No.	Year	Temp. Range, K	Reported Error, %	Name and Specimen Designation	Composition (weight percent), Specifications and Remarks
1	253	1961	273-1873	0. 8-1. 2	Ferroniobium	58. 55 Nb, 17. 09 Fe, 10. 91 Si, 7. 40 Ti, 3. 34 Al, 1. 17 Zr, 0. 53 Cr, 0. 042 P, 0. 011 Cu, 0. 011 S.

DATA TABLE NO. 158 SPECIFIC HEAT OF NIOBIUM + IRON + ΣX_i $Nb + Fe + \Sigma X_i$

[Temperature, T, K; Specific Heat, C_p, Cal $g^{-1} K^{-1}$]

T	C_p
CURVE 1	
273	9.023×10^{-2}
300	9.313
400	1.005×10^{-1}
500	1.055
600	1.096
700	1.132
800	1.166
900	1.198
1000	1.230
1100	1.261
1200	1.291
1300	1.321
1400	1.352
1500	1.382
1600	1.411
1700	1.441
1800	1.471
1873	1.492

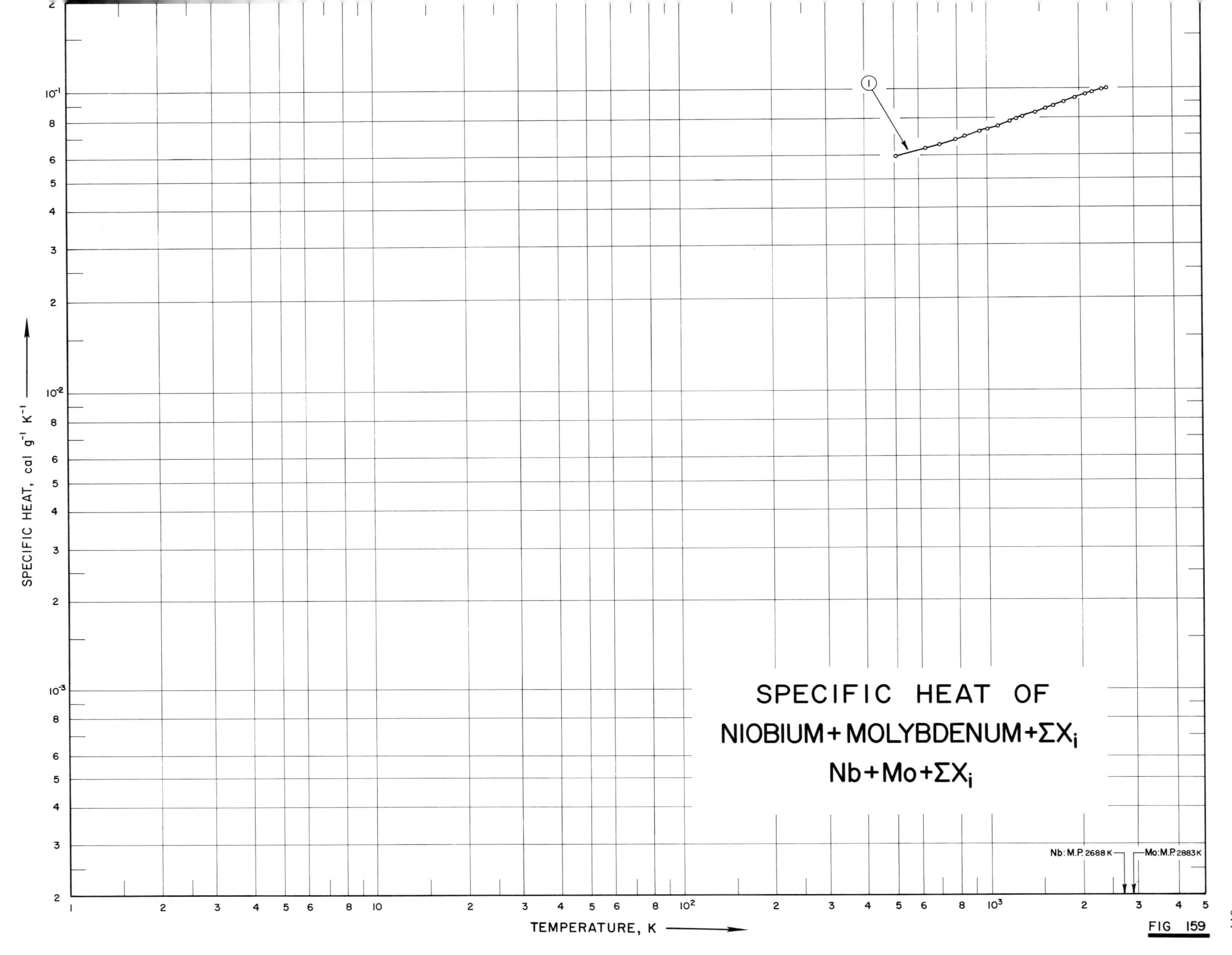

SPECIFIC HEAT OF
NIOBIUM+MOLYBDENUM+ΣXi
Nb+Mo+ΣXi
Nb: M.P. 2688 K
Mo: M.P. 2883 K
SPECIFIC HEAT, cal g⁻¹ K⁻¹
TEMPERATURE, K

FIG 159

SPECIFICATION TABLE NO. 159 SPECIFIC HEAT OF NIOBIUM + MOLYBDENUM + ΣX_i Nb + Mo + ΣX_i

[For Data Reported in Figure and Table No. 159]

Curve No.	Ref. No.	Year	Temp. Range, K	Reported Error, %	Name and Specimen Designation	Composition (weight percent), Specifications and Remarks
1	232	1963	505-2469	±5		Bal Nb, 5.03 Mo, 5.02 V, 1.13 Zr, 0.028 C, 0.0136 N_2, 0.0093 O_2; sample supplied by the Westinghouse Electric Co.; density = 538 lb ft^{-3}.

DATA TABLE NO. 159 SPECIFIC HEAT OF NIOBIUM + MOLYBDENUM + ΣX_i Nb + Mo + ΣX_i

[Temperature, T, K; Specific Heat, C_p, Cal $g^{-1} K^{-1}$]

T	C_p
CURVE 1	
505	6.043×10^{-2}
631	6.410
703	6.614
791	6.858
848	7.010
947	7.271
1011	7.434
1090	7.631
1193	7.877
1201	7.898*
1254	8.021
1286	8.094*
1315	8.160
1397	8.341*
1446	8.446
1483	8.525*
1558	8.680
1654	8.870
1789	9.124
1869	9.268*
1947	9.401
2114	9.669
2216	9.821
2372	1.003×10^{-1}
2469	1.016

*Not shown on plot

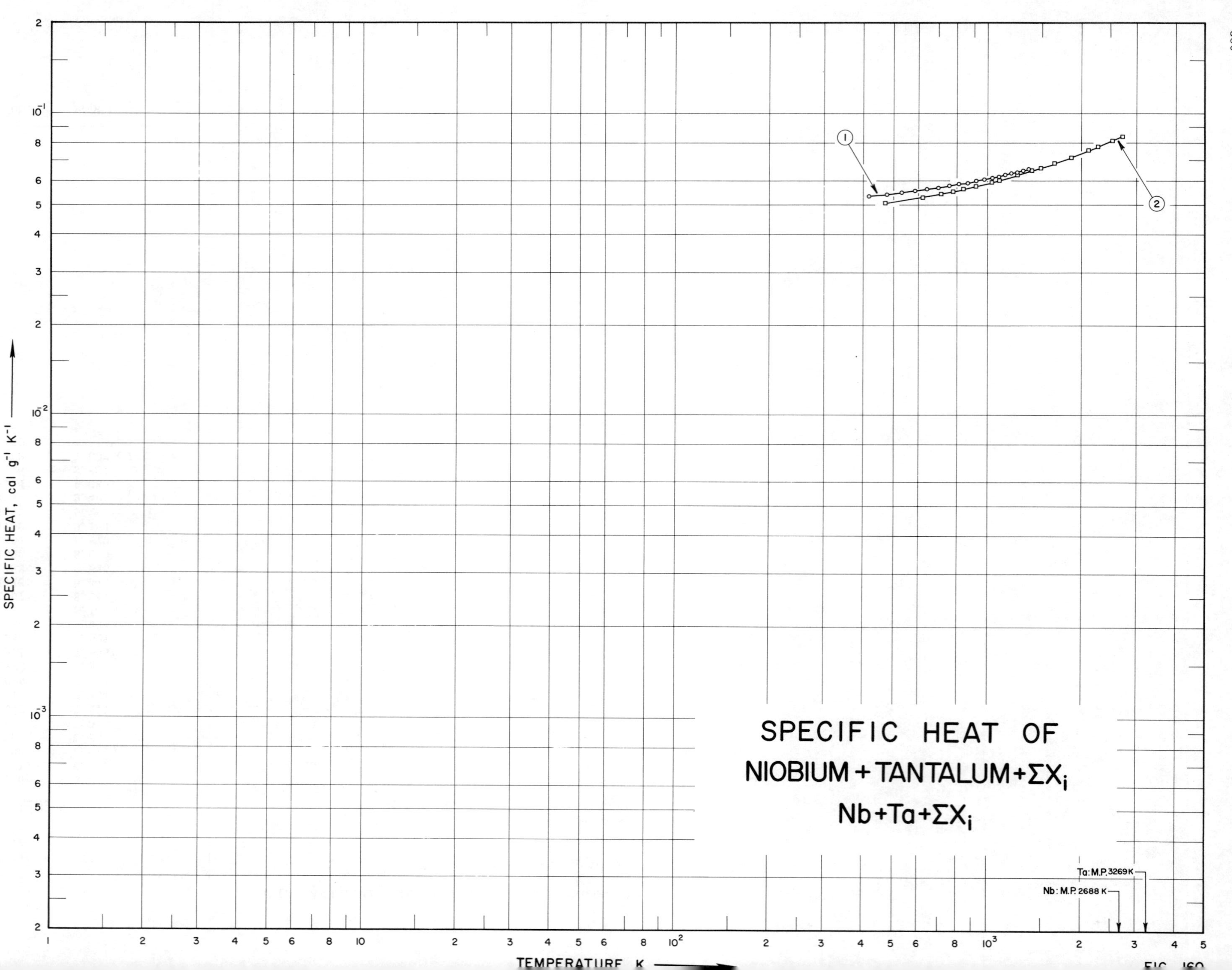
SPECIFIC HEAT OF
NIOBIUM + TANTALUM + ΣX_i
Nb+Ta+ΣX_i
Ta: M.P. 3269 K
Nb: M.P. 2688 K
SPECIFIC HEAT, cal g^{-1} K^{-1}

SPECIFICATION TABLE NO. 160 SPECIFIC HEAT OF NIOBIUM + TANTALUM + ΣX_i Nb + Ta + ΣX_i

[For Data Reported in Figure and Table No. 160]

Curve No.	Ref. No.	Year	Temp. Range, K	Reported Error, %	Name and Specimen Designation	Composition (weight percent), Specifications and Remarks
1	254	1961	422-1364		FS-82B Alloy	Bal Nb, 33 Ta, 0.7-1 Zr; heat treated.
2	232	1963	472-2705	±5		Bal Nb; 27.84 Ta, 10.4 W, 0.92 Zr, 0.01 Si, 0.009 Ni, 0.007 Fe, 0.005 Ti, 0.004 C, 0.005 O_2, 0.002 N_2; sample supplied by the Fansteel Metallurgical Corp; density = 669 lb ft^{-3}.

DATA TABLE NO. 160 SPECIFIC HEAT OF NIOBIUM + TANTALUM + ΣX_i Nb + Ta + ΣX_i

[Temperature, T, K; Specific Heat, C_p, Cal $g^{-1} K^{-1}$]

T	C_p
CURVE 1	
422	5.376×10^{-2}
478	5.447
533	5.518
589	5.589
644	5.660
700	5.731
755	5.802
811	5.880
866	5.944
922	6.015
978	6.086
1033	6.158
1089	6.229
1144	6.300
1200	6.371
1255	6.442
1310	6.513
1366	6.584
CURVE 2	
472	5.094×10^{-2}
621	5.323
711	5.462
780	5.566
844	5.664
853	5.678*
915	5.773
920	5.780*
1035	5.955
1093	6.042
1202	6.208*
1255	6.289
1308	6.367*
1396	6.500
1428	6.548*
1481	6.629
1558	6.744*
1650	6.881
1799	7.103*
1864	7.199
1994	7.393*
2114	7.570
2264	7.790
2422	8.023*
CURVE 2 (cont.)	
2514	8.157×10^{-2}
2705	8.436

* Not shown on plot

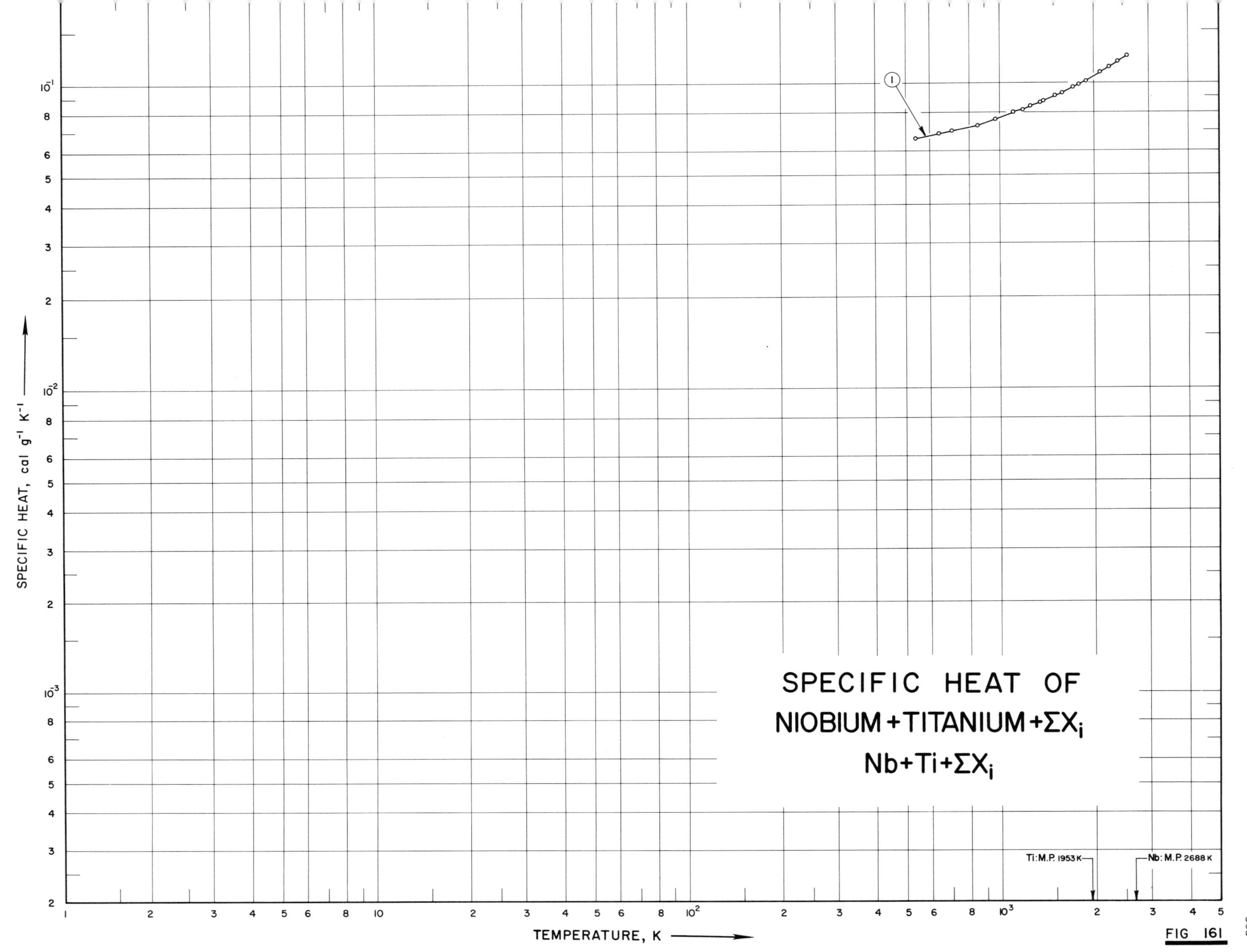
SPECIFIC HEAT OF
NIOBIUM + TITANIUM + ΣX_i
Nb+Ti+ΣX_i
SPECIFIC HEAT, cal g⁻¹ K⁻¹
TEMPERATURE, K
Ti: M.P. 1953 K
Nb: M.P. 2688 K
1

FIG 161

SPECIFICATION TABLE NO. 161 SPECIFIC HEAT OF NIOBIUM + TITANIUM + ΣX_i Nb + Ti + ΣX_i

[For Data Reported in Figure and Table No. 161]

Curve No.	Ref. No.	Year	Temp. Range, K	Reported Error, %	Name and Specimen Designation	Composition (weight percent), Specifications and Remarks
1	232	1963	542-2560	5.0		Bal Nb, 10.0 Ti, 4.9 Zr, 0.0014 C, 0.0244 O_2, 0.0024 N_2, 0.0014 H_2; sample supplied by DuPont; density = 485 lb ft^{-3}.

DATA TABLE NO. 161 SPECIFIC HEAT OF NIOBIUM + TITANIUM + ΣX_i Nb + Ti + ΣX_i

[Temperature, T, K; Specific Heat, C_p, Cal g^{-1} K^{-1}]

T	C_p
CURVE 1	
542	6.630 x 10^{-2}
641	6.864
705	7.018
721	7.057*
853	7.383
856	7.391*
966	7.668
971	7.681*
1119	8.066
1181	8.229
1264	8.453
1285	8.509*
1364	8.726
1391	8.802
1414	8.863*
1509	9.130
1566	9.293*
1590	9.361
1724	9.752
1797	9.967
1883	1.023 x 10^{-1}
2105	1.091
2236	1.132
2397	1.184
2561	1.239

* Not shown on plot

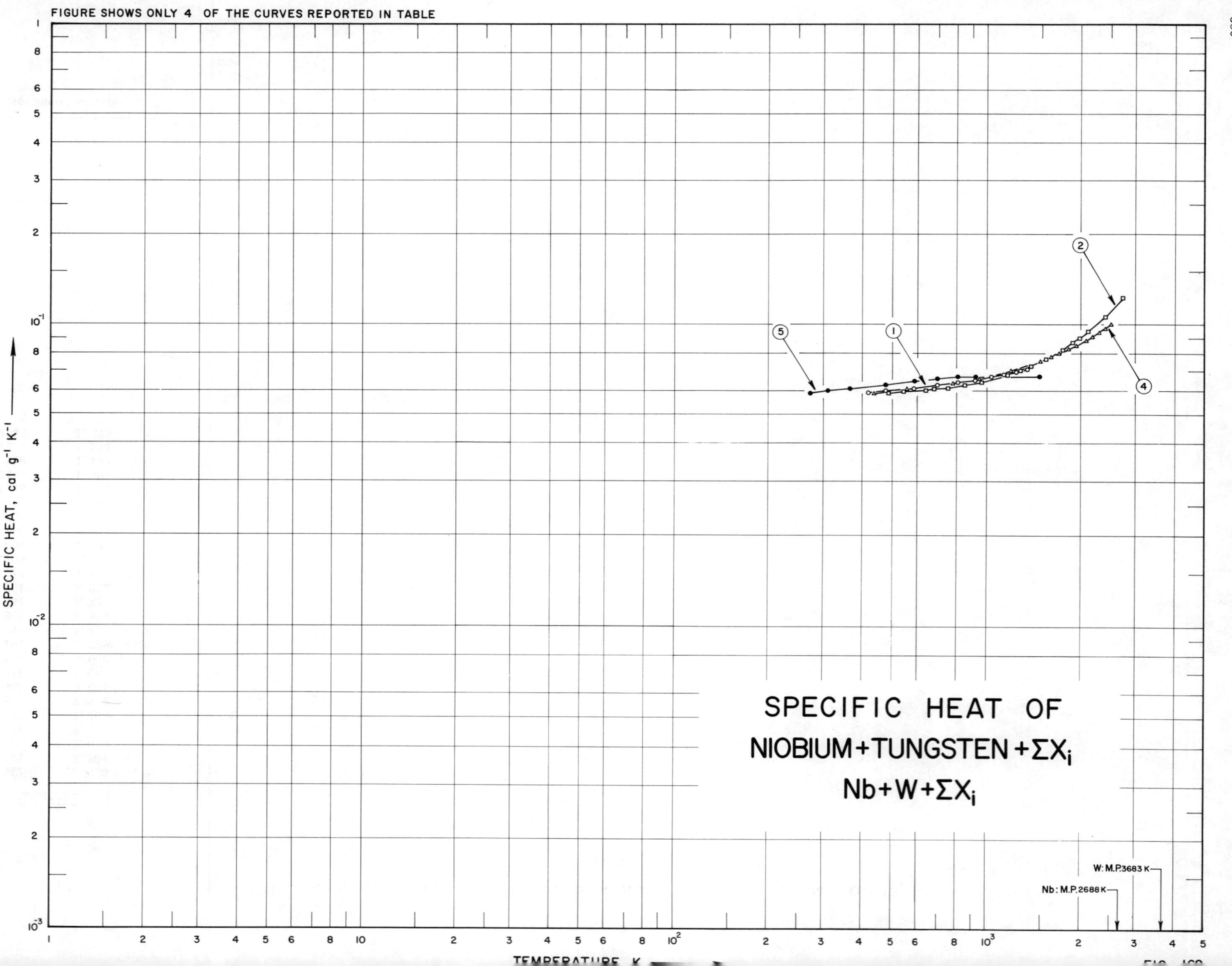

FIGURE SHOWS ONLY 4 OF THE CURVES REPORTED IN TABLE
SPECIFIC HEAT OF
NIOBIUM + TUNGSTEN + ΣXi
Nb + W + ΣXi
SPECIFIC HEAT, cal g⁻¹ K⁻¹
W: M.P. 3683 K
Nb: M.P. 2688 K
1
2
4
5

SPECIFICATION TABLE NO. 162 SPECIFIC HEAT OF NIOBIUM + TUNGSTEN + ΣX_i Nb + W + ΣX_i

[For Data Reported in Figure and Table No. 162]

Curve No.	Ref. No.	Year	Temp. Range, K	Reported Error, %	Name and Specimen Designation	Composition (weight percent), Specifications and Remarks
1	254	1961	422-1367		F-48	Bal Nb, 13.8 W, 4.8 Mo, 0.90 Zr, 0.041 C, 0.036 O, 0.017 N; sample supplied by the General Electric Co.; heat treated.
2	232	1963	487-2744	±5		Bal Nb, 15.3 W, 5.26 Mo, 1.08 Zr, 0.034 C, 0.0211 N_2, 0.0167 O_2, 0.0061 H_2.
3	232	1963	549-2572	±5		Bal Nb, 9.93 W, 2.58 Zr, 0.002 C, 0.012 O_2, 0.006 N_2, 0.0009 H_2; sample supplied by the Haynes Stellite Co.; density = 572 lb ft^{-3}.
4	232	1963	435-2513	±5		Bal Nb, 9.7 W, 0.88 Zr, 0.0810 C, 0.0052 O_2, 0.0033 N_2, 0.0004 H_2; sample supplied by DuPont Co.; density = 564 lb ft^{-3}.
5	255	1963	273-1477	4	CB-752	87.5 Nb, 10.0 W, 2.5 Zr.

DATA TABLE NO. 162 SPECIFIC HEAT OF NIOBIUM + TUNGSTEN + ΣX_i Nb + W + ΣX_i

[Temperature, T, K; Specific Heat, C_p, Cal $g^{-1} K^{-1}$]

T	C_p
CURVE 1	
422	5.939 x 10^{-2}
778	6.006
533	6.073*
589	6.141
644	6.208*
700	6.275
755	6.343*
811	6.410
866	6.477*
922	6.544
978	6.612*
1033	6.679
1089	6.746*
1144	6.814
1200	6.881*
1255	6.948
1311	7.015*
1366	7.083
CURVE 2	
487	5.932 x 10^{-2}
540	5.964
641	6.040
685	6.079
754	6.148
850	6.262
860	6.274*
966	6.422
975	6.437*
1128	6.691*
1172	6.775
1254	6.939*
1263	6.957
1284	7.002
1357	7.163*
1399	7.261
1418	7.307*
1485	7.473*
1562	7.674
1593	7.759*
1753	8.224
1811	8.409*
1891	8.671
1980	8.977
CURVE 2 (cont.)	
2122	9.498 x 10^{-2}
2189	9.757*
2416	1.071 x 10^{-1}
2555	1.134*
2744	1.226
CURVE 3*	
549	5.987 x 10^{-2}
640	6.026
701	6.063
754	6.104
844	6.190
870	6.219
973	6.350
1000	6.390
1151	6.643
1181	6.700
1266	6.874
1278	6.900
1356	7.080
1394	7.173
1422	7.245
1483	7.408
1575	7.671
1589	7.711
1676	7.984
1804	8.421
1933	8.903
2111	9.640
2241	1.023 x 10^{-1}
2386	1.094
2572	1.193
CURVE 4	
435	5.942 x 10^{-2}
554	6.071
676	6.218*
782	6.355
843	6.439*
947	6.589
1028	6.712*
1144	6.901*
1198	6.992
CURVE 4 (cont.)	
1291	7.155 x 10^{-2}*
1403	7.362*
1483	7.517
1620	7.796
1707	7.982
1828	8.251
1946	8.528
2078	8.849
2180	9.111
2297	9.420
2416	9.749
2514	1.003 x 10^{-1}
CURVE 5	
273	5.9 x 10^{-2}
311	6.0
366	6.1
478	6.3
589	6.5
700	6.6
811	6.7
922	6.7
1033	6.7*
1144	6.7*
1255	6.7*
1366	6.7*
1478	6.7

* Not shown on plot

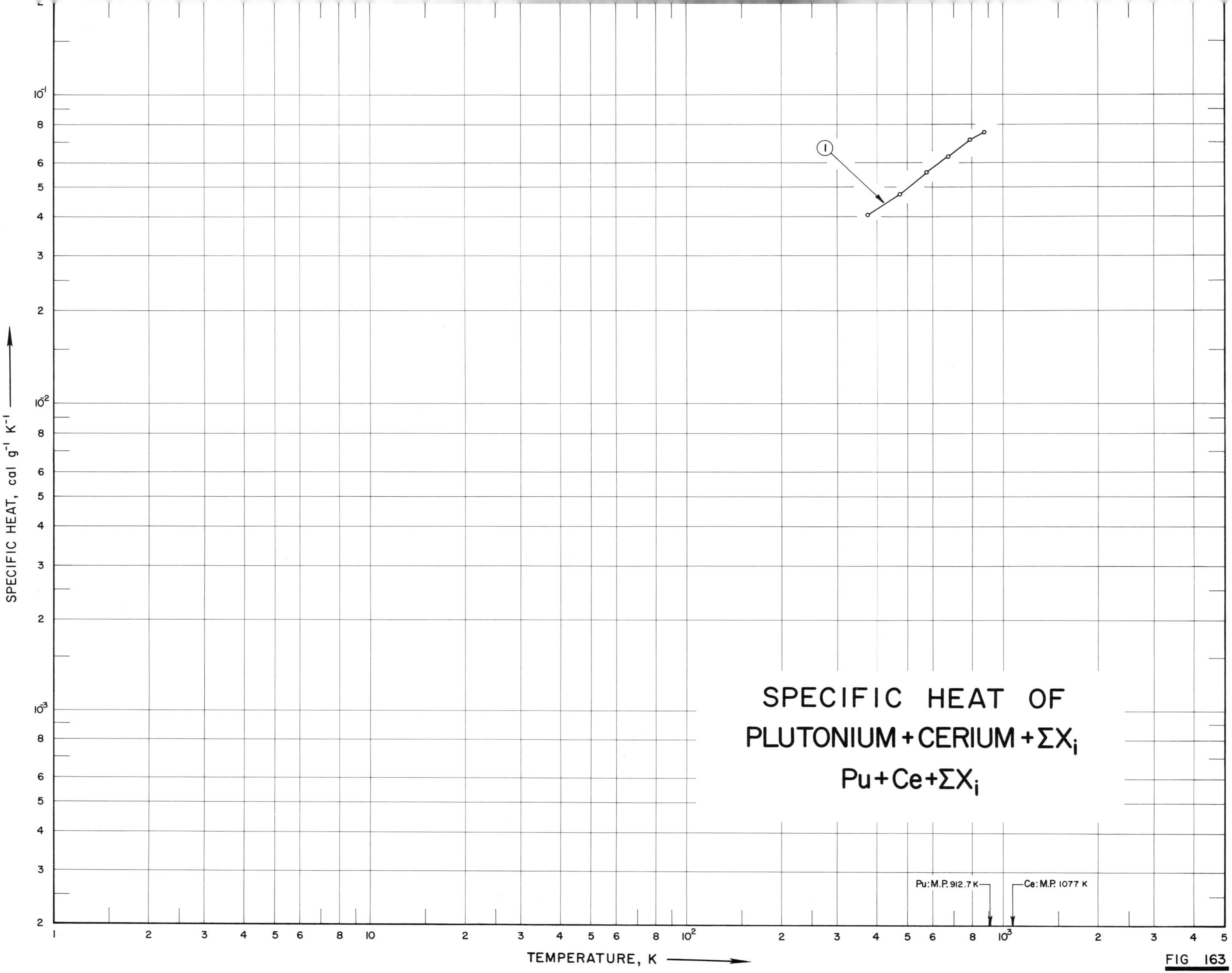

SPECIFIC HEAT OF PLUTONIUM + CERIUM + ΣX_i

$Pu + Ce + \Sigma X_i$

SPECIFIC HEAT, cal g^{-1} K^{-1}

TEMPERATURE, K

Pu: M.P. 912.7 K

Ce: M.P. 1077 K

1

FIG 163

SPECIFICATION TABLE NO. 163 SPECIFIC HEAT OF PLUTONIUM + CERIUM + ΣX_i Pu + Ce + ΣX_i

[For Data Reported in Figure and Table No. 163]

Curve No.	Ref. No.	Year	Temp. Range, K	Reported Error, %	Name and Specimen Designation	Composition (weight percent), Specifications and Remarks
1	256	1963	373-873		Eutectic Alloy	

DATA TABLE NO. 163 SPECIFIC HEAT OF PLUTONIUM + CERIUM + ΣX_i Pu + Ce + ΣX_i

[Temperature, T, K; Specific Heat, C_p, Cal $g^{-1} K^{-1}$]

T	C_p
CURVE 1	
373	4.05×10^{-2}
473	4.75
573	5.60
673	6.30
788	7.15
873	7.55

*Not shown on plot

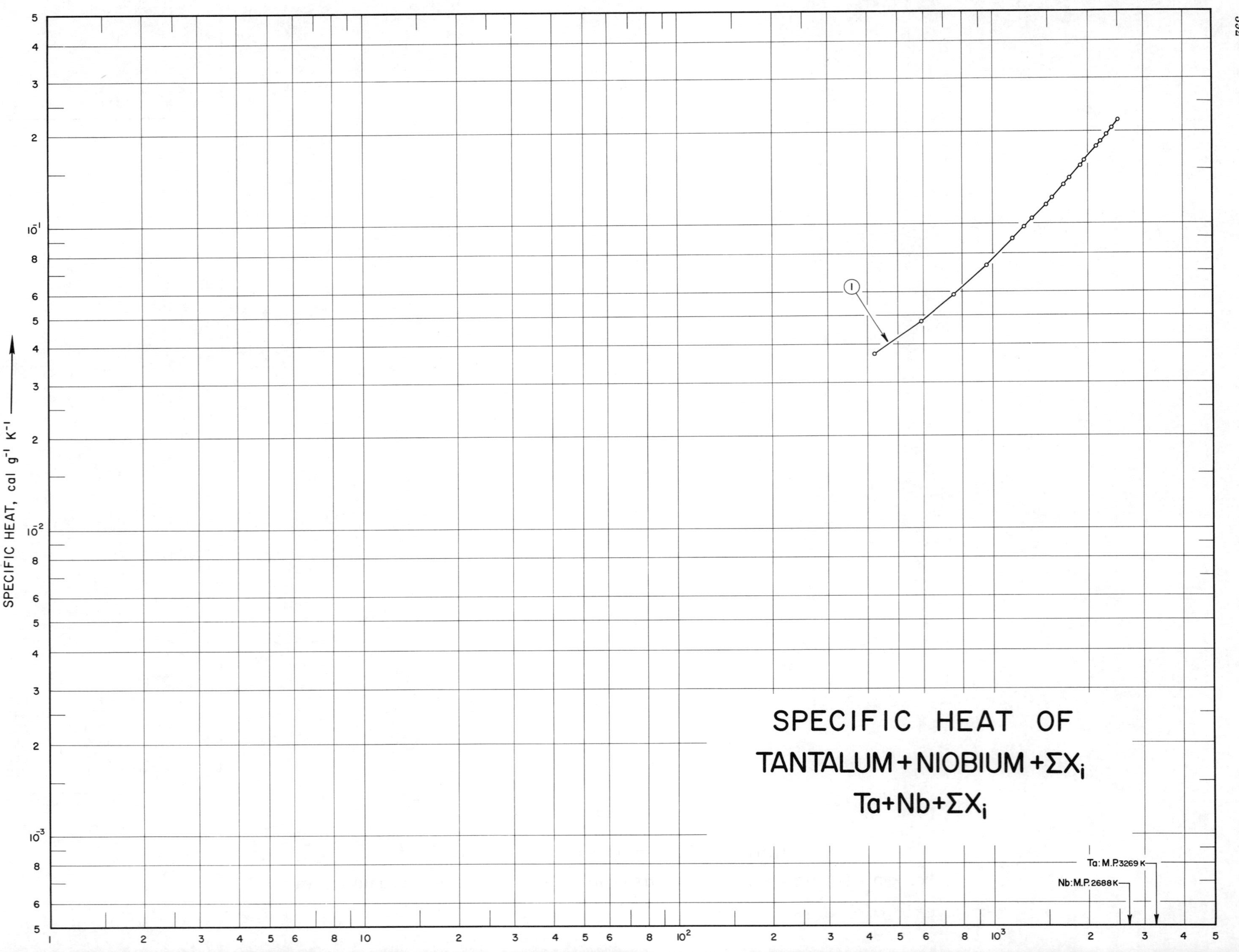

SPECIFIC HEAT OF
TANTALUM + NIOBIUM + ΣXi
Ta+Nb+ΣXi
Ta: M.P. 3269 K
Nb: M.P. 2688 K
1
SPECIFIC HEAT, cal g⁻¹ K⁻¹

SPECIFICATION TABLE NO. 164 SPECIFIC HEAT OF TANTALUM + NIOBIUM + ΣX_i Ta + Nb + ΣX_i

[For Data Reported in Figure and Table No. 164]

Curve No.	Ref. No.	Year	Temp. Range, K	Reported Error, %	Name and Specimen Designation	Composition (weight percent), Specifications and Remarks
1	232	1963	422-2509	±5		Bal Ta, 30.3 Nb, 7.47 V, 0.09 C, 0.015 O_2, 0.0065 N_2; sample supplied by Wah Chang Corp.

DATA TABLE NO. 164 SPECIFIC HEAT OF TANTALUM + NIOBIUM + ΣX_i Ta + Nb + ΣX_i

[Temperature, T, K; Specific Heat, C_p, Cal g^{-1} K^{-1}]

T	C_p
CURVE 1	
422	3.702 x 10^{-2}
593	4.781
755	5.866
958	7.322
965	7.372*
1166	8.919
1272	9.769
1344	1.037 x 10^{-1}
1480	1.153
1555	1.218
1689	1.339
1764	1.409
1901	1.540
1971	1.608
2141	1.781
2216	1.859
2308	1.956
2400	2.055
2508	2.175

*Not shown on plot

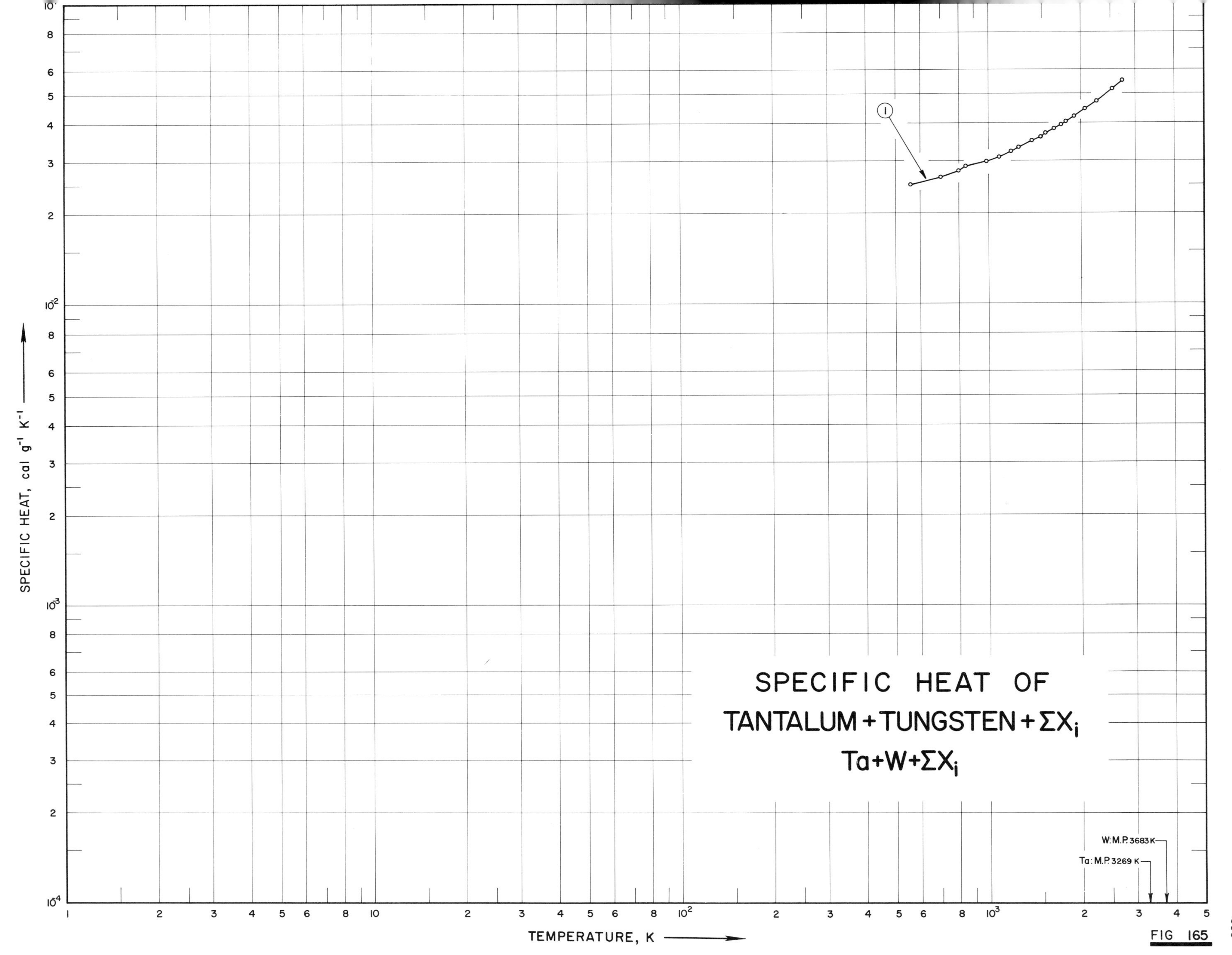

SPECIFIC HEAT OF
TANTALUM + TUNGSTEN + ΣX_i
Ta+W+ΣX_i
W: M.P. 3683 K
Ta: M.P. 3269 K
1
TEMPERATURE, K
SPECIFIC HEAT, cal g^{-1} K^{-1}
FIG 165

SPECIFICATION TABLE NO. 165 SPECIFIC HEAT OF TANTALUM + TUNGSTEN + ΣX_i Ta + W + ΣX_i

[For Data Reported in Figure and Table No. 165]

Curve No.	Ref. No.	Year	Temp. Range, K	Reported Error, %	Name and Specimen Designation	Composition (weight percent), Specifications and Remarks
1	232	1963	561-2733	±5		Bal Ta, 9.0 W, 2.2 Hf, 0.0041 C, 0.004 O_2, 0.0023 N_2; sample supplied by the Westinghouse Corp; density = 1058 lb ft^{-3}.

DATA TABLE NO. 165 SPECIFIC HEAT OF TANTALUM + TUNGSTEN + ΣX_i $Ta + W + \Sigma X_i$

[Temperature, T, K; Specific Heat, C_p, Cal g^{-1} K^{-1}]

T	C_p
CURVE 1	
561	2.507 x 10^{-2}
705	2.662
804	2.773
845	2.884
990	2.989
1079	3.096
1188	3.230
1261	3.324
1314	3.392*
1397	3.500
1483	3.614
1547	3.701
1556	3.713*
1646	3.838
1729	3.955
1798	4.054
1908	4.215
2072	4.462
2253	4.744
2264	4.762*
2541	5.217
2733	5.545

*Not shown on plot

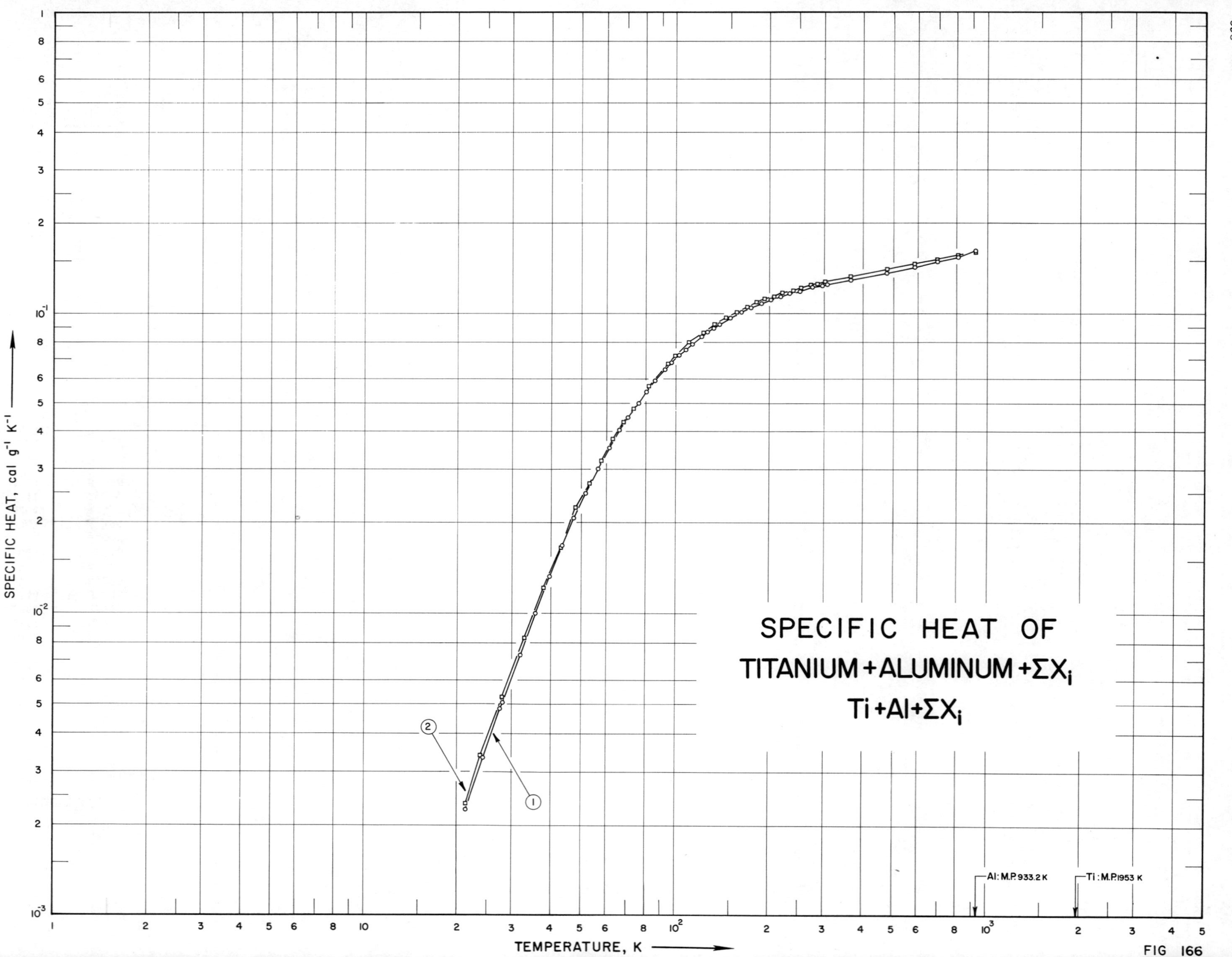

SPECIFIC HEAT OF
TITANIUM + ALUMINUM + ΣX_i
Ti + Al + ΣX_i
Al: M.P. 933.2 K
Ti: M.P. 1953 K
TEMPERATURE, K
SPECIFIC HEAT, cal g^{-1} K^{-1}
1
2

FIG 166

SPECIFICATION TABLE NO. 166 SPECIFIC HEAT OF TITANIUM + ALUMINUM + ΣX_i Ti + Al + ΣX_i

[For Data Reported in Figure and Table No. 166]

Curve No.	Ref. No.	Year	Temp. Range, K	Reported Error, %	Name and Specimen Designation	Composition (weight percent), Specifications and Remarks
1	257	1961	21-922	<2.0		4.4 Al, 3.0 Mo, 1.0 V, 0.1 Fe, 0.03 C, 0.011 N_2, 0.0057 H_2; solution heat treated at 1655 F and then aged at 925 F for 12 hrs.
2	257	1961	21-922	<2.0		5.89 Al, 3.87 V, 0.15 Fe, 0.02 C, 0.015 N_2, 0.005 H_2; sample supplied by the Mallory-Sharon Metals Corp; solution heat treated at 1700 F for 20 min; oil quenched; then aged at 900 F for 4 hrs and cooled in air.

DATA TABLE NO. 166 SPECIFIC HEAT OF TITANIUM + ALUMINUM + ΣX_i Ti + Al + ΣX_i

[Temperature, T, K; Specific Heat, C_p, Cal $g^{-1} K^{-1}$]

T	C_p
CURVE 1	
Series 1	
277.22	1.242 x 10^{-1}
280.09	1.244*
283.43	1.250*
286.89	1.248*
291.30	1.252*
296.79	1.258*
302.25	1.264*
Series 2	
202.11	1.123 x 10^{-1}
207.55	1.134*
212.78	1.143*
218.43	1.154
220.83	1.159*
226.51	1.167*
232.34	1.177
238.32	1.186*
244.45	1.196*
250.52	1.203
256.56	1.212*
262.55	1.221*
268.52	1.228*
274.44	1.237*
296.54	1.261*
301.81	1.267*
307.46	1.272
Series 3	
70.35	4.473 x 10^{-2}
76.05	5.039
80.82	5.498
85.93	5.981
92.61	6.537
97.41	6.900
102.51	7.273
107.56	7.616
112.81	7.956
121.36	8.458
127.03	8.766
132.53	9.036
138.07	9.297
CURVE 1 (cont.)	
Series 3 (cont.)	
143.85	9.542 x 10^{-2}*
150.02	9.786
156.39	1.002 x 10^{-1}*
162.64	1.023
168.77	1.042*
175.12	1.060
181.70	1.078*
188.18	1.094
192.88	1.105
297.21	1.260
300.01	1.265*
303.01	1.267*
306.01	1.271*
Series 4	
21.40	2.27 x 10^{-3}
24.30	3.35
28.00	5.09
31.83	7.32
35.70	1.017 x 10^{-2}
39.64	1.339
43.50	1.691
47.35	2.076
51.70	2.520
56.61	3.026
61.52	3.554
66.32	4.062
Series 5	
21.39	2.28 x 10^{-3}*
24.32	3.36*
27.60	4.85
Series 6	
366.48	1.312 x 10^{-1}
477.59	1.381
588.71	1.445
699.82	1.507
810.93	1.568
922.04	1.629
CURVE 2	
Series 1	
202.26	1.146 x 10^{-1}*
207.85	1.158
214.10	1.170*
220.48	1.181
226.72	1.191*
233.23	1.202*
239.85	1.212
246.45	1.222*
252.99	1.232*
259.49	1.241
265.95	1.249*
272.36	1.258
Series 2	
82.34	5.708 x 10^{-2}
86.81	6.132*
94.99	6.803
100.22	7.201
105.65	7.588*
112.70	8.061
118.09	8.396*
123.49	8.709
129.02	9.004*
134.58	9.281
140.16	9.543*
146.02	9.792
151.74	1.002 x 10^{-1}*
157.56	1.023
163.46	1.043*
171.48	1.070
177.14	1.087*
183.12	1.102
189.02	1.117*
194.84	1.132
200.60	1.144*
Series 3	
21.26	2.37 x 10^{-3}
23.95	3.40
27.99	5.29
32.90	8.33
CURVE 2 (cont.)	
Series 3 (cont.)	
37.97	1.224 x 10^{-2}
43.00	1.672
48.09	2.242
53.14	2.692
58.03	3.214
63.27	3.782
68.57	4.334
73.67	4.846
79.04	5.383*
Series 4	
21.47	2.42 x 10^{-3}*
24.21	3.50*
27.22	4.87*
Series 5	
276.01	1.261 x 10^{-1}*
281.50	1.267*
287.32	1.273
293.26	1.280*
298.86	1.286*
303.87	1.290
Series 6	
366.48	1.348 x 10^{-1}
477.59	1.421
588.71	1.478
699.82	1.529
810.93	1.576
922.04	1.621

* Not shown on plot

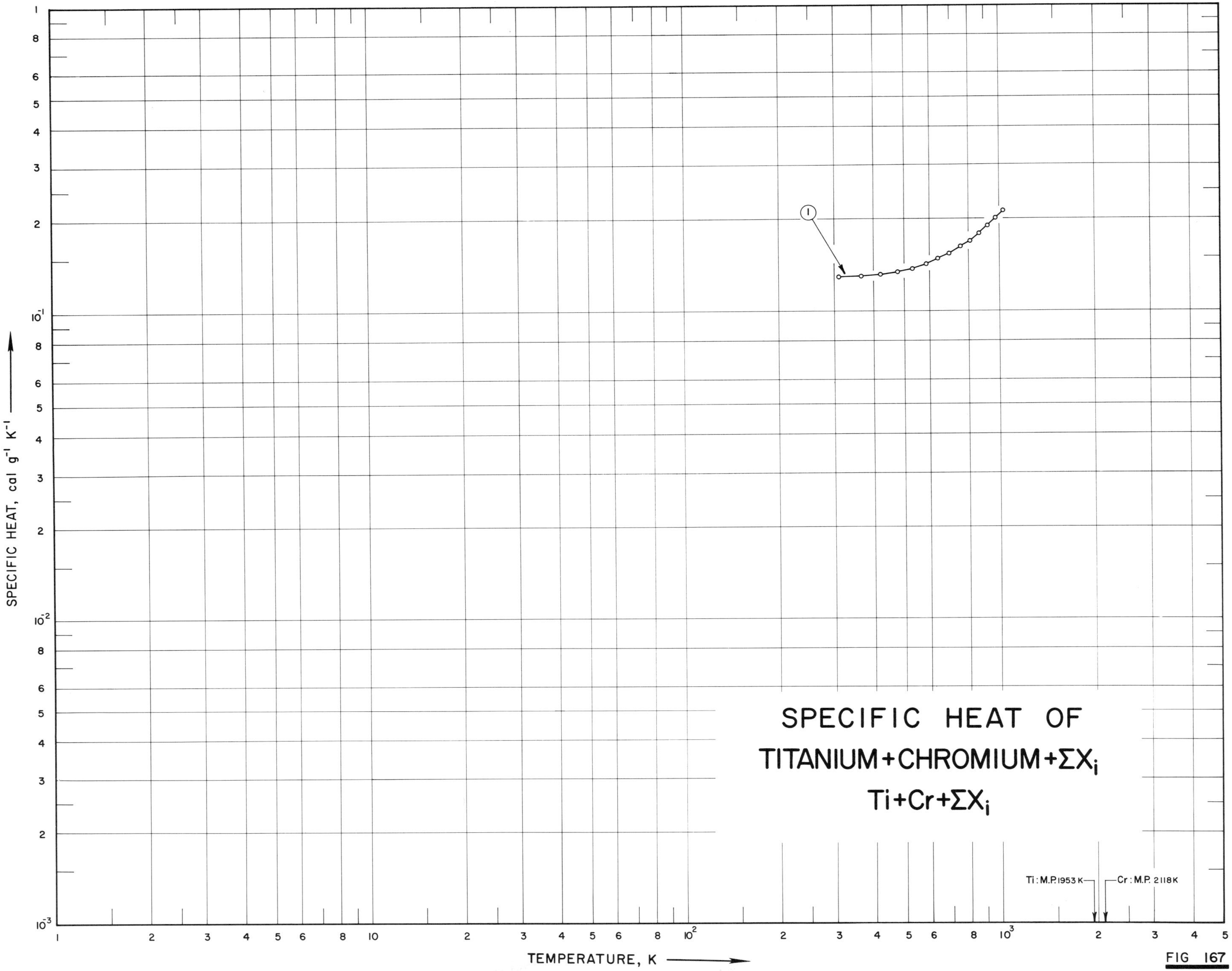

SPECIFIC HEAT OF
TITANIUM + CHROMIUM + ΣX_i
Ti+Cr+ΣX_i

SPECIFIC HEAT, cal g^{-1} K^{-1}

TEMPERATURE, K

Ti : M.P. 1953 K

Cr : M.P. 2118 K

1

FIG 167

SPECIFICATION TABLE NO. 167 SPECIFIC HEAT OF TITANIUM + CHROMIUM + ΣX_i Ti + Cr + ΣX_i

[For Data Reported in Figure and Table No. 167]

Curve No.	Ref. No.	Year	Temp. Range, K	Reported Error, %	Name and Specimen Designation	Composition (weight percent), Specifications and Remarks
1	135	1956	311-1033			95.65 Ti, 2.71 Cr, 1.40 Fe, 0.105 O_2, 0.076 N_2, 0.05 C, 0.0092 H_2.

DATA TABLE NO. 167 SPECIFIC HEAT OF TITANIUM + CHROMIUM + ΣX_i Ti + Cr + ΣX_i

[Temperature, T, K; Specific Heat, C_p, Cal $g^{-1} K^{-1}$]

T	C_p
CURVE 1	
311	1.300 x 10^{-1}
366	1.308
422	1.325
478	1.352
533	1.387
589	1.435
644	1.490
700	1.556
755	1.630
811	1.715
866	1.809
922	1.912
978	2.026
1033	2.149

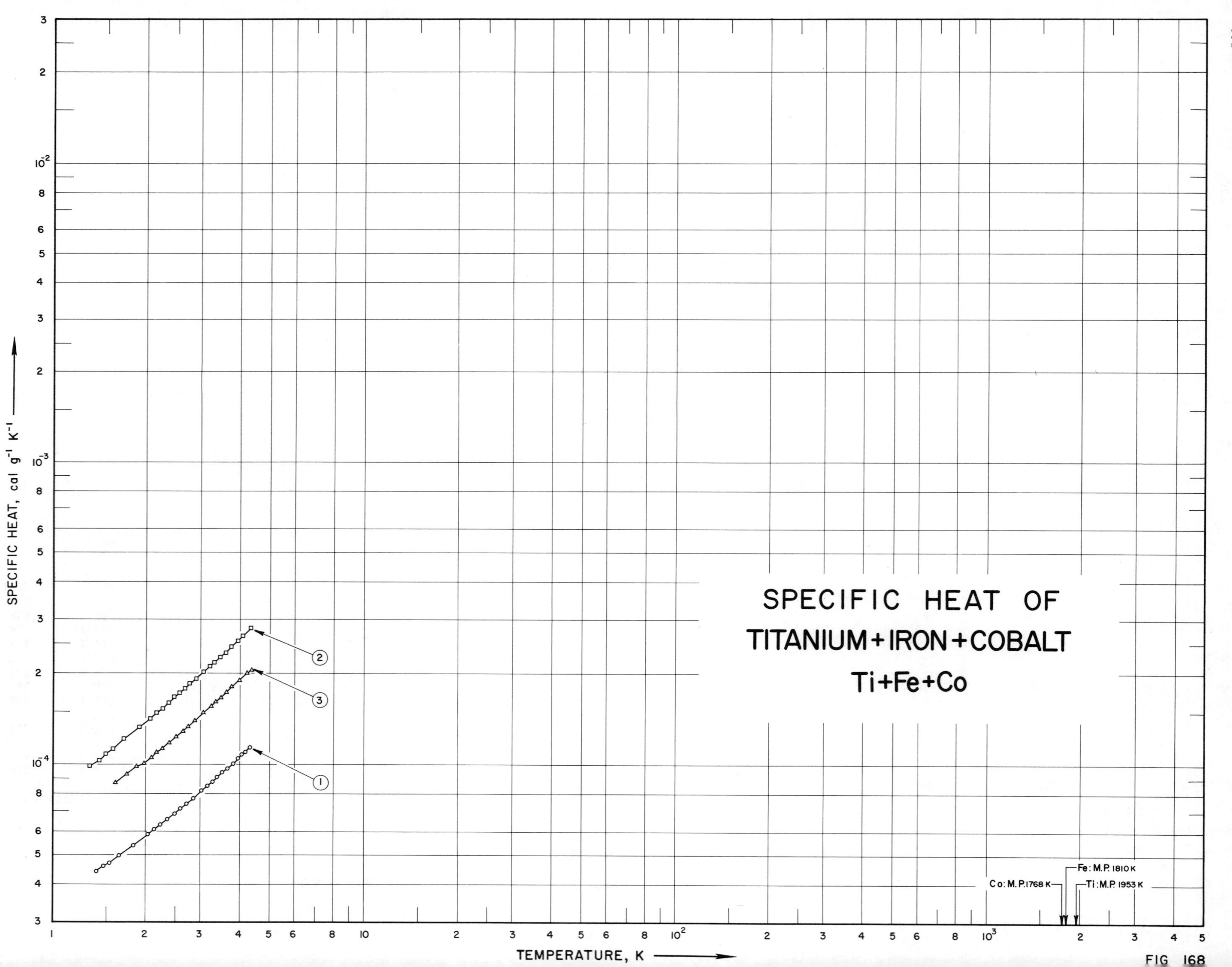
SPECIFIC HEAT OF
TITANIUM+IRON+COBALT
Ti+Fe+Co
Fe: M.P. 1810 K
Ti: M.P. 1953 K
Co: M.P. 1768 K
TEMPERATURE, K
SPECIFIC HEAT, cal g⁻¹ K⁻¹

FIG 168

SPECIFICATION TABLE NO. 168 SPECIFIC HEAT OF TITANIUM + IRON + COBALT Ti + Fe + Co

[For Data Reported in Figure and Table No. 168]

Curve No.	Ref. No.	Year	Temp. Range, K	Reported Error, %	Name and Specimen Designation	Composition (weight percent), Specifications and Remarks
1	405	1960	1.4-4.3		Ti_4Fe_3Co	44.36 Ti, 41.99 Fe, 13.65 Co; body centered cubic crystal structure, single phase; annealed 24 hrs at 900 C and 72 hrs at 1100 C; quenched in vacuum; etched with 3% HF, 3% HCl, and 94% M_2O for about 5 to 10 sec.
2	405	1960	1.3-4.3		Ti_4FeCo	44.84 Ti, 41.38 Co, 13.77 Fe; body centered cubic crystal structure, 1% impurity phase; same as above.
3	405	1960	1.6-4.4		Ti_2FeCo	44.85 Ti, 27.55 Fe, 27.59 Co; body centered cubic crystal, single phase; same as above.

DATA TABLE NO. 168 SPECIFIC HEAT OF TITANIUM + IRON + COBALT Ti + Fe + Co

[Temperature, T, K; Specific Heat, C_p, Cal $g^{-1} K^{-1}$]

T	C_p
CURVE 1	
1.3951	4.408 x 10^{-5}
1.4734	4.590
1.5389	4.709
1.6578	4.994
1.8251	5.385
2.0345	5.885
2.1345	6.093
2.2390	6.307
2.3554	6.589
2.4800	6.852
2.5928	7.153
2.7070	7.414
2.8431	7.708
3.0230	8.179
3.1718	8.516
3.2788	8.764
3.3997	9.087
3.5272	9.411
3.6652	9.699
3.8263	1.016 x 10^{-4}
3.9639	1.051
4.0674	1.083
4.1855	1.113
4.3183	1.145
CURVE 2	
1.3240	9.895 x 10^{-5}
1.3704	1.008 x 10^{-4}*
1.4231	1.037
1.4941	1.086
1.5756	1.136
1.7177	1.223
1.9069	1.333
2.0704	1.420
2.1723	1.483
2.2637	1.535
2.3643	1.604
2.4738	1.675
2.5730	1.720
2.6629	1.787
2.7700	1.852
2.9083	1.922
3.0697	2.028

T	C_p
CURVE 2 (cont.)	
3.2054	2.110 x 10^{-4}
3.3211	2.174
3.4580	2.258
3.6038	2.338
3.7565	2.444
3.9363	2.568
4.0994	2.654
4.2213	2.726*
4.3412	2.818
CURVE 3	
1.6016	8.731 x 10^{-5}
1.7490	9.284
1.8709	9.850
1.9870	1.013 x 10^{-4}
2.0921	1.061
2.1749	1.100
2.2648	1.139
2.3762	1.186
2.5046	1.248
2.6271	1.296
2.7447	1.341
2.8737	1.413
3.0560	1.489
3.2315	1.568
3.3507	1.621
3.4837	1.668
3.6248	1.737
3.7699	1.821
3.8870	1.852*
3.9795	1.896
4.0925	1.951*
4.2246	2.018
4.3616	2.050

*Not shown on plot

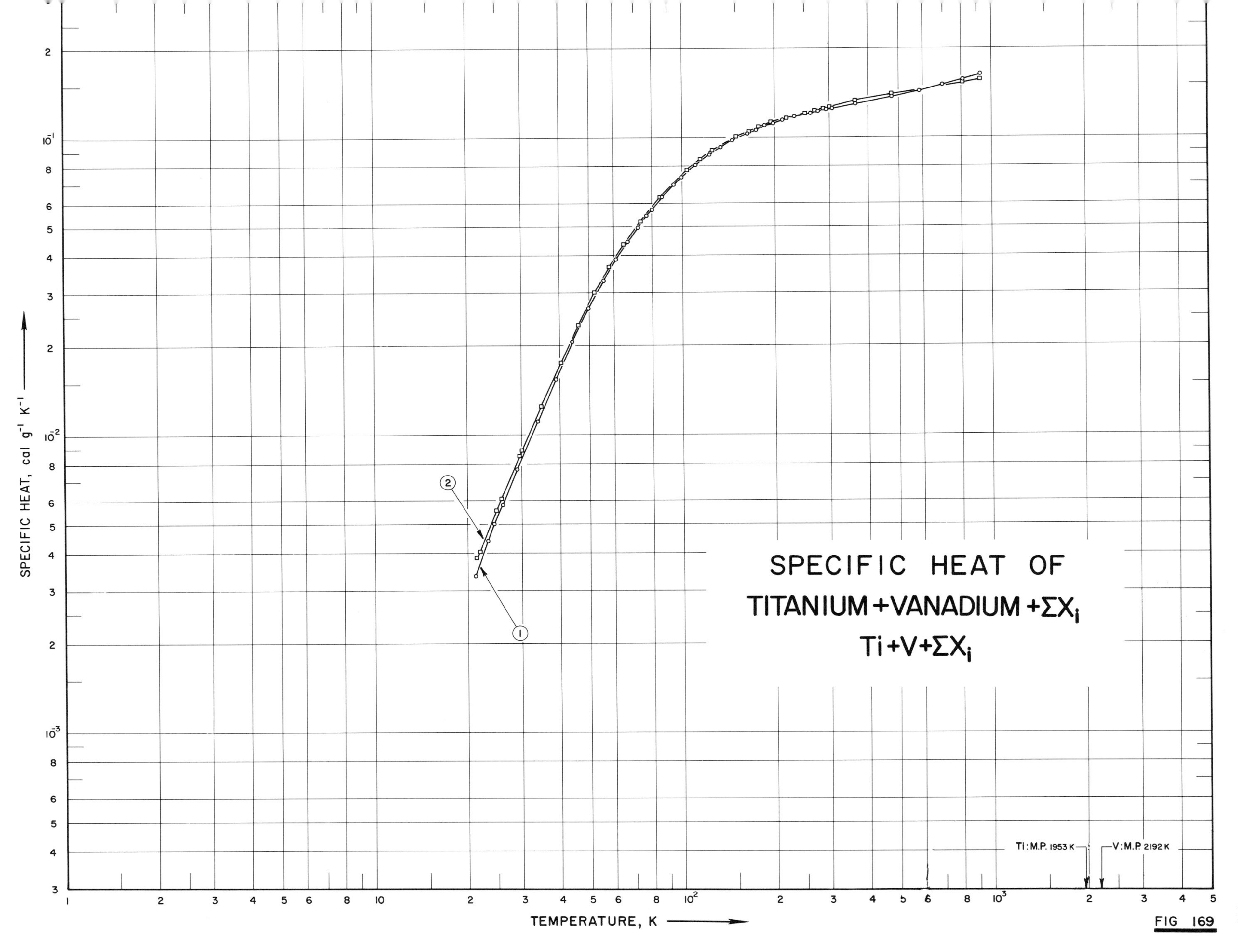
SPECIFIC HEAT OF
TITANIUM + VANADIUM + ΣXi
Ti + V + ΣXi
Ti: M.P. 1953 K
V: M.P. 2192 K
SPECIFIC HEAT, cal g⁻¹ K⁻¹
TEMPERATURE, K
1
2

FIG 169

SPECIFICATION TABLE NO. 169 SPECIFIC HEAT OF TITANIUM + VANADIUM + ΣX_i Ti + V + ΣX_i

[For Data Reported in Figure and Table No. 169]

Curve No.	Ref. No.	Year	Temp. Range, K	Reported Error, %	Name and Specimen Designation	Composition (weight percent), Specifications and Remarks
1	257	1961	21-922	<2.0		13.9 V, 10.4 Cr, 3.5 Al, 0.25 Fe, 0.04 C, 0.025 N_2, 0.0114 H_2; solution treated at 1450 F for 20 min; air cooled; aged at 900 F for 60 hrs; air cooled.
2	257	1961	21-922	<2.0		14.95 V, 2.75 Al, 0.21 Fe, 0.03 C, 0.015 N_2, 0.0066 H_2; solution heat treated at 1410 F for 30 min; aged at 990 F for 4 hrs.

DATA TABLE NO. 169 SPECIFIC HEAT OF TITANIUM + VANADIUM + ΣX_i $Ti + V + \Sigma X_i$

[Temperature, T, K; Specific Heat, C_p, Cal $g^{-1} K^{-1}$]

T	C_p
CURVE 1	
Series 1	
276.15	1.234 x 10^{-1}
282.02	1.240*
288.39	1.246*
294.95	1.254
301.57	1.259*
306.75	1.264
Series 2	
79.91	5.741 x 10^{-2}
86.20	6.325
94.21	6.961
99.71	7.357
105.56	7.754*
111.41	8.123
117.09	8.453*
122.98	8.779
128.75	9.069*
134.43	9.332
140.38	9.583*
146.39	9.824
157.35	1.021 x 10^{-1}*
163.12	1.039
169.02	1.056*
174.98	1.073
180.86	1.087*
186.64	1.101
192.34	1.113*
198.15	1.125
Series 3	
21.21	3.36 x 10^{-3}
24.51	5.03
29.01	7.68
33.82	1.123 x 10^{-2}
38.83	1.556
44.09	2.064
49.90	2.664
55.69	3.278
61.25	3.875
66.72	4.450
CURVE 1 (cont.)	
Series 3 (cont.)	
72.05	4.974 x 10^{-2}
76.97	5.453
Series 4	
21.18	3.43 x 10^{-3}*
23.33	4.43
26.00	5.84
Series 5	
201.39	1.135 x 10^{-1}*
206.97	1.145*
213.17	1.156
219.48	1.166*
225.73	1.176*
231.93	1.185
238.08	1.193*
244.19	1.202*
262.30	1.222
267.29	1.227*
272.42	1.234*
Series 6	
366.48	1.306 x 10^{-1}
477.59	1.377
588.71	1.443
699.82	1.509
810.93	1.573
922.04	1.637
CURVE 2	
Series 1	
68.08	4.679 x 10^{-2}*
73.62	5.229
79.51	5.797
85.20	6.328
93.14	6.972*
98.72	7.380*
104.48	7.774
CURVE 2 (cont.)	
110.17	8.139 x 10^{-1}*
115.81	8.476
121.48	8.790*
126.97	9.074
Series 2	
21.87	4.05 x 10^{-3}
25.76	6.10
30.11	8.91
Series 3	
21.44	3.86 x 10^{-3}
24.80	5.57
29.53	8.53
34.84	1.266 x 10^{-2}
40.33	1.760
46.19	2.347
52.41	3.002
58.56	3.665
64.93	4.357
70.97	4.967*
140.21	9.675
145.75	9.896
151.24	1.010 x 10^{-1}
159.05	1.036*
165.13	1.054
171.32	1.071*
177.55	1.088
184.06	1.103*
190.58	1.117*
196.57	1.130
201.96	1.141*
276.27	1.243*
282.54	1.249*
289.05	1.255
295.73	1.261*
302.69	1.267
Series 4	
201.32	1.139 x 10^{-1}*
207.49	1.150*
213.92	1.161*
CURVE 2 (cont.)	
Series 4 (cont.)	
220.65	1.172 x 10^{-1}
227.68	1.182*
234.71	1.192*
252.35	1.215
258.76	1.223*
264.28	1.230*
270.20	1.236
274.38	1.242*
Series 5	
366.48	1.330 x 10^{-1}
477.59	1.403
588.71	1.455*
699.82	1.499*
810.93	1.538
922.04	1.575

*Not shown on plot

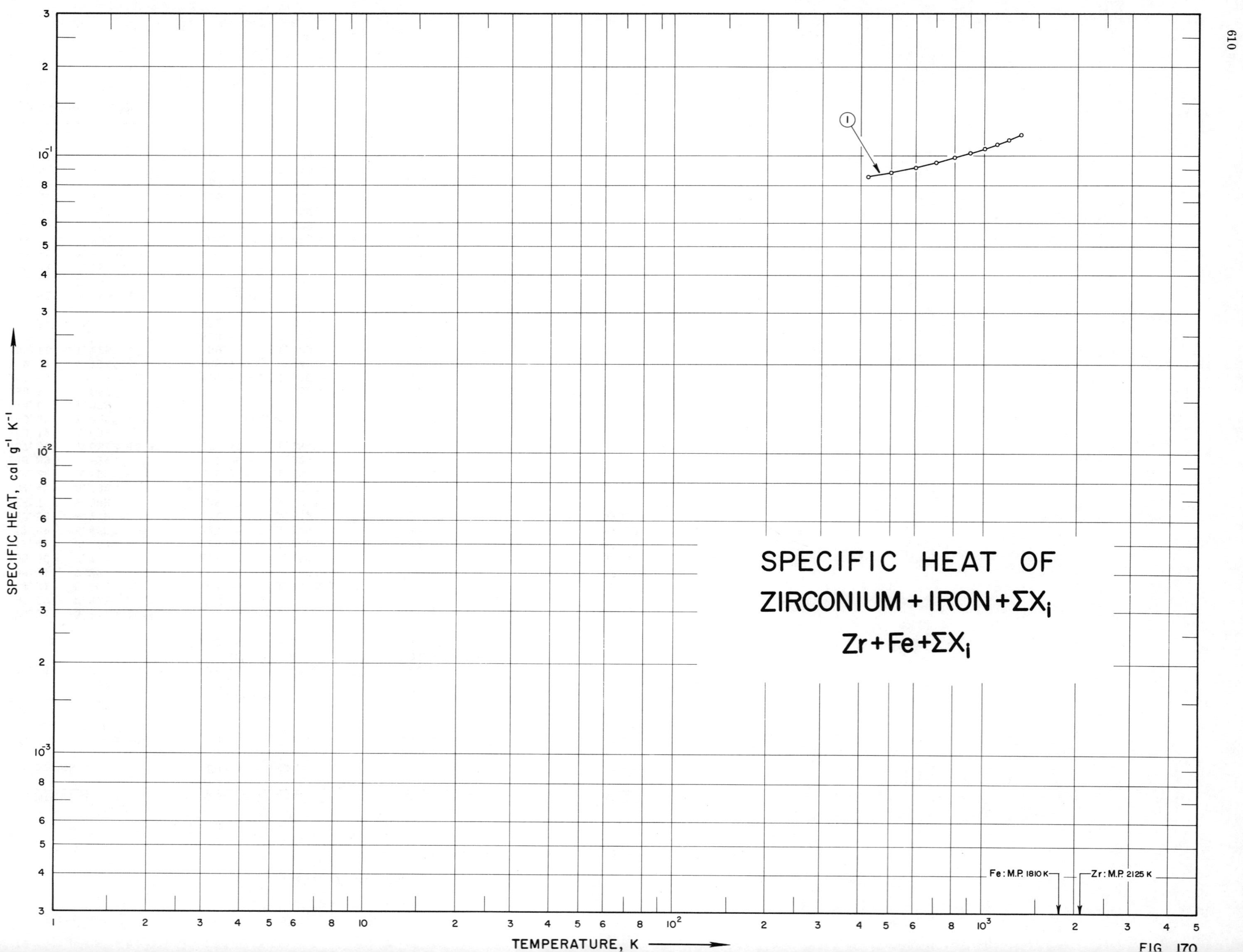

SPECIFIC HEAT OF
ZIRCONIUM + IRON + ΣX_i
Zr + Fe + ΣX_i
1
Fe: M.P. 1810 K
Zr: M.P. 2125 K
TEMPERATURE, K
SPECIFIC HEAT, cal g^{-1} K^{-1}
FIG 170

SPECIFICATION TABLE NO. 170 SPECIFIC HEAT OF ZIRCONIUM + IRON + ΣX_i Zr + Fe + ΣX_i

[For Data Reported in Figure and Table No. 170]

Curve No.	Ref. No.	Year	Temp. Range, K	Reported Error, %	Name and Specimen Designation	Composition (weight percent), Specifications and Remarks
1	**406**	1952	423-1323	±5		1 Fe, 1 Hf, 0.04 Mg, <0.04 Ba, <0.04 Cd, 0.02 Cu, 0.02 Mn, 0.02 Ni, 0.01 Si, 0.0004 Ca, 0.0004 Ti, 0.0002 Cr, 0.0002 Pb, <0.0002 Sn, <0.0002 V, <0.0028 all others.

DATA TABLE NO. 170 SPECIFIC HEAT OF ZIRCONIUM + IRON + ΣX_i $Zr + Fe + \Sigma X_i$

[Temperature, T, K; Specific Heat, C_p, Cal $g^{-1} K^{-1}$]

T	C_p
CURVE 1	
423	8.523×10^{-2}
500	8.800
600	9.160
700	9.520
800	9.880
900	1.024×10^{-1}
1000	1.060
1100	1.096
1200	1.132
1300	1.168*
1323	1.176

*Not shown on plot

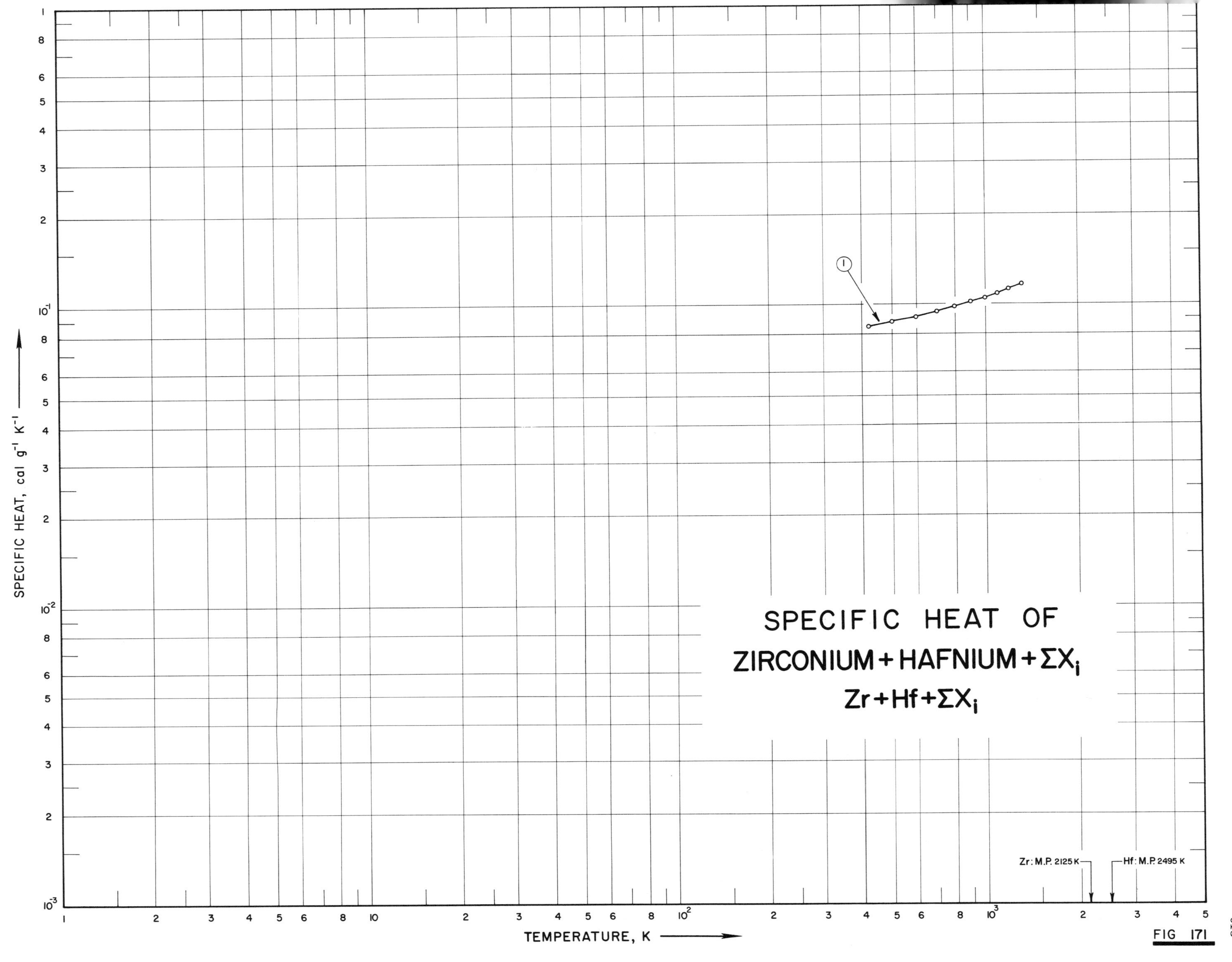

SPECIFIC HEAT OF
ZIRCONIUM + HAFNIUM + ΣX_i
Zr + Hf + ΣX_i
SPECIFIC HEAT, cal g⁻¹ K⁻¹
TEMPERATURE, K
Zr: M.P. 2125 K
Hf: M.P. 2495 K
1

FIG 171

SPECIFICATION TABLE NO. 171 SPECIFIC HEAT OF ZIRCONIUM + HAFNIUM + ΣX_i Zr + Hf + ΣX_i

[For Data Reported in Figure and Table No. 171]

Curve No.	Ref. No.	Year	Temp. Range, K	Reported Error, %	Name and Specimen Designation	Composition (weight percent), Specifications and Remarks
1	406	1952	423-1323	±5		1 Hf, 1 Fe, 0.04 Mg, <0.04 Ba, <0.04 Cd, 0.02 Cu, 0.02 Mn, 0.02 Ni, 0.01 Si, 0.0004 Ca, 0.0004 Ti, 0.0002 Cr, 0.0002 Pb, <0.0002 Sn, <0.0002 V, <0.0028 all others.

DATA TABLE NO. 171 SPECIFIC HEAT OF ZIRCONIUM + HAFNIUM + ΣX_i $Zr + Hf + \Sigma X_i$

[Temperature, T, K; Specific Heat, C_p, Cal $g^{-1} K^{-1}$]

T	C_p
CURVE 1	
423	8.523×10^{-2}
500	8.800
600	9.160
700	9.520
800	9.880
900	1.024×10^{-1}
1000	1.060
1100	1.096
1200	1.132
1300	1.168*
1323	1.176

* Not shown on plot

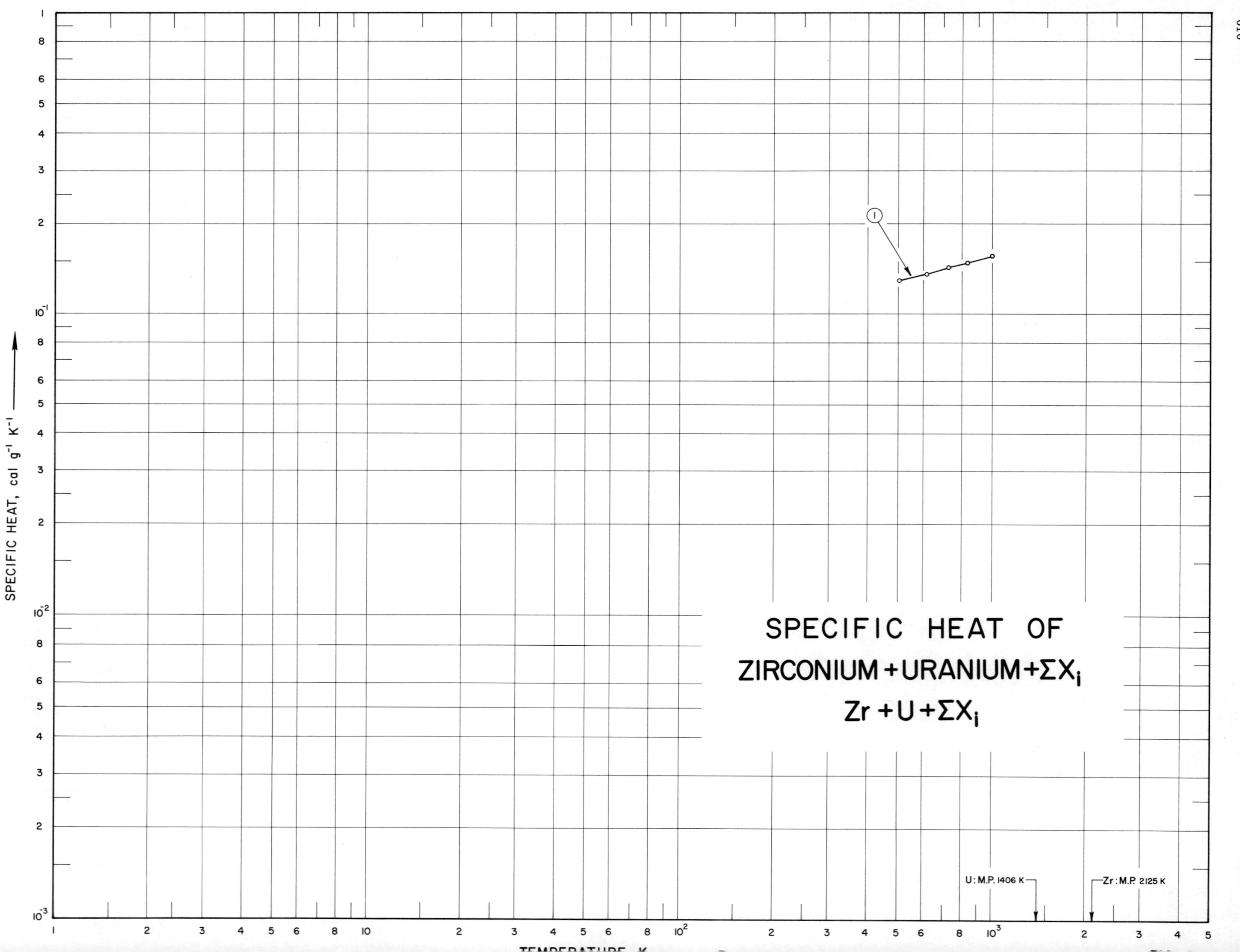

SPECIFIC HEAT OF
ZIRCONIUM + URANIUM + ΣX_i
Zr + U + ΣX_i
1
U: M.P. 1406 K
Zr: M.P. 2125 K
SPECIFIC HEAT, cal g^{-1} K^{-1}

SPECIFICATION TABLE NO. 172 SPECIFIC HEAT OF ZIRCONIUM + URANIUM + ΣX_i Zr + U + ΣX_i

[For Data Reported in Figure and Table No. 172]

Curve No.	Ref. No.	Year	Temp. Range, K	Reported Error, %	Name and Specimen Designation	Composition (weight percent), Specifications and Remarks
1	242	1963	505-1006	±2. 0	Hydrided	~87. 92 Zr, 10. 58 U, 1. 5 H_2; measured under hydrogen atmosphere; density = 383 lb ft^{-3}.

DATA TABLE NO. 172 SPECIFIC HEAT OF ZIRCONIUM + URANIUM + ΣX_i Zr + U + ΣX_i

[Temperature, T, K; Specific Heat, C_p, Cal $g^{-1} K^{-1}$]

T	C_p
CURVE 1	
506	1.31×10^{-1}
617	1.37
728	1.43
839	1.48
1006	1.57

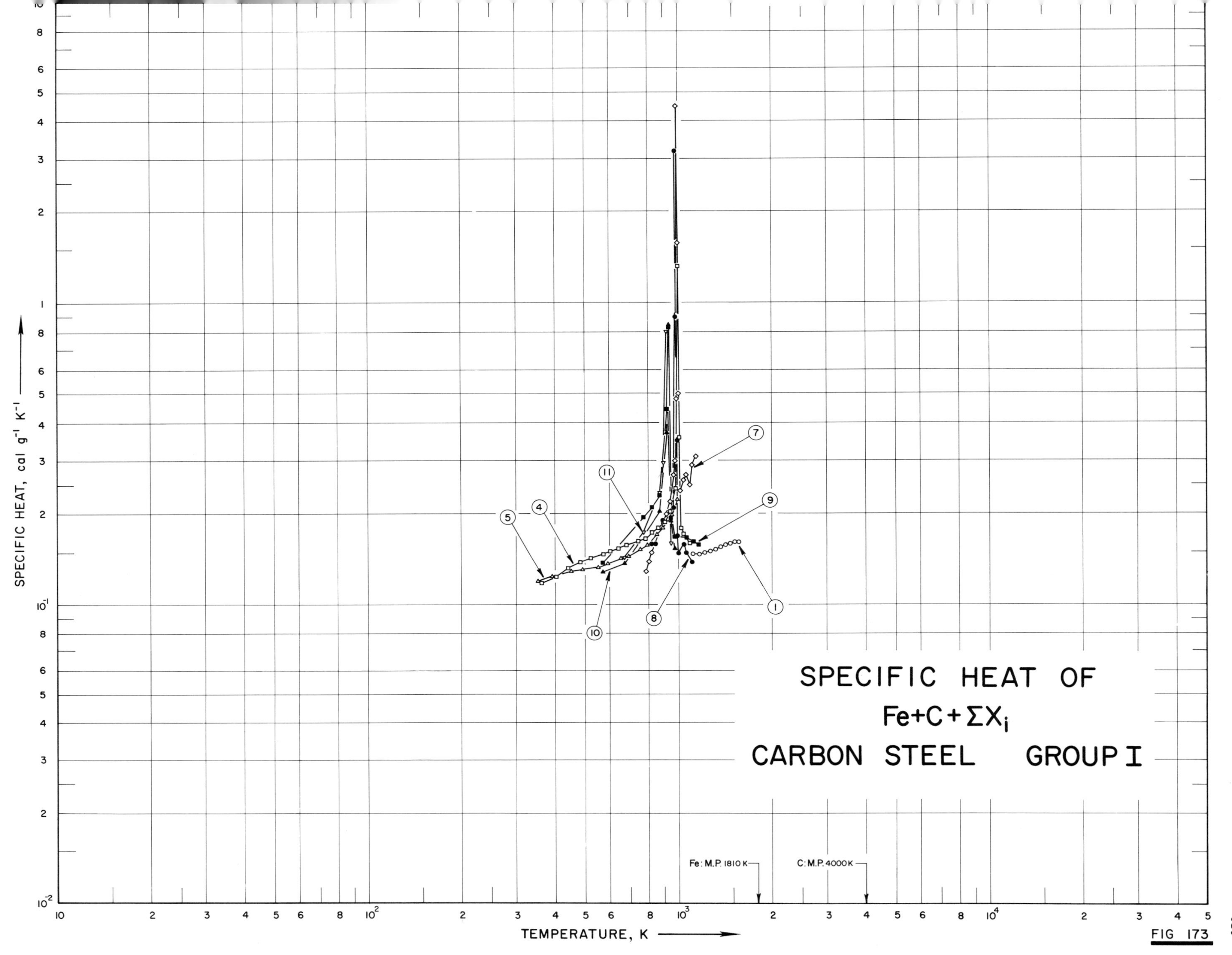
SPECIFIC HEAT OF
Fe+C+ΣXi
CARBON STEEL GROUP I
SPECIFIC HEAT, cal g⁻¹ K⁻¹
TEMPERATURE, K
Fe: M.P. 1810 K
C: M.P. 4000 K
1
4
5
7
8
9
10
11

FIG 173

SPECIFICATION TABLE NO. 173 SPECIFIC HEAT OF IRON + CARBON + ΣX_i (C < 2.00), CARBON STEEL GROUP I

[For Data Reported in Figure and Table No. 173]

Curve No.	Ref. No.	Year	Temp. Range, K	Reported Error, %	Name and Specimen Designation	Composition (weight percent), Specifications and Remarks
1	104	1946	1123-1573	4.0	Carbon steel No. 7	0.8 C, 0.32 Mn, 0.13 Ni, 0.13 Si, 0.11 Cr, 0.07 Cu, 0.021 As, <0.01 Mo, 0.009 S, 0.008 P and 0.004 Al; annealed at 800 C; density = 490 lb ft^{-3} at 15 C.
2	104	1946	1173-1523	4.0	Carbon steel No. 1	0.85 C, 0.31 Mn, and 0.15 Si.
3	104	1946	1123-1523	4.0	Carbon steel No. 8	1.22 C, 0.35 Mn, 0.16 Si, 0.13 Ni, 0.11 Cr, 0.077 Cu, 0.025 As, 0.015 S, 0.01 Mo, 0.009 P, and 0.006 Al.
4	83	1954	363-1143			0.87 C; furnace cooled from the homogenizing temperature of 1100 C.
5	33	1957	353-993	≤0.9	Steel A	>99.147 Fe, 0.77 C, 0.021 S, <0.01 Mn, <0.005 Si, <0.002 P, 0.01-0.09 Cu, 0.001-0.009 Ni; free cooled.
6	33	1957	353-993	≤0.9	Steel B	Same as above; slow cooled.
7	407	1961	785-1142		Steel B	1.2 C, 0.21 Mn, 0.115 Si, 0.02 Cr, 0.023 P, 0.016 S, and 0.01 each Mo and Ni; annealed.
8	407	1961	820-1117		Steel C	1.53 C, 0.25 Mn, 0.067 Si, 0.021 P, and 0.018 S; annealed.
9	408	1940	573-1223			0.67 C, 0.31 Mn, 0.078 Si, 0.025 S, and 0.012 P.
10	408	1940	573-1223			0.97 C, 0.18 Mn, 0.12 Si, 0.028 S, and 0.018 P.
11	408	1940	573-1223			1.21 C, 0.25 Mn, 0.18 Si, 0.038 P, and 0.021 S.
12	408	1940	573-1273			0.81 C, 0.39 Si, 0.32 Mn, 0.008 P, and 0.008 S.

DATA TABLE NO. 173 SPECIFIC HEAT OF IRON + CARBON + ΣX_i ($C < 2.00$), CARBON STEEL GROUP I

[Temperature, T, K; Specific Heat, C_p, Cal g^{-1} K^{-1}]

T	C_p
CURVE 1	
1123	1.48 x 10^{-1}
1173	1.48
1223	1.50
1273	1.52
1323	1.54
1373	1.57
1423	1.59
1473	1.61
1523	1.62
1573	1.62
CURVE 2*	
1173	1.48 x 10^{-1}
1223	1.47
1273	1.48
1323	1.51
1373	1.54
1423	1.58
1473	1.61
1523	1.64
CURVE 3*	
1123	1.48 x 10^{-1}
1173	1.48
1223	1.49
1273	1.51
1323	1.53
1373	1.54
1423	1.56
1473	1.58
1523	1.60
CURVE 4	
Series 1	
363	1.1879 x 10^{-1}
383	1.2210*
403	1.2422
423	1.2798*
443	1.3295
463	1.3712*
483	1.3974
503	1.4089*
CURVE 4 (cont.)	
Series 1 (cont.)	
523	1.4328 x 10^{-1}
543	1.4554*
563	1.4778
583	1.4954*
603	1.5145
623	1.5308*
643	1.5480
663	1.5636*
683	1.5831
703	1.6195*
743	1.6261
763	1.6579*
783	1.6690
803	1.7007*
823	1.7347
843	1.7671*
863	1.8094
883	1.8424*
903	1.8867
923	1.9399*
943	2.0347
963	2.0836*
983	2.4297
1003	1.32545 x 10^{0}
1008	3.5812 x 10^{-1}
1023	1.7976
1043	1.7234
1063	1.6263
1083	1.6117
Series 2*	
1023	1.5802 x 10^{-1}
1043	1.5747
1063	1.5588
1083	1.5980
1103	1.6364
1123	1.6659
1143	1.6171
CURVE 5	
353	1.204 x 10^{-1}
373	1.226*
CURVE 5 (cont.)	
393	1.244 x 10^{-1}
413	1.261*
433	1.271*
453	1.297
473	1.313*
493	1.322
513	1.327*
533	1.333*
553	1.346
573	1.363*
593	1.378
613	1.398*
633	1.408*
653	1.434
673	1.443*
693	1.466
713	1.487*
733	1.514*
753	1.541
773	1.571*
783	1.581*
793	1.590
803	1.626*
813	1.641*
823	1.648*
833	1.658*
843	1.672*
853	1.724
863	1.737*
873	1.748*
883	1.761*
893	1.808
903	1.828*
913	1.847*
923	1.882*
933	1.930*
943	1.966*
953	1.991
963	2.068*
973	2.097*
983	2.164*
993	2.258
CURVE 6*	
353	1.205 x 10^{-1}
373	1.226
393	1.241
413	1.257
433	1.271
453	1.285
473	1.307
493	1.320
513	1.328
533	1.334
553	1.344
573	1.352
593	1.368
613	1.386
633	1.405
653	1.423
673	1.439
693	1.458
713	1.484
733	1.504
753	1.530
773	1.555
783	1.572
793	1.586
803	1.603
813	1.618
823	1.638
833	1.655
843	1.674
853	1.692
863	1.717
873	1.741
883	1.767
893	1.791
903	1.822
913	1.849
923	1.885
933	1.918
943	1.950
953	1.980
963	2.036
973	2.093
983	2.155
993	2.241
CURVE 7	
785	1.3 x 10^{-1}
801	1.4
820	1.5
844	1.7*
868	1.8*
892	1.9*
917	2.0
941	2.2
966	2.7
978	3.0
986	4.8
991	4.52 x 10^{0}
996	1.58
1001	5.0 x 10^{-1}*
1008	2.5
1016	2.4
1041	2.6
1066	2.7
1091	2.5
1117	2.9
1142	3.1
CURVE 8	
820	1.6 x 10^{-1}
844	1.6
868	1.8*
893	1.9
917	2.0*
941	2.0*
966	2.1
981	9.0
985	3.21 x 10^{0}
988	8.9 x 10^{-1}*
991	3.5
996	1.7
1003	1.5
1016	1.5*
1041	1.6
1066	1.5
1091	1.4*
1117	1.4
CURVE 9	
573	1.38 x 10^{-1}
CURVE 9 (cont.)	
673	1.58 x 10^{-1}*
773	1.95
823	2.10
873	2.31
923	4.45
933	8.38
948	1.95
973	1.69
1073	1.68
1173	1.63
1223	1.59
CURVE 10	
573	1.29 x 10^{-1}
673	1.38
773	1.68*
873	2.05
923	3.70
938	8.43
948	1.90
973	1.55
1073	1.50*
1173	1.51*
1223	1.51*
CURVE 11	
573	1.31 x 10^{-1}*
673	1.44
773	1.74
873	2.35
898	2.96
923	8.07
948	1.60
973	1.55*
1073	1.53*
1173	1.52*
1223	1.52*
CURVE 12*	
573	1.470 x 10^{-1}
673	1.530
773	1.565
873	1.590

* Not shown on plot

DATA TABLE NO. 173 (continued)

T	C_p
CURVE 12 (cont.)*	
973	1.603×10^{-1}
1073	1.617
1173	1.630
1273	1.642

*Not shown on plot

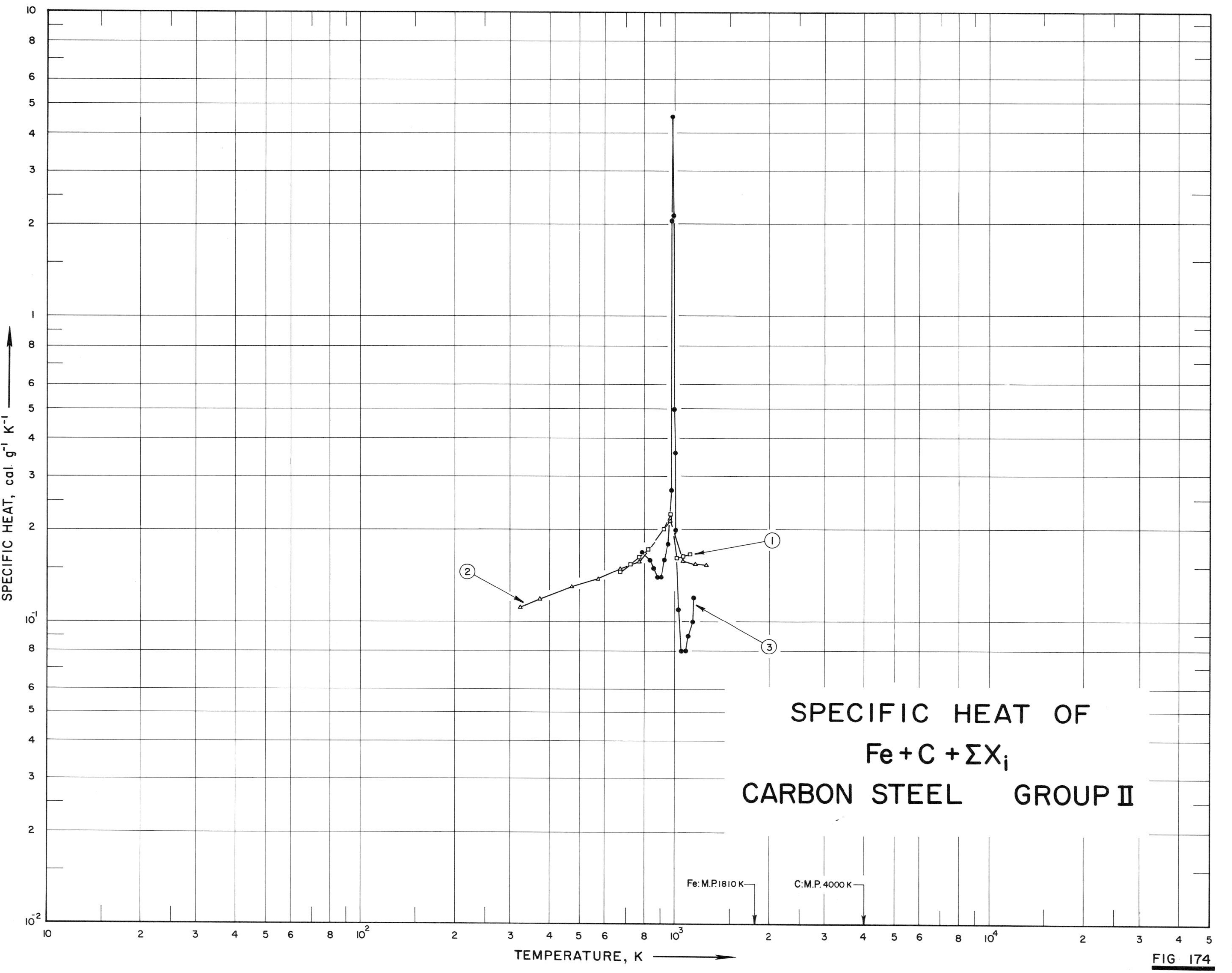
SPECIFIC HEAT OF
Fe + C + ΣX$_i$
CARBON STEEL GROUP II
SPECIFIC HEAT, cal. g^{-1} K^{-1}
TEMPERATURE, K
Fe: M.P. 1810 K
C: M.P. 4000 K
1
2
3

FIG 174

SPECIFICATION TABLE NO. 174 SPECIFIC HEAT OF IRON + CARBON + ΣX_i (C < 2.00), CARBON STEEL GROUP II

[For Data Reported in Figure and Table No. 174]

Curve No.	Ref. No.	Year	Temp. Range, K	Reported Error, %	Name and Specimen Designation	Composition (weight percent), Specifications and Remarks
1	54	1954	673-1123		Eutectoid steel	0.79 C, 0.51 Mo, 0.19 Si, 0.12 Mn, 0.005 each S and P.
2	15	1959	323-1273		Steel U-8	0.75-0.84 C, 0.2-0.4 Mn, 0.15-0.35 Si, 0.25 Ni, 0.25 Cu, 0.2 Cr, 0.035 P, and 0.03 S.
3	407	1961	785-1155		Hyper Eutectoid Steel A	0.79 C, 0.64 Mn, 0.091 Si, 0.038 S, 0.031 P, and 0.01 each Cr, Mo, and Ni; annealed.

DATA TABLE NO. 174 SPECIFIC HEAT OF IRON + CARBON + ΣX_i (C < 2.00), CARBON STEEL GROUP II

[Temperature, T, K; Specific Heat, C_p, Cal g^{-1} K^{-1}]

T	C_p
CURVE 1	
673	1.470×10^{-1}
723	1.551
773	1.636
823	1.738
873	1.861*
923	2.015
973	2.265
1023	1.624
1073	1.649
1123	1.675
CURVE 2	
323	1.12×10^{-1}
373	1.18
473	1.30
573	1.38
673	1.49
773	1.58
873	1.86*
973	2.15
1073	1.59
1173	1.56
1273	1.54
CURVE 3	
785	1.7×10^{-1}
808	1.7*
832	1.6
856	1.5
880	1.4
905	1.4
929	1.6
953	1.8
978	2.7
983	2.06×10^{0}
986	4.55
991	2.15
998	5.0×10^{-1}
CURVE 3 (cont.)	
1003	3.6×10^{-1}
1016	2.0
1028	1.1
1053	8.0×10^{-2}
1079	8.0
1104	9.0
1130	1.0×10^{-1}
1155	1.2

* Not shown on plot

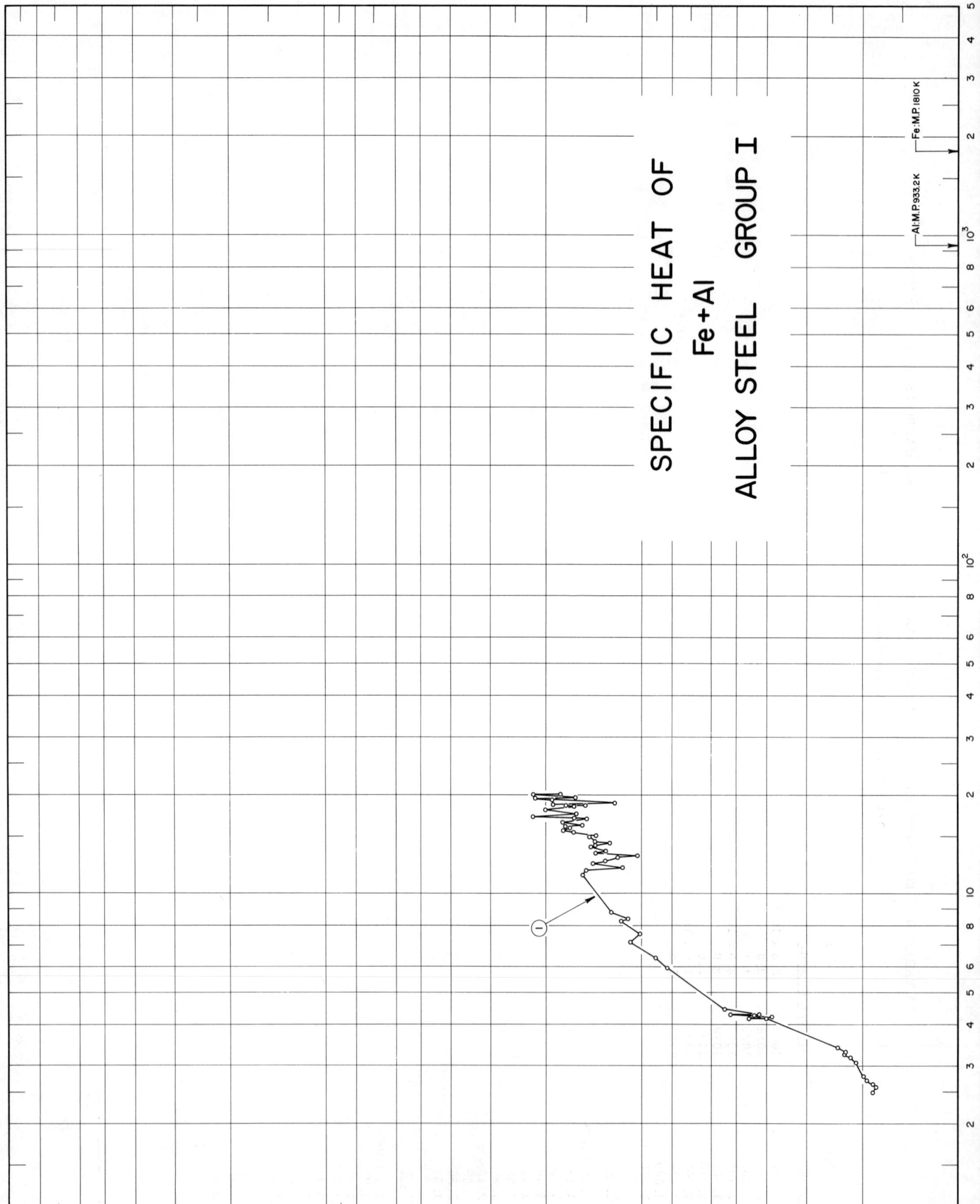

SPECIFIC HEAT OF Fe + Al ALLOY STEEL GROUP I

SPECIFICATION TABLE NO. 175 SPECIFIC HEAT OF IRON + ALUMINUM, Fe + Al, ALLOY STEEL GROUP I

[For Data Reported in Figure and Table No. 175]

Curve No.	Ref. No.	Year	Temp. Range, K	Reported Error, %	Name and Specimen Designation	Composition (weight percent), Specifications and Remarks
1	409	1961	2.5-20		$Fe_{63}Al_{37}$	77.20 Fe and 22.80 Al.

DATA TABLE NO. 175 SPECIFIC HEAT OF IRON + ALUMINUM, Fe + Al, ALLOY STEEL GROUP I

[Temperature, T, K; Specific Heat, C_p, Cal $g^{-1}K^{-1}$]

T	C_p	T	C_p
CURVE 1		CURVE 1 (cont.)	
2.492	1.868 x 10^{-4}	14.444	1.413 x 10^{-3}
2.518	1.843*	14.661	1.459*
2.522	1.847*	14.865	1.472
2.587	1.820	15.058	1.397
2.631	1.858	15.246	1.517*
2.661	1.877*	15.439	1.649
2.707	1.930	15.614	1.767
2.771	1.989	15.806	1.685
2.824	1.994*	16.010	1.745
3.068	2.096	16.172	1.541
3.113	2.104*	16.376	1.541*
3.183	2.185	16.560	1.771
3.256	2.277	16.763	1.523*
3.306	2.261	16.800	1.494
3.358	2.295*	16.913	1.636*
3.400	2.397	17.000	1.632
4.18	4.030	17.018	1.703*
4.18	4.561	17.049	1.745*
4.22	3.852	17.154	2.210
4.25	4.428	17.231	1.754*
4.26	4.207*	17.262	1.720*
4.27	5.247	17.369	1.621*
4.29	4.207	17.530	1.618
4.46	5.469	17.646	1.745*
5.94	8.325	17.920	1.791*
6.37	9.011	18.079	2.013
7.12	1.089 x 10^{-3}	18.195	1.771*
7.17	1.078*	18.338	1.824*
7.55	1.012	18.419	1.665*
8.25	1.169	18.495	1.630
8.38	1.109	18.596	1.738
8.79	1.251	18.604	1.508
11.46	1.539	18.730	1.902
11.77	1.506	18.795	1.658*
12.038	1.156	18.868	1.229
12.233	1.426	18.935	1.862*
12.409	1.393*	19.109	1.789*
12.601	1.304	19.268	1.920
12.862	1.191	19.418	2.041*
13.165	1.034	19.514	2.172
13.356	1.406	19.609	1.623
13.570	1.313	20.532	2.194
13.803	1.457	20.628	2.092*
14.022	1.404	20.714	1.800
14.266	1.269		

* Not shown on plot

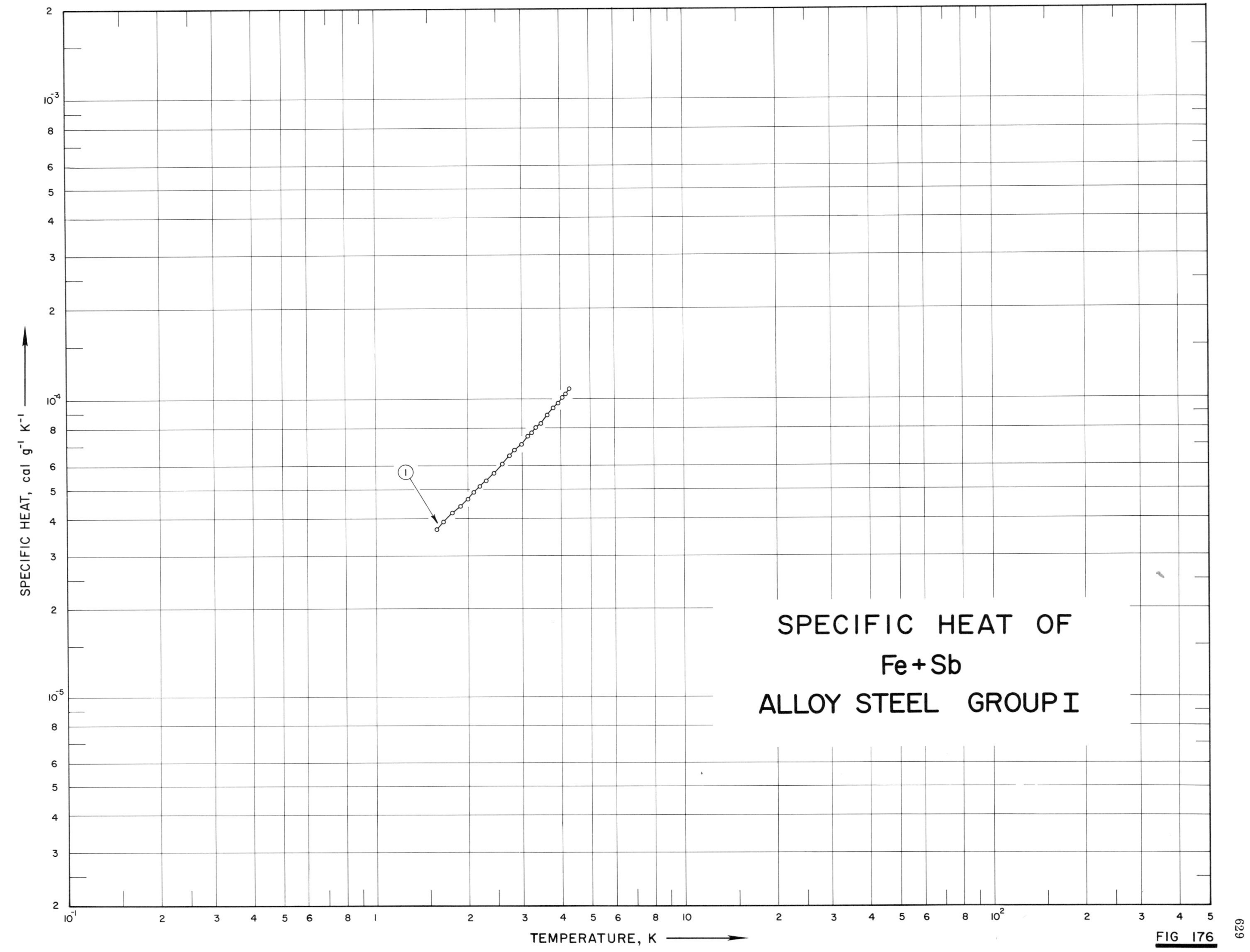
SPECIFIC HEAT OF
Fe + Sb
ALLOY STEEL GROUP I
SPECIFIC HEAT, cal g⁻¹ K⁻¹
TEMPERATURE, K
1

FIG 176

SPECIFICATION TABLE NO. 176 SPECIFIC HEAT OF IRON + ANTIMONY, Fe + Sb, ALLOY STEEL GROUP I

[For Data Reported in Figure and Table No. 176]

Curve No.	Ref. No.	Year	Temp. Range, K	Reported Error, %	Name and Specimen Designation	Composition (weight percent), Specifications and Remarks
1	349	1962	1.6-4.3	≤2	$Fe_{93.5}Sb_{6.5}$	88.81 Fe and 11.13 Sb; annealed under He + 8% H_2 gas atmosphere at 1100 C for 72 hrs; etched with 2% HNO_3.

DATA TABLE NO. 176 SPECIFIC HEAT OF IRON + ANTIMONY, Fe + Sb, ALLOY STEEL GROUP I

[Temperature, T,K; Specific Heat, C_p, Cal $g^{-1}K^{-1}$]

T	C_p
CURVE 1	
1.585	3.713×10^{-5}
1.674	3.937
1.789	4.211
1.895	4.448
1.896	4.435*
2.004	4.687
2.096	4.942
2.189	5.154
2.303	5.401
2.434	5.745
2.593	6.138
2.736	6.535
2.857	6.816
3.009	7.142
3.135	7.564
3.228	7.786
3.338	8.096
3.474	8.363
3.635	8.958
3.805	9.402
3.949	9.762
4.063	1.023×10^{-4}
4.179	1.056
4.305	1.090

* Not shown on plot

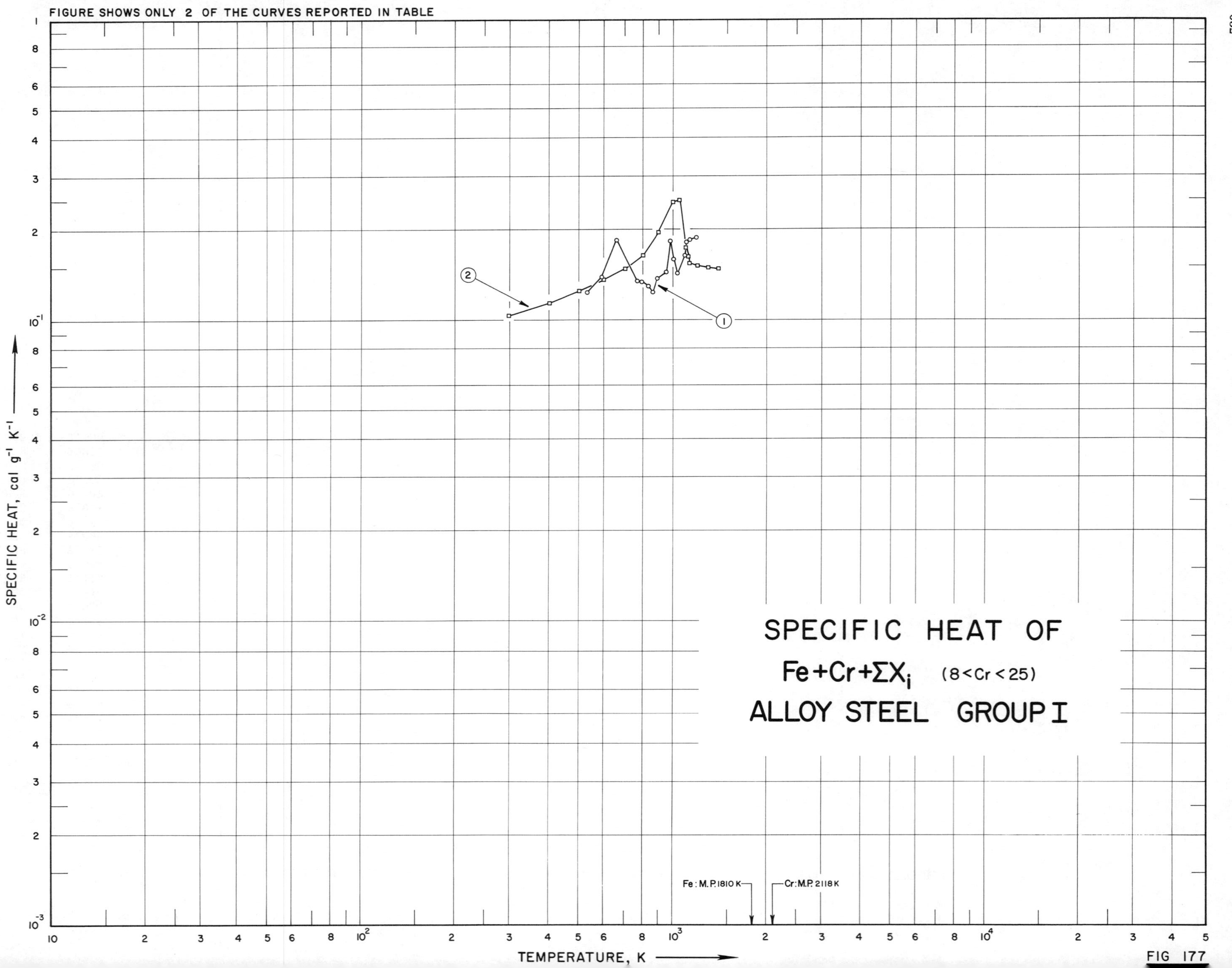
FIGURE SHOWS ONLY 2 OF THE CURVES REPORTED IN TABLE
SPECIFIC HEAT OF
Fe+Cr+ΣX$_i$ (8<Cr<25)
ALLOY STEEL GROUP I
SPECIFIC HEAT, cal g^{-1} K^{-1}
TEMPERATURE, K
Fe: M.P. 1810 K
Cr: M.P. 2118 K
1
2

FIG 177

SPECIFICATION TABLE NO. 177 SPECIFIC HEAT OF IRON + CHROMIUM, Fe + Cr ($8 \leq Cr < 25$), ALLOY STEEL GROUP I

[For Data Reported in Figure and Table No. 177]

Curve No.	Ref. No.	Year	Temp. Range, K	Reported Error, %	Name and Specimen Designation	Composition (weight percent), Specifications and Remarks
1	426	1955	533-1195	≤5.0		Main constituent is Fe and 8.05 Cr.
2	222	1959	298-1400	±0.5	Sample No. 9 Cr	91.2 Fe and 8.8 Cr; homogenized 4 days at 1350 C under helium atmosphere; air cooled to room temperature.
3	320	1959	1.8-4.3			Top of specimen: 78.65 Fe, 21.14 Cr, 0.18 Si, 0.088 O_2, 0.0006 Al, and 0.00014 H_2, bottom of specimen: 78.66 Fe, 21.12 Cr, 0.16 Si, 0.088 O_2, 0.0004 Al, and 0.00014 H_2; induction melted from electrolytic materials of Cr and Fe; alloy kept at molten state 3 min for homogenization; annealed 3 days at 1170 C under 92 He-8H_2 gas mixture.

DATA TABLE NO. 177 SPECIFIC HEAT OF IRON + CHROMIUM, Fe + Cr (8 ≤ Cr < 25), ALLOY STEEL GROUP I

[Temperature, T, K; Specific Heat, C_p, Cal $g^{-1}K^{-1}$]

T	C_p
CURVE 1	
533	1.250 x 10^{-1}
593	1.410
662	1.860
768	1.360
795	1.350
833	1.300
853	1.270*
862	1.250
893	1.380
951	1.450
984	1.840
1006	1.610
1018	1.510*
1039	1.440
1095	1.650
1118	1.830
1139	1.860
1195	1.890
CURVE 2	
298.15	1.046 x 10^{-1}
400	1.158
500	1.266
600	1.375
700	1.496
800	1.669
900	1.977
1000	2.486
1050	2.520
1100	1.763
1121	1.647
1131	1.550
1200	1.533
1300	1.508
1400	1.483
CURVE 3*	
1.820	3.392 x 10^{-5}
1.874	3.499
1.929	3.617
2.010	3.745
2.093	3.912
2.200	4.101
CURVE 3 (cont.)*	
2.348	4.369 x 10^{-5}
2.562	4.793
2.756	5.184
2.928	5.551
3.062	5.813
3.149	5.930
3.247	6.203
3.365	6.439
3.495	6.803
3.650	7.005
3.778	7.340
3.879	7.753
3.990	7.946
4.118	8.352
4.267	8.324

*Not shown on plot

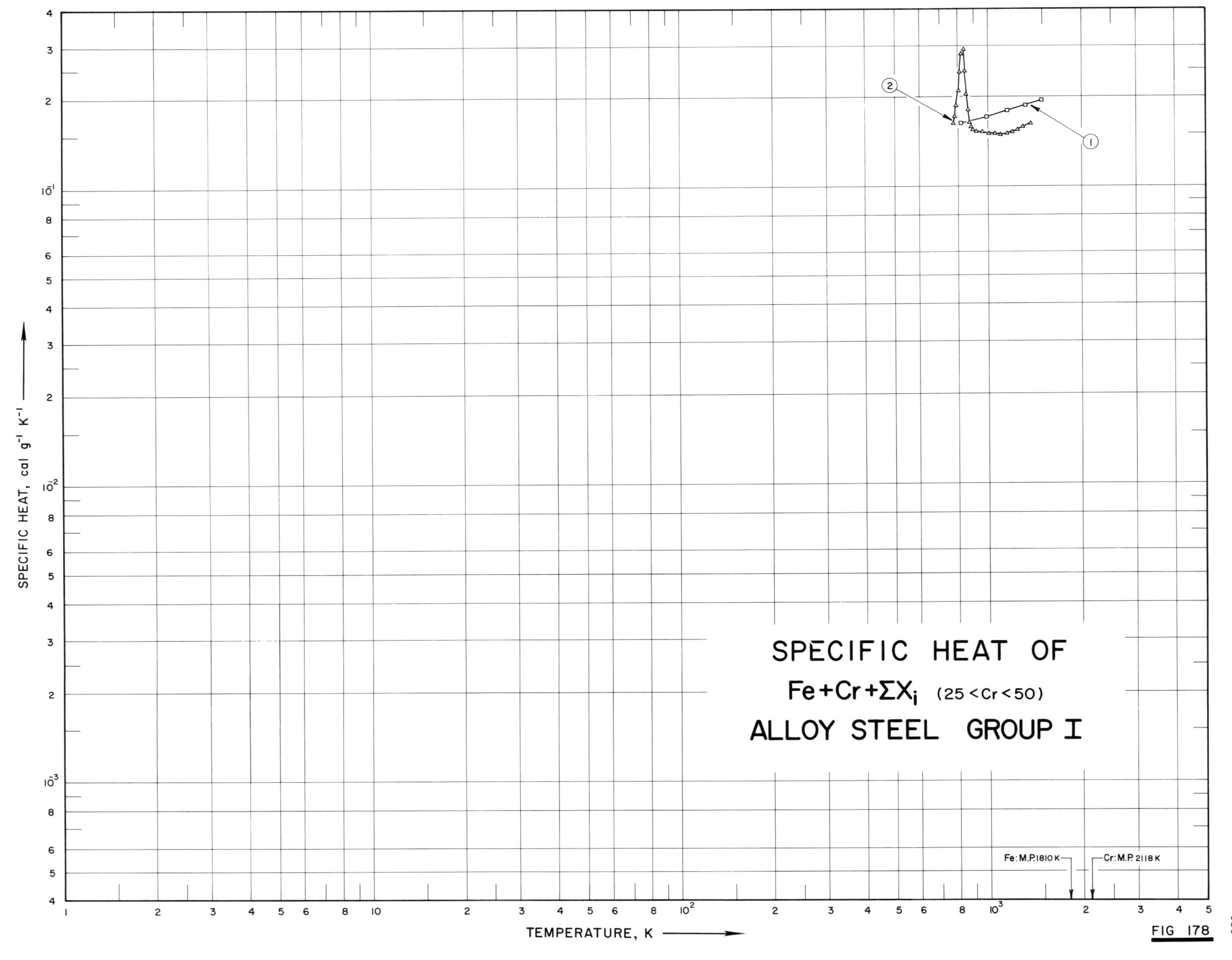
SPECIFIC HEAT OF
Fe+Cr+ΣXi (25<Cr<50)
ALLOY STEEL GROUP I
Fe: M.P. 1810 K
Cr: M.P. 2118 K
SPECIFIC HEAT, cal g⁻¹ K⁻¹
TEMPERATURE, K
1
2

FIG 178

SPECIFICATION TABLE NO. 178 SPECIFIC HEAT OF IRON + CHROMIUM, Fe + Cr ($25 \leq Cr < 50$), ALLOY STEEL GROUP I

[For Data Reported in Figure and Table No. 178]

Curve No.	Ref. No.	Year	Temp. Range, K	Reported Error, %	Name and Specimen Designation	Composition (weight percent), Specifications and Remarks
1	245	1958	811-1478	3.0	Stainless steel 446	70.55 Fe, 27.61 Cr, 0.086 C, and 0.01 Mo, after test: 70.59 Fe, 27.64 Cr, 0.066 C, and 0.01 Mo; density = 473.5 lb ft^{-3}.
2	130	1958	768-1368	±3.0		44.0 Cr.

DATA TABLE NO. 178 SPECIFIC HEAT OF IRON + CHROMIUM, Fe + Cr ($25 \leq Cr < 50$), ALLOY STEEL GROUP I

[Temperature, T, K; Specific Heat, C_p, Cal $g^{-1}K^{-1}$]

T	C_p	T	C_p
CURVE 1		CURVE 2 (cont.)	
811	1.640 x 10^{-1}	1148	1.520 x 10^{-1}
978	1.720	1158	1.522*
1144	1.800	1168	1.530*
1311	1.880	1178	1.534*
1478	1.960	1188	1.539*
		1198	1.544
CURVE 2		1208	1.549*
		1218	1.554*
768	1.647 x 10^{-1}	1228	1.558*
778	1.733	1238	1.565*
788	1.898	1248	1.573
798	2.118	1258	1.577*
808	2.457	1268	1.582*
818	2.825	1278	1.587*
828	2.913	1288	1.594*
838	2.471	1298	1.601
848	2.065	1308	1.606*
858	1.821	1318	1.611*
868	1.666	1328	1.616*
878	1.604	1338	1.620*
888	1.575	1348	1.625*
898	1.558*	1358	1.630*
908	1.551*	1368	1.635
918	1.544		
928	1.539*		
938	1.537*		
948	1.532		
958	1.530*		
968	1.527*		
978	1.527*		
988	1.527*		
998	1.525*		
1008	1.522		
1018	1.522*		
1028	1.522*		
1038	1.522*		
1048	1.522*		
1058	1.522		
1068	1.520*		
1078	1.518*		
1088	1.515*		
1098	1.513		
1108	1.513*		
1118	1.513*		
1128	1.513*		
1138	1.515*		

* Not shown on plot

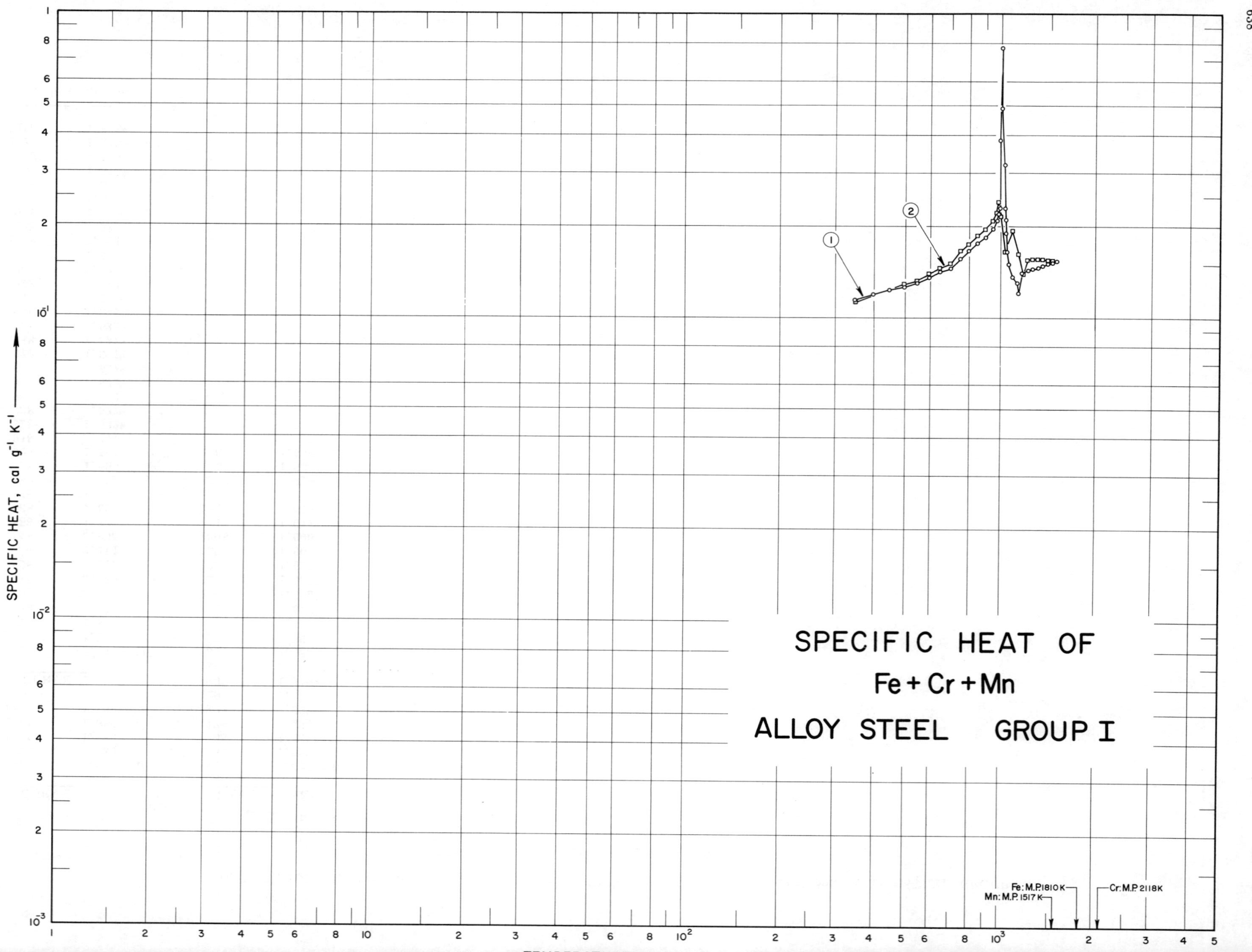

SPECIFIC HEAT OF
Fe + Cr + Mn
ALLOY STEEL GROUP I
Fe: M.P. 1810 K
Mn: M.P. 1517 K
Cr: M.P. 2118 K
1
2
SPECIFIC HEAT, cal g⁻¹ K⁻¹

SPECIFICATION TABLE NO. 179 SPECIFIC HEAT OF IRON + CHROMIUM + MANGANESE, Fe + Cr + Mn, ALLOY STEEL GROUP I

[For Data Reported in Figure and Table No. 179]

Curve No.	Ref. No.	Year	Temp. Range, K	Reported Error, %	Name and Specimen Designation	Composition (weight percent), Specifications and Remarks
1	104	1946	348-1523	2-4	Alloy Steel No. 20	0.88 Cr, 0.59 Mn, 0.35 C, 0.26 Ni, 0.21 Si, 0.20 Mo, 0.12 Cu, 0.039 As, 0.031 S, 0.028 P, and 0.004 Al; annealed at 860 C; then reheated to 640 C and furnace cooled; density (15 C) = 489 lb ft^{-3}.
2	104	1946	348-1523	2-4	High Alloy Steel No. 16	12.95 Cr, 0.25 Mn, 0.17 Si, 0.14 Ni, 0.13 C, 0.060 Cu, 0.034 Al, 0.024 S, 0.018 P, 0.015 As, and 0.012 V; heated at 960 C in air; heated at 750 C for 2 hrs; air cooled; density (15 C) = 482 lb ft^{-3}.

DATA TABLE NO. 179 SPECIFIC HEAT OF IRON + CHROMIUM + MANGANESE, Fe + Cr + Mn, ALLOY STEEL GROUP I

[Temperature, T, K; Specific Heat, C_p, Cal $g^{-1}K^{-1}$]

T	C_p
CURVE 1	
Series I	
348	1.14 x 10^{-1}
398	1.19
448	1.23
498	1.26
548	1.30
598	1.36
648	1.42
698	1.45
748	1.57
798	1.66
848	1.76
898	1.84
948	1.97
998	3.86
1048	2.11
1098	1.36
Series II	
978	2.1 x 10^{-1}
988	2.2
998	2.3
1008	4.9
1018	7.7
1028	3.2
1038	2.3
1048	1.9
1058	1.65
1068	1.50
1128	1.31
1138	1.2*
1148	1.2
Series III	
1123	1.39 x 10^{-1}
1173	1.41
1223	1.43
1273	1.45
1323	1.46
1373	1.48
1423	1.51
1473	1.52
1523	1.54

T	C_p
CURVE 2	
Series I	
348	1.13 x 10^{-1}
398	1.19*
448	1.23*
498	1.27
548	1.32
598	1.38
648	1.45
698	1.51
748	1.65
798	1.74
848	1.86
898	1.95
948	2.09
998	2.16
1048	1.65
1098	1.93
1148	1.60
Series II	
978	2.2 x 10^{-1}
988	2.4
998	2.2*
1008	1.9*
1018	1.8*
1078	1.5*
1088	1.5*
1098	2.1*
1108	2.3*
1118	1.9*
1188	1.4
Series III	
1173	1.53 x 10^{-1}*
1223	1.55
1273	1.56
1323	1.56
1373	1.56
1423	1.55
1473	1.55
1523	1.55

*Not shown on plot

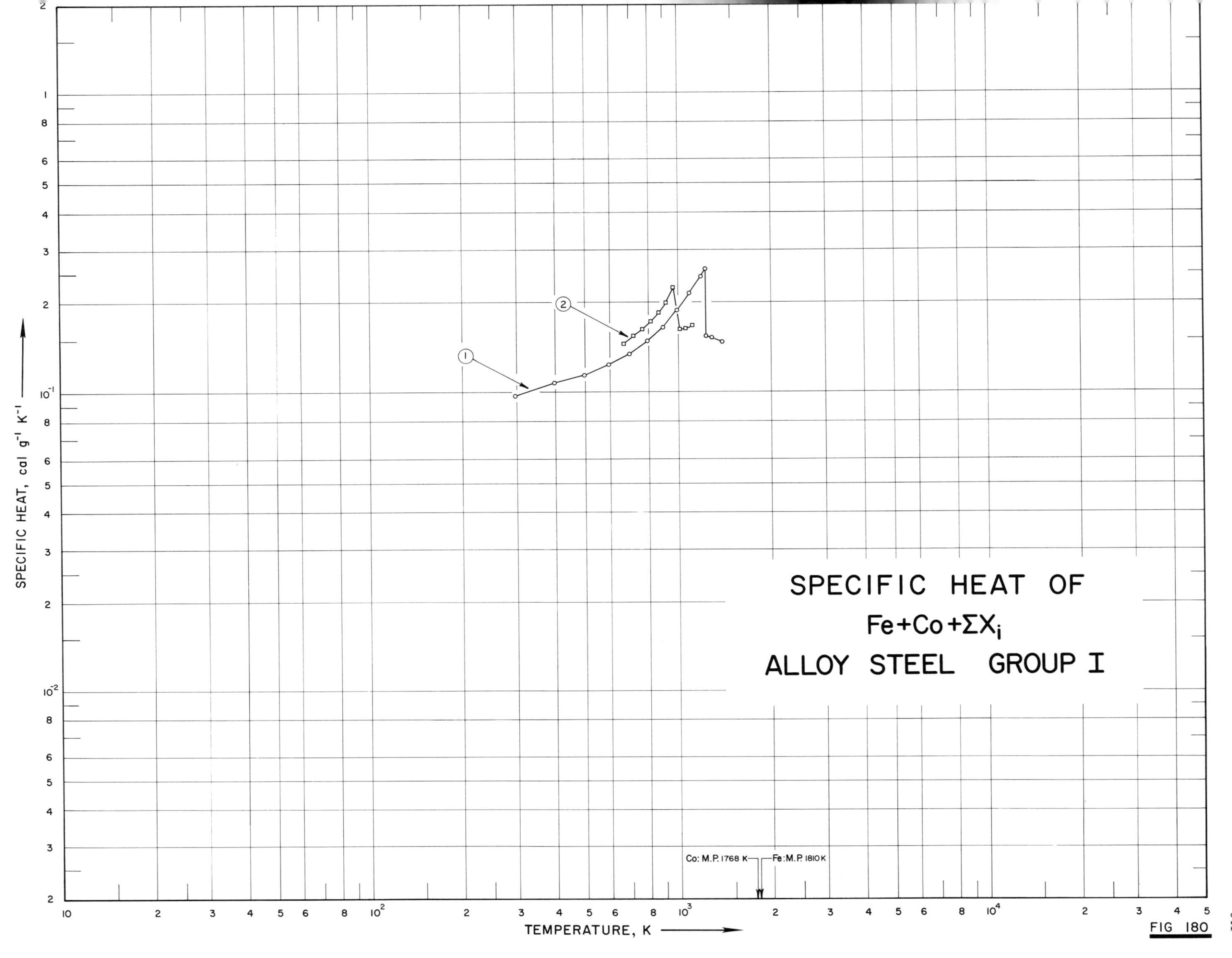
SPECIFIC HEAT OF
Fe+Co+ΣX_i
ALLOY STEEL GROUP I
SPECIFIC HEAT, cal g^-1 K^-1
TEMPERATURE, K
Co: M.P. 1768 K
Fe: M.P. 1810 K
1
2

FIG 180

SPECIFICATION TABLE NO. 180 SPECIFIC HEAT OF IRON + COBALT + ΣX_i, Fe + Co + ΣX_i, ALLOY STEEL GROUP I

[For Data Reported in Figure and Table No. 180]

Curve No.	Ref. No.	Year	Temp. Range, K	Reported Error, %	Name and Specimen Designation	Composition (weight percent), Specifications and Remarks
1	222	1959	299-1400	±0.5	Sample No. 32 Co	67.9 Fe and 32.1 Co; homogenized for 4 days at 1350 C under helium atmosphere; air cooled to room temperature.
2	54	1954	673-1123		Eutectoid steel	1.91 Co, 0.79 C, 0.22 Si, 0.12 Mn, 0.014 S, and 0.005 P; pearlitic; annealed at 900 C for 20 hrs; hammer-cogged to 1.75 in. square bullets from 1120 C; rolled to 1 in. rounds from 1040 C.

DATA TABLE NO. 180 SPECIFIC HEAT OF IRON + COBALT + ΣX_i, Fe + Co + ΣX_i, ALLOY STEEL GROUP I

[Temperature, T, K; Specific Heat, C_p, Cal $g^{-1}K^{-1}$]

T	C_p
CURVE 1	
298.15	9.824×10^{-2}
400	1.079×10^{-1}
500	1.158
600	1.252
700	1.361
800	1.495
900	1.667
1000	1.896
1100	2.165
1200	2.460
(α) 1243	2.593
(γ) 1248	1.555
1300	1.530
1400	1.484
CURVE 2	
673.15	1.470×10^{-1}
723.15	1.551
773.15	1.636
823.15	1.738
873.15	1.861
923.15	2.015
973.15	2.265
1023.15	1.624
1073.15	1.649
1123.15	1.675

*Not shown on plot

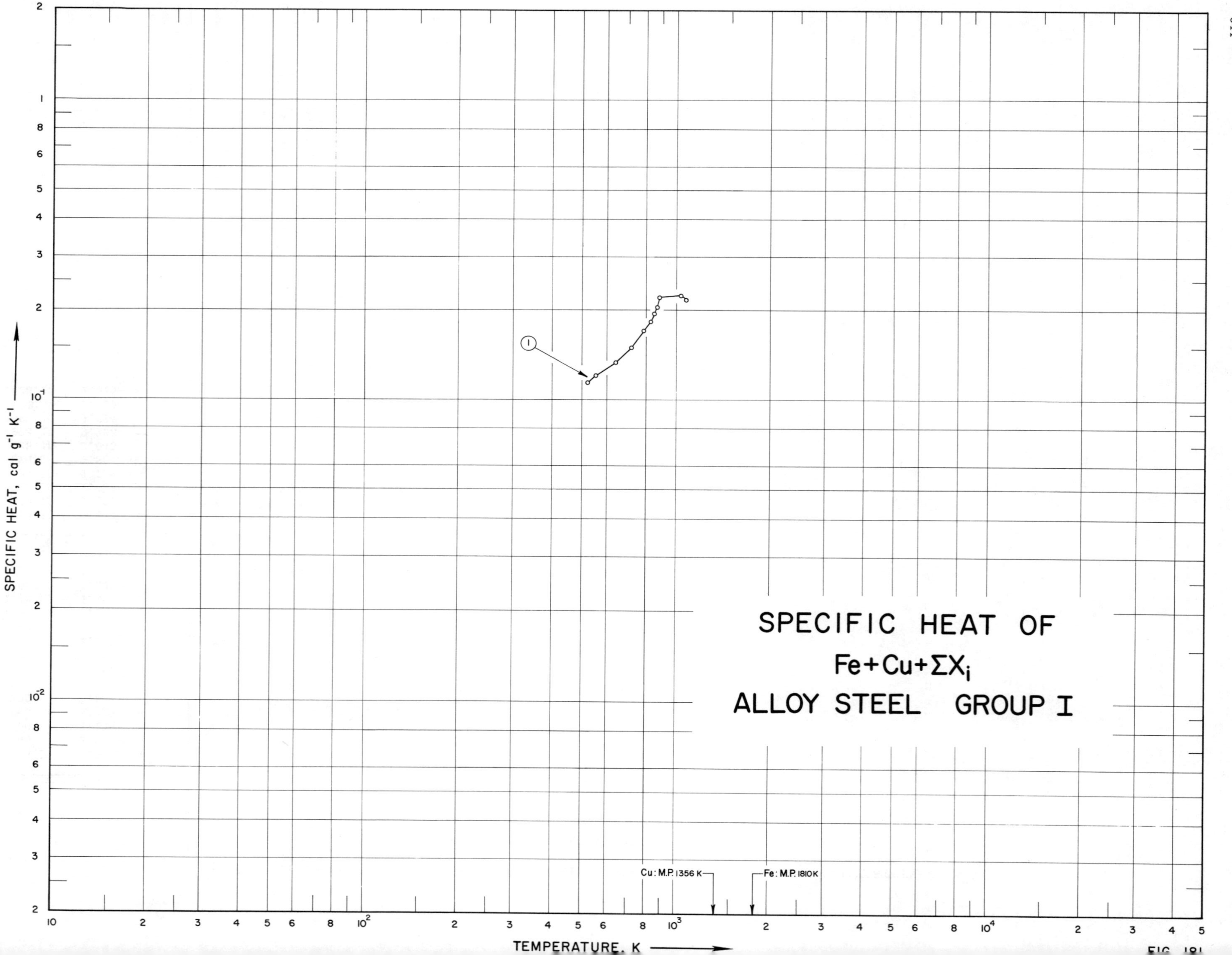
SPECIFIC HEAT OF
$Fe+Cu+\Sigma X_i$
ALLOY STEEL GROUP I
Cu: M.P. 1356 K
Fe: M.P. 1810K
1
TEMPERATURE, K
SPECIFIC HEAT, cal g^{-1} K^{-1}

SPECIFICATION TABLE NO. 181 SPECIFIC HEAT OF IRON + COPPER + ΣX_i, Fe + Cu + ΣX_i, ALLOY STEEL GROUP I

[For Data Reported in Figure and Table No. 181]

Curve No.	Ref. No.	Year	Temp. Range, K	Reported Error, %	Name and Specimen Designation	Composition (weight percent), Specifications and Remarks
1	427	1955	528-1095	≤5.0		~50.0 Cu.

DATA TABLE NO. 181 SPECIFIC HEAT OF IRON + COPPER + ΣX_i, Fe + Cu + ΣX_i, ALLOY STEEL GROUP I

[Temperature, T, K; Specific Heat, C_p, Cal $g^{-1}K^{-1}$]

T	C_p
CURVE 1	
528	1.160×10^{-1}
563	1.220
651	1.340
728	1.500
795	1.720
840	1.840
862	1.960
884	2.060
895	2.220
1051	2.240
1095	2.170

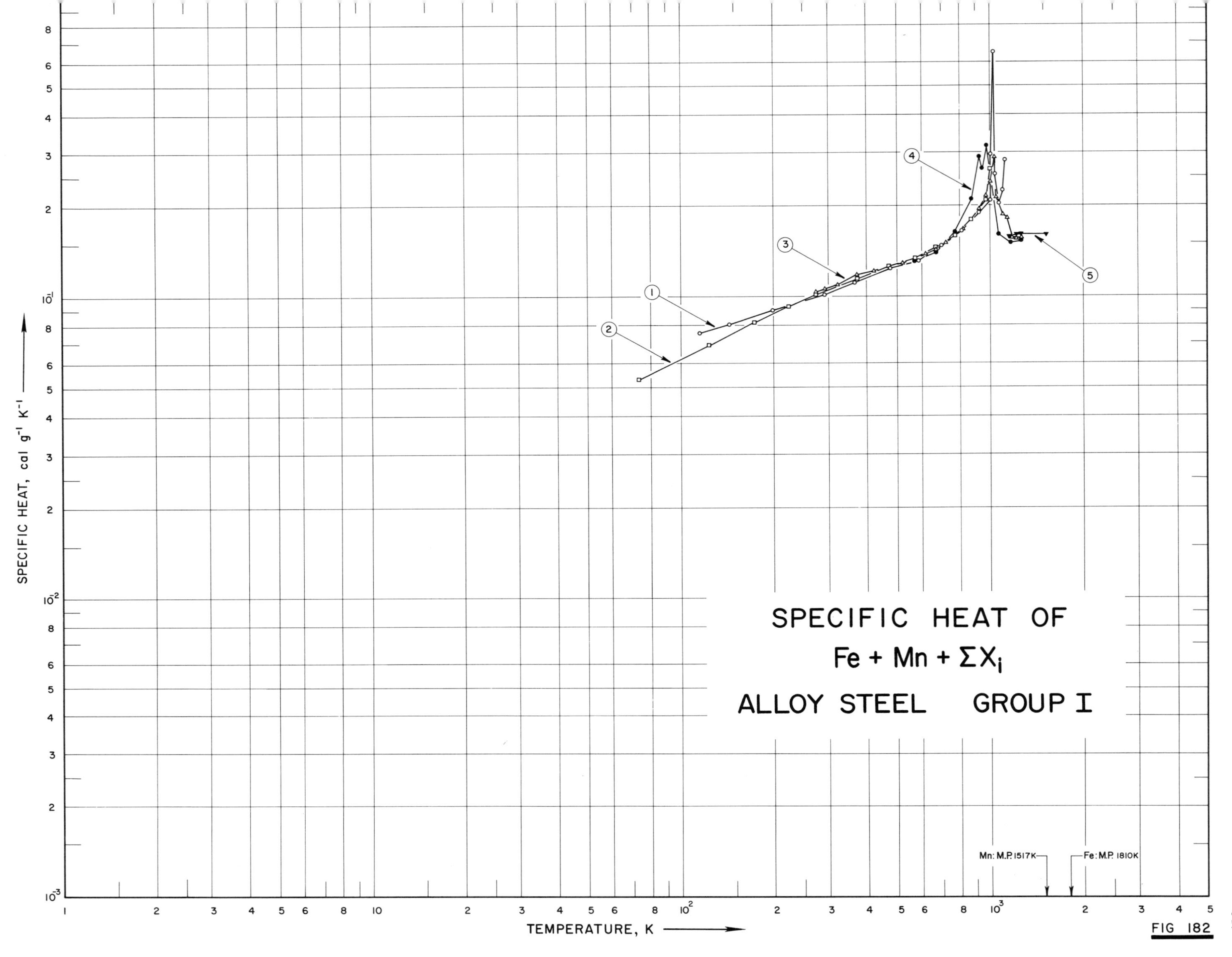
SPECIFIC HEAT OF
Fe + Mn + ΣX_i
ALLOY STEEL GROUP I
SPECIFIC HEAT, cal g^{-1} K^{-1}
TEMPERATURE, K
Mn: M.P. 1517K
Fe: M.P. 1810K
1
2
3
4
5

FIG 182

SPECIFICATION TABLE NO. 182 SPECIFIC HEAT OF IRON + MANGANESE + ΣX_i, Fe + Mn + ΣX_i, ALLOY STEEL GROUP I

[For Data Reported in Figure and Table No. 182]

Curve No.	Ref. No.	Year	Temp. Range, K	Reported Error, %	Name and Specimen Designation	Composition (weight percent), Specifications and Remarks
1	10	1958	116-1122		Hot rolled SAE 1010 steel	Nominal composition: 0.3-0.6 Mn, 0.08-0.13 C, ≤0.05 S, ≤0.04 P, and ≤0.01 Si; sample supplied by the U.S. Steel Corp; sealed in helium in a capsule; density (32 F) = 490 lb ft^{-3}.
2	243	1954	73-1123		Mild Steel SAE 1010	0.3-0.6 Mn, 0.08-0.13 C, <0.05 S and <0.04 P.
3	428	1957	273-1273		Mild Steel	0.61 Mn, 0.2 Si, 0.13 C, 0.12 Ni and 0.01 Cr; density = 489 lb ft^{-3}.
4	408	1940	573-1273			0.53 Mn, 0.15 C, 0.045 P, 0.038 S and 0.004 Si.
5	104	1946	1173-1523	4.0	Carbon Steel No. 1	0.38 Mn, 0.08 Cu, 0.06 C, 0.055 Ni, 0.039 As, 0.035 S, 0.03 Mo, 0.022 Cr, 0.017 P, 0.01 Si, and 0.001 Al; annealed at 930 C; density (15 C) = 491 lb ft^{-3}.
6	104	1946	1173-1573	4.0	Carbon Steel No. 2	0.31 Mn, 0.08 C, 0.08 Si, 0.07 Ni, 0.05 S, 0.045 Cr, 0.032 As, 0.029 P, 0.02 Mo, and 0.002 Al; annealed at 930 C; density (15 C) = 490 lb ft^{-3}.

DATA TABLE NO. 182 SPECIFIC HEAT OF IRON + MANGANESE + ΣX_i, Fe + Mn + ΣX_i, ALLOY STEEL GROUP I

[Temperature, T, K; Specific Heat, C_p, Cal $g^{-1}K^{-1}$]

T	C_p
CURVE 1	
116	7.600 x 10^{-2}
144	8.100
200	9.000
293	1.020 x 10^{-1}
366	1.120
478	1.240
589	1.320
700	1.480
811	1.660
922	1.900
1005	2.080
1005	2.960
1033	6.500
1044	2.550
1075	2.030
1103	2.260
1122	2.830
CURVE 2	
73	5.3 x 10^{-2}
123	6.9
173	8.2
223	9.3
273	1.02 x 10^{-1}
373	1.15
473	1.26
573	1.34
673	1.45
773	1.59
873	1.79
973	2.09
1003	2.63
1033	6.50*
1043	2.62*
1078	2.03*
1103	2.26*
1123	2.83*
CURVE 3	
273	1.039 x 10^{-1}
293	1.061
323	1.109
373	1.180
CURVE 3 (cont.)	
423	1.221 x 10^{-1}
473	1.260*
523	1.291
573	1.331*
623	1.381
673	1.429
723	1.501
773	1.582*
823	1.671
873	1.802*
923	1.953
973	2.151
1013	2.414
1036	2.916
1053	2.130
1073	1.991*
1113	1.871
1141	1.821
1179	1.570
1223	1.561
1273	1.570
CURVE 4	
573	1.32 x 10^{-1}
673	1.40
773	1.63
873	2.10
923	2.90
943	2.65
978	3.16
1073	1.61
1173	1.52
1273	1.53
CURVE 5	
1173	1.57 x 10^{-1}
1223	1.59
1273	1.60
1323	1.60*
1373	1.60*
1423	1.60*
1473	1.60*
1523	1.60
CURVE 6*	
1173	1.56 x 10^{-1}
1223	1.56
1273	1.57
1323	1.57
1373	1.58
1423	1.58
1473	1.59
1523	1.59
1573	1.59

* Not shown on plot

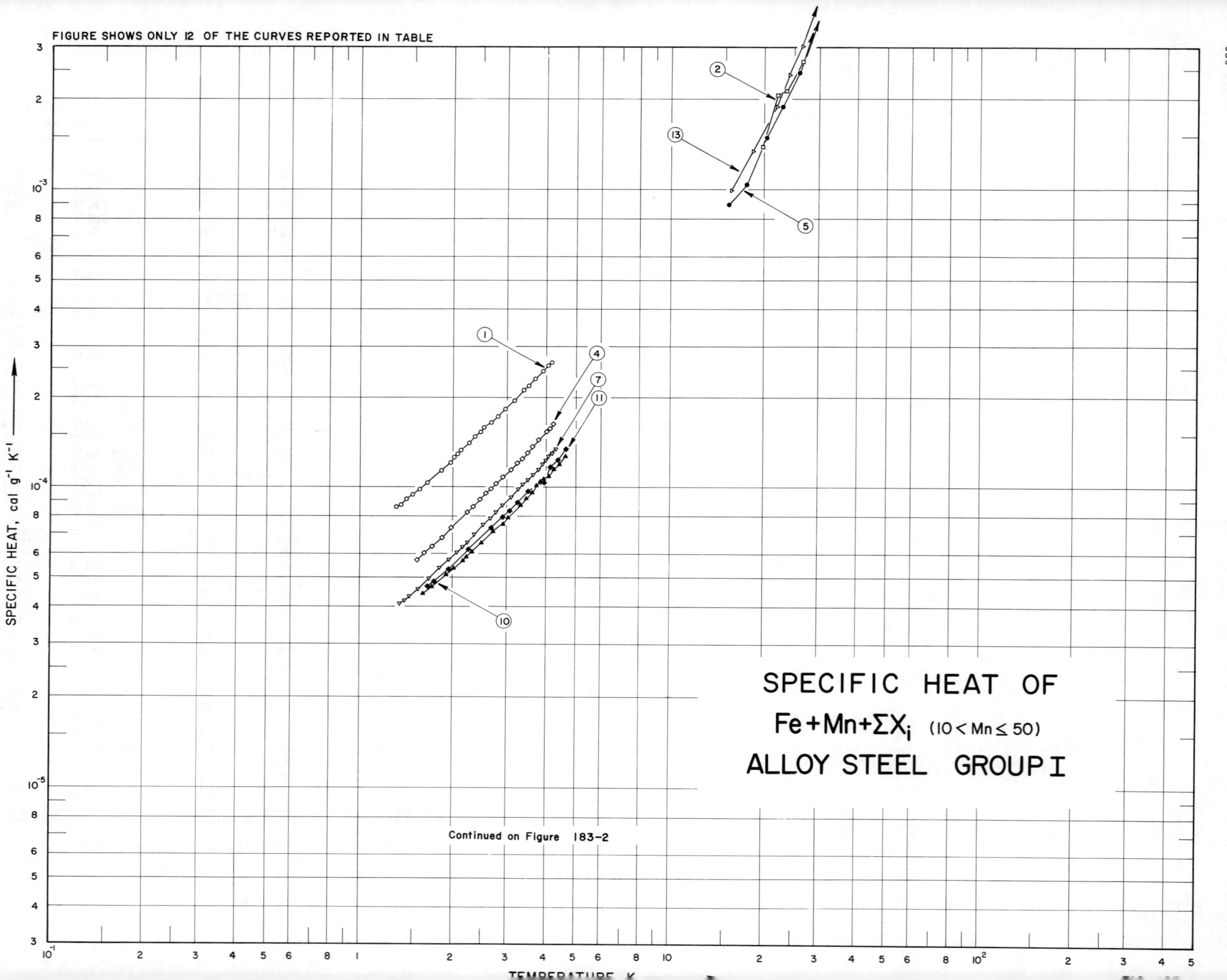
FIGURE SHOWS ONLY 12 OF THE CURVES REPORTED IN TABLE
SPECIFIC HEAT OF
Fe+Mn+ΣXi (10 < Mn ≤ 50)
ALLOY STEEL GROUP I
Continued on Figure 183-2
SPECIFIC HEAT, cal g⁻¹ K⁻¹
TEMPERATURE, K
1
2
4
5
7
10
11
13

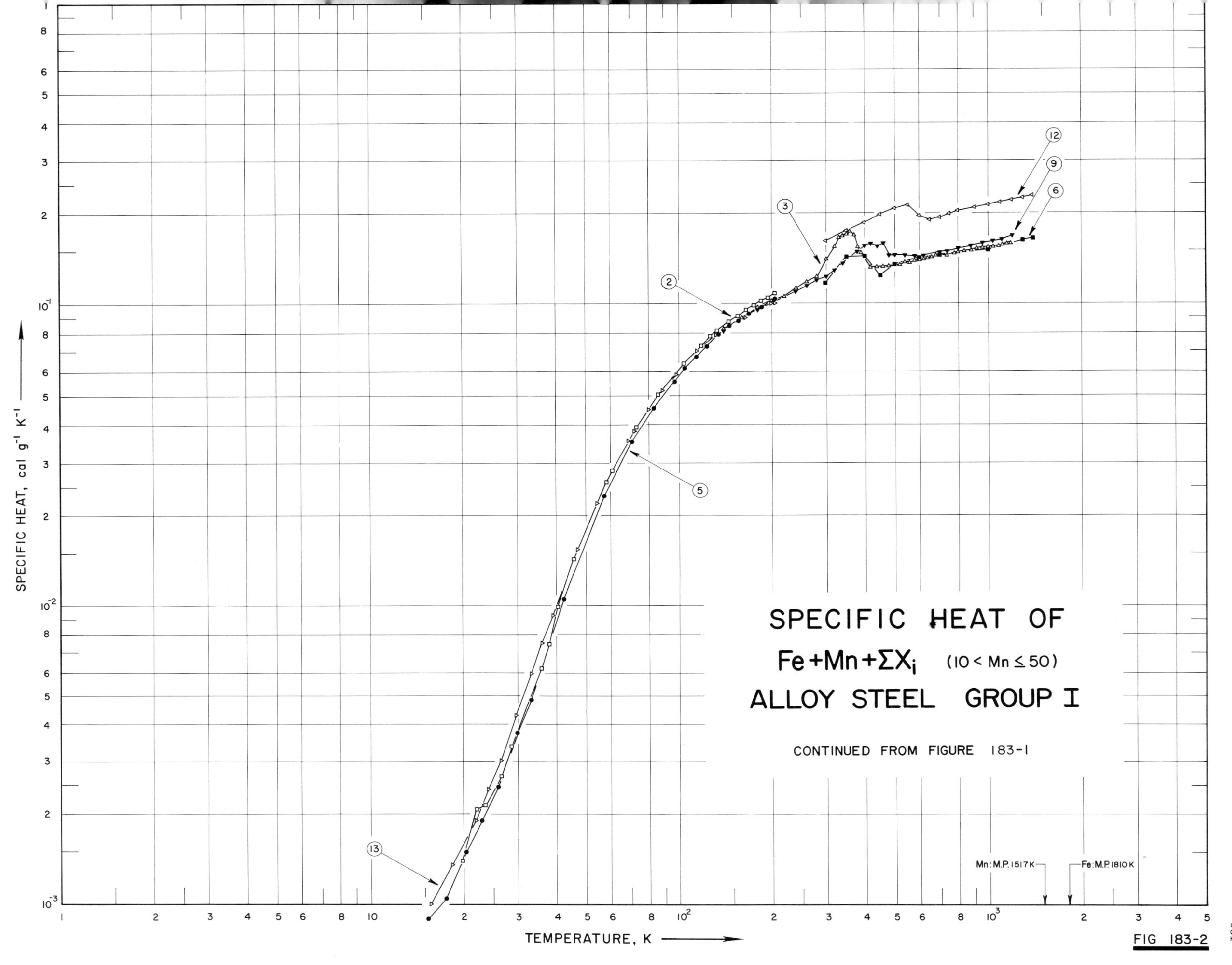
SPECIFIC HEAT OF
Fe+Mn+ΣXi (10 < Mn ≤ 50)
ALLOY STEEL GROUP I
CONTINUED FROM FIGURE 183-1
SPECIFIC HEAT, cal g⁻¹ K⁻¹
TEMPERATURE, K
Mn: M.P. 1517 K
Fe: M.P. 1810 K
2
3
5
6
9
12
13
FIG 183-2

SPECIFICATION TABLE NO. 183 SPECIFIC HEAT OF IRON + MANGANESE + ΣX_i, Fe + Mn + ΣX_i ($10 < Mn \leq 50$), ALLOY STEEL GROUP I

[For Data Reported in Figure and Table No. 183]

Curve No.	Ref. No.	Year	Temp. Range, K	Reported Error, %	Name and Specimen Designation	Composition (weight percent), Specifications and Remarks
1	349	1962	1.3-4.1	≤2.0		83.7 Fe, 11.0 Mn, and 5.3 C; annealed at 1100 C for 72 hrs in He + 8% H_2 gas atmosphere; etched in 1-3% HNO_3.
2	55	1930	20-205			80.6 Fe and 19.4 Mn.
3	310	1962	140-1180			79.005 Fe, 20.55 Mn, 0.39 Si, and 0.055 C.
4	349	1962	1.5-4.2	≤2.0		76.4 Fe, 20.0 Mn, and 3.6 C; annealed at 1100 C for 72 hrs, under He + 8% H_2 gas atmosphere; etched in 1-3% HNO_3.
5	55	1930	15-216	1.5		70 Fe and 30 Mn.
6	222	1959	298-1400	0.5	Sample No. 30 Mn	70.0 Fe and 30.0 Mn; homogenized for 4 days at 1350 C under helium atmosphere; air cooled to room temperature.
7	349	1962	1.3-4.3	≤2.0		66.4 Fe, 30.0 Mn, and 3.6 C; annealed at 1100 C for 72 hrs, under He + 8% H_2 gas atmosphere; etched in 1-3% HNO_3.
8	310	1962	140-1140			65.935 Fe, 32.8 Mn, 1.21 Si, and 0.055 C.
9	310	1962	140-1240			59.165 Fe, 40.4 Mn, 0.41 Si, and 0.025 C.
10	297	1959	1.6-4.6			55.4 Fe and 44.6 Mn; induction melted.
11	297	1959	1.7-4.6			55.0 Fe, 44.1 Mn and 0.9 C; induction melted.
12	222	1959	298-1400	0.5	Sample No. 50 Mn	51.6 Fe and 48.4 Mn; homogenized for 4 days at 1350 C under helium atmosphere; air cooled to room temperature.
13	55	1930	16-205	1.5		50.0 Fe and 50.0 Mn.
14	388	1956	2.2-4.1	7.0		90.0 Fe and 10.0 Mn.

DATA TABLE NO. 183 SPECIFIC HEAT OF IRON + MANGANESE + ΣX_i, Fe + Mn + ΣX_i ($10 < Mn \le 50$), ALLOY STEEL GROUP I

[Temperature, T, K; Specific Heat, C_p, Cal $g^{-1}K^{-1}$]

T	C_p
CURVE 1	
1.315	8.682 x 10^{-5}
1.359	8.816
1.416	9.245
1.476	9.500
1.552	9.902
1.648	1.052 x 10^{-4}
1.825	1.157
1.957	1.227
2.016	1.274
2.059	1.302
2.111	1.341
2.168	1.367*
2.246	1.423
2.349	1.494
2.449	1.555
2.513	1.600
2.572	1.614*
2.651	1.672
2.786	1.756
2.942	1.857
3.131	1.973
3.366	2.141
3.498	2.218
3.672	2.338
3.880	2.479
4.032	2.583
4.138	2.666
CURVE 2	
19.95	1.401 x 10^{-3}
22.00	2.078
23.50	2.147
26.50	2.695
28.63	3.388
35.80	6.240
37.90	7.467
40.55	9.892
45.70	1.438 x 10^{-2}
58.30	2.592
60.80	2.839
73.00	3.954
85.90	5.054
104.47	6.423
118.63	7.348
CURVE 2 (cont.)	
127.47	7.857 x 10^{-2}
133.88	8.193
146.25	8.785
155.54	9.177
166.41	9.599
176.25	9.939
186.22	1.025 x 10^{-1}
195.84	1.058
205.40	1.086
CURVE 3	
140	8.309 x 10^{-2}
160	9.308
180	9.766
200	1.032 x 10^{-1}
220	1.062
240	1.135
260	1.190
280	1.234
300	1.413
320	1.568
330	1.670
340	1.688
350	1.718
360	1.743
370	1.692
380	1.551
390	1.482
400	1.435*
420	1.324
440	1.328
460	1.332
480	1.332
500	1.349*
520	1.358
540	1.379
560	1.375
580	1.392
600	1.396
620	1.418
640	1.422
660	1.435
680	1.452
700	1.452
CURVE 3 (cont.)	
720	1.448 x 10^{-1}*
740	1.452
760	1.456*
780	1.469
800	1.486
820	1.495*
840	1.503
860	1.508*
880	1.508*
900	1.521*
920	1.525
940	1.525*
960	1.542
980	1.546
1000	1.546
1020	1.546
1040	1.551
1060	1.555*
1080	1.563
1100	1.568*
1120	1.576
1140	1.576*
1160	1.589
1180	1.589
CURVE 4	
1.521	5.769 x 10^{-5}
1.604	6.094
1.711	6.414
1.837	6.872
1.969	7.375
2.216	8.348
2.320	8.695
2.439	9.193
2.540	9.607
2.633	9.965
2.738	1.031 x 10^{-4}
2.882	1.091
3.068	1.160
3.219	1.223
3.331	1.266
3.451	1.320
3.596	1.377
3.756	1.460
CURVE 4 (cont.)	
3.987	1.553 x 10^{-4}
4.093	1.597
4.199	1.653
CURVE 5	
15.32	9.012 x 10^{-4}
17.66	1.055 x 10^{-3}
20.33	1.509
22.97	1.902
25.80	2.476
29.97	3.766
33.22	4.869
42.45	1.059 x 10^{-2}
57.57	2.327
70.80	3.543
83.00	4.560
97.00	5.582
105.25	6.172
114.50	6.759
124.90	7.345
135.60	7.954
147.07	8.486
157.33	8.857
169.57	9.322
176.12	9.498*
186.94	9.786
196.61	1.004 x 10^{-1}*
215.94	1.049
CURVE 6	
298.15	1.175 x 10^{-1}
350	1.439
400	1.447
450	1.241
500	1.354
600	1.421
700	1.450
800	1.479*
900	1.508*
1000	1.537
1100	1.566*
1200	1.595*
1300	1.624
1400	1.653
CURVE 7	
1.347	4.121 x 10^{-5}
1.389	4.230
1.447	4.365
1.533	4.608
1.670	4.993
1.817	5.439
1.935	5.776
2.054	6.147
2.140	6.380
2.219	6.613
2.342	7.047
2.493	7.532
2.625	7.945
2.749	8.328
2.887	8.765
3.070	9.343
3.231	9.871
3.348	1.027 x 10^{-4}
3.471	1.070
3.602	1.116
3.751	1.165
3.865	1.211
3.950	1.240
4.047	1.277
4.159	1.315
4.278	1.361
CURVE 8*	
140	8.224 x 10^{-2}
160	8.995
180	9.637
200	1.019 x 10^{-1}
220	1.079
240	1.139
260	1.191
280	1.229
300	1.264
320	1.306
340	1.311
360	1.328
380	1.328
400	1.332
420	1.341
440	1.345
460	1.354
CURVE 8 (cont.)*	
480	1.371 x 10^{-1}
500	1.388
520	1.388
540	1.392
560	1.388
580	1.396
600	1.396
620	1.396
640	1.405
660	1.405
680	1.422
700	1.435
720	1.452
740	1.452
760	1.456
780	1.461
800	1.469
820	1.482
840	1.499
860	1.516
880	1.516
900	1.512
920	1.525
940	1.525
960	1.533
980	1.533
1000	1.538
1020	1.542
1040	1.542
1060	1.559
1080	1.559
1100	1.568
1120	1.593
1140	1.606
CURVE 9	
140	8.181 x 10^{-2}
160	8.823*
180	9.551
200	1.024 x 10^{-1}*
220	1.062*
240	1.105
260	1.156
280	1.204
300	1.234

* Not shown on plot

DATA TABLE NO. 183 (continued)

CURVE 9 (cont.)

T	C_p
320	1.298 x 10^{-1}
340	1.371
360	1.443*
380	1.491
400	1.563
420	1.589
440	1.563
460	1.589
480	1.456
500	1.456
520	1.465*
540	1.456
560	1.448*
580	1.443
600	1.439*
620	1.439
640	1.443*
660	1.448*
680	1.461*
700	1.478
720	1.482*
740	1.491
760	1.486*
780	1.499*
800	1.525
820	1.529*
840	1.546*
860	1.555*
880	1.563
900	1.559*
920	1.568*
940	1.589*
960	1.593
980	1.581*
1000	1.602*
1020	1.619*
1040	1.619
1060	1.628*
1080	1.623*
1100	1.632*
1120	1.636
1140	1.645*
1160	1.658*
1180	1.666*
1200	1.679
1220	1.705*
1240	1.739*

CURVE 10

T	C_p
1.598	4.457 x 10^{-5}
1.710	4.717
1.891	5.154
2.011	5.446
2.038	5.437*
2.145	5.765
2.215	5.949
2.298	6.186
2.470	6.608
2.499	6.685*
2.680	7.179
2.884	7.638
2.993	7.997
3.294	8.817
3.434	9.317
3.587	9.719
3.686	1.027 x 10^{-4}
3.800	1.069
3.918	1.079
4.061	1.113
4.224	1.173
4.404	1.218
4.603	1.288

CURVE 11

T	C_p
1.660	4.732 x 10^{-5}
1.735	4.891
1.939	5.375
1.997	5.409*
2.246	6.255
2.656	7.357
2.888	8.000
3.039	8.420
3.232	8.953
3.498	9.790
3.649	1.020 x 10^{-4}*
3.913	1.111
3.930	1.106*
3.989	1.127*
4.106	1.178
4.173	1.184*
4.297	1.214*
4.339	1.255
4.364	1.237*
4.631	1.361

CURVE 12

T	C_p
298.15	1.625 x 10^{-1}
350	1.753
400	1.872
450	1.983
500	2.089
550	2.133
600	1.970
650	1.906
700	1.946
750	1.988
800	2.057
900	2.099
1000	2.141
1100	2.185
1200	2.227
1300	2.269
1400	2.311

CURVE 13

T	C_p
15.65	1.016 x 10^{-3}
18.37	1.356
21.85	1.919
24.13	2.438
26.30	3.032
29.63	4.303
33.20	5.999
35.90	7.535
38.80	9.328
46.93	1.551 x 10^{-2}
54.20	2.205
68.95	3.561
71.90	3.833
79.75	4.502
88.65	5.218
98.05	5.899
106.25	6.437*
115.80	7.033
123.17	7.461*
132.66	7.941
141.90	8.229*
157.57	8.754*
163.94	9.029
173.13	9.289*
183.41	9.570*
190.52	9.781*
198.04	1.002 x 10^{-1}
205.46	1.019

CURVE 14

T	C_p
2.244	1.657 x 10^{-4}
2.305	1.712*
2.364	1.745
2.418	1.782*
2.472	1.806
2.525	1.864*
2.597	1.890
2.710	1.991
2.813	2.039
2.930	2.140
3.031	2.222
3.119	2.294*
3.197	2.337
3.285	2.402*
3.390	2.482
3.508	2.563*
3.705	2.717
3.850	2.751*
3.983	2.929
4.101	3.010

* Not shown on plot

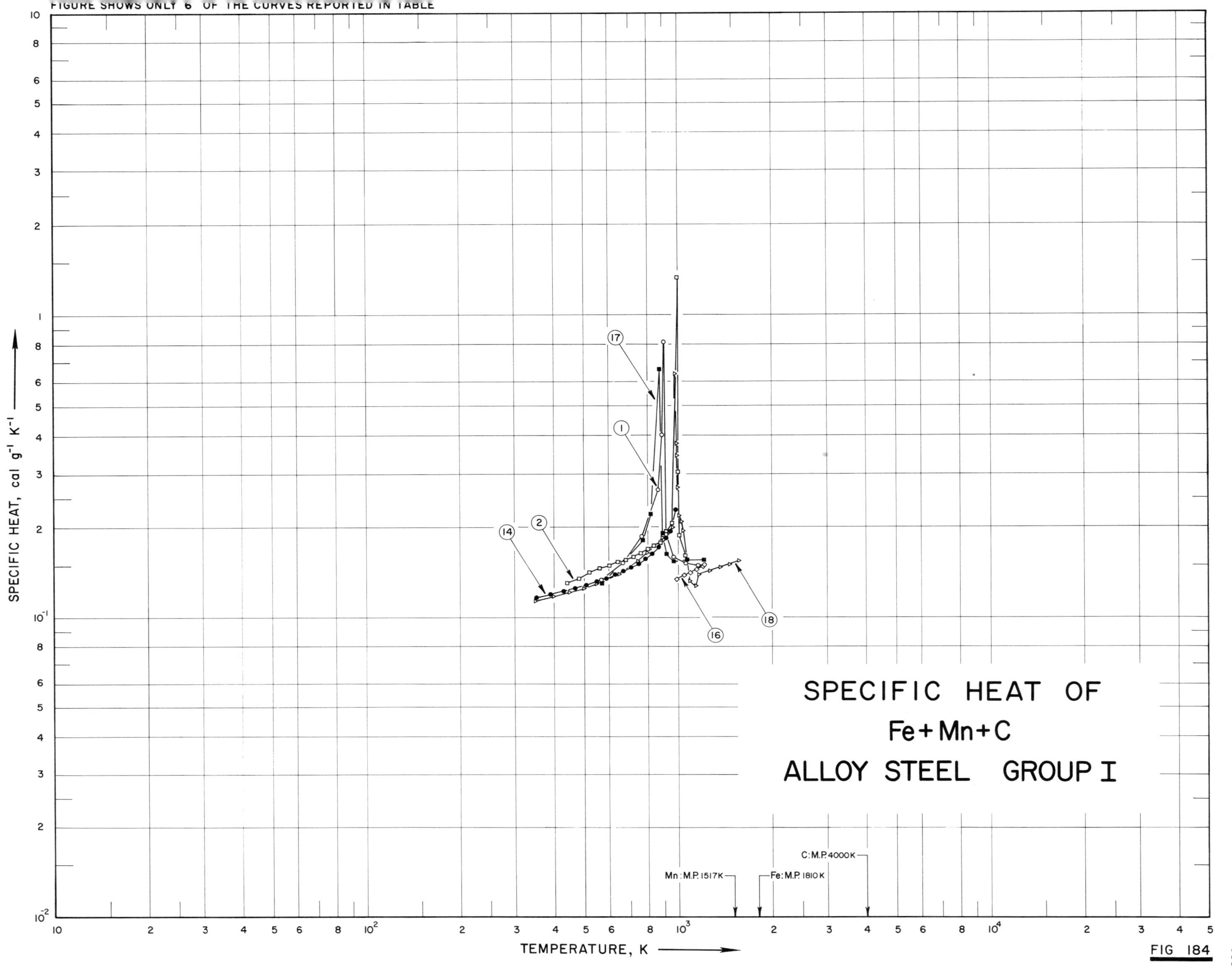
FIGURE SHOWS ONLY 6 OF THE CURVES REPORTED IN TABLE
SPECIFIC HEAT OF
Fe+Mn+C
ALLOY STEEL GROUP I
SPECIFIC HEAT, cal g⁻¹ K⁻¹
TEMPERATURE, K
Mn: M.P. 1517 K
Fe: M.P. 1810 K
C: M.P. 4000 K
17
1
2
14
16
18

FIG 184

SPECIFICATION TABLE NO. 184 SPECIFIC HEAT OF IRON + MANGANESE + CARBON, Fe + Mn + C, ALLOY STEEL GROUP I

[For Data Reported in Figure and Table No. 184]

Curve No.	Ref. No.	Year	Temp. Range, K	Reported Error, %	Name and Specimen Designation	Composition (weight percent), Specifications and Remarks
1	408	1940	573-1223			0.72 Mn, 0.5 C, 0.30 Si, 0.035 P, and 0.03 S.
2	83	1954	443-1063		Steel T-261	1.0 Mn and 0.75 C.
3	83	1954	343-1063		Steel T-262	Same as above.
4	83	1954	343-1063		Steel T-310	Same as above.
5	83	1954	363-1063		Steel T-311	Same as above.
6	83	1954	363-1063		Steel T-270	Same as above.
7	83	1954	343-1063		Steel T-278	Same as above.
8	83	1954	403-1083		Steel T-279	Same as above.
9	83	1954	443-1063		Steel Pearlite	Same as above.
10	83	1954	428-1123		Steel Pearlite	Same as above.
11	83	1954	363-1123		Steel Pearlite	Same as above.
12	83	1954	1008-1083		Steel Austenite	Same as above.
13	83	1954	1023-1123		Steel Austenite	Same as above.
14	33	1957	353-993	≤0.9	Steel B	97.969 Fe, 1.03 Mn, 0.97 C, 0.024 S, $<$0.005 Si, and $<$0.002 P; free cooled.
15	33	1957	353-993	≤0.9	Steel B	Same as above; slow cooled.
16	33	1957	993-1218	$<$0.9	Austenite	
17	408	1940	573-1223			1.04 Mn, 0.33 C, 0.11 P, 0.1 Si, and 0.05 S.
18	104	1946	348-1148	2.0	Alloy Steel No. 4	1.51 Mn, 0.23 C, 0.12 Si, 0.105 Cu, 0.06 Cr, 0.038 B, 0.037 P, 0.04 Ni, 0.033 As, 0.025 Mo, and 0.015 Al; annealed at 860 C; density (15 C) = 489 lb ft^{-3}.
19	54	1954	673-1123		Eutectoid Steel	0.25-1.85 Mn, 0.79-0.80 C, 0.22 Si, 0.011-0.02 P, and 0.011-0.016 S; pearlite.

DATA TABLE NO. 184 SPECIFIC HEAT OF IRON + MANGANESE + CARBON, Fe + Mn + C, ALLOY STEEL GROUP I

[Temperature, T,K; Specific Heat, C_p, Cal $g^{-1}K^{-1}$]

T	C_p
CURVE 1	
573	1.34 x 10^{-1}
673	1.53
773	1.88
873	2.68
898	4.05
913	8.22
923	1.90
973	1.60
1073	1.53
1173	1.50
1223	1.51
CURVE 2	
443	1.3134 x 10^{-1}
463	1.3299*
483	1.3587
503	1.3914*
523	1.4249
543	1.4499*
563	1.4731
583	1.4882*
603	1.5060
623	1.5220*
643	1.5403
663	1.5518*
683	1.5693
703	1.5852*
723	1.6070
743	1.6351*
763	1.6652
783	1.6809*
803	1.7048
823	1.7186*
843	1.7449
863	1.7731*
883	1.7982
903	1.8559*
923	1.9340
943	1.9895*
963	2.0742
983	2.1815*
1003	1.34745 x 10^{0}
1008	3.0630 x 10^{-1}
1023	1.8924
CURVE 2 (cont.)	
1043	1.6232 x 10^{-1}*
1063	1.6222
CURVE 3*	
343	1.1277 x 10^{-1}
363	1.1455
383	1.1946
403	1.2342
423	1.2752
443	1.3207
463	1.3443
483	1.3725
503	1.4031
523	1.4321
543	1.4555
563	1.4781
583	1.4906
603	1.5113
623	1.5274
643	1.5434
663	1.5574
683	1.5752
703	1.5945
723	1.6195
743	1.6417
763	1.6781
783	1.6955
803	1.7255
823	1.7361
843	1.7631
863	1.7943
883	1.8180
903	1.8594
923	1.9157
943	1.9936
963	2.0867
983	2.2153
1003	1.38031 x 10^{0}
1008	3.0209 x 10^{-1}
1023	1.7761
1043	1.5787
1063	1.5805
CURVE 4*	
343	1.1277 x 10^{-1}
363	1.1588
383	1.2197
403	1.2636
423	1.2925
443	1.3268
463	1.3596
483	1.3921
503	1.4149
523	1.4256
543	1.4520
563	1.4738
583	1.4946
603	1.5118
623	1.5290
643	1.5495
663	1.5622
683	1.5770
703	1.5953
723	1.6184
743	1.6438
763	1.6713
783	1.6881
803	1.7183
823	1.7364
843	1.7680
863	1.8034
883	1.8280
903	1.8696
923	1.9095
943	1.9832
963	2.0713
983	2.2009
1003	1.36609 x 10^{0}
1008	3.1225 x 10^{-1}
1023	1.9170
1043	1.6099
1063	1.6013
CURVE 5*	
363	1.1449 x 10^{-1}
383	1.1951
403	1.2355
423	1.2727
CURVE 5 (cont.)*	
443	1.3130 x 10^{-1}
463	1.3365
483	1.3713
503	1.3988
523	1.4257
543	1.4513
563	1.4727
583	1.4886
603	1.5077
623	1.5264
643	1.5429
663	1.5547
683	1.5724
703	1.5903
723	1.6146
743	1.6250
763	1.6667
783	1.6829
803	1.7138
823	1.7328
843	1.7628
863	1.7934
883	1.8199
903	1.8519
923	1.9031
943	1.9727
963	2.0655
983	2.1899
1003	1.37727 x 10^{0}
1008	3.0719 x 10^{-1}
1023	1.5521
1043	1.5872
1063	1.5889
CURVE 6*	
363	1.1450 x 10^{-1}
383	1.1926
403	1.2415
423	1.2917
443	1.3405
463	1.3632
483	1.3845
503	1.4081
523	1.4347
CURVE 6 (cont.)*	
663	1.5488 x 10^{-1}
683	1.5679
703	1.5858
723	1.6046
743	1.6250
763	1.6589
783	1.6805
803	1.7133
823	1.7439
843	1.7704
863	1.8044
883	1.8299
903	1.8629
923	1.9035
943	1.0567
963	2.0401
983	2.1647
1003	1.35617 x 10^{0}
1008	3.0061 x 10^{-1}
1023	1.7571
1043	1.5942
1063	1.5877
CURVE 7*	
343	1.1372 x 10^{-1}
363	1.1711
383	1.2298
403	1.2604
423	1.2867
443	1.3233
463	1.3521
483	1.3787
503	1.4119
523	1.4430
543	1.4599
563	1.4697
583	1.5012
603	1.5299
623	1.5494
643	1.5735
663	1.5901
683	1.6139
703	1.6303
723	1.6552
CURVE 7 (cont.)*	
743	1.6742 x 10^{-1}
763	1.7025
783	1.7273
803	1.7518
823	1.7803
843	1.8211
863	1.8648
883	1.9043
903	1.9439
923	2.0014
943	2.0611
963	2.1369
983	2.2466
1003	1.32727 x 10^{0}
1008	3.2741 x 10^{-1}
1023	1.9467
1043	1.6734
1063	1.6623
CURVE 8*	
403	1.2414 x 10^{-1}
423	1.2663
443	1.2965
463	1.3299
483	1.3538
503	1.3887
523	1.4213
543	1.4447
563	1.4704
583	1.4882
603	1.5060
623	1.5207
643	1.5406
663	1.5563
683	1.5697
703	1.5916
723	1.6099
743	1.6386
763	1.6705
783	1.6858
803	1.7181
823	1.7429
843	1.7823
863	1.8219

* Not shown on plot

DATA TABLE NO. 184 (continued)

CURVE 8 (cont.)*

T	C_p
883	1.8560 x 10^{-1}
903	1.8994
923	1.9535
943	2.0164
963	2.0922
983	2.2139
1003	1.35526 x 10^{0}
1008	3.1856 x 10^{-1}
1023	1.8675
1043	1.6115
1063	1.5856
1083	1.5865

CURVE 9*

T	C_p
443	1.3028 x 10^{-1}
463	1.3119
483	1.3416
503	1.3735
523	1.4065
543	1.4338
563	1.4618
583	1.4737
603	1.4972
623	1.5147
643	1.5343
663	1.5545
683	1.5724
703	1.5964
723	1.6024
743	1.6231
763	1.6505
783	1.6666
803	1.6953
823	1.7215
843	1.7547
863	1.8089
883	1.8444
903	1.8918
923	1.9484
943	2.0146
963	2.1004
1043	1.6163
1063	1.6011

CURVE 10*

T	C_p
428	1.2948 x 10^{-1}
438	1.3083
448	1.3291
458	1.3262
468	1.3388
478	1.3387
503	1.3919
523	1.4167
543	1.4348
563	1.4529
583	1.4698
603	1.4913
623	1.5065
643	1.5256
663	1.5407
683	1.5670
703	1.5861
723	1.6082
743	1.6307
763	1.6614
783	1.6871
803	1.7177
823	1.7451
843	1.7853
863	1.8300
883	1.8635
903	1.9123
943	2.0347
963	2.0958
983	2.2257
1043	1.5973
1063	1.5713
1083	1.5713
1103	1.6574
1123	1.6377

CURVE 11*

T	C_p
363	1.2288 x 10^{-1}
383	1.2044
403	1.2544
423	1.2833
438	1.2998
448	1.3187

CURVE 11 (cont.)*

T	C_p
458	1.3233 x 10^{-1}
468	1.3376
483	1.3579
503	1.3808
523	1.4094
543	1.4298
563	1.4526
583	1.4684
603	1.4901
623	1.5106
643	1.5311
663	1.5411
683	1.5616
703	1.5797
723	1.6007
743	1.6258
763	1.6554
783	1.6762
803	1.7117
823	1.7387
843	1.7790
863	1.8143
883	1.8528
903	1.9034
923	1.9682
943	2.0392
963	2.1217
1043	1.6097
1063	1.5969
1083	1.5969
1103	1.6210
1123	1.6238

CURVE 12*

T	C_p
1008	1.5641 x 10^{-1}
1023	1.5680
1043	1.5797
1063	1.5822
1083	1.5830

CURVE 13*

T	C_p
1023	1.5733 x 10^{-1}
1043	1.5883
1063	1.5892
1083	1.5831
1103	1.5921
1123	1.6043

CURVE 14

T	C_p
353	1.175 x 10^{-1}
373	1.192*
393	1.205
413	1.226*
433	1.239
453	1.256*
473	1.271
493	1.287*
513	1.296
533	1.318*
553	1.338
573	1.357*
593	1.370
613	1.388*
633	1.405
653	1.426*
673	1.448
693	1.458*
713	1.484
733	1.506*
753	1.520
773	1.556*
783	1.578*
793	1.588
803	1.603*
813	1.622*
823	1.645*
833	1.657
843	1.676*
853	1.689*
863	1.724*
873	1.735
883	1.743*
893	1.775*
903	1.798*

CURVE 14 (cont.)

T	C_p
913	1.823 x 10^{-1}*
923	1.850
933	1.909*
943	1.927*
953	1.950
963	1.992*
973	2.041*
983	2.143*
993	2.303

CURVE 15*

T	C_p
353	1.165 x 10^{-1}
373	1.185
393	1.202
413	1.220
433	1.238
453	1.241
473	1.263
493	1.285
513	1.296
533	1.318
553	1.337
573	1.356
593	1.368
613	1.388
633	1.406
653	1.426
673	1.442
693	1.454
713	1.479
733	1.506
753	1.528
773	1.556
783	1.573
793	1.587
803	1.603
813	1.623
823	1.640
833	1.658
843	1.678
853	1.697
863	1.720
873	1.741

CURVE 15 (cont.)*

T	C_p
883	1.767 x 10^{-1}
893	1.793
903	1.833
913	1.872
923	1.885
933	1.925
943	1.957
953	2.011
963	2.021
973	2.105
983	2.183
993	2.294

CURVE 16

T	C_p
993	1.366 x 10^{-1}
1003	1.371*
1013	1.376*
1023	1.382*
1033	1.387*
1038	1.390*
1043	1.393*
1053	1.398
1063	1.403*
1073	1.409*
1083	1.415*
1093	1.420
1103	1.426*
1113	1.431*
1123	1.437*
1133	1.444*
1143	1.449*
1153	1.455
1163	1.459*
1173	1.467*
1198	1.486*
1208	1.488*
1218	1.496

* Not shown on plot

DATA TABLE NO. 184 (continued)

T	C_p
CURVE 17	
573	1.30 x 10^{-1}
673	1.43*
773	1.82
823	2.22
883	6.63
898	1.92
923	1.64
973	1.55
1073	1.57
1173	1.57*
1223	1.57
CURVE 18	
Series I	
348	1.14 x 10^{-1}
398	1.18
448	1.22
498	1.26
548	1.30
598	1.35*
648	1.41
698	1.47*
748	1.55
798	1.66
848	1.77*
898	1.86
948	2.00*
998	3.46
1048	1.96
1098	1.33
1148	1.28
Series II	
928	1.9 x 10^{-1}*
938	1.9*
948	1.9*
958	2.0*
968	2.0
978	2.2*
988	6.4
998	3.8
1008	2.7
1018	2.2
1028	2.1
1038	2.1*

T	C_p
CURVE 18 (cont.)	
1048	2.0 x 10^{-1}*
1058	1.8*
1068	1.6*
Series III	
1173	1.40 x 10^{-1}
1223	1.42*
1273	1.44
1323	1.46*
1373	1.48
1423	1.50*
1473	1.51
1523	1.53*
1573	1.55
CURVE 19*	
673	1.470 x 10^{-1}
723	1.551
773	1.636
823	1.738
873	1.861
923	2.015
973	2.265
1023	1.624
1073	1.649
1123	1.675

* Not shown on plot

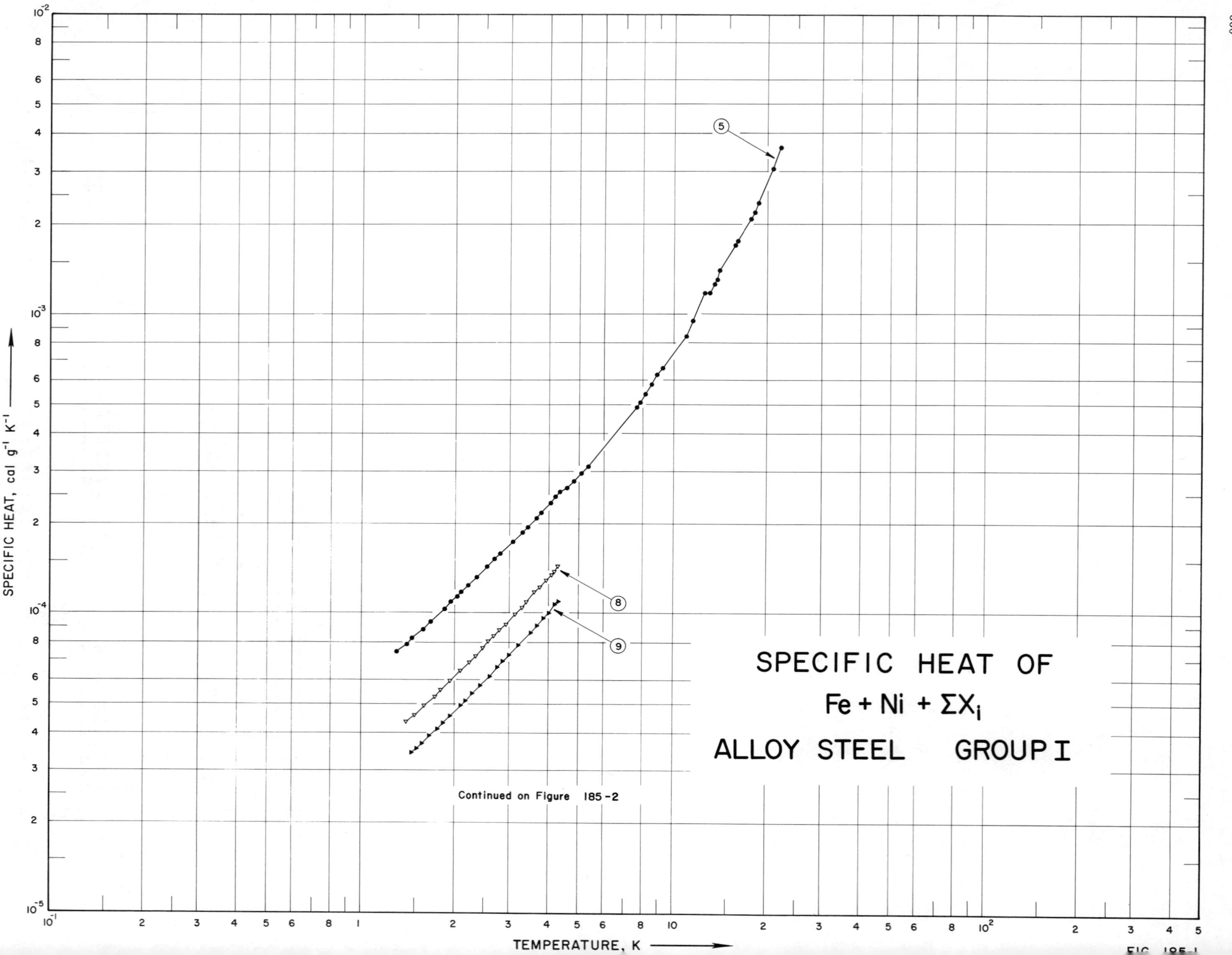
SPECIFIC HEAT OF
Fe + Ni + ΣX$_i$
ALLOY STEEL GROUP I
Continued on Figure 185-2
TEMPERATURE, K
SPECIFIC HEAT, cal g^{-1} K^{-1}
5
8
9

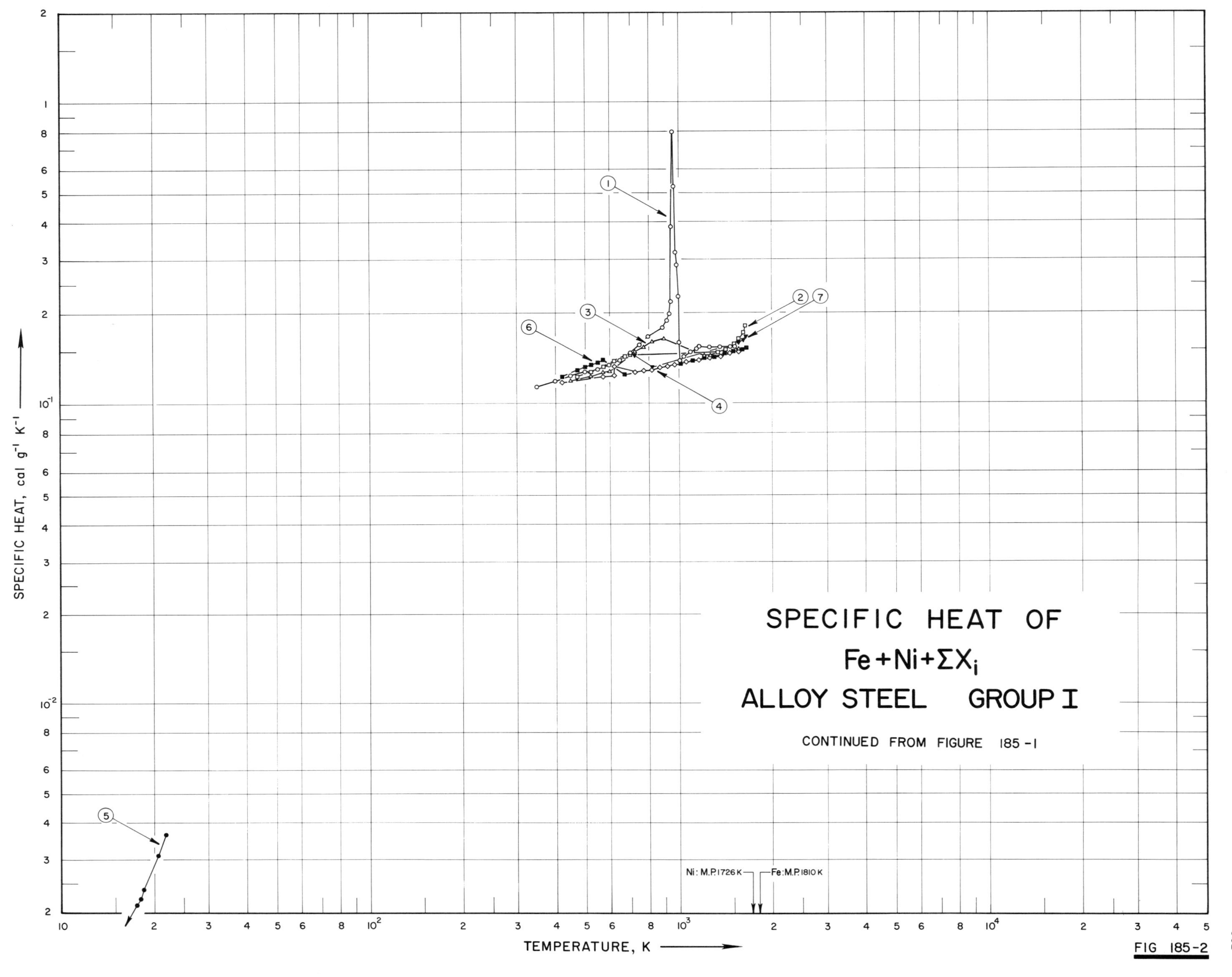
SPECIFIC HEAT OF
$Fe+Ni+\Sigma X_i$
ALLOY STEEL GROUP I
CONTINUED FROM FIGURE 185-1
SPECIFIC HEAT, cal g^{-1} K^{-1}
TEMPERATURE, K
Ni: M.P. 1726 K
Fe: M.P. 1810 K
1
2
3
4
5
6
7

FIG 185-2

SPECIFICATION TABLE NO. 185 SPECIFIC HEAT OF IRON + NICKEL + ΣX_i, Fe + Ni + ΣX_i, ALLOY STEEL GROUP I

[For Data Reported in Figure and Table No. 185]

Curve No.	Ref. No.	Year	Temp. Range, K	Reported Error, %	Name and Specimen Designation	Composition (weight percent), Specifications and Remarks
1	104	1946	348-1523	2-4	Alloy Steel No. 9	3.47 Ni, 0.55 Mn, 0.325 C, 0.18 Si, 0.17 Cr, 0.086 Cu, 0.04 Mo, 0.034 S, 0.032 P, 0.023 As, 0.01 V and 0.006 Al; annealed at 860 C; density (15 C) = 490 lb ft^{-3}.
2	236	1940	473-1648			90.9 Fe, 9.1 Ni; prepared from electrolytically deposited raw material; vacuum melted; heated 5 hrs at 1100 C and cooled slowly.
3	236	1940	448-1573			80.7 Fe and 19.3 Ni.
4	236	1940	423-1573			70.5 Fe and 29.5 Ni.
5	410	1965		0.5-4.0	Invar	64.6 Fe, 35.3 Ni, and 0.05 Co; sample supplied by Carpenter Steel Co.
6	236	1940	423-1673			61.0 Fe and 36.0 Ni.
7	236	1940	473-1673			50.98 Fe and 49.02 Ni.
8	349	1962	1.4-4.3	≤2	$Ni_{40}Fe_{60}$	58.95 Fe and 40.97 Ni; annealed under He + 8% H_2 gas atmosphere at 1100 C for 72 hrs; etched with 30 ml HNO_3 and 20 ml CH_3COOH.
9	349	1962	1.5-4.3	≤2	$Ni_{45}Fe_{55}$	54.09 Fe and 45.77 Ni; same as above.

DATA TABLE NO. 185 SPECIFIC HEAT OF IRON + NICKEL + ΣX_i, Fe + Ni + ΣX_i, ALLOY STEEL GROUP I

[Temperature, T,K; Specific Heat, C_p, Cal $g^{-1}K^{-1}$]

T	C_p
CURVE 1	
Series I	
348	1.15 x 10^{-1}
398	1.20
448	1.25
498	1.28
548	1.31
598	1.36
648	1.41
698	1.48
748	1.58
798	1.68
848	1.79*
898	1.89
948	3.91
998	2.28
1048	1.44
1098	1.49
1148	1.53
Series II	
893	1.8 x 10^{-1}
918	1.9
928	1.9*
938	2.0
948	2.2
958	8.0
968	5.3
978	3.2
988	2.9
998	2.2*
1008	1.6
1018	1.4
Series III	
1173	1.56 x 10^{-1}
1223	1.55*
1273	1.54
1323	1.53*
1373	1.53
1423	1.54*
1473	1.56
1523	1.57
CURVE 2	
473	1.231 x 10^{-1}
523	1.284
573	1.330
623	1.391
673	1.445
698	1.471
1348	1.48
1373	1.50*
1423	1.52
1473	1.55*
1523	1.58
1573	1.66
1623	1.73
1648	1.82
CURVE 3	
448	1.212 x 10^{-1}
473	1.225*
523	1.252
573	1.280
598	1.293
673	1.440*
723	1.501
773	1.562
823	1.623
898	1.654
1223	1.46
1273	1.46
1373	1.48*
1473	1.50*
1573	1.53
CURVE 4	
423	1.178 x 10^{-1}
473	1.220*
523	1.262*
573	1.304*
623	1.346
573	1.232
623	1.250
673	1.267*
723	1.283
CURVE 4 (cont.)	
773	1.300 x 10^{-1}
823	1.315
873	1.330
923	1.345
973	1.360
1023	1.370*
1073	1.387
1123	1.400*
1173	1.413
1223	1.425*
1273	1.437
1323	1.448*
1373	1.459
1423	1.470*
1473	1.480
1523	1.491*
1573	1.499
CURVE 5	
Series I	
1.462	8.300 x 10^{-5}
1.526	8.632*
1.579	8.947
1.626	9.212*
1.668	9.438
1.690	9.531*
1.856	1.047 x 10^{-4}
1.948	1.103
2.027	1.147
2.097	1.182
2.155	1.217*
2.208	1.256
2.264	1.274*
2.352	1.335
2.445	1.403*
2.536	1.456
2.611	1.506*
2.679	1.538
2.733	1.558*
2.794	1.604
3.070	1.765
3.286	1.897
3.288	1.895*
CURVE 5 (cont.)*	
3.345	1.929 x 10^{-4}*
3.405	1.972
3.471	2.009*
3.634	2.110
3.706	2.164*
3.773	2.207
3.834	2.214*
3.884	2.270*
4.043	2.388
4.140	2.448*
4.198	2.501
4.264	2.545*
4.330	2.592
Series II	
1.317	7.489 x 10^{-5}
1.411	7.909
1.492	8.426*
1.587	8.947*
1.678	9.485*
1.769	1.004 x 10^{-4}*
1.848	1.029*
1.934	1.092*
2.032	1.137*
2.120	1.189*
2.197	1.219*
2.385	1.373*
2.479	1.405*
2.563	1.455*
2.647	1.517*
2.748	1.563*
2.847	1.621*
2.958	1.749*
3.082	1.767*
3.199	1.826*
3.310	1.903*
3.412	1.967*
3.504	2.022*
3.575	2.075*
3.651	2.134*
3.718	2.178*
3.816	2.240*
3.909	2.316*
3.995	2.377*
CURVE 5 (cont.)*	
Series III	
4.346	2.512 x 10^{-4}*
4.462	2.609*
4.569	2.663
4.690	2.750*
4.792	2.812
4.906	2.867*
5.059	2.982
5.321	3.156
7.607	4.978
7.773	5.196
7.943	5.402*
8.090	5.528
8.451	5.939
8.819	6.385
9.029	6.548*
9.198	6.691
10.83	8.598
11.10	9.086*
11.50	9.657
11.66	9.846*
12.54	1.200 x 10^{-3}
12.93	1.203
13.43	1.287
13.68	1.334
13.95	1.433
14.32	1.451*
15.65	1.736
15.88	1.793
16.07	1.809*
16.25	1.855*
17.61	2.133
17.85	2.272*
18.05	2.240
18.26	2.335*
18.53	2.406
20.42	3.052*
20.65	3.101
21.31	3.276*
21.82	3.651
CURVE 6	
423	1.234 x 10^{-1}
448	1.268*
473	1.301
498	1.332
523	1.362
548	1.390
573	1.416
673	1.261
723	1.278*
773	1.295*
823	1.312*
873	1.328*
923	1.343*
973	1.358*
1023	1.374
1073	1.388*
1123	1.403
1173	1.417*
1223	1.431
1273	1.444*
1323	1.457
1373	1.470*
1423	1.482
1473	1.494*
1523	1.506
1573	1.517*
1623	1.528
1673	1.539
CURVE 7	
473	1.243 x 10^{-1}*
523	1.291*
573	1.338*
623	1.383*
673	1.429*
723	1.469
823	1.332
873	1.336*
923	1.343*
973	1.352*
1023	1.362*
1073	1.375*
1123	1.390*
1173	1.406*

* Not shown on plot

DATA TABLE NO. 185 (continued)

T	C_p
CURVE 7 (cont.)	
1223	1.425×10^{-1}*
1273	1.446*
1323	1.468*
1373	1.493*
1423	1.520*
1473	1.548*
1523	1.580*
1573	1.612
1623	1.647
1673	1.683
CURVE 8	
1.419	4.399×10^{-5}
1.493	4.611
1.606	4.966
1.723	5.323
1.815	5.634
1.832	5.685*
1.947	6.025
1.976	6.123*
2.087	6.502
2.151	6.661*
2.235	6.942
2.345	7.268
2.470	7.710
2.571	8.152
2.664	8.488
2.787	8.839
2.924	9.254
3.125	1.001×10^{-4}
3.293	1.064
3.375	1.102
3.471	1.133*
3.597	1.189
3.750	1.239
3.937	1.315
4.079	1.373
4.153	1.414
4.281	1.459
CURVE 9	
1.465	3.467×10^{-5}
1.512	3.574
1.575	3.714
1.664	3.931

T	C_p
CURVE 9 (cont.)	
1.761	4.152×10^{-1}
1.846	4.369
1.936	4.589
1.956	4.628*
2.094	4.986
2.173	5.174
2.279	5.455
2.420	5.785
2.594	6.248
2.740	6.677
2.853	6.983
2.994	7.312
3.205	7.895
3.511	8.711
3.658	9.164
3.847	9.749
3.987	1.018×10^{-4}
4.077	1.049*
4.173	1.081
4.282	1.117

* Not shown on plot

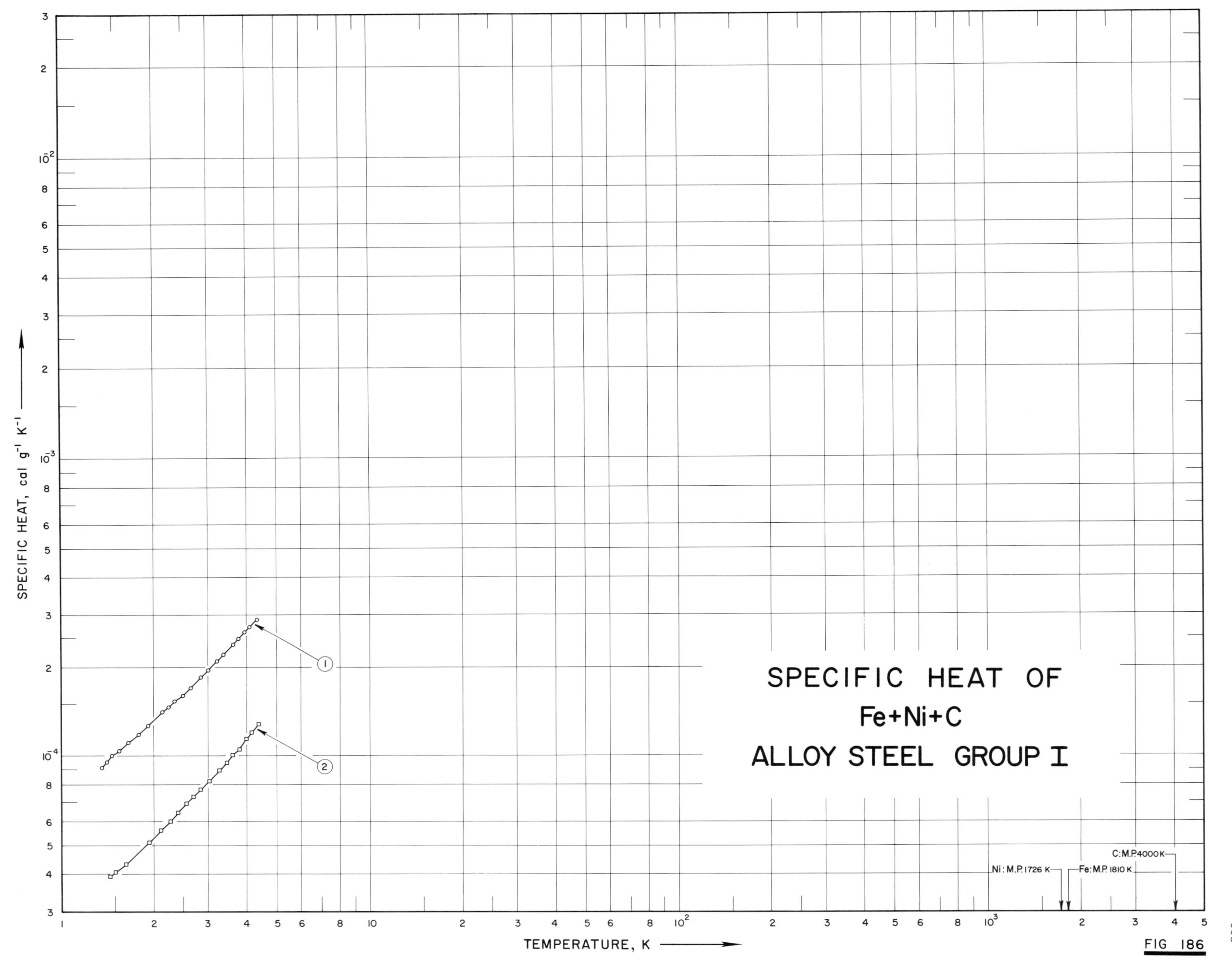

SPECIFIC HEAT OF
Fe+Ni+C
ALLOY STEEL GROUP I
SPECIFIC HEAT, cal g⁻¹ K⁻¹
TEMPERATURE, K
1
2
Ni: M.P. 1726 K
Fe: M.P. 1810 K
C: M.P. 4000 K

FIG 186

SPECIFICATION TABLE NO. 186 SPECIFIC HEAT OF IRON + NICKEL + CARBON, Fe + Ni + C, ALLOY STEEL GROUP I

[For Data Reported in Figure and Table No. 186]

Curve No.	Ref. No.	Year	Temp. Range, K	Reported Error, %	Name and Specimen Designation	Composition (weight percent), Specifications and Remarks
1	349	1962	1.4-4.3	≤2	$Ni_{28}Fe_{65.4}C_{6.6}$	65.4 Fe, 28.0 Ni, and 6.6 C; annealed under vacuum at 1100 C for 72 hrs; etched with 30 ml HNO_3 and 20 ml CH_3COOH.
2	349	1962	1.4-4.4	≤2	$Ni_{37.4}Fe_{56}C_{6.6}$	56.0 Fe, 37.4 Ni, and 6.6 C; same as above.

DATA TABLE NO. 186 SPECIFIC HEAT OF IRON + NICKEL + CARBON, Fe + Ni + C, ALLOY STEEL GROUP I

[Temperature, T, K; Specific Heat, C_p, Cal $g^{-1}K^{-1}$]

T	C_p
CURVE 1	
1.363	9.134×10^{-5}
1.417	9.558
1.473	9.995
1.553	1.041×10^{-4}
1.662	1.112
1.792	1.188
1.924	1.271
2.142	1.403
2.233	1.469
2.332	1.538
2.438	1.617
2.533	1.674*
2.633	1.713
2.741	1.783*
2.859	1.866
3.015	1.971
3.213	2.106
3.374	2.212
3.496	2.297*
3.625	2.378
3.775	2.492
3.949	2.621
4.087	2.731
4.202	2.795*
4.322	2.896
CURVE 2	
1.442	3.946×10^{-5}
1.515	4.054
1.628	4.335
1.930	5.113
2.112	5.607
2.263	6.008
2.398	6.432
2.553	6.945
2.692	7.299
2.837	7.682
3.025	8.222
3.260	8.944
3.456	9.518
3.602	1.005×10^{-4}
3.781	1.065
3.991	1.147
4.158	1.215
4.267	1.252*
4.379	1.277

*Not shown on plot

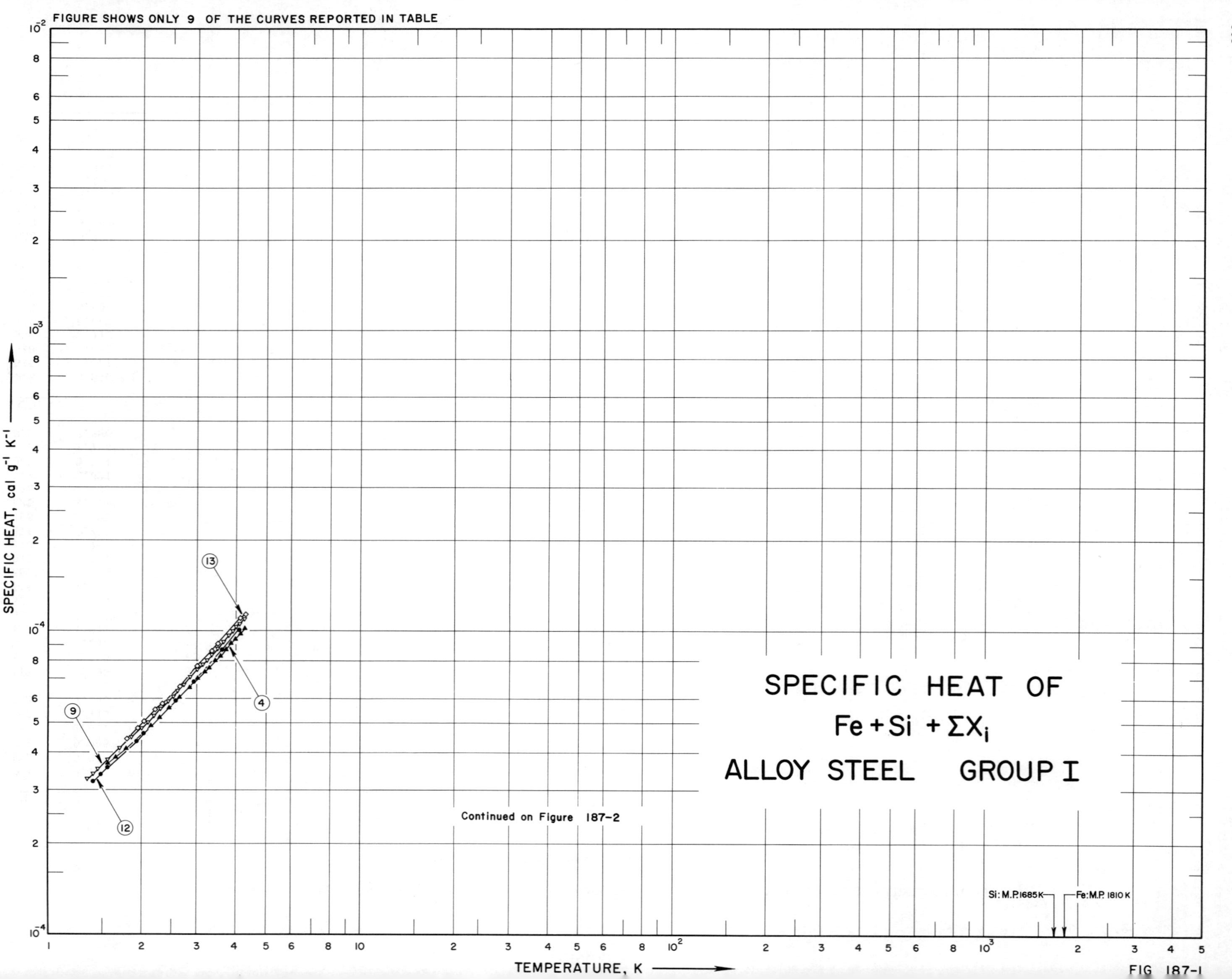

FIGURE SHOWS ONLY 9 OF THE CURVES REPORTED IN TABLE
SPECIFIC HEAT OF
Fe + Si + ΣXi
ALLOY STEEL GROUP I
Continued on Figure 187-2
Si: M.P. 1685 K
Fe: M.P. 1810 K
SPECIFIC HEAT, cal g⁻¹ K⁻¹
TEMPERATURE, K
13
9
4
12
FIG 187-1

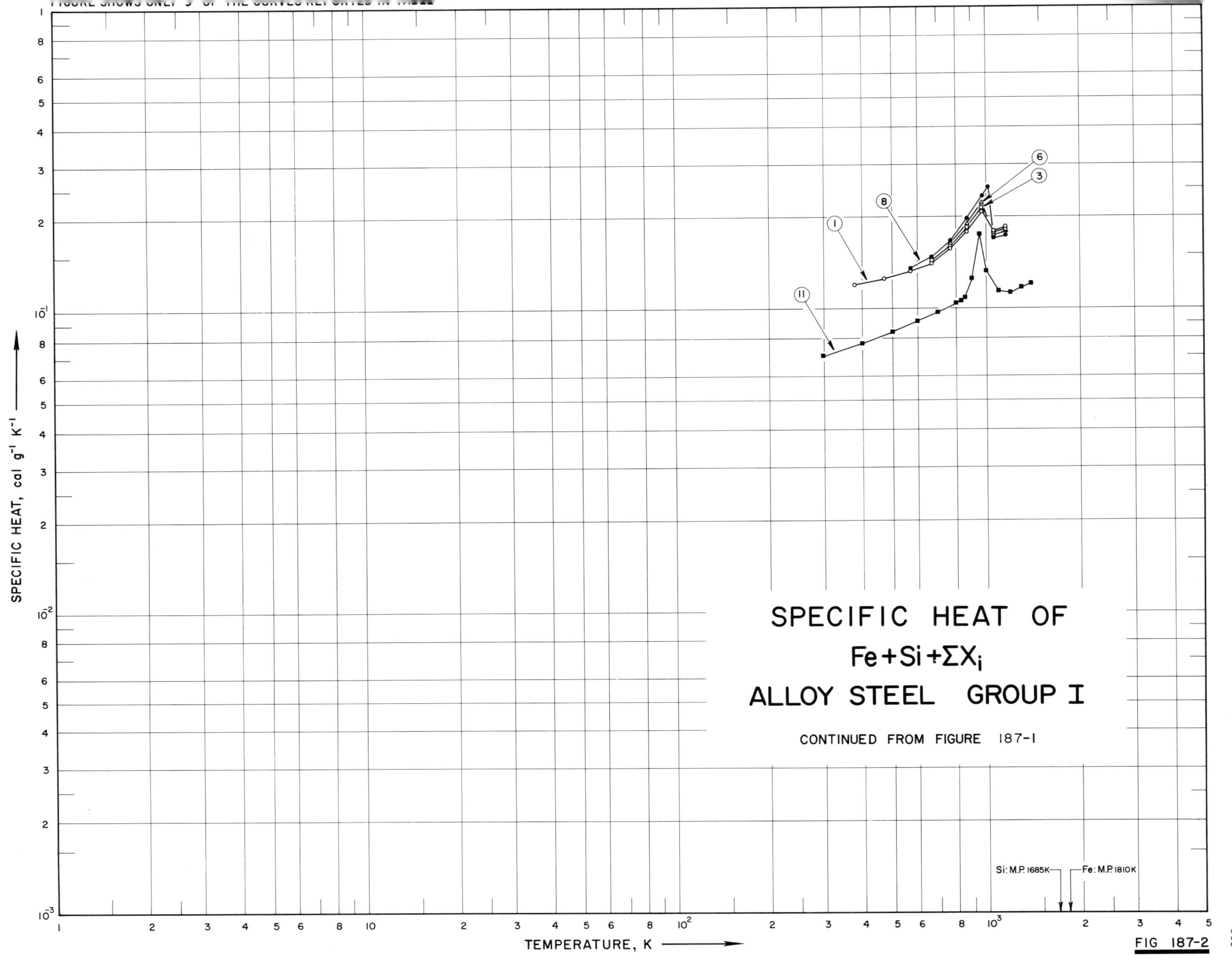

SPECIFIC HEAT OF
Fe+Si+ΣXi
ALLOY STEEL GROUP I
CONTINUED FROM FIGURE 187-1
SPECIFIC HEAT, cal g-1 K-1
TEMPERATURE, K
Si: M.P. 1685K
Fe: M.P. 1810K
1
3
6
8
11

FIG 187-2

SPECIFICATION TABLE NO. 187 SPECIFIC HEAT OF IRON + SILICON + ΣX_i, Fe + Si + ΣX_i, ALLOY STEEL GROUP I

[For Data Reported in Figure and Table No. 187]

Curve No.	Ref. No.	Year	Temp. Range, K	Reported Error, %	Name and Specimen Designation	Composition (weight percent), Specifications and Remarks
1	411	1956	373-1173			1.0 Si, 0.25 Mn, 0.07 C, 0.026 S, and 0.024 P; specimen soaked isothermally in furnace for 1 hr prior to drop.
2	411	1956	373-1173			1.23 Si, 0.29 Mn, 0.09 C, 0.047 P, and 0.029 S.
3	411	1956	373-1173			1.80 Si, 0.32 Mn, 0.09 C, 0.038 P, 0.023 S, and 0.01 Al.
4	349	1962	1.6-4.3	≤2.0	$Fe_{96}Si_4$	97.94 Fe and 2.01 Si; annealed under vacuum at 1200 C for 3 days and at 900 C for 2 days; etched with 1-3% HNO_3.
5	411	1956	373-1173			2.2 Si.
6	411	1956	373-1173			2.78 Si, 0.35 Mn, 0.09 C, 0.06 Al, 0.034 P, and 0.023 S.
7	411	1956	373-1173			3.94 Si, 0.27 Mn, 0.09 Al, 0.08 C, 0.027 P, and 0.008 S.
8	411	1956	373-1173			4.28 Si, 0.08 Mn, 0.06 C, 0.05 Al, 0.012 P, and 0.006 S.
9	349	1962	1.3-4.3	≤2.0	$Fe_{92}Si_8$	95.86 Fe and 4.11 Si; annealed under vacuum at 1200 C for 72 hrs; etched with 1-3% HNO_3.
10	411	1956	373-1173			4.38 Si, 0.20 Mn, 0.07 C, 0.05 Al, 0.015 P, and 0.008 S.
11	222	1959	298-1400		$Fe_{88}Si_{22}$	93.6 Fe and 6.4 Si; homogenized for 4 days at 1350 C under helium atmosphere; air cooled to room temperature.
12	349	1962	1.4-4.3	≤2.0	$Ni_{85}Si_{15}$	92.42 Fe and 7.42 Si; annealed under vacuum at 1200 C for 72 hrs; etched with 1-3% HNO_3.
13	349	1962	1.8-4.3	≤2.0	$Fe_{75}Si_{25}$	85.76 Fe and 14.00 Si; same as above.

DATA TABLE NO. 187 SPECIFIC HEAT OF IRON + SILICON + ΣX_i, Fe + Si + ΣX_i, ALLOY STEEL GROUP I

[Temperature, T, K; Specific Heat, C_p, Cal, $g^{-1}K^{-1}$]

T	C_p
CURVE 1	
373	1.203×10^{-1}
473	1.264
573	1.324
673	1.419
773	1.574
873	1.796
973	2.089
1073	1.812
1173	1.871
CURVE 2*	
373	1.203×10^{-1}
473	1.264
573	1.324
673	1.419
773	1.58
873	1.81
973	2.194
1073	1.80
1173	1.86
CURVE 3	
373	1.204×10^{-1}*
473	1.269*
573	1.33*
673	1.44
773	1.60
873	1.85
973	2.14
1073	1.79
1173	1.84
CURVE 4	
1.571	3.678×10^{-5}
1.658	3.857
1.782	4.138
2.145	4.915
2.288	5.244
2.449	5.627
2.636	6.069
2.840	6.567
3.028	7.024
CURVE 4 (cont.)	
3.180	7.392×10^{-5}
3.289	7.637
3.423	8.000
3.565	8.336
3.721	8.752
3.857	9.136
3.976	9.437
4.107	9.807
4.261	1.023×10^{-4}
CURVE 5*	
373	1.205×10^{-1}
473	1.270
573	1.34
673	1.45
773	1.62
873	1.89
973	2.20
1073	1.77
1173	1.82
CURVE 6	
373	1.20×10^{-1}*
473	1.27*
573	1.34*
673	1.46*
773	1.63
873	1.91
973	2.24
1073	1.76
1173	1.80
CURVE 7*	
373	1.209×10^{-1}
473	1.278
573	1.352
673	1.474
773	1.660
873	1.96
973	2.32
1073	1.74
1173	1.77
CURVE 8	
373	1.21×10^{-1}*
473	1.28*
573	1.36
673	1.48
773	1.67
873	1.98
973	2.37
1023	2.53
1073	1.72
1173	1.75
CURVE 9	
1.340	3.271×10^{-5}
1.406	3.403
1.457	3.539
1.567	3.793
1.714	4.143
1.866	4.495
2.010	4.817
2.102	5.043
2.188	5.262
2.300	5.564
2.423	5.881
2.524	6.134
2.619	6.391
2.739	6.699
2.868	7.073
3.013	7.477
3.139	7.760
3.236	8.034
3.356	8.355
3.502	8.759
3.653	9.197
3.796	9.659
3.911	9.968
4.009	1.029×10^{-4}
4.127	1.066
4.250	1.099
CURVE 10*	
373	1.21×10^{-1}
473	1.28
573	1.36
673	1.48
773	1.67
873	1.98
973	2.37
1023	2.53
1073	1.72
1173	1.75
CURVE 11	
298.15	7.045×10^{-2}
400	7.744
500	8.420
600	9.107
700	9.783
800	1.047×10^{-1}
830	1.068
856	1.083
900	1.262
950	1.767
1000	1.327
1100	1.145
1200	1.136
1300	1.176
1400	1.214
CURVE 12	
1.404	3.214×10^{-5}
1.477	3.381
1.565	3.566
1.677	3.821*
1.797	4.088*
1.924	4.366
2.034	4.611
2.108	4.869*
2.192	5.038*
2.300	5.277*
2.444	5.606*
2.574	5.936
2.678	6.147*
2.798	6.469*
CURVE 12 (cont.)	
2.941	6.841×10^{-5}
3.101	7.237*
3.234	7.699*
3.347	7.878*
3.488	8.237*
3.645	8.693
3.791	9.092*
3.902	9.359*
4.004	9.678*
4.127	1.005×10^{-4}
4.258	1.035*
CURVE 13	
1.810	4.446×10^{-5}
1.912	4.625*
1.959	4.833
2.099	5.061
2.215	5.539
2.345	5.787
2.508	6.047*
2.669	6.564
2.816	7.087*
3.005	7.693
3.219	8.169*
3.372	8.631
3.506	9.070
3.682	9.426*
3.880	1.005×10^{-4}*
4.038	1.051*
4.174	1.102
4.316	1.135

* Not shown on plot

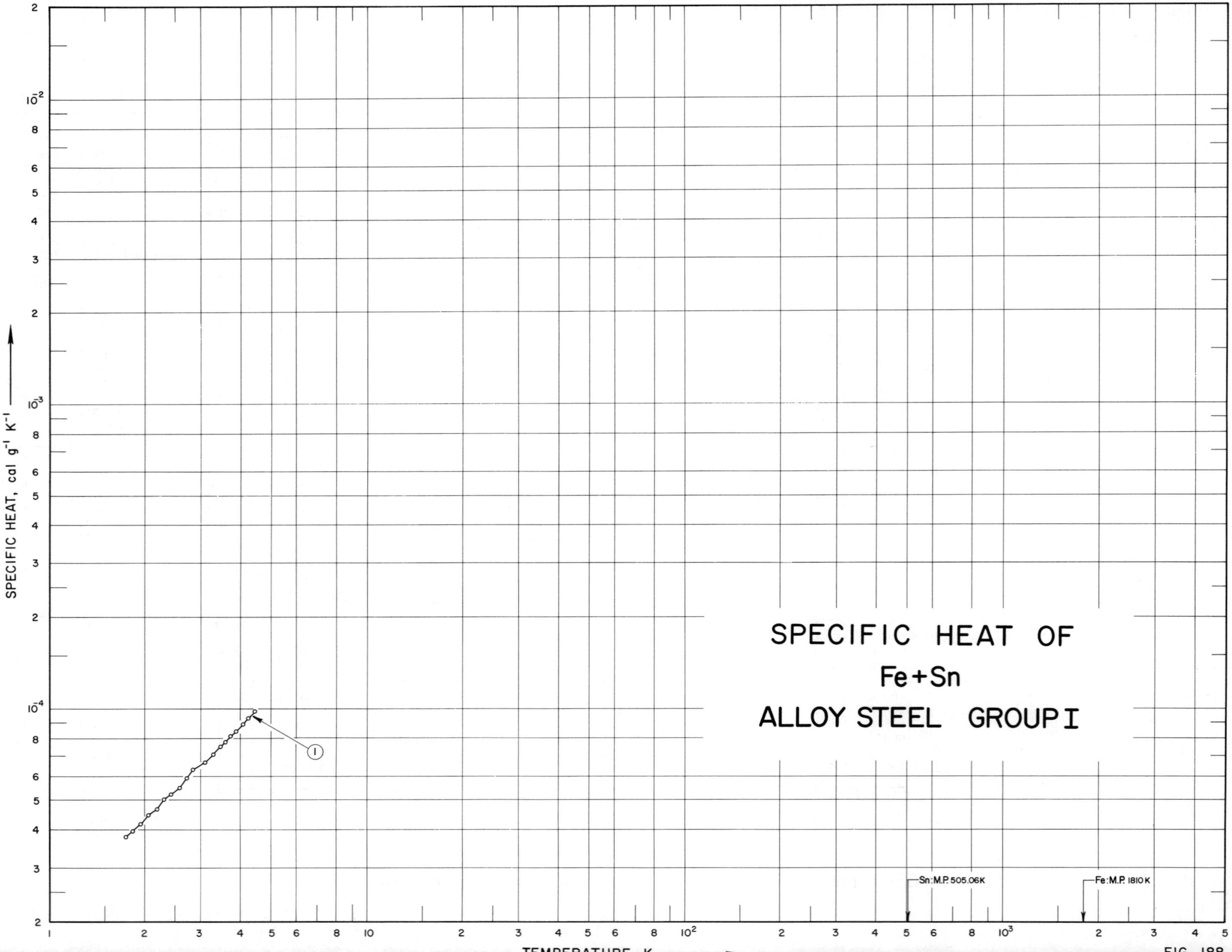

SPECIFIC HEAT OF
Fe+Sn
ALLOY STEEL GROUP I

SPECIFIC HEAT, cal g^{-1} K^{-1}

TEMPERATURE, K

Sn: M.P. 505.06 K

Fe: M.P. 1810 K

FIG. 188

SPECIFICATION TABLE NO. 188 SPECIFIC HEAT OF IRON + TIN, Fe + Sn, ALLOY STEEL GROUP I

[For Data Reported in Figure and Table No. 188]

Curve No.	Ref. No.	Year	Temp. Range, K	Reported Error, %	Name and Specimen Designation	Composition (weight percent), Specifications and Remarks
1	349	1962	1.8–4.5	≤ 2	$Fe_{92}Sn_8$	84.56 Fe and 15.41 Sn.

DATA TABLE NO. 188 SPECIFIC HEAT OF IRON + TIN, Fe + Sn, ALLOY STEEL GROUP I

[Temperature, T, K; Specific Heat, C_p, Cal, $g^{-1}K^{-1}$]

T	C_p
CURVE 1	
1.755	3.796 x 10^{-5}
1.849	3.959
1.953	4.184
2.063	4.465
2.071	4.439*
2.199	4.662
2.318	5.024
2.425	5.236
2.577	5.483
2.721	5.915
2.817	6.071*
2.932	6.301
3.106	6.675
3.300	7.097
3.464	7.518
3.594	7.799
3.722	8.166
3.885	8.450
4.085	8.935
4.237	9.359
4.339	9.543*
4.457	9.808

* Not shown on plot

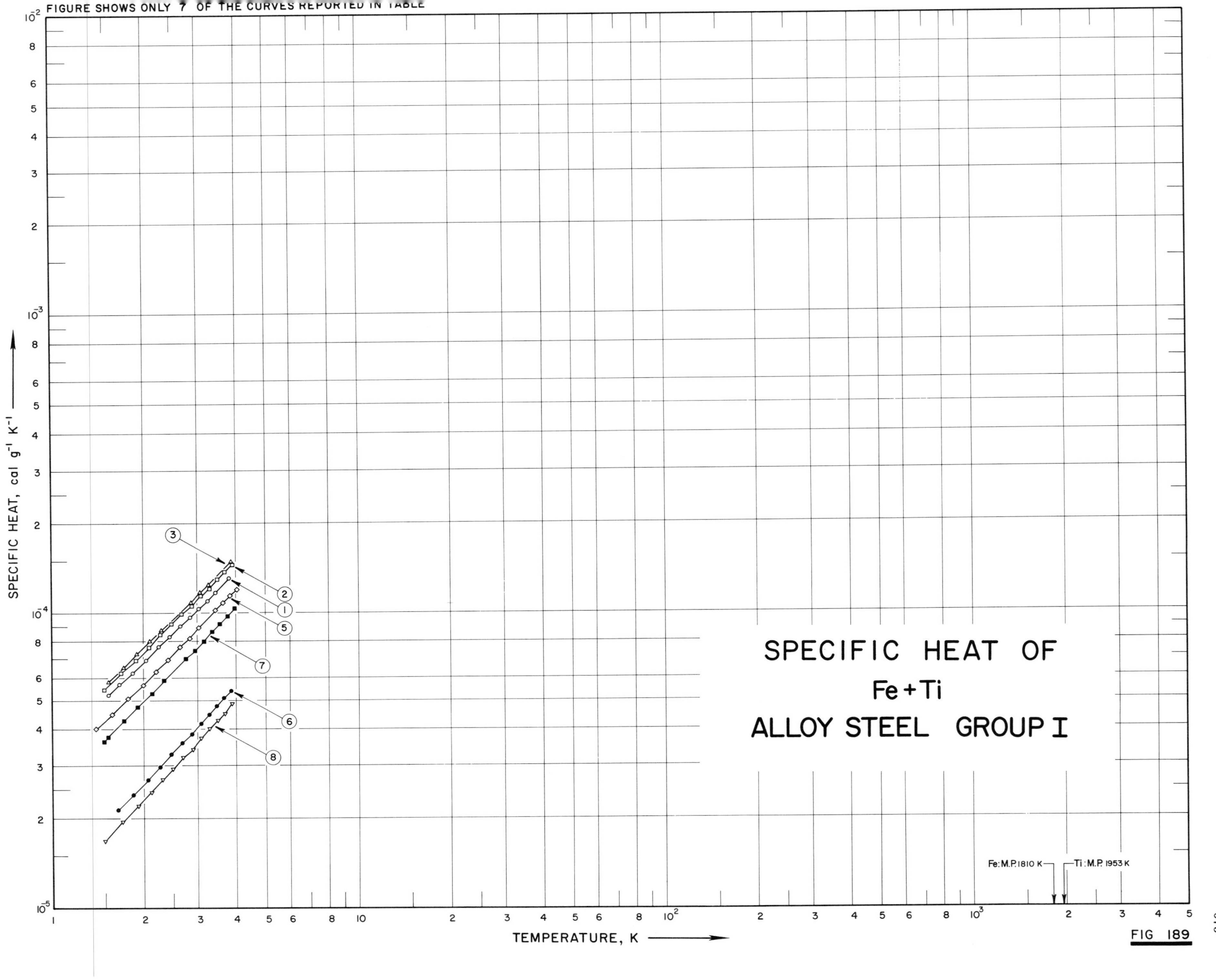
FIGURE SHOWS ONLY 7 OF THE CURVES REPORTED IN TABLE
SPECIFIC HEAT OF
Fe + Ti
ALLOY STEEL GROUP I
SPECIFIC HEAT, cal g⁻¹ K⁻¹
TEMPERATURE, K
Fe: M.P. 1810 K
Ti: M.P. 1953 K
1
2
3
5
6
7
8

FIG 189

SPECIFICATION TABLE NO. 189 SPECIFIC HEAT OF IRON + TITANIUM, Fe + Ti, ALLOY STEEL GROUP I

[For Data Reported in Figure and Table No. 189]

Curve No.	Ref. No.	Year	Temp. Range, K	Reported Error, %	Name and Specimen Designation	Composition (weight percent), Specifications and Remarks
1	240	1962	1.5-3.8		$TiFe_2$	Prepared from titanium: 99.98 Ti, 0.005 Si, 0.0001 Al, 0.0002 Mg, 0.001 Mn, 0.0005 Cr, 0.0003 Fe, 0.0001 Ni, 0.0005 Ca, 0.001 C, 0.002 O_2, and 0.002 N_2; sample supplied by Foote Mineral Co.; Iron: 99.95 Fe, 0.024 C, 0.001-0.005 Mn, 0.0023 O_2, 0.0004 N_2, 0.007 Si, 0.005 Ni, 0.006 Sn, 0.001-0.004 Mo, 0.003-0.006 Co, 0.001-0.006 Al, 0.001-0.003 Cu, and 0.001 Pb; sample supplied by Crucible Steel Co.; alloy prepared with levitation apparatus which uses magnetic field to support raw materials; under argon atmosphere; washed with CCl_4.
2	240	1962	1.5-3.9		Same as above	Same as above.
3	240	1962	1.5-3.9		Same as above	Same as above.
4	240	1962	1.8-4.0		Same as above	Same as above.
5	240	1962	1.4-4.0		Same as above	Same as above.
6	240	1962	1.7-3.9		Same as above	Same as above.
7	240	1962	1.5-4.0		Same as above	Same as above.
8	240	1962	1.5-3.9		Same as above	Same as above.

DATA TABLE NO. 189 SPECIFIC HEAT OF IRON + TITANIUM, Fe + Ti, ALLOY STEEL GROUP I

[Temperature, T,K; Specific Heat, C_p, Cal, $g^{-1}K^{-1}$]

T	C_p
CURVE 1	
1.543	5.251×10^{-5}
1.674	5.728
1.846	6.266
2.045	6.937
2.247	7.657
2.445	8.300
2.649	9.048
2.842	9.658
3.038	1.033×10^{-4}
3.234	1.104
3.428	1.177
3.815	1.316
CURVE 2	
1.489	5.479×10^{-5}
1.686	6.266
1.881	6.936
2.078	7.685
2.273	8.447
2.470	9.162
2.672	9.938
2.875	1.064×10^{-4}
3.078	1.147
3.283	1.219
3.490	1.298
3.696	1.376
3.902	1.467
CURVE 3	
1.543	5.825×10^{-5}
1.727	6.557
1.914	7.269
2.102	7.996
2.291	8.735
2.480	9.387*
2.672	1.019×10^{-4}*
2.867	1.087
3.065	1.174
3.264	1.244
3.464	1.327*
3.667	1.405*
3.873	1.486

T	C_p
CURVE 4*	
1.771	6.544×10^{-5}
1.987	7.329
2.196	8.203
2.400	8.992
2.601	9.757
2.803	1.049×10^{-4}
3.005	1.125
3.207	1.210
3.411	1.287
3.613	1.369
3.815	1.449
4.017	1.534
CURVE 5	
1.418	4.037×10^{-5}
1.577	4.493
1.776	5.081
1.984	5.677
2.192	6.337
2.401	6.948
2.625	7.665
2.829	8.177
3.036	8.923
3.445	1.021×10^{-4}
3.644	1.082
3.839	1.147
4.032	1.202
CURVE 6	
1.656	2.143×10^{-5}
1.851	2.402
2.060	2.682
2.264	2.998
2.469	3.284
2.672	3.593
2.874	3.846
3.076	4.185
3.274	4.482
3.470	4.789
3.665	5.111
3.862	5.433

T	C_p
CURVE 7	
1.481	3.629×10^{-5}
1.538	3.752
1.726	4.267
1.923	4.755
2.126	5.300
2.335	5.874
2.750	6.980
2.954	7.440
3.158	7.981
3.361	8.600
3.565	9.180
3.767	9.758
3.972	1.040×10^{-4}
CURVE 8	
1.484	1.679×10^{-5}
1.709	1.951
1.922	2.207
2.117	2.453
2.302	2.700
2.488	2.922
2.690	3.202
2.891	3.407
3.088	3.746
3.287	4.007
3.487	4.268
3.688	4.508
3.888	4.876

* Not shown on plot

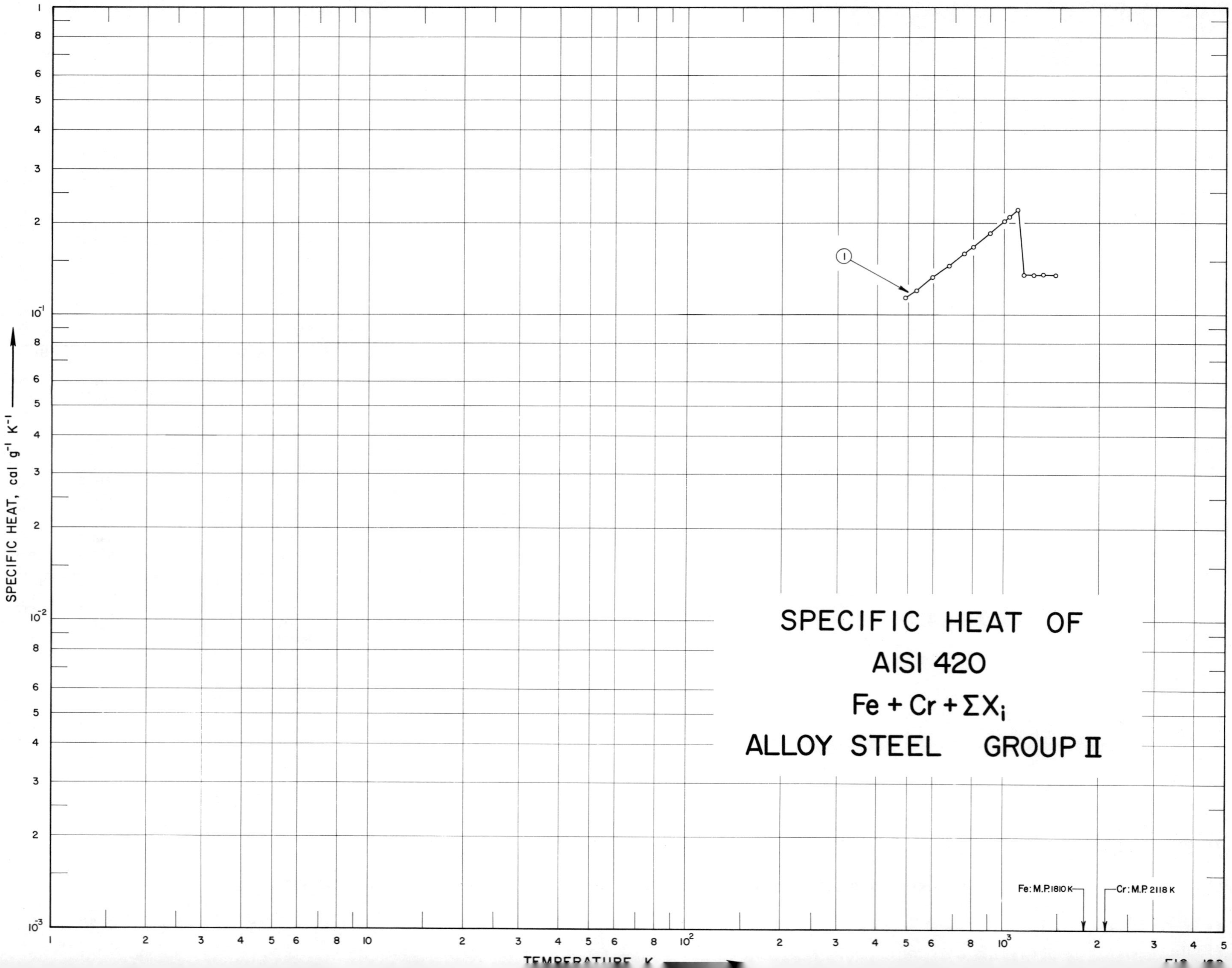

SPECIFIC HEAT OF
AISI 420
$Fe + Cr + \Sigma X_i$
ALLOY STEEL GROUP II

Fe: M.P. 1810 K
Cr: M.P. 2118 K

SPECIFIC HEAT, cal g^{-1} K^{-1}

TEMPERATURE, K

SPECIFICATION TABLE NO. 190 SPECIFIC HEAT OF AISI 420, Fe + Cr + ΣX_i, ALLOY STEEL GROUP II

[For Data Reported in Figure and Table No. 190]

Curve No.	Ref. No.	Year	Temp. Range, K	Reported Error, %	Name and Specimen Designation	Composition (weight percent), Specifications and Remarks
1	146	1961	493-1471	3.0	AISI 420	84.999 Fe, 13.1 Cr, 0.5 Ni, 0.48 Mn, 0.41 Si, 0.3 C, 0.12 Cu, 0.06 Mo, 0.02 P and 0.011 S; under helium atmosphere; density = 481 lb ft^{-3}.

DATA TABLE NO. 190 SPECIFIC HEAT OF AISI 420, $Fe + Cr + \Sigma X_i$, ALLOY STEEL GROUP II

[Temperature, T, K; Specific Heat, C_p, Cal $g^{-1}K^{-1}$]

T	C_p
CURVE 1	
493	1.144 x 10^{-1}
531	1.211
599	1.328
675	1.458
754	1.594
801	1.676
909	1.860
928	1.893*
1005	2.026
1046	2.097*
1047	2.098
1116	2.217
1149	2.273*
1173	1.36
1251	1.36
1272	1.36*
1345	1.36
1468	1.36*
1471	1.36

* Not shown on plot

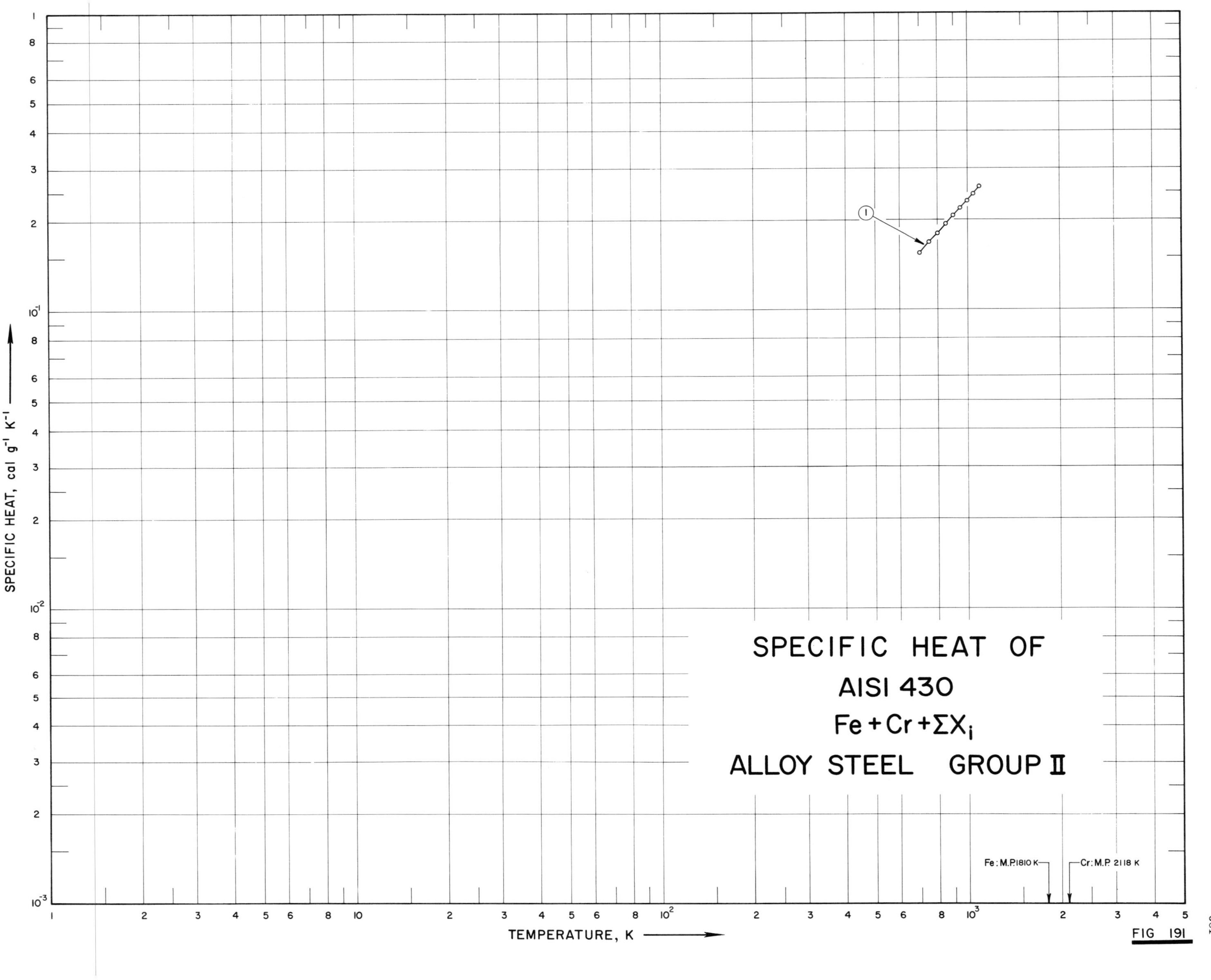
SPECIFIC HEAT OF
AISI 430
Fe + Cr + ΣXi
ALLOY STEEL GROUP II
SPECIFIC HEAT, cal g⁻¹ K⁻¹
TEMPERATURE, K
Fe: M.P. 1810 K
Cr: M.P. 2118 K
1

FIG 191

SPECIFICATION TABLE NO. 191 SPECIFIC HEAT OF AISI 430, $Fe + Cr + \Sigma X_i$, ALLOY STEEL GROUP II

[For Data Reported in Figure and Table No. 191]

Curve No.	Ref. No.	Year	Temp. Range, K	Reported Error, %	Name and Specimen Designation	Composition (weight percent), Specifications and Remarks
1	45	1955	700-1100		AISI 430	

DATA TABLE NO. 191 SPECIFIC HEAT OF AISI 430, $Fe + Cr + \Sigma X_i$, ALLOY STEEL GROUP II

[Temperature, T, K; Specific Heat, C_p, Cal $g^{-1}K^{-1}$]

T	C_p
CURVE 1	
700	1.55 x 10^{-1}
750	1.68
800	1.80
850	1.93
900	2.06
950	2.19
1000	2.32
1050	2.46
1100	2.59

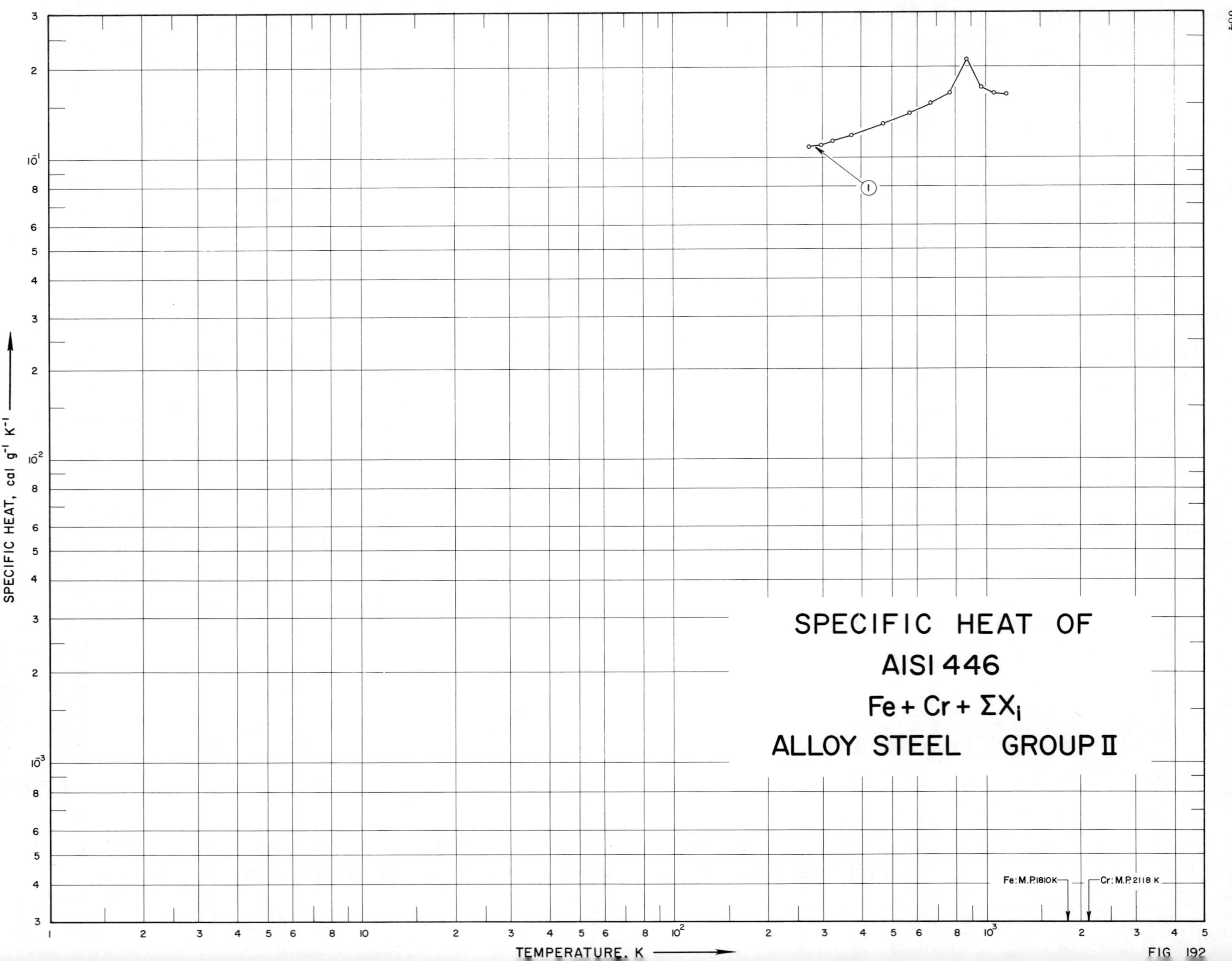
SPECIFIC HEAT OF
AISI 446
Fe + Cr + ΣXi
ALLOY STEEL GROUP II
1
Fe: M.P. 1810 K
Cr: M.P. 2118 K
SPECIFIC HEAT, cal g⁻¹ K⁻¹
TEMPERATURE, K

FIG 192

SPECIFICATION TABLE NO. 192 SPECIFIC HEAT OF AISI 446, Fe + Cr + ΣX_i, ALLOY STEEL GROUP II

[For Data Reported in Figure and Table No. 192]

Curve No.	Ref. No.	Year	Temp. Range, K	Reported Error, %	Name and Specimen Designation	Composition (weight percent), Specifications and Remarks
1	248	1960	273-1173	±0.3	AISI 446	25.58 Cr, 0.68 Si, 0.42 Mn, 0.32 Ni, 0.23 C, 0.019 P and 0.016 S.

DATA TABLE NO. 192 SPECIFIC HEAT OF AISI 446, $Fe + Cr + \Sigma X_i$, ALLOY STEEL GROUP II

[Temperature, T, K; Specific Heat, C_p, Cal $g^{-1}K^{-1}$]

T	C_p
CURVE 1	
273	1.08×10^{-1}
298	1.10
323	1.13
373	1.18
473	1.29
573	1.40
673	1.51
773	1.63
873	2.10
973	1.70
1073	1.63
1173	1.62

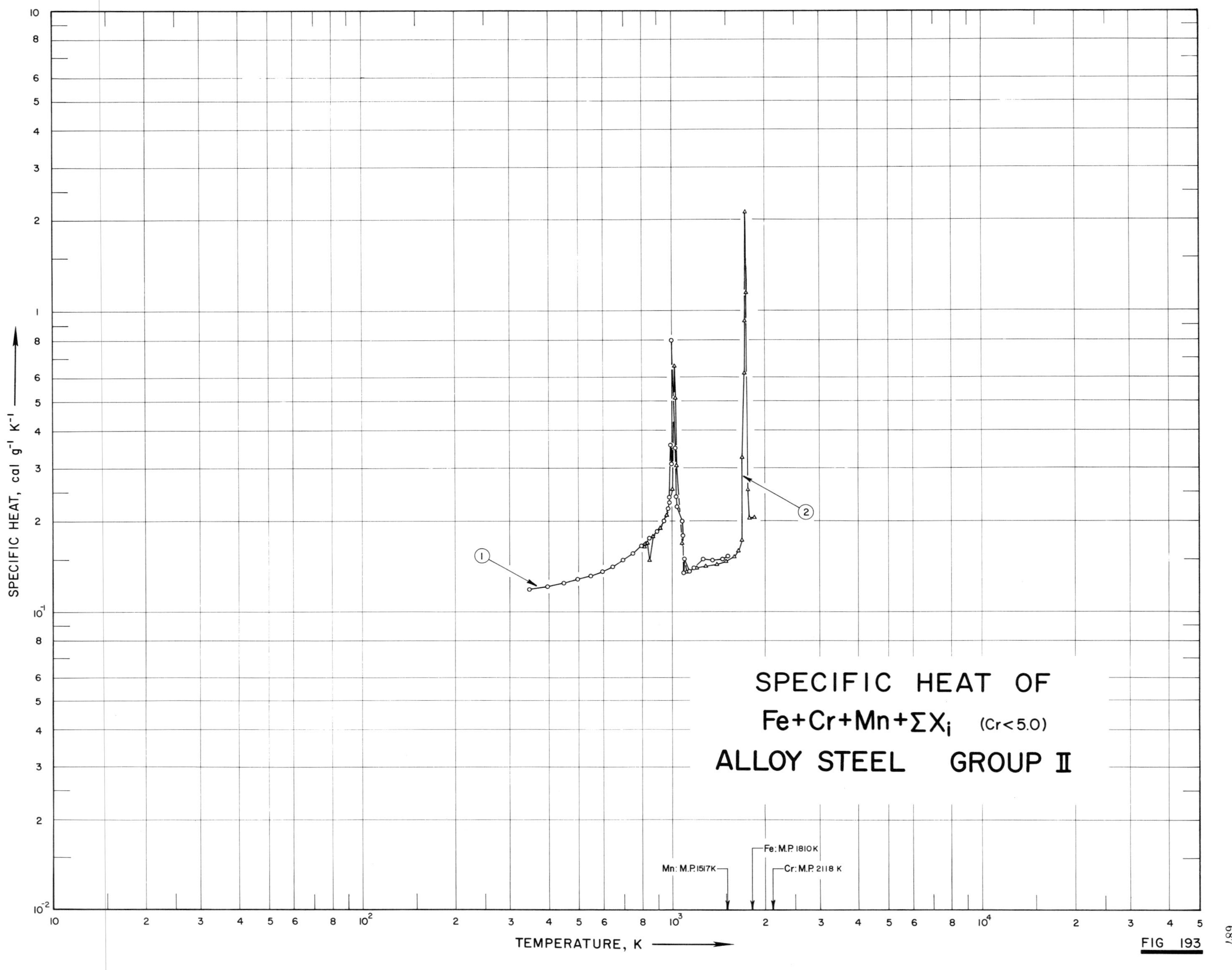
SPECIFIC HEAT OF
Fe+Cr+Mn+ΣXi (Cr<5.0)
ALLOY STEEL GROUP II
SPECIFIC HEAT, cal g⁻¹ K⁻¹
TEMPERATURE, K
Mn: M.P. 1517K
Fe: M.P. 1810K
Cr: M.P. 2118 K
1
2

FIG 193

SPECIFICATION TABLE NO. 193 SPECIFIC HEAT OF Fe + Cr + Mn + ΣX_i (Cr < 5.0), ALLOY STEEL GROUP II

[For Data Reported in Figure and Table No. 193]

Curve No.	Ref. No.	Year	Temp. Range, K	Reported Error, %	Name and Specimen Designation	Composition (weight percent), Specifications and Remarks
1	104	1946	348-1523	2.0	Alloy Steel No. 19	1.09 Cr, 0.69 Mn, 0.315 C, 0.2 Si, 0.073 Ni, 0.066 Cu, 0.039 P, 0.036 S, 0.028 As, 0.012 Mo and 0.005 Al; annealed at 860 C; density (15 C) = 488 lb ft^{-3}.
2	130	1958	818-1868		Alloy Steel No. 19	1.09 Cr, 0.69 Mn, 0.315 C, 0.2 Si, 0.073 Ni, 0.066 Cu, 0.039 P, 0.036 S, 0.028 As, 0.012 Mo and 0.005 Al; heated at a constant rate of 40 watts, up to 1210 C during one day; left at 1000 C overnight.

DATA TABLE NO. 193 SPECIFIC HEAT OF $Fe + Cr + Mn + \Sigma X_i$ ($Cr < 5.0$), ALLOY STEEL GROUP II

[Temperature, T, K; Specific Heat, C_p, Cal $g^{-1}K^{-1}$]

T	C_p
CURVE 1	
Series I	
348	1.18 x 10^{-1}
398	1.22
448	1.25
498	1.28
548	1.32
598	1.37
648	1.42
698	1.49
748	1.57
798	1.66
848	1.77
898	1.85
948	2.00
998	3.58
1048	2.23
1098	1.35
1148	1.37
Series II	
978	2.2 x 10^{-1}
988	2.3
998	2.4
1008	3.1
1018	8.0
1028	3.5
1038	2.4
1088	2.0
1098	1.8
1108	1.5
1188	1.4
Series III	
1173	1.48 x 10^{-1}*
1223	1.49*
1273	1.50
1323	1.49
1373	1.48
1423	1.48
1473	1.50
1523	1.53

T	C_p
CURVE 2	
818	1.666 x 10^{-1}
828	1.690*
838	1.714
848	1.496
858	1.759*
868	1.785
878	1.793*
888	1.836*
898	1.862*
908	1.891*
918	1.919
928	1.948*
938	1.979*
948	2.015*
958	2.051*
968	2.091
978	2.141*
988	2.206*
998	2.292*
1008	2.412*
1018	2.569*
1028	6.606
1038	5.160
1048	3.052
1058	2.180*
1068	2.063*
1078	1.704
1088	1.508*
1098	1.415*
1108	1.386*
1118	1.379*
1128	1.377
1138	1.379*
1148	1.381*
1158	1.384*
1168	1.386*
1178	1.389*
1188	1.391*
1198	1.396*
1208	1.401*
1218	1.405
1228	1.410*
1238	1.413*
1248	1.415*
1258	1.417*

T	C_p
CURVE 2 (cont.)	
1268	1.420 x 10^{-1}*
1278	1.420*
1288	1.422*
1298	1.424
1308	1.424*
1318	1.424*
1328	1.427*
1338	1.427*
1348	1.429*
1358	1.429*
1368	1.432*
1378	1.432*
1388	1.432*
1398	1.434
1408	1.436*
1418	1.439*
1428	1.441*
1438	1.444*
1448	1.448*
1458	1.453*
1468	1.458*
1478	1.465*
1488	1.472*
1498	1.479
1508	1.487*
1518	1.491*
1528	1.499*
1538	1.503*
1548	1.508*
1558	1.511*
1568	1.513*
1578	1.518*
1588	1.525*
1598	1.532*
1608	1.542
1618	1.554*
1628	1.568*
1638	1.587*
1648	1.613
1658	1.640*
1668	1.671*
1678	1.702*
1688	1.745
1698	3.262
1708	4.104*

T	C_p
CURVE 2 (cont.)	
1718	4.328 x 10^{-1}*
1728	6.238
1738	9.312
1748	2.123 x 10^{0}
1758	1.393*
1768	1.153
1778	2.533 x 10^{-1}
1788	2.046
1798	2.046*
1808	2.046*
1818	2.046*
1828	2.046*
1838	2.046*
1848	2.046*
1858	2.046*
1868	2.046

* Not shown on plot

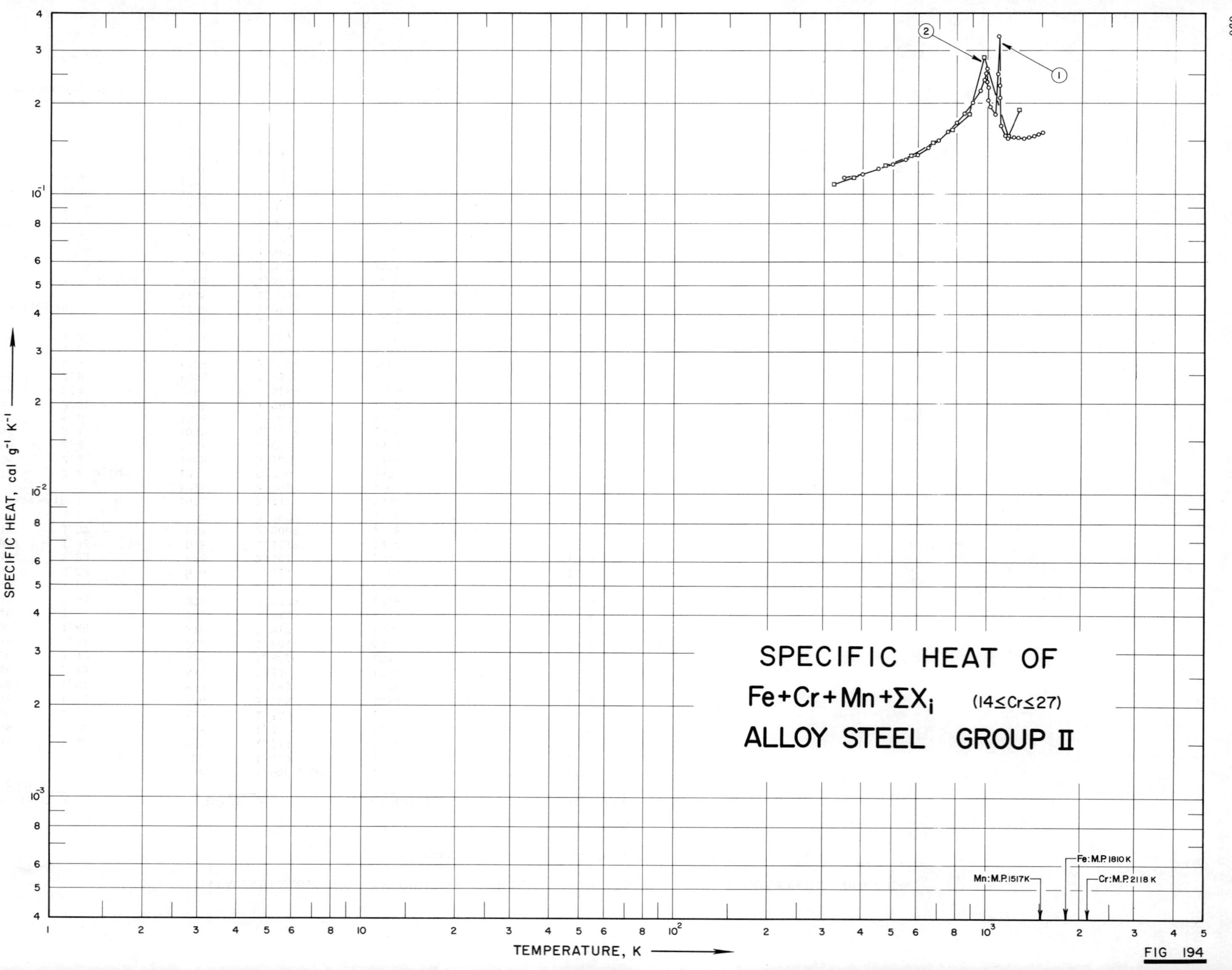
SPECIFIC HEAT OF
Fe+Cr+Mn+ΣXi (14≤Cr≤27)
ALLOY STEEL GROUP II
Mn: M.P. 1517 K
Fe: M.P. 1810 K
Cr: M.P. 2118 K
SPECIFIC HEAT, cal g⁻¹ K⁻¹
TEMPERATURE, K
1
2

FIG 194

SPECIFICATION TABLE NO. 194 SPECIFIC HEAT OF Fe + Cr + Mn + ΣX_i ($14 \leq Cr \leq 27$), ALLOY STEEL GROUP II

[For Data Reported in Figure and Table No. 194]

Curve No.	Ref. No.	Year	Temp. Range, K	Reported Error, %	Name and Specimen Designation	Composition (weight percent), Specifications and Remarks
1	104	1946	348-1523	2.0	High Alloy Steel No. 17	13.69 Cr, 0.28 Mn, 0.27 C, 0.25 W, 0.2 Ni, 0.18 Si, 0.074 Cu, 0.031 Al, 0.022 each S, P, and V, 0.01 Mo and 0.003 As; heated at 960 C in air, 2 hrs at 750 C and air cooled; density = 482 lb ft^{-3} at 15 C.
2	15	1959	323-1273	1.0	Steel 4 Kh13	Nominal composition: 12.0-14.0 Cr, 0.6 Mn, 0.6 Ni, 0.6 Si, 0.35-0.45 C, 0.035 P and 0.03 S.

DATA TABLE NO. 194 SPECIFIC HEAT OF $Fe + Cr + Mn + \Sigma X_i$ ($14 \leq Cr \leq 27$), ALLOY STEEL GROUP II

[Temperature, T, K; Specific Heat, C_p, Cal $g^{-1}K^{-1}$]

T	C_p
CURVE 1	
Series I	
348	1.13 x 10^{-1}
398	1.17
448	1.22
498	1.27
548	1.31
598	1.36
648	1.43
698	1.51
748	1.62
798	1.73
848	1.86
898	2.01
948	2.21
998	2.37
1048	1.87*
1098	2.30
1148	1.57
Series II	
978	2.40 x 10^{-1}
988	2.52
998	2.62
1008	2.27
1018	2.05
1028	1.96
1038	1.89*
1048	1.85*
1058	1.82*
1068	1.84
1078	2.52
1088	3.35
1098	2.09
1118	1.69
Series III	
1173	1.53 x 10^{-1}
1223	1.55
1273	1.54
1323	1.53
1373	1.55
1423	1.56
1473	1.58
1523	1.60

T	C_p
CURVE 2	
323	1.08 x 10^{-1}
373	1.14
473	1.25
573	1.34
673	1.48
773	1.63
873	1.84
973	2.85
1073	1.83*
1173	1.55
1273	1.91

* Not shown on plot

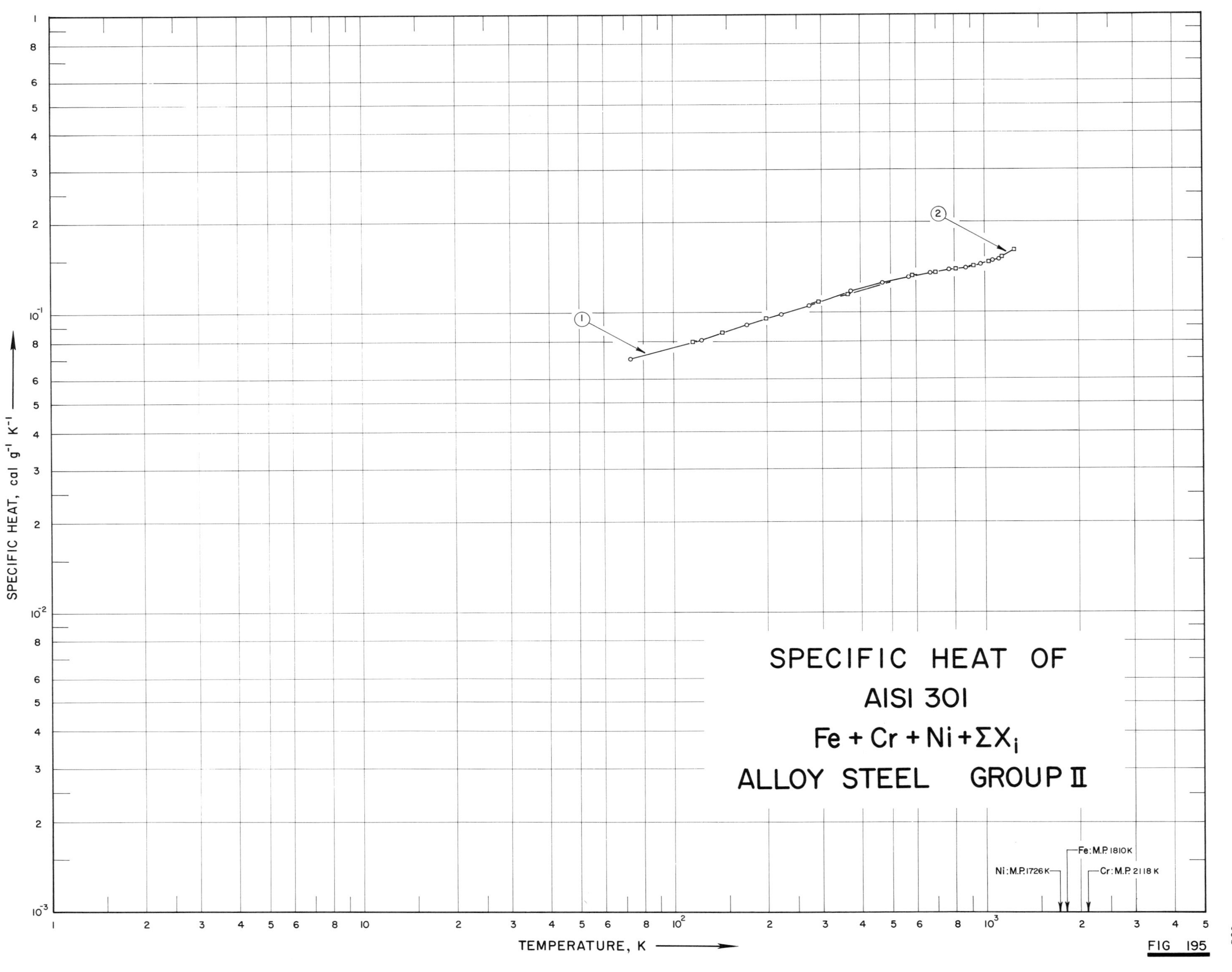
SPECIFIC HEAT OF
AISI 301
Fe + Cr + Ni + ΣX_i
ALLOY STEEL GROUP II
SPECIFIC HEAT, cal g⁻¹ K⁻¹
TEMPERATURE, K
Ni: M.P. 1726 K
Fe: M.P. 1810 K
Cr: M.P. 2118 K
1
2
FIG 195

SPECIFICATION TABLE NO. 195 SPECIFIC HEAT OF AISI 301, Fe + Cr + Ni + ΣX_i, ALLOY STEEL GROUP II

[For Data Reported in Figure and Table No. 195]

Curve No.	Ref. No.	Year	Temp. Range, K	Reported Error, %	Name and Specimen Designation	Composition (weight percent), Specifications and Remarks
1	243	1954	73-1123		AISI 301	Nominal composition: 16-18 Cr and 6-8 Ni.
2	10	1958	116-1255		AISI 301	Nominal composition: 16-18 Cr, 6-8 Ni, $\leq$2.0 Mn, $\leq$1.0 Si, $\leq$0.15 C, $\leq$0.045 P and $\leq$0.03 S; sample supplied by the Republic Steel Corp; specimen sealed in helium capsule; annealed 1 hr at 1900 F; water quenched; density = 495 lb ft^{-3} at 32 F.

DATA TABLE NO. 195 AISI 301, $Fe + Cr + Ni + \Sigma X_i$, ALLOY STEEL GROUP II

[Temperature, T, K; Specific Heat, C_p, Cal $g^{-1} K^{-1}$]

T	C_p
CURVE 1	
73	7.0×10^{-2}
123	8.1
173	9.1
223	9.9
273	1.07×10^{-1}
373	1.18
473	1.27
573	1.32
673	1.36
773	1.39
873	1.42
973	1.45
1073	1.49
1123	1.52
CURVE 2	
116	8.00×10^{-2}
144	8.60
200	9.60
293	1.09×10^{-1}
366	1.17
478	1.27*
589	1.33
700	1.37
811	1.40
922	1.43
1033	1.48
1144	1.54
1255	1.62

* Not shown on plot

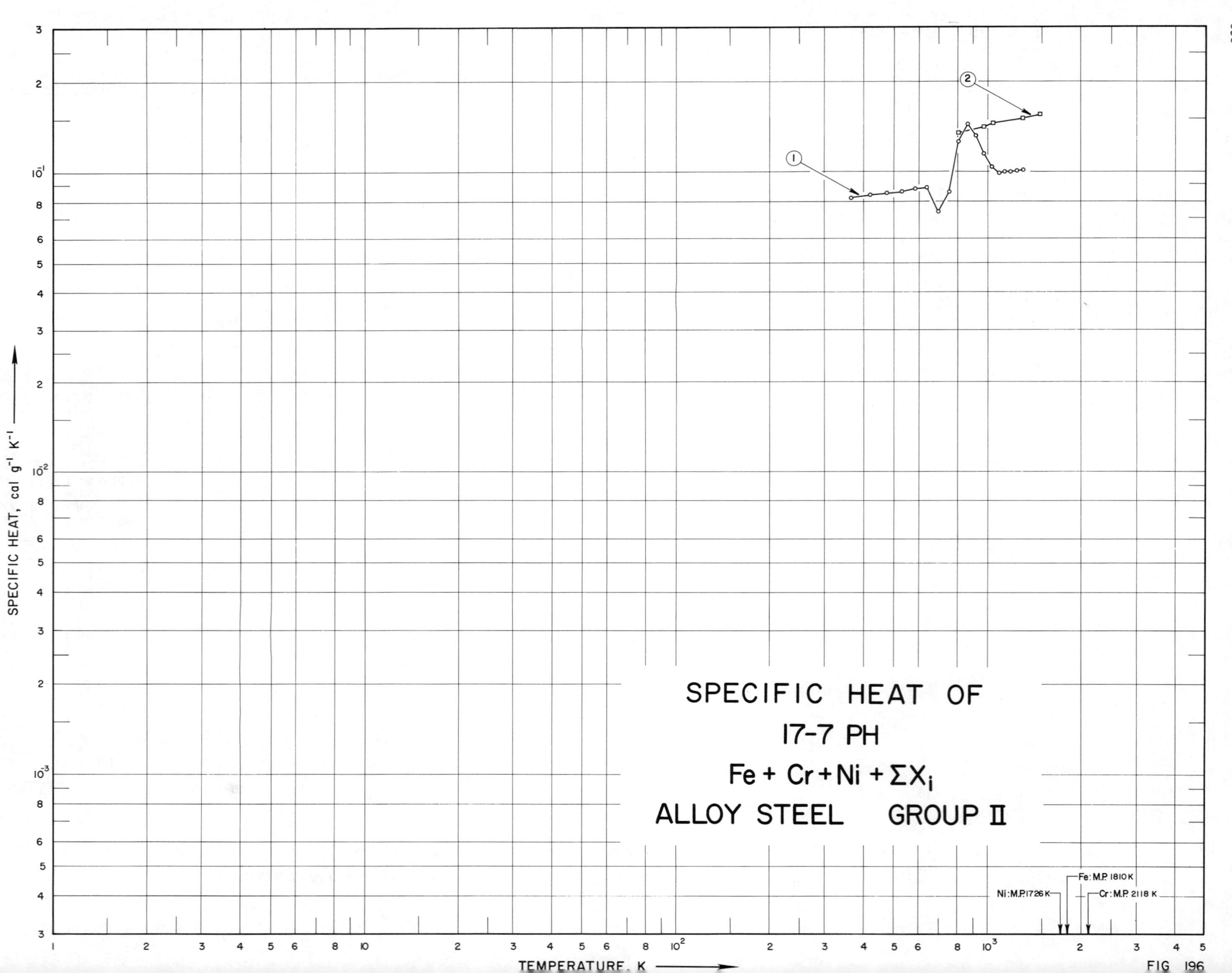

SPECIFIC HEAT OF
17-7 PH
Fe + Cr + Ni + ΣXi
ALLOY STEEL GROUP II
SPECIFIC HEAT, cal g-1 K-1
TEMPERATURE, K
Fe: M.P. 1810 K
Ni: M.P. 1726 K
Cr: M.P. 2118 K
1
2
FIG 196

SPECIFICATION TABLE NO. 196 SPECIFIC HEAT OF 17-7 PH, Fe + Cr + Ni + ΣX_i, ALLOY STEEL GROUP II

[For Data Reported in Figure and Table No. 196]

Curve No.	Ref. No.	Year	Temp. Range, K	Reported Error, %	Name and Specimen Designation	Composition (weight percent), Specifications and Remarks
1	249	1959	366-1311	5-10	17-7 PH	16.99 Cr, 7.26 Ni, 1.25 Al, 0.85 Mn, 0.49 Si, 0.069 C, 0.026 P, and 0.012 S; heated to 1400 F for 1.5 hrs; air cooled; heated to 1050 F for 1.5 hrs; air cooled.
2	245	1958	811-1478		17-7 PH	As received: 72.21 Fe, 17.30 Cr, 7.06 Ni, 1.11 Al, 0.6 Mn, 0.49 Si and 0.074 C; after test: 72.71 Fe, 17.35 Cr, 7.13 Ni, 1.09 Al, 0.55 Mn, 0.52 Si and 0.074 C; density = 483 lb ft^{-3}.

DATA TABLE NO. 196 SPECIFIC HEAT OF 17-7 PH, $Fe + Cr + Ni + \Sigma X_i$, ALLOY STEEL GROUP II

[Temperature, T, K; Specific Heat, C_p, Cal $g^{-1}K^{-1}$]

T	C_p
CURVE 1	
366	8.2×10^{-2}
422	8.4
478	8.5
533	8.6
589	8.8
644	8.9
700	7.4
755	8.6
811	1.27×10^{-1}
866	1.44
922	1.32
978	1.16
1033	1.04
1089	9.90×10^{-2}
1144	1.00×10^{-1}
1200	1.00
1255	1.01
1311	1.02
CURVE 2	
811	1.36×10^{-1}
978	1.41
1144	1.46
1311	1.51
1478	1.55

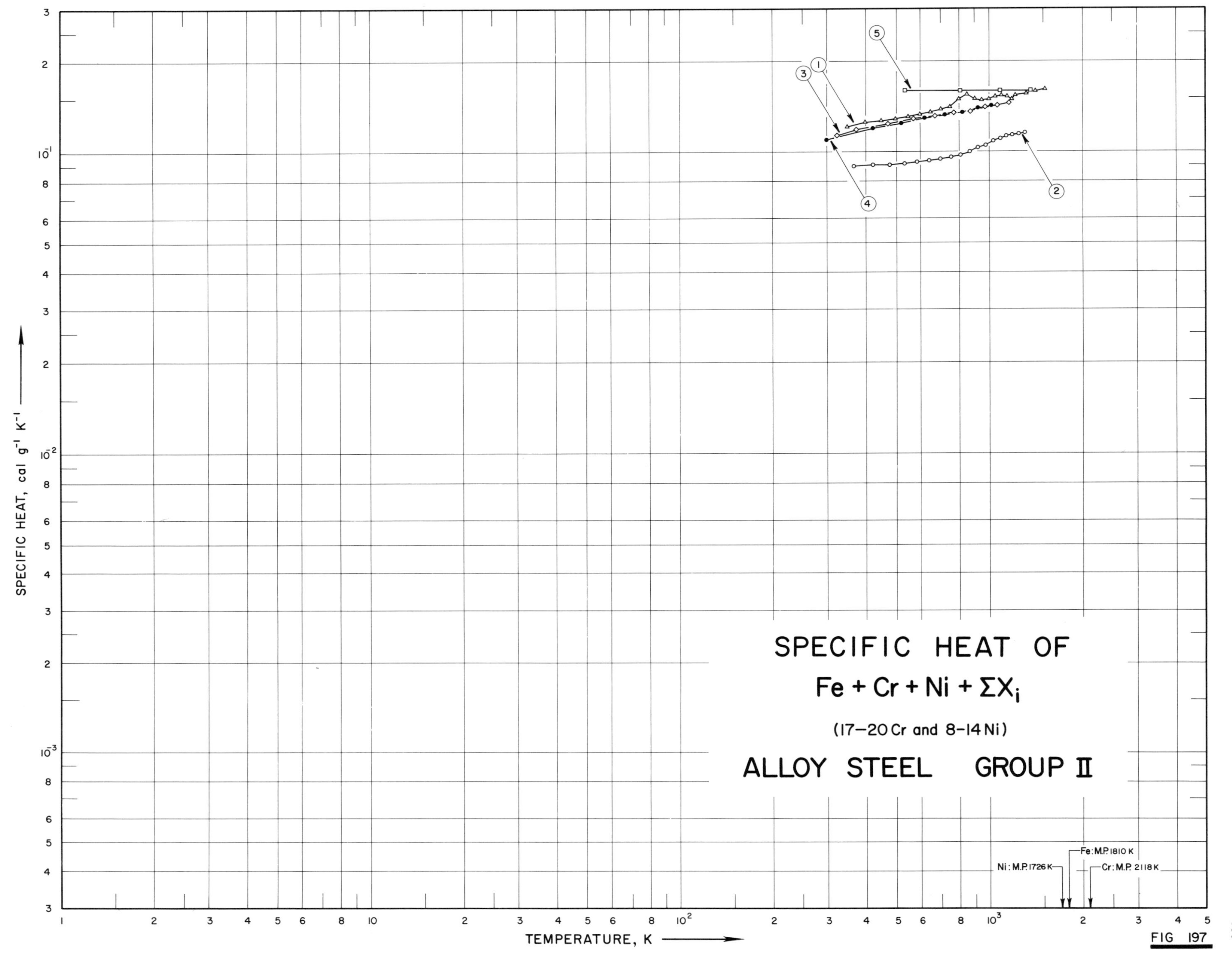
SPECIFIC HEAT OF
Fe + Cr + Ni + ΣXi
(17–20 Cr and 8–14 Ni)
ALLOY STEEL GROUP II
Fe: M.P. 1810 K
Ni: M.P. 1726 K
Cr: M.P. 2118 K
1
2
3
4
5
TEMPERATURE, K
SPECIFIC HEAT, cal g⁻¹ K⁻¹
FIG 197

SPECIFICATION TABLE NO. 197 SPECIFIC HEAT OF Fe + Cr + Ni + ΣX_i (17-20 Cr and 8-14 Ni), ALLOY STEEL GROUP II

[For Data Reported in Figure and Table No. 197]

Curve No.	Ref. No.	Year	Temp. Range, K	Reported Error, %	Name and Specimen Designation	Composition (weight percent), Specifications and Remarks
1	104	1946	348-1523	2-4	High Alloy Steel No. 15	19.11 Cr, 8.14 Ni, 0.68 Si, 0.6 W, 0.37 Mn, 0.08 C, 0.03 Cu, 0.025 As, 0.022 P, 0.011 S and 0.004 Al; heated to 950 C; cooled in water; density = 493 lb ft^{-3} at 15 C.
2	249	1959	366-1311	5-10	AISI 304	18.67 Cr, 9.50 Ni, 1.11 Mn, 0.46 Si, 0.063 C, 0.023 P and 0.017 S; mill annealed condition.
3	15	1959	323-1173	1	Steel Kh18 N9T	Nominal composition: 17.0-20.0 Cr, 8.0-11.0 Ni, 2.0 Mn, 0.8 Si, ≤0.8 Ti, 0.12 C, 0.035 P and 0.03 S.
4	412	1959	298-1073	1	Steel mark 1 x 1 8N9T	Same as above.
5	12	1962	533-1366	≤5	AISI 304	Nominal composition: 18-20 Cr, 8-12 Ni, ≤2 Mn, ≤1.0 Si, ≤0.08 C, ≤0.045 P and ≤0.03 S.

DATA TABLE NO. 197 SPECIFIC HEAT OF $Fe + Cr + Ni + \Sigma X_i$ (17-20 Cr and 8-14 Ni), ALLOY STEEL GROUP II

[Temperature, T, K; Specific Heat, C_p, Cal $g^{-1}K^{-1}$]

T	C_p
CURVE 1	
Series I	
348	1.22 x 10^{-1}
398	1.26
448	1.27
498	1.29
548	1.31
598	1.34
648	1.36
698	1.39
748	1.42
798	1.50
848	1.55
898	1.51
948	1.49
998	1.50
1048	1.53
1098	1.55
1148	1.53
1188	1.5
Series II	
1123	1.58 x 10^{-1}*
1173	1.56*
1223	1.55
1273	1.56*
1323	1.57
1373	1.58*
1423	1.60
1473	1.61*
1523	1.62
CURVE 2	
366	9.0 x 10^{-2}
422	9.1
478	9.1
533	9.2
589	9.3
644	9.4
700	9.5
755	9.65
811	9.80
866	1.01 x 10^{-1}
922	1.04
CURVE 2 (cont.)	
978	1.06 x 10^{-1}
1033	1.09
1089	1.11
1144	1.13
1200	1.14
1255	1.15
1311	1.16
CURVE 3	
323	1.14 x 10^{-1}
373	1.18
473	1.24
573	1.29
673	1.32
773	1.35
873	1.37
973	1.42
1073	1.43
1173	1.46
CURVE 4	
298	1.1 x 10^{-1}
323	1.13*
373	1.18*
423	1.21
473	1.23*
523	1.25
573	1.28*
623	1.30
673	1.31*
723	1.33
773	1.34*
823	1.36
873	1.37*
923	1.40
973	1.42*
1023	1.43
1073	1.42*
CURVE 5	
533	1.600 x 10^{-1}
811	1.600
1089	1.600
1366	1.600

* Not shown on plot

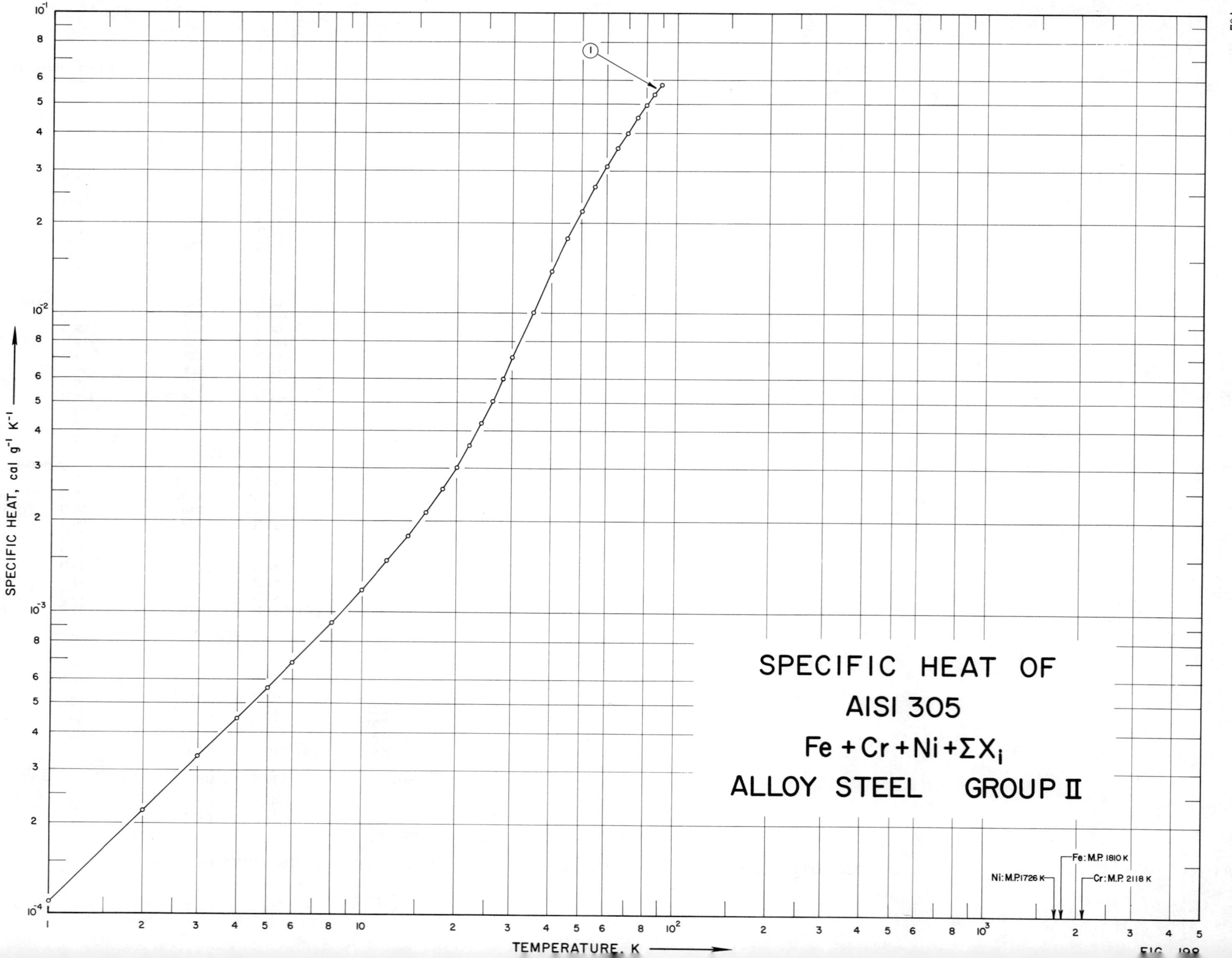
SPECIFIC HEAT OF
AISI 305
Fe + Cr + Ni + ΣXi
ALLOY STEEL GROUP II
Fe: M.P. 1810 K
Ni: M.P. 1726 K
Cr: M.P. 2118 K
TEMPERATURE, K
SPECIFIC HEAT, cal g⁻¹ K⁻¹
1

SPECIFICATION TABLE NO. 198 SPECIFIC HEAT OF AISI 305, $Fe + Cr + Ni + \Sigma X_i$, ALLOY STEEL GROUP II

[For Data Reported in Figure and Table No. 198]

Curve No.	Ref. No.	Year	Temp. Range, K	Reported Error, %	Name and Specimen Designation	Composition (weight percent), Specifications and Remarks
1	413	1965	1-90		AISI 305	17.5 Cr, 11.1 Ni, 0.86 Mn, 0.38 Si, 0.039 C, 0.014 P and 0.01 S.

DATA TABLE NO. 198 SPECIFIC HEAT OF AISI 305, $Fe + Cr + Ni + \Sigma X_i$, ALLOY STEEL GROUP II

[Temperature, T, K; Specific Heat, C_p, Cal $g^{-1}K^{-1}$]

T	C_p
CURVE 1	
1	1.1 x 10^{-4}
2	2.2
3	3.35
4	4.49
5	5.66
6	6.86
8	9.34
10	1.19 x 10^{-3}
12	1.48
14	1.79
16	2.15
18	2.56
20	3.04
22	3.59
24	4.28
26	5.09
28	6.02
30	7.10
35	1.02 x 10^{-2}
40	1.38
45	1.78
50	2.20
55	2.65
60	3.11
65	3.58
70	4.04
75	4.52
80	4.97
85	5.40
90	5.83

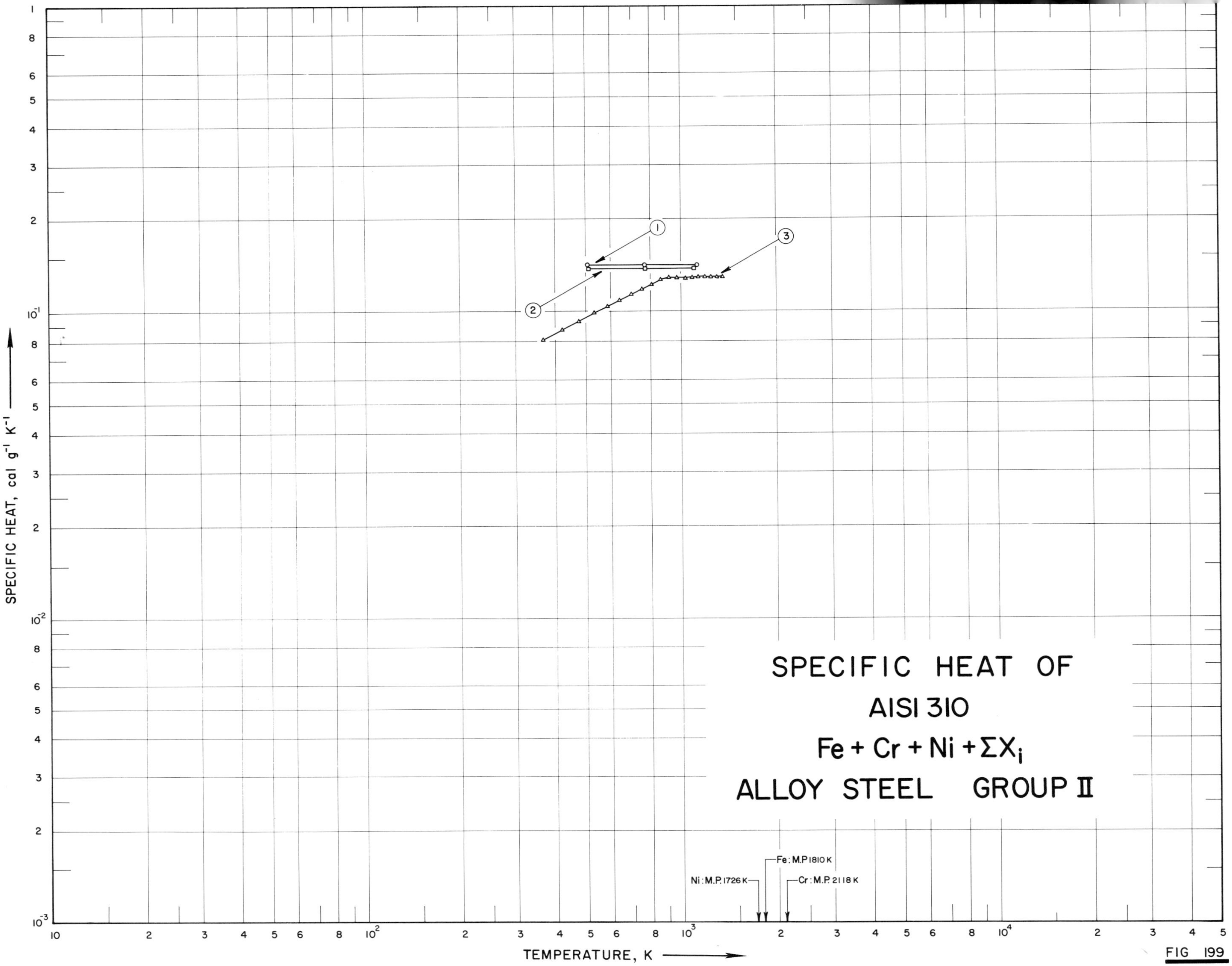

SPECIFIC HEAT OF
AISI 310
$Fe + Cr + Ni + \Sigma X_i$
ALLOY STEEL GROUP II
SPECIFIC HEAT, cal g⁻¹ K⁻¹
TEMPERATURE, K
Ni: M.P. 1726 K
Fe: M.P. 1810 K
Cr: M.P. 2118 K
1
2
3

FIG 199

SPECIFICATION TABLE NO. 199 SPECIFIC HEAT OF AISI 310, Fe + Cr + Ni + ΣX_i, ALLOY STEEL GROUP II

[For Data Reported in Figure and Table No. 199]

Curve No.	Ref. No.	Year	Temp. Range, K	Reported Error, %	Name and Specimen Designation	Composition (weight percent), Specifications and Remarks
1	414	1953	511-1131	±5.0	AISI 310 Heat 64177	24.03 Cr, 16.96 Ni, 0.55 Si, 0.42 Mn, 0.13 C, 0.13 Cu, 0.033 Mo, 0.018 P, 0.01 each Co, Nb, Ta, <0.01 W, Li, Hf, 0.008 S, <0.002 Cd and <0.001 B.
2	414	1953	513-1107	±5.0	AISI 310 Heat 64270	22.3 Cr, 19.14 Ni, 0.5 Mn, 0.43 Si, 0.12 C, 0.1 Cu, 0.06 Nb, 0.042 Mo, 0.025 P, 0.01 Co, <0.01 W and 0.008 S.
3	249	1959	366-1366	5-10	AISI 310	24.90 Cr, 19.63 Ni, 1.6 Mn, 0.42 Si, 0.22 P, 0.036 C and 0.025 S; mill-annealed condition.

DATA TABLE NO. 199 SPECIFIC HEAT OF AISI 310, $Fe + Cr + Ni + \Sigma X_i$, ALLOY STEEL GROUP II

[Temperature, T, K; Specific Heat, C_p, Cal $g^{-1}K^{-1}$]

T	C_p
CURVE 1	
511	1.43 x 10^{-1}
573	1.43*
673	1.43*
773	1.43
873	1.43*
973	1.43*
1073	1.43*
1131	1.43*
CURVE 2	
513	1.39 x 10^{-1}
573	1.39*
673	1.39*
773	1.39
873	1.39*
973	1.39*
1073	1.39*
1107	1.39
CURVE 3	
366	8.2 x 10^{-2}
422	8.8
478	9.4
533	1.00 x 10^{-1}
589	1.055
644	1.11
700	1.16
755	1.20
811	1.24
866	1.28
922	1.30
978	1.30
1033	1.30
1089	1.305
1144	1.31
1200	1.31
1255	1.31
1311	1.31
1366	1.31

* Not shown on plot

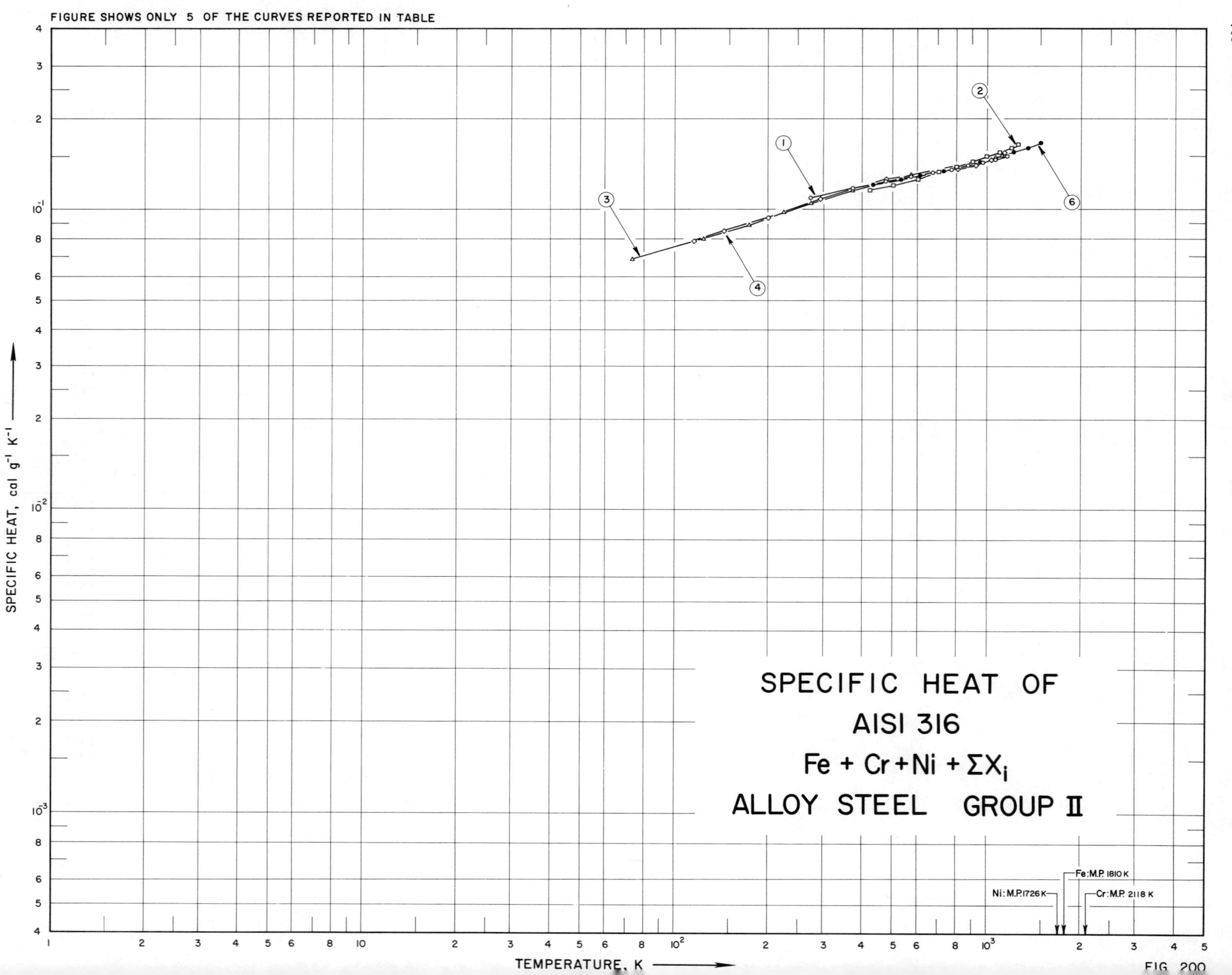
FIGURE SHOWS ONLY 5 OF THE CURVES REPORTED IN TABLE
SPECIFIC HEAT OF
AISI 316
Fe + Cr + Ni + ΣXi
ALLOY STEEL GROUP II
SPECIFIC HEAT, cal g⁻¹ K⁻¹
TEMPERATURE, K
Ni: M.P. 1726 K
Fe: M.P. 1810 K
Cr: M.P. 2118 K
1
2
3
4
6

FIG. 200

SPECIFICATION TABLE NO. 200 SPECIFIC HEAT OF AISI 316, Fe + Cr + Ni + ΣX_i, ALLOY STEEL GROUP II

[For Data Reported in Figure and Table No. 200]

Curve No.	Ref. No.	Year	Temp. Range, K	Reported Error, %	Name and Specimen Designation	Composition (weight percent), Specifications and Remarks
1	150	1957	273-1173		AISI 316	
2	406	1952	423-1273		AISI 316	17.0 Cr, 12.2 Ni, 2.3 Mo, 1.99 Mn, 0.55 Si, 0.12 C, 0.026 P and 0.004 S.
3	243	1958	73-1123		AISI 316	Nominal composition: 16-18 Cr, 10-14 Ni, 2-3 Mo, <2.0 Mn and <0.1 C.
4	10	1958	116-1255		AISI 316	Nominal composition: 16-18 Cr, 10-14 Ni, 2-3 Mo, ≤ 2.0 Mn, ≤ 1.0 Si, ≤ 0.08 C, ≤ 0.045 P and ≤ 0.035 S; sample supplied by the Timken Roller Bearing Co.; sealed in helium capsule; annealed 1 hr at 2000 F; water quenched; density = 496 lb ft^{-3} at 32 F.
5	248	1960	273-1173		AISI 316	17.0 Cr, 12.6 Ni, 2.0 Mo, 1.4 Mn and 0.4 Si; under helium atmosphere.
6	75	1958	433-1500	0.66-2.9	AISI 316	Nominal composition: 16-18 Cr, 10-14 Ni, 2-3 Mo, ≤ 2.0 Mn, ≤ 1.0 Si, ≤ 0.08 C, ≤ 0.045 P and ≤ 0.03 S; helium atmosphere.

DATA TABLE NO. 200 SPECIFIC HEAT OF AISI 316, $Fe + Cr + Ni + \Sigma X_i$, ALLOY STEEL GROUP II

[Temperature, T, K; Specific Heat, C_p, Cal $g^{-1}K^{-1}$]

T	C_p
CURVE 1	
273	1.098 x 10^{-1}
373	1.176
473	1.235
573	1.282
673	1.325
773	1.364
873	1.400
973	1.436
1073	1.470
1173	1.504
CURVE 2	
423	1.171 x 10^{-1}
500	1.215
600	1.272
700	1.329
800	1.386
900	1.443
1000	1.500
1100	1.557
1200	1.614
1273	1.656
CURVE 3	
73	6.9 x 10^{-2}
123	8.0
173	8.9
223	9.8
273	1.05 x 10^{-1}
373	1.17
473	1.25*
573	1.30
673	1.34*
773	1.36*
873	1.39*
973	1.43*
1073	1.48
1123	1.52
CURVE 4	
116	7.9 x 10^{-2}
144	8.5
200	9.4
293	1.08 x 10^{-1}
366	1.16*
478	1.26
589	1.31*
700	1.35
811	1.37
922	1.40
1033	1.47
1144	1.55
1255	1.65*
CURVE 5*	
273	1.100 x 10^{-1}
298	1.122
323	1.142
373	1.177
473	1.234
573	1.282
673	1.325
773	1.364
873	1.402
973	1.438
1073	1.472
1173	1.510
CURVE 6	
433	1.218 x 10^{-1}
445	1.224*
463	1.231*
532	1.261
533	1.261*
550	1.268*
583	1.282*
610	1.294
611	1.294*
675	1.321*
727	1.343
779	1.365*
794	1.372*
795	1.372*
CURVE 6 (cont.)	
796	1.373 x 10^{-1}*
811	1.379*
838	1.390*
887	1.411*
895	1.415*
951	1.438
988	1.454*
1001	1.460*
1089	1.497*
1100	1.502*
1150	1.523*
1197	1.543*
1222	1.554
1279	1.578*
1301	1.587*
1343	1.605*
1366	1.615
1401	1.630*
1451	1.651*
1468	1.658*
1500	1.672

* Not shown on plot

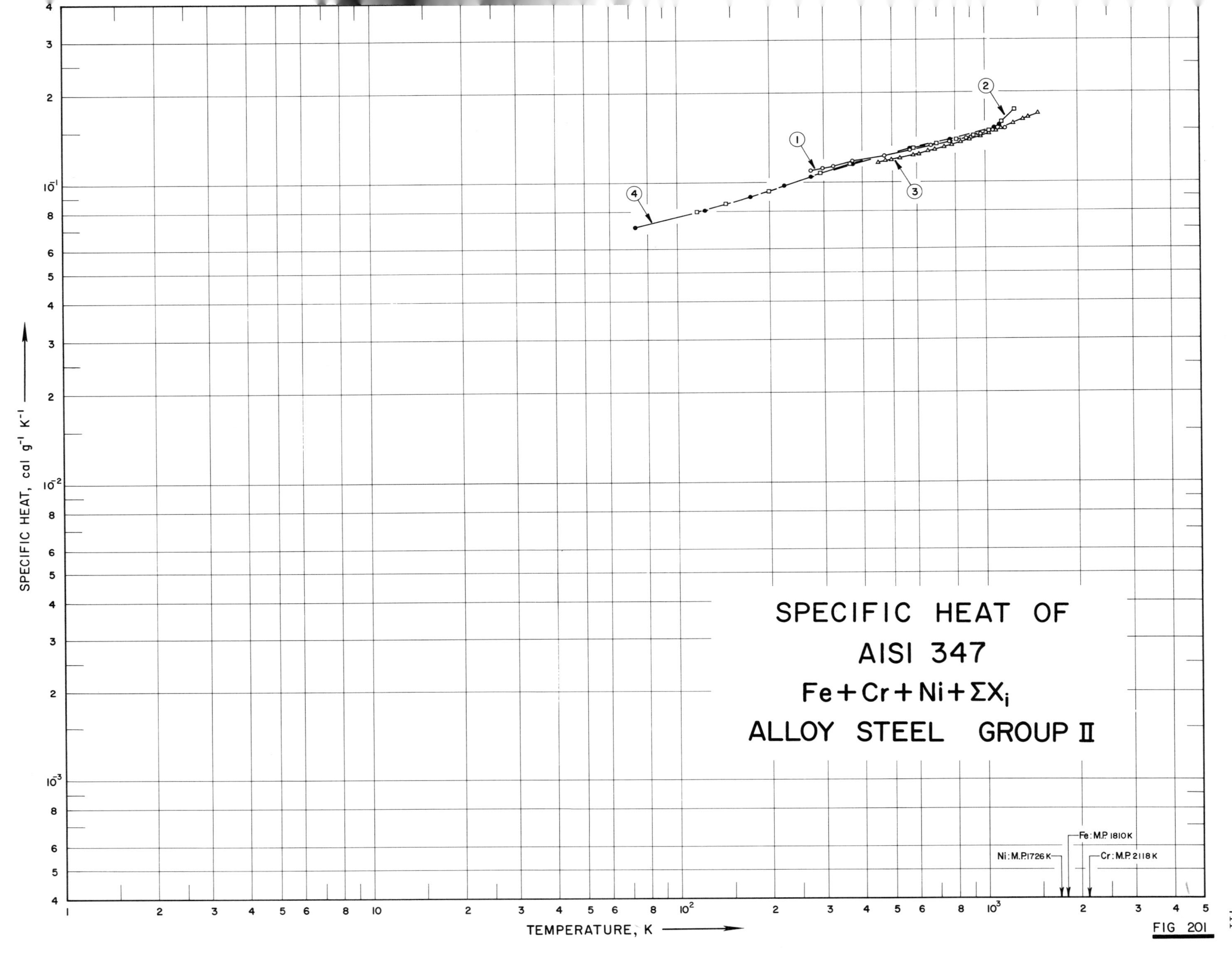
SPECIFIC HEAT OF
AISI 347
Fe+Cr+Ni+ΣXi
ALLOY STEEL GROUP II
SPECIFIC HEAT, cal g⁻¹ K⁻¹
TEMPERATURE, K
Ni: M.P. 1726 K
Fe: M.P. 1810 K
Cr: M.P. 2118 K
1
2
3
4

FIG 201

SPECIFICATION TABLE NO. 201 SPECIFIC HEAT OF AISI 347, Fe + Cr + Ni + ΣX_i, ALLOY STEEL GROUP II

[For Data Reported in Figure and Table No. 201]

Curve No.	Ref. No.	Year	Temp. Range, K	Reported Error, %	Name and Specimen Designation	Composition (weight percent), Specifications and Remarks
1	250	1953	273-1173		AISI 347	18.0 Cr, 11.1 Ni, 1.30 Mn, 0.86 Nb, 0.52 Si and 0.08 C.
2	10	1958	116-1255		AISI 347	Nominal composition: 17-19 Cr, 9-13 Ni, ≤2.0 Mn, ≤1.0 Si, ≤0.08 C, ≤0.047 P, ≤0.03 Si and 10 x C min Nb-Ta; sealed in helium in capsule; annealed 1 hr at 2000 F; water quenched; density = 494 lb ft^{-3} at 32 F.
3	75	1958	451-1133	0.66-2.9	AISI 347	Same as above.
4	243	1954	73-1123		AISI 347	Nominal composition: 17-19 Cr, 9-12 Ni and Nb = 10 X C.

DATA TABLE NO. 201 SPECIFIC HEAT OF AISI 347, $Fe + Cr + Ni + \Sigma X_i$, ALLOY STEEL GROUP II

[Temperature, T, K; Specific Heat, C_p, Cal $g^{-1}K^{-1}$]

T	C_p
CURVE 1	
273	1.104 x 10^{-1}
298	1.123
323	1.142
373	1.176
473	1.233
573	1.283
673	1.326
773	1.370
873	1.410
973	1.448
1073	1.487
1173	1.525
CURVE 2	
166	8.0 x 10^{-2}
144	8.5
200	9.4
293	1.08 x 10^{-1}
366	1.16
478	1.24*
589	1.31
700	1.35
811	1.39
922	1.44
1033	1.49
1144	1.59
1255	1.75
CURVE 3	
451	1.172 x 10^{-1}
454	1.174*
478	1.186
491	1.193*
498	1.196
505	1.200*
507	1.201*
532	1.213
533	1.214*
542	1.219*
585	1.241
597	1.247*
615	1.256
629	1.263*
CURVE 3 (cont.)	
649	1.273 x 10^{-1}*
655	1.276
691	1.295
699	1.299*
699	1.299*
741	1.320
759	1.329*
786	1.343
790	1.345*
811	1.356*
838	1.370
890	1.397
970	1.438
1038	1.472
1086	1.497
1089	1.498*
1133	1.521
1134	1.521*
1140	1.524*
1149	1.529*
1247	1.579
1334	1.623
1366	1.640*
1375	1.644*
1389	1.651
1480	1.698*
1494	1.705
CURVE 4	
73	7.1 x 10^{-2}
123	8.1
173	9.0
223	9.8
273	1.05 x 10^{-1}
373	1.16
473	1.24*
573	1.30
673	1.34*
773	1.38
873	1.42*
973	1.46*
1073	1.52
1123	1.56

* Not shown on plot

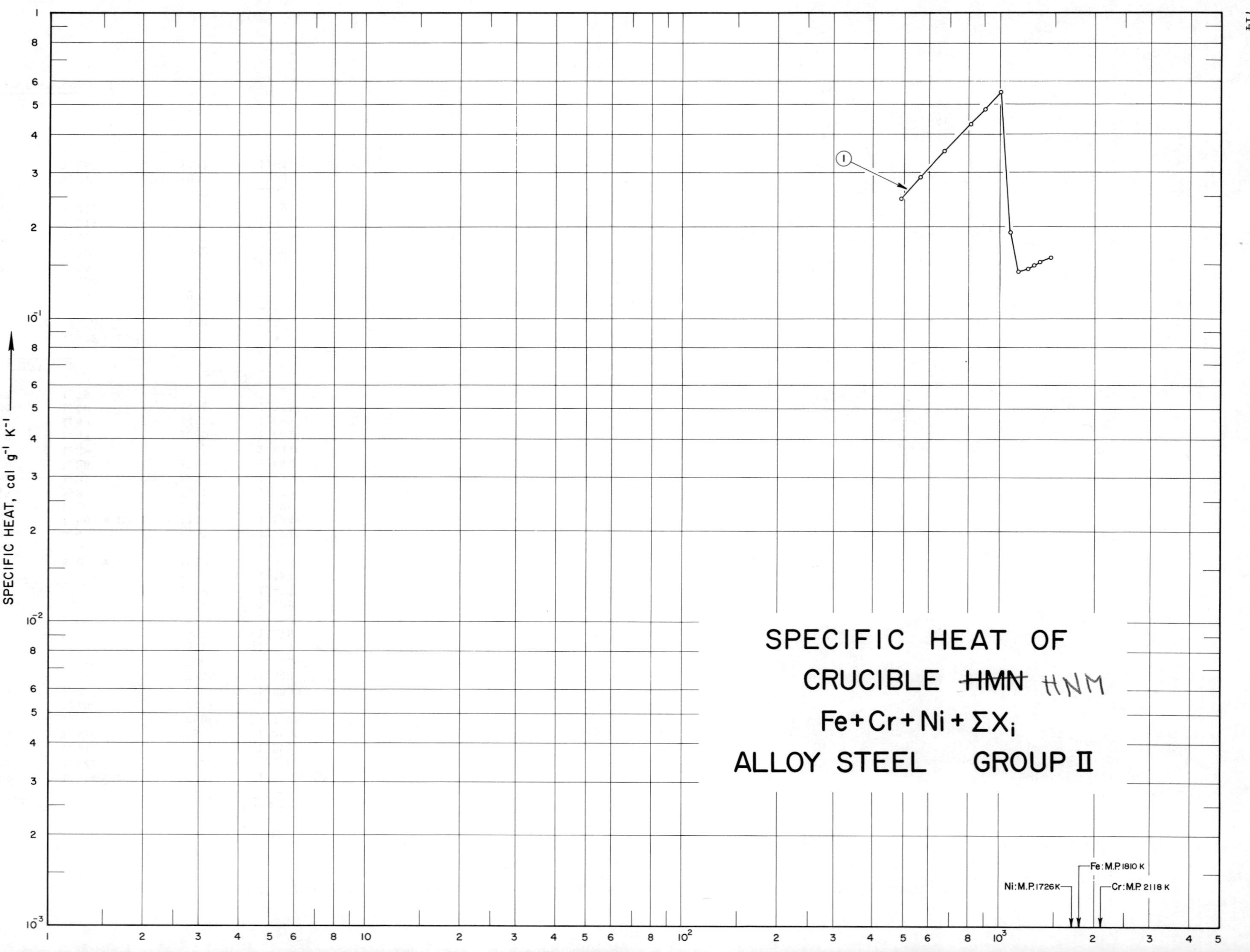
SPECIFIC HEAT OF
CRUCIBLE HMN HNM
Fe+Cr+Ni+ΣXi
ALLOY STEEL GROUP II
SPECIFIC HEAT, cal g⁻¹ K⁻¹
Ni: M.P. 1726 K
Fe: M.P. 1810 K
Cr: M.P. 2118 K
1

SPECIFICATION TABLE NO. 202 SPECIFIC HEAT OF CRUCIBLE ~~HMN~~ HNM, Fe + Cr + Ni + ΣX_i, ALLOY STEEL GROUP II

[For Data Reported in Figure and Table No. 202]

Curve No.	Ref. No.	Year	Temp. Range, K	Reported Error, %	Name and Specimen Designation	Composition (weight percent), Specifications and Remarks
1	146	1961	491-1460	3.0	Crucible ~~HMN~~ HNM	68 Fe, 18.5 Cr, 9.5 Ni, 3.5 Mn, 0.5 Si, 0.3 C, 0.23 P and trace of Mo; helium atmosphere; density = 479 lb ft^{-3}.

DATA TABLE NO. 202 SPECIFIC HEAT OF CRUCIBLE HNM, $Fe + Cr + Ni + \Sigma X_i$, ALLOY STEEL GROUP II

[Temperature, T, K; Specific Heat, C_p, Cal $g^{-1}K^{-1}$]

T	C_p
CURVE 1	
491	2.461×10^{-1}
565	2.884
674	3.505
815	4.315
902	4.812
1020	5.489
1087	1.920
1152	1.425
1231	1.466
1291	1.497
1350	1.528
1460	1.585

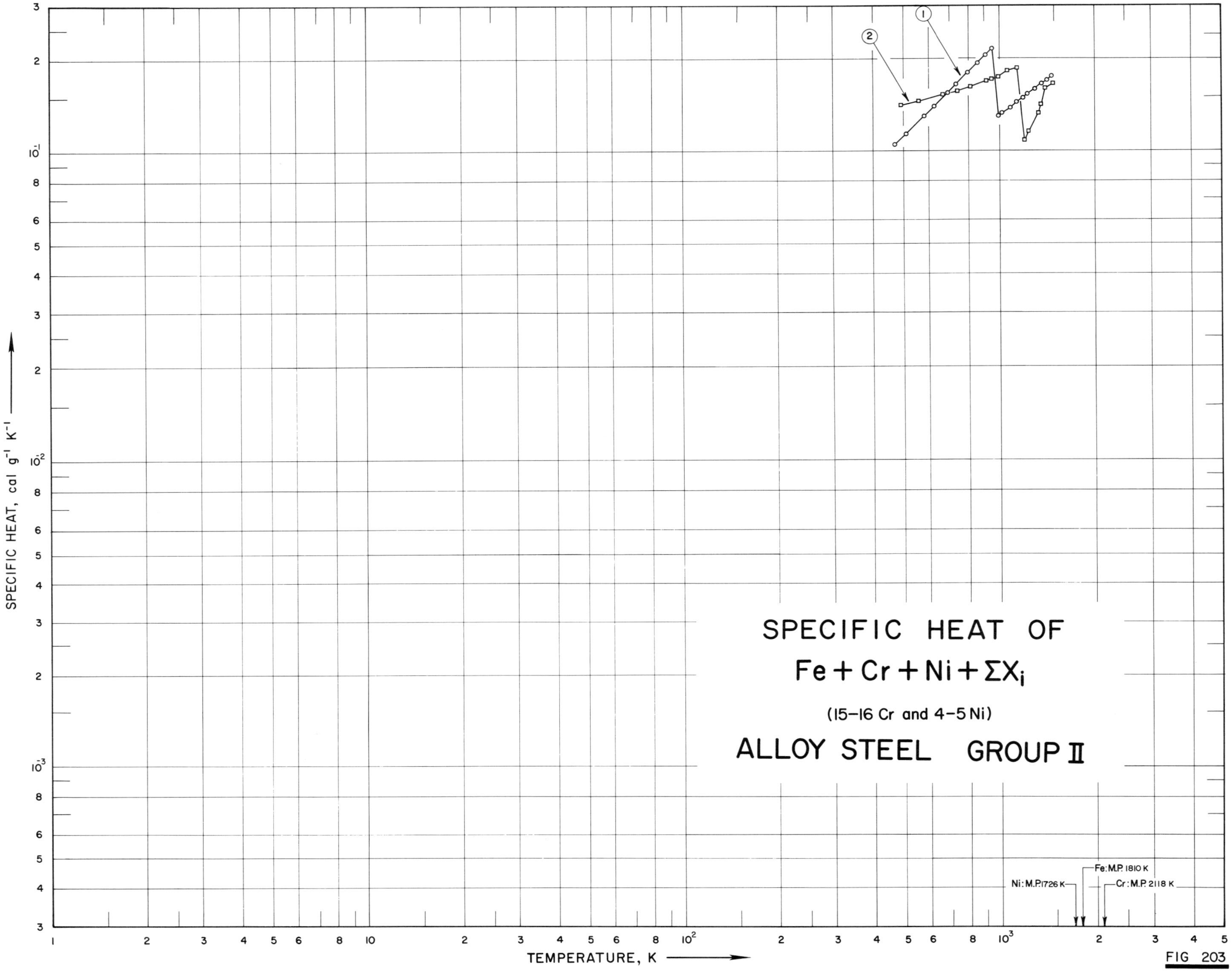

SPECIFIC HEAT OF

$Fe + Cr + Ni + \Sigma X_i$

(15–16 Cr and 4–5 Ni)

ALLOY STEEL GROUP II

FIG 203

SPECIFICATION TABLE NO. 203 SPECIFIC HEAT OF Fe + Cr + Ni + ΣX_i (15-16 Cr and 4-5 Ni), ALLOY STEEL GROUP II

[For Data Reported in Figure and Table No. 203]

Curve No.	Ref. No.	Year	Temp. Range, K	Reported Error, %	Name and Specimen Designation	Composition (weight percent), Specifications and Remarks
1	146	1961	472-1474	3.0	Stainless Steel type 17-4 PH	72.9 Fe, 16.4 Cr, 4.2 Ni, 4.1 Cu, 1.0 Mn, 1.0 Si, 0.3 Nb + Ta, 0.07 C and 0.04 P; helium atmosphere; density = 482 lb ft^{-3}.
2	146	1961	493-1487	3.0	AM355	75.5 Fe, 15.66 Cr, 4.27 Ni, 2.82 Mo, 0.94 Mn, 0.12 C, 0.05 Si and 0.02 P; helium atmosphere; density = 485 lb ft^{-3}.

DATA TABLE NO. 203 SPECIFIC HEAT OF Fe + Cr + Ni + ΣX_i (15-16 Cr and 4-5 Ni), ALLOY STEEL GROUP II

[Temperature, T, K; Specific Heat, C_p, Cal $g^{-1}K^{-1}$]

T	C_p
CURVE 1	
472	1.052×10^{-1}
511	1.142
583	1.308
625	1.405
691	1.558
733	1.653
799	1.805
856	1.938
890	2.016*
909	2.058
954	2.162
996	1.309
1024	1.336
1080	1.389
1143	1.449
1194	1.497
1243	1.543
1248	1.547*
1302	1.599
1372	1.665
1423	1.714
1474	1.762
CURVE 2	
493	1.415×10^{-1}
559	1.459
673	1.533
741	1.578
816	1.627
919	1.696
950	1.715
999	1.747
1070	1.820
1110	1.844*
1146	1.860
1171	1.884*
1207	1.084
1248	1.168
1327	1.332
1408	1.610
1367	1.413
1470	1.624*
1489	1.663

* Not shown on plot

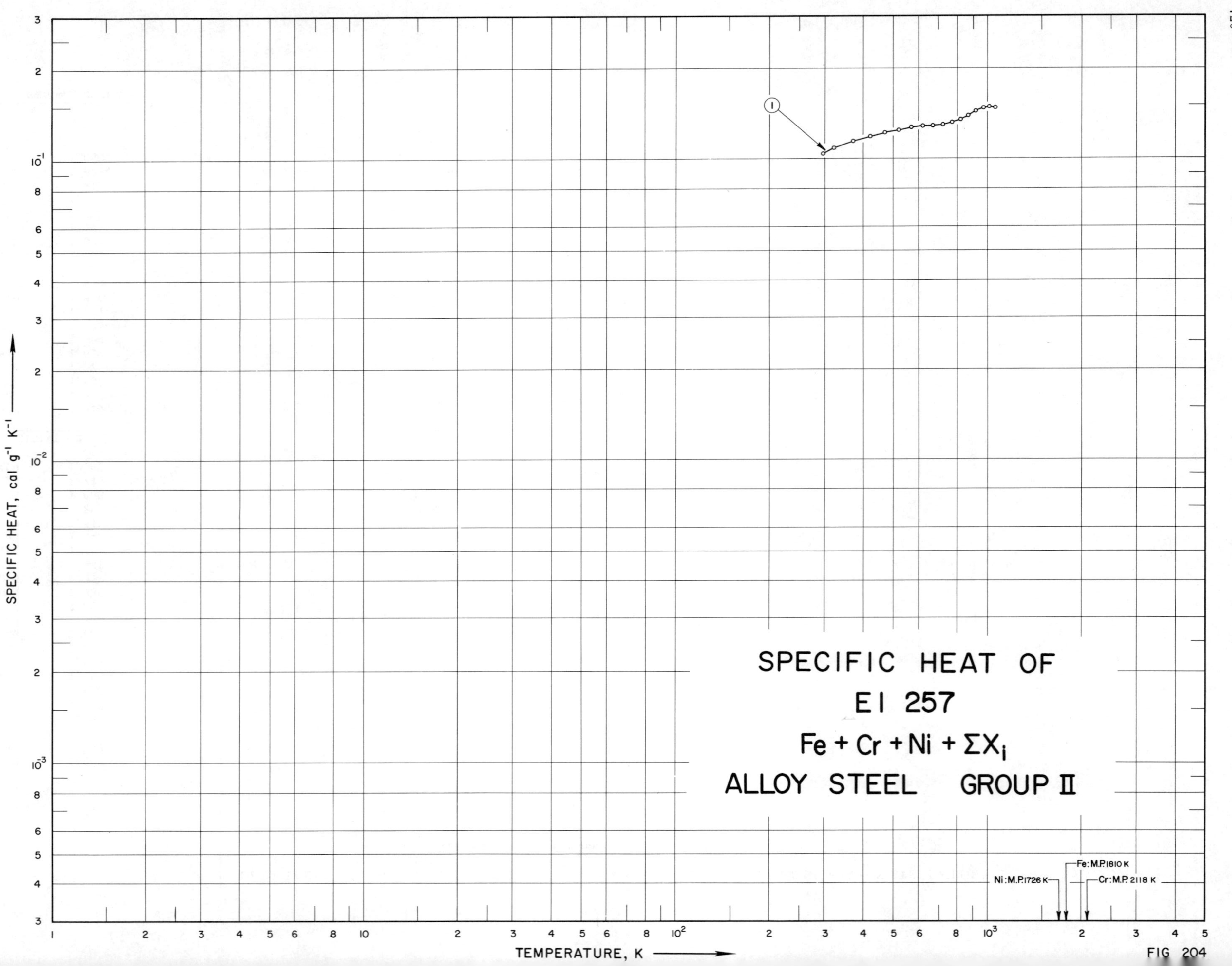
SPECIFIC HEAT OF
E I 257
Fe + Cr + Ni + ΣXi
ALLOY STEEL GROUP II
Fe: M.P. 1810 K
Ni: M.P. 1726 K
Cr: M.P. 2118 K
1
SPECIFIC HEAT, cal g⁻¹ K⁻¹
TEMPERATURE, K
FIG 204

SPECIFICATION TABLE NO. 204 SPECIFIC HEAT OF EI25~~7~~, Fe + Cr + Ni + ΣX_i, ALLOY STEEL GROUP II
75 EI275

[For Data Reported in Figure and Table No. 204]

Curve No.	Ref. No.	Year	Temp. Range, K	Reported Error, %	Name and Specimen Designation	Composition (weight percent), Specifications and Remarks
1	412	1959	298-1073	1. 0	EI275	Nominal composition: 13. 0-15. 0 Cr, 13. 0-15. 0 Ni, 2. 0-2. 75 W, 0. 8 Si, 0. 7 Mn, 0. 4-0. 6 Mo, 0. 035 P, 0. 03 S and 0. 15 C.

DATA TABLE NO. 204 SPECIFIC HEAT OF EI 257, $Fe + Cr + Ni + \Sigma X_i$, ALLOY STEEL GROUP II

[Temperature, T, K; Specific Heat, C_p, Cal $g^{-1}K^{-1}$]

T	C_p
CURVE 1	
298	1.06×10^{-1}
323	1.10
373	1.16
423	1.20
473	1.23
523	1.26
573	1.28
623	1.30
673	1.30
723	1.31
773	1.33
823	1.37
873	1.41
923	1.46
973	1.49
1023	1.50
1073	1.49

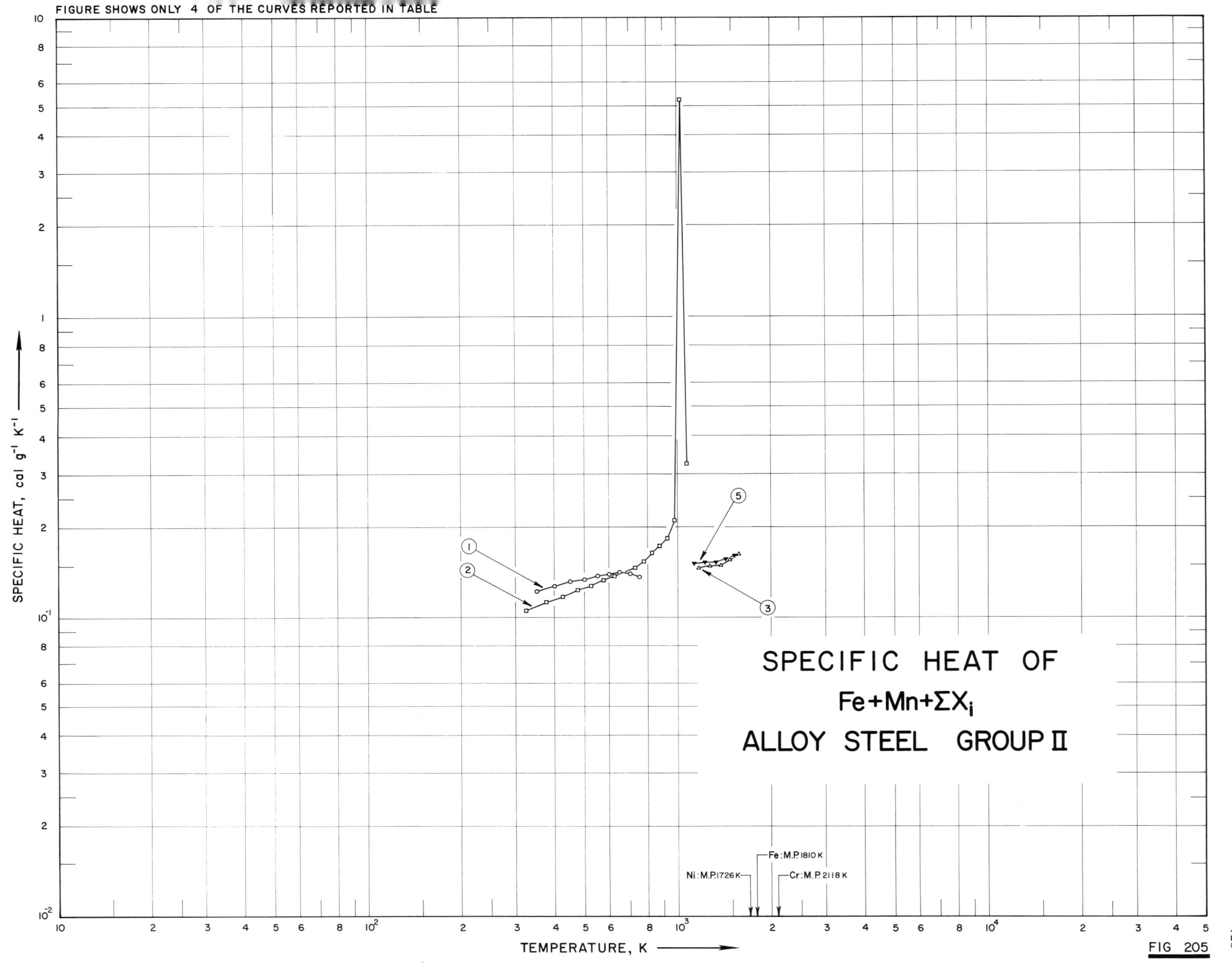
FIGURE SHOWS ONLY 4 OF THE CURVES REPORTED IN TABLE
SPECIFIC HEAT OF
Fe+Mn+ΣXi
ALLOY STEEL GROUP II
SPECIFIC HEAT, cal g-1 K-1
TEMPERATURE, K
Ni: M.P. 1726 K
Fe: M.P. 1810 K
Cr: M.P. 2118 K
1
2
3
5
10
8
6
5
4
3
2
1
10-1
10-2
102
103
104
FIG 205

SPECIFICATION TABLE NO. 205 SPECIFIC HEAT OF Fe + Mn + ΣX_i, ALLOY STEEL GROUP II

[For Data Reported in Figure and Table No. 205]

Curve No.	Ref. No.	Year	Temp. Range, K	Reported Error, %	Name and Specimen Designation	Composition (weight percent), Specifications and Remarks
1	104	1946	348-748		High Alloy Steel No. 13	13.0 Mn, 1.22 C, 0.22 Si, 0.07 Cu, 0.07 Ni, 0.038 As, 0.038 P, 0.03 Cr, 0.01 S and 0.004 Al; heated to 1050 C; cooled in air; density = 491 lb ft^{-3} at 15 C.
2	412	1959	323-1073	1.0	Tempered Steel Mark 12 ~~MX~~ MKh	Nominal composition: 0.4-0.7 Mn, 0.4-0.6 Cr, 0.4-0.6 Mo, 0.3 Ni, 0.15-0.3 Si, 0.25 Cu, 0.09-0.16 C, 0.04 P and 0.04 S.
3	104	1946	1173-1573	4.0	Carbon Steel No. 5	0.643 Mn, 0.415 C, 0.12 Cu, 0.11 Si, 0.063 Ni, 0.033 As, 0.031 P, 0.029 S, and 0.006 Al; annealed at 860 C; density = 490 lb ft^{-3} at 15 C.
4	104	1946	1173-1573	4.0	Carbon Steel No. 6	0.69 Mn, 0.435 C, 0.2 Si, 0.06 Cu, 0.04 Ni, 0.038 S, 0.037 P, 0.03 Cr, 0.024 As, 0.01 Mo and 0.006 Al; annealed at 860 C; density = 489 lb ft^{-3}.
5	104	1946	1123-1573	4.0	Carbon Steel No. 3	0.635 Mn, 0.23 C, 0.13 Cu, 0.11 Si, 0.074 Ni, 0.036 As, 0.034 P, 0.034 S, and 0.01 Al; annealed at 930 C; density = 490 lb ft^{-3} at 15 C.

DATA TABLE NO. 205 SPECIFIC HEAT OF Fe + Mn + ΣX_i, ALLOY STEEL GROUP II

[Temperature, T, K; Specific Heat, C_p, Cal $g^{-1} K^{-1}$]

T	C_p	T	C_p
CURVE 1		CURVE 4*	
348	1.24 x 10^{-1}	1173	1.50 x 10^{-1}
398	1.29	1223	1.50
448	1.33	1273	1.50
498	1.36	1323	1.50
548	1.39	1373	1.52
598	1.41	1423	1.55
648	1.43	1473	1.58
698	1.42	1523	1.61
748	1.38	1573	1.64
CURVE 2		CURVE 5	
323	1.07 x 10^{-1}	1123	1.54 x 10^{-1}
373	1.13	1173	1.55*
423	1.18	1223	1.55
473	1.24	1273	1.55*
523	1.28	1323	1.55
573	1.35	1373	1.56*
623	1.38	1423	1.58
673	1.42*	1473	1.60*
723	1.48	1523	1.63
773	1.56	1573	1.66*
823	1.67		
873	1.75		
923	1.85		
973	2.12		
1023	5.320 x 10^{0}		
1073	3.26 x 10^{-1}		
CURVE 3			
1173	1.48 x 10^{-1}		
1223	1.49*		
1273	1.50		
1323	1.51*		
1373	1.52		
1423	1.54*		
1473	1.58		
1523	1.62*		
1573	1.66		

* Not shown on plot

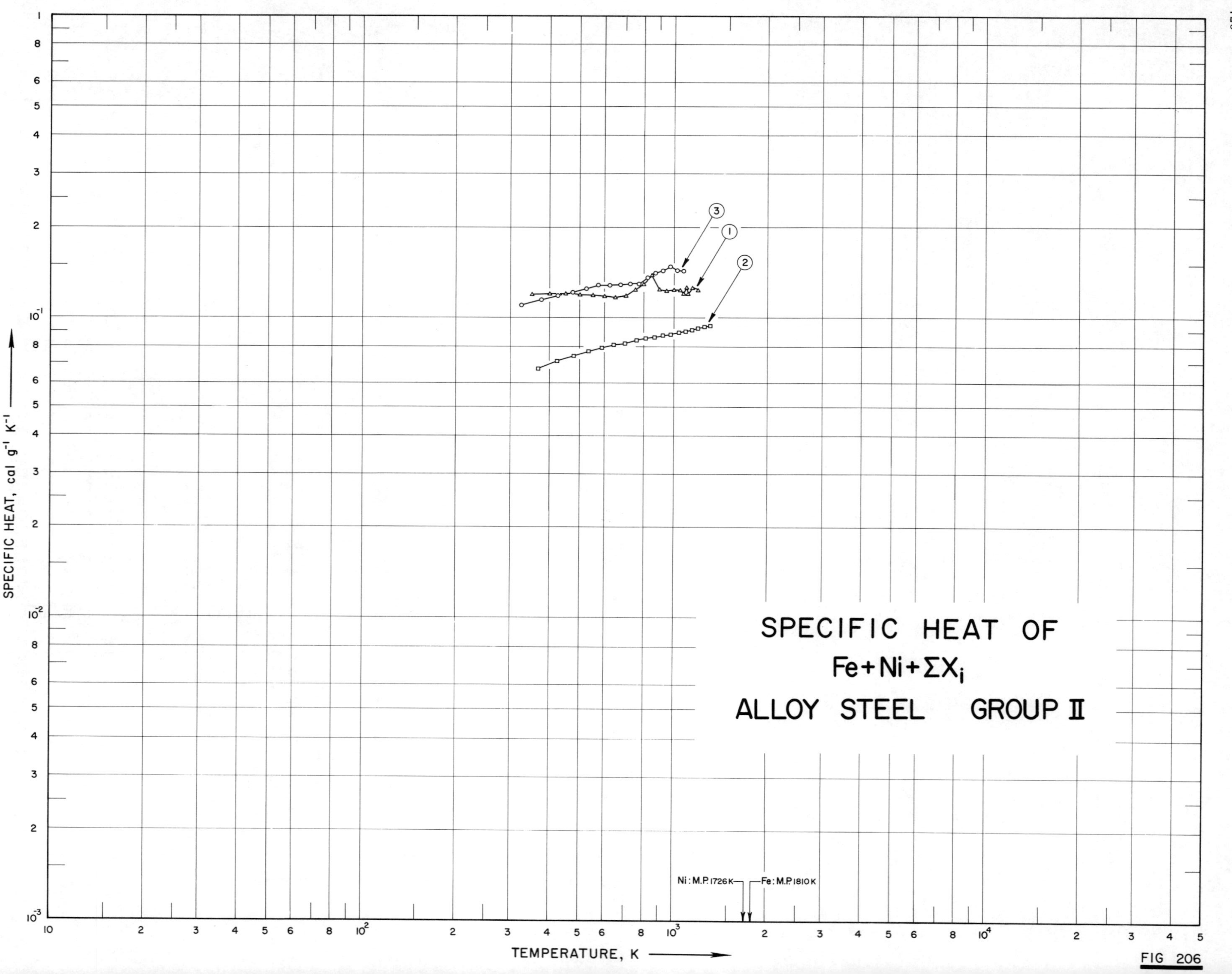
SPECIFIC HEAT OF
Fe+Ni+ΣXi
ALLOY STEEL GROUP II
SPECIFIC HEAT, cal g⁻¹ K⁻¹
TEMPERATURE, K
Ni: M.P. 1726 K
Fe: M.P. 1810 K
1
2
3

FIG 206

SPECIFICATION TABLE NO. 206 SPECIFIC HEAT OF Fe + Ni + ΣX_i, ALLOY STEEL GROUP II

[For Data Reported in Figure and Table No. 206]

Curve No.	Ref. No.	Year	Temp. Range, K	Reported Error, %	Name and Specimen Designation	Composition (weight percent), Specifications and Remarks
1	104	1946	348-1188		High Alloy Steel No. 14	28.37 Ni, 0.89 Mn, 0.28 C, 0.15 Si, 0 03 Cu, 0.027 As, 0.012 Al, 0.009 P and 0.003 S; heated to 950 C; cooled in water; density = 509 lb ft^{-3} at 15 C.
2	249	1955	366-1311	5-10	Incoloy	Nominal composition: 45.0 Fe, 34.0 Ni, 21.0 Cr, 0.5 Cu and 0.05 C; heated to 1975 F for 0.5 hrs; air cooled.
3	252	1964	323-1073	±1.0	OKh16N36V3T EI 855	36.55 Ni, 15.5 Cr, 2.88 W, 0.55 Si, 0.46 Mn, 0.31 Ti, 0.08 C, 0.047 S, and 0.0125 P; quenched in air from 1100 C.

DATA TABLE NO. 206 SPECIFIC HEAT OF Fe + Ni + ΣX_i, ALLOY STEEL GROUP II

[Temperature, T, K; Specific Heat, C_p, Cal $g^{-1}K^{-1}$]

T	C_p
CURVE 1	
Series I	
348	1.19 x 10^{-1}
398	1.20
448	1.20
498	1.19
548	1.19
598	1.18
648	1.17
698	1.19
748	1.24
798	1.29
848	1.30
898	1.25
948	1.23
998	1.24
1048	1.23
1098	1.27
1148	1.26
Series II	
1078	1.2 x 10^{-1}
1088	1.2*
1098	1.2*
1108	1.2
1118	1.4*
1128	1.2*
1188	1.3
CURVE 2	
366	6.7 x 10^{-2}
422	7.1
478	7.4
533	7.7
589	7.9
644	8.1
700	8.2
755	8.4
811	8.5
866	8.6
922	8.7
978	8.8
1033	8.9
CURVE 2 (cont.)	
1089	9.0 x 10^{-2}
1144	9.1
1200	9.2
1255	9.3
1311	9.4
CURVE 3	
323	1.111 x 10^{-1}
373	1.152
423	1.181
473	1.221
523	1.250
573	1.281
623	1.279
673	1.281
723	1.291
773	1.300
823	1.372
873	1.410
923	1.432
973	1.482
1023	1.448
1073	1.441

* Not shown on plot

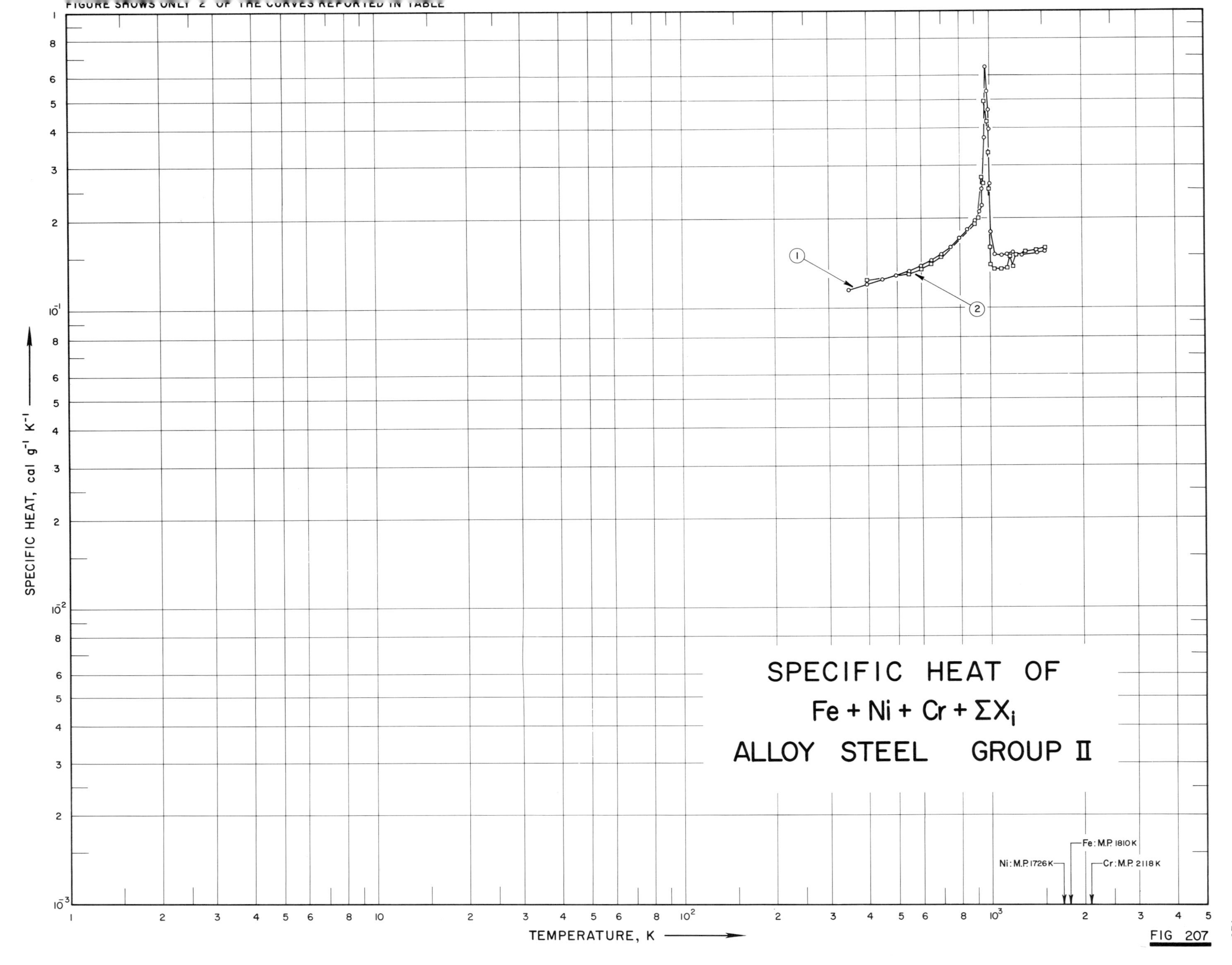
FIGURE SHOWS ONLY 2 OF THE CURVES REPORTED IN TABLE
SPECIFIC HEAT OF
Fe + Ni + Cr + ΣXi
ALLOY STEEL GROUP II
SPECIFIC HEAT, cal g-1 K-1
TEMPERATURE, K
Ni: M.P. 1726 K
Fe: M.P. 1810 K
Cr: M.P. 2118 K
1
2
FIG 207

SPECIFICATION TABLE NO. 207 SPECIFIC HEAT OF Fe + Ni + Cr + ΣX_i, ALLOY STEEL GROUP II

[For Data Reported in Figure and Table No. 207]

Curve No.	Ref. No.	Year	Temp. Range, K	Reported Error, %	Name and Specimen Designation	Composition (weight percent), Specifications and Remarks
1	104	1946	348-1528	2-4	Alloy Steel No. 12	3.53 Ni, 0.78 Cr, 0.55 Mn, 0.39 Mo, 0.34 C, 0.27 Si, 0.05 Cu, 0.037 As, 0.024 P, 0.007 Al, and 0.003 S; annealed at 860 C; reheated to 640 C and cooled in furnace; density = 490 lb ft^{-3} at 15 C.
2	104	1946	348-1523	2-4	Alloy Steel No. 11	3.41 Ni, 0.71 Cr, 0.55 Mn, 0.325 C, 0.25 Si, 0.12 Cu, 0.06 Mo, 0.025 S, 0.023 As, 0.018 P, 0.01 V and 0.008 Al; annealed at 860 C; reheated to 640 C and cooled in furnace; density = 489 lb ft^{-3} at 15 C.
3	104	1946	348-1523	2-4	Alloy Steel No. 10	3.38 Ni, 0.8 Cr, 0.53 Mn, 0.33 C, 0.17 Si, 0.07 Mo, 0.053 Cu, 0.033 S, 0.031 P, 0.028 As, 0.01 V, and 0.008 Al; annealed at 860 C; reheated to 640 C and cooled in furnace; density = 488 lb ft^{-3} at 15 C.

DATA TABLE NO. 207 SPECIFIC HEAT OF $Fe + Ni + Cr + \Sigma X_i$, ALLOY STEEL GROUP II

[Temperature, T, K; Specific Heat, C_p, Cal $g^{-1}K^{-1}$]

CURVE 1

T	C_p
Series I	
348	1.16 x 10^{-1}
398	1.21
448	1.25
498	1.29
548	1.33
598	1.39
648	1.45
698	1.52
748	1.60
798	1.72
848	1.84
898	1.97
948	2.51
998	3.97
1048	1.52
1098	1.50
1148	1.52
1198	1.54
Series II	
928	2.1 x 10^{-1}
938	2.1*
948	2.2
958	2.5
968	3.7
978	6.4
988	5.3
998	4.6
1008	2.6
1018	1.8
Series III	
1178	1.55 x 10^{-1}*
1228	1.53*
1278	1.51
1328	1.51*
1378	1.52*
1428	1.53
1478	1.54*
1528	1.55

CURVE 2

T	C_p
Series I	
348	1.17 x 10^{-1}*
398	1.23
448	1.25*
498	1.28*
548	1.31
598	1.36
648	1.42
698	1.49
748	1.59*
798	1.70*
848	1.82*
898	1.92
948	2.74
998	3.30
1048	1.36
1098	1.36
1148	1.37
1198	1.38
Series II	
928	2.0 x 10^{-1}
938	2.0*
948	2.2*
958	2.6
968	4.9
988	4.2
998	2.5
1008	1.6
1018	1.4
Series III	
1173	1.49 x 10^{-1}
1223	1.51
1273	1.53*
1323	1.55
1373	1.56*
1423	1.57
1473	1.58*
1523	1.59

CURVE 3*

T	C_p
Series I	
348	1.17 x 10^{-1}
398	1.21
448	1.25
498	1.29
548	1.34
598	1.39
648	1.43
698	1.51
748	1.61
798	1.72
848	1.85
898	1.94
948	3.12
998	2.81
1048	1.33
1098	1.36
1148	1.39
1198	1.43
Series II	
928	2.0 x 10^{-1}
938	2.1
948	2.2
958	2.9
968	6.4
978	6.4
988	3.2
998	1.8
1008	1.3
1018	1.3
Series III	
1173	1.50 x 10^{-1}
1223	1.51
1273	1.52
1323	1.53
1373	1.54
1423	1.55
1473	1.56
1523	1.57

* Not shown on plot

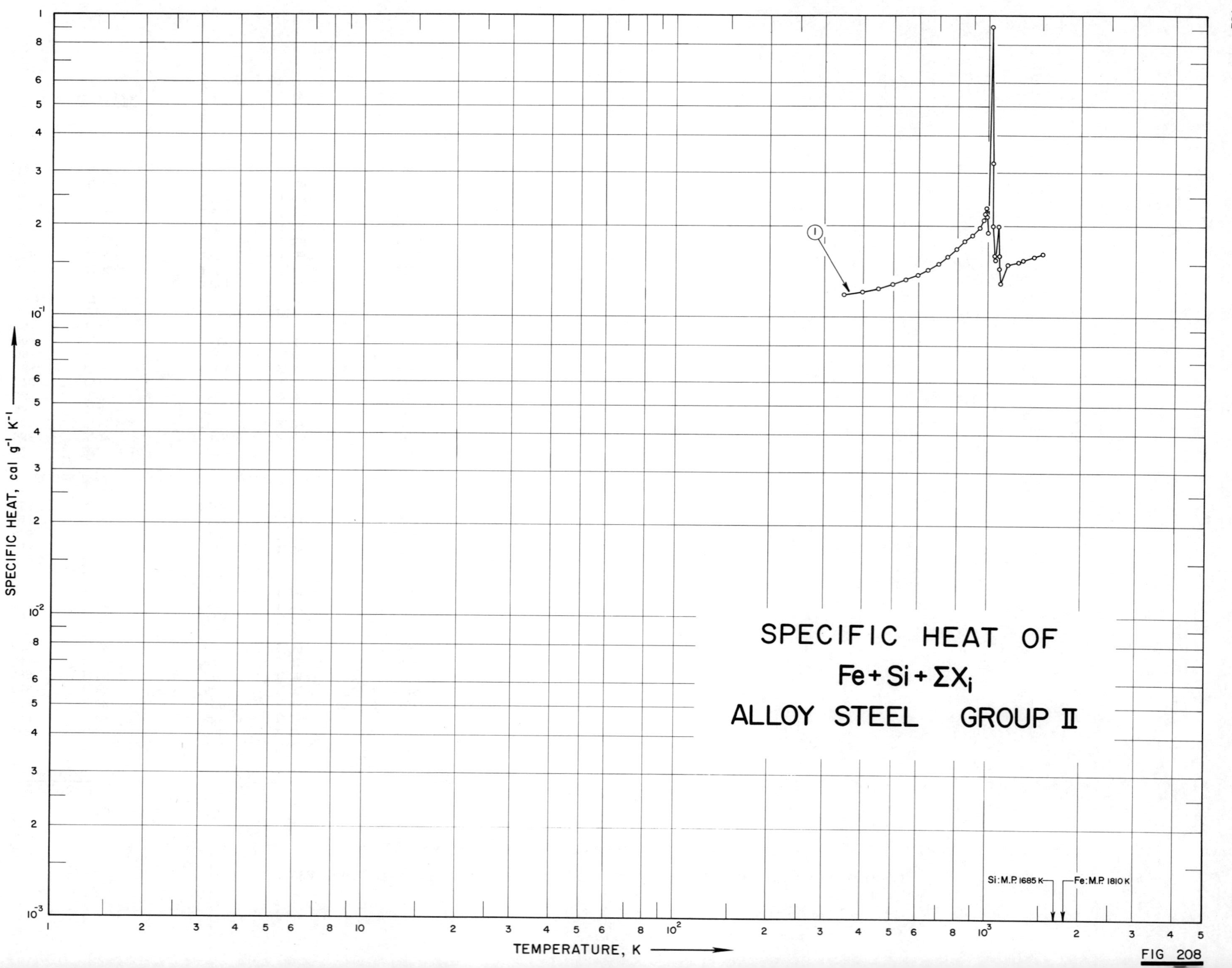

SPECIFIC HEAT OF
Fe + Si + ΣX$_i$
ALLOY STEEL GROUP II
Si: M.P. 1685 K
Fe: M.P. 1810 K
TEMPERATURE, K
SPECIFIC HEAT, cal g^{-1} K^{-1}
1

FIG 208

SPECIFICATION TABLE NO. 208 SPECIFIC HEAT OF Fe + Si + ΣX_i, ALLOY STEEL GROUP II

[For Data Reported in Figure and Table No. 208]

Curve No.	Ref. No.	Year	Temp. Range, K	Reported Error, %	Name and Specimen Designation	Composition (weight percent), Specifications and Remarks
1	104	1946	348-1523	2-4	Alloy Steel No. 21	1.98 Si, 0.9 Mn, 0.637 Cu, 0.485 C, 0.156 Ni, 0.047 S, 0.044 P, 0.04 Cr, 0.029 As, and 0.007 Al; annealed at 930 C; density = 482 lb ft^{-3} at 15 C.

DATA TABLE NO. 208 SPECIFIC HEAT OF $Fe + Si + \Sigma X_i$, ALLOY STEEL GROUP II

[Temperature, T, K; Specific Heat, C_p, Cal $g^{-1} K^{-1}$]

T	C_p
CURVE 1	
Series I	
348	1.19 x 10^{-1}
398	1.22
448	1.25
498	1.29
548	1.33
598	1.38
648	1.44
698	1.51
748	1.59
798	1.68
848	1.79
898	1.87
948	1.98
998	2.16
1048	3.26
1098	1.46
Series III	
978	2.1 x 10^{-1}
988	2.2
998	2.3
1008	2.2*
1018	1.9
1028	1.9*
1038	9.2
1048	2.0
1058	1.6
1068	1.55
1088	2.0
1098	1.6
1108	1.5*
1118	1.3
Series III	
1123	1.48 x 10^{-1}*
1173	1.50
1223	1.51*
1273	1.53
1323	1.55
1373	1.57*
1423	1.59
1473	1.61*
1523	1.63

* Not shown on plot

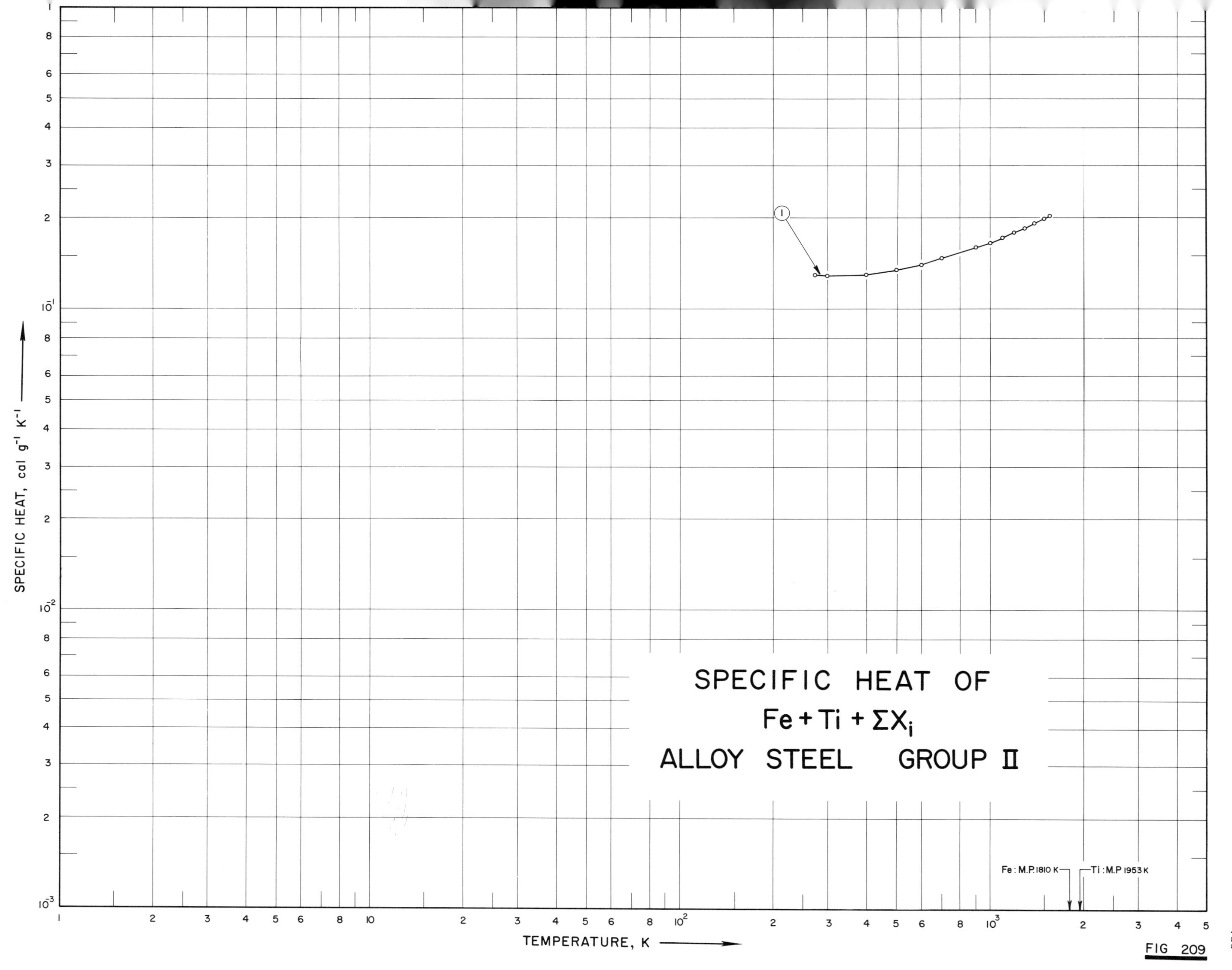
SPECIFIC HEAT OF
Fe + Ti + ΣXi
ALLOY STEEL GROUP II
SPECIFIC HEAT, cal g⁻¹ K⁻¹
TEMPERATURE, K
Fe: M.P. 1810 K
Ti: M.P 1953 K
1

FIG 209

SPECIFICATION TABLE NO. 209 SPECIFIC HEAT OF Fe + Ti + ΣX_i, ALLOY STEEL GROUP II

[For Data Reported in Figure and Table No. 209]

Curve No.	Ref. No.	Year	Temp. Range, K	Reported Error, %	Name and Specimen Designation	Composition (weight percent), Specifications and Remarks
1	253	1961	273-1573	1.2		60.0 Fe, 27.5 Ti, 6.74 Al, 4.3 Si, 0.051 C, 0.025 P, and 0.02 S.

DATA TABLE NO. 209 SPECIFIC HEAT OF $Fe + Ti + \Sigma X_i$, ALLOY STEEL GROUP II

[Temperature, T, K; Specific Heat, C_p, Cal $g^{-1} K^{-1}$]

T	C_p
CURVE 1	
273.15	1.302 x 10^{-1}
300	1.297
400	1.316
500	1.361
600	1.417
700	1.478
900	1.606
1000	1.671
1100	1.737
1200	1.804
1300	1.871
1400	1.938
1500	2.006
1573	2.055

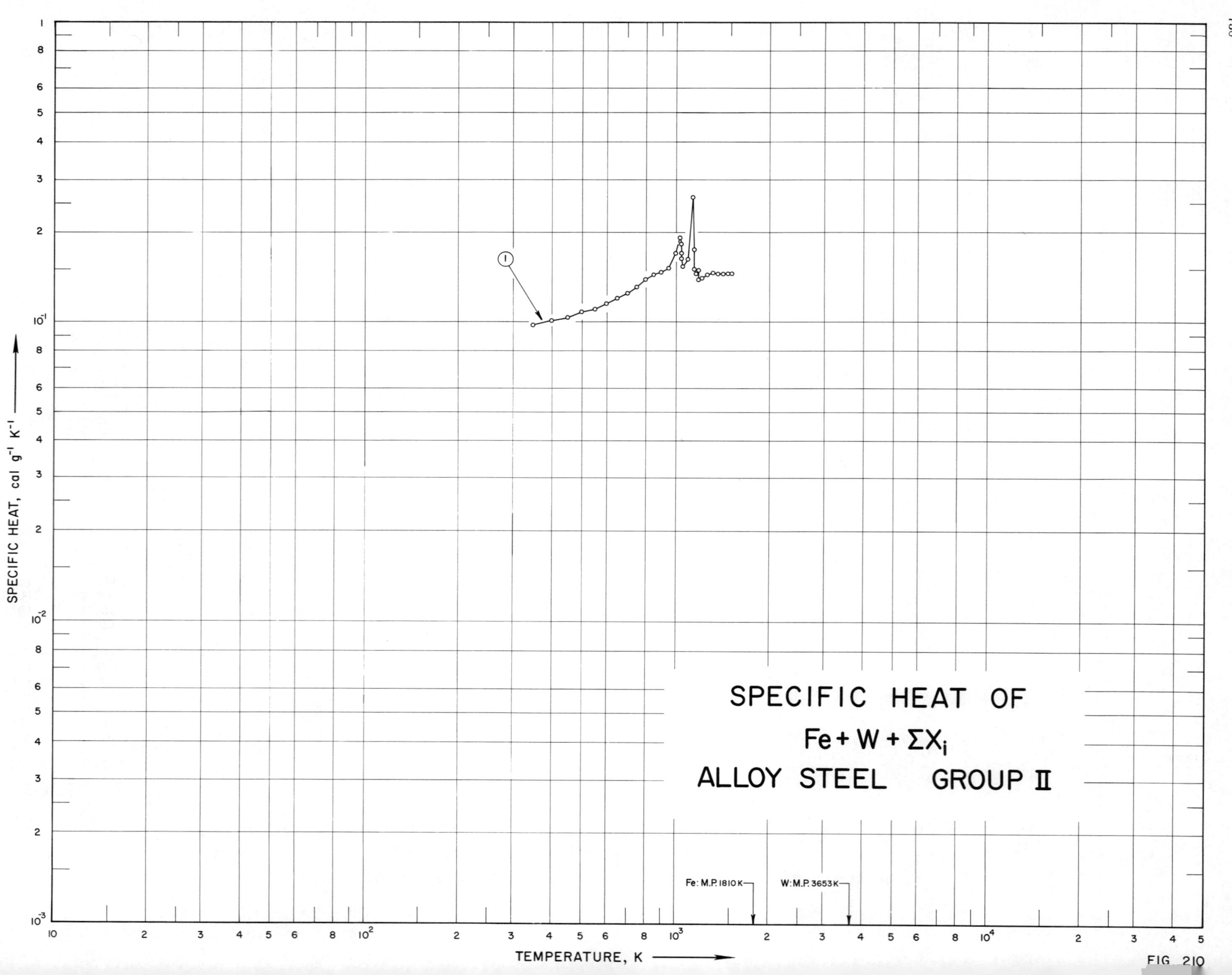
SPECIFIC HEAT OF
Fe + W + ΣXi
ALLOY STEEL GROUP II
SPECIFIC HEAT, cal g⁻¹ K⁻¹
TEMPERATURE, K
Fe: M.P. 1810 K
W: M.P. 3653 K
1

FIG. 210

SPECIFICATION TABLE NO. 210 SPECIFIC HEAT OF Fe + W + ΣX_i, ALLOY STEEL GROUP II

[For Data Reported in Figure and Table No. 210]

Curve No.	Ref. No.	Year	Temp. Range, K	Reported Error, %	Name and Specimen Designation	Composition (weight percent), Specifications and Remarks
1	104	1946	348-1523	2-4	High Alloy Steel No. 18	18.45 W, 4.26 Cr, 1.075 V, 0.715 C, 0.3 Si, 0.25 Mn, 0.067 Ni, 0.064 Cu, 0.035 As, 0.028 S, 0.018 Cr, and 0.004 Al; annealed at 830 C; density = 541 lb ft^{-3} at 15 C.

DATA TABLE NO. 210 SPECIFIC HEAT OF $Fe + W + \Sigma X_i$, ALLOY STEEL GROUP II

[Temperature, T, K; Specific Heat, C_p, Cal $g^{-1}K^{-1}$]

T	C_p
CURVE 1	
Series I	
348	9.8×10^{-2}
398	1.02×10^{-1}
448	1.04
498	1.08
548	1.11
598	1.16
648	1.20
698	1.25
748	1.32
798	1.39
848	1.43
898	1.47
948	1.52
998	1.71
1048	1.71
1098	1.63
1148	1.76
Series II	
1028	1.92×10^{-1}
1038	1.84
1048	1.64
1058	1.54
1068	1.54*
1128	2.62
1138	1.75*
1148	1.50
1158	1.46*
1168	1.46
1178	1.49
Series III	
1173	1.38×10^{-1}
1223	1.40
1273	1.44
1323	1.47
1373	1.46
1423	1.46
1473	1.46
1523	1.46

* Not shown on plot

REFERENCES TO DATA SOURCES

Ref. No.	TPRC No.	
1	24813	Jenkins, R.J. and Parker, W.J., WADD TR 61-95, 1-27, 1961. [AD 268752] [PB 181139]
2	7595	Kok, J.A. and Keesom, W.H., Physica, 4 (9), 835-42, 1937.
3	7514	Pochapsky, T.E., Acta. Met, 1 (6), 747-51, 1953.
4	7827	Giauque, W.F. and Meads, P.F., J. Am. Chem. Soc., 63, 1897-1901, 1941.
5	23058	Hopkins, D.C., Ph.D. Thesis, Univ. Illinois, AROD 2771: 9, 1-82, 1962. [AD 282758]
6	27682	Rorer, D.C., Meyer, H., and Richardson, R.C., 1-29, 1962. [AD 293874]
7	20106	Mit'kina, E.A., At. Energ. (USSR), 7 (2), 163-5, 1959.
8	16061	Martin, D.L., Can. J. Phys., 38, 17-24, 1960.
9	15899	Bell, I.P., U.K. At. Energy Authority, Ind. Group, R&DB (C) TN-127, 1-10, 1959.
10	9736	Lucks, C.F. and Deem, H.W., Am. Soc. Testing Mater. Spec. Tech. Publ. (227), 1-29, 1958.
11	18235	Howse, P.T., Jr., Pears, C.D., and Oglesby, S., Jr., WADD TR 60-657, 1-137, 1961. [AD 260065]
12	26074	Neel, D.S., Pears, C.D., and Oglesby, S., Jr., WADD TR 60-924, N62-12987, 1-216, 1962. [AD 275536]
13	9161	Krauss, F., Z. Metallkunde, 49 (7), 386-92, 1958.
14	16832	Sandenaw, T.A., U.S. At. Energy Comm. LA-2307, 1-19, 1959.
15	19328	Lyusternik, V.E., Pribory i Tekhn. Eksperim., 4, 127-9, 1959.
16	23973	Franck, J.P., Manchester, F.D., and Martin, D.L., Proc. Roy. Soc. (London), 263A, 494-507, 1961.
17	16723	Lehman, G.W., WADD TR 60-581, 1-19, 1960. [AD 247411] [PB 160804]
18	33569	Pawel, R.E., Ph.D. Thesis, Univ. Tennessee, 1-143, 1956.
19	26581	Masuda, Y., Sci. Rept. Res. Inst., Tohoku Univ., A14 (3), 156-64, 1962.
20	10730	Hultgren, R. and Land, C., Trans. Met. Soc. AIME, 215 (1), 165-6, 1959. [AD 216259]
21	6502	Fieldhouse, I.B., Hedge, J.C., Lang, J.I., Takata, A.N., and Waterman, T.E., WADC TR 55-495, Pt. 1, 1-64, 1956. [AD 110404]
22	26103	Clusius, K. and Franzosini, P., Z. Naturforsch., A17 (6), 522-5, 1962.
23	20721	Beaumont, R.H., Chihara, H., and Merrison, J.A., Phil. Mag., 5 (50), 188-91, 1960.
24	6320	Armstrong, L.D. and Grayson-Smith, H., Can. J. Res., A28 (1), 51-9, 1950.
25	6715	Friedberg, S.A., U.S. At. Energy Comm., NP-5668, 1-32, 1954. [AD 63407]
26	7468	Pallister, P.R., J. Iron Steel Inst., 161, 87-90, 1949.
27	1149	Kelley, K.K., J. Chem. Phys., 11, 16-8, 1943.
28	29865	Dench, W.A. and Kubaschewski, O., J. Iron Steel Inst., 201 (2), 140-3, 1963.
29	4677	Awbery, J.H. and Griffiths, E., Proc. Roy. Soc. (London), A174, 1-15, 1940.
30	20619	Lyubimov, A.P. and Belashchenko, D.K., Sbornik Moskov. Inst. Stali, 33, 3-11, 1955.
31	17940	Tret'yakov, Yu.D., Troshkina, V.A., and Khomyakov, K.G., Zh. Neorgan. Khim., 4 (1), 5-12, 1959.
32	10062	Wallace, D.C., Sidles, P.H., and Danielson, G.C., J. Appl. Phys., 31 (1), 168-76, 1960.
33	24279	McElroy, D.L., Ph.D. Thesis, Univ. Tennessee, 1-157, 1957.
34	27879	Anderson, P.D. and Hultgren, R., Trans. Met. Soc. AIME, 224 (4), 842-5, 1962.
35	489	Parkinson, D.H., Simon, F.E., and Spedding, F.H., Proc. Roy. Soc. (London), A207, 137-55, 1951.
36	23623	Berg, J.R., Spedding, F.H., and Daane, A.H., U.S. At. Energy Comm., IS-327, 1-35, 1961.
37	28756	Martin, D.L., Can. J. Phys., 40, 1166-73, 1962.

Ref. No.	TPRC No.	
38	24810	Brooker, J., Paine, R.M., and Stonehouse, A.J., WADD TR 60-889, 1-133, 1961. [AD 265 625]
39	25846	Kogan, V.A. and Lyusternik, V.E., Inzh. Fiz. Zh., Akad. Nauk. Belorussk. SSR, 4 (4), 105-8, 1961.
40	9182	Flubacher, P., Leadbetter, A.J., and Morrison, J.A., Phil. Mag., 4 (39), 273-94, 1959.
41	17357	Gul'tyaev, P.V. and Petrov, A.V., Soviet Phys.-Solid State (English Transl.), 1, 330-4, 1959.
42	37419	Dismukes, J.P., Ekstrom, L., Hockings, E.F., Kudman, I., Lindenbald, N.E., Miller, R.E., Rosi, E.D., and Steigmeier, E.F., 1-100, 1964. [AD 441 794]
43	2122	Hill, R.W. and Parkinson, D.H., Phil. Mag., 43, 309-16, 1952.
44	503	Estermann, I. and Weertman, J.R., J. Chem. Phys., 20, 972-6, 1952.
45	30903	Stull, D.R. and McDonald, R.A., J. Am. Chem. Soc., 77, 5293, 1955.
46	10537	Baker, H., WADC TR 57-194, Pt. II, 1-24, 1957. [AD 131 034]
47	422	Wallace, W.E., Craig, R.S., Saba, W.G., and Sterrett, K.F., U.S. At. Energy Comm., NYO-6334, 1-8, 1957.
48	7471	Craig, R.S., Krier, C.A., Coffer, L.W., Bates, E.A., and Wallace, W.E., J. Am. Chem. Soc., 76, Pt. 6, 238-40, 1954.
49	23142	Mannchen, V.W. and Bornkessel, K., Z. Metallk., 51, 482-5, 1960.
50	19410	Martin, D.L., Proc. Roy. Soc. (London), A254, 444-54, 1960.
51	19285	Martin, D.L., Physica, 25, 1193-9, 1959.
52	6314	Moser, H., Physik. Z., 37 (21), 737-53, 1936.
53	510	Krauss, F. and Warncko, H., Z. Metallk., 46, 61-9, 1955.
54	7115	Hagel, W.C., Pound, G.M., and Mehl, R.F., Carnegie Inst. Tech., 1-103, 1954. [AD 39 272]
55	22751	Eucken, Von A. and Werth, H., Z. Anorg. Allgem. Chem., 188, 152-72, 1930.
56	1884	Busey, R.H. and Giaque, W.F., J. Am. Chem. Soc., 74, 3157-8, 1952.
57	11413	Kelley, K.K., Naylor, B.F., and Shomate, C.H., U. S. Bur. Mines, Tech. Paper 686, 1-34, 1946.
58	36200	Franzosini, P., Losa, C.G., and Clusius, K., Z. Naturforsch., 19a, 1348-53, 1964.
59	7879	Elson, R.G., Grayson-Smith, H., and Wilhelm, J.O., Can. J. Res., A18, 83-9, 1940.
60	197	Armstrong, L.D. and Grayson-Smith, H., Can. J. Res., A27, 9-16, 1949.
61	24797	Lazareva, L.S., Kantor, P.B., and Kandyba, V.V., Fiz. Metal. i Metalloved., 2 (4), 628-9, 1961.
62	27526	Kirillin, V.A., Sheind'lin, A.E., and Chekhovskoi, V.Ya., Intern. J. Heat Mass Transfer, 5, 1-9, 1962.
63	20843	Cape, J.A., Atomics International, AI-6127, 1-12, 1961. [AD 251 317]
64	27207	Lowenthal, G.C., Australian J. Phys., 16 (1), 47-67, 1963.
65	19953	Taylor, R.E. and Finch, R.A., U.S. At. Energy Comm., NAA-SR-6034, 1-32, 1961.
66	7499	Redfield, T.A. and Hill, J.H., U.S. At. Energy Comm., ORNL-1087, 1-9, 1951.
67	31192	Rudkin, R.L., Parker, W.J., and Jenkins, R.J., Temp. Meas. Control Sci. Ind., 3, Pt. 2, 523-34, 1962.
68	27981	Chekhovskoi, V.Ya., Inzh.-Fiz. Zh., Akad. Nauk. Belorussk. SSR, 5 (6), 43-7, 1962.
69	9165	Clusius, K.V. and Franzosini, P., Z. Naturforsch., A14, 99-105, 1959.
70	7602	Jaeger, F.M. and Veenstra, W.A., Rec. Trav. Chim., 53, 677-87, 1934.
71	28530	Boggs, J.H. and Wiebelt, J.A., U.S. At. Energy Comm., TID-5734, 1-91, 1960.
72	37908	Cezairliyan, A., Advances Thermophys. Properties Extrem. Temp. Pres., 3rd ASME Symp., Lafayette, Indiana, 253-63, 1965.
73	18506	Spedding, F.H., McKeown, J.J., and Daane, A.H., J. Phys. Chem., 64 (3), 289-94, 1960.
74	7239	Chou, C., White, D., and Johnston, H.L., Phys. Rev., 109 (2), 788-96, 1958.
75	7689	Fieldhouse, I.B., Hedge, J.C., and Lang, J.I., WADC TR 58-274, 1-79, 1958. [AD 206 892] [PB 151 583]
76	27635	Clusius, K., Franzosini, P., and Piesbergen, U., Z. Naturforsch., 15a, 728-34, 1960.
77	18219	Kaznoff, A.I., Orr, R.L., and Hultgren, R., U.S. Dept. Comm., Office Tech. Serv., 1-83, 1961. [AD 255 273]
78	1004	Hill, R.W. and Smith, P.L., Phil. Mag., 44, 636-44, 1953.
79	22591	Kantor, P.B., Krasovitskaya, R.M., and Kisil, O.M., (English Transl. Fiz. Metal. i Metalloved. USSR, 10 (6), 835-7, 1960), MCL-950/1, 1-6, 1961. [AD 261 792]
80	32715	Kirillin, V.A., Sheind'lin, A.E., Chikhovskoi, V.Ya., and Petrov, V.A., Zh. Fiz., Khim., 37 (10), 2249-56, 1963.

Ref. No.	TPRC No.	Bibliographic Citation
81	27610	Serebryannikov, N. N. and Gel'd, P. V., Izv. Vysshikh. Uchebn. Zavedenii, Tsvetn. Met., 4 (4), 80-6, 1961.
82	11924	Duyckaerts, G., Physica, 6, 401-8, 1939.
83	29379	Picklesimer, M. L., Ph. D. Thesis, Univ. Tennessee, 1-106, 1954.
84	38430	Hoeven, Van der, B. J. C., Jr. and Keesom, P. H., Phys. Rev., 137 (1A), 103-7, 1965.
85	38436	O'Neal, H. R. and Phillips, N. E., Phys. Rev., 137 (3A), 748-59, 1965.
86	32530	Lounasmaa, O. V., Phys. Rev., 133 (2A), 502-9, 1964.
87	32500	Lounasmaa, O. V., Phys. Rev., 133 (1A), 211-8, 1964.
88	28959	Guenther, R. A., M. S. Thesis, Ill. Inst. Tech., ANL-6594, 1-30, 1962.
89	22753	Lounasmaa, O. V. and Roach, P. R., Phys. Rev., 128, 622-6, 1962.
90	7662	DeSorbo, W., J. Phys. Chem., 62, 965-7, 1958.
91	36201	Franzosini, P. and Clusius, K., Z. Naturforsch., 19a, 1430-1, 1964.
92	1591	Johnston, H. L., Hersh, H. N., and Kerr, E. C., J. Am. Chem. Soc., 73, 1112-7, 1951.
93	5278	Jaeger, F. M., Rosenbohm, E., and Zuithoff, A. J., Rec. Trav. Chim., 59, 831-56, 1940.
94	1840	Clusius, K. and Schachinger, L., Z. Naturforsch., A7, 185-91, 1952.
95	1005	Skochdopole, R. E., Griffel, M., and Spedding, F. H., J. Chem. Phys., 23, 2258-63, 1955.
96	1761	Geballe, T. H. and Giauque, W. F., J. A. Chem. Soc., 74, 2368-9, 1952.
97	37348	Westrum, E. F., Jr. (author), McClaine, L. A. (editor), ASD TDR 62-204, Pt. 3, 189-96, 1964. [AD 601 424]
98	8184	Smith, W. T., Jr., Oliver, G. D., and Cobble, J. W., J. Am. Chem. Soc., 75, 5785-6, 1953.
99	28519	Jaeger, F. M. and Rosenbohm, E., Physica, 6, 1123-5, 1939.
100	871	Strittmater, R. C. and Danielson, G. C., U. S. At. Energy Comm., ISC-666, 1-27, 1955. [AD 96 968]
101	31341	Butler, C. P. and Inn, E. C. Y., USNRDL-TR-235, 1-30, 1958. [AD 200 857]
102	27596	Yurkov, V. A. and Ivoninskaya, L. A., Izv. Vysshikh Uchebn. Zavedenii, Fiz., (1), 138-43, 1962.
103	7811	Meads, P. F., Forsythe, W. R., and Giauque, W. F., J. Am. Chem. Soc., 63, 1902-5, 1941.
104	9350	Awbery, J. H., Challoner, A. R., and Pallister, P. R., J. Iron Steel Inst. (London), 154 (2), 83-111, 1946.
105	6381	Trice, J. B., Neely, J. J., and Teeter, C. E., Jr., U. S. At. Energy Comm., NEPA 816-SCR-28, 1-12, 1948.
106	10119	Sims, C. T., Craighead, C. M., Jaffee, R. I., Gideon, D. N., Kleinschmidt, W. W., Nexsen, W. E., Jr., Gaines, G. B., Todd, F. C., Peet, C. S., Rosenbaum, D. M., Runck, R. J., and Campbell, I. E., WADC TR 54-371, 1-79, 1956. [AD 97 301] [PB 121 653]
107	340	Clusius, K. and Losa, C. G., Z. Naturforsch., A10, 545-51, 1955.
108	23964	Clusius, K. and Piesbergen, U., Z. Naturforsch., A14 (1), 23-7, 1959.
109	8151	Roberts, L. M., Proc. Phys. Soc. (London), B70, 434-5, 1957.
110	7408	Desorbo, W., J. Chem. Phys., 21 (7), 1144-8, 1953.
111	29948	Shanks, H. R., Maycock, P. D., Sidles, P. H., and Danielson, G. C., Phys. Rev., 130 (5), 1743-8, 1963.
112	21087	Kantor, P. B., Kisel, O. M., and Fomichev, E. M., Ukrain. Fiz. Zhur., 5, 358-62, 1960.
113	19411	Martin, D. L., Proc. Roy. Soc. (London), A254, 433-43, 1960.
114	8409	Sterrett, K. F. and Wallace, W. E., J. Am. Chem. Soc., 80, 3176-7, 1958.
115	20973	Hildenbrand, D. L., Theard, L. P., and Potter, N. D., AERO-U-1606, 1-49, 1962. [AD 273 792]
116	28434	Hoch, M. and Johnston, H. L., J. Phys. Chem., 65 (5), 855-60, 1961.
117	31542	Kraftmakher, Ya. A., Zh. Prikl. Mekhan. i Tekhn. Fiz., (2), 158-60, 1963.
118	8684	Jennings, L. D., Stanton, R. M., and Spedding, F. H., J. Chem. Phys., 27 (4), 909-13, 1957.
119	20739	Wallace, D. C., Phys. Rev., 120 (1), 84-8, 1960.
120	326	Griffel, M. and Skochdopole, R. E., J. Am. Chem. Soc., 75 (21), 5250-1, 1953.
121	20568	Eichelberger, J. F., U. S. At. Energy Comm., MLM-1133, 1-31, 1962.
122	8419	Dempsey, E. and Kay, A. E., J. Inst. Metals, 86, 379-84, 1958.
123	8420	Dean, D. J., Kay, A. E., and Loasby, R. G., J. Inst. Metals, 86, 464, 1958.

Ref. No.	TPRC No.	Bibliographic Citation
124	32435	Kay, A.E. and Loasby, R.G., Phil. Mag., 9 (97), 37-49, 1964.
125	37902	Conway, J.B. and Hein, R.A., Advances Thermophys. Properties Extreme Temp. Pres., 3rd ASME Symp., Lafayette, Indiana, 131-7, 1965.
126	24076	Jennings, L.D., Hill, E., and Spedding, F.H., J. Chem. Phys., 34, 2082-9, 1961.
127	17597	Golutvin, Yu.M., Zhur. Fiz. Khim., 33, 1798-805, 1959.
128	1271	Aven, M.H., Craig, R.S., Waite, T.R., and Wallace, W.E., Phys. Rev., 102, 1263-4, 1956.
129	24340	Holland, L.R., J. Appl. Phys., 34 (8), 2350-7, 1963.
130	7170	Backhurst, I., J. Iron Steel Inst. (London), 189, 124-34, 1958.
131	23749	Stalinski, B. and Bieganski, Z., Roczniki Chem., 35, 273-83, 1961.
132	1877	Kelley, K.K., Ind. Eng. Chem., 36, 865-6, 1944.
133	10139	Rea, J.A., M.S. Thesis, Oklahoma Agricultural Mechanical College, 1-52, 1956. [AD 94417]
134	250	Scott, J.L., U.S. At. Energy Comm., ORNL-2328, 1-122, 1957. [AD 138 838]
135	131	Loewen, E.G., Trans. Am. Soc. Mech. Engrs., 78, 667-70, 1956.
136	532	Kothen, C.W. and Johnston, H.L., J. Am. Chem. Soc., 75, 3101-2, 1953.
137	26916	Sheindlin, A.E. and Chekhovskoi, V.Ya., Proc. Acad. Sci. USSR, Phys. Chem. Sect. (English Transl.), 142 (6), 184-6, 1962.
138	26919	Kirillin, V.A., Sheindlin, A.E., Chekhovskoi, V.Ya., and Petrov, V.A., Doklady Akad. Nauk. SSSR, 144 (2), 390-1, 1962.
139	26001	Kirillin, V.A., Sheindlin, A.E., and Chekhovskoi, V.Ya., Teploenerg., 9 (2), 63-6, 1962.
140	31541	Kraftmakher, Ya.A., Zh. Prikl. Mekhan. i Tekhn. Fiz., (5), 176-50, 1962.
141	14238	Flotow, H.E. and Lohr, H.R., J. Phys. Chem., 64, 904-6, 1960.
142	1741	Moore, G.E. and Kelley, K.K., J. Am. Chem. Soc., 69, 2105-7, 1947.
143	1584	North, J.M., At. Energy Res. Establ. (G. Brit.), Rep., M/R 1016, 1-9, 1956.
144	807	Ginnings, D.C. and Corruccini, R.J., J. Res. Natl. Bur. Std., 39, 309-16, 1947.
145	33559	Long, E.A., Jones, W.M., and Gordon, J., U.S. At. Energy Comm., A-329, MDDC-609, 1-13, 1942.
146	24812	Fieldhouse, I.B. and Lang, J.I., WADD TR 60-904, 1-119, 1961. [AD 268304]
147	22075	Golutvin, Yu.M. and Kozlovskaya, T.M., Zh. Fiz. Khim., 36 (2), 362-4, 1962.
148	28734	Bieganski, Z. and Stalinski, B., Bull. Acad. Polon. Sci., Ser. Sci. Chim., 9 (5), 367-72, 1961.
149	37425	Gertstein, B.C., Mullay, J., Phillips, E., Miller, R.E., and Spedding, F.H., J. Chem. Phys., 41 (3), 883-9, 1964.
150	7000	Douglas, T.B. and Victor, A.C., WADC TR 57-374, Pt. 2, 1-75, 1957. [AD 150128]
151	10322	Lounasmaa, O. V., Phys. Rev., 126 (4), 1352-6, 1962.
152	31102	Kochetkova, N.M. and Rezukhina, T.N., Semiconductor Materials Conference, Moscow, 26-8, 1961.
153	918	Desorbo, W., Acta Met., 1 (5), 503-7, 1953.
154	6560	Lucks, C.F. and Deem, H.W., WADC TR 55-496, 1-65, 1956. [AD 97185]
155	342	Clusius, K. and Schachinger, L., Z. Angew. Phys., 4 (12), 442-4, 1952.
156	29954	Mitacek, P., Jr. and Aston, J.G., J. Am. Chem. Soc., 85, 137-41, 1963.
157	27993	Weller, W.W. and Kelley, K.K., U.S. Bur. Mines, Rep. Invest. 5984, 1-3, 1962.
158	16682	Jennings, L.D., Miller, R.E., and Spedding, F.H., J. Chem. Phys., 33 (6), 1849-52, 1960.
159	1363	Coughlin, J.P. and King, E.G., J. Am. Chem. Soc., 72, 2262-5, 1950.
160	36669	Parker, R., Trans. Met. Soc., AIME, 233, 1545-9, 1965.
161	17553	Wolcott, N.M., Bull. Inst. Intern. Froid, Annexe, 286-9, 1955.
162	14232	Wise, S.S., Margrave, J.L., and Altman, R.L., J. Phys. Chem., 64, 915-7, 1960.
163	22370	Jaeger, F.M. and Rosenbohm, E., Proc. Acad. Sci. Amsterdam, 34, 85-99, 1931.
164	22461	Jaeger, F.M. and Rosenbohm, E., Rec. Trav. Chim., 51, 1-46, 1932.
165	6251	Jaeger, F.M. and Poppema, T.J., Rec. Trav. Chim., 55 (6), 492-517, 1936.
166	1721	Clusius, K. and Schachinger, L., Z. Naturforsch, 2A (2), 90-7, 1947.
167	6323	Kok, J.A. and Keesom, W.H., Physica, 3 (9), 1035-45, 1936.

Ref. No.	TPRC No.	Bibliographic Citation
168	712	Clusius, K., Losa, C.G., and Franzosini, P., Z. Naturforsch, 12A, 34-8, 1957.
169	29972	Kendall, W.B., Orr, R.L., and Hultgren, R., J. Chem. Eng. Data, Pt. 1, 7, 516-8, 1962.
170	20920	Eastman, E.D. and Rodebush, W.H., J. Am. Chem. Soc., 40, 489-500, 1918.
171	6387	Carpenter, L.G. and Steward, C.J., Phil. Mag., 27 (184), 551-64, 1939.
172	897	Krier, C.A., Ph.D. Thesis, Univ. of Pittsburg, 1-126, 1954.
173	7276	Roberts, L.M., Proc. Phys. Soc. (London), 70 (452B), 744-52, 1957.
174	26732	Lemmon, A.W., Jr., Deem, H.W., Eldridge, E.A., Hall, E.H., Matolich, J., Jr., and Walling, J.F., NASA CR-54017, BATT-4673-Final, 1-66, 1963.
175	1223	Horowitz, M. and Daunt, J.G., Phys. Rev., 91, 1099-106, 1953.
176	863	Dauphinee, T.M., Martin, D.L., and Preston-Thomas, H., Proc. Roy. Soc. (London), 233A, 214-22, 1955.
177	21831	Anderson, C.T., J. Am. Chem. Soc., 52, 2301-4, 1930.
178	30243	Pearlman, N., Ph.D. Thesis, Purdue Univ., 1-104, 1952.
179	21779	Eastman, E.D., Williams, A.M., and Young, T.F., J. Am. Chem. Soc., 46, 1178-83, 1924.
180	6420	Jaeger, F.M., Rosenbohm, J.E., and Veenstra, W.A., Prok. K. Akad. Wet. Amsterdam, 36, 291-8, 1933.
181	6330	Keesom, W.H. and Kok, J.A., Physica, 1, 770-8, 1934.
182	6245	Bronson, H.L. and Wilson, A.J.C., Can. J. Res., 14A (10), 181-93, 1936.
183	6246	Bronson, H.L., Hewson, E.W., and Wilson, A.J.C., Can. J. Res., 14A (10), 194-99, 1936.
184	34308	Martin, D.L., Phys. Rev., 141 (2), 576-82, 1966.
185	1482	Ginnings, D.C., Douglas, T.B., and Ball, A.F., J. Res. Natl. Bur. Stds., 45, 23-33, 1950.
186	21823	Clusius, K. and Vaughen, J.V., J. Am. Chem. Soc., 52, 4686-99, 1930.
187	22754	Seekamp, H., Z. Anorg. Allgem. Chem., 195, 345-65, 1931.
188	6385	Keesom, W.H. and Kok, J.A., Physica, 1, 595-608, 1934.
189	21774	Rodebush, W.H., J. Am. Chem. Soc., 45, 1413-6, 1923.
190	22849	Fritz, L., Z. Physik Chem., 110, 343-62, 1924.
191	22378	Keesom, W.H. and Ende, Van der, J.N., Proc. Acad. Sci. Amsterdam, 35, 143-55, 1932.
192	22375	Jaeger, F.M. and Bottema, J.A., Proc. Acad. Sci. Amsterdam, 35, 347-52, 1932.
193	1376	Bartenev, G.M., Zhur. Tekh. Fiz., 17, 1321-4, 1947.
194	1190	Webb, F.J. and Wilks, J., Proc. Roy. Soc. (London), 230A, 549-59, 1955.
195	490	Corak, W.S. and Satterthwaite, C.B., Phys. Rev., 102, 662-6, 1956.
196	29573	Heffan, H., M.S. Thesis, Univ. California, 1-17, 1958.
197	24027	Yaqub, M., Cryogenics, 1 (3), 166-70, 1961.
198	6415	Keesom, W.H. and van Laer, P.H., Physica, 5 (3), 193-201, 1938.
199	6797	Jaeger, F.M., Rosenbohm, E., and Fonteyne, R., Rec. Trav. Chim., 55 (7/8), 615-54, 1936.
200	21224	Clusius, K. and Franzosini, P., Z. Physik. Chem. (Frankfurt), 16 (3/6), 194-202, 1958.
201	21407	Magnus, A. and Holzmann, H., Ann. Physik [5], 3, 585-613, 1929.
202	1813	Silvidi, A.A. and Daunt, J.G., Phys. Rev., 77, 125-9, 1950.
203	468	Waite, T.R., University Microfilm. 18264, 1-79, 1957.
204	8451	White, D., Chou, C., and Johnston, H.L., Phys. Rev., [2], 109 (3), 797-802, 1958.
205	34604	Lounasmaa, O. V., Phys. Rev., 143 (2), 399-405, 1966.
206	22855	Behrens, W.U. and Drucker, C., Z. Physik. Chem., 113, 79-110, 1924.
207	7025	Clusius, K. and Harteck, P., Z. Physik. Chem., 134, 243-63, 1928.
208	6329	Poppema, T.J. and Jaeger, F.M., Proc. K. Akad. Wet. Amsterdam, 38, 510-20, 1935.
209	1705	Smith, P.L., Phil. Mag., 46, 744-50, 1955.
210	23965	Eichenauer, W. and Schulze, M., Z. Naturforsch., A14 (1), 28-32, 1959.
211	9136	Srinivasan, T.M., Proc. Indian Acad. Sci., 49, 61-5, 1959.
212	8454	Roberts, L.M., Proc. Phys. Soc. (London), B70, Pt. 8, 738-43, 1957.
213	22861	Zeidler, W., Z. Physik Chem., 123, 383-404, 1926.

Ref. No.	TPRC No.	Bibliographic Citation
214	21799	Dixon, A. L. and Rodebush, W. A., J. Am. Chem. Soc., 49, 11162-74, 1927.
215	22369	Jaeger, F. M. and Rosenbohm, E., Proc. Acad. Sci. (Amsterdam), 34, 808-22, 1931.
216	19226	Seidel, G. and Keesom, P. H., Phys. Rev., 112 (4), 1083-8, 1958.
217	1708	Griffel, M., Skochdopole, R. E., and Spedding, F. H., Phys. Rev., [2], 93, 657-61, 1954.
218	19921	Seidel, G. M., Ph. D. Thesis, Purdue Univ., Univ. Microfilms, Mic 58-7994, 1-100, 1958.
219	248	Gerstein, B. C., Griffel, M., Jennings, L. D., Miller, R. E., Skochdopole, R. E., and Spedding, F. H., J. Chem. Phys., 27 (2), 394-9, 1957.
220	32459	Kempen, van H., Miedama, A. R., and Huiskamp, W. J., Physica, 30 (1), 229-36, 1964.
221	27515	Gunther, P., Ann. Physik., 51, 828-46, 1916.
222	10729	Kendall, W. B., Orr, R. L., and Hultgren, R., AFOSR-TN-59-524, 1-14, 1959. [AD 216 258]
223	19947	Gaumer, R. E., Ph. D. Thesis, Ohio State Univ., 1-118, Univ. Microfilms, Mic 60-1182, 1-123, 1959.
224	32871	Braun, von M. and Kohlhaas, R., Z. Naturforsch, [A], 19 (5), 663-4, 1964.
225	31842	Panin, V. E. and Zenkova, E. K., Soviet Phys. Doklady, 4, 1368-71, 1960.
226	371	DeSorbo, W., Acta Met., 3 (3), 227-31, 1955.
227	27439	Day, G. F. and Hultgren, R., J. Phys. Chem., 66 (8), 1532-4, 1962.
228	16062	Martin, D. L., Can. J. Phys., 38, 25-31, 1960.
229	28518	Kono, H., J. Phys. Soc. Japan, 13, 1444-51, 1958.
230	14256	Titman, J. M., Proc. Phys. Soc., 77, Pt. 3, 807-10, 1961.
231	16590	Fieldhouse, I. B., Lang, J. I., and Blau, H. H., WADC TR 59-744, 4, 1-78, 1960. [AD 249 166] [PB 171 390]
232	25961	Hedge, J. C., Kostenko, C., and Lang, J. I., ASD-TDR-63-597, 1-128, 1963. [AD 424 375]
233	20083	Land, C. C., M. S. Thesis, Univ. of California, 1-25, 1957.
234	29988	Gangopadhyay, A. K. and Margrave, J. L., J. Chem. Eng. Data, 8 (2), 204-5, 1963.
235	28346	Aoyama, S. and Kanda, E., J. Chem. Soc. Japan, 62, 312-5, 1941.
236	2968	Zuithoff, A. J., Rec. Trav. Chim., 59, 131-60, 1940.
237	26008	Pears, C. D., ASD-TDR 62-765, 1-420, 1963. [AD 298 061]
238	34310	Isaacs, L. L. and Massalski, T. B., Phys. Rev., 141 (2), 634-7, 1966.
239	280	Passagalia, E., Ph. D. Thesis, Univ. of Penn., Univ. Microfilms Publ. No. 13417, 1955.
240	32845	Massena, C. W., Ph. D. Thesis, Univ. of Pittsburg, 86, 1962.
241	31965	Deem, H. W. and Eldridge, E. A., BMI-1644, UC-25, TID-4500, V1-V2, 1963.
242	27792	Hedge, J. C., Lang, J. I., Howe, E., and Elliot, K., Armour Res. Foundation, Air Force Special Weapons Center, TDR 63-17, 1-19, 1963. [AD 412 5867]
243	6565	Lucks, C. F., Matolich, J., and Van Valzor, J. A., U. S. A. F. TR 6145, 1-71, 1954. [AD 95 406]
244	24512	Krentsis, R. P., Gel'd, P. V., and Serebrennikov, N. N., Izvestiya, VUZ, Chernaya Met., 5-11, 1960.
245	6970	Fieldhouse, I. B., Hedge, J. C. Lang J. I. and Waterman, T. E., WADC-TR-57-487, 1-79, 1958. [AD 150 954] [PB 131 718]
246	24547	McDonald R. A. and Stull, D. R., J. Chem. Eng. Data, 6, 609-10, 1961.
247	8114	Evans, J. P. and Mardon, P. G., 1-4, 1958. [AD 210 812-L]
248	20286	Douglas, T. B. and Victor, A. C., Natl. Bur. Std., J. Res., 65C (1), 65-9, 1961.
249	16743	Venturi, R. and Seibel, R. D., DRI Rept. 1023, 1-31, 1959.
250	7412	Douglas, T. B. and Dever, J. L., USAEC, NBS-2302, 1-23, 1953.
251	677	Douglas, T. B. and Dever, J. L., Natl. Bur. Std., J. Res., 54 (1), 15-9, 1955.
252	26219	Neimark, B. E., Lyusternik, V. E., Anichkina, E. Yu., and Bykova, T. I., High Temp., 1 (1), 9-12, 1963.
253	23704	Serebrennikov, N. N., Gel'd, P. V., and Krentsis, R. P., Izv. Vysshikh Uchebn. Zavedenii, Tsvetn. Met., 4 (1), 82-7, 1961.
254	33326	Neff, C. W., Frank, R. G., and Luft, L., ASD TR 61-392, 2, 1-342, 1961. [AD 267 844]
255	31325	Bewley, J. G., ASD-TDR-63-201, 1-14, 1963. [AD 402 066]
256	28663	Wittenberg, L. J., USAEC, MLM-1162, TID-4500 (30th Ed.), 1-22, 1963.
257	29805	Ziegler, W. T. and Mullins, J. C., Georgia Inst. Technology, Final Rept., Proj. No. A-504, 1-54, 1961.

Ref. No.	TPRC No.	Bibliographic Citation
258	22905	Lounasmaa, O.V., Phys. Rev., 128 (3), 1136-9, 1962.
259	1329	Krier, C.A., Craig, R.S., and Wallace, W.E., J. Phys. Chem., 61, 522-9, 1957.
260	21830	Anderson, C.T., J. Am. Chem. Soc., 52, 2296-300, 1930.
261	30170	Maier, C.G. and Anderson, C.T., J. Chem. Phys., 2, 513-27, 1934.
262	6795	Quinney, H. and Taylor, G.I., Proc. Roy. Soc. (London), 163A (913), 157-81, 1937.
263	6448	Avramescu, A., Z. Tech. Physik, 20 (7), 213-17, 1939.
264	7169	Mäder, H., Metall, 5 (1), 1-5, 1951.
265	19230	Phillips, N. E., Phys. Rev., 114 (3), 676-85, 1959.
266	23058	Hopkins, D.C., Ph.D. Thesis, Univ. of Illinois, 1-82, 1962. [AD 282 758]
267	21394	Günther, P., Ann. Physik, 63, 476-80, 1920.
268	31762	Umino, S., Sci. Repts., Tohoku Impl. Univ., 15 (Series 1), 597-617, 1926.
269	33729	Ahlers, G., Phys. Rev., 145 (2), 419-23, 1966.
270	9862	Lewis, E.J., Phys. Rev., 34, 1575-87, 1929.
271	19099	Vernotte, P. and Jeufroy, A., Compt. Rend., 192, 612-4, 1931.
272	10323	Jaeger, F.M. and Rosenbohm, E., Rec. Trav. Chim., 53, 451-63, 1934.
273	7573	Losana, L., Alluminio, 8 (2), 67-75, 1939.
274	22865	Drucker, C., Z. Physik Chem., 130, 673-90, 1927.
275	21834	Anderson, C.T., J. Am. Chem. Soc., 52, 2720-3, 1930.
276	22368	Keesom, W.H. and Van den Ende, J.N., Proc. Acad. Sci. Amsterdam, 33, 243-54, 1930; (correction: 34, 210-11, 1931).
277	22446	Carpenter, L.G. and Harle, T.F., Proc. Roy. Soc. (London), 136A, 243-50, 1932.
278	6393	Bronson, H.L. and MacHattie, L.E., Can. J. Res., 16A (9), 177-82, 1938.
279	4342	Kubaschewski, O. and Schrag, G., Z. Electrochem., 46, 675-80, 1940.
280	16018	Bell, H. and Hultgren, R., Univ. California, Tech. Rept. No. 1, 1-17, 1960. [AD 242 479] [PB 149 842]
281	14646	Schneider, A. and Hilmer, O., Z. Anorg. Allgem., 286, 97-117, 1956.
282	24325	Gruneisen, E. and Goens, E., Z. Physik, 26, 250-73, 1924.
283	22871	Lange, F. and Simon, F., Z. Phys. Chem., 134, 374-80, 1928.
284	307	Smith, P.L. and Wolcott, N.M., Phil. Mag., 1 (8), 854-65, 1956.
285	9822	McKeown, J.J., Ph.D. Thesis, Iowa State College, 1-113, 1958.
286	25870	Lemmon, A.W., Jr., Deem, H.W., Eldridge, E.A., Hall, E.H., Matolich, J., Jr., and Walling, J.F., NASA-CR-54018 BATT-4673-T7, 1-34, 1964.
287	12111	Anderson, C.T., J. Am. Chem. Soc., 59, 488-91, 1937.
288	1886	Estermann, I., Friedberg, S.A., and Goldman, J.E., Phys. Rev., 87, 582-8, 1952.
289	21420	Klinkhardt, H., Ann. Physik, 84, 167-200, 1927.
290	14596	Jaeger, F.M., Rosenbohm, E. and Bottema, J.A., Proc. Acad. Sci. Amsterdam, 35, 772-9, 1932.
291	6379	Dockerty, S.M., Can. J. Res., 9, 84-93, 1933.
292	6240	Dockerty, S.M., Can. J. Res., A15 (4), 59-66, 1937.
293	28346	Aoyama, S. and Kanda, E., J. Chem. Soc. Japan, 62, 312-15, 1941.
294	1525	Brown, A., Zemansky, M.W., and Boorse, H.A., Phys. Rev., 86, 134-5, 1952.
295	17559	Eder, F.X., Bull. Inst. Intern. Froid. Annexe, 2, 137-41, 1955.
296	32573	Corak, W.S., Ph.D. Thesis, Univ. of Pittsburgh, 1-92, 1955.
297	19923	Cheng, C.-H., Ph.D. Thesis, Dept. of Metallurgy, Univ. of Illinois, 1-103, 1959.
298	32681	Chang, Y.-S.A., Ph.D. Thesis, Univ. of California, 1-79, 1963.
299	32968	Phillips, N. E., Phys. Rev., 134 (2A), A385-91, 1964.
300	34916	Ahlers, G., Rev. Sci. Instr., 37 (4), 477-80, 1966.
301	39107	Dennison, D.H., Gschneidner, K.A., Jr., and Daane, A.H., J. Chem. Phys., 44 (11), 4273-82, 1966.
302	10409	Christescu, S. and Simon, F., Z. Physik Chem., B25, 273-82, 1934.

Ref. No.	TPRC No.	Bibliographic Citation
303	32919	Piesbergen, U., Z. Naturforsch, A18 (2), 141-7, 1963.
304	34510	Sommelet, P. and Orr, R. L., J. Chem. Eng. Data, 11 (1), 64-5, 1966.
305	7453	Corak, W. S., Garfunkel, M. P., Satterthwaite, C. B., and Wexler, A., Phys. Rev., 98 (2), 1699-1707, 1955.
306	32743	Franzosini, P. and Clusius, K., Z. Naturforsch, 18a (12), 1243-6, 1963.
307	32260	Hawkins, D. T., Onillon, M., and Orr, R. L., J. Chem. Eng. Data, 8 (4), 628-9, 1963.
308	7242	Wolcott, N. M., Phil. Mag., 8 (2), 1246-54, 1957.
309	20978	Wohler, L. and Jochum, N., Z. Phys. Chem. (Leipzig), 167A (3), 169-79, 1933.
310	30108	Kohlhaas, R. and Braun, M., Archiv. fur das Eisenhuttenwesen, 34 (5), 391-9, 1963.
311	21789	Rodebush, W. H. and Mickalek, J. C., J. Am. Chem. Soc., 47, 2117-21, 1925.
312	22636	Umino, S., Sci. Rept., Tohoku Impl. Univ., 15, 331-69, 1926.
313	9938	Umino, S., Sci. Rept., Tohoku Impl. Univ., 18, 91-107, 1929.
314	11805	Simon, F. and Swain, R. C., Z. Physik Chem., 28B, 189-98, 1935.
315	6206	Honda, K. and Tokunaga, M., Sci. Rept., Tohoku Impl. Univ., 23 (1), 816-34, 1935.
316	7668	Jaeger, F. M., Rosenbohm, E., and Zuithoff, A. J., Rec. Trav. Chim., 57, 1313-40, 1938.
317	1903	Duyckaerts, G., Mem. Soc. Roy. Sci. Liege, 6, 193-329, 1945.
318	1234	Pallister, P. R., J. Iron Steel Inst., 178, 346-8, 1954.
319	9156	Valentiner, S., Optik, 15 (6), 343-57, 1958.
320	19922	Wei, C.-T., Ph.D. Thesis, Univ. Illinois, Univ. Microfilms Publ. 59-2063, 1-89, 1959.
321	6310	Jaeger, F. M., Bottema, J. A., and Rosenbohm, E., Proc. K. Akad. Wet. Amsterdam, 39 (8), 921-7, 1936.
322	536	Horowitz, M., Silvidi, A. A., Malaker, S. F., and Daunt, J. G., Phys. Rev., 88, 1182-6, 1952.
323	4654	Douglas, T. B. and Dever, J. L., J. Am. Chem. Soc., 76, 4824-6, 1954.
324	39226	Bonnerot, J., Rev. Physique Appliquee, 1 (1), 61-7, 1966.
325	7511	Cabbage, A. M., AECD-3240, 1-10, 1950.
326	16656	Yaggee, F. L. and Untermyer, S., ANL-4458, 1-27, 1950.
327	920	Bates, A. G. and Smith, D. J., U. S. At. Energy Comm. Rept. K729, 1-31, 1951.
328	10357	Douglas, T. B., Epstein, L. F., Dever, J. L., and Howland, W. H., J. Am. Chem. Soc., 77, 2144-50, 1955.
329	298	Logan, J. K., Clement, J. R., and Jeffers, H. R., Phys. Rev., 105, 1435-7, 1957.
330	1113	Shomate, C. H., J. Chem. Phys., 13, 326-8, 1945.
331	28295	Naylor, B. F., J. Chem. Phys., 13, 329-32, 1945.
332	1309	Booth, G. L., Hoare, F. E., and Murphy, B. T., Proc. Phys. Soc. (London), 68B, 830-2, 1955.
333	21397	Simon, F., Ann. Physik, 68, 241-80, 1922.
334	22846	Simon, F., Z. Physik Chem., 107, 279-84, 1923.
335	22306	Carpenter, L. G. and Stoodley, L. G., Phil. Mag., 10 (7), 249-65, 1930.
336	28517	Pickard, G. L. and Simon, F. E., Proc. Phys. Soc. (London), 61, 1-9, 1948.
337	1292	Douglas, T. B., Ball, A. F., and Ginnings, D. C., J. Res. NBS, 46, 334-8, 1951.
338	1230	Busey, R. H. and Giauque, W. F., J. Am. Chem. Soc., 75, 806-9, 1953.
339	22316	Cooper, D. and Langstroth, G. O., Phys. Rev., 33, 243-8, 1929.
340	29598	Kothen, C. W., Ph.D. Thesis, Ohio State Univ., Univ. Microfilms Publ. 52-23697, 1-89, 1952.
341	33660	Rorer, D. C., Ph.D. Thesis, Duke Univ., Univ. Microfilms Inc. 64-8562, 1-124, 1964.
342	19090	Lapp, Ch., Compt. Rend., 186, 1104-6, 1928.
343	7373	Grew, K. E., Proc. Roy. Soc., A145, 509-22, 1934.
344	6212	Keesom, W. H. and Clark, C. W., Physica, 2, 513-20, 1935.
345	6479	Clusius, K. and Goldmann, J., Z. Phys. Chem., B31 (4), 256-62, 1936.
346	6238	Ewert, M., Proc. K. Akad. Wet. Amsterdam, 39 (7), 833-8, 1936.
347	6405	Sykes, C. and Wilkinson, H., Proc. Roy. Soc., 50 (5), 834-51, 1938.
348	6434	Persoz, B., Compt. Rend., 208 (21), 1632-4, 1939.
349	29759	Gupta, K. P., Ph.D. Thesis, Univ. Illinois, Univ. Microfilms Publ. 62-6148, 1-132, 1962.

Ref. No.	TPRC No.	
350	24836	Schmidt, E.O. and Leidenfrost, W., ASME Second Symp. on Thermophysical Properties, 178-84, 1962.
351	911	Brown, A., Zemansky, M.W., and Boorse, H.A., Phys. Rev., 92, 52-8, 1953.
352	37179	Leupold, H.A. and Boorse, H.A., Phys. Rev., 134 (5A), A1322-8, 1964.
353	1846	Douglas, T.B., Ball, A.F., Ginnings, D.C., and Davis, W.D., J. Am. Chem. Soc., 74, 2472-8, 1952.
354	9460	Kurti, N. and Safrata, R.S., Phil. Mag., 3 (31), 780-3, 1958.
355	24162	Gordon, J.E., Dempsey, C.W. and Soller, T., Phys. Rev., 124 (3), 724-5, 1961.
356	32512	Lien, W.H. and Phillips, N.E., Phys. Rev., 133 (5A), A1370-7, 1964.
357	21852	DeVries, T. and Dobry, L.F., J. Am. Chem. Soc., 54, 3258-61, 1932.
358	6248	Anderson, C.T., J. Am. Chem. Soc., 59 (6), 1036-7, 1937.
359	29729	Kalishevich, G.I., Gel'd, P.V., and Krentsis, R.P., Russ. J. Phys. Chem., 39 (12), 1602-3, 1965.
360	20966	Keesom, W.H. and Kok, J.A., Proc. Akad. Sci. Amsterdam, 35, 301-6, 1932.
361	1037	Dauphinee, T.M., McDonald, D.K.C., and Preston-Thomas, H., Proc. Roy. Soc. (London), 221A, 267-76, 1954.
362	1518	Parkinson, D.H. and Quarrington, J.E., Proc. Phys. Soc. (London), 68A, 762-3, 1955.
363	8047	Sterrett, K.F., Ph.D. Thesis, Univ. of Pittsburgh, Univ. Microfilms Publ. 22865, 1-119, 1957.
364	11101	Kelley, K.K., J. Chem. Phys., 8, 316-22, 1940.
365	28520	Keesom, W.H. and Desirant, M., Physica, 8, 273-88, 1941.
366	9686	Clusius, K. and Losa, C.G., Z. Naturforsch, A10, 939-43, 1955.
367	11869	Slansky, C.M. and Coulter, L.V., J. Am. Chem. Soc., 61, 564, 1939.
368	6394	Hicks, J.F.G., Jr., J. Am. Chem. Soc., 60 (5), 1000-4, 1938.
369	6232	Keesom, W.H. and Kok, J.A., Physica, 1, 175-81, 1934.
370	1318	Snider, J.L. and Nicol, J., Phys. Rev., 105, 1242-6, 1957.
371	32860	Bryant, C.A., Ph.D. Thesis, Purdue Univ., Univ. Microfilms Publ. 61-2464, 1-119, 1961.
372	1271	Aven, M., Craig, R.S., Waite, T.R., and Wallace, W.E., Phys. Rev., 102, 1263-4, 1956.
373	15368	Anderson, C.T., J. Am. Chem. Soc., 58, 564-6, 1936.
374	1235	Corak, W.S., Goodman, B.B., Satterthwaite, C.B., and Wexler, A., Phys. Rev., 96, 1442-4, 1954.
375	22614	Lounasmaa, O.V., Phys. Rev., 129, 2460-4, 1963.
376	424	Zabetakis, M.G., Ph.D. Thesis, Univ. of Pittsburgh, Univ. Microfilms Publ. 16530, 1-66, 1956.
377	10453	Zimmerman, J.E. and Crane, L.T., Phys. Rev., 126 (2), 513-6, 1962.
378	1478	Todd, S.S., J. Am. Chem. Soc., 72, 2914-5, 1950.
379	2023	Skinner, G.B. and Johnston, H.L., J. Am. Chem. Soc., 73, 4549-51, 1951.
380	39692	Lounasmaa, O.V. and Sundstrom, L.J., Phys. Rev., 150 (2), 399-412, 1966.
381	409	Johnston, W.V., Ph.D. Thesis, Univ. of Pittsburgh, Univ. Microfilms Publ. 15089, 1-106, 1956.
382	10440	Khomyakov, K.G., Kholler, V.A., and Slavhova, G.K., Vestnik Moskov Univ. Ser. Mat. Mekhan. Astron. Fiz. i Khim., 13 (4), 223-30, 1958.
383	21310	Saba, W.G. and Wallace, W.E., J. Chem. Phys., 35, 689-92, 1961.
384	29687	Brooks, C.R., Ph.D. Thesis, Univ. Tennessee, Univ. Microfilms Publ. 62-3711, 1-256, 1962.
385	38688	Slick, P.I., Massena, C.W., and Craig, R.S., J. Chem. Phys., 43 (8), 2788-94, 1965.
386	31040	Ho, J.C., O'Neal, H.R., and Phillips, N.E., Rev. Sci. Instr., 34 (7), 782-3, 1963.
387	20462	Keesom, W.H. and Kurrelmeyer, B., Physica, 7, 1003-24, 1940.
388	40676	Guthrie, G.L., Ph.D. Thesis, Carnegie Inst. of Tech., 1-126, 1956.
389	39449	Yee, R. and Zimmerman, G.O., J. Appl. Phys., 37 (9), 3577-80, 1966.
390	39735	Veal, B.W., Jr., M.S. Thesis, Univ. of Pittsburgh, 1-104, 1962.
391	40607	Chang, Y.A. and Hultgren, R., J. Chem. Eng. Data, 12 (1), 98-9, 1967.
392	26447	Orr, R.L., Giraud, H.J., and Hultgren, R., ASM Trans. Quart., 55 (3), 853-7, 1962.
393	32472	DeNobel, J. and DuChatenier, F.J., Physica, 29 (11), 1231-2, 1963.
394	19936	Galli, G., Ph.D. Thesis, Univ. Pittsburgh, Univ. Microfilms Publ. 59-4448, 1-77, 1959.

Ref. No.	TPRC No.	
429	28616	Skinner, G.B. Ohio State University Ph.D. Thesis, Univ. Microfilms Publ. 51-25473, 115pp. 1951
395	39887	Arledge, T.L., Jr., M.S. Thesis, Univ. Tennessee, 1-86, 1963.
396	17231	Wollam, J.S. and Wallace, W.E., J. Phys. Chem. Solids, 13, 212-20, 1960.
397	31333	Proshina, Z.V. and Rezukhina, T.N., Russ. J. Phys. Chem., 36 (1), 76-7, 1962.
398	7403	Hoare, F.E., Matthews, J.C., and Walling, J.C., Proc. Roy. Soc. (London), A216 (1127), 502-15, 1953.
399	32927	Ziegler, W.T. and Mullins, J.C., Cryogenics, 4 (1), 39-40, 1964.
400	40631	Jelinek, F.J., Shickell, W.D., and Gerstein, B.C., J. Phys. Chem. Solids, 28 (2), 267-70, 1967.
401	23357	Hake, R.R., Phys. Rev., 123, 1986-94, 1961.
402	31690	Proshina, Z.V. and Rezukhina, T.N., Russ. J. Phys. Chem., 36 (8), 941-2, 1962.
403	39917	Hoare, F.E. and Wheeler, J.C.C., Phys. Letters, 23 (7), 402-3, 1966.
404	39473	Klar, E., Mallya, R.M., and Craig, R.S., J. Chem. Phys., 45 (3), 934-6, 1966.
405	31065	Starke, E.A., Jr., M.S. Thesis, Univ. Illinois, 1-48, 1961.
406	1712	Redmond, R.F. and Lones, J., ORNL-1342, 3-20, 1952. [AD 3 665]
407	23739	Gregory, B. and Bray, H.J., Metallurgia, 63, 276-8, 1961.
408	33186	Bartenev, G.M., J. Tech. Phys. (USSR), 10, 1074-84, 1940.
409	39743	Bernard, H.W., M.S. Thesis, Penn. State Univ., 1-59, 1961.
410	38509	Burford, J.C. and Graham, G.M., Can. J. Phys., 43 (10), 1915-7, 1965.
411	3919	Gel'd, P.V., Kuprovskii, B.B., and Serebrennikov, N.N., Teploenergetika (USSR), 3 (6), 45-51, 1956.
412	21240	Liusternik, V.E., Phys. Metals Metallog. (USSR), 7 (3), 40-3, 1959.
413	35093	DuChatenier, F.J., Boerstoel, B.M., and DeNobel, J., Physica, 31 (7), 1061-2, 1965.
414	26610	Powers, W.C. and Blalock, G.C., CF-53-9-98, 1-9, 1953.
415	7366	Craig, R.S., Satterthwaite, C.B., and Wallace, W.E., USAEC Publ., NYO-946, 1-28, 1952.
416	8143	Holden, R.B. and Kopelman, B., USAEC, SEP-128, 1-17, 1953.
417	41263	Huffstuller, M.C., Jr., Ph.D. Thesis, Univ. California, 1-52, 1961.
418	32370	Kussman, Von A. and Wollenberger, H., Z. Metallkunde, 50, 94-100, 1959.
419	38419	Brewer, D.F., Howe, D.R., and Turrell, B.G., Phys. Letters, 13 (3), 204-5, 1964.
420	40967	Proctor, W., Scurlock, R.G., and Wray, E.M., Proc. Phys. Soc. (London), 90 (3), 697-705, 1967.
421	6335	Mendelssohn, K. and Moore, J.R., Proc. Roy. Soc. (London), 151A, 334-41, 1935.
422	44070	Lounasmaa, O.V. and Sundstrom, L.J., Phys. Rev., 158 (3), 591-600, 1967.
423	30698	Pollitzer, F., Z. Electrochem., 17, 5-14, 1911.
424	35142	Taylor, W.A., McCollum, D.C., Passenheim, B.C., and White, H.W., Phys. Rev., 161 (3), 652-5, 1967.
425	19184	Evans, J.P. and Mardon, P.G., J. Phys. Chem. Solids, 10, 311-3, 1959.
426	36776	Sandenaw, T.A., J. Phys. Chem. Solids, 26 (6), 1075-8, 1965.
427	20619	Lyubimov, A.P. and Belashchenko, D.K., Sbor. Moskovskogo Inst. Stali, 33, 3-11, 1955.
428	7294	Pallister, P.R., J. Iron Steel Inst., 185, 474-82, 1957.

Material Index

MATERIAL INDEX TO SPECIFIC HEAT COMPANION VOLUMES 4, 5, AND 6

Material Name	Vol.	Page
Magnesium aluminate (see Magnesium dialuminum tetraoxide)		
Magnesium metaaluminate (see Magnesium dialuminum tetraoxide)		
Magnesium dialuminum tetraoxide ($MgAl_2O_4$)	5	1479
Magnesium aluminum silicate (see Dimagnesium tetraaluminum pentasilicon 18-oxide)		
Dimagnesium tetraaluminum pentasilicon 18-oxide ($Mg_2Al_4Si_5O_{18}$)	5	1503
Magnesium borides:		
MgB_2	5	345
MgB_4	5	348
Magnesium diboride (MgB_2)	5	345
Magnesium tetraboride (MgB_4)	5	348
Magnesium cadmium alloys:		
MgCd	4	294
$MgCd_3$	4	300
Mg_3Cd	4	297
Magnesium chlorides:		
$MgCl_2$	5	838
$MgCl_2 \cdot H_2O$	5	841
$MgCl_2 \cdot 2H_2O$	5	844
$MgCl_2 \cdot 4H_2O$	5	847
$MgCl_2 \cdot 6H_2O$	5	850
Magnesium dichloride ($MgCl_2$)	5	838
Magnesium dichloride monohydrate ($MgCl_2 \cdot H_2O$)	5	841
Magnesium dichloride dihydrate ($MgCl_2 \cdot 2H_2O$)	5	844
Magnesium dichloride tetrahydrate ($MgCl_2 \cdot 4H_2O$)	5	847
Magnesium dichloride hexahydrate ($MgCl_2 \cdot 6H_2O$)	5	850
Magnesium chromite (see Magnesium dichromium tetraoxide)		
Magnesium dichromium tetraoxide ($MgCr_2O_4$)	5	1482
Magnesium ferrites:		
$MgFe_2O_4$	5	1485
$Mg_xFe_yO_4$	5	1488
Magnesium difluoride (MgF_2)	5	956
Magnesium germanide (see Dimagnesium germanide)		
Dimagnesium germanide (Mg_2Ge)	5	481
Magnesium diiron tetraoxide ($MgFe_2O_4$)	5	1485
Magnesium iron tetraoxide, nonstoichiometric ($Mg_xFe_yO_4$)	5	1488
Magnesium molybdate (see Magnesium molybdenum tetraoxide)		
Magnesium molybdenum tetraoxide ($MgMoO_4$)	5	1491
Magnesium nitride (see Trimagnesium dinitride)		
Trimagnesium dinitride (Mg_3N_2)	5	1084
Magnesium oxide (MgO)	5	140
Magnesium silicates:		
$MgSiO_3$	5	1494
Mg_2SiO_4	5	1497
$Mg_3Si_4O_{11} \cdot H_2O$	5	1500
Magnesium silicon trioxide ($MgSiO_3$)	5	1497
Dimagnesium silicon tetraoxide (Mg_2SiO_4)	5	1497
Trimagnesium tetrasilicon undecaoxide monohydrate ($Mg_3Si_4O_{11} \cdot H_2O$)	5	1500
Magnesium titanates:		
$MgTiO_3$	5	1506
$MgTi_2O_5$	5	1509
Mg_2TiO_4	5	1512
Magnesium dititanate (see Magnesium dititanium pentaoxide)		
Magnesium metatitanate (see Magnesium titanium trioxide)		
Dimagnesium titanate (see Dimagnesium titanium tetraoxide)		
Magnesium titanium trioxide ($MgTiO_3$)	5	1506
Magnesium dititanium pentaoxide ($MgTi_2O_5$)	5	1509
Dimagnesium titanium tetraoxide (Mg_2TiO_4)	5	1512
Magnesium tungstate (see Magnesium tungsten tetraoxide)		
Magnesium tungsten tetraoxide ($MgWO_4$)	5	1515
Magnesium vanadates:		
MgV_2O_6	5	1518
$Mg_2V_2O_7$	5	1521

Material Name	Vol.	Page
Magnesium metavanadate (see Magnesium divanadium hexaoxide)		
Magnesium pyrovanadate (see Magnesium divanadium hexaoxide)		
Magnesium divanadium hexaoxide (MgV_2O_6)	5	1518
Dimagnesium divanadium heptaoxide ($Mg_2V_2O_7$)	5	1521
Magnesium wolframate (see Magnesium tunsten tetraoxide)		
Manganese	4	127
Manganese, electrolytic	4	127
Manganese + Aluminum	4	372
Manganese + Copper	4	377
Manganese + Nickel	4	380
Manganese aluminum carbide (see Trimanganese aluminum carbide)		
Trimanganese aluminum carbide (Mn_3AlC)	5	427
Manganese carbide (see Trimanganese carbide)		
Trimanganese carbide (Mn_3C)	5	433
Manganese carbonate ($MnCO_3$)	5	1121
Manganese chlorides:		
$MnCl_2$	5	853
$MnCl_2 \cdot 4H_2O$	5	856
Manganese dichloride ($MnCl_2$)	5	853
Manganese dichloride tetrahydrate (see Manganous dichloride tetrahydrate)		
Manganese difluoride (MnF_2)	5	959
Manganese oxides:		
MnO	5	145
MnO_2	5	148
Mn_2O_3	5	151
Mn_3O_4	5	154
Manganese binoxide (see Manganese dioxide)		
Manganese dioxide (MnO_2)	5	148
Manganese monoxide (MnO)	5	145
Manganese peroxide (see Manganese dioxide)		
Manganese protoxide (see Manganese monoxide)		
Manganese sesquioxide (Mn_2O_3)	5	151
Dimanganese trioxide (see Manganese sesquioxide)		
Trimanganese tetraoxide (Mn_3O_4)	5	154
Manganese (ic) oxide (see Manganese sesquioxide)		
Manganese (ous) chloride (see Manganese dichloride)		
Manganese (ous) fluoride (see Manganese difluoride)		
Manganese (ous) oxide (see Manganese monoxide)		
Manganese (ous) sulfide (see Manganese sulfide)		
Manganese selenide (see Manganous selenide)		
Manganese silicate (see Manganese silicon trioxide)		
Manganese silicides:		
Mn_3Si	5	586
$MnSi_X$ (nonstoichiometric)	5	589
Trimanganese silicide (Mn_3Si)	5	586
Manganese silicide, nonstoichiometric ($MnSi_X$)	5	589
Manganese silicon trioxide ($MnSiO_3$)	5	1524
Manganese sulfide (MnS)	5	684
Manganese monosulfide (see Manganese sulfide)		
Manganese telluride (see Manganous telluride)		
Manganese zinc carbide (see Trimanganese zinc carbide)		
Trimanganese zinc carbide (Mn_3ZnC)	5	430
Manganin	4	338
Manganomanganic oxide (see Trimanganese tetraoxide)		
Manganous dichloride tetrahydrate ($MnCl_2 \cdot 4H_2O$)	5	856
Manganous selenide ($MnSe$)	5	539
Manganous telluride ($MnTe$)	5	732
Marsh gas (see Methane)		
Mercuric oxide [see Mercury (ic) oxide]		

QUEEN MARY
COLLEGE
LIBRARY